Remote Sensing Handbook, Volume IV (Six Volume Set)

Volume IV of the Six Volume *Remote Sensing Handbook*, Second Edition, is focused on the use of remote sensing in forestry, biodiversity, ecology, land use and land cover, and global terrestrial carbon mapping and monitoring. It discusses remote sensing studies of multi-scale habitat modeling, forest informatics, tree and stand height studies, land cover and land use (LCLU) change mapping, forest biomass and carbon modeling and mapping, and advanced image analysis methods and advances in land remote sensing using optical, radar, LiDAR, and hyperspectral remote sensing. This thoroughly revised and updated volume draws on the expertise of a diverse array of leading international authorities in remote sensing and provides an essential resource for researchers at all levels interested in using remote sensing. It integrates discussions of remote sensing principles, data, methods, development, applications, and scientific and social context.

FEATURES

- Provides the most up-to-date comprehensive coverage of remote sensing science for forests, biodiversity, land cover and land use change (LCLUC), biomass, and carbon.
- Discusses and analyzes data from old and new generations of satellites and sensors spread across 60 years.
- Extensive forestry, LCLUC studies, biomass, and carbon using optical, radar, LiDAR, and hyperspectral data.
- Includes numerous case studies on advances and applications at local, regional, and global scales.
- Introduces advanced methods in remote sensing, such as machine learning, cloud computing, and AI.
- Highlights scientific achievements over the last decade and provides guidance for future developments.

This volume is an excellent resource for the entire remote sensing and GIS community. Academics, researchers, undergraduate and graduate students, as well as practitioners, decision makers, and policymakers, will benefit from the expertise of the professionals featured in this book and their extensive knowledge of new and emerging trends.

Remote Sensing Handbook, Volume IV (Six Volume Set)

Forests, Biodiversity, Ecology, LULC, and Carbon

Second Edition

Edited by Prasad S. Thenkabail, PhD

CRC Press is an imprint of the
Taylor & Francis Group, an **informa** business

Designed cover image: © Prasad S. Thenkabail

Second edition published 2025
by CRC Press
2385 NW Executive Center Drive, Suite 320, Boca Raton FL 33431

and by CRC Press
4 Park Square, Milton Park, Abingdon, Oxon, OX14 4RN

CRC Press is an imprint of Taylor & Francis Group, LLC

© 2025 selection and editorial matter, Prasad S. Thenkabail; individual chapters, the contributors

First edition published by CRC Press 2016

Reasonable efforts have been made to publish reliable data and information, but the author and publisher cannot assume responsibility for the validity of all materials or the consequences of their use. The authors and publishers have attempted to trace the copyright holders of all material reproduced in this publication and apologize to copyright holders if permission to publish in this form has not been obtained. If any copyright material has not been acknowledged please write and let us know so we may rectify in any future reprint.

Except as permitted under U.S. Copyright Law, no part of this book may be reprinted, reproduced, transmitted, or utilized in any form by any electronic, mechanical, or other means, now known or hereafter invented, including photocopying, microfilming, and recording, or in any information storage or retrieval system, without written permission from the publishers.

For permission to photocopy or use material electronically from this work, access www.copyright.com or contact the Copyright Clearance Center, Inc. (CCC), 222 Rosewood Drive, Danvers, MA 01923, 978-750-8400. For works that are not available on CCC please contact mpkbookspermissions@tandf.co.uk

Trademark notice: Product or corporate names may be trademarks or registered trademarks and are used only for identification and explanation without intent to infringe.

Library of Congress Cataloging-in-Publication Data
Names: Thenkabail, Prasad Srinivasa, 1958– editor.
Title: Remote sensing handbook / edited by Prasad S. Thenkabail ; foreword by Compton J. Tucker.
Description: Second edition. | Boca Raton, FL : CRC Press, 2025. | Includes bibliographical references and index. | Contents: v. 1. Remotely sensed data characterization, classification, and accuracies—v. 2. Image processing, change detection, GIS and spatial data analysis—v. 3. Agriculture, food security, rangelands, vegetation, phenology, and soils—v. 4. Forests, biodiversity, ecology, LULC, and carbon—v. 5. Water, hydrology, floods, snow and ice, wetlands, and water productivity—v. 6. Droughts, disasters, pollution, and urban mapping.
Identifiers: LCCN 2024029377 (print) | LCCN 2024029378 (ebook) | ISBN 9781032890951 (hbk ; v. 1) | ISBN 9781032890968 (pbk ; v. 1) | ISBN 9781032890975 (hbk ; v. 2) | ISBN 9781032890982 (pbk ; v. 2) | ISBN 9781032891019 (hbk ; v. 3) | ISBN 9781032891026 (pbk ; v. 3) | ISBN 9781032891033 (hbk ; v. 4) | ISBN 9781032891040 (pbk ; v. 4) | ISBN 9781032891453 (hbk ; v. 5) | ISBN 9781032891477 (pbk ; v. 5) | ISBN 9781032891484 (hbk ; v. 6) | ISBN 9781032891507 (pbk ; v. 6)
Subjects: LCSH: Remote sensing—Handbooks, manuals, etc.
Classification: LCC G70.4 .R4573 2025 (print) | LCC G70.4 (ebook) | DDC 621.36/780285—dc23/eng/20240722
LC record available at https://lccn.loc.gov/2024029377
LC ebook record available at https://lccn.loc.gov/2024029378

ISBN: 978-1-032-89103-3 (hbk)
ISBN: 978-1-032-89104-0 (pbk)
ISBN: 978-1-003-54117-2 (ebk)

DOI: 10.1201/9781003541172

Typeset in Times
by Apex CoVantage, LLC

Contents

Foreword by Compton J. Tucker xiii
Preface xxi
About the Editor xxix
List of Contributors xxxiii
Acknowledgments xxxvii

PART I Forests

Chapter 1 Characterizing Tropical Forests with Multispectral Imagery 3

E.H. Helmer, Nicholas R. Goodwin, Valéry Gond, Carlos M. Souza Jr., and Gregory P. Asner

Acronyms and Definitions 3
1.1 Introduction 3
1.2 Multispectral Imagery and REDD+ 4
    1.2.1 Greenhouse Gas Inventories and Forest Carbon Offsets 4
    1.2.2 The Roles of Multispectral Imagery 5
1.3 Characteristics of Multispectral Image Types 6
1.4 Preprocessing Imagery to Address Clouds 10
    1.4.1 Cloud Screening 10
    1.4.2 Filling Cloud and Scan-line Gaps 15
1.5 Forest Biomass, Degradation, Regrowth Rates From Multispectral Imagery 16
    1.5.1 Tropical Forest Biomass from High-Resolution Multispectral Imagery 16
    1.5.2 The Biomass, Age, and Rates of Biomass Accumulation in Forest Regrowth 17
    1.5.3 Limitations to Mapping Forest Biomass or Age with One Multispectral Image Epoch 18
    1.5.4 Detecting Tropical Forest Degradation with Multispectral Imagery 21
1.6 Mapping Tropical Forest Types with Multispectral Imagery 24
    1.6.1 Forest Types as Strata for REDD+ and Other C Accounting 24
    1.6.2 High-Resolution Multispectral Imagery for Mapping Finely Scaled Habitats 24
    1.6.3 Remote Tree Species Identification and Forest Type Mapping 26
    1.6.4 Mapping Tropical Forest Types with Medium-Resolution Imagery 26
    1.6.5 Species Richness, Endemism and Functional Traits, and Multispectral Imagery 28
    1.6.6 Tropical Forest Type Mapping at Coarse Spatial Scale 29
    1.6.7 Tropical Forest Type Mapping and Image Spatial Resolution ... 29
1.7 Monitoring Effects of Global Change on Tropical Forests 30
    1.7.1 Progress in Monitoring Tropical Forests at Subcontinental to Global Scales 30

		1.7.2	The Feedbacks among Tropical Forest Disturbance, Drought, and Fire 31
		1.7.3	Storm-Related and Other Tree Mortality 32
	1.8	Summary and Conclusions 32
	1.9	Acknowledgments 33

Chapter 2 Remote Sensing of Forests from LiDAR and Radar 47

Juha Hyyppä, Xiaowei Yu, Mika Karjalainen, Xinlian Liang,
Anttoni Jaakkola, Mike Wulder, Markus Hollaus, Joanne C. White,
Mikko Vastaranta, Jiri Pyörälä, Tuomas Yrttimaa, Ninni Saarinen,
Josef Taher, Juho-Pekka Virtanen, Leena Matikainen, Yunsheng Wang,
Eetu Puttonen, Mariana Campos, Matti Hyyppä, Kirsi Karila,
Harri Kaartinen, Matti Vaaja, Ville Kankare, Antero Kukko,
Markus Holopainen, Hannu Hyyppä, Masato Katoh, and Eric Hyyppä

Acronyms and Definitions 47
2.1 Introduction 48
2.2 Conventional Practices for Acquisition of Forest Resource Information 49
  2.2.1 Forest Inventory 49
  2.2.2 Forest Measurements 50
2.3 General Features of Laser Scanning/LiDAR and Radar 52
2.4 Obtaining 3D Data from Forestry 52
  2.4.1 Space-borne LiDAR 54
  2.4.2 Space-borne Synthetic Aperture Radar (SAR) 55
  2.4.3 Airborne Laser Scanning, Airborne LiDAR 58
  2.4.4 Terrestrial Laser Scanning 58
  2.4.5 Mobile Laser Scanning 59
2.5 Processing 3D Data into Forest Information 60
  2.5.1 DTM Processing 61
  2.5.2 DSM Processing and Canopy Height Model 62
  2.5.3 Point Height Metrics 62
  2.5.4 Approaches for Obtaining Forest Data from Point Clouds 62
  2.5.5 Individual Tree Detection or Locating with ALS 65
  2.5.6 Individual Tree Height Derivation 66
  2.5.7 Diameter and Stem Curve Derivation 67
2.6 Future Operational Possibilities 69
  2.6.1 The Concept and Utility of the LiDAR Plots 69
  2.6.2 Improving Large-Area Mapping of Forest Attributes Using Satellite Radar 71
  2.6.3 Precision Forestry—Toward Individual Tree Inventories and Change Detection 72
  2.6.4 Detection of Forest Cuttings 76
  2.6.5 Automizing Field Inventories 77
  2.6.6 Harvester Laser Scanning 80
  2.6.7 Permanent Laser Scanning Systems for Continuous Forest Monitoring 80
2.7 Summary 82
2.8 Acknowledgment 84

Chapter 3	Forest Biophysical and Biochemical Properties from Hyperspectral and LiDAR Remote Sensing 96	

Gregory P. Asner, Susan L. Ustin, Philip Townsend, and Roberta E. Martin

- Acronyms and Definitions 96
- 3.1 Introduction 96
- 3.2 HSI and LiDAR Data 97
  - 3.2.1 HSI Data Sources 97
  - 3.2.2 LiDAR Data Sources 98
  - 3.2.3 Data Quality 99
- 3.3 HSI Remote Sensing of Forests 100
  - 3.3.1 Biophysical Properties 100
  - 3.3.2 Biochemical Properties 102
  - 3.3.3 Canopy Physiology 105
- 3.4 LiDAR Remote Sensing of Forests 108
  - 3.4.1 Canopy Structure and Biomass 108
  - 3.4.2 Light Penetration 110
- 3.5 Integrating HSI and LiDAR 111
  - 3.5.1 Benefits of Data Fusion 111
- 3.6 Conclusions 113

Chapter 4	Optical Remote Sensing of Tree and Stand Heights 125	

Sylvie Durrieu, Cédric Vega, Marc Bouvier, Frédéric Gosselin, Jean-Pierre Renaud, Laurent Saint-André, Anouk Schleich, and Maxime Soma

- Acronyms and Definitions 125
- 4.1 Introduction 126
- 4.2 Why Measure Tree Heights? 127
  - 4.2.1 The Determinants of Heights 127
  - 4.2.2 Importance of Tree Height Distribution for Forest Management and Ecology 127
  - 4.2.3 Limitations of Field Measurements and How Remote Sensing Can Help Meet Information Requirements 129
- 4.3 Two Promising Optical Remote Sensing Techniques for Tree Height Measurements: LiDAR and Digital Photogrammetry 135
  - 4.3.1 LiDAR 135
  - 4.3.2 3D Modeling of the Canopy by Digital Photogrammetry 143
  - 4.3.3 Comparison of ALS and Photogrammetric Products 147
- 4.4 Assessing Height Characteristics at the Stand Level 150
  - 4.4.1 General Presentation of Area-based Approaches 150
  - 4.4.2 Area-based Model Implementation 152
  - 4.4.3 Model Extrapolation and Inferences for Large-Area Inventories 157
- 4.5 Approaches for Individual Tree Height Assessment 158
  - 4.5.1 Raster-based Approaches 159
  - 4.5.2 Point-based Approaches 162
- 4.6 Conclusion and Perspectives 165
- 4.7 Acknowledgments 167

PART II Biodiversity

Chapter 5 Biodiversity of the World: A Study from Space 191

Thomas W. Gillespie, Morgan Rogers, Chelsea Robinson, and Duccio Rocchini

Acronyms and Definitions 191
5.1 Introduction 191
5.2 Measuring Biodiversity from Space 192
    5.2.1 Mapping Species and Vegetation Types 192
    5.2.2 Mapping Individual Trees 193
    5.2.3 Mapping Animals from Space 194
    5.2.4 Mapping Species Assemblages 194
5.3 Modeling Biodiversity from Space 194
    5.3.1 Species Distribution Modeling 194
    5.3.2 Land Cover and Diversity 196
    5.3.3 Spectral Indices and Diversity 197
    5.3.4 Multiple Sensors and Diversity 198
5.4 Monitoring Biodiversity from Space 198
    5.4.1 Remote Sensing of Protected Areas 199
    5.4.2 Remote Sensing of Urban Areas 200
5.5 Spaceborne Sensors and Biodiversity 201
    5.5.1 Spectral Sensors and Biodiversity 201
    5.5.2 Radar Sensors and Biodiversity 202
    5.5.3 LiDAR Sensors and Biodiversity 203
    5.5.4 Ideal Biodiversity Satellites and Sensors 203
5.6 Conclusions 204

Chapter 6 Multi-Scale Habitat Mapping and Monitoring Using Satellite
Data and Advanced Image Analysis Techniques 211

*Stefan Lang, Christina Corbane, Palma Blonda, Kyle Pipkins,
and Michael Forster*

Acronyms and Definitions 211
6.1 Introduction—The Policy Framework 212
    6.1.1 Monitoring Global Change 212
    6.1.2 Biodiversity and Related Policies 214
    6.1.3 Mapping the State of Ecosystems 216
    6.1.4 The EU Habitats Directive 216
6.2 Satellite Sensor Capabilities 217
    6.2.1 Spatial Resolution—What Detail Can Be Mapped? 219
    6.2.2 Spectral Resolution—Plant and Plant Feature Discrimination 222
    6.2.3 Active Systems—Radar and LiDAR 222
    6.2.4 Revisiting Time—Phenology 225
    6.2.5 Advanced Image Analysis Techniques 226
6.3 EO-Based Biodiversity and Habitat Mapping 227
    6.3.1 Land Cover, Habitats, and Indicators 228
    6.3.2 Distinguishing between and within Broad Habitat Categories 229
6.4 Observing Quality, Pressures, and Changes 233
    6.4.1 Measuring Habitat Quality 233
    6.4.2 Identifying Pressures and Changes 234
6.5 Toward a Biodiversity Monitoring Service 236

PART III Ecology

Chapter 7 Ecological Characterization of Vegetation Using Multi-Sensor Remote Sensing in the Solar Reflective Spectrum 249

Conghe Song, Jing Ming Chen, Taehee Hwang, Alemu Gonsamo, Holly Croft, Quanfa Zhang, Matthew Dannenberg, Yulong Zhang, Christopher Hakkenberg, and Junxiang Li

Acronyms and Definitions 249
7.1 Introduction 250
7.2 A Brief History of Key Optical Sensors for Vegetation Mapping 250
    7.2.1 The NOAA/AVHRR Program 251
    7.2.2 MODIS and MISR 251
    7.2.3 Suomi NPP/VIIRS 253
    7.2.4 The Landsat Program 253
    7.2.5 The SPOT Program 257
    7.2.6 Commercial High-Resolution Satellite Era 259
    7.2.7 Future Direction of Optical Remote Sensing 260
7.3 Optical Remote Sensing of Vegetation Structure 260
    7.3.1 Vegetation Cover 261
    7.3.2 Forest Successional Stages 263
    7.3.3 Remote Sensing of Leaf Area Index and Clumping Index 265
    7.3.4 Biomass 270
    7.3.5 Uncertainties, Errors, and Accuracy 274
7.4 Optical Remote Sensing of Vegetation Functions 275
    7.4.1 Vegetation Phenology 275
    7.4.2 Fraction of Absorbed Photosynthetically Active Radiation 279
    7.4.3 Leaf Chlorophyll Content 282
    7.4.4 Light Use Efficiency 285
    7.4.5 Gross Primary Productivity (GPP)/Net Primary Productivity (NPP) 288
    7.4.6 Uncertainties, Errors, and Accuracy 291
7.5 Future Directions 292
7.6 Acknowledgment 292

PART IV Land Use/Land Cover

Chapter 8 Land Cover Change Detection 311

John Rogan and Nathan Mietkiewicz

Acronyms and Definitions 311
8.1 Introduction 311
8.2 Land Cover Change Detection and Monitoring—Theory and Practice 313
8.3 Trends in Land Cover Change Detection and Monitoring 315
    8.3.1 Historical Trends—Eight Epochs 315
    8.3.2 Cause of Land Cover Change 316
8.4 Land Cover Change Detection Approaches 318
    8.4.1 Monotemporal Change Detection—Products for Real Time and Specific Disturbance Types 318

		8.4.2	Bitemporal Change Detection—Map Comparison and Disturbance Analysis 319
		8.4.3	Temporal Trend Analysis—Automation and Big Data 322
		8.4.4	Comparison of Several Automated Change Detection Approaches 323
	8.5	Accuracy Assessment—Beyond Statistics 327
	8.6	Massachusetts Case Study—CLASlite 328
		8.6.1	CLASlite Results 330
		8.6.2	Deforestation and Disturbance Mapping 332
		8.6.3	Gardner, Massachusetts, Forest Change 332
		8.6.4	2011 Tornado Disturbance 333
	8.7	Knowledge Gaps and Future Directions 333
	8.8	Acknowledgments 336

Chapter 9 Land Use and Land Cover Mapping and Monitoring with Radar Remote Sensing 343

Zhixin Qi, Anthony Gar-On Yeh, Xia Li, and Qianwen Lv

Acronyms and Definitions 343
9.1 Introduction 343
9.2 Radar System Parameters and Development 346
9.3 Radar System Parameter Consideration for LULC Mapping 351
9.4 Classification of Radar Imagery 355
9.4.1 Image Preprocessing 355
9.4.2 Feature Extraction and Selection 355
9.4.3 Selection of Classifiers 362
9.5 Change Detection Methods for Radar Imagery 364
9.5.1 Unsupervised Change Detection Methods 364
9.5.2 Combining Unsupervised Change Detection and Post-classification Comparison 368
9.6 Applications of Radar Imagery in LULC Mapping and Monitoring 369
9.6.1 LULC Classification and Change Detection 369
9.6.2 Forestry Inventory and Mapping 372
9.6.3 Crop and Vegetation Identification 374
9.6.4 Application on Urban Environment 377
9.6.5 Snow and Ice Mapping 379
9.6.6 Flood Detection and Monitoring 380
9.6.7 Other Applications 380
9.7 Future Developments 380

PART V Carbon

Chapter 10 Global Carbon Budgets and the Role of Remote Sensing 395

R.A. Houghton

Acronyms and Definitions 395
10.1 The Global Carbon Budget 395
    10.1.1 The Contemporary Carbon Budget 396
    10.1.2 A History of Carbon Cycle Research 397

Contents

	10.1.3	Sources and Sinks of Carbon from Land 398
	10.1.4	A Bookkeeping Model 398
	10.1.5	Spatial Analyses 400
10.2	Land Use and Land Cover Change (LULCC), Disturbances, and Recovery 401
	10.2.1	Use of Satellite Data 402
10.3	The Policy Realm: Issues Inherent in Estimating the Flux of Carbon from LULCC, with an Example Using RED, REDD, and REDD+ 408
	10.3.1	Definitions 408
	10.3.2	Assigning a Carbon Density to the Areas Deforested 409
	10.3.3	Committed versus Actual Emissions (Legacy Effects)....................409
	10.3.4	Gross and Net Emissions of Carbon from LULCC 410
	10.3.5	Initial Conditions... 410
	10.3.6	Full Carbon Accounting 411
	10.3.7	Accuracy and Precision 412
	10.3.8	Attribution 412
	10.3.9	Uncertainties 412
10.4	The Residual Terrestrial Sink... 413
	10.4.1	The Orbiting Carbon Observatory (OCO) 413
	10.4.2	Satellite Monitoring of Vegetation Activity (Greenness)................. 414
10.5	Conclusions.. 415
10.6	Acknowledgment 416

Chapter 11 Aboveground Terrestrial Biomass and Carbon Stock Estimations from Multi-Sensor Remote Sensing 423

Wenge Ni-Meister

Acronyms and Definitions... 423
11.1 Introduction 424
   11.1.1 Importance of the Terrestrial Ecosystem Carbon and Carbon Changes Estimates...424
   11.1.2 Importance of Tropical Rainforests in Carbon Storage 424
   11.1.3 Summary of Methods Used to Estimate Terrestrial Biomass and Carbon Stocks 426
   11.1.4 The Role of Remote Sensing in Terrestrial Ecosystem Carbon Estimates 427
   11.1.5 Specific Topics Covered in This Chapter 427
11.2 Conventional Methods of Carbon Stocks Estimates 427
   11.2.1 Biome Average Methods 428
   11.2.2 Allometric Biomass Methods.. 429
11.3 Remote Sensing Data... 432
   11.3.1 Passive Optical Remote Sensing Data.. 432
   11.3.2 Radar Data... 433
   11.3.3 LiDAR Data 434
11.4 Research Approaches/Methods 435
11.5 Remote Sensing Based Aboveground Biomass Estimates 436
   11.5.1 Optical Remote Sensing 436
   11.5.2 Radar 438
   11.5.3 LiDAR 439
   11.5.4 Multi-Sensor Fusion 441
11.6 Summary 445

	11.7	Conclusions and Future Directions	447
	11.8	Acknowledgment	448

PART VI Summary and Synthesis of Volume IV

Chapter 12 Forests, Biodiversity, Ecology, LULC, and Carbon 457

Prasad S. Thenkabail

		Acronyms and Definitions	457
	12.1	Tropical Forest Characterization Using Multi-Spectral Imagery	458
	12.2	LiDAR and Radar for Forest Informatics	461
	12.3	Hyperspectral Imager (HSI) and LiDAR Data in the Study of Forest Biophysical, Biochemical, and Structural Properties	463
	12.4	Tree and Stand Heights from Optical Remote Sensing	465
	12.5	Study of Biodiversity from Space	466
	12.6	Multi-Scale Habitat Mapping and Monitoring Using Satellite Data and Advanced Image Analysis Techniques	468
	12.7	Ecological Characterization of Vegetation Using Multi-Sensor Remote Sensing	470
	12.8	Land Cover Change Detection	472
	12.9	Radar Remote Sensing in Land Use and Land Cover Mapping and Change Detection	475
	12.10	Global Carbon Budgets and Remote Sensing	477
	12.11	Remote Sensing of Global Terrestrial Carbon	479
	12.12	Acknowledgments	484

Index 491

Foreword

Satellite remote sensing has progressed tremendously since the first Landsat was launched on June 23, 1972. Since the 1970s, satellite remote sensing and associated airborne and *in situ* measurements have resulted in geophysical observations for understanding our planet through time. These observations have also led to improvements in numerical simulation models of the coupled atmosphere-land-ocean systems at increasing accuracies and predictive capabilities. This was made possible by data assimilation of satellite geophysical variables into simulation models, to update model variables with more current information. The same observations document the Earth's climate and have driven consensus that *Homo sapiens* are changing our climate through greenhouse gas emissions.

These accomplishments are the work of many scientists from a host of countries and a dedicated cadre of engineers who build and operate the instruments and satellites that collect geophysical observation data from satellites, all working toward the goal of improving our understanding of the Earth. This edition of the *Remote Sensing Handbook* (Second Edition, Volumes I, II, III, IV, V, and VI) is a compendium of information for many research areas of the Earth System that have contributed to our substantial progress since the 1970s. The remote sensing community is now using multiple sources of satellite and *in situ* data to advance our studies of Planet Earth. In the following paragraphs, I will illustrate how valuable and pivotal satellite remote sensing has been in climate system study since the 1970s. The chapters in the *Remote Sensing Handbook* provide other specific studies on land, water, and other applications using Earth observation data of the past 60+ years.

The Landsat system of Earth-observing satellites led the way in pioneering sustained observations of our planet. From 1972 to the present, at least one and frequently two Landsat satellites have been in operation (Wulder et al. 2022; Irons et al. 2012). Starting with the launch of the first NOAA-NASA Polar Orbiting Environmental Satellites NOAA-6 in 1978, improved imaging of land, clouds, and oceans and atmospheric soundings of temperature were accomplished. The NOAA system of polar-orbiting meteorological satellites has continued uninterrupted since that time, providing vital observations for numerical weather prediction. These same satellites are also responsible for the remarkable records of sea surface temperature and land vegetation index from the Advanced Very High-Resolution Radiometers (AVHRR) that now span more than 46 years as of 2024, although no one anticipated valuable climate records from these instruments before the launch of NOAA-6 in 1978 (Cracknell 2001). AVHRR instruments are expected to remain in operation on the European MetOps satellites into 2026 and possibly beyond.

The successes of data from the AVHRR led to the Moderate Resolution Imaging Spectroradiometer (MODIS) instruments on NASA's Earth Observing System (EOS) of satellite platforms that improved substantially upon the AVHRR. The first of the EOS platforms, Terra, was launched in 2000 and the second of these platforms, Aqua, was launched in 2002. Both platforms are nearing their operational end of life, and many of the climate data records from MODIS will be continued with the Visible Infrared Imaging Suite (VIIRS) instrument on NOAA's Joint Polar Satellite System (JPSS) meteorological satellites. The first of these missions, the NPOES Preparation Project, was launched in 2012 with the first VIIRS instrument that currently is operating along with similar instruments on JPSS-1 (launched in 2017) and JPSS-2 (launched in 2022). However, unlike the morning/afternoon overpasses of MODIS, the VIIRS instruments are all in an afternoon overpass orbit. One of the strengths of the MODIS observations was morning and afternoon data from identical instruments.

Continuity of observations is crucial for advancing our understanding of the Earth's climate system. Many scientists feel the crucial climate observations provided by remote sensing satellites are among the most important satellite measurements because they contribute to documenting the current state of our climate and how it is evolving. These key satellite observations of our climate are second in importance only to the polar orbiting and geostationary satellites needed for numerical weather prediction that provide natural disaster alerts.

The current state of the art for remote sensing is to combine different satellite observations in a complementary fashion for what is being studied. Climate study is an example of using disparate observations from multiple satellites coupled with *in situ* data to determine if climate change is occurring, where it is occurring, and identify the various component processes responsible.

1. **Planet warming quantified by satellite radar altimetry:** Remotely sensed climate observations provide the data to understand our planet and identify forces and drivers that influence climate. The primary sea level climate observations come from radar altimetry that started in late 1992 with Topex-Poseidon and has been continued by Jason-1, Jason-2, Jason-3, and Sentinel-6 to provide an uninterrupted record of global sea level. Changes in global sea level provide unequivocal evidence that our planet is warming, cooling, or staying at the same temperature. Radar altimetry from 1992 to date has shown global sea level increases of ~3.5 mm/yr, hence our planet is warming (Figure 0.1). Sea level rise has two components, ocean thermal expansion and ice melt from the ice sheets of Greenland and Antarctica, and to a lesser extent for glacier concentrations in places like the Gulf of Alaska and Patagonia. The combination of GRACE and GRACE Follow-On gravity measurements quantifies the ice mass losses of Greenland and Antarctica to a high degree of accuracy. Combining the gravity data with the flotilla of almost 4,000 Argo floats provides the temperature data with depth necessary to quantify ocean temperatures and isolate the thermal component of sea level rise.

2. **Our Sun is remarkably stable in total solar irradiance**. Observations of total solar irradiance have been made from satellites since 1979 and show that total solar irradiance has varied only ±1 part in 500 over the past 35 years, establishing that our Sun is not to blame for global warming (Figure 0.2).

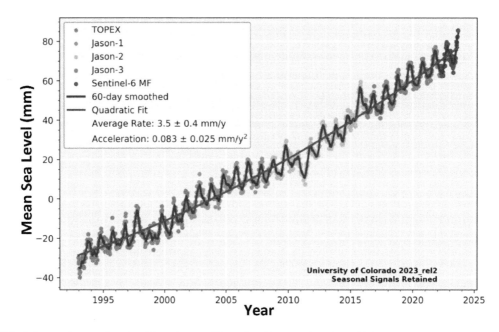

FIGURE 0.1 Seasonal sea level from five satellite radar altimeters from late 1992 to the present. Sea level is the unequivocal indicator of the Earth's climate—when sea level rises, the planet is warming; when sea level falls, the planet is cooling (Nerem 2018 updated to 2023; https://sealevel.colorado.edu/data/total-sea-level-change).

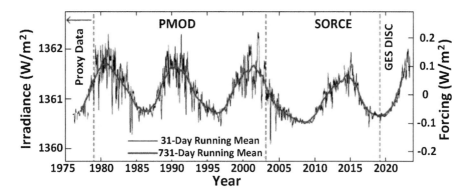

FIGURE 0.2 The Sun is not to blame for global warming, based on total solar irradiance observations from satellites. The few watts per m² solar irradiance variations covary with the sunspot cycle. The luminosity of the Sun varies 0.2% over the course of the 11-year solar and sunspot cycle. The SORCE TSI data set continues these important observations with improved accuracy on the order of ±0.035 (Kopp et al. 2024) and from https://lasp.colorado.edu/sorce/data/tsi-data/.

3. **Determining ice sheet contributions to sea level rise**. Since 2002 gravity observations from the Gravity Recovery and Climate Experiment Satellite (GRACE) mission and the GRACE Follow-On mission have been measured. GRACE data quantify ice mass changes from the Antarctic and Greenland ice sheets that constitute 98% of the ice mass on land (Luthcke et al. 2013). GRACE data are truly remarkable—their retrieval of variations in the Earth's gravity field is quantitatively and directly linked to mass variations. With GRACE data we are able for the first time to determine the mass balance with time of the Antarctic and Greenland ice sheets and concentrations of glaciers on land. GRACE data show sea level rise is 60% explained by ice sheet mass loss (Figure 0.3). GRACE data have many other uses, such as changes in ground water storage. See: <http://www.csr.utexas.edu/grace/>.

4. **40% sea level rise explained by thermal expansion in the planet's oceans measured by *in situ* ~ 3700 Argo drifting floats**. The other contributor to sea level rise is the thermal expansion or "steric" component of our planet's oceans. Documenting this necessitates using diving and drifting floats or buoys in the Argo network to record temperature with depth (Romerich et al. 2009; Figure 0.4). Argo floats are deployed from ships; they then submerge, descend slowly to a depth of 1000 m, recording temperature, pressure, and salinity as they descend. At 1000 m, they drift for ten days, continuing their measurements of temperature and salinity. After ten days, they slowly descend to 3000 m and then ascend to the surface, all the time recording their measurements. At the surface, each float transmits all the data collected on the most recent excursion to a geostationary satellite and then descend again to repeat this process.

Argo temperature data show that 40% of sea level rise results from warming and thermal expansion of our oceans. Combining radar altimeter data, GRACE and GRACE Follow-On data, and Argo data provide confirmation of sea level rise and show what is responsible for it and in what proportions. With total solar irradiance being near-constant, what is driving global warming can be determined. Analysis of surface *in situ* air temperature coupled with lower tropospheric air temperature and stratospheric temperature data from remote sensing infrared and microwave sounders show that the surface and near-surface are warming, while the stratosphere is cooling. This is an unequivocal confirmation that greenhouse gases are warming the planet.

Combining sea level radar altimetry, GRACE, and GRACE Follow-On use gravity data to quantify ice sheet mass losses, and the use of Argo floats to measure ocean

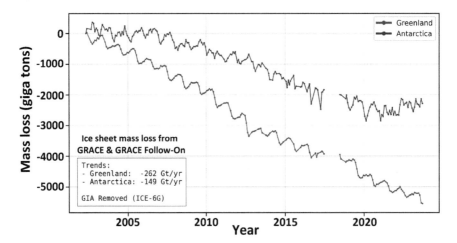

FIGURE 0.3 Sixty percent sea level rise explained by mass balance of ice melting measured by GRACE and GRACE Follow-On satellites. Ice mass variations from 2003 to 2023 for the Antarctica and Greenland ice sheets using gravity data (Croteau et al. 2021 updated to 2023). The Antarctic and Greenland Ice Sheets constitute 98% of the Earth's land ice.

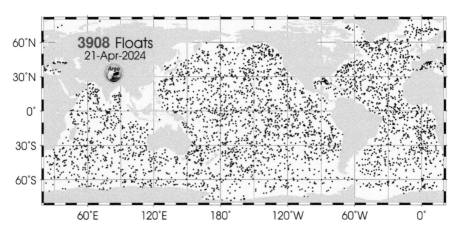

FIGURE 0.4 Forty percent sea level rise explained by thermal expansion in the planet's oceans measured *in situ* by ~ 3908 drifting floats that were in operation on March 25, 2024. These floats provide the data needed to document thermal expansion of the oceans (Roemmich & the Argo Float Team 2009 updated to 2024 and <http://www.argo.ucsd.edu/>).

temperatures with depth enables reconciliation of sea level increases with mass loss of ice sheets and ocean thermal expansion. The ice and steric expansion explains 95% of sea level rise (Figure 0.5).

5. **The global carbon cycle**. Many scientists are actively working to study the Earth's carbon cycle, and there are several chapters in this *Remote Sensing Handbook* (Volumes I–VI) on various components under study.

Carbon cycles through reservoirs on the Earth's surface in plants and soils, exists in the atmosphere as gases such as carbon dioxide (CO_2), and exists in ocean water in phytoplankton and marine sediments. CO_2 is released into the atmosphere from the combustion of fossil fuels, by land

Foreword xvii

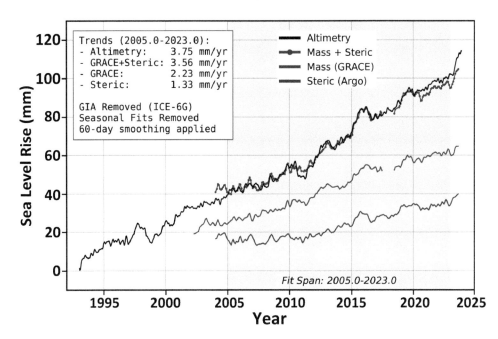

FIGURE 0.5 Sea level rise with the gravity ice mass loss and the Argo thermal expansion quantities added to the plot of global mean sea level. The GRACE and GRACE Follow-On ice sheet gravity term and Argo thermal expansion terms together explain 95% of sea level rise (Croteau et al. 2021 updated to 2023).

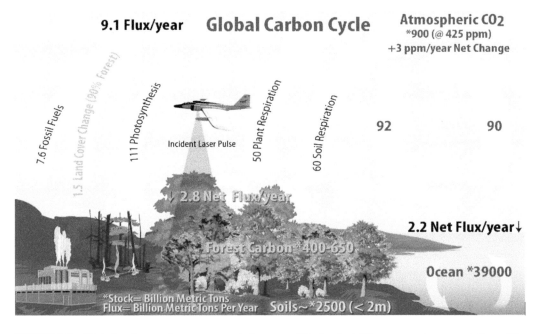

FIGURE 0.6 Global carbon cycle measurements from a multitude of satellite sensors. A representation of the global carbon cycle showing our best estimates of carbon fluxes and carbon reservoirs as of 2024. A series of satellite observations are needed simultaneously to understand the carbon cycle and its role in the Earth's climate system (Cias et al. 2014 updated to 2023). The major unknowns in the global carbon cycle are fluxes between different reservoirs, oceanic gross primary production, carbon in soils, and the carbon in woody vegetation.

cover changes on the Earth's surface, by the respiration of green plants, and by the decomposition of carbon in dead vegetation and in soils, including carbon in permafrost.

Land gross primary production is a MODIS product that has extended into the VIIRS era (Running et al. 2004; Roman et al. 2024). MODIS data also provide burned area and CO_2 emissions from wildfire (Giglio et al. 2016). Oceanic gross primary production will be provided by the Plankton, Aerosol, Cloud, ocean Ecosystem (PACE) satellite that was launched in early 2024 (Gorman et al. 2019). This complements the GPP land portion of the carbon cycle and will enable global gross primary production to be determined by MODIS-VIIRS and PACE.

Furthermore, Harmonized Landsat-8, Landsat-9, and Sentinel-2 30 m data (HLS) provide multispectral time series data at 30 m with a revisit frequency of three days at the equator (Crawford et al. 2023; Masek et al. 2018). This will enable time series improvements in spatial detail to 30 m from the 250 m scale of MODIS. The revisit time of Sentinel-2 with 10 m data is five days at the equator, which is a major improvement from 30 m. Multispectral time series observations are the basis for providing gross primary production estimates on land that are also used for food security (Claverie et al. 2018).

Refinements in satellite multispectral spatial resolution to the 50 cm to 3–4 m scale provided by commercial satellite data have enabled tree carbon to be determined from large areas of trees outside of forests. NASA has started using commercial satellite data to complement MODIS, Landsat, and other observations. One of the uses for Planet 3–4 m and Maxar < 1 m data have been for mapping trees outside of forests (Brandt et al. 2020; Reiner et al. 2023; Tucker et al. 2023). Tucker et al. (2023) mapped 10 billion trees at the 50 cm scale over 10 million km^2 and converted them into carbon at the tree level with allometry. The value of Planet and Maxar (formerly Digital Globe) data allows carbon studies to be extended into areas with discrete trees, and Huang et al. (2024) has successfully mapped one tree species across the entire Sahelian and Sudanian Zones of Africa.

The height of trees is an important measurement to determine their carbon content. For areas of contiguous tree crowns, GEDI and ICESat laser altimetry (Magruder et al. 2024) coupled with Landsat and Sentinel-2 observations, enable improved estimates of carbon in these forests (Claverie et al. 2018).

The key to closing several uncertainties in the carbon cycle is to quantify fluxes among the various components. Passive CO_2 retrieval methods from the Greenhouse Gases Observing Satellite (GOSAT) (Noël et al. 2021) and the Orbiting Carbon Observatory-2 (OCO-2) (Jacobs et al. 2024) are inadequate to provide this. Passive methods are not possible at night, in all seasons, and require specific Sun-target-sensor viewing perspectives and conditions. A recent development of the Aerosol and Carbon dioxide Detection Lidar (ACDL) instrument (Dai et al. 2023) by our Chinese colleagues offers a ten-fold coverage improvement in CO_2 retrievals over those provided by OCO-2 and 20-fold coverage improvement over GOSAT. The reported uncertainty of ACDL is on the order of ±0.6 ppm.

Understanding the carbon cycle requires a "full court press" of satellite and *in situ* observations because all these observations must be made at the same time. Many of these measurements have been made over the past 30–40 years, but new measurements are needed to quantify carbon storage in vegetation, to quantify CO_2 fluxes, to quantify land respiration, and improve numerical carbon models. Similar work needs to be performed for the role of clouds and aerosols in climate and to improve our understanding of the global hydrological cycle.

The remote sensing community has made tremendous progress over the last six decades, as captured in various chapters of the *Remote Sensing Handbook* (Second Edition, Volumes I–VI). Handbook chapters provide comprehensive understanding of land and water studies through detailed methods, approaches, algorithms, synthesis, and key references. Every type of remote sensing data obtained from systems such as optical, radar, LiDAR, hyperspectral, and hyperspatial are presented and discussed in different chapters. Chapters in this volume address remote sensing data characteristics, within and between sensor calibrations, classification methods, and accuracies, taking a wide array of remote sensing data from a wide array of platforms over the last five decades. Volume I also brings in new remote sensing technologies, such as radio occultation and reflectometry from the global navigation satellite system or GPS satellites, crowdsourcing, drones, cloud computing, artificial intelligence, machine learning, hyperspectral, radar, and remote sensing law. The chapters in the *Remote Sensing Handbook* are written by leading remote sensing scientists of the world and ably edited by Dr. Prasad S. Thenkabail, Senior Scientist (ST), at the U.S. Geological Survey (USGS) in Flagstaff, Arizona. The importance and value of the *Remote Sensing Handbook* is clearly demonstrated by the need for this second edition. The *Remote Sensing Handbook* (First Edition, Volumes I–III) was published in 2014, and now, after ten years, the *Remote Sensing Handbook* (Second Edition, Volumes I–VI), with 91 chapters and nearly 3500 pages, will be published. It is certainly monumental work in remote sensing science, and for this I want to compliment Dr. Prasad Thenkabail. Remote sensing is now important to many scientific disciplines beyond our community, and I recommend the *Remote Sensing Handbook* (Second Edition, six volumes) to not only remote sensors but to the entire scientific community.

We can look forward in the coming decades to improving our quantitative understanding of the global carbon cycle, understanding the interaction of clouds and aerosols in our radiation budget, and understanding the global hydrological cycle.

by Compton J. Tucker
Satellite Remote Sensing Beyond 2025
NASA/Goddard Space Flight Center
Earth Science Division
Greenbelt, Maryland 20771 USA

REFERENCES

Brandt, M., Tucker, C.J., Kariryaa, A., et al. 2020. An unexpectedly large count of trees in the West African Sahara and Sahel. *Nature* 587: 78–82. http://doi.org/10.1038/s41586-020-2824-5

Ciais, P., et al. 2014. Current systematic carbon-cycle observations and the need for implementing a policy-relevant carbon observing system. *Biogeosciences* 11(13): 3547–3602.

Claverie, M., Ju, J, Masek, J.G., Dungan, J.L., Vermote, E.F., Roger, J.-C., Skakun, S.V., et al. 2018. The Harmonized Landsat and Sentinel-2 surface reflectance data set. *Remote Sensing of Environment* 219: 145–161. http://doi.org/10.1016/j.rse.2018.09.002

Cracknell, A. 2001. The exciting and totally unanticipated success of the AVHRR in applications for which it was never intended. *Advances in Space Research* 28: 233–240. http://doi.org/10.1016/S0273-1177(01)00349-0

Crawford, C.J., Roy, D.P., Arab, S., Barnes, C., Vermote, E., Hulley, G., et al. 2023. The 50-year Landsat collection 2 archive. *Science of Remote Sensing* 8(2023): 100103. ISSN 2666-0172. https://doi.org/10.1016/j.srs.2023.100103. https://www.sciencedirect.com/science/article/pii/S2666017223000287

Croteau, M.J., Sabaka, T.J., Loomis, B.D. 2021. GRACE fast mascons from spherical harmonics and a regularization design trade study. *Journal of Geophysical Research—Solid Earth* 126: e2021JB022113. https://doi.org/10.1029/2021JB022113.10.1029/2021JB022113

Dai, G., Wu, S., Sun, K., Long, W., Liu, J., Chen, W. 2023. Aerosol and Carbon Dioxide Detection Lidar (ACDL) overview. *Presentation at ESA-JAXA EarthCare Workshop*, November 2023. https://doi.org/10.5194/egusphere-2023-2182

Giglio, L., Schroeder, W., Justice, C.O. 2016. The collection 6 MODIS active fire detection algorithm and fire products. *Remote Sensing of Environment* 178: 31–41. http://doi.org/10.1016/j.rse.2016.02.054

Gorman, E.T., Kubalak, D.A., Patel, D., Dress, A., Mott, D.B., Meister, G., Werdell, P.J. 2019. The NASA Plankton, Aerosol, Cloud, ocean Ecosystem (PACE) mission: An emerging era of global, hyperspectral Earth system remote sensing. *Sensors, Systems, and Next Generation Satellites* XXIII: 11151. http://doi.org/10.1117/12.2537146

Huang, K. et al. 2024. Mapping every adult baobab (Adansonia digitata L.) across the Sahel to uncover the coexistence with rural livelihoods. *Nature Ecology and Evolution*. http://doi.org/10.21203/rs.3.rs-3243009/v1

Irons, J.R., Dwyer, J.L., Barsi, J.A. 2012. The next Landsat satellite: The Landsat data continuity mission. *Remote Sensing of Environment* 122: 11–21. http://doi.org/10.1016/j.rse.2011.08.026

Jacobs, N., et al. 2024. The importance of digital elevation model accuracy in X_{CO2} retrievals: Improving the Orbiting Carbon Observatory-2 Atmospheric Carbon Observations from Space version 11 retrieval product. *Atmospheric Measurement Techniques* 17(5): 1375–1401. http://doi.org/10.5194/amt-17-1375-2024

Kopp, G., Nèmec, N.E., Shapiro, A. 2024. Correlations between total and spectral solar irradiance variations. *Astrophysical Journal* 964(1). http://doi.org/10.3847/1538-4357/ad24e5

Magruder, L.A., Farrell, S.L., Neuenschwander, A., Duncanson, L., Csatho, B., Kacimi, S., et al. 2024. Monitoring Earth's climate variables with satellite laser altimetry. *Nature Reviews of Earth and Environment* 5(2):120–136. http://doi.org/10.1038/s43017-023-00508-8

Masek, J., Ju, J., Roger, J.-C., Skakun, S., Claverie, M., Dungan, J. 2018. Harmonized Landsat/Sentinel-2 products for land monitoring. *IGARSS 2018—2018 IEEE International Geoscience and Remote Sensing Symposium*, Valencia, Spain, 2018, pp. 8163–8165. http://doi.org/10.1109/IGARSS.2018.8517760

Nerem, R.S., Beckley, B.D., Fasullo, J.T., Mitchum, G.T. 2018. Climate-change–driven accelerated sea-level rise detected in the altimeter era. *Proceeding of the National Academy of Sciences* 115(9): 2022–2025. http://doi.org/10.1073/pnas.1717312115

Noël, S., et al. 2021. XCO_2 retrieval for GOSAT and GOSAT-2 based on the FOCAL algorithm. *Atmospheric Measurement Techniques* 14(5): 3837–3869. http://doi.org/10.5194/amt-14-3837-2021

Reiner, F., et al. 2023. More than one quarter of Africa's tree cover is found outside areas previously classified as forest. *Nature Communications*. http://doi.org/10.1038/s41467-023-37880-4

Roemmich, D., The Argo Steering Team. 2009. Argo—The challenge of continuing 10 years of progress. *Oceanography* 22(3): 46–55.

Román, M., et al. 2024. Continuity between NASA MODIS collection 6.1 and VIIRS collection 2 land products. *Remote Sensing of Environment* 302. http://doi.org/10.1016/j.rse.2023.113963

Running, S.W., Nemani, R.R., Heinsch, F.A., Zhao, M.S., Reeves, M., Hashimoto, H. 2004. A continuous satellite-derived measure of global terrestrial primary production. *Bioscience* 54(6): 547–560. http://doi.org/10.1641/0006-3568

Tucker, C., Brandt, M., Hiernaux, P., Kariryaa, A., et al. 2023. Sub-continental-scale carbon stocks of individual trees in African drylands. *Nature* 615: 80–86. http://doi.org/10.1038/s41586-022-05653-6

Wulder, M.A., Roy, D.P., Radeloff, V.C., Loveland, T.R., Anderson, M.C., Johnson, D.M., et al. 2022. Fifty years of Landsat science and impacts. *Remote Sensing of Environment* 280(2022): 113195. ISSN 0034-4257. https://doi.org/10.1016/j.rse.2022.113195. https://www.sciencedirect.com/science/article/pii/S0034425722003054

Preface

The overarching goal of this six-volume, 91-chapter, and approximately 3500-page *Remote Sensing Handbook* (Second Edition, Vols. I–VI) was to capture and provide the most comprehensive, state-of-the-art remote sensing science and technology development and advancement in the last 60+ years, by clearly demonstrating the: (1) scientific advances, (2) methodological advances, and (3) societal benefits achieved during this period, as well as to provide a vision of what is to come in the years ahead. The book volumes are, to date and to my best knowledge, the most comprehensive documentation of the scientific and methodological advances that have taken place in understanding remote sensing data, methods, and a wide array of applications. Written by 300+ leading global experts in the area, each chapter: (1) focuses on a specific topic (e.g., data, methods, and a specific set of applications), (2) reviews existing state-of-the-art knowledge, (3) highlights the advances made, and (4) provides guidance for areas requiring future development. Chapters in the book cover a wide array of subject matters of remote sensing applications. The *Remote Sensing Handbook* (Second Edition, Vols. I–VI) is planned as reference material for a broad spectrum of remote sensing scientists to understand the fundamentals as well as the latest advancements and the wide array of applications, such as for land and water resource practitioners, natural and environmental practitioners, professors, students, and decision-makers.

Special features of the six-volume *Remote Sensing Handbook* (Second Edition) include:

1. Participation of an outstanding group of remote sensing experts, an unparalleled team of writers for such a book project.
2. Exhaustive coverage of a wide array of remote sensing science: data, methods, and applications.
3. Each chapter being led by a luminary and most chapters written by writing teams, which further enriched the chapters.
4. Broadening the scope of the book to make it ideal for expert practitioners as well as students.
5. A global team of writers, global geographic coverage of study areas, a wide array of satellites and sensors; and
6. Plenty of color illustrations.

Chapters in the book have covered remote sensing:

State-of-the-art on satellites, sensors, science, technology, and applications
Methods and techniques
A wide array of applications, such as land and water applications, natural resources management, and environmental issues
Scientific achievements and advancements of the aforementioned over the last 60+ years
Societal benefits
Knowledge gaps
Future possibilities in the 21st century

Great advances have taken place over the last 60+ years in the study of Planet Earth from remote sensing, especially using data gathered from a multitude of Earth Observation (EO) satellites launched by various governments as well as private entities. A large part of the initial remote sensing technology was developed and tested during the two world wars. In the 1950s remote sensing slowly began its foray into civilian applications. But, during the years of the Cold War both civilian and military remote sensing applications increased swiftly. But it was also an age when remote sensing was the domain of very few top experts who often had multiple skills in engineering, science, and computer technology. From the 1960s onwards, there have been many governmental agencies that have initiated civilian remote sensing. The National Aeronautics and Space Administration (NASA) of the USA has been at the forefront of many of these efforts. Others who have provided leadership in civilian remote sensing include, but are not limited to, the European Space Agency (ESA), the Indian Space Research Organization (ISRO), the Centre national d'études spatiales (CNES) of France, the Canadian Space Agency (CSA), the Japan Aerospace Exploration Agency (JAXA), the German Aerospace Center (DLR), the China National Space Administration (CNSA), the United Kingdom Space Agency (UKSA), and the Instituto Nacional de Pesquisas Espaciais (INPE) of Brazil. Many private entities, such as Planet Labs PBC, have launched and operate satellites. These government and private agencies and enterprises have launched, and continue to launch and operate, a wide array of satellites and sensors that capture data of the Earth in various regions of the electromagnetic spectrum and in various spatial, radiometric, and temporal resolutions, routinely and repeatedly. However, the real thrust for remote sensing advancement came during the last decade of the 20th century and the beginning of the 21st century. These initiatives included the launch of a series of new generation EO satellites to gather data more frequently and routinely, the release of pathfinder datasets, free web-enabling of the data by many agencies (e.g., USGS released the entire Landsat archives as well as real-time acquisitions of the world for free by making them web accessible), and providing processed data to users (e.g., the Harmonized Landsat and Sentinel-2 or HLS data, surface reflectance products of MODIS). Other efforts, like Google Earth, made remote sensing more popular and brought in a new platform for easy visualization and navigation of remote sensing data. Advances in computer hardware and software made it possible to handle big data. Crowdsourcing, web access, cloud computing such as in the Google Earth Engine (GEE) platform, machine learning, deep learning, coding, artificial intelligence, mobile apps, and mobile platforms (e.g., drones) added a new dimension to how remote sensing data is used. Integration with global positioning systems (GPS) and global navigation satellite systems (GNSS) and the inclusion of digital secondary data (e.g., digital elevation, precipitation, temperature) in analysis has made remote sensing much more powerful. Collectively, these initiatives provided new vision in making remote sensing data more popular, widely understood, and increasingly used for diverse applications, hitherto considered difficult. The availability of free archival data when combined with more recent acquisitions has also enabled quantitative studies of change over space and time. The *Remote Sensing Handbook* (Vols. I–VI) is targeted to capture these vast advances in data, methods, and applications, so a remote sensing student, scientist, or a professional practitioner will have the most comprehensive, all-encompassing reference material in one place.

Modern-day remote sensing technology, science, and applications are growing exponentially. This growth is a result of a combination of factors that include: (1) advances and innovations in data capture, access, processing, computing, and delivery (e.g., big data analytics, harmonized and normalized data, inter-sensor relationships, web enabling of data, cloud computing such as in GEE, crowdsourcing, mobile apps, machine learning, deep learning, coding in Python and Java Script, and artificial intelligence); (2) increasing the number of satellites and sensors gathering data of the planet, repeatedly and routinely, in various portions of the electromagnetic spectrum as well as in an array of spatial, radiometric, and temporal resolutions; (3) efforts at integrating data from multiple satellites and sensors (e.g., Sentinels with Landsat); (4) advances in data normalization, standardization, and harmonization (e.g., delivery of data in surface reflectance, inter-sensor calibration); (5) methods and techniques for handling very large data volumes (e.g., global mosaics); (6) quantum leap in computer hardware and software capabilities (e.g., the ability to process several terabytes of data); (7) innovation in methods, approaches, and techniques leading to sophisticated algorithms (e.g., spectral matching techniques, neural network perceptron); and (8) the development of new spectral indices to quantify and study specific land and water parameters (e.g., hyperspectral vegetation indices, or HVIs). As a result of these developments, remote sensing science is today very mature and is widely used in virtually every discipline of Earth Sciences for quantifying, mapping, modeling, and monitoring Planet Earth. Such rapid advances are captured in a number of remote sensing and Earth Science journals. However, students, scientists, and practitioners of remote sensing science and applications have significant difficulty in gathering a complete understanding of various developments and advances that have taken place because of their vastness spread across the last 60+ years. Thereby, the chapters in the *Remote Sensing Handbook* are designed to give a whole picture of the scientific and technological advances of the last 60+ years.

Today the science, art, and technology of remote sensing is truly ubiquitous and increasingly part of everyone's everyday life, often without even the user knowing it. Whether looking at your own home or farm (e.g., the following figure), helping you navigate when you drive, visualizing a phenomenon occurring in a distant part of the world (e.g., the following figure), monitoring events such as droughts and floods, reporting weather, detecting and monitoring troop movements or nuclear sites, studying deforestation, assessing biomass carbon, addressing disasters like earthquakes or tsunamis, and a host of other applications (e.g., precision farming, crop productivity, water productivity, deforestation, desertification, water resources management), remote sensing plays a key role. Already, many new innovations are taking place. Companies such as Planet Labs PBC and Skybox are capturing very high spatial resolution imagery and even videos from space using a large number of microsatellite (CubeSat) constellations. Planet Labs also will soon launch

FIGURE 0.7 Google Earth can be used to seamlessly navigate and precisely locate any place on Earth, often with very high spatial resolution data (VHRI; sub-meter to 5 m) from satellites such as IKONOS, Quickbird, and Geoeye (Note: the image is from one of the VHRI). Here, the editor-in-chief (EiC) (Thenkabail) of this *Remote Sensing Handbook* (Vols. I–VI) located his village home and surroundings that have land cover such as secondary rainforests, lowland paddy farms, areca nut plantations, coconut plantations, minor roads, walking routes, open grazing lands, and minor streams (typically, first and second order) (Note: land cover based on ground knowledge of the EiC). The first primary school attended by the EiC is located precisely. Precise coordinates (1345 39.22 Northern latitude, 75 06 56.03 Eastern longitude) of Thenkabail's village house and the date of image acquisition (March 1, 2014). Google Earth images are used for visualization as well as for numerous science applications, such as accuracy assessment, reconnaissance, determining land cover, and establishing land use for various ground surveys.

hyperspectral Tanager satellites. There are others (e.g., Pixxel, India) who have launched and continue to launch constellations of hyperspectral or other sensors. China is constantly putting a wide array of satellites into orbit. Just as the smart phone and social media connected the world, remote sensing is making the world our backyard (e.g., the following figure). No place goes unobserved, and no event gets reported without an image. True liberation for any technology and science comes when it is widely used by common people who often have no idea about how it all comes together but do intuitively understand the information provided. That is already happening (e.g., how we use smart phones is significantly driven by satellite data–driven maps and GPS-driven locations). These developments make it clear that not only do we need to understand the state of the art, but we also have a vision of where the future of remote sensing is headed. Thereby, in a nutshell, the goal of the *Remote Sensing Handbook* (Vols. I–VI) is to cover the developments and advancement of six distinct eras (listed next) in terms of data characterization and processing as well as myriad land and water applications:

Pre-civilian remote sensing era of the pre-1950s: World War I and II when remote sensing was a military tool;
Technology demonstration era of the 1950s and 1960s: Sputnik-I and NOAA AVHRR era of the 1950s and 1960s;
Landsat era of the 1970s: when the first truly operational land remote sensing satellite (Earth Resources Technology Satellite, or ERTS, later re-named Landsat) was launched and operated;
Earth observation era of the 1980s and 1990s: when a number of space agencies began launching and operating satellites (e.g., Landsat 4,5 by the USA; SPOT-1,2 by France; IRS-1a,1b by India)
Earth observation and new Millennium era of the 2000s: when data dissemination to users became as important as launching, operating, and capturing data (e.g., MODIS terra/acqua, Landsat-8, Resourcesat);
21st-Century era starting in the 2010s: when new generation micro/nano satellites or CubeSats (e.g., Planet Labs PBC, Skybox) and hyperspectral satellite sensors (e.g., Tanager-1, DESIS, PRISMA, EnMAP, upcoming NASA SBG) add to increasing constellations of multi-agency sensors (e.g., Sentinels; Landsat-8,9; upcoming Landsat-Next).

Motivation to take up editing the six-volume *Remote Sensing Handbook* (Second Edition) wasn't easy. It is a daunting work and requires extraordinary commitment over two to three years. After repeated requests from Ms. Irma Shagla-Britton, Manager and Leader for Remote Sensing and GIS books of Taylor and Francis/CRC Press, and considerable thought, I finally agreed to take the challenge in 2022. Having earlier edited the three-volume *Remote Sensing Handbook*, published in November 2015, I was pleased that the books were of considerable demand for a second edition. This was enough motivation. Further, I wanted to do something significant at this stage of my career that would make considerable contribution to the global remote sensing community. When I edited the first edition during 2012–2014, I was still recovering from colon cancer surgery and chemotherapy. But this second edition is a celebration of my complete recovery from the dreaded disease. I have not only fully recovered but have never felt so completely full of health and vigor. This naturally gave me sufficient energy and enthusiasm to back my motivation to edit this monumental six-volume *Remote Sensing Handbook*. At least for me this is the *magnum opus* that I feel proud to have accomplished and feel confident of the immense value for students, scientists, and professional practitioners of remote sensing who are interested in a standard reference on the subject. They will find this six-volume *Remote Sensing Handbook*: "Complete and comprehensive coverage of the state-of-art remote sensing, capturing the advances that have taken place over last 60+ years, which will set the stage for a vision for the future."

Above all, I am indebted to some 300+ authors and co-authors of the chapters who have spent so much of their creative energy to work on the chapters, deliver them in time, and patiently address all edits and comments. These are amongst the very best remote sensing scientists from around the world, extremely busy people who made time for the book project and have made outstanding contributions. I went back to everyone who contributed to the *Remote Sensing Handbook* (First Edition, three volumes) published in 2015 and requested that they revise their chapters. Most of the lead authors of the chapters agreed to revise, which was reassuring. However, some were not available, some retired, and some declined for other reasons. In such cases I adopted two strategies: (1) invite a few new chapter authors to make up for this gap and (2) update the chapters myself in other cases. I am convinced this strategy worked well to capture the latest information and maintain the integrity of every chapter. What was also important was to ensure the latest advances in remote sensing science were adequately covered. Authors of the chapters amazed me with their commitment and attention to detail. First, the quality of each of the chapters was of the highest standard. Second, with very few exceptions, chapters were delivered on time. Third, edited chapters were revised thoroughly and returned on time. Fourth, all my requests on various formatting and quality enhancements were addressed. My heartfelt gratitude to these great authors for their dedication to quality science. It has been my great honor and privilege to work with these dedicated legends. Indeed, I call them my "heroes" in a true sense. These are highly accomplished, renowned, pioneering scientists of the highest merit in remote sensing science, and I am ever grateful to have their time, effort, enthusiasm, and outstanding intellectual contributions. I am indebted to their kindness and generosity. In the end we had 300+ authors writing 91 chapters.

Overall, the *Remote Sensing Handbook* (Vols. I–VI) took about two years, from the time book chapters and authors were identified to the final publication of the book. The six volumes of the *Remote Sensing Handbook* were designed in such a way that a reader can have all six volumes as a standard reference or have individual volumes to study specific subject areas. The six volumes are:

Remote Sensing Handbook, Second Edition, Vol. I
Volume I: *Sensors, Data Normalization, Harmonization, Cloud Computing, and Accuracies—9781032890951*

Remote Sensing Handbook, Second Edition, Vol. II
Volume II: *Image Processing, Change Detection, GIS, and Spatial Data Analysis—9781032890975*

Remote Sensing Handbook; Second Edition, Vol. III
Volume III: *Agriculture, Food Security, Rangelands, Vegetation, Phenology, and Soils—9781032891019*

Remote Sensing Handbook; Second Edition, Vol. IV
Volume IV: *Forests, Biodiversity, Ecology, LULC, and Carbon—9781032891033*

Remote Sensing Handbook; Second Edition, Vol. V
Volume V: *Water Resources: Hydrology, Floods, Snow and Ice, Wetlands, and Water Productivity—9781032891453*

Remote Sensing Handbook; Second Edition, Vol. VI
Volume VI: *Droughts, Disasters, Pollution, and Urban Mapping—9781032891484*

There are 18, 17, 17, 12, 13, and 14 chapters, respectively, in the six volumes.

A wide array of topics are covered in the six volumes.

The topics covered in Volume I include: (1) satellites and sensors; (2) global navigation satellite systems (GNSS); (3) remote sensing fundamentals; (4) data normalization, harmonization, and standardization; (5) vegetation indices and their within and across sensor calibration; (6) crowdsourcing; (7) cloud computing; (8) Google Earth Engine–supported remote sensing; (9) accuracy assessments; and (10) remote sensing law.

The topics covered in Volume II include: (1) digital image processing fundamentals and advances; (2) digital image classifications for applications such as urban, land use, and land cover; (3) hyperspectral image processing methods and approaches; (4) thermal infrared image processing principles and practices; (5) image segmentation; (6) object-based image analysis (OBIA), including geospatial data integration techniques in OBIA; (7) image segmentation in specific applications like land use and land cover; (8) LiDAR digital image processing, (9) change detection; and (10) integrating geographic information systems (GIS) with remote sensing in geoprocessing workflows, democratization of GIS data and tools, fronters of GIScience, and GIS and remote sensing policies.

The topics covered in Volume III include: (1) vegetation and biomass, (2) agricultural croplands, (3) rangelands, (4) phenology and food security, and (5) soils.

The topics covered in Volume IV include: (1) forests, (2) biodiversity, (3) ecology, (4) land use/land cover, and (5) carbon. Under each of these broad topics, there are one or more chapters.

Volume V focuses on hydrology, water resources, ice, wetlands, and crop water productivity. The chapters are broadly classified into: (1) geomorphology, (2) hydrology and water resources, (3) floods, (4) wetlands, (5) crop water use and productivity, and snow and ice.

Volume VI focuses on water resources, disasters, and urban remote sensing. The chapters are broadly classified into: (1) droughts and drylands, (2) disasters, (3) volcanoes, (4) fires, and (5) nightlights.

There are many ways to use the *Remote Sensing Handbook* (Second Edition, six volumes). A lot of thought went into organizing the volumes and chapters, so you will see a "flow" from chapter to chapter and from volume to volume. As you read through the chapters, you will see how they are interconnected and how reading all of them provides you with greater in-depth understanding. You will also realize, as someone deeply interested in one of the topics, that you will have greater interest in one volume. Having all six volumes as reference material is ideal for any remote sensing expert, practitioner, or student. However, you can also refer to individual volumes based on your interest. We have also made great attempts to ensure that chapters are self-contained. That way, you can focus on a chapter and read it through, without having to be overly dependent on other chapters. Taking this perspective, there is some (~5–10%) material that may be repeated across chapters. This is done deliberately. For example, when you are reading a chapter on LiDAR or radar, you don't want to go all the way back to another chapter to understand the characteristics of these data. Similarly, certain indices (e.g., vegetation condition index, or VCI, and temperature condition index, or TCI) that are defined in one chapter (e.g., on drought) may be repeated in another chapter (also on drought). Such minor overlaps are helpful to readers to avoid going back to another chapter to understand a phenomenon, index, or characteristic of a sensor. However, if you want a lot of details of these sensors, indices, phenomenon, then you will have to read the appropriate chapter where there is in-depth coverage of the topic.

Each volume has a summary chapter (the last chapter of each volume). The summary chapter can be read two ways: (1) as the last chapter to recapture the main points of each of the previous chapters or (2) as an initial overview to get the first feeling for what is in the volume, before diving in to read each chapter in detail. I suggest readers do it both ways: read it first before reading chapters in detail to gather an idea of what to expect in each chapter and then read it at the end to recapture what was read in each of the previous chapters.

It has been a great honor as well as a humbling experience to edit the *Remote Sensing Handbook* (Vols. I–VI). I truly enjoyed the effort, albeit I felt overwhelmed at times with never-ending work. What an honor to work with luminaries in their field of expertise. I learned a lot from them and am very grateful for their support, encouragement, and deep insights. Also, it has been a pleasure working with the outstanding professionals at Taylor & Francis Group/CRC Press. There is no greater joy than being immersed in the pursuit of excellence, knowledge gain, and knowledge capture. At the same time, I am happy it is over. If there is a third edition a decade or so from now, it will be taken up by someone else (individually or as a team) and certainly not me!

I expect the book to be a standard reference of immense value to any student, scientist, professional, or practical practitioner of remote sensing. Any book that has the privilege of 300+ truly outstanding and dedicated remote sensing scientists ought to be a *magnum opus* deserving to be a standard reference on the subject.

Dr. Prasad S. Thenkabail, PhD
Editor-in-Chief (EiC)
Remote Sensing Handbook (Second Edition, Volumes I, II, III, IV, V, VI)

Volume 1: Sensors, Data Normalization, Harmonization, Cloud Computing, and Accuracies
Volume 2: Image Processing, Change Detection, GIS, and Spatial Data Analysis
Volume 3: Agriculture, Food Security, Rangelands, Vegetation, Phenology, and Soils
Volume 4: Forests, Biodiversity, Ecology, LULC, and Carbon
Volume 5: Water Resources: Hydrology, Floods, Snow and Ice, Wetlands, and Water Productivity
Volume 6: Droughts, Disasters, Pollution, and Urban Mapping

About the Editor

Dr. Prasad S. Thenkabail, PhD, is a senior scientist with the United States Geological Survey (USGS), specializing in remote sensing science for agriculture, water, and food security. He is a world-recognized expert in remote sensing science with multiple major contributions in the field sustained for 40+ years. Dr. Thenkabail has conducted pioneering research in hyperspectral remote sensing of vegetation, global croplands mapping for water and food security, and crop water productivity. His work on hyperspectral remote sensing of agriculture and vegetation are widely cited. His papers on hyperspectral remote sensing are the first of their kind and, collectively, they have: (1) determined optimal hyperspectral narrowbands (OHNBs) in the study of agricultural crops; (2) established hyperspectral vegetation indices (HVIs) to model and map biophysical and biochemical quantities of crops; (3) created a framework and sample data for the global hyperspectral imaging spectral libraries of crops (GHISA); (4) developed methods and techniques of overcoming Hughes's phenomenon; (5) demonstrated the strengths of hyperspectral narrowband (HNB) data in advancing classification accuracies relative to multispectral broadband (MBB) data; (6) showed advances one can make in modeling biophysical and biochemical quantities of crops using HNB and HVI data relative to MBB data; and (7) created a body of work in understanding, processing, and utilizing HNB and HVI data in agricultural cropland studies. This body of work has become a widely referred reference worldwide. In studies of global croplands for food and water security, he has led the release of the world's first 30-m Landsat satellite–derived global cropland extent product at 30 m (GCEP30; https://www.usgs.gov/apps/croplands/app/map) (Thenkabail et al., 2021; https://lpdaac.usgs.gov/news/release-of-gfsad-30-meter-cropland-extent-products/) and Landsat-derived global rainfed and irrigated area product at 30 m (LGRIP30; https://lpdaac.usgs.gov/products/lgrip30v001/) (Teluguntla and Thenkabail et al., 2023). He earlier led production of the world's first global irrigated area map (https://lpdaac.usgs.gov/products/lgrip30v001/ and https://lpdaac.usgs.gov/products/gfsad1kcmv001/) using multi-sensor satellite data that led to crops. The global cropland datasets using satellite remote sensing demonstrates a "paradigm shift" in global cropland mapping using remote sensing through big data analytics, machine learning, and petabyte-scale cloud computing on the Google Earth Engine (GEE). The LGRIP30 and GCEP30 products are released through NASA's LP DAAC and published in USGS's professional paper 1868 (Thenkabail et al., 2021). He has been principal investigator of many projects over the years, including the NASA-funded global food security support analysis data in the 30-m (GFSAD) project (www.usgs.gov/wgsc/gfsad30).

Thenkabail's career scientific achievements can be gauged by successfully making the list of the world's top 1% of scientists per the Stanford study ranking the world's scientists from across 22 scientific fields and 176 sub-fields. The fields were based on deep analysis evaluating the 1996–2023 SCOPUS data from Elsevier of about 10 million scientists (Ioannidis, 2023; Ioannidis et al., 2020). Dr. Thenkabail was recognized as Fellow of the American Society of Photogrammetry and Remote Sensing (ASPRS) in 2023. He has published over 150 peer-reviewed scientific papers and edited 15 books. His scientific papers have won several awards over the years, demonstrating world-class research of the highest quality. These include: the 2023 Talbert Abrams Grand Award, the highest scientific paper award of the ASPRS (with Itiya Aneece); the 2015 ASPRS ERDAS award for best scientific paper in remote sensing (with Michael Marshall); the 2008 John I. Davidson ASPRS President's Award for practical papers (with Pardhasaradhi Teluguntla); and the 1994 Autometric Award for outstanding paper in remote sensing (with Dr. Andy Ward).

Dr. Thenkabail's contributions to series of leading edited books places him as a world leader in remote sensing science. There are three seminal book-sets with a total of 13 volumes that he edited, and which have demonstrated his major contributions as an internationally acclaimed remote sensing scientist. These are: (1) *Remote Sensing Handbook* (Second Edition, six-volume book-set, 2024) with 91 chapters and nearly 3000 pages and for which he is the sole editor, (2) *Remote Sensing Handbook* (First Edition, three-volume book-set, 2015) with 82 chapters and 2304 pages and for which he is the sole editor, and (3)

Hyperspectral Remote Sensing of Vegetation (four-volume book-set, 2018) with 50 chapters and 1632 pages that he edited as the chief editor (co-editors: Prof. John Lyon and Prof. Alfredo Huete).

Dr. Thenkabail is at the center of rendering scientific service to the world's remote sensing community over his long period of service. This includes serving as Editor-in-Chief (2011–present) of *Remote Sensing Open Access Journal*; Associate Editor (2017–present) of *Photogrammetric Engineering and Remote Sensing (PE&RS)*; Editorial Advisory Board (2016–present) of the International Society of Photogrammetry and Remote Sensing (ISPRS), and Editorial Board Member (2007–2017) of *Remote Sensing of Environment* (RSE).

The USGS and NASA selected him as one of the three international members on the Landsat Science Team (2006–2011). He is an advisory board member of the online library collection to support the United Nations Sustainable Development Goals (UN SDGs) and is currently a scientist for the NASA and ISRO (Indian Space Research Organization) Professional Engineer and Scientist Exchange Program (PESEP) for 2022–2024. He was the chair of the International Society of Photogrammetry and Remote Sensing (ISPRS) Working Group WG VIII/7 (land cover and its dynamics) from 2013 to 2016; played a vital role for USGS as Global Coordinator, Agricultural Societal Beneficial Area (SBA), Committee for Earth Observation (CEOS) (2010–2013), during which he co-wrote the global food security case study for the CEOS *Earth Observation Handbook* (EOS), Special Edition for the UN Conference on Sustainable Development, presented in Rio de Janeiro, Brazil; and was the co-lead (2007–2011) of IEEE's "Water for the World" initiative, a non-profit effort funded by IEEE that worked in coordination with the Group on Earth Observations (GEO) in its GEO Water and GEO Agriculture initiatives.

Dr. Thenkabail worked as Postdoctoral Researcher and Research Faculty at the Center for Earth Observation (YCEO), Yale University (1997–2003) and led remote sensing programs in three international organizations:

- International Water Management Institute (IWMI), 2003–2008
- International Center for Integrated Mountain Development (ICIMOD), 1995–1997
- International Institute of Tropical Agriculture (IITA), 1992–1995

He began his scientific career as a scientist (1986–1988) working for the National Remote Sensing Agency (NRSA) (now renamed the National Remote Sensing Center, or NRSC), Indian Space Research Organization (ISRO), Department of State, Government of India.

Dr. Thenkabail's work experience spans over 25 countries, including East Asia (China), Southeast Asia (Cambodia, Indonesia, Myanmar, Thailand, Vietnam), the Middle East (Israel, Syria), North America (United States, Canada), South America (Brazil), Central Asia (Uzbekistan), South Asia (Bangladesh, India, Nepal, Sri Lanka), West Africa (Republic of Benin, Burkina Faso, Cameroon, Central African Republic, Cote d'Ivoire, Gambia, Ghana, Mali, Nigeria, Senegal, Togo), and Southern Africa (Mozambique, South Africa). Dr. Thenkabail is regularly invited as keynote speaker or invited speaker at major international conferences and at other important national and international forums every year.

Dr. Thenkabail obtained his PhD in agricultural engineering from the Ohio State University, USA in 1992 and has a master's degree in hydraulics and water resources engineering and a bachelor's degree in civil engineering (both from India). He has 168 publications, including 15 books; 175+ peer-reviewed journal articles, book chapters, and professional papers/monographs; and 15+ significant major global and regional data releases.

REFERENCES

Scientific papers

https://scholar.google.com/citations?user=9IO5Y7YAAAAJ&hl=en

USGS Professional Paper, Data & Product Gateways, Interactive Viewers

Ioannidis, J.P.A. (2023). October 2023 data-update for "Updated science-wide author databases of standardized citation indicators". *Elsevier Data Repository*, 6, https://doi.org/10.17632/btchxktzyw.6.

Ioannidis, J.P.A., Boyack, K.W., and Baas, J. (2020). Updated science-wide author databases of standardized citation indicators. *PLoS Biology,* 18(10), p. e3000918, https://doi.org/10.1371/journal.pbio.3000918.

Teluguntla, P., Thenkabail, P., Oliphant, A., Gumma, M., Aneece, I., Foley, D., and McCormick, R. (2023). Landsat-Derived Global Rainfed and Irrigated-Cropland Product @ 30-m (LGRIP30) of the World (GFSADLGRIP30WORLD). *The Land Processes Distributed Active Archive Center (LP DAAC) of NASA and USGS.* p. 103. https://lpdaac.usgs.gov/news/release-of-lgrip30-data-product/ (download data, documents).

Thenkabail, P.S., Teluguntla, P.G., Xiong, J., Oliphant, A., Congalton, R.G., Ozdogan, M., Gumma, M.K., Tilton, J.C., Giri, C., Milesi, C., Phalke, A., Massey, R., Yadav, K., Sankey, T., Zhong, Y., Aneece, I., and Foley, D. (2021). Global Cropland-Extent Product at 30-m Resolution (GCEP30) Derived from Landsat Satellite Time-Series Data for the Year 2015 Using Multiple Machine-Learning Algorithms on Google Earth Engine Cloud: U.S. *Geological Survey Professional Paper 1868*, 63 p., https://doi.org/10.3133/pp1868 (research paper). https://lpdaac.usgs.gov/news/release-of-gfsad-30-meter-cropland-extent-products/ (download data, documents). https://www.usgs.gov/apps/croplands/app/map (view data interactively).

Books

Remote Sensing Handbook (Second Edition, Six Volumes, 2024)

Thenkabail, Prasad. 2024. *Remote Sensing Handbook (Second Edition, Six Volume Book-set)*, Volume I: Sensors, Data Normalization, Harmonization, Cloud Computing, and Accuracies. Taylor and Francis Inc./CRC Press, Boca Raton, London, New York. 978-1-032-89095-1—CAT# T132478. Print ISBN: 9781032890951. eBook ISBN: 9781003541141. Pp. 581.

Thenkabail, Prasad. 2024. *Remote Sensing Handbook (Second Edition, Six Volume Book-set)*, Volume II: Image Processing, Change Detection, GIS, and Spatial Data Analysis. Taylor and Francis Inc./CRC Press, Boca Raton, London, New York. 978-1-032-89097-5—CAT# T133208. Print ISBN: 9781032890975. eBook ISBN: 9781003541158. Pp. 464.

Thenkabail, Prasad. 2024. *Remote Sensing Handbook (Second Edition, Six Volume Book-set)*, Volume III: Agriculture, Food Security, Rangelands, Vegetation, Phenology, and Soils. Taylor and Francis Inc./ CRC Press, Boca Raton, London, New York. 978-1-032-89101-9—CAT# T133213. Print ISBN: 9781032891019; eBook ISBN: 9781003541165. Pp. 788.

Thenkabail, Prasad. 2024. *Remote Sensing Handbook (Second Edition, Six Volume Book-set)*, Volume IV: Forests, Biodiversity, Ecology, LULC, and Carbon. Taylor and Francis Inc./CRC Press, Boca Raton, London, New York. 978-1-032-89103-3—CAT# T133215. Print ISBN: 9781032891033. eBook ISBN: 9781003541172. Pp. 501.

Thenkabail, Prasad. 2024. *Remote Sensing Handbook (Second Edition, Six Volume Book-set)*, Volume V: Water, Hydrology, Floods, Snow and Ice, Wetlands, and Water Productivity. Taylor and Francis Inc./CRC Press, Boca Raton, London, New York. 978-1-032-89145-3—CAT# T133261. Print ISBN: 9781032891453. eBook ISBN: 9781003541400. Pp. 516.

Thenkabail, Prasad. *Remote Sensing Handbook (Second Edition, Six Volume Book-set)*, Volume VI: Droughts, Disasters, Pollution, and Urban Mapping. Taylor and Francis Inc./CRC Press, Boca Raton, London, New York. 978-1-032-89148-4—CAT# T133267. Print ISBN: 9781032891484; eBook ISBN: 9781003541417. Pp. 467.

Hyperspectral Remote Sensing of Vegetation (Second Edition, Four Volumes, 2018)

Thenkabail, P.S., Lyon, G.J., and Huete, A. (Editors) (2018). Book Title: *Hyperspectral Remote Sensing of Vegetation* (Second Edition, Four Volume-set).

Volume I Title: *Fundamentals, Sensor Systems, Spectral Libraries, and Data Mining for Vegetation*. Publisher: CRC Press- Taylor and Francis group, Boca Raton, London, New York. Pp. 449, Hardback ID: 9781138058545; eBook ID: 9781315164151.

Volume II Title: *Hyperspectral Indices and Image Classifications for Agriculture and Vegetation*. Publisher: CRC Press- Taylor and Francis group, Boca Raton, London, New York. Pp. 296. Hardback ID: 9781138066038; eBook ID: 9781315159331.

Volume III Title: *Biophysical and Biochemical Characterization and Plant Species Studies*. Publisher: CRC Press- Taylor and Francis group, Boca Raton, London, New York. Pp. 348. Hardback: 9781138364714; eBook ID: 9780429431180.

Volume IV Title: *Advanced Applications in Remote Sensing of Agricultural Crops and Natural Vegetation*. Publisher: CRC Press- Taylor and Francis group, Boca Raton, London, New York. Pp. 386. Hardback: 9781138364769; eBook ID: 9780429431166.

Remote Sensing Handbook (First Edition, Three Volumes, 2015)

Thenkabail, P.S., (Editor-in-Chief), 2015. "Remote Sensing Handbook"

Volume I: *Remotely Sensed Data Characterization, Classification, and Accuracies*. Taylor and Francis Inc./ CRC Press, Boca Raton, London, New York. ISBN 9781482217865—CAT# K22125. Print ISBN: 978-1-4822-1786-5; eBook ISBN: 978-1-4822-1787-2. Pp. 678.

Volume II: *Land Resources Monitoring, Modeling, and Mapping with Remote Sensing*. Taylor and Francis Inc./ CRC Press, Boca Raton, London, New York. ISBN 9781482217957—CAT# K22130. Pp. 849.

Volume III: *Remote Sensing of Water Resources, Disasters, and Urban Studies*. Taylor and Francis Inc./CRC Press, Boca Raton, London, New York. ISBN 9781482217919—CAT# K22128. Pp. 673.

Hyperspectral Remote Sensing of Vegetation (First Edition, Single Volume, 2013)

Thenkabail, P.S., Lyon, G.J., and Huete, A. (Editors) (2012). Book entitled: *Hyperspectral Remote Sensing of Vegetation*. CRC Press/Taylor and Francis group, Boca Raton, London, New York. Pp. 781 (80+ pages in color). http://www.crcpress.com/product/isbn/9781439845370

Remote Sensing of Global Croplands for Food Security (First Edition, Single Volume, 2009)

Thenkabail, P., Lyon, G.J., Turral, H., and Biradar, C.M. (Editors) (2009). Book entitled: *Remote Sensing of Global Croplands for Food Security*. CRC Press/Taylor and Francis Group, Boca Raton, London, New York. Pp. 556 (48 pages in color). Published in June, 2009.

FIGURE Snap shots of the Editor-in-Chief's work and life.

Contributors

Gregory P. Asner
Department of Global Ecology
Carnegie Institution for Science
Stanford, CA, USA
and
Center for Global Discovery and Conservation Science
Arizona State UniversityHilo, HI, USA

Palma Blonda
Italian National Research Council CNR-Institute of Intelligent Systems for Automation ISSIA
Bari, Italy

Marc Bouvier
UMR TETIS Irstea-Cirad-AgroParisTech/ENGREF
Maison de la Télédétection
Montpellier, France

Mariana Campos
PhD in Photogrammetry
Senior Researcher at Finnish Geospatial Research Institute FGI
National Land Survey of Finland
Brazil

Jing Ming Chen
Professor
Department of Geography and Planning
University of Toronto
100 St. George St., Room 5047
Toronto, Ontario, Canada M5S 3G3

Christina Corbane
Irstea—UMR TETIS
Montpellier, France
and
European Commission—Joint Research Centre
Institute for the Protection and Security of the Citizen (IPSC)
Global Security and Crisis Management (GlobeSEC) Unit
Italy

Holly Croft
Department of Geography and Program in Planning
University of Toronto
Toronto, Ontario, Canada

Matthew Dannenberg
Department of Geography
University of North Carolina at Chapel Hill
Chapel Hill, NC, USA

Sylvie Durrieu
Irstea—UMR TETIS
 Irstea-Cirad-AgroParisTech/ENGREF
Maison de la Télédétection
Montpellier, France

Michael Forster
Technische Universität Berlin (TU)
Geoinformation in Environmental Planning
EB 5
Strasse des 17. Juni 145
10623 Berlin

Thomas W. Gillespie
Professor
University of California, Los Angeles
Los Angeles, CA, USA

Valéry Gond
Forest Ecosystems Goods and Services,
 Agricultural Research for Development
Campus de Baillarguet
Montpellier Cedex, France

Alemu Gonsamo
Department of Geography and Program in Planning
University of Toronto
Toronto, Ontario, Canada

Nicholas R. Goodwin
Research Fellow-Lidar Specialist
School of the Environment
Science Division, Department of Science, IT, Innovation and the Arts (DSITIA)
Brisbane, Queensland, Australia

Frédéric Gosselin
Irstea
Domaine des Barres
Nogent sur Vernisson, France

Christopher Hakkenberg
Curriculum for the Environment and Ecology
University of North Carolina at Chapel Hill
Chapel Hill, NC, USA

E.H. Helmer
International Institute of Tropical Forestry,
 USDA Forest Service
Jardín Botánico Sur
Río Piedras, Puerto Rico, USA

Markus Hollaus
Senior Scientist Dipl.-Ing. Dr.techn
E120-07-Forschungsbereich Photogrammetrie
Wiedner Hauptstraße 8
1040 Wien
Raumnummer DC 02 D30

Markus Holopainen
Professor
Department of Forest Sciences
University of Helsinki
Viikinkaari 1, Biocentre 3
00790 Helsinki
Finland

R.A. Houghton
Woods Hole Research Center
Falmouth, MA, USA

Taehee Hwang
Department of Geography
Indiana University
Bloomington, IN, USA

Eric Hyyppä
National Land Survey of Finland
Finnish Geospatial Research Institute
Espoo, Finland

Hannu Hyyppä
Aalto University
School of Engineering
Espoo, Finland

Juha Hyyppä
National Land Survey of Finland
Finnish Geospatial Research Institute
Espoo, Finland

Matti Hyyppä
National Land Survey of Finland
Finnish Geospatial Research
 Institute
Espoo, Finland

Anttoni Jaakkola
Cruise
San Francisco, USA

Harri Kaartinen
National Land Survey of Finland
Finnish Geospatial Research Institute
Espoo, Finland

Ville Kankare
University of Turku
Department of Geography and Geology
Turku, Finland

Kirsi Karila
National Land Survey of Finland
Finnish Geospatial Research Institute, Espoo,
 Finland

Mika Karjalainen
National Land Survey of Finland
Finnish Geospatial Research Institute
Espoo, Finland

Masato Katoh
Shinshu University
Faculty of Agriculture

Antero Kukko
National Land Survey of Finland
Finnish Geospatial Research Institute
Espoo, Finland

Stefan Lang
Department of Geoinformatics—Z_GIS
Paris Lodron University Salzburg
Salzburg, Austria

Juxiang Li
College of Ecology and Environmental
 Sciences
East China Normal University
Shanghai, P. R. China

Contributors

Xia Li
School of Geographic Sciences
Key Lab. of Geographic Information Science
 (Ministry of Education)
East China Normal University
Shanghai, P.R. China

Xinlian Liang
Wuhan University
State Key Laboratory of Information
 Engineering in Surveying
Mapping and Remote Sensing
Wuhan, China

Qianwen Lv
School of Geography and Planning
Sun Yat-sen University
Guangzhou, P. R. China

Roberta E. Martin
Center for Global Discovery and Conservation
 Science
Arizona State University
Hilo, HI, USA

Leena Matikainen
National Land Survey of Finland
Finnish Geospatial Research Institute
Espoo, Finland

Nathan Mietkiewicz
Graduate School of Geography
Clark University
Worcester, MA, USA

Wenge Ni-Meister
Department of Geography
Hunter College
New York, NY, USA
and
Doctoral Program in Earth and
 Environmental Sciences,
 Graduate Center
The City University of New York
New York, NY, USA

Kyle Pipkins
Geoinformation in Environmental Planning Lab
Technische Universität Berlin
Berlin, Germany

Eetu Puttonen
National Land Survey of Finland
Finnish Geospatial Research Institute
Espoo, Finland

Jiri Pyörälä
National Land Survey of Finland
Finnish Geospatial Research Institute
Espoo, Finland

Jean-Pierre Renaud
Office National des Forêts
Nancy, France

Zhixin Qi
School of Geography and Planning
Sun Yat-sen University
Guangzhou, P. R. China

Chelsea Robinson
Department of Geography
University of California, Los Angeles
Los Angeles, CA, USA

Duccio Rocchini
BIOME Lab, Department of Biological,
 Geological and Environmental Sciences
Alma Mater Studiorum University of Bologna
Bologna, Italy
and
Faculty of Environmental Sciences
Department of Spatial Sciences
Czech University of Life Sciences Prague
Praha—Suchdol, Czech Republic

John Rogan
Graduate School of Geography
Clark University
Worcester, MA, USA

Morgan Rogers
Department of Urban Planning, Luskin School
 of Public Affairs
University of California, Los Angeles
Los Angeles, CA, USA

Ninni Saarinen
Laurent Saint-André
INRA UR BEF and CIRAD UMR
 Eco&Sols

Unité Biogéochimie des Ecosystèmes Forestiers
Champenoux, France

Conghe Song
Department of Geography
University of North Carolina at Chapel Hill
Chapel Hill, NC, USA

Carlos M. Souza Jr.
Instituto do Homen e Meio Ambiente (Imazon)
Rua Domingos Marreiros
Belém, Brazil

Josef Taher
Prasad Thenkabail
US Geological Survey (USGS)
Flagstaff, AZ, USA

Philip Townsend
Department of Forest and Wildlife Ecology
University of Wisconsin
Madison, WI, USA

Susan L. Ustin
Department of Land, Air, and Water Resources
University of California
Davis, CA, USA

Matti Vaaja
Aalto University
School of Engineering
Espoo, Finland

Mikko Vastaranta
University of Eastern Finland
School of Forest Sciences
Joensuu, Finland

Cédric Vega
Institut National de l'Information Geographique et Forestière, Laboratoire de l'Inventaire Forestier Nancy, France
and
Institut Français de Pondichéry
Pondicherry, India

Juho-Pekka Virtanen

Yunsheng Wang

Joanne C. White
Natural Resources Canada
Canadian Forest Service
Victoria, British Columbia, Canada

Mike Wulder
Natural Resources Canada
Canadian Forest Service
Victoria, British Columbia, Canada

Anthony Gar-On Yeh
Department of Urban Planning and Design
The University of Hong Kong
P. R. China

Tuomas Yrttimaa
University of Eastern Finland
School of Forest Sciences
Joensuu, Finland

Xiaowei Yu
National Land Survey of Finland,
Finnish Geospatial Research Institute,
Espoo, Finland

Quanfa Zhang
Key Laboratory of Aquatic Botany and Watershed Ecology
Wuhan Botanical Garden
Chinese Academy of Sciences
Wuhan, P. R. China

Yulong Zhang
Department of Geography
University of North Carolina at Chapel Hill
Chapel Hill, NC, USA
and
Key Laboratory of Aquatic Botany and Watershed Ecology
Wuhan Botanical Garden Chinese Academy of Sciences
Wuhan, P. R. China

Acknowledgments

The *Remote Sensing Handbook* (Second Edition, Vols. I–VI) brought together a galaxy of highly accomplished, renowned remote sensing scientists, professionals, and legends from around the world. The lead authors were chosen by me after careful review of their accomplishments and sustained publication record over the years. The chapters in the second edition were written/revised over a period of two years. All chapters were edited and revised.

Gathering such a galaxy of authors was the biggest challenge. These are all extremely busy people, and committing to a book project that requires substantial workload is never easy. However, almost all of those whom I requested agreed to write a chapter specific to their area of specialization, and only a few I had to convince to make time. The quality of the chapters should convince readers why these authors are such highly rated professionals and why they are so successful and accomplished in their field of expertise. They not only wrote very high-quality chapters, but delivered them on time, addressed any editorial comments in a timely manner without complaints, and were extremely humble and helpful. Their commitment to quality science is what makes them special. I am truly honored to have worked with such great professionals.

I would like to mention the names of everyone who contributed and made the *Remote Sensing Handbook* (Second Edition, Vols. I–VI) possible. In the end, we had 91 chapters, a little over 3000 pages, and a little over 300+ authors. My gratitude goes to each one of them. These are many of the globally well-known "who's who" in remote sensing science. A list of all authors is provided next. The names of the authors are organized chronologically for each volume and the chapters within them. Each lead author of the chapter is in bold. **The names of the 400+ authors who contributed to six volumes are as follows:**

Volume I: Sensors, Data Normalization, Harmonization, Cloud Computing, and Accuracies: 18 chapters written by 53 authors (Editor-in-chief: Prasad S. Thenkabail):

Drs. Sudhanshu S. Panda, Mahesh Rao, Prasad S. Thenkabail, Debasmita Misra, and James P. Fitzerald; **Mohinder S. Grewal**; **Kegen Yu**, Chris Rizos, and Andrew Dempster; **D. Myszor**, O. Antemijczuk, M. Grygierek, M. Wierzchanowski, and K.A. Cyran; **Natascha Oppelt** and Arnab Muhuri; **Philippe M. Teillet**; **Philippe M. Teillet** and Gyanesh Chander; **Rudiger Gens** and Jordi Cristóbal Rosselló; **Aolin Jia**, Dongdong Wang; **Tomoaki Miura**, Kenta Obata, Hiroki Yoshioka, and Alfredo Huete; **Michael D. Steven**, Timothy J. Malthus, and Frédéric Baret; **Fabio Dell'Acqua** and Silvio Dell'Acqua; **Ramanathan Sugumaran**, James W. Hegeman, Vivek B. Sardeshmukh, and Marc P. Armstrong; **Lizhe Wang**, Jining Yan, Yan Ma, Xiaohui Huang, Jiabao Li, Sheng Wang, Haixu He, Ao Long, and Xiaohan Zhang; **John E. Bailey** and Josh Williams; **Russell G. Congalton**; **P.J. Blount**; **Prasad S. Thenkabail**.

Volume II: Image Processing, Change Detection, GIS, and Spatial Data Analysis: 17 chapters written by 64 authors (Editor-in-chief: Prasad S. Thenkabail):

Sunil Narumalani and Paul Merani; **Mutlu Ozdogan**; **Soe W. Myint**, Victor Mesev, Dale Quattrochi, and Elizabeth A. Wentz; **Jun Li**, Paolo Gamba, and Antonio Plaza; **Qian Du**, Chiranjibi Shah, Hongjun Su, and Wei Li; **Claudia Kuenzer**, Philipp Reiners, Jianzhong Zhang, and Stefan Dech; **Mohammad D. Hossain** and Dongmei Chen; **Thomas Blaschke**, Maggi Kelly, Helena Merschdorf; **Stefan Lang** and Dirk Tiede; **James C. Tilton**, Selim Aksoy, and Yuliya Tarabalka; **Shih-Hong Chio**, Tzu-Yi Chuang, Pai-Hui Hsu, Jen-Jer Jaw, Shih-Yuan Lin, Yu-Ching Lin, Tee-Ann Teo, Fuan Tsai, Yi-Hsing Tseng, Cheng-Kai Wang, Chi-Kuei Wang, Miao Wang, and Ming-Der Yang; **Guiying Li**, Mingxing Zhou, Ming Zhang, and Dengsheng Lu; **Jason A. Tullis**, David P. Lanter, Aryabrata Basu, Jackson D. Cothren, Xuan Shi, W. Fredrick Limp, Rachel F. Linck, Sean G. Young, Jason Davis, and Tareefa S. Alsumaiti; **Gaurav Sinha**, Barry J. Kronenfeld, and Jeffrey

C. Brunskill; **May Yuan**; **Stefan Lang**, Stefan Kienberger, Michael Hagenlocher, Lena Pernkopf; **Prasad S. Thenkabail**.

Volume III: Agriculture, Food Security, Rangelands, Vegetation, Phenology, and Soils: 17 chapters written by 110 authors (Editor-in-chief: Prasad S. Thenkabail):
  Alfredo Huete, Guillermo Ponce-Campos, Yongguang Zhang, Natalia Restrepo-Coupe, and Xuanlong Ma; **Juan Quiros-Vargas**, Bastian Siegmann, Juliane Bendig, Laura Verena Junker-Frohn, Christoph Jedmowski, David Herrera, and Uwe Rascher; **Frédéric Baret**; **Lea Hallik**, Egidijus Šarauskis, Ruchita Ingle, Indrė Bručienė, Vilma Naujokienė, and Kristina Lekavičienė; **Clement Atzberger** and Markus Immitzer; **Agnès Bégué**, Damien Arvor, Camille Lelong, Elodie Vintrou, and Margareth Simoes; **Pardhasaradhi Teluguntla**, Prasad S. Thenkabail, Jun Xiong, Murali Krishna Gumma, Chandra Giri, Cristina Milesi, Mutlu Ozdogan, Russell G. Congalton, James Tilton, Temuulen Tsagaan Sankey, Richard Massey, Aparna Phalke, and Kamini Yadav; **Yuxin Miao**, David J. Mulla, and Yanbo Huang; **Baojuan Zheng**, James B. Campbell, Guy Serbin, Craig S.T. Daughtry, Heather McNairn, and Anna Pacheco; **Prasad S. Thenkabail**, Itiya Aneece, Pardhasaradhi Teluguntla, Richa Upadhyay, Asfa Siddiqui, Justin George Kalambukattu, Suresh Kumar, Murali Krishna Gumma, and Venkateswarlu Dheeravath; **Matthew C. Reeves**, Robert Washington-Allen, Jay Angerer, Raymond Hunt, Wasantha Kulawardhana, Lalit Kumar, Tatiana Loboda, Thomas Loveland, Graciela Metternicht, Douglas Ramsey, Joanne V. Hall, Trenton Benedict, Pedro Millikan, Angus Retallack, Arjan J.H. Meddens, William K. Smith, and Wen Zhang; **E. Raymond Hunt Jr.**, Cuizhen Wang, D. Terrance Booth, Samuel E. Cox, Lalit Kumar, and Matthew C. Reeves; **Lalit Kumar**, Priyakant Sinha, Jesslyn F. Brown, R. Douglas Ramsey, Matthew Rigge, Carson A. Stam, Alexander J. Hernandez, E. Raymond Hunt Jr., and Matt Reeves; **Molly E. Brown**, Kirsten de Beurs, and Kathryn Grace; **José A.M. Demattê**, Cristine L.S. Morgan, Sabine Chabrillat, Rodnei Rizzo, Marston H.D. Franceschini, Fabrício da S. Terra, Gustavo M. Vasques, Johanna Wetterlind, Henrique Bellinaso, and Letícia G. Vogel; **E. Ben-Dor**, J.A.M. Demattê; **Prasad S. Thenkabail**.

Volume IV: Forests, Biodiversity, Ecology, LULC, and Carbon: 12 chapters written by 71 authors (Editor-in-chief: Prasad S. Thenkabail)
  E.H. Helmer, Nicholas R. Goodwin, Valéry Gond, Carlos M. Souza Jr., and Gregory P. Asner; **Juha Hyyppä**, Xiaowei Yu, Mika Karjalainen, Xinlian Liang, Anttoni Jaakkola, Mike Wulder, Markus Hollaus, Joanne C. White, Mikko Vastaranta, Jiri Pyörälä, Tuomas Yrttimaa, Ninni Saarinen, Josef Taher, Juho-Pekka Virtanen, Leena Matikainen, Yunsheng Wang, Eetu Puttonen, Mariana Campos, Matti Hyyppä, Kirsi Karila, Harri Kaartinen, Matti Vaaja, Ville Kankare, Antero Kukko, Markus Holopainen, Hannu Hyyppä, Masato Katoh, and Eric Hyyppä; **Gregory P. Asner**, Susan L. Ustin, Philip A. Townsend, and Roberta E. Martin; **Sylvie Durrieu**, Cédric Vega, Marc Bouvier, Frédéric Gosselin, Jean-Pierre Renaud, and Laurent Saint-André; **Thomas W. Gillespie**, Morgan Rogers, Chelsea Robinson, and Duccio Rocchini; **Stefan Lang**, Christina Corbane, Palma Blonda, Kyle Pipkins, and Michael Förster; **Conghe Song**, Jing Ming Chen, Taehee Hwang, Alemu Gonsamo, Holly Croft, Quanfa Zhang, Matthew Dannenberg, Yulong Zhang, Christopher Hakkenberg, and Juxiang Li; **John Rogan** and Nathan Mietkiewicz; **Zhixin Qi**, Anthony Gar-On Yeh, Xia Li, and Qianwen Lv; **R.A. Houghton**; **Wenge Ni-Meister**; **Prasad S. Thenkabail**.

Volume V: Water Resources: Hydrology, Floods, Snow and Ice, Wetlands, and Water Productivity: 13 chapters written by 60 authors (Editor-in-chief: Prasad S. Thenkabail)
  James B. Campbell and Lynn M. Resler; **Sadiq I. Khan**, Ni-Bin Chang, Yang Hong, Xianwu Xue, and Yu Zhang; **Santhosh Kumar Seelan**; **Allan S. Arnesen**, Frederico T. Genofre, Marcelo P. Curtarelli, and Matheus Z. Francisco; **Allan S. Arnesen**, Frederico T. Genofre, Marcelo P. Curtarelli, and Matheus Z. Francisco; **Sandro Martinis**, Claudia Kuenzer, and André Twele; **Le Wang**, Jing Miao, and Ying Lu; **Chandra Giri**; **D.R. Mishra**, X. Yan, S. Ghosh, C. Hladik,

J.L. O'Connell, and H.J. Cho; **Murali Krishna Gumma**, Prasad S. Thenkabail, Pranay Panjala, Pardhasaradhi Teluguntla, Birhanu Zemadim Birhanu, and Mangi Lal Jat; **Trent W. Biggs**, Pamela Nagler, Anderson Ruhoff, Triantafyllia Petsini, Michael Marshall, George P. Petropoulos, Camila Abe, and Edward P. Glenn; **Antônio Teixeira**, Janice Leivas, Celina Takemura, Edson Patto, Edlene Garçon, Inajá Sousa, André Quintão, Prasad Thenkabail, and Ana Azevedo; **Hongjie Xie**, Tiangang Liang, Xianwei Wang, Guoqing Zhang, Xiaodong Huang, and Xiongxin Xiao; **Prasad. S. Thenkabail**.

Volume VI: Droughts, Disasters, Pollution, and Urban Mapping: 14 chapters written by 53 authors (Editor-in-chief: Prasad S. Thenkabail)
  Felix Kogan and Wei Guo; **F. Rembold**, M. Meroni, O. Rojas, C. Atzberger, F. Ham, and E. Fillol; **Brian D. Wardlow**, Martha A. Anderson, Tsegaye Tadesse, Mark S. Svoboda, Brian Fuchs, Chris R. Hain, Wade T. Crow, and Matt Rodell; **Jinyoung Rhee**, Jungho Im, and Seonyoung Park; **Marion Stellmes**, Ruth Sonnenschein, Achim Röder, Thomas Udelhoven, Gabriel del Barrio, and Joachim Hill; **Norman Kerle**; **Stefan Lang**, Petra Füreder, Olaf Kranz, Brittany Card, Shadrock Roberts, Andreas Papp; **Robert Wright**; **Krishna Prasad Vadrevu** and Kristofer Lasko; **Anupma Prakash**, Claudia Kuenzer, Santosh K. Panda, Anushree Badola, and Christine F. Waigl; **Hasi Bagana**, Chaomin Chena, and Yoshiki Yamagata; **Yoshiki Yamagata**, Daisuke Murakami, Hajime Seya, and Takahiro Yoshida; **Qingling Zhang**, Noam Levin, Christos Chalkias, Husi Letu, and Di Liu; **Prasad S. Thenkabail**.
  The authors not only delivered excellent chapters, but they also provided valuable insights and inputs for me in many ways throughout the book project.
  I was delighted when **Dr. Compton J. Tucker**, Senior Earth Scientist, Earth Sciences Division, Science and Exploration Directorate, NASA Goddard Space Flight Center (GSFC) agreed to write the foreword for the book. For anyone practicing remote sensing, Dr. Tucker needs no introduction. He has been a "godfather" of remote sensing and has inspired a generation of remote sensing scientists. I have been a student of his without ever really being one. I mean, I have not been his student in the classroom, but I have followed his legendary work throughout my career. I remember reading his highly cited paper (now with citations nearing 7700!):

- Tucker, C.J. (1979) 'Red and Photographic Infrared Linear Combinations for Monitoring Vegetation,' *Remote Sensing of Environment,* **8(2)**, 127–150.

I first read this paper in 1986 when I had just joined the National Remote Sensing Agency (NRSA; now NRSC), Indian Space Research Organization (ISRO). Dr. Tucker's pioneering works have been a guiding light for me ever since. In 1975, after getting his PhD from Colorado State University, Dr. Tucker joined NASA GSFC as a postdoctoral fellow and became a full-time NASA employee in 1977. Ever since, he has conducted several path-finding studies. He has used NOAA AVHRR, MODIS, SPOT Vegetation, and Landsat satellite data for studying deforestation, habitat fragmentation, desert boundary determination, ecologically coupled diseases, terrestrial primary production, glacier extent, and how climate affects global vegetation. He has authored or coauthored more than 280 journal articles that have been cited more than 93,000 times, he is an adjunct professor at the University of Maryland and a consulting scholar at the University of Pennsylvania's Museum of Archaeology and Anthropology, and he has appeared in more than 20 radio and TV programs. He is a fellow of the American Geophysical Union and has been awarded several medals and honors, including NASA's Exceptional Scientific Achievement Medal, the Pecora Award from the US Geological Survey, the National Air and Space Museum Trophy, the Henry Shaw Medal from the Missouri Botanical Garden, the Galathea Medal from the Royal Danish Geographical Society, and the Vega Medal from the Swedish Society of Anthropology and Geography. He was the NASA representative to the US Global Change Research Program from 2006 to 2009. He is instrumental in releasing the AVHRR 33-year (1982–2014) Global Inventory Monitoring and Modeling Studies

(GIMMS) data. I strongly recommend that everyone reads his excellent foreword before reading this book. In the foreword, Dr. Tucker demonstrates the importance of data from Earth Observation (EO) sensors from orbiting satellites to maintaining a reliable and consistent climate record. Dr. Tucker further highlights the importance of continued measurements of these variables of our planet in the new millennium through new, improved, and innovative EO sensors from sun synchronous and/or geostationary satellites.

I want to acknowledge with thanks for the encouragement and support received by my US Geological Survey (USGS) colleagues. I would like to mention the late Mr. Edwin Pfeifer, Dr. Susan Benjamin (my director at the Western Geographic Science Center), Dr. Dennis Dye, Mr. Larry Gaffney, Mr. David F. Penisten, Ms. Emily A. Yamamoto, Mr. Dario D. Garcia, Mr. Miguel Velasco, Dr. Chandra Giri, Dr. Terrance Slonecker, Dr. Jonathan Smith, Timothy Newman, and Zhouting Wu. Of course, my dear colleagues at USGS, Dr. Pardhasaradhi Teluguntla, Dr. Itiya Aneece, Mr. Adam Oliphant, and Mr. Daniel Foley, have helped me in numerous ways. I am ever grateful for their support and significant contributions to my growth and this body of work. Throughout my career, there have been many postdoctoral-level scientists who have worked with me closely and contributed to my scientific growth in different ways. They include Dr. Murali Krishna Gumma, Head of Remote Sensing at the International Crops Research Institute for the Semi-Arid Tropics; Dr. Jun Xiong, Geo ML ≠ ML with GeoData, Climate Corp.; Dr. Michael Marshall, Associate Professor, University of Twente, Netherlands; Dr. Isabella Mariotto, Former USGS postdoctoral researcher; Dr. Chandrashekar Biradar, Country Director, India for World Agroforestry; and numerous others. I am thankful for their contributions. I know I am missing many names: too numerous to mention them all, but my gratitude for them is the same as the names I have mentioned here.

There is a very special person I am very thankful for: the late Dr. Thomas Loveland. I first met Dr. Loveland at USGS, Sioux Falls for an interview to work for him as a scientist in the late 1990s when I was still at Yale University. But even though I was selected, I was not able to join him, as I was not a citizen of the United States at that time and working for USGS required that. He has been my mentor and pillar of strength for over two decades, particularly during my Landsat Science Team days (2006–2011) and later once I joined USGS in 2008. I have watched him conduct Landsat Science Team meetings with great professionalism, insight, and creativity. I remember him telling my PhD advisor on me being hired at USGS: "we don't make mistakes!" During my USGS days, he was someone I could ask for guidance and seek advice from, and he would always be there to respond with kindness and understanding. Above all he shared his helpful insights. It is sad that we lost him too early. I pray for his soul. Thank you, Tom, for your kindness and generosity.

Over the years, there have been numerous people who have come into my professional life who have helped me grow. It is a tribute to their guidance, insights, and blessings that I am here today. In this regard I need to mention a few names out of gratitude: (1) Prof. G. Ranganna, my master's thesis advisor at the National Institute of Technology (NIT) in Surathkal, Karnataka, India. Prof. Ranganna is 92 years old (2024), and to this day he is my guiding light on how to conduct oneself with fairness and dignity in professional and personal conduct. Prof. Ranganna's trait of selflessly caring for his students throughout his life is something that influenced me to follow. (2) Prof. E.J. James, former Director of the Center for Water Resources Development and Management (CWRDM) in Calicut, Kerala, India. Prof. James was my master's thesis advisor in India, and his dynamic personality in professional and personal matters had an influence on me. Dr. James always went out of his way to help his students despite his busy schedule. (3) The late Dr. Andrew Ward, my PhD advisor at the Ohio State University in Columbus, Ohio. He funded my PhD studies in the United States through grants. Through him I learned how to write scientific papers and how to become a thorough professional. He was a tough task maker, your worst critic (to help you grow), but also a perfectionist who helped you grow as a peerless professional and, above all, a very kind human being at the core. He would write you long memos on flaws in your research but then help you out of it by making you work double the time! To make you work harder, he would tell you, "You won't get my sympathy." Then when you accomplished your goals, he would tell you, "you have paid back for your scholarship many times over!" (4) Dr. John G. Lyon, also my PhD advisor at the Ohio State University. He was a peerless motivator who encouraged you to believe in yourself. (5) Dr. Thiruvengadachari, Scientist at the National Remote Sensing Agency (NRSA), which is now the National Remote Sensing Center (NRSC) in India. He was my first boss at the Indian Space Research Organization (ISRO), and through him I learned the initial steps in remote sensing science. I was just 25 years old then and had joined NRSA after earning my Master of Engineering (hydraulics and water resources) and Bachelor of Engineering (civil engineering) degrees. The first day in the office Dr. Thiruvengadachari asked me how much remote sensing I knew. I told him "zero" and instantly thought he would ask me to leave the room. But his response was "very good!" and he gave me a manual on remote sensing from Purdue University's Laboratory for Applications of Remote Sensing (LARS) to study. Those were the days before there was formal training in remote sensing at universities. So, my remote sensing lessons began by working practically on projects, and one of our first projects was "drought monitoring for India using NOAA AVHRR data." This was an intense period of learning the fundamentals of remote sensing science for me by practicing on a daily basis. Data came in 9 mm tapes and was read on massive computing systems, image processing was done mostly during night shifts by booking time on a centralized computer, fieldwork was conducted using false color composite (FCC) outputs and topographic maps (there was no GPS), geographic information system (GIS) was in its infancy, a lot of calculations were done using calculators, and we had just started working in IBM 286 computers with floppy disks. So, when I decided to resign my NRSA job and go to the United States to do my PhD, Dr. Thiruvengadachari told me, "Prasad, I am losing my right hand, but you can't miss this opportunity." Those initial wonderful days of learning from Dr. Thiruvengadachari will remain etched in my memory. I am also thankful to my very good friend Shri C.J. Jagadeesha, who was my colleague at NRSA/NRSC, ISRO. He was a friend who encouraged me to grow as a remote sensing scientist through our endless rambling discussions over tea in Iranian restaurants outside NRSA those days and elsewhere.

I am ever grateful to my former professors at the Ohio State University: the late Prof. Carolyn Merry, Dr. Duane Marble, and Dr. Michael Demers. They have taught and/or encouraged, inspired, and given me opportunities at the right time. The opportunity to work for six years at the Center for Earth Observation at Yale University (YCEO) was incredibly important. I am thankful to Prof. Ronald G. Smith, Director of YCEO, for the opportunity, guidance, and kindness. At YCEO I advanced myself as a remote sensing scientist. The opportunities I got working for the International Institute of Tropical Agriculture (IITA), based in Nigeria, and the International Water Management Institute (IWMI), based in Sri Lanka, where I worked on remote sensing science pertaining to a number of applications such as agriculture, water, wetlands, food security, sustainability, climate, natural resources management, environmental issues, droughts, and biodiversity were extremely important in my growth as a remote sensing scientist—especially from the point of view of understanding the real issues on the ground in real-life situations. Finding solutions and applying one's theoretical understanding to practical problems and seeing them work has its own nirvana.

As it is clear from the previous paragraphs, it is of great importance to have guiding pillars of light at crucial stages of your education. It is how you become who you are in the end, how you grow and make your own contributions. I am so blessed to have had these wonderful guiding lights come into my professional life at the right moments of my career (which also influenced me positively in my personal life). From that firm foundation, I could build on what I learned and through the confidence of knowledge and accomplishments pursue my passion for science and do several significant pioneering studies throughout my career.

I mention all these people out of gratitude for my ability today to edit this monumental *Remote Sensing Handbook* (Second Edition, Vols. I–VI).

I am very thankful to Ms. Irma Shagla-Britton, Manager and Leader for Remote Sensing and GIS books at Taylor & Francis/CRC Press. Without her consistent encouragement to take on this responsibility of editing the *Remote Sensing Handbook*, especially in trusting me to accomplish this momentous work over so many other renowned experts, I would never have gotten to work on this in the first place. Thank you, Irma. Sometimes you need to ask several times before one can say yes to something!

I am very grateful to my wife (Sharmila Prasad), my daughter (Spandana Thenkabail), and my son-in-law (Tejas Mayekar) for their usual unconditional understanding, love, and support. My wife and daughter have always been pillars of my life, now joined by my equally loving son-in-law. I learned the values of hard work and dedication from my revered parents. This work wouldn't come through without their life of sacrifices to educate their children and their silent blessings. My father's emphasis on education and sending me to the best schools to study despite our family's very modest income, as well as my mother's endless hard work, are my guiding light and inspiration. Of course, there are many, many others to be thankful for, but there are too many to mention here. Finally, it must be noted that a work of this magnitude, editing the monumental *Remote Sensing Handbook* (Second Edition, Vols. I–VI), continuing from the three-volume first edition, requires almighty blessings. I firmly believe nothing happens without the powers of the universe blessing you and providing needed energy, strength, health, and intelligence. To that infinite power my humble submission of everlasting gratefulness.

It has been my deep honor and great privilege to have edited the *Remote Sensing Handbook* (Second Edition, Vols. I–VI) after having edited the three-volume first edition that was published in 2015. Now, after nearly ten years, we will have a six-volume second edition in 2024. A huge thanks to all the authors, the publisher, my family and friends, and everyone else who made this huge task possible.

Dr. Prasad S. Thenkabail, PhD
Editor-in-Chief
Remote Sensing Handbook (Second Edition, Volumes I, II, III, IV, V, VI)

Volume 1: Sensors, Data Normalization, Harmonization, Cloud Computing, and Accuracies
Volume 2: Image Processing, Change Detection, GIS, and Spatial Data Analysis
Volume 3: Agriculture, Food Security, Rangelands, Vegetation, Phenology, and Soils
Volume 4: Forests, Biodiversity, Ecology, LULC, and Carbon
Volume 5: Water Resources: Hydrology, Floods, Snow and Ice, Wetlands, and Water Productivity
Volume 6: Droughts, Disasters, Pollution, and Urban Mapping

Part I

Forests

1 Characterizing Tropical Forests with Multispectral Imagery

E.H. Helmer, Nicholas R. Goodwin, Valéry Gond, Carlos M. Souza Jr., and Gregory P. Asner

ACRONYMS AND DEFINITIONS

ALI	Advanced Land Imager
ASTER	Advanced spaceborne thermal emission and reflection radiometer
AVHRR	Advanced very high resolution radiometer
DEM	Digital Elevation Model
EOS	Earth Observing System on board Aqua satellite
ETM+	Enhanced Thematic Mapper+
GHG	Greenhouse gas
GLAS	Geoscience Laser Altimeter System
IKONOS	A commercial earth observation satellite, typically, collecting sub-meter to 5 m data
IRS	Indian Remote Sensing Satellites
LiDAR	Light detection and ranging
MERIS	Medium-resolution imaging spectrometer
MODIS	Moderate-resolution imaging spectroradiometer
MSS	Multi-spectral scanner
NDVI	Normalized Difference Vegetation Index
NIR	Near-infrared
NOAA	National Oceanic and Atmospheric Administration
OLI	Operational land imager
REDD	Reducing Emissions from Deforestation in Developing Countries
SPOT	Satellite Pour l'Observation de la Terre, French Earth Observing Satellites
STARFM	Spatial and temporal adaptive reflectance fusion model
SWIR	Shortwave infrared
UNFCCC	United Nations Framework Convention on Climate Change
USDA	United States Department of Agriculture

1.1 INTRODUCTION

Tropical forests abound with regional and local endemic species and house at least half of the species on Earth, while covering less than 7% of its land (Lahssini et al., 2022; Bolívar-Santamaría et al., 2021; Bullock et al., 2020; Wilson, 1988; Gentry, 1988; as cited in Skole and Tucker, 1993). Their clearing, burning, draining, and harvesting can make slopes dangerously unstable, degrade water resources, change local climate, or release to the atmosphere as greenhouse gases (GHGs) the organic carbon (C) that they store in their biomass and soils. These forest disturbances accounted for 19% or more of annual human-caused emissions of CO_2 to the atmosphere from the years 2000–2010, and that level is more than the global transportation sector, which accounted for 14% of these emissions. Forest regrowth from disturbances removes about half of the CO_2 emissions coming from the forest disturbances (Suab et al., 2024; Bullock et al., 2020; Houghton, 2013; IPCC, 2014). Another GHG of concern when considering tropical forests is N_2O released from forest fires.

Tropical forests (including subtropical forests) occur where hard frosts are absent at sea level (Holdridge, 1967), which means low latitudes, and where the dominant plants are trees, including palm trees, tall woody bamboos, and tree ferns. They include former agricultural or other lands that are now undergoing forest succession (Zhe et al., 2024; Adrah et al., 2022; Lahssini et al., 2022; Faber-Langendoen et al., 2012). They receive from < 1000 MM yr^{-1} of precipitation to more than ten times that much as rainfall or fog condensation. Whether dry or humid, tropical forests have far more species diversity than temperate or boreal forests, and their role in Earth's atmospheric GHG budgets is large.

Multispectral satellite imagery, i.e., remotely sensed imagery with discrete bands ranging from visible to shortwave infrared wavelengths, is the timeliest and most accessible remotely sensed data for monitoring these forests (Pletcher et al., 2024; Doughty et al., 2023; Ngo et al., 2023; Rana et al., 2023; Bolívar-Santamaría et al., 2021; Cross et al., 2018; Erinjery et al., 2018; Zaki et al., 2017). Given this relevance, we summarize here how multispectral imagery can help characterize tropical forest attributes of widespread interest, particularly attributes that are relevant to GHG emissions inventories and other forest C accounting: forest type, age, structure, and disturbance type or intensity; the storage, degradation, and accumulation of C in aboveground live tree biomass (AGLB, in Mg dry weight ha^{-1}); the feedback between tropical forest degradation and climate; and cloud screening and gap-filling in imagery. In this chapter, the term *biomass* without further specification is referring to AGLB.

1.2 MULTISPECTRAL IMAGERY AND REDD+

1.2.1 GREENHOUSE GAS INVENTORIES AND FOREST CARBON OFFSETS

Multispectral satellite imagery can provide crucial data to inventories of forest GHG sinks and sources. Inventories of GHGs that have forest components include national inventories for negotiations related to the United Nations Framework Convention on Climate Change (UNFCCC). The UNFCCC now includes a vision of compensating countries for Reducing greenhouse gas Emissions to the atmosphere from Deforestation, Degradation, sustainable management of forests, or conservation or enhancement of forest C stocks in developing countries (known as REDD+). Inventories of GHG emissions for the UNFCCC Clean Development Mechanism (CDM) may also include forests, and there are other forest C offset programs.

Programs like REDD+ could help moderate Earth's climate. They could also help conserve tropical forests and raise local incomes, as long as countries make these latter goals a priority in REDD+ planning. Compensation in REDD+ is for organic C stored in forest AGLB, dead wood, belowground live biomass, soil organic matter, or litter, as long as the stored C is "produced" by avoiding GHG emissions, such as avoiding deforestation or the degradation of forest C stores (Avoided Deforestation Partners, 2013; Bhavsar et al., 2024; Costa et al., 2023; Lahssini et al., 2022; Staal et al., 2020; Cross et al., 2019).

In forest C offsets, avoided emissions are estimated as the difference between net GHG emissions that would have occurred without implementing change (the *baseline case* or *business as usual scenario*) and actual net emissions that are reduced from what they would have been without the management change (the *project case*). Logging, burning, and fragmentation are examples of disturbances that degrade forest C stores. Replacing conventional logging with reduced impact logging reduces associated C emissions and is an example of avoided C emissions. For subnational projects such as those developed under voluntary carbon markets or the CDM, *leakage* must also be subtracted. *Leakage* refers to net emissions that a carbon offset project displaces from its location to elsewhere. Examples are deforestation or removals of roundwood or fuelwood in a forest not far from the forest where such activities have ceased for forest C credits (Njomaba et al., 2024; Ngo et al., 2023; Bullock et al., 2020; Ganivet and Bloomberg, 2019; Cross et al., 2018).

Many countries and organizations have officially proposed that forest C stored by enrichment planting, or by forest growth or regrowth on lands that were not forest before 1990, should also be explicitly eligible for REDD+ compensation (Parker et al., 2009). These latter activities, afforestation and reforestation, already dominate forest projects developed under the CDM.

1.2.2 The Roles of Multispectral Imagery

The United Nations Intergovernmental Panel on Climate Change (IPCC) provides guidelines for GHG emissions inventories, including for forest land (IPCC, 2006). Expanded methods based on these guidelines include those from the Verified Carbon Standard program (http://www.v-c-s.org). Summaries of these guidelines for communities seeking to certify carbon credits for voluntary carbon markets are also available (e.g., Vickers et al., 2012). For each stratum of each land use considered, changes in C stocks are estimated on an annual basis as the net of changes in the following C pools (in Mg C yr^{-1}) (Equation 2.3, IPCC, 2006):

$$\Delta C_{LU} = \Delta C_{AB\text{-}C} + \Delta C_{BB\text{-}C} + \Delta C_{DW\text{-}C} + \Delta C_{LI\text{-}C} + \Delta C_{SO\text{-}C} + \Delta C_{HW\text{-}C} \quad (1.1)$$

Where,
ΔC_{LU} = carbon stock changes for a land-use stratum, e.g., a forest stratum, in Mg C yr^{-1}
And $\Delta C_{SUBSCRIPT}$ represents carbon stock changes for a given pool.
Subscripts denote the following carbon pools in units of Mg C yr^{-1}:
AB-C = aboveground live biomass carbon
BG-C = belowground biomass carbon
DW-C = dead wood carbon
LI-C = litter carbon
SO-C = soil organic carbon
HW-C = harvested wood carbon

For forest GHG inventories for REDD+ and other programs, multispectral satellite imagery can be used to estimate some of the key variables for Equation 1:

1. Areas of forest strata (e.g., forest types, disturbance/degradation classes, or management);
2. Baseline and ongoing rates of change in the areas of forest strata;
3. The AGLB and rates of C accumulation in young forests;
4. Point estimates of forest C pools in AGLB with fine resolution imagery to supplement ground plot data;
5. Potentially, forest AGLB if shown to be accurate for a given landscape; and
6. Potentially, GHG emission factors for forest disturbances if spectral indices of disturbance intensity can be calibrated to correlate well with associated GHG emissions and remaining C pools.

Monitoring forest extent over large scales is also crucial to this forest C accounting, and multispectral satellite imagery is the best data for this purpose, but this topic is covered in other chapters of this volume. Other chapters also cover multispectral image fusion with radar to map forest AGLB (e.g., Saatchi et al., 2011) or estimation of tropical forest biomass with airborne LiDAR (e.g., Asner et al., 2012). Multi-angular image data can also improve forest age mapping (Braswell et al., 2003).

When using the "stock-difference" method (IPCC, 2006) to quantify the parameters in Equation 1, the total C pool for each time period is estimated by multiplying the spatial density of C by the area (in ha) of the forest stratum. The change in the C pool is estimated as the difference in C pools between two time periods divided by the elapsed time in years (please see Equation 2.5 in IPCC, 2006). In addition, in Equation 1 belowground biomass is usually estimated as a fraction of aboveground biomass with default values by ecological zone, region, or country. Also, when the type of land use is forest, litter can often be ignored.

The average spatial density of carbon in live biomass, in Mg C ha^{-1}, is estimated from the average spatial density of the dry weight of live biomass (in Mg ha^{-1}) multiplied by the C fraction of dry weight biomass. Typically this C fraction is about 50% of dry weight mass. The IPCC (2006) has published default values for average C fraction of dry weight wood biomass by ecological zone. Dry weight is estimated with equations that relate the size of the trees growing in a forest to their dry weight, mainly as gauged by tree stem diameter and height. Then the estimated dry weights of all trees in a known area are summed. Species-specific or regional equations are sometimes available (Zhe et al., 2024; Rana et al., 2023; Adrah et al., 2022; Bolívar-Santamaría et al., 2021; Koyama et al., 2019).

1.3 CHARACTERISTICS OF MULTISPECTRAL IMAGE TYPES

Multispectral satellite imagery is available at spatial resolutions ranging from high (< 5 m) to medium (5–100 m) to coarse (> 100 m) (e.g., Table 1.1). The data usually include reflective bands covering the visible (blue, green, red) and near infrared (NIR) wavelengths of the electromagnetic

TABLE 1.1
Multispectral Satellite Imagery Most Commonly Used to Characterize Tropical Forests

Satellite Repeat/Revisit1 Cycle, Scene Size/Swath Width Quantization	Band	Wavelength (μm)	Distributed Spatial Resolution (m)	Approximate Active Dates (Year-Month-Day)
High Resolution (<5 m)				
PlanetScope Dove CubeSat Constellations (Bands for SuperDove shown here)	Coastal Blue	0.431–0.452	3	Dove 2014-06-19 to present SuperDove 2022-01-13 to present
Daily	Blue	0.465–0.515	3	
4 km × 8 km Dove-C	Green I	0.513–0.549	3	
24 km × 16 km Dove-R				
32 km × 20 km SuperDove				
12 bits rescaled to 16 bits	Green	0.547–0.583	3	
	Yellow	0.600–0.620	3	
	Red	0.650–0.680	3	
	RedEdge	0,697–0.713	3	
	NIR	0.845–0.885	3	
IKONOS	Panchromatic	0.45–0.90	1	1999-09-24 to 2015-03-31
3–5 day Revisit	1-Blue	0.445–0.516	4	
11 × 11 km Scenes	2-Green	0.506–0.595	4	
11 bits	3-Red	0.632–0.698	4	
	4-Near Infrared	0.757–0.853	4	
Quickbird	Panchromatic	0.45–0.90	0.6	2001-10-18 to 2014-12-17
2–6 days Revisit	1-Blue	0.45–0.52	2.4	
18 × 18 km Scenes	2-Green	0.52–0.60		
11 bits	3-Red	0.63–0.69		
	4-Near Infrared	0.76–0.90		

TABLE 1.1 *(Continued)*
Multispectral Satellite Imagery Most Commonly Used to Characterize Tropical Forests

Satellite Repeat/Revisit1 Cycle, Scene Size/Swath Width Quantization	Band	Wavelength (μm)	Distributed Spatial Resolution (m)	Approximate Active Dates (Year-Month-Day)
Medium Resolution (5 to 100 m) with High Resolution Panchromatic				
SPOT 4 HRVIR; SPOT 5 HRG	Panchromatic	0.51–0.73	2.5	SPOT 4: 1998-03-24 to 2013-07
2–3 days Revisit	Panchromatic	0.51–0.73	5	SPOT 5: 2002-05-04 to March 2015
60 × 60 km	Green	0.50–0.59	10	
8 bits	Red	0.61–0.68	10	
	Near Infrared	0.78–0.89	10	
	Shortwave Infrared	1.58–1.75	20	
Medium Resolution (5 to 100 m) with High Resolution Panchromatic				
SPOT 1,2,3 HRV	Panchromatic	0.51–0.73	10	SPOT 1: 1986-02-22 to 1990-09
1 to 3 days Revisit	Green	0.50–0.59	20	SPOT 2: 1990-01-22 to 2009-07-16
60 km × 60 km	Red	0.61–0.68	20	SPOT 3: 1993-09-26 to 1996-11-14
8 bits	Near Infrared	0.78–0.89	20	
Medium Resolution (5 to 100 m)				
Landsat MSS 1,2,3 (4,5)	4 (1)-Blue-Green	0.5–0.6	602	Landsat 1: 1972-07-23 to 1978-01-06
16 days Repeat	5 (2)-Red	0.6–0.7	602	Landsat 2: 1975-01-22 to 1982-02-25
170 × 185 km	6 (3)-Near Infrared	0.7–0.8	602	Landsat 3: 1978-03-05 to 1983-03-31
4 bits	7 (4)-Near Infrared	0.8–1.1	602	
Medium Resolution (5 to 100 m)				
Landsat 4 TM, 5 TM, 7 ETM+	1-Blue	0.45–0.52	30	Landsat 4: 1982-07-17 to 1993-12-14
16 days Repeat	2-Green	0.52–0.60	30	Landsat 5: 1984-03-01 to 2013-01
170 × 183 km	3-Red	0.63–0.69	30	Landsat 7: 1999-04-15 to 2022-04-06
8 bits	4-Near Infrared	0.76–0.90	30	
	5-Shortwave Infrared	1.55–1.75	30	
	6-Thermal (2 ETM+ bands)	10.40–12.50	L4,5 1203 (30) L7 603 (30)	
	7-Shortwave Infrared	2.08–2.35	30	

(Continued)

TABLE 1.1 *(Continued)*
Multispectral Satellite Imagery Most Commonly Used to Characterize Tropical Forests

Satellite Repeat/Revisit1 Cycle, Scene Size/Swath Width Quantization	Band	Wavelength (μm)	Distributed Spatial Resolution (m)	Approximate Active Dates (Year-Month-Day)
	8-Panchromatic (L7 only)	0.52–0.90	15	
Landsat 8–9 Operational Land Imager (OLI)	1-Coastal aerosol	0.433–0.453	30	Landsat 8 2013-02-11 to present
16 days Repeat	2-Blue	0.450–0.515	30	Landsat 9 2021-09-27 to present
170 × 183 km	3-Green	0.525–0.600	30	
12 bits	4-Red	0.630–0.680	30	
	5-Near Infrared	0.845–0.885	30	
	6-SWIR 1	1.560–1.660	30	
	7-SWIR 2	2.100–2.300	30	
	8-Panchromatic	0.500–0.680	15	
	9-Cirrus	1.360–1.390	30	
	10-Thermal Infrared 1	10.60–11.19	1003 (30)	
	11-Thermal Infrared 2	11.50–12.51	1003 (30)	
Copernicus Sentinel-2 Multi-Spectral Instrument (MSI)	1-Ultra Blue (Coastal and Aerosol)	Central wavelength 0.443	60	Sentinel-2A 2015-06-23 to present
10 Days Repeat	2-Blue	Central wavelength 0.49	10	Sentinel-2B 2017-03-07 to present
290 km	3-Green	Central wavelength 0.56	10	
12 bits	4-Red	Central wavelength 0.665	10	
	5-Vegetation Red Edge	Central wavelength 0.705	20	
	6-Vegetation Red Edge	Central wavelength 0.74	20	
	7-Vegetation Red Edge	Central wavelength 0.783	20	
	8-Near Infrared	Central wavelength 0.842	10	
	9-Vegetation Red Edge	Central wavelength 0.865	20	

(Continued)

TABLE 1.1 *(Continued)*
Multispectral Satellite Imagery Most Commonly Used to Characterize Tropical Forests

Satellite Repeat/Revisit1 Cycle, Scene Size/Swath Width Quantization	Band	Wavelength (μm)	Distributed Spatial Resolution (m)	Approximate Active Dates (Year-Month-Day)
	10-Water vapor Short Wave Infrared (SWIR)	Central wavelength 0.94	60	
	11-Cirrus Short Wave Infrared (SWIR)	Central wavelength 0.137	60	
	12-Short Wave Infrared (SWIR)	Central wavelength 0.161	20	
	13-Short Wave Infrared (SWIR)	Central wavelength 0.219	20	
Coarse Resolution (>100 m)				
Terra/Aqua MODIS4 (7 of 36 bands are shown)	1	0.620–0.670	250	Terra (EOS AM): 1999-08-12 to December 2025
1 day Revisit	2	0.841–0.876	250	Aqua (EOS PM): 2002-05-04 to August 2026
2330 km Swath Width	3	0.459–0.479	500	
12 bits	4	0.545–0.565	500	
	5	1.230–1.250	500	
	6	1.628–1.652	500	
	7	2.105–2.155	500	
SPOT 4,5 Vegetation 1, 24	0-Blue	0.43–0.47	1150	Aboard SPOT 4: 1998-03-24 to 2013-07
1 day Revisit	2-Red	0.61–0.68	1150	Aboard SPOT 5: 2002-05-04 to May 2014
2250 km Swath Width	3-Near Infrared	0.78–0.89	1150	
10 bits	SWIR -Shortwave Infrared	1.58–1.75	1150	

1. Revisit cycles change with latitude.
2. The original MSS pixel size of 79 × 57 m is now resampled to 60 m.
3. Thermal infrared Landsat bands are now resampled to 30 m.
4. For coarse-resolution sensors, resolution given is at nadir.

spectrum. Several sensors also include shortwave infrared (SWIR) bands (e.g., Landsat Thematic Mapper [TM] and subsequent Landsat sensors; Sentinel-2; the sensors aboard the fourth and fifth missions of Satellite Pour l'Observation de la Terre (SPOT 4 HRVIR, SPOT 5 HRG and the SPOT 4 and 5 Vegetation instruments); the Moderate Resolution Imaging Spectroradiometer (MODIS) the Advanced Wide Field Sensor (AWiFS); and the Infrared Multispectral Scanner Camera (IRMSS) aboard the China-Brazil Earth Resources Satellite series (CBERS). Sentinel-2, the most recent PlanetScope imagery, and the planned Landsat Next mission include a red edge band.

Satellite launches in the years 1998–1999 greatly increased the amount of imagery available for monitoring tropical forests. These launches brought (1) the first public source of high spatial resolution imagery (IKONOS, with < 5-m pixels); (2) the first medium-resolution imagery (5- to 100-m pixels) with some degree of consistent global data collection (Landsat 7); (3) the first medium-resolution imagery with fine-resolution panchromatic bands of 2.5–15 m (SPOT 4 and Landsat 7, respectively); and (4) the first coarse-resolution imagery (>100-m pixels) distributed with higher-level preprocessing, like atmospheric correction and cloud-minimized compositing (MODIS and SPOT Vegetation). Before IKONOS, remotely sensed reference data had to come from air photos that in many places were outdated and costly to obtain. Since IKONOS, many sources of high-resolution imagery have become available.

The next big advances in tropical forest monitoring with satellite imagery came in 2005–2008, when (1) Google, Inc. and the producers of high-resolution imagery, such as Quickbird and IKONOS, made high-resolution data viewable on Google Earth for many sites, making reference data free and accessible for subsets of project areas; and (2) the Brazilian National Institute for Space Research (INPE) and the US Geological Survey (USGS) began to freely distribute Landsat and other imagery with medium spatial resolution, making long, dense time series of medium-resolution imagery available over large areas (Suab et al., 2024; Rana et al., 2023; Adrah et al., 2022; Staal et al., 2020; Cross et al., 2019; Wang et al., 2019; Zaki et al., 2017).

Other sources of multispectral imagery for monitoring tropical forests over large areas that are not shown in Table 1.1, mainly to highlight them here, include the Japan-US Advanced Spaceborne Thermal Emission and Reflection Radiometer (ASTER) (aboard Terra) (Pletcher et al., 2024; Suab et al., 2024; Lahssini et al., 2022; Ferreira et al., 2019; Erinjery et al., 2018; Zaki et al., 2017). In addition to 15-m visible to near infrared (VNIR) bands, it has several SWIR and thermal bands with 30- to 90-m spatial resolution. Data for Brazil, China, and nearby areas are also available from CBERS. The series of CBERS satellites, 1, 2, and 2B, collected panchromatic to SWIR images with medium spatial resolution (20- to 80-m, 113- to 120-km swath width), and red and NIR images with coarse spatial resolution (260-m, 890-km swath width) from 1999 to 2010 and missions to collect with medium resolution multispectral imagery with a five-day revisit cycle are scheduled. In the Indian Resources Satellite (IRS) series, the Wide Field Sensor has a 740-km swath width, 188-m spatial resolution, and red and NIR bands. More recently, the IRS-P6 satellite carries the AWiFS instrument. AWiFS has 60-m pixels for green through SWIR bands, a 740-km swath width, a five-day revisit cycle, and a SWIR band, combining advantages of imagery with medium and coarse spatial resolution. The later of the IRS series sensors include data from LISS (Linear Imaging Self Scanner) with multispectral imagery with a 23.5-m spatial resolution. Ground stations receiving data from early CBERS and the IRS satellite series did not cover all of the tropics.

1.4 PREPROCESSING IMAGERY TO ADDRESS CLOUDS

1.4.1 CLOUD SCREENING

We begin with cloud and cloud shadow screening, as this step is crucial in the image processing chain for characterizing tropical forests. Clouds and their shadows obscure the ground and contaminate temporal trends in reflectance. Automated systems for processing large archives of satellite imagery are becoming more common for natural resource applications and must screen clouds.

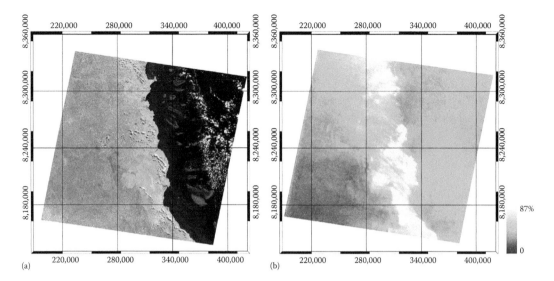

FIGURE 1.1 Illustration of cloud distribution spatially and temporally over tropical forests of north Queensland: (1) Landsat image (RGB: 542, Path/Row: 96/71, and date: 20070702) and (2) percentage of observations classified as cloud between 1986 and 2012 ($n = 445$). Note: high cloud fractions were not included in calculations.

Clouds are composed of condensed water vapor that forms water droplets and scatters VNIR light, reducing direct illumination on the surface below and forming a cloud shadow. In multispectral satellite imagery, clouds are characterized by a high albedo (Choi and Bindschadler, 2004), while their shadows have lower reflectance than surrounding pixels. The easiest solution to cloud contamination is to restrict analyses to cloud-free imagery, which may only include dry season imagery for tropical and coastal environments due to frequent cloud cover. Alternatively, methods to screen cloud- and shadow-contaminated pixels can increase the number of observations available (Figure 1.1). Increasing the number of available observations in a time series may also improve the detection of land surface change and reflectance trends.

Manual and semi-automated approaches to cloud screening are undesirable for processing large numbers of images due to the time-consuming nature of the work, which may depend not only on analyst experience but also on image contrast. Several automated approaches have been developed, but separating cloud and shadow from the land surface is not necessarily straightforward given the diversity of land surfaces coupled with large variations in cloud and shadow optical properties (Zhe et al., 2024; Koyama et al., 2019; Zaki et al., 2017; Zhu et al., 2012b; Goodwin et al., 2013; Lyapustin et al., 2008). A summary of current approaches to cloud and shadow screening for Landsat TM/ETM+, SPOT, and MODIS sensors follows. Cloud screening and gap-filling methods have also been developed for high-resolution PlanetScope (Wang et al., 2021, 2022), Landsat Multispectral Scanner (MSS), and Sentinel-2 imagery (e.g., Skakun et al., 2022; Zhu and Helmer, 2018).

1.4.1.1 Landsat TM and OLI Imagery

The Landsat TM/ETM+ archives of countries with receiving stations now contain up to three decades of imagery (1984 to present), with varying levels of cloud and cloud shadow contained in the archive of images. The USGS is working with other countries to consolidate these archives through consistent processing and distribution through its website (landsat.usgs.gov). Image preprocessing by the Landsat program has included the Automatic Cloud Cover Assessment (ACCA) algorithms for both Landsat-5 TM and Landsat-7 ETM+ missions, which use optical and thermal (ETM+ only) bands to identify clouds (Irish, 2006). It is designed for reporting the percentage of cloud cover over scenes rather than producing per-pixel masks. Further modifications have also been tested for application to Landsat 8 imagery (Scaramuzza et al., 2011), which includes a new cirrus band (1.360–1.390 μm) that is sensitive to aerosol loadings and should improve cloud detection. ACCA is designed to limit the impacts of cloud and scene variability on thresholding. The ETM+ ACCA incorporates two passes: one to conservatively estimate "certain" cloud at the pixel level with a series of spectral and thermal tests. The result is then used to derive scene-based thermal thresholds for the second pass. The error in scene-averaged cloud amount was estimated to be around 5% (Irish et al., 2006). Scaramuzza et al. (2011) validated the per-pixel classification of the ETM+ ACCA (pass 1) and found a 79.9% agreement between the reference and ACCA at the pixel scale. Using a subset of the same reference set, Oreopoulos et al. (2011) evaluated both per-pixel ACCA masks and a cloud detection algorithm modified from the MODIS Luo-Trishchenko-Khlopenkov algorithm (Luo et al., 2008). Both ACCA and the modified LTK showed greater than 90% agreement with the reference, although like ACCA, the LTK had limited ability to detect thin cirrus clouds. Furthermore, ACCA has been used as the starting point for further cloud masking (Njomaba et al., 2024; Ngo et al., 2023; Cross et al., 2019; Laurin et al., 2019; Zaki et al., 2017; Choi and Bindshadler, 2004; Roy et al., 2010; Scaramuzza et al., 2011).

Earlier studies have shown that several approaches work well for classifying clouds and cloud shadows over particular paths/rows. One approach is image differencing based on image pairs (Lahssini et al., 2022; Laurin et al., 2019; Wang, 1999), while other studies have empirically defined thresholds for cloud brightness and coldness in one or more spectral/thermal bands, including, for example, Landsat TM Bands 1 and 6 (Martinuzzi et al., 2007); Bands 3 and 6 (Huang et al., 2010); Bands 1, 3, 4, and 5 (Oreopoulos et al., 2011); and Bands 1, 4, 5, and 6 (Helmer et al., 2012). The application of these methods to a range of paths/rows around the globe, however, remains untested and may encounter issues due to spectral similarities among the wide range of combinations of land surfaces and cloud/cloud shadows (Zhe et al., 2024; Dupuis et al., 2020; Laurin et al., 2019; Zaki et al., 2017).

The automated method that Huang et al. (2010) developed to allow forest change detection in cloud-contaminated imagery considers brightness and temperature thresholds for clouds that are self-calibrated against forest pixels. It requires a digital elevation model (DEM) to normalize top of atmosphere brightness temperature values and helps to project cloud shadow on the land surface. Published validation data for this method is currently limited to four US images with forest and would benefit from further calibration/validation.

Additional approaches included function of mask (Fmask) (Zhu and Woodcock, 2012) and a time series approach by Goodwin and Collett (2014) (Figure 1.2). Fmask integrates existing algorithms and metrics with optical and thermal bands to separate contaminated pixels from land surface pixels. Fmask also considers contextual information for mapping potential cloud shadow using a flood-fill operation applied to the near-infrared band. Cloud shadows are then identified by linking clouds and their shadows with solar/sensor geometry and cloud height inferred from the thermal Landsat TM Band 6. The results were validated with a global dataset and were a significant improvement to ACCA with Fmask achieving overall user and producer accuracies of 96%, 89%, and 92%, respectively, compared to 85%, 92%, and 72%, respectively, for ACCA.

The time series method uses temporal change to detect cloud and cloud shadow (Goodwin et al., 2013). It smooths pixel time series of land surface reflectance using minimum and median filters and then locates outliers with multi-temporal image differencing. Seeded region grow is applied to the difference layer using a watershed region grow algorithm to map clusters of change pixels, with clumps smaller than 5 pixels removed to minimize classification speckle. This has the effect of increasing the cloud/shadow detection rate whilst restricting commission errors; smaller magnitudes of change associated with cloud/cloud shadows are only mapped if they are in the neighborhood of larger changes. Morphological dilation operations were applied to map a larger spatial extent of the cloud and cloud shadow, while shadows were translated along the image plane in the reverse solar azimuth direction to assess the overlap with clouds and confirm the object is

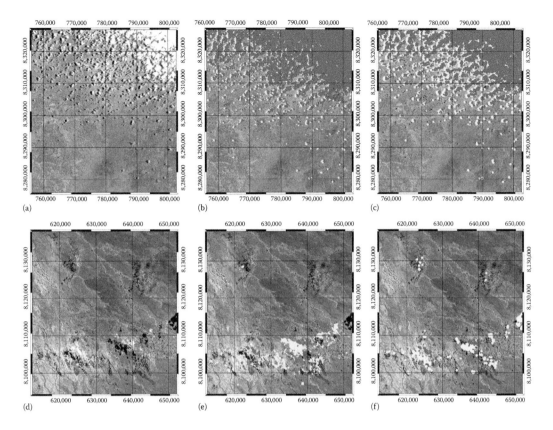

FIGURE 1.2. Examples of Fmask and time series approaches to cloud and cloud shadow screening: (a) to (c) well-detected cumulus cloud and cloud shadow (RGB: 542, Path/Row: 97/71, and date: 19981020), and (d) to (f) a complex example where both methods miss sections of cirrus cloud (RGB: 542, Path/Row: 98/72, and date: 20010418).

a shadow. A comparison with Fmask showed the time series method could screen more cloud and cloud shadow than Fmask across Queensland, Australia (cloud and cloud shadow producers' accuracies were eight and 12 percentage points higher, respectively).

Several trade-offs exist between these two automated approaches to cloud and shadow screening. The time series method might detect more cloud and cloud shadow, yet Fmask is more computationally efficient and practical for individual images. At present, the time series method is processed using entire time series for each Landsat path/row. For operational systems processing many images, the computational overhead of the time series approach could be worthwhile, as it can detect more cloud/shadow contamination. Locations with few cloud-free observations per year and high land use change are also less desirable for a time series method. In the absence of an atmospheric aerosol correction, pixels contaminated by smoke and haze are more likely to be classified as cloud by the time series method. Neither the Fmask nor the time series method, nor previous attempts, adequately map high-level, semi-transparent cirrus clouds (Figure 1.2 (d)–(f)). New methods for Landsat 8 will likely detect more clouds with the new band sensitive to cirrus clouds. Both Fmask and the time series methods are highly configurable, allowing calibration for a localized region or a wider application. Fmask has been calibrated using a global reference set, while the time series approach was calibrated and tested mainly for northeastern Australian conditions.

Although both methods have high accuracy, further improvements could be made, particularly to screening cloud shadow. Removing the dependency of a link between cloud and shadow would be a considerable advancement, as clouds are often missed, especially thin clouds, or under/over mapped, causing the shadow test to fail. Furthermore, adding thermal information to the time-series method has the potential to remove commission errors where bright surfaces such as exposed soil are falsely classified as cloud. Both methods use a series of rules to classify cloud and shadow and have the flexibility to add new algorithms and criteria to improve the detection of contaminated pixels. There are many new cloud and cloud shadow masking methods, including ones that use deep learning or online computing platforms, and new reference datasets are now available to test them (Skakun et al., 2022; Hermosilla et al., 2023; Yin et al., 2020). Detecting thin clouds has also progressed. The automatic time-series analysis (ATSA) method (Zhu and Helmer, 2018), for example, uses time series of the haze optimized transformation (Zhang et al., 2002) to detect thin clouds, clouds in imagery without a thermal band, and associated shadows, an approach adopted by Fmask 4.0 (Qiu et al., 2019).

1.4.1.2 SPOT, MSS, and Sentinel-2 Imagery

The spatial and spectral characteristics of SPOT (Satellite Pour l'Observation de la Terre) and Sentinel-2 imagery have similarities to Landsat imagery, with the first satellite launched in 1986 (SPOT 1), and similar methods for screening cloud and cloud shadows may be useful. A difference is that SPOT and Sentinel-2, along with older MSS imagery, lack a thermal band which has been useful in discriminating clouds (e.g., ACCA). Many cloud-screening methods have been developed for Sentinel-2, some of which were compared in Skakun et al. (2022). ATSA was tested for Landsat 4 MSS, Landsat 8 OLI, and Sentinel-2 images. Although a limited number of studies have been published on screening cloud and cloud shadow from SPOT data, SPOT is a commercially operated sensor, and unlike Landsat TM/ETM+ and MODIS, scenes are typically purchased/tasked with limited cloud cover or would otherwise prove cost prohibitive for many vegetation applications (Costa et al., 2023; Bolívar-Santamaría et al., 2021; Ganivet and Bloomberg, 2019; Berveglieri et al., 2018; Cross et al., 2018). The New South Wales government of Australia, for example, acquired 1850 images between 2004 and 2012, of which only 313 contained cloud with the maximum cloud cover values < 10% (Fisher, 2014).

Le Hégarat-Mascle and André (2009) used a Markov Random Field framework that assumes clouds are connected objects, solar/sensor geometry is known, and shadow has a similar shape to its corresponding cloud (excluding the influence of topography). Potential cloud pixels were identified using a relationship between green and SWIR bands; shadows were located using cloud shape, orientation of shadow relative to cloud, and SWIR band reflectance, removing objects not part of a cloud-shadow pair. The method was applied to 39 SPOT 4 HRVIR (High Resolution Visible IR) images over West Africa with encouraging results. However, when applying this method Fisher (2014) found commission errors, as bright surfaces were frequently matched to dark surfaces that were not cloud contaminated. They suggest first masking vegetation and water bodies, then locating marker pixels for clouds and shadows in the green-SWIR space and NIR bands, respectively, then growing objects with the watershed transform. Sensor/solar geometry and object size are also used to match clouds with their shadows (Doughty et al., 2023; Rana et al., 2023).

1.4.1.3 MODIS Imagery

MODIS has a standard cloud product, in contrast to SPOT or, until recently, Landsat and Sentinel-2, which includes information on whether a pixel is clear from cloud/shadow contamination. The cloud mask is based on several per-pixel spectral tests and is produced at 250-m and 1-km spatial resolutions (Strabala, 2005). A validation with active ground-based LiDAR/radar sensors showed a < 10% agreement with the MODIS cloud mask (Ackerman et al., 2008).

Recent research has found that time series information can improve cloud detection in MODIS imagery (Costa et al., 2023; Zaki et al., 2017; Hilker et al., 2012; Lyapustin et al., 2008). The cloud screening method in MAIAC (Multi-Angle Implementation of Atmospheric Correction), for example, uses a dynamic clear skies reference image and covariance calculations, in addition to spectral and thermal tests, to locate clouds over land (Lyapustin et al., 2008). In a tropical Amazonian environment, Hilker et al. (2012) demonstrated that this method was better at detecting clouds and increasing the number of useable pixels than the standard product (MYD09GA), which translated into more accurate patterns in NDVI.

1.4.2 Filling Cloud and Scan-Line Gaps

Cloud and cloud shadow screening removes contaminated pixels from analyses but leaves missing data in the imagery and derived products. The scan-line correction error affecting Landsat 7 post-2003 also leaves gaps approximating 20% of affected images (USGS, 2003). Data gaps in maps are aesthetically unappealing and the derivation of statistics more difficult. As a result, approaches have been developed to fill data gaps, including temporal compositing and fusing imagery from two different sensors.

A range of temporal compositing algorithms have been developed to minimize cloud contamination and noise (Bhavsar et al., 2024; Koyama et al., 2019; Dennison et al., 2007; Flood, 2013). Compositing involves analyzing band/metric values across a date range, with an algorithm deciding the pixel value most likely to be cloud/noise free. The choice of algorithm may vary depending on the application and land-cover type. Compositing algorithms have generally been applied to high temporal frequency data such as MODIS and AVHRR; however, methods for compositing imagery with a lower temporal resolution have also been developed. For example, the MOD 13 products use the maximum value compositing (MVC) algorithm with NDVI as the metric in 16-day and monthly composites of MODIS imagery (Strabala, 2005). Landsat has similarly been composited using a parametric weighting scheme (Griffiths et al., 2013). The result is an image that ideally is free from noise or cloud that can be used as a product itself or the corresponding pixels used to infill data gaps.

The fusion or blending of MODIS and Landsat offers another approach to predict image pixel values within data gaps. These methods integrate medium spatial resolution Landsat with temporal trends in reflectance (e.g., seasonality) captured by the higher temporal frequency of MODIS. Roy et al. (2008) integrated the MODIS Bidirectional Reflectance Distribution Function (BRDF)/albedo product and Landsat data to model Landsat reflectance. They found that infrared bands were more accurately predicted than visible wavelengths, probably in response to greater atmospheric effects at shorter wavelengths. The Spatial and Temporal Adaptive Reflectance Fusion Model (STARFM) requires a MODIS-Landsat image pair captured on the same day, plus a MODIS image on the prediction date, and applies spatial weighting to account for reflectance outliers (Gao et al., 2006). Further algorithm development has produced an Enhanced STARFM (ESTARFM) method that was found to improve predictions in heterogeneous landscapes (Zhu et al., 2010). However, there are known limitations with blending or fusing Landsat and MODIS imagery. Solutions involving MODIS will only work post-2000, when imagery was first captured, and potentially 2002 onwards, where stable BRDF predictions are needed (Roy et al., 2008). Furthermore, Emelyanova et al. (2013) found that land-cover type and temporal and spatial variances impact the fusion of MODIS and Landsat as well as the choice of algorithm. Where the temporal variance of MODIS is considerably less than the spatial variance of Landsat, blending may not improve predictions (Zhe et al., 2024; Adrah et al., 2022; Cross et al., 2019).

Gap filling using Landsat imagery alone has also been performed. Helmer and Ruefenacht (2005) developed a method for predicting Landsat values using two Landsat images for change detection. This method develops a relationship between uncontaminated pixels in an image pair with regression tree models, and it then applies these models to predict the values in areas with missing data in the target image. Additional images are used in the same way to predict pixels in remaining cloud gaps. Langner et al. (2014) segment such pairwise predictive models according to forest type. Approaches using geostatistics have also been developed. Pringle et al. (2009) use an image before and after the target image in geostatistical interpolation to predict values in Landsat 7 SLC-off imagery. Based on their results they recommend images captured within weeks, rather than months, of each other to limit temporal variance in a tropical savanna environment. Zhu et al. (2012) also use geostatistics with encouraging results to predict missing Landsat 7 SLC-off data based on the Geostatistical Neighborhood Similar Pixel Interpolator (GNSPI).

A potential limitation with gap filling is the introduction of image noise or artifacts. This is because of differences in vegetation phenology, illumination, and atmospheric effects, as gap-filled imagery contains data from multiple dates and/or sensors. These effects can be minimized by atmospheric and illumination corrections, as well as methods that seek to balance the distribution of pixel values such as histogram matching, linear regression, or regression trees (Ngo et al., 2023; Bullock et al., 2020; Koyama et al., 2019; Helmer and Ruefenacht, 2007).

1.5 FOREST BIOMASS, DEGRADATION, REGROWTH RATES FROM MULTISPECTRAL IMAGERY

Studies have used multispectral imagery to map or estimate some key inputs to the variables in Equation 1 (Section 1.2.2) for forests: forest AGLB (in Mg dry weight ha^{-1}), rates of C accumulation in reforesting lands (in Mg dry weight ha^{-1}yr^{-1}); and area or intensity of forest degradation or disturbance (in ha). In addition, multispectral imagery is the most common satellite imagery for mapping tropical forest types, which we discuss in Section 1.6, and AGLB estimates are often more precise and accurate if stratified by forest type (Pletcher et al., 2024; Costa et al., 2023; Adrah et al., 2022; Dupuis et al., 2020; Koyama et al., 2019; Laurin et al., 2019; Wang et al., 2019).

In this section we first review work that uses the spectral and textural information in multispectral imagery of high spatial resolution to estimate tropical forest AGLB. We then discuss how the spectral information inherent to multiyear image time series has high sensitivity to the height, AGLB, and age of forests that have established since about ten years before the start of an image sequence (so as early as ten years before 1972 for Landsat data), which we refer to here as *young forests*, allowing estimates of biomass and C accumulation rates in reforested lands. Next, we discuss how multispectral imagery from a single epoch of medium to coarse spatial resolution imagery has limited sensitivity to tropical forest age or biomass (Suab et al., 2024; Zhe et al., 2024; Bullock et al., 2020; Wang et al., 2019). Section 1.5.3 focuses on detecting tropical forest degradation at pixel and subpixel scales.

1.5.1 TROPICAL FOREST BIOMASS FROM HIGH-RESOLUTION MULTISPECTRAL IMAGERY

When considering forest structure mapping, multispectral imagery of high spatial resolution, with pixels ≤ 5-m, is distinct from imagery with medium spatial resolution because the spatial patterns of dominant and codominant tree crowns are visible. The possibility of detecting tree crown size suggests a way to estimate AGLB by allometry between stem diameters, used to estimate AGLB, and crown size (Zhe et al., 2024; Rana et al., 2023; Dupuis et al., 2020; Koyama et al., 2019; Laurin et al., 2019; Wang et al., 2019; Asner et al., 2002; Couteron et al., 2005; Palace et al., 2008). Automated crown delineation in these images is more accurate than manual means, but both methods overestimate the area of large crowns and underestimate the frequency of understory and codominant trees (Asner et al., 2002; Palace et al., 2008), such that biomass estimates from crown delineation alone require adjustments.

A new technique, however, predicts the biomass of high-biomass tropical forests with stand-level spatial patterns of tree crowns in images with ~1-m or finer pixels. The new method first applies two-dimensional Fourier transforms to subsets (*samples*) of high-resolution panchromatic images, from which it produces a dataset with a row for each sample of imagery and columns that bin the outputs from the transform so that the columns in each row together form a proxy for the distribution of crown sizes discerned, or "apparent" in each image sample (Njomaba et al., 2024; Ngo et al., 2023; Lahssini et al., 2022; Staal et al., 2020; Koyama et al., 2019). Principal components transformation

(PCA) of this matrix yields axes that serve as predictors in regression models of stand structural parameters, like basal area, AGLB, or "apparent" dominant crown size (calculated by inversion) (Couteron et al., 2005; Barbier et al., 2010; Ploton et al., 2011). Ploton et al. (2011) predicted forest biomass ranging from ~100 to over 600 Mg ha^{-1} in Western Ghats, India, with IKONOS image extracts downloaded from Google Earth Pro (0.6–0.7 m resolution). Their model explained 75% of the variability in forest biomass. They estimated that the relative uncertainty in AGLB estimates that was due to the remote sensing technique, of < 15%, was similar to uncertainties associated with estimating forest AGLB with LiDAR. With this new technique, AGLB estimates from high-resolution imagery on Google Earth could supplement ground or LiDAR-based surveys. The resulting increase in the number and density of AGLB estimates for forests should better characterize the landscape-scale spatial variability in AGLB and increase the precision of forest C pool estimates.

Related to the aforementioned work on AGLB are studies that have characterized how gradients in the spatial patterns of tropical forest canopies correspond with climate. These gradients are apparent in high-resolution imagery, and future changes in these patterns could reflect and help monitor effects of global climate change (Bhavsar et al., 2024; Costa et al., 2023; Koyama et al., 2019; Barbier et al., 2010; Palace et al., 2008; Malhi and Román-Cuesta, 2008). Barbier et al. (2010), for example, showed how dominant crown size and canopy size heterogeneity change with climate and substrate across Amazonia.

1.5.2 The Biomass, Age, and Rates of Biomass Accumulation in Forest Regrowth

With a long time series of medium-resolution multispectral images such as Landsat, key variables for GHG inventories (and forest C accounting for REDD+) can be mapped and estimated for young tropical forests, including area, age, height, AGLB, and rates of biomass accumulation. Where an image time series spans the age range of young forests, its spectral data can precisely estimate age, which is needed to estimate biomass accumulation rates and can also help estimate the height or AGLB of these forests (Suab et al., 2024; Adrah et al., 2022; Staal et al., 2020). Helmer et al. (2009) estimated a landscape-level rate of AGLB accumulation in Amazonian secondary forest by regressing forest biomass estimates from the Geoscience Laser Altimeter System (GLAS) (Figure 1.3) against remotely sensed forest age (R-square=0.60). The estimated landscape-level biomass accumulation rate of 8.4 Mg ha^{-1}yr^{-1} agreed well with ground-based studies. Forest age was mapped with an algorithm that automatically processed a time series of Landsat MSS and TM imagery (1975–2003) with self-calibrated thresholds that detect when secondary forests established on previously cleared land. The technique mapped the extent of old-growth forest and age of secondary forest with an overall accuracy of 88%. With the time series, tropical secondary forest > 28 years old was accurately distinguished from old-growth forest, even though it was spectrally indistinct in the most recent Landsat scenes. This older secondary forest clearly stored less C than the old-growth forest, being shorter and having much smaller average canopy diameters than nearby old growth.

Forest height and AGLB are strongly related, and the height or AGLB of young forests can be mapped with long time series of Landsat images in tropical (Helmer et al., 2010) and temperate (Li et al., 2011; Pflugmacher et al., 2012; Ahmed et al., 2014) regions. With a regression tree model based on the spectral data from all of the images in a time series of cloud-gap-filled Landsat imagery (1984–2005 with 1- to 5-year intervals), Helmer et al. (2010) mapped the height (RMSE = 0.9 m, R-square = 0.84, range 0.6–7 m) and foliage height profiles of tropical semi-evergreen forest (Figure 1.4). In contrast with mapping the height of old forests, local-scale spatial variability in young forest structure was mapped, because within-patch differences in disturbance intensity and type, and subsequent forest recovery rate, were reflected in the spectral data from the multiyear image stack (Costa et al., 2023; Bolívar-Santamaría et al., 2021; Ferreira et al., 2019). This study also mapped forest disturbance type, age, and wetland forest type, with an overall accuracy of 88%, with a decision tree model of the entire time series of cloud-minimized composite images to better understand avian habitat. As a result, the classification distinguished different agents of forest disturbance,

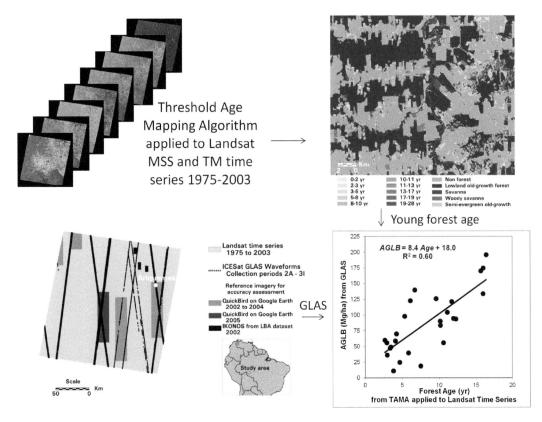

FIGURE 1.3 The average age of secondary forest pixels, as estimated from automatic processing of a time series of Landsat MSS, TM, and ETM+ imagery, in the 150-m window surrounding GLAS waveform centers explained 60% of the variance in GLAS-estimated canopy height and biomass (Aboveground Live Biomass, AGLB, in Mg ha^{-1}yr^{-1} dry weight). The standard error of the slope and intercept are 1.4 and 13.2, respectively, for 26 observations.

including classes of cleared forests and forests affected by escaped fire, and allowed estimation of rates of forest regrowth. Forest age, vertical structure, and disturbance type explained differences in woody species composition, including the abundance of forage species for an endangered Neotropical migrant bird, the Kirtland's warbler (*Dendroica kirtlandii*).

1.5.3 Limitations to Mapping Forest Biomass or Age with One Multispectral Image Epoch

1.5.3.1 Tropical Forest Biomass with One Image Epoch

Forest biomass mapping with multispectral imagery empirically predicts the AGLB of forested pixels with models that relate forest AGLB or height, from ground plots or LiDAR, to spectral bands, spectral indices, or spectral texture variables (Doughty et al., 2023; Bolívar-Santamaría et al., 2021; Bullock et al., 2020; Laurin et al., 2019). It remains a challenge (Song, 2013). Forest AGLB is usually estimated in units of Mg dry weight ha^{-1} (see Section 1.2). As more data on stand species composition and species-specific wood densities become available, maps of C storage in forest biomass, as in Asner et al. (2013) and Mitchard et al. (2014), rather than forest biomass itself, may become more common.

Medium to coarse spatial resolution imagery from one epoch is not that sensitive to small changes in the AGLB or C storage in aboveground biomass of dense tropical forests. (By *epoch* we mean

Characterizing Tropical Forests with Multispectral Imagery

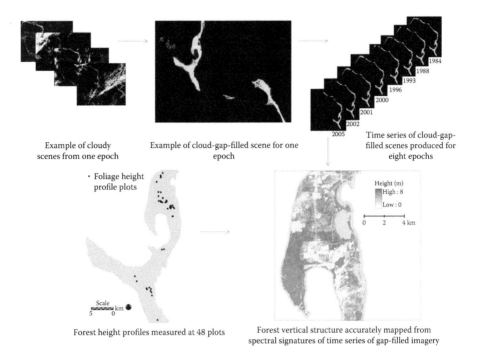

FIGURE 1.4 Tropical dry forest height and foliage height profiles were mapped from a time series of gap-filled Landsat and ALI imagery on the island of Eleuthera, in the Bahamas, substituting time for vertical canopy space. The time series was also used to map forest disturbance type and age

imagery from one date, one gap-filled or composite image composed of imagery from one to several years, or multiseason imagery from one year). This limited sensitivity appears in biomass mapping models as high per-pixel uncertainty that can manifest itself in several ways (Bhavsar et al., 2024; Cross et al., 2019; Cross et al., 2018; Zaki et al., 2017):

1. Mapping models may explain a minority of variance in reference data (i.e., regressions of predicted vs. observed values have low coefficients of determination or R-squared values of less than 0.50) (e.g., Oza et al., 1996 for volume of Indian deciduous forest; Steininger, 2000 for Bolivian sites; Wijaya et al., 2010 in Indonesia);
2. Mapping models may both underestimate AGLB at high-biomass sites and overestimate AGLB where biomass is low (e.g., Baccini et al., 2008 for tropical Africa; Blackard et al., 2008 for the US, including Puerto Rico; Wijaya et al., 2010);
3. Spectral responses to AGLB may saturate at relatively low levels of around 175 Mg C ha^{-1}. For example, studies indicate that stand-level multispectral responses saturate at 150–170 Mg ha^{-1} for study sites in Brazilian Amazonia (Steininger, 2000; Lu, 2005), ~ 180 Mg C ha^{-1} in Panama (Asner et al., 2013), and 175 Mg ha^{-1} across Uganda (Avitabile et al., 2012). These saturation levels may be half or less of the biomass of the most structurally complex or old-growth tropical forests in humid lowlands. In many landscapes the relationship between multispectral data and tropical forest AGLB may saturate at even lower levels; and
4. Continental to global-scale mapping models may not capture gradients in AGLB and C pools that stem from differences in forest allometry and average wood density (Mitchard et al., 2014).

Despite per-pixel uncertainties, estimates of the total forest biomass may be accurate when pixels are summed over large areas that have a wide range of AGLB (Costa et al., 2023; Dupuis et al., 2020; Staal et al., 2020; Wang et al., 2019). This result could happen when the average biomass of pixels covering a large area approaches the mean of the ground or LiDAR data used to estimate the mapping model. Estimates of total forest AGLB across tropical landscapes can also be accurate if the landscapes that have few forest patches with AGLB that exceeds the levels where spectral response becomes saturated (e.g., Avitabile et al., 2012).

Texture variables from SPOT 5 imagery may improve mapping models of AGLB, because SPOT 5 imagery has finer spatial resolutions of 10–20 m compared with many other image sources with medium spatial resolution (Table 1.1), but results may still have relative errors of around 20% (Castillo-Santiago et al., 2010). Exceptions may include Asian bamboo forests (Xu et al., 2011) or low-biomass tropical forests.

Mapping models of tropical forest AGLB or height that rely on multispectral imagery benefit from added predictors. Example predictors that may improve models include topography, forest type, climate, soils, geology, or indicators of disturbance, like tree canopy cover (Costa et al., 2023; Doughty et al., 2023; Bullock et al., 2020; Berveglieri et al., 2018; Helmer and Lefsky, 2006; Saatchi et al., 2007; Blackard et al., 2008; Asner et al., 2009; Wijaya et al., 2010, Lefsky, 2010). After including these predictors in mapping models, the variability in the biomass mapped for undisturbed forests may reflect more of the variability in AGLB that stems from regional to landscape-scale environmental gradients in attributes like rainfall. Maps of these spatial patterns may be useful, but they may not reveal much local-scale AGLB variation.

1.5.3.2 Tropical Forest Age with One Image Epoch

As with AGLB, multispectral imagery has limited sensitivity to increasing forest age. Many studies show that spectral indices that contrast the mid-infrared bands with the near infrared or visible bands are the most sensitive indices to tropical forest age, height, and AGLB (e.g., Njomaba et al., 2024; Ngo et al., 2023; Bullock et al., 2020; Koyama et al., 2019; Cross et al., 2018; Boyd et al., 1996; Helmer et al., 2000; Steininger, 2000; Thenkabail et al., 2003; Helmer et al., 2010). For example, with Landsat TM or ETM+ data, these indices include the NIR/SWIR ratio, the tasseled cap wetness index (Crist and Cicone, 1984; Huang et al., 2002), the Wetness Brightness Difference Index (WBDI) (Helmer et al., 2009), and the Normalized Difference Moisture Index (NDMI) (also referred to as the Normalized Difference Structure Index and the Normalized Difference Infrared Index). The WBDI and NDMI are calculated as:

$$\text{WBDI} = \text{TC Wetness} - \text{TC Brightness} \tag{1.2}$$

$$\text{NDMI} = (\text{NIR}_{b4} - \text{SWIR}_{b5})/(\text{NIR}_{b4} + \text{SWIR}_{b5}) \tag{1.3}$$

However, lowland humid tropical forests recovering from previous clearing may become spectrally indistinct from mature forests within 15–20 years (Boyd et al., 1996; Steininger, 2000), though slower-growing tropical forests, like montane or dry forests, can remain spectrally distinct longer (Helmer et al., 2000; Vieira et al., 2003). Only a handful of forest age classes can be reliably distinguished in single-date multispectral imagery. Age differences are blurred by differences in disturbance type and intensity that affect regrowth rates and related spectral responses during forest succession (Nelson et al., 2000; Arroyo-Mora et al., 2005; Thenkabail et al., 2004; Foody and Hill 1996), although age explains more variability in rates of forest regrowth than does disturbance type (Helmer et al., 2010; Omeja et al., 2012).

Recently logged forest has less biomass than old-growth forest, but it may become spectrally indistinct from mature forest within a year or two (Asner et al., 2004a), which is another case in which the forest canopy recovers faster than forest AGLB. In a study in Sabah, Malaysia, conventional logging reduced forest biomass by 67% but reduced impact logging by 44% (Pinard and Putz, 1996). In moist forests of Amazonia, AGLB decreased by only 11–15% after reduced impact logging (Miller et al., 2011).

The youngest regenerating forest patches in landscapes usually do not dominate pixels as large as those of coarse spatial resolution imagery like MODIS. The outcome is that maps from such imagery have high error rates for secondary tropical forest (Adrah et al., 2022; Dupuis et al., 2020; Staal et al., 2020; Laurin et al., 2019), but see Lucas et al. (2002). When modeling pixel fractional cover of one or more young forest classes vs. nonforest vs. old forest with MODIS, for example, secondary forest is modeled with the most bias and the least precision (Braswell et al., 2003; Tottrup et al., 2007). In Amazonia, the model R-square values for fraction of secondary forest cover were 0.35 for MODIS data alone and 0.61 for MODIS plus MISR data. At the spatial resolution of 1.1 km, corresponding to most of the MISR bands, resulting maps overestimated secondary forest area by 26%. Converting fractional secondary forest cover to discrete classes underestimated secondary forest area by 43% (Braswell et al., 2003). Similarly Carreiras et al. (2006) concluded that the errors for decision tree classification of secondary forest with SPOT 4 Vegetation across Amazonia were unacceptably high.

1.5.4 Detecting Tropical Forest Degradation with Multispectral Imagery

Tropical forests suffer anthropogenic pressures that perturb their structure and ecological functioning (Vitousek et al., 1994). Human activities that disturb them range from plant collecting and human habitation to total deforestation. Many of these forest disturbances can occur at fine spatial scales of less than 5 to tens of meters, including forest fire (Aragão and Shimabukuro, 2010), recent logging (Costa et al., 2023; Lahssini et al., 2022; Cross et al., 2019; Ferreira et al., 2019; Berveglieri et al., 2018; Asner et al., 2005; Sist and Ferreira, 2007), road networks (Laporte et al., 2007; Laurance et al., 2009), mining (Peterson and Heemskerk, 2001), and expanding agricultural frontiers (Dubreuil et al., 2012). These human impacts appear like small isolated objects within an ocean of greenness (Souza et al., 2003). They appear as points (logging gaps), lines (roads, trails), both points and lines (logging decks plus skid trails), and with mining areas both bare soil and pooled water are present.

Although these disturbances can be small, medium-resolution remote sensing techniques can detect and quantify them within homogeneous forest cover (Gond et al., 2004). Compared with fine-scale imagery, images with pixels of 5–30 m have lower or no cost while more frequently covering larger areas of tropical forest. Consequently, medium-resolution imagery constitutes an excellent tool for assessing logging activities in tropical forests across large scales (Asner et al., 2005). Much work to detect finely scaled disturbances of tropical forests uses pixel-level spectra (Section 1.5.3.1) Other work models subpixel spectra to derive continuous variables for monitoring fine-scale disturbances, focusing on degradation of forest C storage for REDD+ programs and ecosystem models (Section 1.5.3.2).

1.5.4.1 Detecting Fine-Scale Forest Degradation at the Pixel Level

Detecting small canopy gaps and skid trails that have been open for less than six months is possible in French Guiana with SPOT 5 HRG (High Resolution Geometric) images (Gond and Guitet, 2009). The technique developed is based on the local contrast between a photosynthetically active surface (the forest) and one with no or little photosynthetic activity (the gap itself). Using the three main channels dedicated to vegetation identification (Red [0.61–0.68µm], near-infrared [0.79–0.89µm], and short-wave infrared [1.58–1.75µm] wavelengths) the contrast between forests and gap is increased enough to be accurately depicted. The detection of an undisturbed forest pixel is made by multiple thresholds on the different reflectances. The advantage of standard remotely sensed data like SPOT 4/5 or Landsat 5/7/8 is the possibility to detect automatically the focused object (Pithon et al., 2013). The automatic processing makes the system operational for tropical forest management and depends only on image availability.

1.5.4.1.1 Road and Trail Detection

Road and trail detection is also a challenge for tropical forest management. Opening, active, and abandoned road and trail networks are a permanent landmark of tropical forest openness and degradation (Costa et al., 2023; Lahssini et al., 2022; Cross et al., 2019; Zaki et al., 2017; Laurance et al.,

2009). Documenting this dynamic is possible with the 30 years of medium-resolution radiometer archives (Landsat and SPOT). In 2007 Laporte et al. (2007) photo interpreted Landsat imagery to map the road and trail network across the forests of Central Africa to show which forest areas are endangered by logging activity. When displaying red, near-infrared, and short-wave infrared channels in red, green, and blue, active roads and trails are "brown"; abandoned roads and trails are "green," and intact tropical forests are "dark green" (de Wasseige et al., 2004). To process automatically the archives for large areas, Bourbier et al. (2013) proposed a method for using Landsat archives to allow tropical forest managers to visualize the road and trail network dynamism at local (concession) or national scales.

1.5.4.1.2 Mining Detection

Detecting mining activity is slightly different. In general, detecting legal mining is not a real challenge because bare surfaces are prominent and easily mapped. When mining is illegal in tropical forests, however, the bare surface is much smaller and difficult to detect (Almeida-Filho and Shimabukuro, 2002). The additional difficulty comes from the mobility of the illegal miners. A recent abandoned mining site is detectable, but the miners have left. Detecting active mining sites where miners are illegally working is most critical to managers. To map active mining sites in French Guiana, an automatic system using SPOT 5 imagery from a local reception station has been operational since 2008 (Gond et al., unpublished). The system is based on detecting turbid waters resulting from debris washing. Again, the object "turbid water" sharply contrasts with its environment, as with tropical forest vs. bare soil. Using red, near-infrared, and short-wave infrared channels, turbid water is detected by multiple thresholds on reflectances. So far the operational system has processed over 1230 SPOT 5 images to ensure regular coverage in space and time of illegal mining activity in French Guiana (Joubert et al., 2012).

1.5.4.2 Detecting Forest Degradation at the Subpixel Level with Spectral Mixture Analysis

Forest degradation in the context of REDD+ can be defined as a persistent reduction in carbon stocks or canopy cover caused by sustained or high-impact disturbance (Bhavsar et al., 2024; Njomaba et al., 2024; Ngo et al., 2023; Bullock et al., 2020; Koyama et al., 2019; Laurin et al., 2019; Wang et al., 2019; Berveglieri et al., 2018; Cross et al., 2018; Erinjery et al., 2018; Zaki et al., 2017). As a result, forest degradation is often expressed as a complex, three-dimensional change in forest structure related to the introduction of areas of bare soil, piles of dead vegetation created by the residues and collateral damage of removed trees and other plants, and areas with standing dead or damaged tree trunks associated with partial tree fall. Burned forests also leave surface fire scars, indicated by patches of charred vegetation and bare ground (Cochrane et al., 1999; Alencar et al., 2011). Much of tropical forest degradation occurring around the world is driven by selective logging and fires that escape into forests from neighboring clearings. At the multispectral sensor resolution of Landsat, SPOT, and MODIS, it is expected that forest degradation will be expressed in varying combinations of green vegetation (GV), soil, non-photosynthetic vegetation (NPV) and shade within image pixels.

Spectral mixture analysis (SMA) models can be used to decompose the mixture of GV, NPV, soil, and shade reflectances into component fractions known as endmembers (Adrah et al., 2022; Lahssini et al., 2022; Cross et al., 2019; Wang et al., 2019; Berveglieri et al., 2018; Cross et al., 2018; Erinjery et al., 2018; Zaki et al., 2017; Adams et al., 1995). SMA has been extensively used throughout the world's tropical forests to detect and map forest degradation (Asner et al., 2009a). For example, subpixel fractional cover of soils derived from SMA was used to detect and map logging infrastructure, including log landings and logging roads (Souza and Barreto, 2000), while NPV fraction improved the detection of burned forests and of logging damage areas (Cochrane and Souza, 1998; Cochrane et al., 1999). GV and shade enhances the detection of canopy gaps created by tree fall (Asner et al., 2004c) and forest fires (Morton et al., 2011).

SMA models usually assume that the image spectra are formed by a linear combination of n pure spectra, or endmembers (Adams et al., 1995), such that:

$$R_b = \sum_{i=1}^{n} F_i R_{i,b} + \varepsilon_b \quad (1.4)$$

for

$$\sum_{i=1}^{n} F_i = 1 \quad (1.5)$$

where R_b is the reflectance in band b, $R_{i,b}$ is the reflectance for endmember i in band b, F_i the fraction of endmember i, and ε_b is the residual error for each band. The SMA model error is estimated for each image pixel by computing the RMS error, given by:

$$\text{RMS} = \sum \left[n^{-1} \sum_{b=1}^{n} \varepsilon_b \right]^{1/2} \quad (1.6)$$

As mentioned, in the case of degraded forests, the expected endmembers are GV, NPV, soil, and shade fractions. Including a cloud endmember is also possible, which improves the detection and masking of clouds when mapping forest degradation over large areas with long time series of imagery in the Amazon region (Souza et al., 2013). To calibrate the model, the endmembers can be obtained directly from the images (Small, 2004) or from reflectance spectra acquired in the field with a handheld spectrometer (Roberts et al., 2002). The advantage of obtaining endmembers directly from images is that spatial and radiometric calibration between field and sensor observations is not required. SMA can be automated to make this technique useful for mapping and monitoring large tropical forest regions. A Monte Carlo unmixing technique using reference endmember bundles was proposed for that purpose (Bateson et al., 2000), as well as generic endmember spectral libraries (Souza et al., 2013).

1.5.4.3 Interpreting and Combining Subpixel Endmember Fractions and Derived Indices

SMA fractions can be combined into indices to further accentuate areas of forest degradation. For example, the Normalized Difference Fraction Index (NDFI) was developed to enhance the detection of forest degradation by combining the detection capability of individual fractions (Souza et al., 2005). NDFI values range from −1 to 1. For intact forests, NDFI values are expected to be high (i.e., about 1) due to the combination of high GV_{shade} (i.e., high GV and canopy shade) and low NPV and soil values. As forest becomes degraded, the NPV and soil fractions are expected to increase, lowering NDFI values relative to intact forest. Bare soil areas will produce an NDFI value of −1 because of the absence of GV.

Another approach to SMA allows for uncertainty in the endmember reflectance spectra used for decomposing each pixel into constituent cover types. Referred to as endmember bundles (Bateson et al., 2000), SMA with spectral endmember variability provides a means to estimate GV, NPV, soil, and shade fractions with quantified uncertainty in each image pixel. Using a Monte Carlo approach, Asner and Heidebrecht (2002) developed automated SMA procedures that have subsequently been used to map forest degradation due to logging or understory fire in a wide variety of tropical regions (e.g., Alencar et al., 2011; Allnutt et al., 2013; Bryan et al., 2013; Carlson et al., 2012).

Several mapping algorithms based on spatial and contextual classifiers, decision trees, and change detection have also been applied to SMA results to better map forest degradation using Landsat, SPOT, and MODIS imagery. These techniques are discussed elsewhere (Souza Jr. and Siqueira, 2013; Asner et al., 2009b). Additionally, large-area mapping and estimates of forest degradation in the Amazon region have also been conducted using these techniques (Suab et al., 2024; Bolívar-Santamaría et al., 2021; Ferreira et al., 2019; Wang et al., 2019; Berveglieri et al., 2018; Cross et al., 2018; Erinjery et al., 2018; Zaki et al., 2017; Asner et al., 2005; Souza et al., 2013).

1.6 MAPPING TROPICAL FOREST TYPES WITH MULTISPECTRAL IMAGERY

1.6.1 Forest Types as Strata for REDD+ and Other C Accounting

Maps of forest type are critical to tropical forest management, including for REDD+ and other GHG inventories. When estimating tropical forest AGLB and other C stores with existing inventory ground plots or LiDAR data, the estimates are generally stratified by forest type (Zhe et al., 2024; Bullock et al., 2020; Helmer et al., 2009; Asner, 2009; Salimon et al., 2011). When designing forest inventories or LiDAR surveys, stratifying sample locations by forest type improves the efficiency of the sample design (Wertz-Kanounnikoff, 2008), including stratification with types defined by disturbance history (Salk et al., 2013). Stratification by topography or geology may also be important (Ferry et al., 2010; Laumonier et al., 2010) if forest type does not inherently account for related variability in AGLB. An informative review and synthesis of LiDAR sample design as it relates to forest parameter estimation over large forest areas is available in Wulder et al. (2012). Another important role of forest type maps based on multispectral satellite imagery is that they are often used to account for the distributions of species and habitats when planning representative reserve systems. For this reason forest type maps are also useful to identify where deforestation or wood harvesting is "leaking" to forests that are critical to conserve but that store less C than forest areas being targeted in REDD+ or carbon offset projects.

Most satellite image-based maps of tropical forest types map classes of forest *formations*. Vegetation formations are defined by growth form and physiognomy. At the simplest level, forest formations may distinguish among closed, open, and wetland forests. More detailed formations may distinguish among forests with different leaf forms or phenology (e.g., deciduous vs. evergreen, broad-leaved vs. needle-leaved, or descriptors that imply a suite of physiognomic characteristics, such as "dry," "montane," or "cloud" forests. More detailed than forest formations are forest *associations*, which distinguish among tree species assemblages. For example, in Figure 1.5, which we discuss in Section 1.6.4, the upper-level headings for forests are forest formations. The subheadings under each forest formation are forest associations.

1.6.2 High-Resolution Multispectral Imagery for Mapping Finely Scaled Habitats

High-resolution imagery makes excellent reference data for calibrating classification and mapping models based on imagery with coarser spatial resolution, but using it as the primary basis for mapping forest types has several disadvantages. In high-resolution imagery, the within-stand spectral variability of forest types can be large, varying within tree crowns, for example, such that digital classifications at the pixel scale cannot distinguish many forest types (Zhe et al., 2024; Lahssini et al., 2022; Cross et al., 2019; Wang et al., 2019). Also, these images cover relatively small areas, making them inefficient for mapping forest types over large areas (Nagendra and Rocchini, 2008). Existing archives of high-resolution imagery also lack short-wave infrared bands (SWIR), which are important in vegetation mapping. Because Landsat ETM+ data has SWIR bands, for example, Thenkabail et al. (2003) found that three floristic tropical forest classes were more distinct in ETM+ data than in IKONOS imagery. WorldView-3, however, will have eight SWIR bands collected at a spatial resolution of 3.7 m.

Yet satellite imagery with high spatial resolution can aid in mapping finely scaled habitats or habitat characteristics. Example habitats are edges or linear features: riparian areas (Nagendra and Rocchini, 2008), roadsides or other corridors, or strands of vegetation types along coastlines. Habitats with high mechanical, chemical, or moisture stress can also be finely scaled. Example stresses are fast-draining substrates where microtopography strongly affects vegetation, like substrates of limestone (Martinuzzi et al., 2008) or sand, or substrates that are also semi-toxic, like serpentines. High winds or drier climate, as in savanna ecotones, also lead to finely scaled habitats.

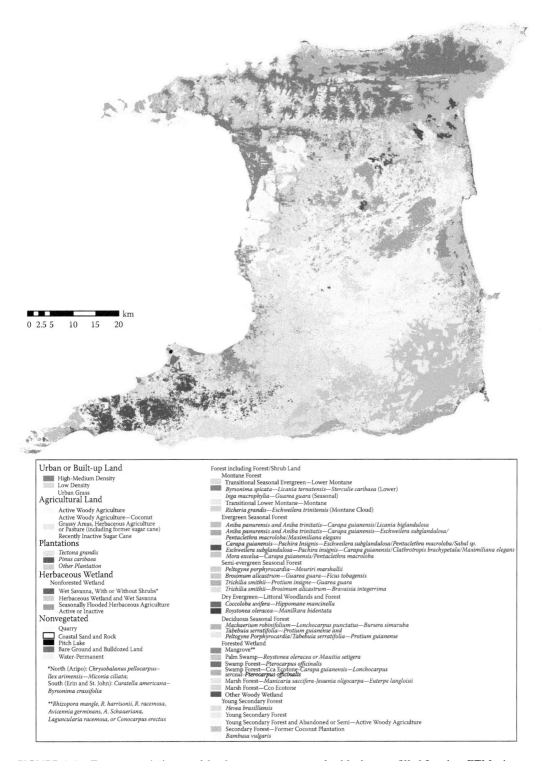

FIGURE 1.5 Forest associations and land cover were mapped with the gap-filled Landsat ETM+ imagery, centered around the year 2007, plus synthetic multiseason imagery developed from three gap-filled TM images from the 1980s that were from the mid- to late dry season, including from severe drought.

Savanna ecosystems, for example, range in tree cover from grassland to forest, which is why we mention them here. Tree cover may change over meters, and high-resolution imagery may be most effective for habitat mapping. Boggs (2010) applied object-oriented classification to 4-m multispectral IKONOS imagery to map tree cover patterns in the Mozambique savanna.

In Namibia, tree clusters and grass patches are distinguishable with object-oriented or pixel-level classifications of pan-sharpened Quickbird imagery (0.6-m pixels). In contrast, 10-m multispectral SPOT 5 pixels, though pan-sharpened to 2.5 m, required object-oriented classification (Gibbes et al., 2010).

Object-oriented classification of medium-resolution imagery can indeed sometimes substitute for high-resolution imagery when it can discern finer-scale features of interest that are missed with pixel-level classifications. Jamaica Newman et al. (2011) found that object-oriented classification of medium-resolution imagery led to better characterization of roads and forest fragmentation metrics than pixel-level classification did. Object-oriented classification of ASTER data can map savanna habitats in northwest Australia, and it was also more accurate than pixel-level classification (Whiteside et al., 2011). Longer-wave infrared bands were resampled to the 15-m resolution of the visible and near-infrared bands.

1.6.3 Remote Tree Species Identification and Forest Type Mapping

Many tropical tree species can be identified by photo interpretation of high-resolution satellite imagery or air photos. With tree crowns in tropical forest often reaching > 10 m in diameter, subcrown features are visible. In subtropical to warm temperate forests of east central Queensland, Australia, Tickle et al. (2006) correctly identified dominant tree species in most of 150 air photo plots with stereo color air photos of scale 1:4000 (~2–m resolution). With these data they categorized the air photo plots into five genus groups (Bhavsar et al., 2024; Costa et al., 2023; Adrah et al., 2022; Dupuis et al., 2020; Koyama et al., 2019; Zaki et al., 2017).

In the moist forests of Panama, Garzón-López et al. (2013) found that visual analysis of high-resolution color air photos (0.13 m pixels) can reveal the spatial distributions of some tropical forest canopy trees. Of 50 common canopy species on a 50-ha plot, 22% had crowns that were distinct in the photos. Of four species tested, interpreters found 40% of the stems that were recorded in field surveys; the resulting maps accurately showed the species' spatial patterns. Sánchez-Azofeifa et al. (2011) concluded that 2.4-m multispectral Quickbird imagery can reveal the spatial distribution and clusters of a species that is conspicuous when flowering, though immature or nonflowering individuals are often missed.

In French Guiana, Trichon and Julien (2006) found that 12 of the 15 most common canopy species or species groups were identifiable, with an accuracy of 87%, in color air photos ranging in scale from 1:1500 to 1:18000 (~0.75- to 4-m pixels). Twenty to 25% of trees with dbh >= 10 cm, and all trees with dbh >= 20 cm, were visible in the photos. For ten taxa from old-growth Ecuadorian Amazon forest representing a range of crown structures, González-Orozco et al. (2010) found that photo interpretation of large-scale air photos with a dichotomous key correctly identified individuals at a rate of > 70% for three of the taxa and > 50% for two of them.

That photo interpreters can identify many of the dominant species in tropical forest canopies in high-resolution imagery suggests that, given field-based knowledge of the composition and distribution of tree floristic classes (i.e., tree species associations), which are defined by dominant tree species, floristic types of tropical forest can be identified in high-resolution multispectral imagery. Consequently, reference data from photos interpreting high-resolution multispectral imagery can supplement field data as a source of training and validation data for mapping tropical tree communities with satellite imagery (Helmer et al., 2012).

1.6.4 Mapping Tropical Forest Types with Medium-Resolution Imagery

In mapping tropical forest types with multispectral imagery, spectral similarity among forest classes is a major challenge. Disturbance, differences in topographic illumination, artifacts from filling cloud and other data gaps or from scene mosaicking, all increase class signature variability and consequently increase signature overlap among classes. Secondary forest in a humid montane zone, for example, may be spectrally similar to shade coffee or old-growth forest on highly illuminated slopes. When on a shaded slope that same secondary montane forest is spectrally similar to old-growth forest in a less productive zone at higher altitudes (Helmer et al., 2000). Yet digital classifications of multispectral imagery can map many different forest types with some additions: (1) ancillary geographic data; (2) multiseason or multiyear imagery or derived phenology; and (3) pixels for training classification models that represent the variability in environmental and image conditions.

Digital maps of environmental data like topography, climate, or geology help distinguish spectrally similar forest types. With Landsat TM/ETM+, linear discriminant function classifications have incorporated ancillary data via post-classification rules based on topography to map eucalyptus forest types (Skidmore, 1989); adding topographic bands to spectral bands to map land cover and forest physiognomic types (Zhe et al., 2024; Lahssini et al., 2022; Cross et al., 2019; Ferreira et al., 2019; Cross et al., 2018; Elumnoh and Shrestha, 2000; Helmer et al., 2002; Gottlicher et al., 2009) or distinguish among tree floristic classes (Foody and Cutler, 2003; Salovaara et al., 2005); and classifying imagery by geoclimatic zone (Helmer et al., 2002). Image smoothing or segmentation can improve these classifications by reducing within-class spectral variation (Tottrup, 2004; Thessler et al., 2008).

Tree associations or other floristic classes can be separable with multispectral imagery within an ecological zone, particularly if topographic bands are included. With TM/ETM+ and 18–127 plots, studies have separated three to nine floristic classes within lowland evergreen forest in Central Africa, Amazonia, Borneo, and Costa Rica (Zhe et al., 2024; Costa et al., 2023; Staal et al., 2020; Laurin et al., 2019; Zaki et al., 2017; Thenkabail et al., 2003; Salovaara et al., 2005; Foody and Cutler, 2003; Thessler et al., 2005; Sesnie et al., 2010). Chust et al. (2006) mapped nine floristic subclasses with ETM+ data, elevation, and geographic position over a broad environment across central Panama. With Landsat TM data Wittmann et al. (2002) mapped three structural classes of Amazonian várzea forests that corresponded to four associations: early successional low várzea, late secondary and climax low várzea (two associations), and climax high várzea. These studies use spectral data from a single image date and consider only forest; cloudy areas were mapped as such.

When mapping many classes, machine learning classifications more effectively incorporate ancillary environmental data, including date bands for gap-filled images. They also do not assume that class spectral distributions are parametric, and they typically outperform linear classifications. Combining ancillary data and machine learning classification permits classifications that distinguish many forest and land-cover types, even with noisy, cloud gap-filled imagery. Examples with TM/ETM+ include decision tree classifications of one or two seasons of cloud gap-filled Landsat plus ancillary data to map tropical forest physiognomic types and land cover (Kennaway and Helmer, 2007; Helmer et al., 2008; Kennaway et al., 2008). Sesnie et al. (2008) mapped land cover, agriculture type, floristic classes of lowland old-growth forest and three higher-elevation classes based on a map of life zones (*sensu* Holdridge, 1967) with a relatively cloud-free image for each of two scenes. To map tree floristic classes of lowland through montane tropical forest types and land cover in Trinidad and Tobago, Helmer et al. (2012) applied decision tree classification to recent cloud gap-filled Landsat imagery stacked with decades-old, gap-filled synthetic multiseason imagery from droughts (Figure 1.5).

Mapping many physiognomic or floristic classes of tropical forest, as in the aforementioned studies requires (1) thousands of training and testing pixels representing the environmental and spectral range of each class, including the range of pixel dates where gap-filled imagery was used (Helmer and Ruefenacht, 2007); (2) a band that represents the date of the source image for each pixel in the composite image (a *date band*); and (3) a machine learning classification model. The extensive training data needed are rarely available from field plots. But analysts can learn to identify many physiognomic and floristic classes in remotely sensed imagery given field-based knowledge of general distributions, particularly given free viewing of high-resolution imagery online and Landsat image archives, allowing almost unlimited reference data collection.

Helmer et al. (2012) found that all mono- and bi-dominant tree floristic classes and many other tree communities in Trinidad and Tobago could be distinguished in reference imagery from nearby associations by (1) unique canopy structure in high-resolution imagery or (2) distinct or unique phenology on specific dates of either high- or medium-resolution reference imagery. For example, distinct canopy structure at high resolution distinguished *Mora excelsa* forests, littoral associations (frequent palms in one; prostrate stems in the other); *Pterocarpus officinalis* swamps, palm swamps, mangroves, and stands of bamboo (*Bambusa vulgaris*), abandoned coconut (*Cocos nucifera*), teak (*Tectona grandis*), pine (*Pinus caribaea*), and Brazilian rubber (*Hevea brasiliensis*). Phenology, including characteristics like flowering, deciduousness, leaf flushes, or inundation, helped to distinguish seven forest associations in high-resolution reference imagery and four associations in phenologically unique Landsat reference scenes. With this knowledge and reference imagery, thousands of training data pixels could be collected.

Including multiseason imagery in classification models of course resolution imagery also improves spectral distinction among tropical forest types (Bohlman et al., 1998; Tottrup, 2004). What is exciting is that we can now think beyond multiseason imagery to multiyear imagery that captures climate or weather extremes or disturbance history. Helmer et al. (2012) found that adding bands from cloud gap-filled TM imagery from a severe drought that occurred 20 years earlier than the most recent imagery used in the stack of data for classification contributed to the largest increases in accuracy when mapping forest associations in Trinidad. Mapping accuracy of seasonal associations benefited the most. Accuracy improved by 14–21% for deciduous, 7–36% for semi-evergreen, and 3–11% for seasonal evergreen associations, and by 5–8% for secondary forest and woody agriculture. Multiyear multispectral imagery that displays different flood stages helps distinguish between upland and periodically flooded tropical forests (Helmer et al., 2009) and among tropical forested wetland types (and can reflect differences in secondary forest species composition by mapping disturbance type, as mentioned) (Helmer et al., 2010). In Amazonia, de Carvahlo et al. (2013) determined the life cycle length of native bamboo patches with multiyear TM/ETM+ data (Costa et al., 2023; Bolívar-Santamaría et al., 2021; Ganivet and Bloomberg, 2019; Zaki et al., 2017).

1.6.5 Species Richness, Endemism and Functional Traits, and Multispectral Imagery

The tree species richness of tropical forests increases with some of the same variables that influence forest reflectance in multispectral satellite imagery (Suab et al., 2024; Rana et al., 2023; Dupuis et al., 2020; Koyama et al., 2019; Zaki et al., 2017). Richness increases with forest height (among lowland forests with strong edaphic differences), soil fertility (after accounting for rainfall), canopy turnover, and time since catastrophic disturbance; richness decreases with dry season length, latitude, and altitude (Givnish, 1999). We know from forest ground plots that tree species richness also increases with secondary forest age (Wittmann et al., 2002; Chazdon et al., 2007; Helmer et al., 2008). Consequently, over gradients that span from dry to humid, multispectral bands and indices related to vegetation greenness, structure, or disturbance may correlate with species richness. And, in fact, studies have documented such relationships with single-date Landsat TM or ETM+, or Sentinel-2 imagery (Foody and Cutler, 2003; Nagendra et al., 2010; Hernández-Stefanoni et al., 2011). Single-date multispectral data are unlikely, however, to be sensitive to differences in species richness along short environmental gradients.

An important consideration in biodiversity conservation is that species richness alone does not define conservation value: representation across as many native ecosystems and species as possible is important. Many less productive tropical forest types with less tree species richness, like cloud forests, or forests on harsh or drying soils, like those on ultramafic or limestone substrates or ombrotrophic sands, have greater endemic and native species richness. In addition, land-use history and forest structure, species composition and functional attributes, and the relationships among these factors are often strongly related to geoclimatic variables across landscapes. As a result, relative basal area of endemic, native, and introduced tropical tree species counts, for example, have been reliably mapped with models based on decadal and seasonal time series of multispectral imagery, topography, climate, and geology (Helmer et al., 2018). Tropical forest community-level functional traits, functional groups, or functional diversity have also been mapped with multispectral imagery plus geoclimate and topography data (Helmer et al., 2018; Aguirre-Gutiérrez et al., 2021).

1.6.6 Tropical Forest Type Mapping at Coarse Spatial Scale

In tropical regions extending over large areas, multiseason data from monthly, annual, or multiyear composites of imagery with coarse spatial resolution have supported large-area mapping of tropical forest formations with even linear classification methods (Zhe et al., 2024; Bullock et al., 2020; Koyama et al., 2019; Gond et al., 2011, 2013; Joshi et al., 2006; Pennec et al., 2011; Verhegghen et al., 2012). For example, Gond et al. (2011) mapped five classes of forest canopy openness across the French Guiana with an unsupervised classification of an annual composite image of SPOT 4 Vegetation data. Across Central Africa, Gond et al. (2013) mapped 14 forest formations with one year of 8- and 16-day MODIS image composites. The forest formations were based on leaf phenology and canopy openness. With one year of NDVI composite images from the Indian Resource Satellite (IRS 1C) Wide Field Sensor (WiFS) across India, Joshi et al. (2006) mapped 14 forest formations. The formations were labeled by phenology and climatic class (e.g., tropical dry deciduous forest, tropical moist deciduous forest, and so on). Verhegghen et al. (2012) applied unsupervised classification to seasonal and annual composites of MERIS (Medium Resolution Imaging Spectrometer) and SPOT 4 Vegetation data for the Congo basin, producing a map with six forest classes that were based on leaf phenology, canopy openness, and elevation class. Producer and user accuracy for forest classes in the latter two studies were mostly between 80% and 100%.

Combining ancillary data, monthly image composites of imagery with coarse spatial resolution but high temporal resolution, and decision tree classification, has permitted forest classifications at subcontinental to global scales or has distinguished many more forest formations. Decision tree classification of monthly composites of imagery with coarse spatial resolution, and mosaics of such composites, is also used to map tropical forests over large areas. Examples of such large-area maps based on MODIS image composites are of tropical forest ecoregion (Muchoney et al., 2000), biome (Friedl et al., 2002), or forest formation (Carreiras et al., 2006). With decision tree classification of dry season MODIS image composites, Portillo-Quintero and Sánchez-Azofeifa (2010) mapped the extent of two classes of tropical dry forests (tropical dry forest and forests in tropical grasslands, savannas, and shrublands), for the mainland Neotropics plus the Greater Antilles. Overall accuracy was 82%. The importance of this latter work is that global land-cover maps often misclassify dry tropical forests as some other land cover.

1.6.7 Tropical Forest Type Mapping and Image Spatial Resolution

Without question, multiseason data greatly improves the number of different physiognomic or floristic classes of tropical forest that can be mapped with multispectral satellite imagery (Njomaba et al., 2024; Ngo et al., 2023; Bullock et al., 2020; Koyama et al., 2019). Monthly image composites or derived phenology metrics, as are possible with coarse-resolution imagery, are optimal. Joshi et al. (2006) qualitatively compared their WiFS-based map of forest types across India with a forest map of the country based on Linear Imaging Self Scanner data, which has a pixel resolution of 23.5 m but a 24-day repeat cycle. They concluded that the five-day revisit cycle of WiFS, which allowed them to incorporate 12 monthly image composites, yielded better information on forest types and other vegetation and land-cover classes, even though WiFS has a spatial resolution of 188 m.

However, tropical forest types can change greatly over small areas, and spatial resolutions coarser than 100–200 m are too coarse to distinguish important differences in forest types in many places. In tropical islands, for example, forest floristic and physiognomic types that are critical to distinguish for conservation planning would be poorly delineated. Medium-resolution imagery with a shorter revisit cycle would greatly improve prospects for mapping tropical forest types with multispectral imagery. This could be more easily accomplished, for example, if AWiFS data, with its 56-m spatial resolution and five-day revisit cycle, were available for all of the tropics, or if the Landsat program had of a constellation of at least four satellites.

In addition, past disturbances affect forest physiognomy and species composition, and some forest classes may only become spectrally distinct during periodic drought and flooding. Consequently, forest type mapping can also benefit when older satellite imagery or long image time series are incorporated into forest type mapping, as in Helmer et al. (2010, 2012).

Finally, to distinguish tropical forest types on small mountains or small islands; along coastlines, rivers, and other linear features; or in other finely scaled landscapes, high-resolution imagery will be needed.

1.7 MONITORING EFFECTS OF GLOBAL CHANGE ON TROPICAL FORESTS

1.7.1 Progress in Monitoring Tropical Forests at Subcontinental to Global Scales

Tropical forest mapping with coarse-resolution imagery in optical remote sensing is very constrained by cloud cover. Helpfully, its high temporal frequency of acquisition balances the handicaps of cloud-contaminated pixels (McCallum et al., 2006). Historical long time series from NOAA-AVHRR paved the way for this research (Tucker et al., 1985; Townshend et al., 1991). Indeed, the spectral capacities from visible to short-wave infrared of these sensors motivated many applications and technological developments. The identification of tropical forest patterns has improved over time (Holben, 1986; Mayaux et al., 1998; DeFries et al., 2000) and benefits from a large panel of vegetation indices for evaluating photosynthetic activity (Bhavsar et al., 2024; Costa et al., 2023; Lahssini et al., 2022; Staal et al., 2020; Koyama et al., 2019; Cross et al., 2018; Rouse et al., 1974; Huete, 1988; Pinty and Verstraete, 1992; Qi et al., 1994; Gao, 1996).

At the end of the 1990s the experiences gained from these applications led to new sensors adapted to land surface observation, including SPOT-Vegetation (March 1998) and TERRA-MODIS (December 1999) (Friedl et al., 2010). Spatial resolutions were improved from 1.1 km (NOAA-AVHRR) to 1.0 km (Vegetation), 0.3 km (MERIS), and 0.5/0.25 km (MODIS). Geo-location was improved. Specific spectral bands dedicated to vegetation were implemented. New sensor technology was developed, such as the push-broom system on Vegetation, which avoids large swath distortions. After 15 years of feedback, we may now measure the added value of these sensors.

Research to characterize tropical forests at subcontinental to global scales has become more accurate and precise (Mayaux et al., 2004; Vancutsem et al., 2009) by taking phenology into account (Xiao et al., 2006; Myneni et al., 2007; Doughty and Goulden, 2008; Park, 2009; Brando et al., 2010). Repetitive observation and long temporal archives make possible land-surface observation on 8-, 10-, or 16-day time periods and allows phenology studies to take advantage of both high spectral quality and high observation frequency (Verhegghen et al., 2012 for MERIS and Vegetation; Gond et al., 2013 for MODIS). In addition, there are more forest attributes being characterized, including forest edges (to delimit forest patches and more accurately estimate forest areas) (Mayaux et al., 2013; Verhegghen et al., 2012), aboveground biomass (Costa et al., 2023; Lahssini et al., 2022; Staal et al., 2020; Laurin et al., 2019; Wang et al., 2019; Malhi et al., 2006; Saatchi et al., 2007; Baccini et al., 2008), deforestation and forest degradation (Doughty et al., 2023; Ngo et al., 2023; Dupuis et al., 2020; Staal et al., 2020; Cross et al., 2019; Ferreira et al., 2019; Cross et al., 2018; Achard et al., 2002; Duveiller et al., 2008; Hansen et al., 2008; Baccini et al., 2012; Desclée et al., 2013), and climate change impacts (Phillips et al., 2009; Lewis et al., 2011; Samanta et al., 2010).

Sensor capabilities and computer capacities now allow production of global-scale land-cover maps (Bartholomé and Belward, 2005, for Vegetation; Friedl et al., 2002 and Hansen et al., 2008, for MODIS; Bontemps et al. 2012, for MERIS), which have greatly improved our knowledge of land surface cover in comparison with previous views obtained from NOAA-AVHRR (Loveland and Belward, 1997; DeFries and Townshend, 1994).

Tropical forest characterizations with multispectral imagery have now begun to address a real challenge: that of monitoring and understanding climate change impacts on the biosphere (Gibson et al., 2011). Tropical forests are particularly threatened by global temperature increases and the possibility of modified rainfall regimes (Zelazowski et al., 2011). These changes will influence vegetation spatial distribution (Parmesan and Yohe, 2003), forest functioning (Nemani et al., 2003), and carbon storage capacity (Stephens et al., 2007), which may in turn affect climate. In this context monitoring tropical forests with coarse-resolution satellite imagery is of prime importance to understanding biological processes and managing forest resilience. Zhao and Running (2010), for example, showed that large-scale droughts have decreased net primary productivity in the Southern Hemisphere, including tropical Asia and South America. As we discuss next, however, some critical remote sensing problems still need to be addressed before we can effectively monitor some important effects of droughts on tropical forests.

1.7.2 The Feedbacks among Tropical Forest Disturbance, Drought, and Fire

Multispectral imagery can help characterize the positive feedback among tropical forest disturbance, fire, and climate. First, tropical forest clearing dries nearby forest, and multispectral imagery can detect forest clearing. In Amazonia, for example, Briant et al. (2010) delineated forest boundaries with MODIS multispectral bands and found that as the forest becomes more fragmented, drops in MODIS-based indices related to canopy moisture extend farther into intact forest, and that the old forest in more fragmented landscapes has lower canopy moisture to begin with. Second, forest cover data also reveal that forests desiccated by fragmentation and other disturbance are more susceptible to fire. Armenteras et al. (2013) used forest fragmentation indices from forest cover maps, along with active fire data from MODIS, which uses MODIS thermal bands to show that forest fires increase in extent and frequency with fragmentation. Logging also increases forest vulnerability to fire (Uhl and Buschbacher, 1985; Woods, 1989), and as outlined earlier logging can be detected with medium-resolution multispectral imagery. In Ghana, Dwomoh et al. (2019) also found that forest degradation increases tropical forest susceptibility to fire.

A third aspect of the disturbance-fire-climate feedback is that drought magnifies the association between disturbance and fire (Siegert et al., 2001; Alencar et al., 2006). In Amazonia, fire scars mapped with Landsat occurred mostly within 1 km of clearings during normal dry seasons but extended to 4 km from clearings during drought years (Alencar et al., 2006). Some of these studies relied on Landsat imagery to quantify forest fragmentation, because of its finer spatial resolution, or radar imagery to map fire scars, to avoid clouds.

Amazonian droughts are likely to become more common and severe with climate change. During droughts, reduced forest growth and increased tree mortality cause intact forests to shift from a net sink to a net source of CO_2 to the atmosphere (Lewis et al., 2011). However, monitoring drought effects that are spectrally subtle, like increased tropical tree mortality or changes in phenology, remains a challenge. For example, studies have found that vegetation greenness may increase, decrease, or show no change during drought. The increases could stem from decreased cloud cover, leaf flushes related to increased sunlight, decreased canopy shadow from increased mortality of the tallest trees, an artifact of seasonal variations in sun-sensor geometry, or all of these factors, and despite observation frequency, cloud and smoke contamination in pixels still obscures trends in vegetation greenness (Anderson et al., 2010; Asner and Alencar, 2010; Samanta et al., 2010; Morton et al., 2014). Though Amazon forest greenness may increase during severe drought, photosynthesis may decline (Yang et al., 2018). Asner et al. (2004b) suggest that metrics from hyperspectral imagery may be better suited to resolve drought effects on tropical forests because they are sensitive to canopy leaf water content and light use efficiency. A challenge, then, is to develop a system that, despite cloud and smoke contamination, integrates these different sensors to continuously monitor the feedback between forest fragmentation, logging, fire, and climate.

1.7.3 Storm-Related and Other Tree Mortality

Storm intensities and frequencies, and associated severe wind and rain events, are likely increasing from climate change (Keellings and Hernández Ayala, 2019; Feng et al., 2023). Besides direct tree death from forest clearing, logging, drought, and fire, storms and associated landslides, flooding, and subsequent fire can cause canopy damage or tree mortality that is detectable with multispectral satellite imagery (Feng et al., 2020; Hall et al., 2020; Yu and Gao, 2020; Pascual et al., 2022; Emmert et al., 2023; Ping et al., 2023). Among forest types and along environmental gradients that multispectral imagery can identify, average tree mortality rates can also vary. Image bands, indices, and phenology metrics from multispectral imagery, including from past decades, were among the top predictors of tree mortality across Puerto Rico, where dry and more deciduous forests, older forests, and forests less impacted by hurricanes had lower mortality rates (Helmer et al., 2023). Tree canopy cover and changes in tree canopy cover or vegetation indices can be post-hurricane predictors of tropical tree mortality (Helmer et al., 2023) or biomass loss (Hall et al., 2020). Fire is more widespread after hurricanes (Helmer et al., 2010) and increases tropical tree mortality. The patchy nature of storm-related mortality, and the speed with which canopy greenness can recover, means that imagery with finer spatial or temporal resolution may more precisely gauge-related tree mortality (Emmert et al., 2023; Ping et al., 2023).

1.8 SUMMARY AND CONCLUSIONS

Across spatial scales, increased image access and data usability are the main factors driving an explosion of progress in characterizing tropical forests with multispectral satellite imagery (Bhavsar et al., 2024; Njomaba et al., 2024; Pletcher et al., 2024; Suab et al., 2024; Zhe et al., 2024; Costa et al., 2023; Doughty et al., 2023; Ngo et al., 2023; Rana et al., 2023; Adrah et al., 2022; Lahssini et al., 2022; Bolívar-Santamaría et al., 2021; Bullock et al., 2020; Dupuis et al., 2020; Staal et al., 2020; Cross et al., 2019; Ferreira et al., 2019; Ganivet and Bloomberg, 2019; Koyama et al., 2019; Laurin et al., 2019; Wang et al., 2019; Berveglieri et al., 2018; Cross et al., 2018; Erinjery et al., 2018; Zaki et al., 2017). The menu of pre-processed image products of the second generation of high-frequency Earth Observation satellite sensors, MODIS and SPOT Vegetation, along with their improved spatial and spectral resolution, led to a wider group of users applying multispectral imagery across larger areas and in more diverse ways. Products like cloud-screened composites of Earth surface reflectance, vegetation indices, quality flags, fire flags, and land cover have enabled efforts to map tropical forest productivity, type, phenology, moisture status, and biomass and to study the effects of climate change on tropical forests, particularly feedback among drought, fire, and deforestation.

At the scale of medium-resolution imagery, free access to Landsat, and in some cases free access to SPOT imagery, has spawned many new applications that rely on dozens, hundreds, or thousands of scenes, including scenes with scan-line gaps or scenes previously considered too cloudy to bother with. Cloud- and gap-filled Landsat imagery and image time series are now used to automatically detect forest clearing, partial disturbance, or regrowth; quantify degradation of tropical forest C storage; map the age, structure, biomass, height, and disturbance type of secondary tropical forests; automatically and more precisely mask clouds and cloud shadows in imagery; and create detailed maps of forest types in these often cloudy landscapes. Characterizing tropical forest phenology at medium and even high resolution will now be possible for many places, given recent additions to image preprocessing, like atmospheric correction, reduced repeat cycles for available medium- and high-resolution image sources, and harmonized Landsat 8 and Sentinel-2 (HLS) imagery (Claverie et al., 2018) will allow better insights and monitoring of tropical forest ecosystem seasonal patterns of productivity, biomass, and phenological diversity (e.g., Wang et al., 2020, 2024; Zhu et al., 2021; Medeiros et al., 2022). Many of these automated applications build on the experiences gained from the high-frequency, coarse spatial resolution imagery, and all of them are relevant to REDD+ monitoring, reporting, and verification.

At fine spatial scales, free viewing and low-cost printing of georeferenced high-resolution imagery via online tools like Google Earth and Bing supplement field data for training and testing the aforementioned products that are based on medium- and coarse-resolution imagery. In addition, scientists have used image products from Google Earth to estimate tropical forest biomass directly. New commercial sensors that produce multispectral satellite imagery with spatial resolutions ≤ 0.5 m should also allow more disturbance types and tropical tree communities to be remotely identifiable.

1.9 ACKNOWLEDGMENTS

Thanks to John Armston and Ariel Lugo for their invaluable comments on this text. This research was conducted in cooperation with the University of Puerto Rico and the USDA Forest Service Rocky Mountain Research Station.

REFERENCES

Achard, F., Eva, H. D., Stibig, H.-J., Mayaux, P., Gallego, J., Richards, T., and Malingreau, J.-P. 2002. Determination of deforestation rates of the world's humid tropical forests. *Science*, 297, 999–1002.

Ackerman, S., Holz, R., Frey, R., Eloranta, E., Maddux, B., and McGill, M. 2008. Cloud detection with MODIS. Part II: Validation. *Journal of Atmospheric and Oceanic Technology*, 25, 1073–1086.

Adams, J. B., Sabol, D. E., Kapos, V., Almeida Filho, R., Roberts, D. A., Smith, M. O., and Gillespie, A. R. 1995. Classification of multispectral images based on fractions of endmembers: Application to landcover change in the Brazilian Amazon. *Remote Sensing of Environment*, 52, 137–154.

Adrah, E., Wan Mohd Jaafar, W. S., Omar, H., Bajaj, S., Leite, R. V., Mazlan, S. M., Silva, C. A., Chel Gee Ooi, M., Mohd Said, M. N., Abdul Maulud, K. N., et al. 2022. Analyzing canopy height patterns and environmental landscape drivers in tropical forests using NASA's GEDI spaceborne LiDAR. *Remote Sensing*, 14, 3172. https://doi.org/10.3390/rs14133172

Aguirre-Gutiérrez, J., Rifai, S., Shenkin, A., Oliveras, I., Bentley, L. P., Svátek, M., Girardin, C. A., Both, S., Riutta, T., Berenguer, E., and Kissling, W. D. 2021. Pantropical modelling of canopy functional traits using Sentinel-2 remote sensing data. *Remote Sensing of Environment*, 252, 112122.

Ahmed, O. S., Franklin, S. E., and Wulder, M. A. 2014. Interpretation of forest disturbance using a time series of Landsat imagery and canopy structure from airborne lidar. *Canadian Journal of Remote Sensing*, 39, 521–542.

Alencar, A., Asner, G. P., Knapp, D., and Zarin, D. 2011. Temporal variability of forest fires in eastern Amazonia. *Ecological Applications*, 21, 2397–2412.

Alencar, A., Nepstad, D., and Diaz, M. C. V. 2006. Forest understory fire in the Brazilian Amazon in ENSO and non-ENSO years: Area burned and committed carbon emissions. *Earth Interactions*, 10, 1–17.

Allnutt, T. F., Asner, G. P., Golden, C. D., and Powell, G. V. 2013. Mapping recent deforestation and forest disturbance in northeastern Madagascar. *Tropical Conservation Science*, 6, 1–15.

Almeida-Filho, R., and Shimabukuro, Y. E. 2002. Digital processing of a Landsat-TM time series for mapping and monitoring degraded areas caused by independent gold miners, Roraima State, Brazilian Amazon. *Remote Sensing of Environment*, 79, 42–50.

Anderson, L. O., Malhi, Y., Aragão, L. E., Ladle, R., Arai, E., Barbier, N., and Phillips, O. 2010. Remote sensing detection of droughts in Amazonian forest canopies. *New Phytologist*, 187, 733–750.

Aragão, L. E., and Shimabukuro, Y. E. 2010. The incidence of fire in Amazonian forests with implications for REDD. *Science*, 328, 1275–1278.

Armenteras, D., González, T. M., and Retana, J. 2013. Forest fragmentation and edge influence on fire occurrence and intensity under different management types in Amazon forests. *Biological Conservation*, 159, 73–79.

Arroyo-Mora, J. P., Sánchez-Azofeifa, G. A., Kalacska, M. E., Rivard, B., Calvo-Alvarado, J. C., and Janzen, D. H. 2005. Secondary forest detection in a Neotropical dry forest landscape using Landsat 7 ETM+ and IKONOS Imagery1. *Biotropica*, 37, 497–507.

Asner, G. P. 2009. Tropical forest carbon assessment: Integrating satellite and airborne mapping approaches. *Environmental Research Letters*, 4, 034009.

Asner, G. P., and Alencar, A. 2010. Drought impacts on the Amazon forest: The remote sensing perspective. *New Phytologist*, 187, 569–578.

Asner, G. P., and Heidebrecht, K. B. 2002. Spectral unmixing of vegetation, soil and dry carbon cover in arid regions: Comparing multispectral and hyperspectral observations. *International Journal of Remote Sensing*, 23, 3939–3958.

Asner, G. P., Keller, M., Pereira, Jr., R., Zweede, J. C., and Silva, J. N. 2004a. Canopy damage and recovery after selective logging in Amazonia: Field and satellite studies. *Ecological Applications*, 14, 280–298.

Asner, G. P., Keller, M., and Silva, J. N. 2004b. Spatial and temporal dynamics of forest canopy gaps following selective logging in the eastern Amazon. *Global Change Biology*, 10, 765–783.

Asner, G. P., Knapp, D. E., Balaji, A., and Páez-Acosta, G. 2009a. Automated mapping of tropical deforestation and forest degradation: CLASlite. *Journal of Applied Remote Sensing*, 3, 033543–033543–033524.

Asner, G. P., Knapp, D. E., Broadbent, E. N., Oliveira, P. J., Keller, M., and Silva, J. N. 2005. Selective logging in the Brazilian Amazon. *Science*, 310, 480–482.

Asner, G. P., Mascaro, J., Anderson, C., Knapp, D. E., Martin, R. E., Kennedy-Bowdoin, T., van Breugel, M., et al. 2013. High-fidelity national carbon mapping for resource management and REDD+. *CarbonBalance and Management*, 8, 1–14.

Asner, G. P., Mascaro, J., Muller-Landau, H. C., Vieilledent, G., Vaudry, R., Rasamoelina, M., Hall, J. S., et al. 2012. A universal airborne LiDAR approach for tropical forest carbon mapping. *Oecologia*, 168, 1147–1160.

Asner, G. P., Nepstad, D., Cardinot, G., and Ray, D. 2004c. Drought stress and carbon uptake in an Amazon forest measured with spaceborne imaging spectroscopy. *Proceedings of the National Academy of Sciences of the United States of America*, 101, 6039–6044.

Asner, G. P., Palace, M., Keller, M., Pereira, R., Silva, J. N., and Zweede, J. C. 2002. Estimating canopy structure in an Amazon forest from laser range finder and IKONOS satellite observations 1. *Biotropica*, 34, 483–492.

Asner, G. P., Rudel, T. K., Aide, T. M., Defries, R., and Emerson, R. 2009b. A contemporary assessment of change in humid tropical forests. *Conservation Biology*, 23, 1386–1395.

Avitabile, V., Baccini, A., Friedl, M. A., and Schmullius, C. 2012. Capabilities and limitations of Landsat and land cover data for aboveground woody biomass estimation of Uganda. *Remote Sensing of Environment*, 117, 366–380.

Avoided Deforestation Partners. 2013. Approved VCS methodology VM0007, version 1.4, REDD methodology modules (REDD-MF). In *Sectoral Scope 14*, edited by T. A. W. International. Richmond, CA: Verified Carbon Standard.

Baccini, A., Goetz, S., Walker, W., Laporte, N., Sun, M., Sulla-Menashe, D., Hackler, J., et al. 2012. Estimated carbon dioxide emissions from tropical deforestation improved by carbon-density maps. *Nature Climate Change*, 2, 182–185.

Baccini, A., Laporte, N., Goetz, S., Sun, M., and Dong, H. 2008. A first map of tropical Africa's above-ground biomass derived from satellite imagery. *Environmental Research Letters*, 3, 045011.

Barbier, N., Couteron, P., Proisy, C., Malhi, Y., and Gastellu-Etchegorry, J. P. 2010. The variation of apparent crown size and canopy heterogeneity across lowland Amazonian forests. *Global Ecology and Biogeography*, 19, 72–84.

Bartholomé, E., and Belward, A. 2005. GLC2000: A new approach to global land cover mapping from Earth observation data. *International Journal of Remote Sensing*, 26, 1959–1977.

Bateson, C. A., Asner, G. P., and Wessman, C. A. 2000. Endmember bundles: A new approach to incorporating endmember variability into spectral mixture analysis. *Geoscience and Remote Sensing, IEEE Transactions on*, 38, 1083–1094.

Berveglieri, A., Imai, N. N., Tommaselli, A. M. G., Casagrande, B., and Honkavaara, E. 2018. Successional stages and their evolution in tropical forests using multi-temporal photogrammetric surface models and superpixels. *ISPRS Journal of Photogrammetry and Remote Sensing*, 146, 548–558. ISSN 0924-2716. https://doi.org/10.1016/j.isprsjprs.2018.11.002. https://www.sciencedirect.com/science/article/pii/S0924271618302983

Bhavsar, D., Das, A. K., Chakraborty, K., et al. 2024. Above ground biomass mapping of tropical forest of tripura using EOS-04 and ALOS-2 PALSAR-2 SAR data. *Journal of the Indian Society of Remote Sensing*. https://doi.org/10.1007/s12524-024-01838-w

Blackard, J., Finco, M., Helmer, E., Holden, G., Hoppus, M., Jacobs, D., Lister, A., et al. 2008. Mapping US forest biomass using nationwide forest inventory data and moderate resolution information. *Remote Sensing of Environment*, 112, 1658–1677.

Boggs, G. 2010. Assessment of SPOT 5 and QuickBird remotely sensed imagery for mapping tree cover in savannas. *International Journal of Applied Earth Observation and Geoinformation*, 12, 217–224.

Bohlman, S. A., Adams, J. B., Smith, M. O., and Peterson, D. L. 1998. Seasonal foliage changes in the eastern Amazon basin detected from Landsat Thematic Mapper satellite images. *Biotropica*, 30, 13–19.

Bolívar-Santamaría, S., and Reu, B. 2021. Detection and characterization of agroforestry systems in the Colombian Andes using sentinel-2 imagery. *Agroforestry Systems*, 95, 499–514.https://doi.org/10.1007/s10457-021-00597-8

Bontemps, S., Arino, O., Bicheron, P., Carsten Brockmann, C., Leroy, M., Vancutsem, C., and Defourny, P. 2012. Operational service demonstration for global land-cover mapping. The GlobCover and GlobCorine experiences for 2005 and 2009. Pp. 243–264 in *Remote Sensing of Land Use and Land Cover: Principles and Applications*, edited by C. Giri. Terre Haute, IN: CRC Press.

Bourbier, L., Cornu, G., Pennec, A., Brognoli, C., and Gond, V. 2013. Large-scale estimation of forest canopy opening using remote sensing in Central Africa. *Bios et Forets des Tropiques*, 3–9.

Boyd, D. S., Foody, G. M., Curran, P., Lucas, R., and Honzak, M. 1996. An assessment of radiance in Landsat TM middle and thermal infrared wavebands for the detection of tropical forest regeneration. *International Journal of Remote Sensing*, 17, 249–261.

Brando, P. M., Goetz, S. J., Baccini, A., Nepstad, D. C., Beck, P. S., and Christman, M. C. 2010. Seasonal and interannual variability of climate and vegetation indices across the Amazon. *Proceedings of the National Academy of Sciences*, 107, 14685–14690.

Braswell, B., Hagen, S., Frolking, S., and Salas, W. 2003. A multivariable approach for mapping sub-pixel land cover distributions using MISR and MODIS: Application in the Brazilian Amazon region. *Remote Sensing of Environment*, 87, 243–256.

Briant, G., Gond, V., and Laurance, S. G. 2010. Habitat fragmentation and the desiccation of forest canopies: A case study from eastern Amazonia. *Biological Conservation*, 143, 2763–2769.

Bryan, J. E., Shearman, P. L., Asner, G. P., Knapp, D. E., Aoro, G., and Lokes, B. 2013. Extreme differences in forest degradation in Borneo: Comparing practices in Sarawak, Sabah, and Brunei. *PLoS ONE*, 8.

Bullock, E. L., Woodcock, C. E., and Olofsson, P. 2020. Monitoring tropical forest degradation using spectral unmixing and Landsat time series analysis. *Remote Sensing of Environment*, 238, 110968. ISSN 0034-4257. https://doi.org/10.1016/j.rse.2018.11.011. https://www.sciencedirect.com/science/article/pii/S0034425718305200

Carlson, K. M., Curran, L. M., Ratnasari, D., Pittman, A. M., Soares-Filho, B. S., Asner, G. P., Trigg, S. N., et al. 2012. Committed carbon emissions, deforestation, and community land conversion from oil palm plantation expansion in West Kalimantan, Indonesia. *Proceedings of the National Academy of Sciences*, 109, 7559–7564.

Carreiras, J., Pereira, J., Campagnolo, M. L., and Shimabukuro, Y. E. 2006. Assessing the extent of agriculture/pasture and secondary succession forest in the Brazilian Legal Amazon using SPOT VEGETATION data. *Remote Sensing of Environment*, 101, 283–298.

Castillo-Santiago, M. A., Ricker, M., and de Jong, B. H. 2010. Estimation of tropical forest structure from SPOT-5 satellite images. *International Journal of Remote Sensing*, 31, 2767–2782.

Chazdon, R. L., Letcher, S. G., Van Breugel, M., Martínez-Ramos, M., Bongers, F., and Finegan, B. 2007. Rates of change in tree communities of secondary Neotropical forests following major disturbances. *Philosophical Transactions of the Royal Society B: Biological Sciences*, 362, 273–289.

Choi, H., and Bindschadler, R. 2004. Cloud detection in Landsat imagery of ice sheets using shadow matching technique and automatic normalized difference snow index threshold value decision. *Remote Sensing of Environment*, 91, 237–242.

Chust, G., Chave, J., Condit, R., Aguilar, S., Lao, S., and Pérez, R. 2006. Determinants and spatial modeling of tree β-diversity in a tropical forest landscape in Panama. *Journal of Vegetation Science*, 17, 83–92.

Claverie, M., Ju, J., Masek, J. G., Dungan, J. L., Vermote, E. F., Roger, J. C., Skakun, S. V., and Justice, C. 2018. The harmonized Landsat and Sentinel-2 surface reflectance data set. *Remote Sensing of Environment*, 219, 145–161.

Cochrane, M. A., Alencar, A., Schulze, M. D., Souza, C. M., Nepstad, D. C., Lefebvre, P., and Davidson, E. A. 1999. Positive feedbacks in the fire dynamic of closed canopy tropical forests. *Science*, 284, 1832–1835.

Cochrane, M. A., and Souza Jr, C. 1998. Linear mixture model classification of burned forests in the eastern Amazon. *International Journal of Remote Sensing*, 19, 3433–3440.

Costa, A. C. D., Pinto, J. R. R., Miguel, E. P., Xavier, G. D. O., Júnior, B. H. M., and Matricardi, E. A. T. 2023. Artificial intelligence tools and vegetation indices combined to estimate aboveground biomass in tropical forests. *Journal of Applied Remote Sensing*, 17(2), 024512. https://doi.org/10.1117/1.JRS.17.024512

Couteron, P., Pelissier, R., Nicolini, E. A., and Paget, D. 2005. Predicting tropical forest stand structure parameters from Fourier transform of very high-resolution remotely sensed canopy images. *Journal of Applied Ecology*, 42, 1121–1128.

Crist, E. P., and Cicone, R. C. 1984. A physically-based transformation of Thematic Mapper data—the TM Tasseled Cap. *IEEE Transactions on Geoscience and Remote Sensing*, 22, 256–263.

Cross, M. D., Scambos, T. A., Pacifici, F., and Marshall, W. E. 2018. Validating the use of metre-scale multispectral satellite image data for identifying tropical forest tree species. *International Journal of Remote Sensing*, 39(11), 3723–3752. http://doi.org/10.1080/01431161.2018.1448482

Cross, M. D., Scambos, T. A., Pacifici, F., Vargas-Ramirez, O., Moreno-Sanchez, R., and Marshall, W. 2019. Classification of tropical forest tree species using meter-scale image data. *Remote Sensing*, 11(12), 1411. https://doi.org/10.3390/rs11121411

de Carvalho, A. L., Nelson, B. W., Bianchini, M. C., Plagnol, D., Kuplich, T. M., and Daly, D. C. 2013. Bamboo-dominated forests of the southwest Amazon: Detection, spatial extent, life cycle length and flowering waves. *PLoS ONE*, 8, e54852.

DeFries, R., Hansen, M., Townshend, J., Janetos, A., and Loveland, T. 2000. A new global 1-km dataset of percentage tree cover derived from remote sensing. *Global Change Biology*, 6, 247–254.

DeFries, R., and Townshend, J. 1994. NDVI-derived land cover classifications at a global scale. *International Journal of Remote Sensing*, 15, 3567–3586.

Dennison, P. E., Roberts, D. A., and Peterson, S. H. 2007. Spectral shape-based temporal compositing algorithms for MODIS surface reflectance data. *Remote Sensing of Environment*, 109, 510–522.

Desclee, B., Simonetti, D., Mayaux, P., and Achard, F. 2013. Multi-sensor monitoring system for forest cover change assessment in Central Africa. *IEEE Journal on Selected Topics in Applied Earth Observations and Remote Sensing*, 6, 110–120.

de Wasseige, C., and Defourny, P. 2004. Remote sensing of selective logging impact for tropical forest management. *Forest Ecology and Management*, 188, 161–173.

Doughty, C. E., and Goulden, M. L. 2008. Seasonal patterns of tropical forest leaf area index and CO_2 exchange. *Journal of Geophysical Research: Biogeosciences*, 113, 2005–2012.

Doughty, C. E., Keany, J. M., Wiebe, B. C., et al. 2023. Tropical forests are approaching critical temperature thresholds. *Nature*, 621, 105–111. https://doi.org/10.1038/s41586-023-06391-z

Dubreuil, V., Debortoli, N., Funatsu, B., Nédélec, V., and Durieux, L. 2012. Impact of land-cover change in the Southern Amazonia climate: A case study for the region of Alta Floresta, Mato Grosso, Brazil. *Environmental Monitoring and Assessment*, 184, 877–891.

Dupuis, C., Lejeune, P., Michez, A., and Fayolle, A. 2020. How can remote sensing help monitor tropical moist forest degradation?—A systematic review. *Remote Sensing*, 12(7), 1087. https://doi.org/10.3390/rs12071087

Duveiller, G., Defourny, P., Desclée, B., and Mayaux, P. 2008. Deforestation in Central Africa: Estimates at regional, national and landscape levels by advanced processing of systematically-distributed Landsat extracts. *Remote Sensing of Environment*, 112, 1969–1981.

Dwomoh, F. K., Wimberly, M. C., Cochrane, M. A., and Numata, I. 2019. Forest degradation promotes fire during drought in moist tropical forests of Ghana. *Forest Ecology and Management*, 440, 158–168.

Elumnoh, A., and Shrestha, R. P. 2000. Application of DEM data to Landsat image classification: Evaluation in a tropical wet-dry landscape of Thailand. *Photogrammetric Engineering and Remote Sensing*, 66, 297–304.

Emelyanova, I. V., McVicar, T. R., Van Niel, T. G., Li, L. T., and van Dijk, A. I. 2013. Assessing the accuracy of blending Landsat–MODIS surface reflectances in two landscapes with contrasting spatial and temporal dynamics: A framework for algorithm selection. *Remote Sensing of Environment*, 133, 193–209.

Emmert, L., Negrón-Juárez, R. I., Chambers, J. Q., Santos, J. D., Lima, A. J. N., Trumbore, S., and Marra, D. M. 2023. Sensitivity of optical satellites to estimate windthrow tree-mortality in a Central Amazon forest. *Remote Sensing*, 15(16), 4027.

Erinjery, J. J., Singh, M., and Kent, R. 2018. Mapping and assessment of vegetation types in the tropical rainforests of the Western Ghats using multispectral Sentinel-2 and SAR Sentinel-1 satellite imagery. *Remote Sensing of Environment*, 216, 345–354. ISSN 0034-4257. https://doi.org/10.1016/j.rse.2018.07.006. https://www.sciencedirect.com/science/article/pii/S003442571830333X

Faber-Langendoen, D., Keeler-Wolf, T., Meidinger, D., Josse, C., Weakley, A., Tart, D., Navarro, G., et al. 2012. *Classification and description of world formation types*. Hierarchy Revisions Working Group, Federal Geographic Data Committee, FGDC Secretariat, US Geological Survey. Reston, VA, and NatureServe, Arlington, VA.

Feng, Y., Negrón-Juárez, R. I., and Chambers, J. Q. 2020. Remote sensing and statistical analysis of the effects of hurricane María on the forests of Puerto Rico. *Remote Sensing of Environment*, 247, 111940.

Feng, Y., Negrón-Juárez, R. I., Romps, D. M., and Chambers, J. Q. 2023. Amazon windthrow disturbances are likely to increase with storm frequency under global warming. *Nature Communications*, 14(1), 101.

Ferreira, M. P., Wagner, F. H., Aragão, L. E. O. C., Shimabukuro, Y. E., and Filho, C. R. D. S. 2019. Tree species classification in tropical forests using visible to shortwave infrared WorldView-3 images and texture analysis. *ISPRS Journal of Photogrammetry and Remote Sensing*, 149, 119–131. ISSN 0924-2716. https://doi.org/10.1016/j.isprsjprs.2019.01.019. https://www.sciencedirect.com/science/article/pii/S0924271619300280

Ferry, B., Morneau, F., Bontemps, J. D., Blanc, L., and Freycon, V. 2010. Higher treefall rates on slopes and waterlogged soils result in lower stand biomass and productivity in a tropical rain forest. *Journal of Ecology*, 98, 106–116.

Fisher, A. 2014. Cloud and cloud-shadow detection in SPOT5 HRG satellite imagry with automated morphological feature extraction. *Remote Sensing*, 6, 776–800.

Flood, N. 2013. Seasonal composite Landsat TM/ETM+ images using the Medoid (a Multi-Dimensional Median). *Remote Sensing*, 5, 6481–6500.

Foody, G. M., and Cutler, M. E. 2003. Tree biodiversity in protected and logged Bornean tropical rain forests and its measurement by satellite remote sensing. *Journal of Biogeography*, 30, 1053–1066.

Foody, G. M., and Hill, R. 1996. Classification of tropical forest classes from Landsat TM data. *International Journal of Remote Sensing*, 17, 2353–2367.

Foody, G. M., Palubinskas, G., Lucas, R. M., Curran, P. J., and Honzak, M. 1996. Identifying terrestrial carbon sinks: Classification of successional stages in regenerating tropical forest from Landsat TM data. *Remote Sensing of Environment*, 55, 205–216.

Friedl, M. A., McIver, D. K., Hodges, J. C., Zhang, X., Muchoney, D., Strahler, A. H., Woodcock, C. E., et al. 2002. Global land cover mapping from MODIS: Algorithms and early results. *Remote Sensing of Environment*, 83, 287–302.

Friedl, M. A., Sulla-Menashe, D., Tan, B., Schneider, A., Ramankutty, N., Sibley, A., and Huang, X. 2010. MODIS Collection 5 global land cover: Algorithm refinements and characterization of new datasets. *Remote Sensing of Environment*, 114, 168–182.

Ganivet, E., and Bloomberg, M. 2019. Towards rapid assessments of tree species diversity and structure in fragmented tropical forests: A review of perspectives offered by remotely-sensed and field-based data. *Forest Ecology and Management*, 432, 40–53. ISSN 0378-1127. https://doi.org/10.1016/j.foreco.2018.09.003. https://www.sciencedirect.com/science/article/pii/S0378112718307102

Gao, B.-C. 1996. NDWI—a normalized difference water index for remote sensing of vegetation liquid water from space. *Remote Sensing of Environment*, 58, 257–266.

Gao, F., Masek, J. G., Schwaller, M., and Hall, F. 2006. On the blending of the Landsat and MODIS surface reflectance: Predicting daily Landsat. *IEEE Transactions on Geoscience and Remote Sensing*, 44, 2207–2208.

Garzon-Lopez, C. X., Bohlman, S. A., Olff, H., and Jansen, P. A. 2013. Mapping tropical forest trees using high-resolution aerial digital photographs. *Biotropica*, 45, 308–316.

Gentry, A. H. 1988. Tree species richness of upper Amazonian forests. *Proceedings of the National Academy of Sciences*, 85, 156–159.

Gibbes, C., Adhikari, S., Rostant, L., Southworth, J., and Qiu, Y. 2010. Application of object based classification and high resolution satellite imagery for savanna ecosystem analysis. *Remote Sensing*, 2, 2748–2772.

Gibson, L., Lee, T. M., Koh, L. P., Brook, B. W., Gardner, T. A., Barlow, J., Peres, C. A., et al. 2011. Primary forests are irreplaceable for sustaining tropical biodiversity. *Nature*, 478, 378–381.

Givnish, T. J. 1999. On the causes of gradients in tropical tree diversity. *Journal of Ecology*, 87, 193–210.

Gond, V., Bartholomé, E., Ouattara, F., Nonguierma, A., and Bado, L. 2004. Surveillance et cartographie des plans d'eau et des zones humides et inondables en régions arides avec l'instrument VEGETATION embarqué sur SPOT-4. *International Journal of Remote Sensing*, 25, 987–1004.

Gond, V., Fayolle, A., Pennec, A., Cornu, G., Mayaux, P., Camberlin, P., Doumenge, C., et al. 2013. Vegetation structure and greenness in Central Africa from Modis multi-temporal data. *Philosophical Transactions of the Royal Society B: Biological Sciences*, 368.

Gond, V., Freycon, V., Molino, J.-F., Brunaux, O., Ingrassia, F., Joubert, P., Pekel, J.-F., et al. 2011. Broad-scale spatial pattern of forest landscape types in the Guiana Shield. *International Journal of Applied Earth Observation and Geoinformation*, 13, 357–367.

Gond, V., and Guitet, S. 2009. Elaboration d'un diagnostic post-exploitation par télédétection spatiale pour la gestion des forêts de Guyane. *Bois et Forêts des Tropiques*, 299, 5–13.

González-Orozco, C. E., Mulligan, M., Trichon, V., and Jarvis, A. 2010. Taxonomic identification of Amazonian tree crowns from aerial photography. *Applied Vegetation Science*, 13, 510–519.

Goodwin, N. R., and Collett, L. J. 2014. Development of an automated method for mapping fire history captured in Landsat TM and ETM+ time series across Queensland, Australia. *Remote Sensing of Environment*, 148, 206–221.

Goodwin, N. R., Collett, L. J., Denham, R. J., Flood, N., and Tindall, D. 2013. Cloud and cloud shadow screening across Queensland, Australia: An automated method for Landsat TM/ETM+ time series. *Remote Sensing of Environment*, 134, 50–65.

Göttlicher, D., Obregón, A., Homeier, J., Rollenbeck, R., Nauss, T., and Bendix, J. 2009. Land-cover classification in the Andes of southern Ecuador using Landsat ETM+ data as a basis for SVAT modelling. *International Journal of Remote Sensing*, 30, 1867–1886.

Griffiths, P., Kuemmerle, T., Baumann, M., Radeloff, V. C., Abrudan, I. V., Lieskovsky, J., Munteanu, C., et al. 2013. Forest disturbances, forest recovery, and changes in forest types across the Carpathian ecoregion from 1985 to 2010 based on Landsat image composites. *Remote Sensing of Environment*, 151, 72–88.

Gutierrez, G. V., Marcano, H., Ruzycki, T., Wood, T., Anderegg, W., Powers, J., and Helmer, E. 2023. *Aridity and Forest Age Mediate Landscape Scale Patterns of Tropical Forest Resistance to Cyclonic Storms*. Authorea Preprints.

Hall, J., Muscarella, R., Quebbeman, A., Arellano, G., Thompson, J., Zimmerman, J. K., and Uriarte, M. 2020. Hurricane-induced rainfall is a stronger predictor of tropical forest damage in Puerto Rico than maximum wind speeds. *Scientific Reports*, 10(1), 4318.

Hansen, M. C., Roy, D. P., Lindquist, E., Adusei, B., Justice, C. O., and Altstatt, A. 2008. A method for integrating MODIS and Landsat data for systematic monitoring of forest cover and change in the Congo Basin. *Remote Sensing of Environment*, 112, 2495–2513.

Helmer, E. H., Brown, S., and Cohen, W. 2000. Mapping montane tropical forest successional stage and land use with multi-date Landsat imagery. *International Journal of Remote Sensing*, 21, 2163–2183.

Helmer, E. H., Kennaway, T. A., Pedreros, D. H., Clark, M. L., Marcano-Vega, H., Tieszen, L. L., Ruzycki, T. R., et al. 2008. Land cover and forest formation distributions for St. Kitts, Nevis, St. Eustatius, Grenada and Barbados from decision tree classification of cloud-cleared satellite imagery. *Caribbean Journal of Science*, 44, 175–198.

Helmer, E. H., and Lefsky, M. A. 2006. Forest canopy heights in Amazon River basin forests as estimated with the Geoscience Laser Altimeter System (GLAS). Pp. 802–808 in *Monitoring Science and Technology Symposium: Unifying Knowledge for Sustainability in the Western Hemisphere, September 21–25, 2004*, edited by C. Aguirre-Bravo, P. J. Pellicane, D. P. Burns, and S. Draggan. Denver, CO: Proceedings RMRS-P-37CD. Ogden, UT: U.S. Department of Agriculture, Forest Service, Rocky Mountain Research Station, CD-ROM.

Helmer, E. H., Lefsky, M. A., and Roberts, D. A. 2009. Biomass accumulation rates of Amazonian secondary forest and biomass of old-growth forests from Landsat time series and the Geoscience Laser Altimeter System. *Journal of Applied Remote Sensing*, 3, 033505–033505–033531.

Helmer, E. H., Ramos, O., López, T. D. M., Quiñones, M., and Diaz, W. 2002. Mapping forest type and land cover of Puerto Rico, a component of the Caribbean biodiversity hotspot. *Caribbean Journal of Science*, 38, 165–183.

Helmer, E. H., and Ruefenacht, B. 2007b. A comparison of radiometric normalization methods when filling cloud gaps in Landsat imagery. *Canadian Journal of Remote Sensing/Journal Canadien de Teledetection*, 33, 325–340.

Helmer, E. H., Ruzycki, T. S., Benner, J., Voggesser, S. M., Scobie, B. P., Park, C., Fanning, D. W., et al. 2012. Detailed maps of tropical forest types are within reach: Forest tree communities for Trinidad and Tobago mapped with multiseason Landsat and multiseason fine-resolution imagery. *Forest Ecology and Management*, 279, 147–166.

Helmer, E., and Ruefenacht, B. 2005. Cloud-free satellite image mosaics with regression trees and histogram matching. *Photogrammetric Engineering and Remote Sensing*, 71, 1079.

Helmer, E., and Ruefenacht, B. 2007. A comparison of radiometric normalization methods when filling cloud gaps in Landsat imagery. *Canadian Journal of Remote Sensing*, 33, 325–340.

Helmer, E., Ruzycki, T. S., Wunderle, J. M., Vogesser, S., Ruefenacht, B., Kwit, C., Brandeis, T. J., et al. 2010. Mapping tropical dry forest height, foliage height profiles and disturbance type and age with a time series of cloud-cleared Landsat and ALI image mosaics to characterize avian habitat. *Remote Sensing of Environment*, 114, 2457–2473.

Helmer, E. H., Kay, S., Marcano-Vega, H., Powers, J. S., Wood, T. E., Zhu, X., Gwenzi, D., and Ruzycki, T. S. 2023. Multiscale predictors of small tree survival across a heterogeneous tropical landscape. *PLoS One*, 18(3), e0280322.

Helmer, E. H., Ruzycki, T. S., Wilson, B. T., Sherrill, K. R., Lefsky, M. A., Marcano-Vega, H., Brandeis, T. J., Erickson, H. E., and Ruefenacht, B. 2018. Tropical deforestation and recolonization by exotic and native trees: Spatial patterns of tropical forest biomass, functional groups, and species counts and links to stand age, geoclimate, and sustainability goals. *Remote Sensing*, 10(11), 1724.

Hermosilla, T., Francini, S., Nicolau, A. P., Wulder, M. A., White, J. C., Coops, N. C., and Chirici, G. 2023. Clouds and image compositing. Pp. 279–302 in *Cloud-Based Remote Sensing with Google Earth Engine: Fundamentals and Applications*, edited by J. A. Cardille, M. A. Crowley, D. Saah, and N. E. Clinton. Cham, Switzerland: Springer International Publishing.

Hernández-Stefanoni, J. L., Alberto Gallardo-Cruz, J., Meave, J. A., and Dupuy, J. M. 2011. Combining geostatistical models and remotely sensed data to improve tropical tree richness mapping. *Ecological Indicators*, 11, 1046–1056.

Hilker, T., Lyapustin, A. I., Tucker, C. J., Sellers, P. J., Hall, F. G., and Wang, Y. 2012. Remote sensing of tropical ecosystems: Atmospheric correction and cloud masking matter. *Remote Sensing of Environment*, 127, 370–384.

Holben, B. N. 1986. Characteristics of maximum-value composite images from temporal AVHRR data. *International Journal of Remote Sensing*, 7, 1417–1434.

Holdridge, L. R. 1967. *Life Zone Ecology*. San José, Costa Rica: Tropical Science Center.

Houghton, R. A. 2013. The emissions of carbon from deforestation and degradation in the tropics: Past trends and future potential. *Carbon Management*, 4, 539–546.

Huang, C., Thomas, N., Goward, S. N., Masek, J. G., Zhu, Z., Townshend, J. R., and Vogelmann, J. E. 2010. Automated masking of cloud and cloud shadow for forest change analysis using Landsat images. *International Journal of Remote Sensing*, 31, 5449–5464.

Huang, C., Wylie, B., Homer, C., Yang, L., and Zylstra, G. 2002. Derivation of a Tasseled Cap transformation based on Landsat 7 at-satellite reflectance. *International Journal of Remote Sensing*, 23, 1741–1748.

Huete, A. R. 1988. A soil-adjusted vegetation index (SAVI). *Remote Sensing of Environment*, 25, 295–309.

IPCC. 2006. *IPCC Guidelines for National Greenhouse Gas Inventories*. Hayama, Japan: Global Environmental Strategies (IGES).

IPCC. 2014. *Climate Change 2014: Mitigation of Climate Change, Contribution of Working Group III to the Fifth Assessment Report of the Intergovernmental Panel on Climate Change*, edited by O. Edenhofer, R. Pichs-Madruga, Y. Sokona, E. Farahani, S. Kadner, K. Seyboth, A. Adler, et al. Cambridge and New York: Cambridge University Press.

Irish, R. R., Barker, J. L., Goward, S. N., and Arvidson, T. 2006. Characterization of the Landsat-7 ETM+ automated cloud-cover assessment (ACCA) algorithm. *Photogrammetric Engineering and Remote Sensing*, 72, 1179.

Joshi, P. K. K., Roy, P. S., Singh, S., Agrawal, S., and Yadav, D. 2006. Vegetation cover mapping in India using multi-temporal IRS Wide Field Sensor (WiFS) data. *Remote Sensing of Environment*, 103, 190–202.

Joubert, P., Bourgeois, U., Linarés, S., Gond, V., Verger, G., Allo, S., and Coppel, A. 2012. L'observatoire de l'activité minière, un outil adapté à la surveillance de l'environnement. In *XV° symposium de la Société Savante Latino-Américaine de Télédétection et des Systèmes d'Informations Spatiales (SELPER)*. Cayenne, Guyane française, 19–23 November 2012. https://agritrop.cirad.fr/567372/1/document_567372.pdf

Keellings, D., and Hernández Ayala, J. J. 2019. Extreme rainfall associated with Hurricane Maria over Puerto Rico and its connections to climate variability and change. *Geophysical Research Letters*, 46(5), 2964–2973.

Kennaway, T. A., and Helmer, E. H. 2007. The forest types and ages cleared for land development in Puerto Rico. *GIScience & Remote Sensing*, 44, 356–382.

Kennaway, T. A., Helmer, E. H., Lefsky, M. A., Brandeis, T. A., and Sherrill, K. R. 2008. Mapping land cover and estimating forest structure using satellite imagery and coarse resolution lidar in the Virgin Islands. *Journal of Applied Remote Sensing*, 2, 023551–023551–023527.

Koyama, C. N., Watanabe, M., Hayashi, M., Ogawa, T., and Shimada, M. 2019. Mapping the spatial-temporal variability of tropical forests by ALOS-2 L-band SAR big data analysis. *Remote Sensing of Environment*, 233, 111372. ISSN 0034-4257. https://doi.org/10.1016/j.rse.2019.111372. https://www.sciencedirect.com/science/article/pii/S0034425719303918

Lahssini, K., Baghdadi, N., le Maire, G., and Fayad, I. 2022. Influence of GEDI acquisition and processing parameters on canopy height estimates over tropical forests. *Remote Sensing*, 14(24), 6264. https://doi.org/10.3390/rs14246264

Langner, A., Hirata, Y., Saito, H., Sokh, H., Leng, C., Pak, C., and Raši, R. 2014. Spectral normalization of SPOT 4 data to adjust for changing leaf phenology within seasonal forests in Cambodia. *Remote Sensing of Environment*, 143, 122–130.

Laporte, N. T., Stabach, J. A., Grosch, R., Lin, T. S., and Goetz, S. J. 2007. Expansion of industrial logging in Central Africa. *Science*, 316, 1451–1451.

Laumonier, Y., Edin, A., Kanninen, M., and Munandar, A. W. 2010. Landscape-scale variation in the structure and biomass of the hill dipterocarp forest of Sumatra: Implications for carbon stock assessments. *Forest Ecology and Management*, 259, 505–513.

Laurance, W. F., Goosem, M., and Laurance, S. G. 2009. Impacts of roads and linear clearings on tropical forests. *Trends in Ecology & Evolution*, 24, 659–669.

Laurin, G. V., Ding, J., Disney, M., Bartholomeus, H., Herold, M., Papale, D., and Valentini, R. 2019. Tree height in tropical forest as measured by different ground, proximal, and remote sensing instruments, and impacts on above ground biomass estimates. *International Journal of Applied Earth Observation and Geoinformation*, 82, 101899. ISSN 1569-8432. https://doi.org/10.1016/j.jag.2019.101899. https://www.sciencedirect.com/science/article/pii/S0303243419300844

Lefsky, M. A. 2010. A global forest canopy height map from the moderate resolution imaging spectroradiometer and the geoscience laser altimeter system. *Geophysical Research Letters*, 37.

Le Hégarat-Mascle, S., and André, C. 2009. Use of Markov random fields for automatic cloud/shadow detection on high resolution optical images. *ISPRS Journal of Photogrammetry and Remote Sensing*, 64, 351–366.

Lewis, S. L., Brando, P. M., Phillips, O. L., van der Heijden, G. M., and Nepstad, D. 2011. The 2010 amazon drought. *Science*, 331, 554–554.

Li, A., Huang, C., Sun, G., Shi, H., Toney, C., Zhu, Z., Rollins, M. G., et al. 2011. Modeling the height of young forests regenerating from recent disturbances in Mississippi using Landsat and ICESat data. *Remote Sensing of Environment*, 115, 1837–1849.

Li, Z., Ota, T., and Mizoue, N. 2024. Monitoring tropical forest change using tree canopy cover time series obtained from Sentinel-1 and Sentinel-2 data. *International Journal of Digital Earth*, 17(1). http://doi.org/10.1080/17538947.2024.2312222

Loveland, T., and Belward, A. 1997. The IGBP-DIS global 1km land cover data set, DISCover: First results. *International Journal of Remote Sensing*, 18, 3289–3295.

Lu, D. 2005. Aboveground biomass estimation using Landsat TM data in the Brazilian Amazon. *International Journal of Remote Sensing*, 26, 2509–2525.

Lucas, R. M., Xiao, X., Hagen, S., and Frolking, S. 2002. Evaluating TERRA-1 MODIS data for discrimination of tropical secondary forest regeneration stages in the Brazilian Legal Amazon. *Geophysical Research Letters*, 29, 42–41–42–44.

Luo, Y., Trishchenko, A. P., and Khlopenkov, K. V. 2008. Developing clear-sky, cloud and cloud shadow mask for producing clear-sky composites at 250-meter spatial resolution for the seven MODIS land bands over Canada and North America. *Remote Sensing of Environment*, 112, 4167–4185.

Lyapustin, A., Wang, Y., and Frey, R. 2008. An automatic cloud mask algorithm based on time series of MODIS measurements. *Journal of Geophysical Research: Atmospheres*, 113, 1984–2012.

Malhi, Y., and Román-Cuesta, R. M. 2008. Analysis of lacunarity and scales of spatial homogeneity in IKONOS images of Amazonian tropical forest canopies. *Remote Sensing of Environment*, 112, 2074–2087.

Malhi, Y., Wood, D., Baker, T. R., Wright, J., Phillips, O. L., Cochrane, T., Meir, P., et al. 2006. The regional variation of aboveground live biomass in old-growth Amazonian forests. *Global Change Biology*, 12, 1107–1138.

Martinuzzi, S., Gould, W. A., and González, O. M. R. 2007. *Creating Cloud-Free Landsat ETM+ Data Sets in Tropical Landscapes: Cloud and Cloud-Shadow Removal*, edited by F. S. US Department of Agriculture. Río Piedras, Puerto Rico: International Institute of Tropical Forestry.

Martinuzzi, S., Gould, W. A., González, O. M. R., Martínez Robles, A., Calle Maldonado, P., Pérez-Buitrago, N., and Fumero Caban, J. J. 2008. Mapping tropical dry forest habitats integrating Landsat NDVI, Ikonos imagery, and topographic information in the Caribbean Island of Mona. *Revista de Biologia Tropical*, 56, 625–639.

Mayaux, P., Achard, F., and Malingreau, J.-P. 1998. Global tropical forest area measurements derived from coarse resolution satellite imagery: A comparison with other approaches. *Environmental Conservation*, 25, 37–52.

Mayaux, P., Bartholomé, E., Fritz, S., and Belward, A. 2004. A new land-cover map of Africa for the year 2000. *Journal of Biogeography*, 31, 861–877.

Mayaux, P., Pekel, J.-F., Desclée, B., Donnay, F., Lupi, A., Achard, F., Clerici, M., et al. 2013. State and evolution of the African rainforests between 1990 and 2010. *Philosophical Transactions of the Royal Society B: Biological Sciences*, 368.

McCallum, I., Obersteiner, M., Nilsson, S., and Shvidenko, A. 2006. A spatial comparison of four satellite derived 1km global land cover datasets. *International Journal of Applied Earth Observation and Geoinformation*, 8, 246–255.

Medeiros, R., Andrade, J., Ramos, D., Moura, M., Pérez-Marin, A. M., dos Santos, C. A., da Silva, B. B., and Cunha, J. 2022. Remote sensing phenology of the Brazilian caatinga and its environmental drivers. *Remote Sensing*, 14(11), 2637.

Miller, S. D., Goulden, M. L., Hutyra, L. R., Keller, M., Saleska, S. R., Wofsy, S. C., Figueira, A. M. S., et al. 2011. Reduced impact logging minimally alters tropical rainforest carbon and energy exchange. *Proceedings of the National Academy of Sciences*, 108, 19431–19435.

Mitchard, E. T. A., Feldpausch, T. R., Brienen, R. J. W., Lopez-Gonzalez, G., Monteagudo, A., Baker, T. R., Lewis, S. L., et al. 2014. Markedly divergent estimates of Amazon forest carbon density from ground plots and satellites. *Global Ecology and Biogeography*, 23, 935–946.

Morton, D. C., DeFries, R. S., Nagol, J., Souza Jr, C. M., Kasischke, E. S., Hurtt, G. C., and Dubayah, R. 2011. Mapping canopy damage from understory fires in Amazon forests using annual time series of Landsat and MODIS data. *Remote Sensing of Environment*, 115, 1706–1720.

Morton, D. C., Nagol, J., Carabajal, C. C., Rosette, J., Palace, M., Cook, B. D., Vermote, E. F., et al. 2014. Amazon forests maintain consistent canopy structure and greenness during the dry season. *Nature*, 506, 221–224.

Muchoney, D., Borak, J., Chi, H., Friedl, M., Gopal, S., Hodges, J., Morrow, N., and Strahler, A. 2000. Application of the MODIS global supervised classification model to vegetation and land cover mapping of Central America. *International Journal of Remote Sensing*, 21(6–7), 1115–1138.

Myneni, R. B., Yang, W., Nemani, R. R., Huete, A. R., Dickinson, R. E., Knyazikhin, Y., Didan, K., et al. 2007. Large seasonal swings in leaf area of Amazon rainforests. *Proceedings of the National Academy of Sciences*, 104, 4820–4823.

Nagendra, H., and Rocchini, D. 2008. High resolution satellite imagery for tropical biodiversity studies: The devil is in the detail. *Biodiversity and Conservation*, 17, 3431–3442.

Nagendra, H., Rocchini, D., Ghate, R., Sharma, B., and Pareeth, S. 2010. Assessing plant diversity in a dry tropical forest: Comparing the utility of Landsat and IKONOS satellite images. *Remote Sensing*, 2, 478–496.

Nelson, R. F., Kimes, D. S., Salas, W. A., and Routhier, M. 2000. Secondary forest age and tropical forest biomass estimation using Thematic Mapper imagery: Single-year tropical forest age classes, a surrogate for standing biomass, cannot be reliably identified using single-date TM imagery. *Bioscience*, 50, 419–431.

Nemani, R. R., Keeling, C. D., Hashimoto, H., Jolly, W. M., Piper, S. C., Tucker, C. J., Myneni, R. B., et al. 2003. Climate-driven increases in global terrestrial net primary production from 1982 to 1999. *Science*, 300, 1560–1563.

Newman, M. E., McLaren, K. P., and Wilson, B. S. 2011. Comparing the effects of classification techniques on landscape-level assessments: Pixel-based versus object-based classification. *International Journal of Remote Sensing*, 32, 4055–4073.

Ngo, Y.-N., Tong Minh, D. H., Baghdadi, N., and Fayad, I. 2023. Tropical forest top height by GEDI: From sparse coverage to continuous data. *Remote Sensing*, 15(4), 975. https://doi.org/10.3390/rs15040975

Njomaba, E., Ofori, J. N., Guuroh, R. T., Aikins, B. E., Nagbija, R. K., and Surový, P. 2024. Assessing forest species diversity in Ghana's tropical forest using planetscope data. *Remote Sensing*, 16(3), 463. https://doi.org/10.3390/rs16030463

Omeja, P. A., Obua, J., Rwetsiba, A., and Chapman, C. A. 2012. Biomass accumulation in tropical lands with different disturbance histories: Contrasts within one landscape and across regions. *Forest Ecology and Management*, 269, 293–300.

Oreopoulos, L., Wilson, M. J., and Várnai, T. 2011. Implementation on Landsat data of a simple cloud-mask algorithm developed for MODIS land bands. *Geoscience and Remote Sensing Letters, IEEE*, 8, 597–601.

Oza, M., Srivastava, V., and Devaiah, P. 1996. Estimating tree volume in tropical dry deciduous forest from Landsat TM data. *Geocarto International*, 11, 33–39.

Palace, M., Keller, M., Asner, G. P., Hagen, S., and Braswell, B. 2008. Amazon forest structure from IKONOS satellite data and the automated characterization of forest canopy properties. *Biotropica*, 40, 141–150.

Park, S. 2009. Synchronicity between satellite-measured leaf phenology and rainfall regimes in tropical forests. *Photogrammetric Engineering and Remote Sensing*, 75, 1231–1237.

Parker, C., Mitchell, A., Trivedi, M., Mardas, N., and Sosis, K. 2009. The little REDD+ book. *Global Canopy Foundation*, 132.

Parmesan, C., and Yohe, G. 2003. A globally coherent fingerprint of climate change impacts across natural systems. *Nature*, 421, 37–42.

Pascual, A., Tupinambá-Simões, F., Guerra-Hernández, J., and Bravo, F. 2022. High-resolution planet satellite imagery and multi-temporal surveys to predict risk of tree mortality in tropical eucalypt forestry. *Journal of Environmental Management*, 310, 114804.

Pennec, A., Gond, V., and Sabatier, D. 2011. Tropical forest phenology in French Guiana from MODIS time series. *Remote Sensing Letters*, 2, 337–345.

Peterson, G. D., and Heemskerk, M. 2001. Deforestation and forest regeneration following small-scale gold mining in the Amazon: The case of Suriname. *Environmental Conservation*, 28, 117–126.

Pflugmacher, D., Cohen, W. B., and Kennedy, R. E. 2012. Using Landsat-derived disturbance history (1972–2010) to predict current forest structure. *Remote Sensing of Environment*, 122, 146–165.

Phillips, O. L., Aragão, L. E., Lewis, S. L., Fisher, J. B., Lloyd, J., López-González, G., Malhi, Y., et al. 2009. Drought sensitivity of the Amazon rainforest. *Science*, 323, 1344–1347.

Pinard, M. A., and Putz, F. E. 1996. Retaining forest biomass by reducing logging damage. *Biotropica*, 278–295.

Ping, D., Dalagnol, R., Galvão, L. S., Nelson, B., Wagner, F., Schultz, D. M., and Bispo, P. D. C. 2023. Assessing the magnitude of the amazonian forest blowdowns and post-disturbance recovery using Landsat-8 and time series of PlanetScope satellite constellation data. *Remote Sensing*, 15(12), 3196.

Pinty, B., and Verstraete, M. 1992. GEMI: A non-linear index to monitor global vegetation from satellites. *Vegetatio*, 101, 15–20.

Pithon, S., Jubelin, G., Guitet, S., and Gond, V. 2013. A statistical method for detecting logging-related canopy gaps using high-resolution optical remote sensing. *International Journal of Remote Sensing*, 34, 700–711.

Pletcher, E., Smith-Tripp, S., Evans, D., and Schwartz, N. B. 2024. Evaluating global vegetation products for application in heterogeneous forest-savanna landscapes. *International Journal of Remote Sensing*, 45(2), 492–507. http://doi.org/10.1080/01431161.2023.2299278

Ploton, P., Pélissier, R., Proisy, C., Flavenot, T., Barbier, N., Rai, S. N., and Couteron, P. 2011. Assessing aboveground tropical forest biomass using Google Earth canopy images. *Ecological Applications*, 22, 993–1003.

Portillo-Quintero, C. A., and Sánchez-Azofeifa, G. A. 2010. Extent and conservation of tropical dry forests in the Americas. *Biological Conservation*, 143, 144–155.

Pringle, M., Schmidt, M., and Muir, J. 2009. Geostatistical interpolation of SLC-off Landsat ETM+ images. *ISPRS Journal of Photogrammetry and Remote Sensing*, 64, 654–664.

Qi, J., Chehbouni, A., Huete, A., Kerr, Y., and Sorooshian, S. 1994. A modified soil adjusted vegetation index. *Remote Sensing of Environment*, 48, 119–126.

Qiu, S., Zhu, Z., and He, B. 2019. Fmask 4.0: Improved cloud and cloud shadow detection in Landsats 4–8 and Sentinel-2 imagery. *Remote Sensing of Environment*, 231, 111205.

Rana, P., Popescu, S., Tolvanen, A., Gautam, B., Srinivasan, S., and Tokola, T. 2023. Estimation of tropical forest aboveground biomass in Nepal using multiple remotely sensed data and deep learning. *International Journal of Remote Sensing*, 44(17), 5147–5171. http://doi.org/10.1080/01431161.2023.2240508

Roberts, D., Numata, I., Holmes, K., Batista, G., Krug, T., Monteiro, A., Powell, B., et al. 2002. Large area mapping of land-cover change in Rondônia using multitemporal spectral mixture analysis and decision tree classifiers. *Journal of Geophysical Research*, 107, 8073.

Rouse, J., Haas, R., Schell, J., Deering, D., and Harlan, J. 1974. *Monitoring the Vernal Advancement and Retrogradation (Greenwave Effect) of Natural Vegetation*. College Station, TX: Texas A&M University, Remote Sensing Center.

Roy, D. P., Ju, J., Kline, K., Scaramuzza, P. L., Kovalskyy, V., Hansen, M., Loveland, T. R., et al. 2010. Web-enabled Landsat data (WELD): Landsat ETM+ composited mosaics of the conterminous United States. *Remote Sensing of Environment*, 114, 35–49.

Roy, D. P., Ju, J., Lewis, P., Schaaf, C., Gao, F., Hansen, M., and Lindquist, E. 2008. Multi-temporal MODIS-Landsat data fusion for relative radiometric normalization, gap filling, and prediction of Landsat data. *Remote Sensing of Environment*, 112, 3112–3130.

Saatchi, S. S., Harris, N. L., Brown, S., Lefsky, M., Mitchard, E. T., Salas, W., Zutta, B. R., et al. 2011. Benchmark map of forest carbon stocks in tropical regions across three continents. *Proceedings of the National Academy of Sciences*, 108, 9899–9904.

Saatchi, S. S., Houghton, R., Dos Santos Alvala, R., Soares, J., and Yu, Y. 2007. Distribution of aboveground live biomass in the Amazon basin. *Global Change Biology*, 13, 816–837.

Salimon, C. I., Putz, F. E., Menezes-Filho, L., Anderson, A., Silveira, M., Brown, I. F., and Oliveira, L. 2011. Estimating state-wide biomass carbon stocks for a REDD plan in Acre, Brazil. *Forest Ecology and Management*, 262, 555–560.

Salk, C. F., Chazdon, R., and Andersson, K. 2013. Detecting landscape-level changes in tree biomass and biodiversity: Methodological constraints and challenges of plot-based approaches. *Canadian Journal of Forest Research*, 43, 799–808.

Salovaara, K. J., Thessler, S., Malik, R. N., and Tuomisto, H. 2005. Classification of Amazonian primary rain forest vegetation using Landsat ETM+ satellite imagery. *Remote Sensing of Environment*, 97, 39–51.

Samanta, A., Ganguly, S., Hashimoto, H., Devadiga, S., Vermote, E., Knyazikhin, Y., Nemani, R. R., et al. 2010. Amazon forests did not green-up during the 2005 drought. *Geophysical Research Letters*, 37.

Sánchez-Azofeifa, A., Rivard, B., Wright, J., Feng, J.-L., Li, P., Chong, M. M., and Bohlman, S. A. 2011. Estimation of the distribution of Tabebuia guayacan (Bignoniaceae) using high-resolution remote sensing imagery. *Sensors*, 11, 3831–3851.

Scaramuzza, P. L., Bouchard, M. A., and Dwyer, J. L. 2011. Development of the Landsat data continuity mission cloud-cover assessment algorithms. *Geoscience and Remote Sensing, IEEE Transactions on*, 50, 1140–1154.

Sesnie, S. E., Finegan, B., Gessler, P. E., Thessler, S., Bendana, Z. R., and Smith, A. M. 2010. The multispectral separability of Costa Rican rainforest types with support vector machines and Random Forest decision trees. *International Journal of Remote Sensing*, 31, 2885–2909.

Sesnie, S. E., Gessler, P. E., Finegan, B., and Thessler, S. 2008. Integrating Landsat TM and SRTM-DEM derived variables with decision trees for habitat classification and change detection in complex neotropical environments. *Remote Sensing of Environment*, 112, 2145–2159.

Siegert, F., Ruecker, G., Hinrichs, A., and Hoffmann, A. 2001. Increased damage from fires in logged forests during droughts caused by El Nino. *Nature*, 414, 437–440.

Sist, P., and Ferreira, F. N. 2007. Sustainability of reduced-impact logging in the Eastern Amazon. *Forest Ecology and Management*, 243, 199–209.

Skakun, S., Wevers, J., Brockmann, C., Doxani, G., Aleksandrov, M., Batič, M., Frantz, D., Gascon, F., Gómez-Chova, L., Hagolle, O., and López-Puigdollers, D. 2022. Cloud Mask Intercomparison eXercise (CMIX): An evaluation of cloud masking algorithms for Landsat 8 and Sentinel-2. *Remote Sensing of Environment*, 274, 112990.

Skidmore, A. K. 1989. An expert system classifies eucalypt forest types using thematic mapper data and a digital terrain model. *Photogrammetric Engineering and Remote Sensing*, 55, 1449–1464.

Skole, D., and Tucker, C. 1993. Tropical deforestation and habitat fragmentation in the Amazon: Satellite data from 1978 to 1988. *Science*, 260, 1905–1910.

Small, C. 2004. The Landsat ETM+ spectral mixing space. *Remote Sensing of Environment*, 93, 1–17.

Song, C. 2013. Optical remote sensing of forest leaf area index and biomass. *Progress in Physical Geography*, 37, 98–113.

Souza Jr, C. M., and Barreto, P. 2000. An alternative approach for detecting and monitoring selectively logged forests in the Amazon. *International Journal of Remote Sensing*, 21, 173–179.

Souza Jr, C. M., Firestone, L., Silva, L. M., and Roberts, D. 2003. Mapping forest degradation in the Eastern Amazon from SPOT 4 through spectral mixture models. *Remote Sensing of Environment*, 87, 494–506.

Souza Jr, C. M., Roberts, D. A., and Cochrane, M. A. 2005. Combining spectral and spatial information to map canopy damage from selective logging and forest fires. *Remote Sensing of Environment*, 98, 329–343.

Souza Jr, C. M., and Siqueira, J. V. N. 2013. ImgTools: A software for optical remotely sensed data analysis. Pp. 1571–1578 in *Anais XVI Simpósio Brasileiro de Sensoriamento Remoto—SBSR. I. N. d. P. E. (INPE)*. Foz do Iguaçu, PR: Instituto Nacional de Pesquisas Espaciais (INPE).

Souza Jr, C. M., Siqueira, J. V., Sales, M. H., Fonseca, A. V., Ribeiro, J. G., Numata, I., Cochrane, M. A., et al. 2013. Ten-year Landsat classification of deforestation and forest degradation in the Brazilian Amazon. *Remote Sensing*, 5, 5493–5513.

Staal, A., Fetzer, I., Wang-Erlandsson, L., et al. 2020. Hysteresis of tropical forests in the 21st century. *Nature Communications*, 11, 4978. https://doi.org/10.1038/s41467-020-18728-7

Steininger, M. 2000. Satellite estimation of tropical secondary forest above-ground biomass: Data from Brazil and Bolivia. *International Journal of Remote Sensing*, 21, 1139–1157.

Stephens, B. B., Gurney, K. R., Tans, P. P., Sweeney, C., Peters, W., Bruhwiler, L., Ciais, P., et al. 2007. Weak northern and strong tropical land carbon uptake from vertical profiles of atmospheric CO2. *Science*, 316, 1732–1735.

Strabala, K. I. 2005. *MODIS Cloud Mask User's Guide*. P. 32, edited by C. I. f. M. S. Studies. Madison, WI: University of Wisconsin, Madison.

Suab, S. A., Supe, H., Louw, A. S., et al. 2024. Mapping of temporally dynamic tropical forest and plantations canopy height in Borneo Utilizing TanDEM-X InSAR and multi-sensor remote sensing data. *Journal of the Indian Society of Remote Sensing*. https://doi.org/10.1007/s12524-024-01820-6

Thenkabail, P. S., Enclona, E. A., Ashton, M. S., Legg, C., and De Dieu, M. J. 2004. Hyperion, IKONOS, ALI, and ETM+ sensors in the study of African rainforests. *Remote Sensing of Environment*, 90, 23–43.

Thenkabail, P. S., Hall, J., Lin, T., Ashton, M. S., Harris, D., and Enclona, E. A. 2003. Detecting floristic structure and pattern across topographic and moisture gradients in a mixed species Central African forest using IKONOS and Landsat-7 ETM+ images. *International Journal of Applied Earth Observation and Geoinformation*, 4, 255–270.

Thessler, S., Ruokolainen, K., Tuomisto, H., and Tomppo, E. 2005. Mapping gradual landscape-scale floristic changes in Amazonian primary rain forests by combining ordination and remote sensing. *Global Ecology and Biogeography*, 14, 315–325.

Thessler, S., Sesnie, S., Ramos Bendaña, Z. S., Ruokolainen, K., Tomppo, E., and Finegan, B. 2008. Using k-nn and discriminant analyses to classify rain forest types in a Landsat TM image over northern Costa Rica. *Remote Sensing of Environment*, 112, 2485–2494.

Tickle, P., Lee, A., Lucas, R. M., Austin, J., and Witte, C. 2006. Quantifying Australian forest floristics and structure using small footprint LiDAR and large scale aerial photography. *Forest Ecology and Management*, 223, 379–394.

Tottrup, C. 2004. Improving tropical forest mapping using multi-date Landsat TM data and pre-classification image smoothing. *International Journal of Remote Sensing*, 25, 717–730.

Tottrup, C., Rasmussen, M., Samek, J., and Skole, D. 2007. Towards a generic approach for characterizing and mapping tropical secondary forests in the highlands of mainland Southeast Asia. *International Journal of Remote Sensing*, 28, 1263–1284.

Townshend, J., Justice, C., Li, W., Gurney, C., and McManus, J. 1991. Global land cover classification by remote sensing: Present capabilities and future possibilities. *Remote Sensing of Environment*, 35, 243–255.

Trichon, V., and Julien, M.-P. 2006. Tree species identification on large-scale aerial photographs in a tropical rain forest, French Guiana—application for management and conservation. *Forest Ecology and Management*, 225, 51–61.

Tucker, C. J., Goff, T., and Townshend, J. 1985. African land-cover classification using satellite data. *Science*, 227, 369–375.

Uhl, C., and Buschbacher, R. 1985. A disturbing synergism between cattle ranch burning practices and selective tree harvesting in the eastern Amazon. *Biotropica*, 265–268.

USGS. 2003. *Preliminary Assessment of Landsat 7 ETM+ Data Following Scan Line Corrector Malfunction USGS*. Sioux Falls, SD: United States Geological Survey.

Vancutsem, C., Pekel, J.-F., Evrard, C., Malaisse, F., and Defourny, P. 2009. Mapping and characterizing the vegetation types of the Democratic Republic of Congo using SPOT VEGETATION time series. *International Journal of Applied Earth Observation and Geoinformation*, 11, 62–76.

Verhegghen, A., Mayaux, P., De Wasseige, C., and Defourny, P. 2012. Mapping Congo Basin vegetation types from 300 m and 1 km multi-sensor time series for carbon stocks and forest areas estimation. *Biogeosciences*, 9, 5061–5079.

Vickers, B., Trines, E., and Pohnan, E. 2012. Community guidelines for accessing forestry voluntare carbon markets. P. 196, edited by R. o. f. A. a. t. Pacific. Bangkok: Food and Agriculture Organization of the United Nations.

Vieira, I. C. G., de Almeida, A. S., Davidson, E. A., Stone, T. A., Reis de Carvalho, C. J., and Guerrero, J. B. 2003. Classifying successional forests using Landsat spectral properties and ecological characteristics in eastern Amazonia. *Remote Sensing of Environment*, 87, 470–481.

Vitousek, P. M. 1994. Beyond global warming: Ecology and global change. *Ecology*, 75, 1861–1876.

Wang, B. 1999. Automated detection and removal of clouds and their shadows from Landstat TM images. *IEICE Transactions on Information and Systems*, 82, 453–460.

Wang, J., Lee, C. K., Zhu, X., Cao, R., Gu, Y., Wu, S., and Wu, J. 2022. A new object-class based gap-filling method for PlanetScope satellite image time series. *Remote Sensing of Environment*, 280, 113136.

Wang, J., Li, Y., Rahman, M. M., Li, B., Yan, Z., Song, G., Zhao, Y., Wu, J., and Chu, C. 2024. Unraveling the drivers and impacts of leaf phenological diversity in a subtropical forest: A fine-scale analysis using PlanetScope CubeSats. *New Phytologist*.

Wang, J., Yang, D., Chen, S., Zhu, X., Wu, S., Bogonovich, M., Guo, Z., Zhu, Z., and Wu, J. 2021. Automatic cloud and cloud shadow detection in tropical areas for PlanetScope satellite images. *Remote Sensing of Environment*, 264, 112604.

Wang, J., Yang, D., Detto, M., Nelson, B. W., Chen, M., Guan, K., Wu, S., Yan, Z., and Wu, J. 2020. Multi-scale integration of satellite remote sensing improves characterization of dry-season green-up in an Amazon tropical evergreen forest. *Remote Sensing of Environment*, 246, 111865.

Wang, Y., Ziv, G., Adami, M., Mitchard, E., Batterman, S. A., Buermann, W., Marimon, B. S., Junior, B. H. M., Reis, S. M., Rodrigues, D., and Galbraith, D. 2019. Mapping tropical disturbed forests using multi-decadal 30 m optical satellite imagery. *Remote Sensing of Environment*, 221, 474–488. ISSN 0034-4257. https://doi.org/10.1016/j.rse.2018.11.028. https://www.sciencedirect.com/science/article/pii/S0034425718305376

Wertz-Kanounnikoff, S. 2008. *Monitoring Forest Emissions, a Review of Methods*. Bogor, Indonesia: CIFOR.

Whiteside, T. G., Boggs, G. S., and Maier, S. W. 2011. Comparing object-based and pixel-based classifications for mapping savannas. *International Journal of Applied Earth Observation and Geoinformation*, 13, 884–893.

Wijaya, A., Liesenberg, V., and Gloaguen, R. 2010. Retrieval of forest attributes in complex successional forests of Central Indonesia: Modeling and estimation of bitemporal data. *Forest Ecology and Management*, 259, 2315–2326.

Wilson, E. O. 1988. The current state of biological diversity. Pp. 3–18 in *Biodiversity*, edited by E. O. Wilson and F. M. Peters. Washington, DC: National Academy Press.

Wittmann, F., Anhuf, D., and Funk, W. J. 2002. Tree species distribution and community structure of central Amazonian várzea forests by remote-sensing techniques. *Journal of Tropical Ecology*, 18, 805–820.

Woods, P. 1989. Effects of logging, drought, and fire on structure and composition of tropical forests in Sabah, Malaysia. *Biotropica*, 290–298.

Wulder, M. A., White, J. C., Nelson, R. F., Næsset, E., Ørka, H. O., Coops, N. C., Hilker, T., et al. 2012. Lidar sampling for large-area forest characterization: A review. *Remote Sensing of Environment*, 121, 196–209.

Xiao, X., Hagen, S., Zhang, Q., Keller, M., and Moore III, B. 2006. Detecting leaf phenology of seasonally moist tropical forests in South America with multi-temporal MODIS images. *Remote Sensing of Environment*, 103, 465–473.

Xu, X., Du, H., Zhou, G., Ge, H., Shi, Y., Zhou, Y., Fan, W., et al. 2011. Estimation of aboveground carbon stock of Moso bamboo (Phyllostachys heterocycla var. pubescens) forest with a Landsat Thematic Mapper image. *International Journal of Remote Sensing*, 32, 1431–1448.

Yang, J., Tian, H., Pan, S., Chen, G., Zhang, B., and Dangal, S. 2018. Amazon drought and forest response: Largely reduced forest photosynthesis but slightly increased canopy greenness during the extreme drought of 2015/2016. *Global Change Biology*, 24(5), 1919–1934.

Yin, Z., Ling, F., Foody, G. M., Li, X., and Du, Y. 2020. Cloud detection in Landsat-8 imagery in Google Earth Engine based on a deep convolutional neural network. *Remote Sensing Letters*, 11(12), 1181–1190.

Yu, M., and Gao, Q. 2020. Topography, drainage capability, and legacy of drought differentiate tropical ecosystem response to and recovery from major hurricanes. *Environmental Research Letters*, 15(10), 104046.

Zaki, N. A. M., and Latif, Z. A. 2017. Carbon sinks and tropical forest biomass estimation: A review on role of remote sensing in aboveground-biomass modelling. *Geocarto International*, 32(7), 701–716. http://doi.org/10.1080/10106049.2016.1178814

Zelazowski, P., Malhi, Y., Huntingford, C., Sitch, S., and Fisher, J. B. 2011. Changes in the potential distribution of humid tropical forests on a warmer planet. *Philosophical Transactions of the Royal Society A: Mathematical, Physical and Engineering Sciences*, 369, 137–160.

Zhang, Y., Guindon, B., and Cihlar, J. 2002. An image transform to characterize and compensate for spatial variations in thin cloud contamination of Landsat images. *Remote Sensing of Environment*, 82, 173–187. http://dx.doi.org/10.1016/S0034-4257(02)00034-2

Zhao, M., and Running, S. W. 2010. Drought-induced reduction in global terrestrial net primary production from 2000 through 2009. *Science*, 329, 940–943.

Zhu, X., Chen, J., Gao, F., Chen, X., and Masek, J. G. 2010. An enhanced spatial and temporal adaptive reflectance fusion model for complex heterogeneous regions. *Remote Sensing of Environment*, 114, 2610–2623.

Zhu, X., Gao, F., Liu, D., and Chen, J. 2012a. A modified neighborhood similar pixel interpolator approach for removing thick clouds in Landsat images. *Geoscience and Remote Sensing Letters, IEEE*, 9, 521–525.

Zhu, X., and Helmer, E. H. 2018. An automatic method for screening clouds and cloud shadows in optical satellite image time series in cloudy regions. *Remote Sensing of Environment*, 214, 135–153.

Zhu, X., Helmer, E. H., Gwenzi, D., Collin, M., Fleming, S., Tian, J., Marcano-Vega, H., Meléndez-Ackerman, E. J., and Zimmerman, J. K. 2021. Characterization of dry-season phenology in tropical forests by reconstructing cloud-free Landsat time series. *Remote Sensing*, 13(23), 4736.

Zhu, Z., and Woodcock, C. E. 2012. Object-based cloud and cloud shadow detection in Landsat imagery. *Remote Sensing of Environment*, 118, 83–94.

Zhu, Z., Woodcock, C. E., and Olofsson, P. 2012b. Continuous monitoring of forest disturbance using all available Landsat imagery. *Remote Sensing of Environment*, 122, 75–91.

2 Remote Sensing of Forests from LiDAR and Radar

Juha Hyyppä, Xiaowei Yu, Mika Karjalainen, Xinlian Liang, Anttoni Jaakkola, Mike Wulder, Markus Hollaus, Joanne C. White, Mikko Vastaranta, Jiri Pyörälä, Tuomas Yrttimaa, Ninni Saarinen, Josef Taher, Juho-Pekka Virtanen, Leena Matikainen, Yunsheng Wang, Eetu Puttonen, Mariana Campos, Matti Hyyppä, Kirsi Karila, Harri Kaartinen, Matti Vaaja, Ville Kankare, Antero Kukko, Markus Holopainen, Hannu Hyyppä, Masato Katoh, and Eric Hyyppä

ACRONYMS AND DEFINITIONS

AGB	Aboveground biomass
ALOS	Advanced Land Observing Satellite
ALS	Airborne laser scanning
CHM	Canopy height model
COSMO-SkyMed	Constellation of Small Satellites for Mediterranean basin Observation (COSMO)-SkyMed
DEM	Digital Elevation Model
DLR	German Aerospace Center
DTM	Digital terrain model
ERS	European remote sensing satellites
GIS	Geographic Information System
GLAS	Geoscience Laser Altimeter System
GNSS	Global Navigation Satellite Systems
ICESat	Instrument on board the Ice, Cloud, and land Elevation
JERS	Japanese Earth Resources Satellite
LAI	Leaf area index
LiDAR	Light detection and ranging
MLS	Mobile laser scanning
MODIS	Moderate-resolution imaging spectroradiometer
NASA	National Aeronautics and Space Administration
NDVI	Normalized Difference Vegetation Index
PALSAR	Phased Array type L-band Synthetic Aperture Radar
SAR	Synthetic aperture radar
SRTM	Shuttle Radar Topographic Mission
TerraSAR-X	A radar Earth observation satellite with its phased array synthetic aperture radar
TIN	Triangulated irregular network

TLS	Terrestrial laser scanning
UAV	Unmanned Aerial Vehicle

2.1 INTRODUCTION

This chapter is about collecting three-dimensional (3D) information from LiDAR and radar and turning that information into valuable forest informatics. Similar pipelines can be applied for both LiDAR and radar data.

Today, remote sensing processes for forestry are mainly based on point cloud processing or on elevation models (3D techniques). These point cloud data can be provided both by the LiDAR and radar—but also by photogrammetry. Analogously to photogrammetric spatial intersection, a stereo pair of SAR images with different off-nadir angles can be used to calculate the 3D coordinates for corresponding points on the image pair producing point clouds from radar imagery. Also, in SAR interferometry, the pixel-by-pixel phase difference between two complex SAR images acquired from slightly different perspectives can be converted into elevation differences of the terrain/object. Thus, both LiDAR and radar can provide data which can be processed in a similar way either using original points or using surface models in a raster form. From the point clouds you can calculate DTM (Digital Terrain Model), DSM (Digital Surface Model), and CHM/nDSM (Canopy Height Model/normalized Digital Surface Model). The idea is to provide a surface model (DSM) and subtract the ground elevation (DTM) from it in order to get a canopy height. Intensity, coherence (in interferometry-SAR), and texture can be used to improve the estimates in 3D-based inventory.

In general, there is high synergy between LiDAR and radar since they are based on the same measuring principle even though they are using different frequencies/wavelengths.

1. Laser intensity calibration has stemmed from the corresponding work with radar backscatter coefficient determination. In the late 1980s, Finnish and Swedish researchers were developing early versions of scanning LiDAR/radars, and already at that time the radar return versus time was automatically corrected in the hardware (Hallikainen et al. 1993).
2. The LiDAR-derived terrain model is the basic information needed in future radar-based forest inventories. In the future, radar-based forest inventory processing will be mainly done using laser scanning point cloud processing techniques. In large-area forest inventory in the near future, satellite-based radar can cover large areas with relatively high repetition rate, and laser scanning can provide important field reference for satellite data calibration.
3. Both LiDAR and radar are active remote sensing techniques. The major advantages of active remote sensing systems include better penetration of atmosphere; coverage can be obtained at user-specified times, even at night, since it produces its own illumination to the target; images/echoes can be produced from different types of polarized energy; and systems may operate simultaneously in several wavelengths/frequencies and, thus, have multi-wavelength/frequency potential.

There are many previous syntheses that complement this chapter. Starting from the past literature, the reader is referred to known basics of, e.g., LiDAR/laser scanning from, e.g., Hyyppä et al. (2008), and radar from Henderson and Lewis (1998) and Jensen (2000). Forestry applications and processing based on active sensors are also covered in Hyyppä et al. (2008) on using laser scanning data in forestry applications and related algorithms; in Holopainen et al. (2014) on estimation of forest stock and yield with LiDAR; in Nelson (2013) on how did we get there and an early history of forestry LiDAR; in Koch (2010) on status and future using new laser scanning, synthetic aperture radar, and hyperspectral remote sensing data for forest biomass assessment; and in Mallet and Bretar (2009) on full waveform topographic LiDAR. Liang et al. (2016) reviewed the use of terrestrial laser scanning for small-area forest inventories. White et al. (2016) reviewed remote sensing technologies for enhancing forest inventories. Tsitsi (2016) summarized studies on remote sensing of aboveground forest biomass. A similar review was done by Sinha et al. (2015) for radar remote sensing. Lechner et al. (2020) summarized applications of remote sensing to forest ecology and management. Balenović et al. (2021) focused on handheld laser scanning and its possibilities for forest inventory. Guimarães et al. (2020) reviewed drone remote sensing of forests. Tompalski et al. (2021) reviewed estimation of changes in forest attributes, especially using airborne laser scanning. Liang et al. (2022) reviewed close-range remote sensing techniques, especially from the point of view of providing field reference for large-area remote sensing. Fassnacht et al. (2024) analyzed current challenges, considerations, and directions in remote sensing of forests. Use of deep learning methods in forests was reviewed by Wang et al. (2021), Diez et al. (2021), and Hamedianfar et al. (2022). In this handbook, there are many complementary chapters, for example, those dealing with LiDAR processing, tropical rainforests, remote sensing of tree height, terrestrial carbon modeling, and global biomass modeling.

We see that radar and LiDAR are currently changing how operational forest inventory is performed. Remote sensing of forest from LiDAR and radar is the future of forest inventory at local, regional, or national levels. Currently, airborne LiDAR is operationally applied in the Nordic countries when carrying out standwise forest inventories (Fassnacht et al. 2024). It is likely that terrestrial and mobile LiDAR or laser scanning may also soon be used operationally for the collection of detailed field reference data. Radar data may be promising for large-area monitoring applications with new upcoming spaceborne radar missions. Therefore, we planned the content of the chapter in the following way:

- Section 2: Conventional practices to acquire forest resource information
- Section 3: Basics of LiDAR and radar
- Section 4: Obtaining 3D data from LiDAR and radar for forestry
- Section 5: Common processing chain for LiDAR and radar into useful forest informatics.
- Section 6: Operational possibilities with LiDAR and radar.

In this chapter, we aim to demonstrate that LiDAR and radar point clouds are processed in a very similar processing chain, which has been originally developed for airborne LiDAR for stand-level forest inventory in boreal forest area. Additionally, we highlight a number of future techniques that will further challenge current operational inventory systems.

2.2 CONVENTIONAL PRACTICES FOR ACQUISITION OF FOREST RESOURCE INFORMATION

According to data in the Global Forest Resources Assessment 2010 (FRA 2010), the total global forest area is slightly over 4 billion hectares (31% of the total land surface). The five most forest-rich countries are the Russian Federation, Brazil, Canada, the USA, and China. In terms of the ratio of forest cover to total land area, Finland (73% of the land area), Japan, and Sweden (both 69% of the land area) are the world's most extensively forested countries among the industrialized and temperate countries. Forest monitoring via remote sensing plays a crucial role in the assessment, planning, field data collection, image processing, analysis, and modeling for sustainable forest management (SFM). GIS (geographic information systems) are widely employed to manage forest information obtained with remote sensing on a stand polygon basis. The addition of stems by species to each stand polygon could become an important tool for practical forest management operations, such as precise thinning, selective cutting, and harvesting.

2.2.1 Forest Inventory

Forest inventory is carried out to support decision-making by the forest manager or forest owner. Forest resource information is, thus, needed for large-scale strategic planning, operative forest management, and pre-harvest planning (Table 2.1). National forest inventories (NFIs) are examples of inventories undertaken for large-scale strategic planning for gathering information about nationwide forest resources, such as growing stock volume, forest cover, growth and yield, biomass, carbon balance, and large-scale wood procurement potential. In NFIs, it is important to have unbiased estimates and to obtain information also from small strata. The most conventional strategy for NFI is to

TABLE 2.1
Aims and Methods to Collect Forest Resource Information

Aim	Method	Provides a Map	Description
Large-scale strategic planning	National forest inventory (NFI)	No	Nationwide statistics are calculated based on a systematic sample of field plots. Field plots are measured tree by tree.
Operational forest management	Stand-wise field inventory (SWFI)	Yes	Several field plots are measured from every stand. Sample trees are measured from plots.
Pre-harvest planning	SWFI with additional measurements	Yes	SWFI information is double checked by measuring additional plots

use sampling and measure sampling plots at the field. Thus, NFIs do not provide mapping information that is required for operative forest management and preharvest planning. In many countries, information for these purposes has been collected using stand-wise field inventories (SWFI). There are many different methods with varying accuracy to carry out SWFI. In general, visual interpretation of aerial images is combined with field sampling and some rapid measurements. For example, in Finland and Sweden, stands are delineated from aerial photographs and then every stand is visited by a forester. At the stand, basal area is measured using a relascope (angle-count method) from various locations. Then basal area–weighted mean tree diameter and height are measured by caliper and clinometer. Finally, inventory attributes are generalized to stand. Forest management operations are determined using calculated inventory attributes combined with ocular assessments at the site.

During the last 100 years, forest inventory has transferred from the determination of the volume of logs, trees, and stands, and a calculation of the increment and yield toward more but not fully into multifunctional forestry considering wildlife, recreation, watershed management timber, biofuel, climate regulation, air purification, erosion control, and habitat for biodiversity. The multiple use of forest is increasingly considered as different ecosystem services, and from an operational forestry standpoint the information required to support "multiple use" can also satisfy the characterization of "ecosystems services." However, a major focus of forest assessment is still in obtaining accurate information on the volume and growth of trees in forest plots, stands, and large areas. The forest-bound carbon is also an important issue globally. One of the biggest challenges currently in forest inventory research is how to measure and monitor forest biomass and its changes effectively and accurately. Radar and LiDAR provide tools for that.

2.2.2 Forest Measurements

A forest inventory could in principle be based on measuring every tree in a given area, but this is usually not realistic in forestry. As such, the acquisition of forest resource information is typically based on sampling. The most common sample units are a tree and a plot. To obtain usable information related to forest resources, various attributes have to be measured at the tree level. Individual tree measures are summed at the plot level and then plots are used to obtain forest resource information representative of any given area, such as a stand, a woodlot, a county, or a country. Due to this hierarchy, it is highly important to measure individual tree attributes accurately. It should also be pointed out that many times tree attributes are modeled instead of measured, e.g., stem volume is usually modeled based on diameter at breast height (dbh) and height of the tree (h), noting that measurements made at a tree or plot level are often selected by some sampling criteria. Table 2.2 summarizes the main tree attributes from the point of view of forest mensuration. Upper diameter is typically taken from a height of 6 m. Height of the crown base is the height from the ground to the lowest green branch or to the lowest complete living whorl of branches. Basal area is the cross-sectional area defined by the dbh.

Some of them can be directly measured or calculated from these direct measurements, while others need to be predicted through statistical or physical modeling. Traditionally, the following individual tree attributes such as

TABLE 2.2
Attributes of Trees to Be Measured

Attribute	Unit	Typically Expected Accuracy for Measurement*
Height	m	0.5–2 m
Diameter at Breast Height (dbh)	mm	5–10 mm
Upper Diameter (e.g., at height 6 m)	mm	5–10 mm
Height of Crown Base	m	0.2–0.4 m
Species		
Age	yrs	5 yrs
Location	m	0.5–2 m
Basal Area	m²	see diameter accuracy
Volume	m³	10–20%
Biomass	kg/m³	10–20%
Growth	e.g., mm (increment borer)	1 mm

* depends strongly on the use of the data

- Height (and height growth),
- Diameters at different height along the stem (and diameter growth),
- Crown diameter, and
- Tree species

are measured or determined in the field. Diameter is convenient to measure and is one of the directly measurable dimensions from which tree cross-sectional area, surface area, and volume can be computed. Various instruments and methods have been developed for measuring the tree dimension in the field (Husch et al. 1982; Päivinen et al. 1992; Gill et al. 2000; Korhonen et al. 2006; Clark et al. 2000), such as

- Caliper, diameter tape, and optical devices for diameter measurements;
- Level rod, pole, and hypsometers for tree height measurements; and
- Increment borer for diameter growth measurements.

The method used in obtaining the measurements is largely determined by the accuracy required. Past growth of diameter can be obtained from increment borings or cross-section cuts. For some species past height growth may be determined by measuring internodal lengths. Sometimes it is even necessary to fell the tree to obtain more accurate measurements, e.g., to measure the stem volume accurately requires destructive sampling of a tree. Accordingly, direct and indirect methods have been developed for the estimation of such forest attributes. Practically speaking, tree volume is estimated from dbh and possibly together with height and upper diameter for each tree species. The models for stem volume, especially based on diameter information, exist for many commercially important tree species. For example, in Finland, there are volume models v for main tree species based on different inputs (diameter d, height h, diameter at the height of 6 m, d_6)

- $v = f(d)$
- $v = f(d,h)$
- $v = f(d,h,d_6)$

The most accurate and non-destructive way to determine the stem volume of the tree is to use stem curve, i.e., the stem diameter as the function of the tree height. A relascope is often used to measure the basal area of the sample plots if diameters of all the trees within the plot are not measured.

2.3 GENERAL FEATURES OF LASER SCANNING/LIDAR AND RADAR

Active microwave imagery is obtained using instruments and principles that are different from those acquired in the visible, near-, mid-, and thermal infrared portions of the spectrum using passive remote sensing techniques. Therefore, it is necessary to understand the basics of the active microwave systems, such as LiDAR and radars.

Laser scanning (LS) is a *surveying technique* used for mapping topography, vegetation, urban areas, ice, infrastructure, and other targets of interest. LS is often referred to as LiDAR because of the central role of the LiDAR in the system. Also, LS is more commonly used in Europe, whereas LiDAR is used in the US. The basic principle of LiDAR is to use a laser beam to illuminate an object and a photodiode to register the backscatter radiation and to measure the range. More precisely, Airborne Laser Scanning (ALS) is a method based on Light Detection and Ranging (LiDAR) measurements from an aircraft, where the precise position and orientation of the sensor is known, and therefore the position (x, y, z) of the reflecting objects can be determined. In addition to ALS, there is an increasing interest in Terrestrial Laser Scanning (TLS), where the laser scanner is mounted on a tripod or even on a moving platform, i.e., Mobile Laser Scanning (MLS). The output of the laser scanner is then a georeferenced point cloud of LiDAR measurements, including the intensity and possibly waveform information of the returned light. A typical ALS system consists of (1) a laser ranging unit (i.e., LiDAR), (2) an opto-mechanical scanner, (3) a position and orientation unit, and (4) a control, processing, and storage unit. The laser ranging unit can be subdivided into a transmitter, a receiver, and the optics for both units. These components also apply to other types of LS systems, such as MLS. The receiver optics collects the backscattered light and focuses it onto the detector, converting the photons to electrical impulses. The opto-mechanical scanning unit is responsible for the deflection of the transmitted laser beams across the flight track. The type of the applied deflection unit (e.g., oscillating mirror/zig-zag scanning, rotating mirror/line scanning, pushbroom/fiber scanning, Palmer/conical scanning) defines the scan pattern on the ground. A differential GNSS (Global Navigation Satellite System) receiver provides the position of the laser ranging unit. Its orientation is determined by the pitch, roll, and heading of the aircraft, which are measured by an inertial navigation/measurement system (Hyyppä et al., 2008, 2009).

Radar is a similar object-detection system using radio waves to determine the range, altitude, direction, or speed of objects. In remote sensing especially the backscatter strength is used for object recognition/classification. The transmitted energy illuminates an area on the ground. The radar cross-section of the object is defined as the ratio of the backscattered power versus isotropically reflecting the object. The radar backscatter coefficient is defined as the radar cross-section divided by the illuminated area. The radar backscatter coefficient is used to classify targets. With radar operating wavelengths, surface roughness, moisture and biomass, and vegetation structure are major parameters affecting the backscatter. In addition to backscatter, range (e.g., Hallikainen et al. 1993), polarization response, stereoscopy, various incidence angles, and interferometry have been applied for remote sensing of forests.

Table 2.3 gives a short comparison of LiDAR and radar in remote sensing of forests.

2.4 OBTAINING 3D DATA FROM FORESTRY

The focus in this chapter is on techniques capable of providing 2.5D/3D (2.5D stands for 3D surface models) data for forest inventory processing to be depicted in Section 2.5. Short state-of-the-art of the LiDAR/radar techniques providing 3D data covers:

- Space-borne LiDAR
- Space-borne Synthetic Aperture Radar (SAR)
- SAR interferometry

Remote Sensing of Forests from LiDAR and Radar

TABLE 2.3
Comparison of LiDAR and Radar in Remote Sensing of Forests

Characteristics	LiDAR	Radar
Cloud penetration capacity	No	Yes
Coverage can be obtained at user-specified times, even at night	Yes	Yes
May penetrate vegetation, sand, and surface layers of snow	No	Yes
Penetration to forests	Using canopy gaps and small beams	May penetrate through leaves and needles. Usually applied
Images can be produced from different types of polarized energy (HH, HV, VH, VH)	Yes, but not applied	Yes, common approach
May operate simultaneously in several wavelengths (frequencies) and thus has multifrequency potential	Yes, applied currently in bathymetry and hyperspectral LiDARs	Yes
Can produce overlapping images suitable for stereoscopic viewing and radargrammetry	Not applied	Yes
Supports interferometric	Not applied	Yes
Applied wavelengths	Typical wavelengths are between 500 and 1550 nm	The abbreviations associated with the radar frequencies and wavelengths include (K: 18–26.5 GHz, 1.67–1.19 cm; X: 8–12.5 GHz, 3.8–2.4 cm; C: 4–8 GHz, 7.5–3.9 cm; S: 2–4 GHz, 15–7.5 cm; L: 1–2 GHz, 30–15 cm; P: 0.3–1 GHz, 100–30 cm)
Speckle	Since backscatter is not strongly used, the effect of speckle is not well studied. However, the speckle effect is smaller in magnitude as with radars	Strong speckle effect with microwave wavelengths, to remove the speckle, the image is usually processed using several looks, thus, an averaging process takes place.
Foreshortening, layover, and shadowing	Do not exist with LiDAR data, since point clouds are always preferred before backscatter information	Geometric distortions exist in all radar imagery when backscatter is the main data output
Object roughness	Most of the objects are considered rough with LiDAR wavelengths	Incidence angle strongly affects radar backscatter with non-rough surfaces
Biomass measurements	Typically based on point cloud metrics data	Previously based on backscatter data. Radar backscatter increases approximately linearly with increasing biomass until it saturates at a biomass level that depends on the radar frequency. The lower the frequency, the better the penetration.
Soil moisture	Affects intensity	L-band radar penetrates to a maximal depth of approximately 10 cm. Shorter wavelengths penetrate to only 1–3 cm. Multi-temporal radar data can be used to measure soil moisture variation.

- SAR radargrammetry
- Airborne Laser Scanning
- Terrestrial Laser Scanning
- Mobile Laser Scanning

SAR tomography is not covered, since it is still far from practical usability. Since there are many forest attributes to be measured, more focus is on biomass and stem volume, to which all attributes basically correlate. In Section 2.5, there is additional discussion to measure each attribute separately. Table 2.4 gives main characteristics of each of these systems at the general level.

TABLE 2.4
Example Characteristics of the 3D Remote Sensing Systems

Space-borne LiDAR

Potential	Global biomass with remarkably better accuracy than with today's techniques
Possible beam size	Few tens of m, ranging accuracy of 1–3 m (for canopy height)
Challenges	To obtain high power from space, lifetime of the system, coverage, how to monitor when satellite LiDAR provide strips
System providers	NASA with ICESat-2, and with GEDI (NASA 2014) from the International Space Station

Space-borne SAR

Resolution	Resolution up to 1 m from space, typically to 10-km-by-10-km imagery
Frequency	L, S, C, or X-band
Revisit time	Few days
Interferometry	Repeat-pass interferometry/single-pass interferometry
Feasible	Large-area inventories
Characteristics	Ranging accuracy of 3–5 m (for canopy height), cost of the data is typically high compared to data quality
System providers	DLR (TerraSAR X, Tandem X, Tandem-L), ASI (COSMO-SkyMed), ESA (Sentinel-1, Biomass, ROSE-L), Radarsat, SAOCOM, JAXA (ALOS-1, ALOS-4), CONAE (SAOCOM), CSIRO (NOVASAR), ICEye, Capella, NASA/ISRO (NISAR)

Airborne Laser Scanning

Point density	Point density 0.5–1000 pts/m^2
Elevation accuracy	5–30 cm
Planimetric accuracy	10–50 cm
Operating range	Few hundred m to several km
Feasible	Cost-effective for areas larger than 50 km^2
Characteristics	Homogenous point clouds
System providers	Teledyne, Leica, Riegl

Mobile Laser Scanning

Point density	Point density in the range of 100 to tens of thousands pts/m^2
Accuracy	Point accuracy of few centimeters (egg) when collected with good GNSS coverage, alternative positioning acquired with robotic SLAM (Simultaneous Localization and Mapping)
Operating range	
Feasible	
Characteristics	Applicable range of few tens of m
System providers	Collecting large data sets for road environment, collecting homogeneous point clouds for field inventory
	Relatively high variation in the range data, time of acquisition needs to be used in the processing
	Faro, Innoviz, LeddarTech, Hesai, Leica, Riegl, Sick, Teledyne, Trimple, Ouster

Terrestrial Laser Scanning

Point density	Point density in the range of 10000+ pts/m^2 at the 10 m
Accuracy	Distance accuracy of few mm to 1 cm
Operating range	Applicable range of few tens of m
Feasible	Operational scanning range from 1 to several hundred m
Characteristics	Feasible for small areas less than few tens of m distance
System providers	Processing time challenging: image processing techniques applied; small variation in data, e.g., distance variation low, thus surface normal can be calculated
	Faro, Leica, Riegl, Topcon, Trimble, Zoller, and Fröhlich

2.4.1 Space-borne LiDAR

The Geoscience Laser Altimeter System (GLAS) onboard NASA's Ice, Cloud and Land Elevation Satellite (ICESat) was the first spaceborne LiDAR mission providing LiDAR data at a global scale

(Zwally et al. 2002; Schutz et al. 2005). ICESat/GLAS was launched in January 2003 and acquired LiDAR waveform data until October 2009. Several studies also demonstrated the potential of GLAS data to characterize forest structure (e.g., Lefsky et al. 2005, 2007; Rosette et al. 2008; Sun et al. 2008; Ballhorn et al. 2011). Space-borne LiDAR, to provide systematic and widely dispersed measures of vegetation characteristics, is required for robust global biomass estimates, among other information needs (Tsui et al. 2013). NASA's Ice, Cloud and Land Elevation Satellite-2 (ICESat-2), launched in 2018, has been extensively validated in Neuenschwander et al. (2020) in which ICESat-2-derived terrain and canopy heights from 11 months of ICESat-2 data were calibrated using national laser scanning data collected in southern Finland. The terrain heights agreed with the ALS with less than 75 cm. ICESat-2 missed the top 11–13% of canopy heights for coniferous-dominated canopy covers ranging from 40% to 85%. Neuenschwander et al. (2020) also showed that end users interested in canopy height retrievals should avoid the use of weak beam data in boreal forest conditions.

The Global Ecosystem Dynamics Investigation (GEDI) is a full-waveform LiDAR system designed to measure vegetation structure with three 1064 nm lasers. (Dubayah et al. 2020). It is onboard the International Space Station, and thus the latitude of operation is limited between 51.6N and 51.6S. Spatially discrete measurements with a 25 m footprint are acquired along the track every 60 m and 600 m between tracks. GEDI data has been combined with continuous satellite image data, e.g., from Tandem-X to map forest height and biomass (Chen et al. 2021; Choi et al. 2023; Schlund et al. 2023). Performance of Icesat-2 and GEDI for canopy height retrieval has been evaluated by Liu et al. (2021).

2.4.2 Space-borne Synthetic Aperture Radar (SAR)

In the past decades, a remarkable amount of research using SAR data has been conducted in the field of forest inventory, concentrating usually on stand- or plot-level mean stock volume and/or AGB estimation (e.g., Le Toan et al. 1992; Fransson & Israelsson 1999; Wagner et al. 2003; Rauste 2005; Tokola et al. 2007; Holopainen et al. 2010; Solberg et al. 2010). Results have been promising, but not accurate enough for operative forest inventories. In the 2010s SAR satellite images rapidly improved thanks to the very-high-resolution SAR satellites (TerraSAR-X/TanDEM-X, COSMO-SkyMed and Radarsat-2) (Krieger et al. 2007; Torres et al. 2012), enabling more detailed forest mapping. Long time series of archived data availability, high-performance computing, and machine learning methods make global forest monitoring possible.

In general, SAR images contain the following information at the pixel level: (1) radar backscattering intensity, (2) phase of the backscattered signal, and (3) range to target pixel.

1. Radar intensity corresponds to the strength of the backscattered signal compared to the strength of the transmitted signal, and it is a function of the SAR system parameters (such as the wavelength and the polarization of the used electro-magnetic radiation) and target parameters (such as the target area roughness compared to the used radar wavelength and dielectric properties).
2. The phase information in the single-channel SAR data is quasi-random and, therefore, useless for target interpretation. However, phase information is an essential part of multi-polarized data analysis (SAR polarimetry) and SAR interferometry.
3. The range measurement is based on the time-of-flight information of the radar pulse and has typically been neglected in biomass estimation tasks.

The use of intensity and backscattering coefficient information in forest resources mapping has been widely studied over the past few decades. In general, the longer radar wavelengths (L-band, or for airborne sensors also P- and VHF-band) are more suitable for stem volume estimation than the shorter wavelengths of C-band or X-band (Le Toan et al. 1992; Fransson 1999). The reason for this is that the interaction between radar waves and forest structures in the L-band and P-band occurs on the trunks of trees. On the other hand, in the X-band and C-band, the scattering takes place at the top of the forest canopy, on branches and foliage, contributing apparently less to the information related to stem volume. Even though the relationship between radar intensity and stem volume has been well studied, there still remain practical challenges to be overcome due to topography and seasonal variations in weather. Rauste (2005) was able to obtain a correlation coefficient of 0.85 between stem volume and radiometrically normalized L-band JERS-1 data in Finland. However, the L-band intensity appears to saturate at some level of stem volume. Typically, stem volume levels beyond 100–200m^3/ha cannot be observed (Fransson & Israelsson 1999; Rauste 2005) due to saturation. In addition to saturation at higher biomass levels, there are other issues, such as speckle and mixture of surface roughness as well as moisture and biomass, affecting the output of the radar.

The radar pulse illuminates a given surface area that consists of several scattering points. Thus, the returned echo comprises a coherent combination of individual echoes from a large number of points (see Elachi 1987). The result is a single vector representing the amplitude V and phase f ($I \sim V^2$ of the total echo, which is a vector sum of the individual echoes). This variation is called fading or speckle. Thus, an image of a homogeneous surface with constant reflectivity will result in intensity variation from one resolution element to the next. Speckle gives images recorded with radar a grainy texture.

The radar cross-section is defined as the equivalent of a perfectly reflecting area that reflects isotropically (spherically). The backscatter coefficient is defined as the radar cross-section divided by the area illuminated. The radar backscatter coefficient is mainly used to classify target characteristics. Surface roughness, moisture and biomass, and vegetation structure are major environmental parameters within the resolution cell that are responsible for backscattering the incident energy. Surface roughness is the terrain property that strongly influences the strength of the radar backscatter. Co-polarization backscatter toward the sensor results from single reflections from canopy components such as the leaves, stems, branches, and trunk, and these returns are generally very strong (called canopy surface scattering). If the energy is scattered multiple times within a distributed volume, such as a stand of pine trees, this is often called volume scattering. Radar backscatter increases approximately linearly. With higher biomass levels it is hard to separate soil moisture and vegetation backscatter contributions.

Using data from the ERS-1 and ERS-2 tandem mission, it was demonstrated that the phase information can be used to calculate 2D interferometric coherence maps, and this can lead to deriving stem volume and other forest parameters (e.g., Askne et al. 2003; Wagner et al. 2003; Santoro et al. 2007). However, the coherence signal appears to saturate at some point of the biomass, hampering the estimation of high biomass values. Tandem-X coherence data has been used for canopy height estimation (Askne et al. 2013; Olesk et al. 2015; Schlund et al. 2019) and ALOS PALSAR L-band coherence data for boreal forest growing stock volume estimation (Thiel & Schmullius 2014).

Moreover, promising results have been achieved by combining the methods of interferometry and polarimetry, i.e., PolInSAR (Papathanassiou & Cloude 2001), demonstrated in the TanDEM-X satellite. Several data fusion studies of SAR and LiDAR data (scanning or profiling) in forest mapping have been performed (e.g., Hyde et al. 2007; Nelson et al. 2007; Goodenough et al. 2008; Kellndorfer et al. 2010; Banskota et al. 2011; Sun et al. 2008). An overview of using ALS, SAR, and hyperspectral remote sensing data for AGB assessment can be found in Koch (2010). The estimation of AGB solely on the basis of SAR backscatter intensity has proven to be challenging (e.g., Fransson & Israelsson 1999; Rauste 2005; Holopainen et al. 2010). Cross-polarization has been found to be more useful for biomass and stem volume assessment (e.g., Huuva 2023).

The most promising approach to determine forest biomass by radar imaging from space is likely to be via canopy height information (i.e., 3D techniques) similarly to laser scanning. Studies have shown that elevation information extracted from SAR has potential in estimation of forest canopy height even close to ALS data (e.g., Solberg 2010; Perko et al. 2011; Karjalainen et al. 2012). Basically, there are two approaches to extract elevation information from the SAR images: (1) SAR interferometry, i.e., InSAR (Massonnet & Feigl 1998; Rosen et al. 2000), or (2) radargrammetry (Leberl 1979; Toutin & Gray 2000).

2.4.2.1 SAR Interferometry

SAR interferometry (InSAR) is a technique in which the pixel-by-pixel phase difference between two complex SAR images acquired from slightly different perspectives can be converted into elevation differences of the terrain (Massonnet & Feigl 1998; Rosen et al. 2000). When the X- or C-band of the radar is considered, the scattering takes place near the top of the forest canopy (Le Toan et al. 1992). Therefore, if the elevation of the ground surface is known (e.g., a Digital Terrain Model [DTM] is available), then the X- or C-band's interferometric height compared to the ground surface elevation is related to the forest canopy height and accordingly to the stem volume. For forestry applications, simultaneous acquisition of the SAR data used for interferometry is especially advantageous. An example of the use of interferometric data for forest canopy height estimation has been provided by Kellndorfer et al. (2004), who used the C-band's interferometric heights from the Shuttle Radar Topography Mission (SRTM) to estimate the forest canopy height. Similar results using the SRTM X-band data were presented by Solberg et al. (2010), who also estimated the aboveground biomass based on SRTM elevation values. The Tandem-X mission (TDM), launched in 2010, consists of two satellites flying in close formation enabling bistatic acquisition of X-band SAR data. Studies (Solberg et al. 2013; Askne et al. 2013) have demonstrated the use of TDM data retrieval of forest biomass of boreal forests. Karila et al. (2015) extracted and compared the forest parameters utilizing the X-band SAR data from the German single-pass interferometry mission TanDEM-X mission (TDX) and the TerraSAR-X radargrammetry (TSX) mission. TDX data provides stem volume and aboveground biomass estimates at the level of 20–21 RMSE-%, whereas TSX provides RMSE-% at the level of 30%. As a comparison, ALS area-based prediction provides corresponding accuracies of 14–15%. Large-scale forest AGB and stem volume maps have been derived from Tandem-X backscatter and interferometric coherence and height data with 21–25% RMSE (Persson et al. 2017).

In Huuva et al. (2023), site index and canopy age for each studied field plot was predicted by fitting the time series of TanDEM-X-based top heights to established height development curves needed in site index determination. Tandem-X data has also been used in combination with SRTM data and Landsat spectral indices to map boreal forest biomass (Sadeghi et al. 2018). Tandem-X height has been used to estimate tree height in the Canadian Arctic (Antonova et al. 2019), and, in tropical forest, Tandem-X. InSAR height relationship with biomass (Solberg et al. 2017) and tree height (Lei et al. 2021) have also been studied. In addition, a global forest/non-forest map has been derived at 50 m resolution from TDX interferometric data (Martone et al. 2018).

2.4.2.2 SAR Radargrammetry

SAR radargrammetry is an alternative way to InSAR to extract elevation data from radar data. This is based on stereoscopic measurement of SAR images. Analogously to photogrammetric spatial intersection, a stereo pair of SAR images with different off-nadir angles can be used to calculate the 3D coordinates for corresponding points on the image pair. However, contrary to interferometry, radargrammetry is based on the intensity and range values of SAR data and not on the phase information. The foundations for the stereo-viewing capabilities of radar images were recognized already in the 1960s (see, e.g., La Prade 1963). An example of research looking into the mathematical foundations for calculating 3D coordinates and their expected accuracies is the work by Leberl (1979). When the trajectory of a SAR antenna (position and velocity as functions of time) is known accurately enough in relation to the object coordinate system, and when a point target can be clearly identified on two SAR images with different off-nadir angles, the 3D coordinates of the point target can be calculated based on the range information. Typically, the so-called Range-Doppler equation system is used as a sensor model, which describes accurately enough the propagation of electromagnetic radiation from the SAR image pixel to the point target and vice versa (Leberl 1979). The Canadian satellite Radarsat-1 was one of the first SAR satellites to provide images with variable off-nadir angles suitable for radargrammetric processing (Toutin & Gray 2000). The ERS-1 and

ERS-2 satellites of the European Space Agency have also provided suitable stereo-pairs, but with limited stereo overlap areas (Li et al. 2006). However, only a few studies related to the extraction of forest information from radargrammetry have been published, e.g., by Chen et al. (2007). Studies by Perko et al. (2011) and Karjalainen et al. (2012) have revealed the potential of radargrammetric 3D data in forest biomass estimation and change detection. Wittke et al. (2019) estimated forest parameters using radargrammetry TSX data and compared the results with estimations derived from other remote sensing data sources. TSX provided RMSE-% of the level of 30% for stem volume and aboveground biomass. Radargrammetry produces lower results than single-pass interferometry due to lower coherence between the applied image pairs resulting in reduced accuracy for height measurements.

2.4.3 Airborne Laser Scanning, Airborne LiDAR

Airborne Laser Scanning (ALS) is a method based on Light Detection and Ranging (LiDAR) measurements from an aircraft, where the precise position and orientation of the sensor is known, and therefore the point cloud (x, y, z) of the reflecting objects can be determined. The first studies of ALS for forestry purposes included standwise mean height and volume estimation (e.g., Næsset 1997a, 1997b), individual-tree-based height determination and volume estimation (e.g., Hyyppä & Inkinen 1999; Hyyppä et al. 2001), tree species classification (e.g., Hyyppä et al. 2001; Brandtberg et al. 2003; Holmgren & Persson 2004), and measurement of forest growth and detection of harvested trees (e.g., Hyyppä et al. 2003; Yu et al. 2004, 2006). Today, ALS is becoming a standard technique in the mapping and monitoring of forest resources. By using ALS-based inventory, 5–20% error in main forest stand attributes at stand level has been obtained. For overviews on using ALS in forest inventory, see Hyyppä et al. (2008). ALS is a feasible technique also for efficient and accurate aboveground biomass (AGB) retrieval because of its capability of direct measurement of vegetation 3D structure. AGB correlates strongly with canopy height. Popescu et al. (2003), Popescu (2007), van Aardt et al. (2008), and Zhao et al. (2009) showed that AGB can be estimated similarly to other forest attributes by means of ALS metrics. The leaf area index (LAI) has also been used as a predictor of AGB (Koch 2010). Interaction of ALS pulse and forest canopy have also been studied using ALS intensity (Korpela et al. 2010). Forestry applications of ALS has been summarized in Maltamo et al. (2014). Recently the use of single-photon and multispectral ALS has been explored for application in forestry and forest inventory (e.g., White et al. 2021a, 2021b; Queinnec et al. 2022; Irwin et al. 2021; Yu et al. 2017, 2020). In Yu et al. (2020), single-photon LiDAR measurements with SPL100 (Leica/Hexagon) using two flight heights (1900 m and 3800 m) were compared with Optech Titan (400 m) multi-photon airborne laser scanning (ALS) data under leaves-on summer conditions. Yu et al. (2020) found that both data provide forest attribute estimates with comparable accuracy using area-based methods. A similar conclusion was made by Wästlund et al. (2018). White et al. (2021b) demonstrated that leaf-on SPL data were slightly less accurate than leaf-on linear model LiDAR data in capturing elevation heights of forest areas and that leaf-off SPL data were more accurate than the leaf-on linear model LiDAR data. Yu et al. (2017) investigated the potential of multispectral ALS using Optech Titan data for tree species classification in boreal forest, where the objective is to be able to separate three species: pine, spruce, and deciduous. The species of isolated and dominant trees could be classified with an accuracy of 90.5%. The corresponding accuracy was 89.8% for a group of trees, 79.1% for trees next to a larger tree, and 53.9% for trees situated under a larger tree, respectively. The results suggest that Channel 1064 nm contained the highest amount of information for species classification followed by channel 1550 nm and channel 532 nm.

2.4.4 Terrestrial Laser Scanning

Terrestrial laser scanning (TLS), also known as ground-based LiDAR, has been shown to be a promising technique for forest field inventories at tree and plot level. The major advantage of using

FIGURE 2.1 Multispectral point clouds acquired with FGI-developed helicopter-borne multispectral laser scanner. Combined multispectral response and very high density provides adequate tools to classify various tree species.

TLS in forest field inventories lies in its capacity to document the forest in detail. The first commercial TLS system was built by Cyra Technologies (acquired by Leica in 2001) in 1998, and the first papers related to plot-level tree attribute estimation were reported in the early 2000s. Currently, TLS has shown to be feasible for collecting basic tree attributes at the tree and plot level, such as DBH and tree position (Maas et al. 2008; Brolly & Kiraly 2009; Murphy et al. 2010; Liang et al. 2012a; Lovell et al. 2011). By reconstructing stem curve, it is possible to derive high-quality stem volume and biomass estimates comparable in accuracy with the best national allometric models (Liang et al. 2014a). TLS data also permit time series analyses because the entire plot can be documented consecutively over time (Liang et al. 2012b). In Liang et al. (2018a) an international benchmarking of TLS methods for forest measurements was conducted. The results of the benchmarking clearly showed that TLS can provide DBH and stem curve close to practical needs but completeness of stem detection and the accuracy of tree height estimation requires improvements. Wang et al. (2019) further showed that five-scan multi-scan TLS data is feasible for measuring tree heights of trees up to 15–20 m tall; after that, underestimation and errors increase due to limited visibility of tree tops.

2.4.5 Mobile Laser Scanning

Mobile Laser Scanning (MLS) is based on LiDAR measurements from a moving platform, where the precise position and orientation of the sensor is known using a navigation system, similarly to ALS, and therefore the position of the reflecting objects can be determined either from pulse travel time or phase information. An MLS system consists of one or several laser scanners. A navigation system consists of various sensors for positioning and determining the rotation angles of the system, while GNSS and inertial measurement unit (IMU) are the most important parts of the system. Also, other mapping sensors, such as cameras, thermal imagers, and spectrometers, can be incorporated into the MLS systems. In principle, MLS is similar to ALS, whereas the platform is not aircraft. Positioning is solved with GNSS and IMU integration and/or using Simultaneous Localization and Mapping (SLAM).

According to Hyyppä et al. (2020a), MLS can be further divided into: (1) phone-based scanning, (2) vehicle-based scanning, (3) drone-based scanning, and (4) hand-held, backpack, and other personal laser scanning techniques. Phone-based laser scanning has been studied, e.g., in Hyyppä et al. (2017), where Google Tango and Kinect were applied for diameter and stem curve measurements. Mokroš et al. (2021) compared iPad Pro, multi-camera system, hand-held laser scanning, and terrestrial laser scanning. Vehicle-based MLS in forestry is being studied, e.g., in Lin et al. (2010), Holopainen et al. (2013), Liang et al. (2014c), Bienert et al. (2018), Čerňava et al. (2019), Zhao et al. (2018), Wu et al. (2013), and Pierzchała et al. (2018). Hand-held or backpack laser scanning for forestry has also been studied (e.g., Chen et al. 2019; Cabo et al. 2018; Bauwens et al. 2016; Del Perugia et al. 2019; Marselis et al. 2016; Liang et al. 2014b; Liang et al. 2018b; Hyyppä et al. 2020b, 2020c).

Today, drone-based laser scanning has three approaches to measure tree attributes, namely, using (1) a standard above-canopy individual tree approach, (2) an above-canopy individual tree approach in which tree stem is directly measured from the dense point clouds (Hyyppä et al. 2022), and (3) under-canopy drone measuring stems similar to TLS measurements (Hyyppä et al. 2020a).

2.5 PROCESSING 3D DATA INTO FOREST INFORMATION

It is anticipated that many of the future remote sensing processes for forestry will be based on 3D point cloud or elevation data processing, especially those based on LiDAR and radar.

From the 3D data you can calculate DTM (Digital Terrain Model), DSM (Digital Surface Model), CHM/nDSM (Canopy Height Model/normalized Digital Surface Model). Today, most 3D techniques require a good DTM, which typically comes from laser scanning, since radar data with available frequency bands does not provide penetration into the ground floor due to too large of a footprint size and no ranging capacity of the applied radars. In optical wavelength the penetration of the signal through vegetation is lower, but since LiDAR is based on narrow beams finding canopy gaps, the terrain model can be calculated accurately from the LiDAR data. Lower microwave wavelength penetrates the vegetation layer better, but then also the penetration into the ground increases based on dielectric properties of the ground. From that point of view LiDAR data is optimal for topographic mapping. The idea is to provide a surface model (DSM) and subtract the ground elevation (DTM) from it in order to get a canopy height. Intensity, coherence (in interferometry-SAR), and texture can be used to improve the estimates in 3D-based inventory techniques as well as other LiDAR metrics.

We see that there are two kinds of processing needed: forest attribute estimation based on single-time point cloud and use of bi-temporal point clouds for change detection. Additionally, direct measurement of individual tree attributes is the third kind of data processing methodology used, but it can be combined with the first two approaches.

Single-time point cloud processing for forest attribute data collection includes the following steps. DTM is obtained from ALS data or known beforehand (for SAR). CHM is calculated to get tree heights and tree height metrics. Features (point cloud metrics) are calculated from the data, and non-parametric estimation is applied. Non-parametric estimation requires field plots, which are used for teaching the classifier. In addition to the point cloud metrics, other features, such as individual-tree information, texture, waveform LiDAR features, image processing applied to, e.g., DSMs, image-based features (including NDVI), and other channel information and ratios, can be added to improve the prediction. Tree species are predicted also in this phase, and therefore, the system should include features capable of discriminating against species. The optimum output, requested by forest companies, of the process is species-specific height distribution of the trees.

The process includes:

1. DTM generation
2. DSM and CHM generation
3. Derivation of point cloud metric
4. Prediction of forest attributes using non-parametric estimation

The typical process for change detection is to subtract two DSMs from bi-temporal data after they have been shown to match with elevation. After thresholding, which is used to see real changes, and after filtering/smoothing, segmentation of the changed areas can be delineated. The changes can then be compared with the real change of the reference. Prediction of the change can also be done using non-parametric estimation.

Direct measurement of individual tree attributes can be based on detection of individual trees, either tree crowns or stems, measurement of the tree heights, diameters, and stem curves directly from the point cloud data.

In the following sections, these three types of processing are further discussed.

2.5.1 DTM Processing

Removal of low points is an important preprocessing part of DTM and is usually done before ground classification. Low points seem to come below the ground surface, and their origin in forested area may be multiple reflections from trees or the ringing effect (too high return signal entering the receiver). Low points exist also with airborne 3D radar data (Hyyppä 1993). Forested areas may also include buildings, and multiple reflections from windows and ground can cause low points. A single or group of points may cause an anomaly in the correct ground surface if it is not removed from the point cloud in the first step. A point can be defined as being low if all of its neighboring points in a search window are more than a predefined value higher than the point. One or several low point removal filtering processes should be done before the DTM filtering (classification of DTM points).

The second step in DTM calculation is the initial DTM point cloud selection. This step is needed to detect building regions. For example, by selecting the lowest points with an 80-m-by-80-m window will remove all buildings less than 80 m in size. This step can also be done by data pyramids. The first and second steps are not always used, but the user should be careful in such cases.

For the final ground point classification, there are several DTM filtering techniques developed. Mathematical morphology is one applied technology. Using operators such as erosion, dilation, closing, and opening, can produce DTM and DSM. Vosselman (2000) applied a maximum admissible height difference function for a defined distance. Point within the maximum height difference function were included as ground points. Progressive densification strategy starts with step 2 (initial DTM point cloud selection) and then iteratively increases the amount of accepted terrain points. Axelsson (2000) developed a progressive TIN densification method that was implemented into Terrascan software. In Terrascan, laser point clouds are first classified to separate ground points from all other points. The program selects local low points on the ground and makes an initial triangulated model. New laser points are then added to the model iteratively, and the actual ground surface is then described more and more precisely. Maximum building size, iteration angle, and distance parameters determine which points are accepted. Kraus and Pfeifer (1998) developed a DTM algorithm for which laser points between terrain points and non-terrain points were distinguished using an iterative prediction of the DTM and weights attached to each laser point, depending on the vertical distance between the expected DTM level and the corresponding laser point. The method officially goes to category surface-based filtering, in which the starting point is that all given points belong to the terrain class and then the points that do not fit to the surface model are iteratively removed. In the beginning, the method did not use initial DTM, but in order to overcome the limitation in large building areas and in order to speed up the process, initial DTM and data pyramids in a hierarchical framework were employed. The method is implemented in SCOP++ (Kraus & Otepka 2005). Additionally, the filtering can be based on segments, i.e., segment-based filtering. Either object- or feature-based segmentation can be done, and then filtering is performed to each segment separately. Additionally, waveform and intensity can be used to assist in the ground filtering.

A comparison of the filtering techniques used for DTM extraction can be found in a report on ISPRS comparison of filters by Sithole and Vosselman (2004). Selection of the filtering strategy is not a simple process. In practice, the amount of interactive work determines the final quality of the product, but in forest inventory, fully automated DTM calculation is preferred. Examples of commercial software that include DTM generation are REALM, TerraScan, and SCOP++. DTM quality indicators can also be directly calculated from the point cloud and waveform data. Such indicators include, e.g., point density, point spacing, terrain slope, echo width, and estimate of the aboveground biomass estimated with LiDAR data.

Recently, deep learning has enabled DTM extraction from point cloud data. Also, deep learning solutions can be based on raster or point cloud data. Hu and Yuan (2016) proposed a first filtering algorithm based on deep convolutional neural networks (CNN). Since then many new CNN architectures and point-based approaches have been developed.

2.5.2 DSM Processing and Canopy Height Model

Ideally, the first echoes over a forest region comes from the surface model of the canopy, whereas the last echoes form the terrain model. The most frequently used method for the creation of a DSM is, therefore, to take the highest echo within a given neighborhood and interpolate the missing heights. Following the creation of the digital terrain (or elevation) model (DTM or DEM), a canopy height model (CHM) can be calculated by subtracting the height of the ground from the DSM and presented in a raster or TIN height data format.

Airborne LiDAR measurements tend to underestimate tree height (Nelson et al. 1988; Hyyppä & Inkinen 1999; Rönnholm et al. 2004), and the same happens with 3D radars (Hyyppä 1993). The first echo return comes more often from the shoulder of the tree instead of the tree top. Although a laser pulse hits the top, the treetop may not be wide enough to reflect a recordable return signal. On the other hand, dense under vegetation causes overestimation in the DTM. For these reasons, the CHM is typically underestimated. Other factors affecting tree height measurement accuracy are scanning parameters, such as flying height, pulse density, pulse footprint, applied modeling algorithms, scanner properties (e.g., sensitivity, field of view, zenith scan angle, and beam divergence), and structure and density of the tree crown (Holmgren 2003; Hopkinson et al. 2006). With deciduous forests, seasonal aspects have to be recognized. Single-photon and linear-model LiDAR data may have different amounts of low points or high points, which researchers may need to take into consideration for DTM and DSM processing,

2.5.3 Point Height Metrics

The prediction of stand variables is typically based mainly on point height metrics calculated from ALS data. Nelson et al. (1988) divided features related to the height and density, which is the foundation of the area-based technology. Features such as percentiles calculated from a normalized point height distribution, mean point height, densities of the relative heights or percentiles, standard deviation, and coefficient of variation are generally used (Hyyppä & Hyyppä 1999; Næsset 2002). The percentiles are down to the top heights calculated from the vertical distribution of the point heights, that is, the percentile describes the height at which a certain number of cumulative point heights occur. Density-related features are calculated from the proportion of vegetation hits compared with all hits. A hit is seen as a vegetation hit from trees or bushes if it has been reflected from over some threshold limit above ground level. All the features are calculated separately for every echo type. The reason for this is that the sampling between echo types is somewhat different (Korpela et al. 2010). Table 2.5 gives a list of typical point height metrics used.

2.5.4 Approaches for Obtaining Forest Data from Point Clouds

Approaches aimed at obtaining forest and forestry data from point cloud data have been divided into two groups: (1) area-based approaches (ABAs) and (2) individual/single-tree detection approaches (ITDs).

ABA prediction of forest variables is based on the statistical dependency between the variables measured in the field and the predictor features derived from ALS data. The sample unit in the ABA

TABLE 2.5
Typical Point Height Metrics Used in Forest Attribute Derivation

No.	Feature	Explanation
Point Height Metrics		
1	meanH	Mean canopy height calculated as the arithmetic mean of the heights from the point cloud
2	stdH	Standard deviations of heights from the point cloud
3	P	Penetration calculated as a proportion of ground returns to total returns
4	COV	Coefficient of variation
5	H10	10th percentile of canopy height distribution
6	H20	20th percentile of canopy height distribution
7	H30	30th percentile of canopy height distribution
8	H40	40th percentile of canopy height distribution
9	H50	50th percentile of canopy height distribution
10	H60	60th percentile of canopy height distribution
11	H70	70th percentile of canopy height distribution
12	H80	80th percentile of canopy height distribution
13	H90	90th percentile of canopy height distribution
14	maxH	Maximum height
15	D10	10th canopy cover percentile computed as the proportion of returns below 10% of the total height
16	D20	20th canopy cover percentile computed as the proportion of returns below 20% of the total height
17	D30	30th canopy cover percentile computed as the proportion of returns below 30% of the total height
18	D40	40th canopy cover percentile computed as the proportion of returns below 40% of the total height
19	D50	50th canopy cover percentile computed as the proportion of returns below 50% of the total height
20	D60	60th canopy cover percentile computed as the proportion of returns below 60% of the total height
21	D70	70th canopy cover percentile computed as the proportion of returns below 70% of the total height
22	D80	80th canopy cover percentile computed as the proportion of returns below 80% of the total height
23	D90	90th canopy cover percentile computed as the proportion of returns below 90% of the total height

is most often a grid cell, the size of which depends on the size of the field-measured training plot. Stand-level forest inventory results are aggregated by summing and weighting the grid-level predictions inside the stand. When using ITD techniques, individual trees are detected and tree-level variables, such as height and volume, are measured or predicted from the ALS data, i.e., the basic unit is an individual tree. Then the stand-level forest inventory results are aggregated by summing up the treewise data. The ABA does not make use of the neighborhood data of laser returns. On the other hand, ABAs are based on the height and density data acquired by ALS, which are highly correlated with the forest variables. Currently, the ABA is operationally applied in the Nordic countries when carrying out standwise forest inventories. Some 3 Mha of Finnish forests are inventoried every year by applying ABA. White et al. (2013a) report on best practices for using the ABA in a forest management context.

ABA is based on accurate training plot–level data, which should represent the whole population and cover the variations in it as much as possible. The efficient selection of the training plot locations requires pre-knowledge of the inventory area. The statistical relation between the predictors and dependent variables to be defined is modeled using training data. The dependent variables are then predicted for all (other) grid cells without training data typically using non-parametric estimation techniques. If stand-level variables are needed, they are calculated by weighting the grid-level predictions inside the stand to the known stand delineation map.

One of the first tests with ABA in Finland was Suvanto et al. (2005). Regression models were developed using laser height metrics for diameter, height, stem number, basal area, and stem volume of 472 reference plots. The predicted accuracies were 9.5%, 5.3%, 18.1%, 8.3%, and 9.8%, respectively, at stand level. Current forest management planning inventories in Scandinavia require species-specific information for growth projections and simulated bucking. Tree species composition has also a major effect on forest value. Maltamo et al. (2006) added predictor features from aerial photographs and existing stand registers to ALS height metrics, resulting in plot-level volume estimation accuracy from 13% to 16% depending on the predictors used. Similarly, Packalén and Maltamo (2007) used the k-MSN method to impute species-specific stand variables using ALS metrics and aerial photographs to the same dataset as in Suvanto et al. (2005), and the species-specific volume estimates at the stand level were 62.3%, 28.1%, and 32.6% for deciduous tree, Scots pine and Norway spruce, respectively. Thus, there are limitations with current technology, and especially species-specific tree size (height and diameter) distribution information is needed. One possible way is to use more detailed data and ITD type processing.

In addition to NN, k-MSN methods, a random forest (RF) classifier has also been applied in ABA (Yu et al. 2011). RF is a non-parametric regression method in which the prediction is obtained by aggregating regression trees, each constructed using a different random sample of the training data and choosing splits of the trees from among the subsets of the available features, randomly chosen at each node. The samples that are not used in training are called "out-of-bag" observations. They can be used to estimate the feature's importance by randomly permuting out-of-bag data across one feature at a time and then estimating the increment in error due to this permutation. The greater the increment, the more important the feature.

A similar ABA approach can be also used to process point clouds provided by radar imagery. According to the first results obtained, the use of stereo SAR data in the predicting of plot-level forest variables is promising. Karjalainen et al. (2012) obtained a relative error (RMSE%) of 34% for stem volume prediction. For the other forest variables, i.e., the mean basal area, mean diameter at breast height, and mean forest canopy height, the accuracies were, 29%, 19.7%, and 14%, respectively using RF as non-parametric estimation technique. Typically, such a high level of prediction accuracy cannot be obtained using satellite-borne remote sensing at the plot-level data in the boreal forest zone.

Since there are limitations in the ABA and user's need to get better species-specific tree size distribution data, individual tree approaches have been developed. The basic idea is to derive more detailed information of standing trees, which are then used in the prediction of the forest attributes. Thus, area-based prediction can be done using individual tree–based features, as originally proposed in Hyyppä and Inkinen (1999). That was demonstrated in Hyyppä et al. (2012), in which both individual tree–based and point height metrics were used as the inputs for the RF classifier. Individual tree–based features improved the ABA's accuracy significantly since they had very high correlation, e.g., with the reference stem volume. When calculating the importance of the features, most of the individual tree–based features were among the best features confirming that individual tree–based features are applicable in ABA or stand0level inventory in general. When estimating plot-level mean height, the best laser-derived feature was the mean height derived by using the individual tree technique. When estimating DBH, the best laser-derived features were (1) mean canopy height and (2) penetration to the ground, (3) mean tree height (derived from the extracted individual trees), and (4) mid percentiles. For the estimation of stem volume, the best laser-derived feature was the stem volume derived from extracted individual trees, followed by the basal area derived from extracted individual trees. It is possible to easily derive further laser point height metrics and individual tree–based features.

2.5.5 INDIVIDUAL TREE DETECTION OR LOCATING WITH ALS

Individual tree detection (ITD) using airborne laser scanning data was presented in Hyyppä and Inkinen (1999), Hyyppä et al. (2001), and Persson et al. (2002). In Hyyppä and Inkinen (1999), ITD was considered to have four steps: (1) making individual trees or tree groups visible, (2) calculation of normalized height, (3) detecting individual trees or a group of trees from normalized point cloud data, and (4) estimating tree attributes (individual or group of tree attributes), such as volume or basal area, from the features that can be directly measured from ALS point clouds (position, height, crown size, and species).

Originally, most of the approaches for tree detection were based on finding trees from the CHM. Thus, the CHM corresponds to the maximum canopy height of the first pulse data. Hyyppä et al. (2012) utilizes the canopy penetration capability of the last pulse returns with overlapping trees. When trees overlap, the surface model corresponding to the first pulse stays high, whereas with the last pulse, even a small gap results in a drop in elevation, i.e., the trees can be more readily discriminated against. The idea worked for small trees and for interlocked canopies. Better solutions for improving the detection accuracy of individual trees, trunk hits (Rahman & Gorte 2008), and full waveform (Reitberger et al. 2009) have been proposed. Various ITD methods have been internationally benchmarked, see, e.g., the papers of Kaartinen et al. (2012), Wang et al. (2016), Eysn et al. (2015), Vauhkonen et al. (2012), and Parkan (2019). In today's implementations, the tree positions, heights, canopy areas, and crown-related features are directly derived, whereas diameter and volume are predicted by non-parametric estimation (e.g., random forest). For species determination, additional data sources such as RGB images are also applied (Holmgren & Persson 2004; Holmgren et al. 2008).

A major challenge in the ITD process is segmentation. Segmentation of individual trees is complicated due to tree crown adjacency and intersection, differences and changes of tree shape with age, canopy structure, and understory vegetation occurring both within a single tree species and between different species (Parkan 2019). Segmentation methods utilize physical properties of point clouds, such as spacing between trees, point cloud characterizing the tree shape, and intensity. Established clustering techniques, such as K-means, DBSCAN, hierarchical clustering, and mean-shift, have also been applied in segmentation. Initial solution is typically based on preliminary detection of treetops. Combinations of segmentation and clustering are also applied to separate trees inside the segment/group (Hyyppä et al. 2020b).

The most common individual tree detection algorithms include:

1. Raster-based processing of canopy height (Hyyppä & Inkinen 1999)
2. Voxel-based processing of canopy height (Wang et al. 2008)
3. Vector- or point cloud–based delineation of canopy height (Reitberger et al. (2007)
4. Mixed representations (Morsdorf et al. 2003)
5. Detection of stems (Rahman & Gorke 2008)
6. Direct measurement of stem curve from point cloud data (Hyyppä et al. 2022)

The accuracy of tree detection using ALS has been found to vary between 46% and 95% in previous studies (Chang et al. 2013; Ene et al. 2012; Koch et al. 2006; Kwak et al. 2007; Morsdorf et al. 2003; Persson et al. 2002) depending on the difference in forest structure and data used. In a benchmarking study for individual tree detection, Kaartinen et al. (2012) reported detection rates ranging from 25% to 90% with different algorithms and point densities. Wang et al. (2016) benchmarked five individual tree detection algorithms and found that the overall detection rates of dominant and co-dominant trees exceeded 90% and 80%, respectively. The detection rates of the intermediate and suppressed trees ranged from 40% to 80% and up to 30%, respectively. Chang et al. (2013) reported a 77% detection rate for dense and mixed forests with point density of 4.3 points/m². Ene et al. (2012) reported a 46–50% detection rate with point density of circa 10 points/m². Based on

Finnish experiences, the performance of 3D point clouds generated using both LiDAR and photogrammetry were compared for detecting individual trees in the southern Boreal Forest Zone. The compared techniques included drone-based and airborne laser scanning and photogrammetry as well as satellite photogrammetry. Reference dataset consisted of close to 9000 field-measured trees located in 91 sample plots. Overall, individual tree detection accuracy ranged from 39.3% to 69% for all trees and 55.8% to 79.6% for trees taller than 15 m, depending on the data used. Detection rates for individual trees were about 20% units greater using the drone LiDAR data (320 pts/m^2) compared to the ALS data (2.7 pts/m^2) for trees with a diameter less than 25 cm, and about 10% units greater compared to the drone photogrammetry for trees with a diameter less than 15 cm.

2.5.6 Individual Tree Height Derivation

Kaartinen et al. (2012) showed that the variability of point density was negligible for tree height determination when compared to the segmentation accuracy between the methods. In Kaartinen et al. (2012), with the best models, a RMSE of 60–80 cm was obtained for tree height (Figure 2.2). The results with the best automated models were significantly better than those attained when using the manual process. Both underestimation of tree height and standard deviation were decreased in general as the point density increased. The height estimation accuracy was highest for dominant and co-dominant tree layers. According to Wang et al. (2019), ALS-based tree height estimates were robust across all forest stand conditions. The taller the tree, the more reliable the ALS-based tree height. The highest uncertainty took place with suppressed tree layers due to difficulties related to segmentation of such trees. Based on Finnish experiences, RMSE of 2% for height could be reached with national scale 5 pts/m^2 data based on analyses relying on more than 60,000 field-measured trees.

The measurement of tree height using TLS at the plot level has not been thoroughly studied because the visibility of treetops with TLS techniques can be questioned. However, there are past results with TLS showing that tree height is typically underestimated and that the magnitude of estimation error is typically several meters. In Huang et al. (2011), a −0.26 m bias and a 0.76 m RMSE were reported for one plot (212 stems/ha, sparse stand) using the multi-scan approach. In Brolly and Kiraly (2009), a −0.27 m bias with a 1.82 m RMSE and a −2.37 m bias with a 3.25 m RMSE was reported for one more dense plot (753 stems/ha) using the single-scan approach. In Hopkinson et al. (2004), an approximate 1.5 m underestimation of tree heights was reported for two medium-density plots (465 and 661 stems/ha) using the multi-scan approach. Maas et al. (2008) depicted a −0.64 m

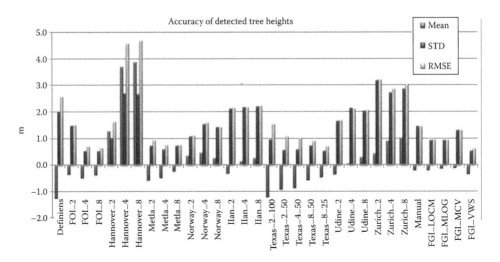

FIGURE 2.2 Tree height comparison from Kaartinen et al. (2012).

bias and a 4.55 m RMSE for nine trees located on four plots (212–410 stems/ha) using the single- and multi-scan approaches. Fleck et al. (2011) concluded a 2.41 m RMSE for 45 selected trees on one plot (392 stems/ha) using multi-scan data. The observation of tree tops from the TLS data is possible on sparse sample plots using many scans, as reported in Huang et al. (2011) and Fleck et al. (2011), but not in dense sample plots. Treetops are most likely shadowed by other parts of the crown in the point cloud, but the use of clearly visible trees in sparse plots could be actually used to calibrate ALS-based tree heights. In the ISPRS/EuroSDR Tree extraction test (Kaartinen & Hyyppä 2008), TLS was able to collect tree height information with a level of 10 cm, but very-high-density scanning was made (both as point cloud density and in the number of scans applied).

One of the advantages of using MLS for plot-level inventories lies in the fact that MLS can see many of the invisible treetops for TLS. Since the MLS platform moves all the time during the data acquisition, the gaps in treetops are more likely visible with MLS than with TLS. In Hyyppä et al. (2020b), the RMSE of 2% for tree height was achieved for easy plots and about 5% for obstructed plots, implying a tree height measurement capacity of 0.5–1 m with the MLS system having a SLAM-based positioning system.

2.5.7 Diameter and Stem Curve Derivation

ALS, TLS, and MLS have been studied to provide accurate diameters at the tree and plot level. This kind of field data collection could also be used as a substitute for field-measured plot-level data in the ABA prediction of forest attributes. The tree stem curve, or stem taper, depicts the tapering of the stem as a function of the height. The tree stem curve holds a significant position in forestry, as it is the key input needed in harvest operations and is used in various ecological studies. If the most important part of the stem curve can be determined, it is also possible to derive biomass and stem volume estimates from the trees without knowledge of conventional allometric models relating, e.g., diameter and height information to volume and biomass. From that point of view, it is surprising that the non-invasive measurement of stem curves using TLS, MLS, and ALS has not been intensively studied. The most popular processing method today for locating trees and determining diameter is to cut a slice of data from the original point cloud and to identify and model tree stems from this layer by point clustering or circle finding (Simonse et al. 2003; Aschoff & Spiecker 2004; Thies et al. 2004; Watt & Donoghue 2005; Maas et al. 2008; Brolly & Kiraly 2009; Tansey et al. 2009; Huang et al. 2011). This assumes that all trees present a clear stem at the same height at which the slice goes through the point cloud. This assumption is typically not valid in most mixed forests having branches at different heights, and nearby branches may be overlapped in the layer. A study of a mixed deciduous stand showed difficulties even in the manual stem detection in a TLS data layer (Hopkinson et al. 2004). Results from studies of different types of forests are highly variable, indicating the need for more research on these topics. In Liang et al. (2012b), the DBH estimation results are reported at the tree level from five plots having bias of 0.16 cm and RMSE of 1.29 cm. In Lindberg (2012), the bias and RMSE of tree-level DBH estimates from six plots were 0.16 cm and 3.8 cm, respectively. Table 2.6 summarizes the accuracy of the plotwise DBH measurements using the single-scan approach. In practice, single-scan, multi-scan, and multi-single-scan techniques can be applied, each having their own pros and cons.

Pioneering TLS work for stem curve measurement and modeling includes nine pine trees studied in Henning and Radtke (2006), a spruce tree in Maas et al. (2008), and two trees, one pine and one spruce, in Liang et al. (2011). The RMSE of the stem curve measurements was 4.7 cm in Maas et al. (2008) utilizing single-scan data. In Liang et al. (2011), the RMSE of the stem curve estimation of the pine tree was 1.3 cm and 1.8 cm with the multi- and single-scan data, respectively; and the RMSE of the curve measurement of the spruce tree was 0.6 cm and 0.6 cm using the multi- and single-scan data, respectively. The first detailed study on the plot-level automatic measuring of the stem curves of different species and different growth stages using TLS was reported in 2014 (Liang et al. 2014a). Twenty-eight trees, 16 pines and 12 spruces, were selected from nine sample

TABLE 2.6
Summary of the Plotwise DBH Estimation from the Single-Scan Methods

	Plot			Result	
	Number	Size	Density (stems/ha)	Bias DBH (cm)	RMSE DBH (cm)
Maas et al. (2008)	3	15 m radius	212–410	−0.67–1.58	1.80–3.25
Brolly and Kiraly (2009)*	1	30 m radius	753	−1.6–0.5	3.4–7.0
Liang and Hyyppä (2013)	5	10 m radius	605–1210	−0.18–0.76	0.74–2.41
Olofsson et al. (2014)	16	20 m radius	358–1042	0.6	2.0–4.2

*Three detection methods were discussed.

plots. The plots were scanned utilizing the multi-scan approach. The trees were felled, and the stem curves were manually measured in the field. For comparison, the stem curve was also manually measured from the point cloud data. The stem curves were automatically measured with a mean bias of 0.15 cm and mean RMSE of 1.13 cm at the tree level. The highest diameters measured were between 50.6% and 74.5% of the total tree height, with a mean of 65.8% for pine trees and 61% for spruce trees. These results showed that TLS data and automated processing have the capability of accurately measuring stem curves of different species and different growth stages. Surprisingly, the automated processing gave clearly more diameter measurements at the upper part of the stems than with manual measurements from the same data. The difficulty of the manual measurement from point cloud data is that the stem edges are difficult to locate when the stem is partly blocked in the data by other branches.

In Liang et al. (2014c), a mobile laser scanning (MLS) system was tested and its implications for forest inventories were discussed. The stem mapping accuracy was 87.5%; the root mean square errors of the DBH estimates and the location were 2.36 cm and 0.28 m, respectively. These results indicate that the MLS system has the potential to accurately map large forest plots and that further research on mapping accuracy and cost–benefit analyses is needed. In Hyyppä et al. (2017) using a Tango smartphone, stem diameters were derived with a RMSE of 0.73 cm compared to tape reference measurements. Mokroš et al. (2021) compared iPad Pro, multi-camera system, hand-held laser scanning, and terrestrial laser scanning and found the iPad Pro acquired 2.6–3.4 cm accuracy for DBH. Gülci et al. (2023) achieved a corresponding accuracy of 2.33 cm for DBH. DBH has been obtained in various MLS, with accuracies ranging from 1 to 4 cm and detection rates varying between 80% and 95% in relatively easy boreal forests (Lin et al. 2010; Holopainen et al. 2013; Liang et al. 2014b, 2014c, 2018b; Bienert et al. 2018; Čerňava et al. 2019; Zhao et al. 2018; Wu et al. 2013; Pierzchała et al. 2018; Chen et al. 2019; Cabo et al. 2018; Bauwens et al. 2016; Del Perugia et al. 2019; Marselis et al. 2016; Hyyppä et al. 2020b, 2020c). It has been found in Hyyppä et al. (2020a, 2020c) that the under-canopy drone system could achieve DBH with RMSE of 2–8% and stem curve of 2–15%. The accuracies are adequate for operational works.

Previously it was not possible to measure stem diameters directly from the ALS or drone point clouds. Rahman and Gorte (2008) demonstrated the value of ALS trunk hits. Jaakkola et al. (2017) tried to measure diameters directly from the point cloud data using a Velodyne-16 sensor from the drone. However, the accuracy of direct DBH measurements was between 5.5 and 6.8 cm due to the large beam of the system. On the other hand, the estimation of DBH from point cloud metrics resulted in an accuracy of 2.6 cm. The technique was further developed in a study by Hyyppä et al. (2022), in which an in-house-developed dual-wavelength laser scanning system was mounted on board a helicopter and collected point cloud data from a total of 1469 trees. The obtained point clouds included many high-quality stem hits, allowing the estimation of the stem curves and stem volumes of individual trees directly from the point clouds. Point density was between 2200 and 3800 returns/m². The scanner was tilted by 15° from the nadir to increase the possibility of recording stem hits. The stem volumes of individual trees were computed by using the measured stem curves and tree heights without any allometric models. In Hyyppä et al. (2022) they were able to estimate the stem curves with a RMSE of 1.7–2.6 cm (6–9%) while detecting 42–71% of the trees. The RMSE of stem volume estimates varied between 0.1 and 0.15 m³ (12–21%). If MLS holds the promise for collecting reference data with an extremely high efficiency, the method proposed in Hyyppä et al. (2022) would allow recording reference individual trees with the speed of flight, i.e., roughly about 500 reference trees per minute.

2.6 FUTURE OPERATIONAL POSSIBILITIES

Today, remote sensing is changing the way we assess forest resources. For a century, foresters conducted the forest inventory. Now, this responsibility is increasingly shifting to remote sensing experts and forestry professionals proficient in handling remote sensing, 3D data, point cloud, and image processing. With the following cases, depicted from the author's works, we aim to anticipate situations where LiDAR and radar remote sensing play a more integral role in forestry practices.

2.6.1 THE CONCEPT AND UTILITY OF THE LIDAR PLOTS

Ground plot-–like measures from airborne laser scanning in many ways resemble field measures and are termed LiDAR plots (Wulder et al. 2012a) and can be applied as a reference for, e.g., satellite-based observations. Ground plots remain invaluable for robust forest characterizations, enabling consistent and reliable measurement of attributes to support forest inventory, mapping, monitoring, modeling, and science. National inventories in many jurisdictions are primarily based upon careful and systematic measurement of ground plots (Kangas & Maltamo 2006). Furthermore, applications that use remotely sensed data to produce forest attribute maps often require ground plot data for building models and validating outcomes. In many jurisdictions, ground plots remain costly to install and as a result are often limited in number and extent. For example, remote locations are difficult to access, such as some northern regions of Canada, further precluding the establishment of ground plots. Another example is locations with non-commercial forests that do not have the requisite economic drivers for maintaining up-to-date plot or inventory data sets. This lack of ground measurements precludes the development of robust large-area forest inventory, mapping, and monitoring applications. As an alternative, Wulder et al. (2012a) have proposed the concept of the LiDAR plot. Airborne scanning LiDAR data have been shown to offer attribute characterizations (especially height-related attributes) that are similar and in some cases more accurate than ground measurements (Næsset 2007; Hyyppä et al. 2008).

The concept of the LiDAR plot is comparable to that of a ground plot with a fixed area. LiDAR plots are an area-based summary of a LiDAR point cloud, whereby descriptive statistics or metrics are generated from the point cloud (e.g., percentiles, mean, standard deviation) and these are used, with a sample of co-located ground measurements, to model forest inventory attributes of interest such as mean height, dominant height, basal area, volume, and biomass (Næsset 2002). Thus, although the LiDAR plot concept requires some amount of traditional ground plots, it enables efficient propagation of the ground plot information over large areas, via airborne measurements. The stability of the empirical relationships between metrics and inventory attributes across many different forest types and structures can be attributed to the nature of the LiDAR data itself: in essence, the LiDAR metrics represent a detailed measurement of all surfaces within a canopy (foliage, branches, and stems). The LiDAR point clouds are generalized on a grid as well as vertically, creating a voxel of information that can be simplified. This voxel-based generalization can be implemented in freely available software (such as LASTools and FUSION) that provides unique metrics that in turn are used for model development. Thus, even when LiDAR data are collected at a lower hit density (i.e., 1 hit/m²) (Jakubowski et al. 2013), or when the vertical structure of the forest is complex (i.e., composed of multiple canopy strata, with a significant understory component) (Vastaranta et al. 2013), meaningful relationships to plot-level forest attributes may still be generated.

LiDAR plots are typically square in shape, and their size is determined in concert with the size of the aforementioned ground plots used for model development. Typical plots sizes are 400 m² or 20 by 20 m (White et al. 2013b), and the plots must be large enough to contain sufficient LiDAR hits, to have a more uniform hit density (Næsset 2002) and to enable reasonable attribute estimates (McGaughey 2013). The full swath of the LiDAR transect (depending on instrument scan angle) may be tessellated into these fixed-area plots, followed by the generation of metrics and estimation of inventory attributes. Figure 2.3 provides a schematic of the LiDAR plot concept.

Note that opportunistically located ground plots (measured for LiDAR plot attribute modeling) and LiDAR transects can be used to improve large-area mapping and monitoring (e.g., Chen et al. 2012a; Mora et al. 2013b; Magnussen & Wulder 2012; Matasci et al. 2018a, 2018b) but may not be appropriate for statistically driven designs for large-area forest inventories (Wulder et al. 2012a). Chen et al. (2012a) used samples of LiDAR plots generated from a national collection of LiDAR transects (Wulder et al. 2012b) as calibration and validation data to support geometric-optical modeling of mean, dominant, and Lorey's height using Landsat imagery. Estimates of vertical forest structure are critical for forest inventory and reporting. Heights were modeled over the area of a single Landsat scene (185 km by 185 km) at a 25 m resolution with average estimation errors (RMSE) of 4.9 m, 4.1 m, and 4.7 m for mean height, dominant height, and Lorey's height, respectively. In this study, the LiDAR plots data were useful for model development, identification of spectral endmembers required for the geometric-optical model, and parameterization of the model's tree variables. Using a different modeling approach and different LiDAR transects in combination with higher-resolution optical imagery (QuickBird), Chen et al. (2012b) obtained an average error (RMSE) of 3.3 m in the estimation of plot-level mean heights. Mora et al. (2013a) used data from LiDAR plots (Wulder et al. 2012b) with samples of very high spatial resolution (VHSR; < 1 m) imagery to achieve estimation errors (RMSE) of 2.3 m for mean stand height.

Obviously, there is promise in the use of LiDAR plot data to enable modeling of forest structural attributes across large areas. What is emerging from the literature is that only modest gains in error reduction are possible when using LiDAR plots with increasingly higher-resolution optical imagery

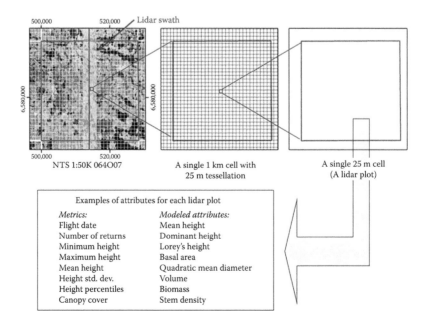

FIGURE 2.3 Schematic of the LiDAR plot concept.

(Mora et al. 2013b). These modest gains in accuracy are offset by the increased level of effort and cost associated with using the higher-resolution imagery. Landsat data is free and readily accessible, with each scene covering a markedly larger area (185 km by 185 km) relative to what is typical for VHSR data (e.g., 10 km by 10 km). Thus, synergies between LiDAR plots and Landsat data offer particular advantages for nations such as Canada, with more than 600 million ha of forested ecosystems that require monitoring and reporting information to be collected in a systematic and transparent manner (Matasci et al. 2018a, 2018b). Deep learning networks (DNNs) have recently challenged traditional machine learning methods. Recently, new types of DNNs have been successfully used in remote sensing data classification (Karila et al. 2023; Bazi et al. 2021; Seely et al. 2023; Murray et al. 2024). For example, deep learning has been utilized to get country-wide estimates of ALS-based forest parameters, such as canopy height, forest canopy density, and cover, from a combination of Sentinel-1 SAR and Sentinel-2 multispectral data with normalized mean absolute errors between 11% and 15% (Becker et al. 2023). Newest versions of deep learning architectures provide potential further improvements in classification accuracy. Most likely the LiDAR plot concept can be further improved using new deep learning architectures. LiDAR plots have demonstrated a unique and valuable role in supporting large-area mapping, monitoring, and modeling for boreal forest ecosystems. The utility has been demonstrated for science and management-related information needs across a broad range of applications. Further, transect installation and application over different forest ecosystems (such as tropical) and to address different management or science questions remains to be undertaken.

2.6.2 Improving Large-Area Mapping of Forest Attributes Using Satellite Radar

Experiences with ALS (Næsset 2002; Hyyppä et al. 2008; Vastaranta et al. 2012; Vastaranta 2012) and digital stereo-photogrammetry (Nurminen et al. 2013; Vastaranta et al. 2014; White et al. 2013b) have shown that precise forest biomass estimations can be achieved using these airborne techniques. The foundation of the high biomass estimation accuracy is their ability to measure the forest canopy height and density. On the other hand, the majority of satellite data-based forest mapping techniques, until now, have only used the intensity information (reflectance values or SAR backscattering coefficient), i.e., 2D information, in estimating forest biomass. Even though in some experiments a good agreement with 2D information and forest biomass have been obtained, the way forward in satellite-based forest mapping appears to be the use of 3D techniques.

In the past decade, commercial SAR satellite data have undergone remarkable progress in terms of spatial resolution, geolocation accuracy, and data availability—mainly thanks to the X-band SAR satellite systems of TerraSAR-X, TanDEM-X, and COSMO-SkyMed. Consequently, there has been a growing interest in using the aforementioned satellites for 3D forest mapping also. In principle, there are two main techniques to extract 3D/elevation data from satellite SAR images, including both SAR interferometry and SAR radargrammetry.

Based on the recent scientific studies in 3D SAR techniques and forest resources mapping, there appears to be potential in these techniques for deriving detailed forest attribute maps over large areas with good temporal resolution. Even though ALS and aerial stereo-photogrammetry provide more accurate estimates for forest attributes compared to 3D SAR techniques, there might be a demand for SAR data, especially for large-area mapping and especially in monitoring of changes in the forest structure. The advantage of 3D SAR techniques compared to 2D estimation techniques is that the data derived using 3D techniques can be easily integrated to existing forest attribute inventory processes. Based on the results of Karila et al. (2015) and Wittke et al. (2019), single-pass interferometry provides RMSE of about 20% for stem volume and biomass, and SAR radargrammetry provides accuracy of about 30%. Such accuracies are better than those provided by 2D optical satellite techniques.

There are many existing or coming satellites working either L-, S-, C-, or X-band SAR. For example, Biomass is an ESA Explorer 7 satellite aiming to take measurements of forest biomass to assess terrestrial carbon stocks and fluxes. The satellite employs a P-band synthetic aperture polarimetric radar operating at 435 MHz. Maps of forest biomass and canopy height are aimed to obtain at a resolution of

200 m. The mission will also have an experimental tomographic phase to provide 3D views of forests. The delayed launch is scheduled for 2024. Sun-synchronous orbit at an altitude of 660 km is planned. Mission life of five years is expected, and the tomographic phase is three months. Fully polarimetric P-band response is obtained. A key question for the future is whether 3D or 2D techniques are applied in operative forest inventories for extremely large areas. The use of intensity information of P-, L-, S-, C-, or X-band, or the difference between them, is a typical 2D technique. 3D techniques include, e.g., the use of height information from canopies. Previous studies (Hyyppä et al. 2000; Karila et al. 2015; Wittke et al. 2019) suggest that 3D techniques are more accurate than 2D techniques.

2.6.3 Precision Forestry—Toward Individual Tree Inventories and Change Detection

Currently, the global forest inventory is about 3B€ annual market, which is mainly based on fieldwork. Inventory has increasingly shifted into using ALS. ALS-based forest inventory has long been operational/commercial in Scandinavia, Baltic countries, Spain, Switzerland, the USA, Canada, Australia, Japan, and New Zealand. Boreal and mountainous forests are easier to measure due to smaller numbers of species and more sparse forest structure. There is also a need to get tree-level information, in which Japan, Scandinavia, and Central Europe are leading the developments.

Both in Finland and Sweden, the countrywide LiDAR data is heavily applied into stand-level (area-based) and large-area/National Forest Inventory (NFI). SLU has implemented country-level inventory maps based on ALS. In Sweden, there exist about 30,000 NFI plots that can be used to train ALS data sets. Since 2020, Finland has moved to the second National Laser Scanning with point density of 5 pts/m^2 (number of transmitted pulses). That allows the identification of log-size individual trees; 90% of trees having DBH higher than 20 cm can be found automatically. Height, DBH, and volume of such trees can be estimated with 2–3%, 15%, and 30% accuracy, respectively, based on Finnish experiences (Hyyppä et al. 2024). FGI has demonstrated the concept with 2.4 billion trees covering 30% of the Finnish territory. In 2026, Finland is starting the 3rd National Laser Scanning with an increased point density, and the objective is to have both individual tree–level and area-based inventories. Figure 2.4 shows an example of the information and visualization system developed for individual tree–level inventories.

Japan has a long history of approximately 10 million ha of conifer plantations under SFM. Timber has been used in important cultural heritage sites since the Asuka and Nara Periods (607–793 AD), including in the oldest existing wooden buildings, castles, temples, and shrines, and forms an important part of the history of the Japanese imperial family, which is one of the longest in the world. Traditional Japanese culture is based on wood and the country's rich natural heritage. The forest survey system in Japan has been taking aerial photographs and recording airborne data every five years since 1947. High-resolution optical images and GIS are widely used for practical forestry. However, forest officers use only remote-sensing data overlaid with GIS forest polygons, and assess boundaries, forest conditions, cutting areas, and damage areas by image interpretation, because the forest resource information extracted from image analysis alone is insufficient. Today, forest officers and landowners require more precise information at the individual tree level on the amount of plantation resources harvested for timber, biomass, and clean energy (Katoh & Gougeon 2012). At present, the coverage of ALS is expanding. Further application of the ALS approach is needed to estimate the number of large trees of each species available for use in traditional wooden buildings (Figure 2.5). ALS data is expensive in Japan and, therefore, there is market even for drone-based remote sensing.

Multi-epoch ALS also allows new methods to analyze the development of the forest (e.g., change detection) (e.g., Hyyppä et al. 2003; Yu et al. 2004; Yu et al. 2006; Hopkinson et al. 2008; Næsset & Nelson 2007; Riofrio et al. 2022, 2023). Multi-epoch data also allows the use of data assimilation, where multi-epoch data could consist of optical and radar satellite data as well as ALS time series (Nyström et al. 2015; Lindgren et al. 2022).

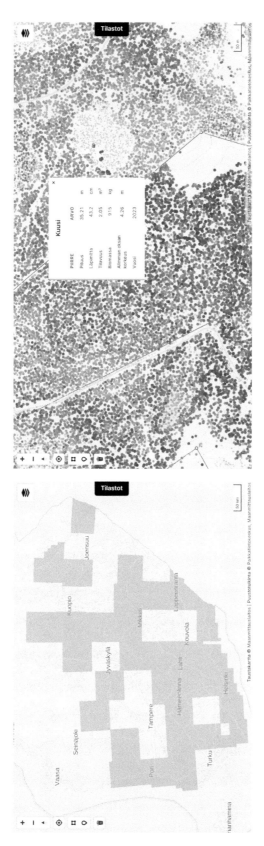

FIGURE 2.4 Example of individual tree-level inventory system metsakanta.com where the first 2.4 billion Finnish trees can be viewed and analyzed. Left: covered area, Right: detail of the calculated data. By clicking a tree in the system, you can see all tree attributes. You can also get summaries of each area coming with one click.

For large areas, multi-epoch ALS data acquisition normally varies in terms of ALS sensors on various platforms, so acquisition times and flying properties (e.g., flying height, scan angles) are used. These facts increase the requirements on the applied algorithms for assessing forest parameters. In general, it is important that for the calculation of the topographic models (i.e., DTM, DSM) and ALS metrics from multi-epoch ALS data, comparable methods will be applied. Therefore, it is strongly recommended that the original 3D ALS point clouds are used as input for current and future analyses. For the study area Vorarlberg, the DSM was calculated based on a land cover–dependent approach (Hollaus et al. 2010), which is robust against varying point densities, acquisition times, and tree species. This approach uses the strengths of different algorithms for generating the final DSM by using surface roughness information to combine two DSMs, which are calculated based on (1) the highest echo within a raster cell and (2) moving least squares interpolation with a plane as functional model (i.e., a tilted regression plane is fitted through the k-nearest neighbors). One of the most important error sources is originating from insufficient geocoding. The experiences in Austria have shown that height differences of stable objects (e.g., roof planes, streets) between the multi-epoch surface models (i.e., DTMs, DSMs) originating from errors in the georeferencing of the individual ALS data sets have to be minimized to ensure that the height differences can be connected to changes of tree heights and consequently to growing stock or biomass changes. To minimize these height differences, least square matching (LSM) of the DSMs from both acquisition times was applied successfully, whereas a 3D shift was sufficient. It could be demonstrated that for streets, roofs, and bare soils the mean height differences of 0.17 m could be reduced to 0.07 m (Hollaus et al. 2013).

For analyzing multi-epoch ALS data, the height differences are the most important measure for forest-monitoring applications. Apart from the described minimization of height differences from stable objects, one can use one reference DTM due to the assumption that the DTM changes within the forest will be negligible. For the study area in Austria the DTM derived from the second ALS data set characterized with higher point density compared to the first one was used as a reference for detecting forest height changes and biomass changes, respectively.

For the district-wide growing stock/biomass estimation the rasterized nDSM (=DSM-DTM) was used as input for the semi-empirical regression model from Hollaus et al. (2009). This method assumes a linear relationship between the growing stock and the ALS-derived canopy volume, stratified according to four canopy height classes to account for height-dependent differences in canopy structure. Each data set was calibrated with the corresponding FI data, and the derived growing stock maps were compared. Finally, the changes of the growing stock maps were detected, whereas the changes were split into exploitation and forest growth. To consider small differences in tree crown representation within each different ALS data set, morphological operations (i.e., open/close) and a minimum mapping area of 10 m² were applied. Finally, the growing stock change map could be limited to the determined forest area, e.g., fulfilling the criteria of the Austrian forest definition (Eysn et al. 2012).

The validation of the growing stock change has shown high agreement between the changes calculated from the sample-plot-based FI data (+43.0 m³/ha) and those derived from the ALS-derived growing stock change map (42.5 m³/ha) and shows the high potential of ALS data for integrating them into operational forest inventories.

Soininen et al. (2022) reported the 20-year boreal forest growth values at the individual tree level obtained with ALS using two totally different sensor systems, namely, Toposys-I Falcon and Riegl VUX-1HA in a test site called Kalkkinen. The point densities were 11 and 1360 points/m² and acquisitions were made in 2000 and 2021, respectively. The growth for individual tree attributes, such as height, DBH, and stem volume, were calculated using direct estimation. The results showed that long-term series growth of height, DBH, and stem volume are possible to record with a high-to-moderate coefficient of determination of 0.90, 0.48, and 0.45 in the best-case scenarios. The respective RMSE values were 0.98 m, 0.02 m, and 0.17 m³. The assumed key errors in growth

FIGURE 2.5 Increased need for ALS at the individual tree level in SFM. Left: Akasawa Forest Reserve, old-growth cypress (*Chamaecyparis obtusa*) forest over 350 years old under SFM for construction timber for the Ise Jingu Shrine in Nagano, Japan. Center: Cutting ceremony for replacing the timber in the Ise Jingu Shrine; note the use of the traditional axe. Right: Ise Jingu Shinto Shrine in Mie, Japan, which uses 10,000 large cypress in 14 wooden buildings, and the timber of which has been replaced every 20 years since 690 AD.

estimates include segmentation errors, matching errors, errors due to two different systems and performance, and due to field reference errors. On the contrary, mean growth values for different tree height classes were reliably obtained.

2.6.4 Detection of Forest Cuttings

In addition to monitoring forest growth, multi-epoch ALS data sets allow the detection and analysis of different types of changes resulting from forest cuttings (e.g., Yu et al. 2004; Arumäe et al. 2020). Cutting changes can include clear-cut areas with no trees or some individual trees left and various forest thinnings. A study was carried out with two ALS data sets from Evo in Finland, anticipating the situation in a few years' time when two high-quality ALS data sets will be available for the entire country. The first data set from 2014 had a point density of 10 points/m^2, and the second one from 2019 was part of the second national laser scanning described earlier (at least 5 transmitted pulses/m^2). Raster DSMs representing the maximum height of each 1-m-by-1-m pixel were created from both data sets. A DTM produced by the National Land Survey of Finland was also available. Forest compartment boundaries and some reference data on cuttings were obtained from the Finnish Forest Centre.

The study included simple change analyses by using individual pixels, tree segments, and forest stands. Tree segments were created with a watershed approach, which was one of the FGI methods presented in Kaartinen et al. (2012). Elementary changes in tree cover can be detected as height changes at the pixel or segment level. Those changes can then be used to determine the state of entire forest stands. Visual analyses revealed that pixel-based change detection gave the most detailed information on changes for further analyses of larger areas. Pixel-based results are not dependent on segmentation and its possible inaccuracies. Pixel-based change detection is also an effective and straightforward approach. Segment-based analysis, however, also has clear benefits because it is related to changes at the individual tree level. For example, the number of remaining trees in a clear-cut area is easy to calculate if tree segments are available. Individual tree segments in a forest stand can also be classified as growing trees, disappeared trees, and uncertain cases (Figure 2.6), which gives further information for change analyses. The results showed that change criteria need special consideration even in the case of simple thresholding methods. For example, the use of the maximum or mean height of a tree segment can result in different results. Difference of maximum heights as a change criterion can lead to unexpected errors if there are inaccuracies in segment boundaries.

A simple thresholding approach gave feasible results in classifying forest stands into clear-cut, thinned, and unchanged stands. The criteria for classification were based on the proportion of pixels with < −3 m height change within the stand. Some thinned stands, where only a small part of the total area had been cut, were not detected. Such cases need more advanced analyses. For example, it is possible to detect small, clear-cut areas separately from the stand-wise classification and highlight those that exist inside stands classified as unchanged.

Experiments were also carried out to test if changes in the height of the ground surface, resulting from cutting operations, can be detected. Such changes are typically small when compared to DSM changes resulting from the cutting of trees. This analysis was based on DTMs created from the two ALS data sets. The results showed various changes on the ground surface, and changes were more typical in clear-cut stands than in thinned or unchanged stands. According to visual evaluation of the input datasets and map data, ground-level changes can include, for example, new trails in the terrain, new or changed ditches, stumps and heaps of tree branches on the ground, and small changes in the height of forest roads, which might have been caused by heavy trucks. Differences in vegetation cover along roads or ditches can also cause changes that become visible when comparing two DTMs. Detailed information on changes in individual trees, ground surface, and forest stands could be utilized, for example, to monitor forest cutting operations and to provide change information for land cover and environmental analyses.

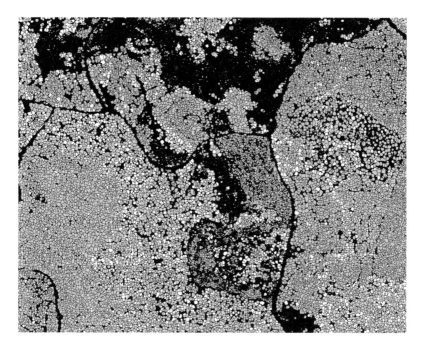

FIGURE 2.6 Growing trees in green (mean height change between 2014 and 2019 positive), disappeared trees in orange (mean height from the ground ≤ 2 m in 2019) and uncertain cases in white. Individual tree segments were created from the 2014 data.

2.6.5 Automizing Field Inventories

Crowdsourcing—Many of the mapping tasks have been performed by state organizations and the mapping has been, therefore, very centralized. This work has been done by trained staff typically having a background in the field of surveying. Since early 2000 it has been possible to map the surroundings by common, non-skilled citizens having, e.g., GNSS receivers, cameras, and smartphones and called by many different terms, such as crowdsourcing, collaboratively contributed geographic information, web-based public participation geographic information system, collaborative mapping, web mapping 2.0, neogeography, wikimapping, and volunteered geographic information. More commonly crowdsourcing is understood as geospatial data collection of voluntary citizens who are untrained in the disciplines of geography, cartography, or related fields such as forestry. A short review of crowdsourcing can be found in Heipke (2010) and Fritz et al. (2009). In the field of forestry, crowdsourcing has been used to assess the condition of city trees. For example, PhillyTreeMap is a web-based application that allows citizens to input tree information of city forests; today, over 144,000 Philadelphian trees are stored in the database.

Field reference data are conventionally collected at the sample plot level by means of manual measurements, which are both labor-intensive and time-consuming. Because of the high costs and laboriousness, the number of tree attributes collected is limited. In practice, some of the most important tree attributes are not even measured or sampled. Automated and more cost-effective techniques are needed to provide plot-level field inventory data. Recently, TLS has been shown to be a promising solution for forest-related studies. There is a huge lack of open data in the field of plot-level or tree-level data to calibrate the ALS estimates. Today's challenge is the creation of large-area forest resource maps with small costs and the highest accuracy possible for various purposes. Since plotwise field data does not openly exist, and TLS processing at the country level produces significant costs, alternative solutions are also being developed. The use of crowdsourcing to get field reference for large-area forest inventory has been recently studied. In Hyyppä et al. (2017), it was shown that locally collected field reference significantly improves forest inventory estimates at boreal forest conditions. Based on these results, Hyyppä et al. (2017) introduced a crowdsourcing concept based on individual tree diameter and stem curve measurements, using Kinect and a mobile phone–embedded 3D sensor (Tango) for ALS-based large-area forest inventory at the individual tree level. An advantage of the ITD approach over ABA in the crowdsourcing concept is that a smaller number of reference trees is needed to get reasonable accuracy, and it is easier for the landowners to measure physically well-established parameters, i.e., the diameter of the tree [cm] instead of the basal area [m^2/ha] of a plot. The disadvantage of the ITD concept, in which suppressed trees are not found, can be overcome by the use of crowdsourcing measuring diameters for the missing trees. Thus, there is synergy with the use of ITD and crowdsourcing.

Crowdsourcing can become more engaging when gamified, a possibility facilitated by consumer-level smartphones and tablets equipped with LiDAR sensors and the capability to provide augmented reality (AR) experiences for users. Generally, the aim of gamification is to enhance the enjoyment, engagement, and motivation of tasks, activities, or processes by leveraging psychological principles that captivate people's interest in games. In a recent study at Tampere University, four AR applications were developed to assess the characteristics and quality of crowdsourced LiDAR data. The study also sought to comprehend user behavior and their experience of gathering data in a forest setting through gamified applications. The findings demonstrate that by incorporating gamification, researchers were not only able to influence the user experience but also the type and quality of the collected data.

Use of mobile laser scanning—There are various forms of mobile laser scanning applied for field reference automatization. In Liang et al. (2014b) a backpack laser scanner for collecting tree attributes was demonstrated. The applied 10-kg FGI system consists of a multi-constellation navigation system and an ultra-high-speed phase-shift laser scanner mounted on a rigid base plate as for a single sensor block. In the data acquisition, the system was tilted by 20° to record the vertical tree stems. That was followed by a large number of backpack and handheld studies. Hyyppä et al. (2020a) continued the work of Liang et al. (2014b). Modifications included pulse-based backpack mobile laser scanner (Riegl VUX-1HA) combined with in-house-developed SLAM (Simultaneous Localization and Mapping), and a novel post-processing algorithm chain that allows one to extract stem curves to individual standing trees. Processing applied an algorithm for scan-line arc extraction, a stem inclination angle correction, and an arc-matching algorithm correcting SLAM drifts. The process was fully based on stem curves. Stem detection accuracy was 100%. The RMSE for stem curves was 1.2 cm (5.1%) and 1.7 cm (6.7%) for the easy and medium plots, respectively. The RMSE for tree heights were 1.8 m (8.7%) and 1.1 m (4.9%) and for the stem volumes were 9.7% and 10.9%, respectively. The new processing chain provided stem volume estimates with better accuracy than previous methods based on MLS data. The accuracy of stem volume estimation was comparable to that provided by TLS approaches in similar forest conditions.

Mini-UAV-based airborne laser scanning data collection has been possible since Jaakkola et al. (2010). Mini-UAVs (< 20 kg) have been previously used for mapping purposes using, for example, aerial images. Zhao et al. (2006) depicted a remote-controlled helicopter supplied with navigation sensors and a laser range finder. In Jaakkola et al. (2010) the first mini-UAV including the laser scanner, intensity recording, spectrometer, thermal camera, and conventional digital camera, was depicted. Mini-UAV laser scanning has since been feasible for small-area and multitemporal surveys. Also, corridor-type applications have been developed for more than a decade now.

Today there is a large number of drone LiDAR sensors that can be used for extracting individual trees. Drone laser scanning now has three approaches to measure tree attributes and getting field reference, namely using:

1. A standard above-canopy individual tree approach (depicted in 15.5.5)
2. An above-canopy individual tree approach in which the tree stem is directly measured from the dense point clouds (Jaakkola et al. 2017; Hyyppä et al. 2022)

3. Under-canopy drone measuring stems similarly to handheld, backpack LS, or TLS measurements (Hyyppä et al. 2020a).

These three drone-based approaches have been studied in Hyyppä et al. (2020b, 2021, under-canopy drone case), Jaakkola et al. (2017, approaches 1 and 2), Hyyppä et al. (2020c, comparison of approaches 1 and 3), and Hyyppä et al. (2022, approach 2). Both Jaakkola et al. (2017) and Hyyppä et al. (2020c) reported DBH and stem volume accuracy at 10% and 20%, respectively, for easy plots for using approach 1. In obstructed plots, the approach resulted in corresponding accuracies of about 20% and 40%. This implies that approach 1 is not the best approach for getting reference measurements at the individual tree level. It should be noticed that in Hyyppä et al. (2020c), the three systems applied were high-quality, expensive sensors, such as for Riegl VQ480U and VUX-1HA on board helicopters and drones. On the contrary, approaches 2 and 3 had promising results. In approach 3 an under-canopy UAV system measured DBH with RMSE of 2–8% (Hyyppä et al. 2020a) and stem volumes of individual trees with a standard error of 10%, equivalent to the error obtained when merging above-canopy UAV laser scanning data with terrestrial point cloud data (Hyyppä et al. 2021). The accuracies obtained with the under-canopy drone laser scanning are very good. It should be understood that the applied sensor and positioning technology determines the quality of the point clouds. The beam divergence is of key importance. In order to get individual tree heights, you should have a rotating laser scanner, as in Hyyppä et al. (2021), or you should have a scanner with a large field of view. Another challenge is the drone system capacity to fly automatically in forests with varying densities. Currently applied technologies are feasible for flying in easy and slightly obstructed forests. In dense forests, it is hard for the system to fly inside the forests and to get adequately good data. Thus, today the best MLS sensor for collecting reference from the dense forests is the handheld laser scanner assuming that the applied sensor is of high quality (high PRF, good SLAM, small beam size). Due to problems in flying autonomously under-canopy and collecting high-quality data from the stem, approach 2 deserves a lot of future work. In general MLS holds the promise for collecting reference data with an extremely high efficiency. It is perhaps ten times faster than using TLS.

The most efficient approach for reference data collection would be to use a drone-based laser scanning system above the canopy. But instead of using standard ITD processes, we need to process data in a new way. In Hyyppä et al. (2022), a high-quality laser scanner, Riegl VUX-1HA, was tilted by 15° from the nadir to increase the possibility of recording stem hits, resulting in high-density point cloud of 2200–3800 echoes/m^2. In Hakula et al. (2023), a three-channel system was used for the same task to get tree species mapped. In Hyyppä et al. (2022), stem hits were used to detect individual trees and to measure the stem curves. About 50% of the stems were found from three test sites covering almost 1500 relatively mature trees. Stem curve and volume accuracy of 6–9% and 12–21%, respectively, was achieved. It should be understood that such trees are candidate trees for forest inventory at the individual tree level. If the future provides bi- or multiwave-length high-quality scanners on drones, they could be used to provide reference, e.g., for the following tree attributes: height, DBH, stem curve, volume, biomass, species, and quality. And the collection speed is several hundred times faster than with TLS. Based on experience gained in FGI, reference data collected with technology depicted in Hyyppä et al. (2022) was tested for producing reference data for individual tree–level forest inventories based on National ALS data and compared to the stand approach where field reference is manually collected. The accuracy of predicted individual tree attributes were equal with both approaches, implying that Hyyppä et al.'s (2022) concept is highly interesting for collecting automatically field reference data. It should be noticed also that all TLS and MLS field reference techniques should be based on automatically getting stem curve and tree height, since that removes the need to use inaccurate national allometric models.

Use of autonomous big data—In Hyyppä et al. (2023), the potential of using autonomous perception data for mapping the stem attributes of roadside trees was demonstrated. It was shown that DBH could be estimated from autonomous perception data with RMSE of 10%, which was slightly lower than the corresponding error obtained from a specially planned mobile mapping survey. The data-processing workflow was optimized by using time-based filtering for stem diameter estimation. Time stamp information was shown to be important for achieving high accuracy in many applications of autonomous perception data. It is expected that reference data can be collected from forest road environments in the same way.

2.6.6 Harvester Laser Scanning

As a result of advancements in mobile laser scanning technologies and associated algorithms, the potential of integrating a laser scanning system to a forest harvester has started attracting interest in recent years. In 2022, Ponsse, one of the world's leading manufacturers of cut-to-length forest machines, introduced a technology concept for a world-first thinning density assistant based on a mobile laser scanner on board a harvester. However, the first system introduced by Ponsse utilized only the number of nearby trees detected by the laser scanner system to optimize the thinning.

In Faitli et al. (2024), a full processing flow was presented for a LiDAR-equipped harvester system starting from the development of a real-time SLAM-based positioning method followed by the use of algorithmic methods to detect tree stems from the acquired MLS point cloud and to estimate their attributes, such as stem curve. In the study, an Ouster OS0–128 (Rev C) scanner was attached to a Ponsse C44 forest harvester while it conducted a thinning operation on multiple boreal forest test sites. As shown in Figure 2.7, the harvester MLS system enabled the detection of 90% of trees with DBH > 20 cm within a 15-m range from the harvester. For trees with a DBH below 20 cm, the detection rate was markedly lower and decreased below 50% for distances above 10 m from the scanner. Within a 15-m range from the harvester, the stem curves of the detected individual trees could be estimated with a bias of −0.1 cm, a RMSE of 3.6 cm, and a median absolute error (MAE) of 1.8 cm. For the DBH, the bias and RMSE were of similar magnitude or slightly lower. The Ouster scanner was characterized by a large beam divergence of 6.1 mrad, necessitating the use of distance-dependent bias correction for the stem diameter. Even with the bias correction, the median absolute deviation of the stem curve estimates was found to rise with increasing distance from the harvester system due to more effective noise in the point cloud, as visualized in Figure 2.7. Even though the SLAM-based positioning method could have been executed in real time, the detection of trees and the estimation of their attributes was conducted offline in a post-processing mode using the full SLAM-corrected MLS point cloud. Future work is still needed to study real-time estimation of tree attributes from harvester MLS data in order to optimize the harvester operation.

A mobile laser scanner on board a harvester can also provide useful features concerning the wood quality–influencing external structures of trees. In Winberg et al. (2023) handheld MLS data features were linked with wood properties measured at a sawmill utilizing state-of-the-art X-ray scanners.

2.6.7 Permanent Laser Scanning Systems for Continuous Forest Monitoring

Detecting and quantifying fine-detail changes in natural targets, like tree organ growth, with at least centimeter-level accuracy poses a state-of-the-art challenge. The complex, inter-linked, dynamics between natural targets and their immediate surroundings create multiple error sources in measurements that make change mapping, prediction, and understanding difficult. Understanding these phenomena requires systematic long-term monitoring with high spatiotemporal resolution.

Permanent laser scanning systems (PLSS) have emerged as a potential alternative for continuous forest monitoring and change detection. A PLSS can be understood as a close-range surveying technique in which a laser scanning system is permanently mounted to continuously monitor natural or human-made surfaces and targets in a fixed scene. PLSS setups are able to detect below centimeter-level changes, deformations, and dynamics through well-registered LiDAR time series by monitoring their target scene from a fixed point of view. Several pioneering PLSS setups already exist with their focus on monitoring of crop growth monitoring (El-Naggar et al. 2021) and boreal forests (Culvenor et al. 2014; Griebel et al. 2015; Campos et al. 2021). The diversity of hardware and

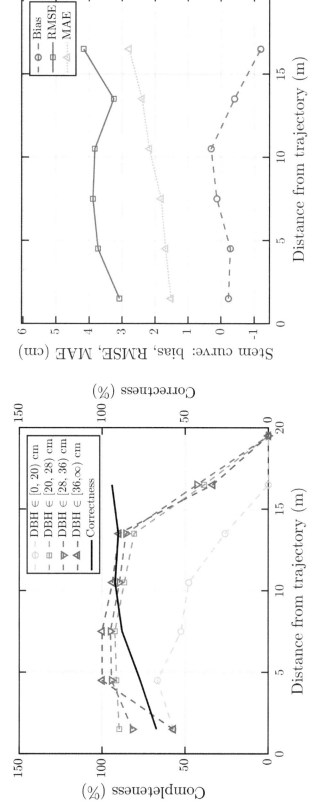

FIGURE 2.7 Left: completeness rate, i.e., recall, of tree stem detection as a function of the distance from the harvester system for trees in four different size categories based on their DBH. The correctness rate, i.e., precision, of tree stem detection is shown with the solid black line. Right: bias, RMSE, and median absolute error (MAE) of stem curve estimates as a function of the distance from the harvester system.

automation options offer many possibilities to design PLSS setups depending on the application and accuracy requirements.

Especially focused on forest monitoring, Culvenor et al. (2014) designed and built an in-situ Monitoring LiDAR (IML) system, VEGNET. The VEGNET IML is a stationary profiling LiDAR unit built with low-cost components. A VEGNET IML is capable of monitoring long-term changes in forest Leaf and Plant Area Index (LAI & PAI, respectively). Griebel et al. (2015) applied a network of three VEGNET units to monitor daily forest canopy dynamics in a dry sclerophyll forest for a two-year period. Portillo-Quintero et al. (2014) used a single VEGNET unit to monitor PAI changes in a boreal mixed-wood forest in Canada for a three-week leaf-fall period. They compared the VEGNET PAI signal with the Moderate Resolution Imaging Spectroradiometer (MODIS) LAI product. They reported a similar decreasing trend between the VEGNET and MODIS signals over the monitoring period with 2–15% difference.

Campos et al. (2021) developed a high-performance PLSS system called a LiDAR Phenology (LiPhe) station that is designed to acquire multiple year-long dense LiDAR time series with high spatial resolution from a boreal forest. Developed by the Finnish Geospatial Research Institute (FGI), the LiPhe station is equipped with a measurement computer for data acquisition automation and a time-of-flight Riegl VZ-2000i scanner (RIEGL Gmbh, Horn, Austria). The scanner is installed inside a weather-protected cover 30 m from the ground near the top of a 35-m tall measurement tower. The tower is part of the Hyytiälä forest research station infrastructure located in central Finland (61°51'N, 24°17'E). The LiPhe station monitoring area comprises more than 4000 trees within a 400 m planar range from the scanner location. The forest stem density in the area is around 625 per ha. The LiPhe system has been fully operational since April 2020. The temporal resolution of the acquired point cloud time series is one scan per hour, and the angular scan resolution is 0.006° and corresponds to a nominal 1-cm distance between two points on a planar surface 100 m away from the system. This dense spatiotemporal time series is able to detect and quantify daily and seasonal phenological changes in boreal tree species over a year. LiPhe station–acquired data have been applied in several applications related to tree growth dynamics and forest structure analyses: circadian movement detection (Puttonen et al. 2019; Junttila et al. 2022), tree height and canopy growth monitoring (Campos et al. 2023; Yrttimaa et al. 2024), phenological change detection and timing (Campos et al. 2021), and biomass assessments (Spadavecchia et al. 2023).

2.7 SUMMARY

In this chapter we presented some highlights of 3D information collection and processing from LiDAR and radar and turning that information into valuable forest informatics. We really see that the processing of 3D data will be more and more based on similar tools that are currently available in the LiDAR community. Some of the most exciting developments anticipated are summarized next.

The concept of the LiDAR plot is comparable to that of a ground plot with a fixed area. LiDAR plots are an area-based summary of ALS point clouds, whereby descriptive statistics or metrics are generated from the point cloud (e.g., percentiles, mean, and standard deviation) and these are used, with a sample of co-located ground measurements, to model forest inventory attributes of interest, such as mean height, dominant height, basal area, volume, and biomass. Modern deep learning approaches are expected to improve the original LiDAR plot concept. The LiDAR plot concept is especially feasible for teaching satellite imagery.

For the calibration of ALS data, a high-quality reference is needed. The last decade has shown rapid development of getting field reference data automatically. Many techniques, such as crowd-sourcing, TLS, and MLS, including smartphone-based scanning, handheld laser scanning, under-canopy drone laser scanning, above-canopy drones measuring tree stems directly from point clouds, and autonomous driving big data are example techniques that provide the high-quality reference required. Many mobile laser scanning techniques provide stem volume estimates with 10–15%

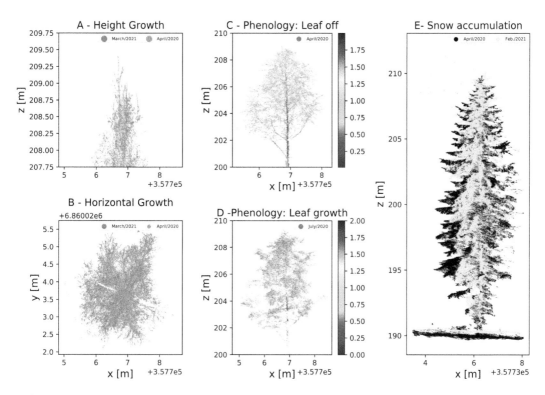

FIGURE 2.8 Different annual dynamics in individual tree structures captured with the permanent laser scanning station (FGI LiPhe). A)–B) Annual height and horizontal growth in a Silver birch (Betula Pendula pub.) canopy. C)–D) Laser backscatter differences in a Silver birch (Betula Pendula pub.) canopy at two different phenological states. E) Effect of snow accumulation to a Norway spruce (*Picea abies*) canopy structure and its immediate surroundings.

errors that is acceptable as a high-quality individual tree reference. All accurate techniques, however, require the derivation of the stem curve. When drone laser scanning data is collected from above the canopy, and standard area-based or individual tree–based techniques (Næsset 2002; Hyyppä & Inkinen 1999) are applied, the accuracy derived for the plot or individual tree level is not adequate enough to calibrate ALS surveys, although it is highly feasible for satellite-based studies. Above-canopy ALS data capable to extract stem curves, presented in Hyyppä et al. (2022), is an interesting approach for getting a high-quality reference. Another interesting new opening is the improved segmentation capacity of very-dense point clouds using new deep learning approaches (e.g., Xiang et al. 2023; Wielgosz et al. 2024).

Based on the recent scientific studies in 3D SAR techniques and forest resources mapping, there appears to be potential in these techniques for deriving detailed forest attribute maps over large areas with good temporal resolution. Even though ALS and aerial stereo-photogrammetry provide more accurate estimates for forest attributes compared to 3D SAR techniques, there might be a demand for 3D SAR data, especially for large-area mapping and especially in monitoring of changes in the forest structure. Single-pass SAR interferometry provides biomass and stem volume with accuracy close to ALS and photogrammetry. The advantage of 3D SAR techniques compared to 2D estimation techniques is that the data derived using 3D techniques can be easily integrated to existing forest attribute inventory processes based on point cloud data.

There is a clear trend toward precision forestry where individual tree detection is applied (Hyyppä & Inkinen 1999) from ALS data. FGI has created a demonstration system, metsakanta. com, where FGI calculated 2.4 billion trees using the National Laser Scanning data with point density of 5 pts/m². That point density is feasible for the identification of log-size individual trees; 90% of trees having DBH higher than 20 cm can be found automatically. Height, DBH, and volume of such trees can be estimated with RMSE of 2–3%, 15%, and 30%, respectively. It is expected that similar precision forest concepts are implemented elsewhere to meet the industrial needs for improved forest data. Multitemporal ALS surveys will complement this data and provide carbon sinks even at the individual tree level.

In addition to these new, exciting developments presented, there are too many new areas that should have been discussed, especially from the point of view of new research potential.

2.8 ACKNOWLEDGMENT

Support from the Research Council of Finland (UNITE Flagship (357908), "Feasibility of Inside-Canopy UAV Laser Scanning for Automated Tree Quality Surveying" (334002), "Capturing structural and functional diversity of trees and tree communities for supporting sustainable use of forests" (348644) are gratefully acknowledged.

REFERENCES

Antonova, S., Thiel, C., Höfle, B., Anders, K., Helm, V., Zwieback, S., Marx, S., and Boike, J. 2019. Estimating tree height from TanDEM-X data at the northwestern Canadian treeline. *Remote Sensing of Environment* 231:111251.

Arumäe, T., Lang, M., and Laarmann, D. 2020. Thinning- and tree-growth-caused changes in canopy cover and stand height and their estimation using low-density bitemporal airborne lidar measurements—a case study in hemi-boreal forests. *European Journal of Remote Sensing* 53(1):113–123.

Aschoff, T., and Spiecker, H. 2004. Algorithms for the automatic detection of trees in laser scanner data. *International Archives of Photogrammetry, Remote Sensing and Spatial Information Sciences* 36:W2.

Askne, J.I., Fransson, J.E., Santoro, M., Soja, M.J., and Ulander, L.M. 2013. Model-based biomass estimation of a hemi-boreal forest from multitemporal TanDEM-X acquisitions. *Remote Sensing* 5(11):5574–5597.

Askne, J.I., Santoro, M., Smith, G., and Fransson, J.E.S. 2003. Multitemporal repeat-pass SAR interferometry of boreal forests. *IEEE Transactions on Geoscience and Remote Sensing* 43(6):1219–1228.

Axelsson, P. 2000. DEM generation from laser scanner data using adaptive TIN models. *International Archives of Photogrammetry and Remote Sensing* 33(Part B4):110–117.

Balenović, I., Liang, X., Jurjević, L., Hyyppä, J., Seletković, A., and Kukko, A. 2021. Hand-held personal laser scanning–current status and perspectives for forest inventory application. *Croatian Journal of Forest Engineering: Journal for Theory and Application of Forestry Engineering* 42(1):165–183.

Ballhorn, U., Jubanski, J., and Siegert, F. 2011. ICESat/GLAS data as a measurement tool for peatland topography and peat swamp forest biomass in Kalimantan, Indonesia. *Remote Sensing* 3(9):1957–1982.

Banskota, A., Wynne, R., Johnson, P., and Emessiene, B. 2011. Synergistic use of very high-frequency radar and discrete-return lidar for estimating biomass in temperate hardwood and mixed forests. *Annals of Forest Science* 68(2):347–356.

Bauwens, S., Bartholomeus, H., Calders, K., and Lejeune, P. 2016. Forest inventory with terrestrial LiDAR: A comparison of static and hand-held mobile laser scanning. *Forests* 7(6):127.

Bazi, Y., Bashmal, L., Rahhal, M.M.A., Dayil, R.A., and Ajlan, N.A. 2021. Vision transformers for remote sensing image classification. *Remote Sensing* 13(3):516.

Becker, A., Russo, S., Puliti, S., Lang, N., Schindler, K., and Wegner, J.D. 2023. Country-wide retrieval of forest structure from optical and SAR satellite imagery with deep ensembles. *ISPRS Journal of Photogrammetry and Remote Sensing* 195:269–286.

Bienert, A., Georgi, L., Kunz, M., Maas, H.G., and Von Oheimb, G. 2018. Comparison and combination of mobile and terrestrial laser scanning for natural forest inventories. *Forests* 9(7):395.

Brandtberg, T., Warner, T., Landenberger, R., and McGraw, J. 2003. Detection and analysis of individual leaf-off tree crowns in small footprint, high sampling density lidar data from the eastern deciduous forest in North America. *Remote Sensing of Environment* 85:290–303.

Brolly, G., and Kiraly, G. 2009. Algorithms for stem mapping by means of terrestrial laser scanning. *Acta Silvatica et Lignaria Hungarica* 5:119–130.

Cabo, C., Del Pozo, S., Rodríguez-Gonzálvez, P., Ordóñez, C., and González-Aguilera, D. 2018. Comparing terrestrial laser scanning (TLS) and wearable laser scanning (WLS) for individual tree modeling at plot level. *Remote Sensing* 10(4):540.

Campos, M.B., Litkey, P., Wang, Y., Chen, Y., Hyyti, H., Hyyppä, J., and Puttonen, E. 2021. A long-term terrestrial laser scanning measurement station to continuously monitor structural and phenological dynamics of boreal forest canopy. *Frontiers in Plant Science* 11:606752.

Campos, M.B., Valve, V., Shcherbacheva, A., Echriti, R., Wang, Y., and Puttonen, E. 2023. Detection of silver birch growth dynamics and timing with dense spatio-temporal LiDAR time-series. *The International Archives of the Photogrammetry, Remote Sensing and Spatial Information Sciences* 48:1715–1722.

Čerňava, J., Mokroš, M., Tuček, J., Antal, M., and Slatkovská, Z. 2019. Processing chain for estimation of tree diameter from GNSS-IMU-based mobile laser scanning data. *Remote Sensing* 11(6):615.

Chang, A., Eo, Y., Kim, Y., and Kim, Y. 2013. Identification of individual tree crowns from LiDAR data using a circle fitting algorithm with local maxima and minima filtering. *Remote Sensing Letters* 4(1):29–37.

Chen, G., Hay, G.J., and St-Onge, B.A. 2012b. GEOBIA framework to estimate forest parameters from LiDAR transects, Quickbird imagery and machine learning: A case study in Quebec, Canada. *International Journal of Applied Earth Observation* 15:28–37.

Chen, G., Wulder, M.A., White, J.C., Hilker, T., and Coops, N.C. 2012a. LiDAR calibration and validation for geometric-optical modeling with Landsat imagery. *Remote Sensing of Environment* 124:384–393.

Chen, H., Cloude, S.R., and White, J.C. 2021. Using GEDI waveforms for improved TanDEM-X forest height mapping: A combined SINC + legendre approach. *Remote Sensing* 13:2882.

Chen, S., Liu, H., Feng, Z., Shen, C., and Chen, P. 2019. Applicability of personal laser scanning in forestry inventory. *PLoS ONE* 14(2):e0211392.

Chen, Y., Shi, P., Deng, L., and Li, J. 2007. Generation of a top-of-canopy Digital Elevation Model (DEM) in tropical rain forest regions using radargrammetry. *International Journal of Remote Sensing* 28(19):4345–4349.

Choi, C., Pardini, M., Armston, J., and Papathanassiou, K.P. 2023. Forest biomass mapping using continuous InSAR and discrete waveform lidar measurements: A TanDEM-X/GEDI test study. *IEEE Journal of Selected Topics in Applied Earth Observations and Remote Sensing* 16:7675–7689.

Clark, N.A., Wynne, R.H., and Schmoldt, D.L. 2000. A review of past research on dendrometers. *Forest Science* 46:570–576.

Culvenor, D.S., Newnham, G.J., Mellor, A., Sims, N.C., and Haywood, A. 2014. Automated in-situ laser scanner for monitoring forest leaf area index. *Sensors* 14(8):14994–15008.

Del Perugia, B., Giannetti, F., Chirici, G., and Travaglini, D. 2019. Influence of scan density on the estimation of single-tree attributes by hand-held mobile laser scanning. *Forests* 10(3):277.

Diez, Y., Kentsch, S., Fukuda, M., Caceres, M.L.L., Moritake, K., and Cabezas, M. 2021. Deep learning in forestry using UAV-acquired RGB data: A practical review. *Remote Sensing* 13(14):2837.

Dubayah, R., Blair, J.B., Goetz, S., Fatoyinbo, L., Hansen, M., Healey, S., Hofton, M., Hurtt, G., Kellner, J., Luthcke, S., and Armston, J. 2020. The global ecosystem dynamics investigation: High-resolution laser ranging of the Earth's forests and topography. *Science of Remote Sensing* 1:100002.

Elachi, C. 1987. *Spaceborne Radar Remote Sensing*. IEEE Press, 255 p.

El-Naggar, A.G., Jolly, B., Hedley, C.B., Horne, D., Roudier, P., and Clothier, B.E. 2021. The use of terrestrial LiDAR to monitor crop growth and account for within-field variability of crop coefficients and water use. *Computers and Electronics in Agriculture* 190:106416.

Ene, L., Næsset, E., and Gobakken, T. 2012. Single tree detection in heterogeneous boreal forests using airborne laser scanning and area-based stem number estimates. *International Journal of Remote Sensing* 33(16):5171–5193.

Eysn, L., Hollaus, M., Lindberg, E., Berger, F., Monnet, J.M., Dalponte, M., Kobal, M., Pellegrini, M., Lingua, E., Mongus, D., and Pfeifer, N. 2015. A benchmark of lidar-based single tree detection methods using heterogeneous forest data from the alpine space. *Forests* 6(5):1721–1747.

Eysn, L., Hollaus, M., Schadauer, K., and Pfeifer, N. 2012. Forest delineation based on airborne LiDAR data. *Remote Sending* 4(3):762–783.

Faitli, T., Hyyppä, E., Hyyti, H., Hakala, T., Kaartinen, H., Kukko, A., Muhojoki, J., and Hyyppä, J. 2024. Enabling forest machinery to measure and map surrounding tree stems during harvesting. Manuscript submitted.

Fassnacht, F.E., White, J.C., Wulder, M.A., and Næsset, E. 2024. Remote sensing in forestry: Current challenges, considerations and directions. *An International Journal of Forest Research* 97(1):11–37.

Fleck, S., Mölder, I., Jacob, M., Gebauer, T., Jungkunst, H.F., and Leuschner, C. 2011. Comparison of conventional eight-point crown projections with LIDAR-based virtual crown projections in a temperate old-growth forest. *Annals of Forest Science* 68:1173–1185.

FRA. 2010. *Global Forest Resources Assessment 2010*. Available online: http://www.fao.org/forestry/fra/fra2010/en/

Fransson, J.E.S. 1999. *Analysis of Synthetic Aperture Radar Images for Forestry Applications. Acta Universitatis Agriculturae Sueciae, Silvestria*, vol. 100, Doctoral Thesis. Swedish University of Agricultural Sciences, Umeå, Sweden.

Fransson, J.E.S., and Israelsson, H. 1999. Estimation of stem volume in boreal forests using ERS-1 C- and JERS-1 L-band SAR data. *International Journal of Remote Sensing* 20(1):123–137.

Fritz, S., McCallum, I., Schill, C., Perger, C., Grillmayer, R., Achard, F., Kraxner, F., and Obersteiner, M. 2009. Geo-Wiki.Org: The use of crowdsourcing to improve global land cover. *Remote Sensing* 1:345–354.

Gill, S.J., Biging, G.S., and Murphy, E.C. 2000. Modeling conifer tree crown radius and estimating canopy cover. *Forest Ecology and Management* 126(3):405–416.

Goodenough, D., Chen, H., Dyk, A., Hobart, G., and Richardson, A. 2008. Data fusion study between polarimetric SAR, hyperspectral and lidar data for forest information. In *IEEE International Symposium on Geoscience and Remote Sensing—IGARSS 2008*. IEEE, pp. 281–284.

Griebel, A., Bennett, L.T., Culvenor, D.S., Newnham, G.J., and Arndt, S.K. 2015. Reliability and limitations of a novel terrestrial laser scanner for daily monitoring of forest canopy dynamics. *Remote Sensing of Environment* 166:205–213.

Guimarães, N., Pádua, L., Marques, P., Silva, N., Peres, E., and Sousa, J.J. 2020. Forestry remote sensing from unmanned aerial vehicles: A review focusing on the data, processing and potentialities. *Remote Sensing* 12(6):1046.

Gülci, S., Yurtseven, H., Akay, A.O., and Akgul, M. 2023. Measuring tree diameter using a LiDAR-equipped smartphone: A comparison of smartphone-and caliper-based DBH. *Environmental Monitoring and Assessment* 195(6):678.

Hakula, A., Ruoppa, L., Lehtomäki, M., Yu, X., Kukko, A., Kaartinen, H., Taher, J., Matikainen, L., Hyyppä, E., Luoma, V., and Holopainen, M. 2023. Individual tree segmentation and species classification using high-density close-range multispectral laser scanning data. *ISPRS Open Journal of Photogrammetry and Remote Sensing* 9:100039.

Hallikainen, M., Hyyppä, J., Haapanen, J., Tares, T., Ahola, P., Pulliainen, J., and Toikka, M. 1993. A helicopter-borne eight-channel ranging scatterometer for remote sensing, Part I: System description. *IEEE Transactions on Geoscience and Remote Sensing* 31(1):161–169.

Hamedianfar, A., Mohamedou, C., Kangas, A., and Vauhkonen, J. 2022. Deep learning for forest inventory and planning: A critical review on the remote sensing approaches so far and prospects for further applications. *Forestry* 95(4):451–465.

Heipke, C. 2010. Crowdsourcing geospatial data. *ISPRS Journal of Photogrammetry and Remote Sensing* 65: 550–557.

Henderson, F., and Lewis, A. 1998. *Principles and Application of Imaging Radar* (Manual of Remote Sensing, Volume 2). 3rd ed. Wiley, 896 p.

Henning, J.G., and Radtke, P.J. 2006. Detailed stem measurements of standing trees from ground-based scanning LiDAR. *Forest Science* 52(1):67–80.

Hollaus, M., Eysn, L., Karel, W., and Pfeifer, N. 2013. *Growing Stock Change Estimation Using Airborne Laser Scanning Data. Silvilaser 2013*, Peking. Paper ID SL2013–060, 8 p.

Hollaus, M., Mandlburger, G., Pfeifer, N., and Mücke, W. 2010. Land cover dependent derivation of digital surface models from airborne laser scanning data. *International Archives of Photogrammetry, Remote Sensing and the Spatial Information Sciences* 39(3):6. PCV 2010, Paris, France.

Hollaus, M., Wagner, W., Schadauer, K., Maier, B., and Gabler, K. 2009. Growing stock estimation for alpine forests in Austria: A robust lidar-based approach. *Canadian Journal of Forest Research* 39(7):1387–1400.

Holmgren, J. 2003. *Estimation of Forest Variables Using Airborne Laser Scanning*, PhD Thesis. Acta Universitatis Agriculturae Sueciae, Silvestria 278, Swedish University of Agricultural Sciences, Umeå, Sweden.

Holmgren, J., and Persson, Å. 2004. Identifying species of individual trees using airborne laser scanning. *Remote Sensing of Environment* 90:415–423.

Holmgren, J., Persson, Å., and Söderman, U. 2008. Species identification of individual trees by combining high resolution LiDAR data with multi-spectral images. *International Journal of Remote Sensing* 29(5):1537–1552.

Holopainen, M., Haapanen, E., Karjalainen, M., Vastaranta, M., Hyyppä, J., Yu, X., Tuominen, S., and Hyyppä, H. 2010. Comparing accuracy of airborne laser scanning and TerraSAR-X radar images in the estimation of plot-level forest variables. *Remote Sensing* 2(2):432–445.

Holopainen, M., Kankare, V., Vastaranta, M., Liang, X., Lin, Y., Vaaja, M., Yu, X., Hyyppä, J., Hyyppä, H., Kaartinen, H., and Kukko, A. 2013. Tree mapping using airborne, terrestrial and mobile laser scanning–A case study in a heterogeneous urban forest. *Urban Forestry & Urban Greening* 12(4):546–553.

Holopainen, M., Vastaranta, M., Liang, X., Hyyppä, J., Jaakkola, A., and Kankare, V. 2014. Estimation of forest stock and yield using LiDAR data. In *Remote Sensing of Natural Resources*. Edited by Guangxing Wang and Qihao Weng. CRC Press.

Hopkinson, C., Chasmer, L., and Hall, R.J. 2008. The uncertainty in conifer plantation growth prediction from multi-temporal lidar datasets. *Remote Sensing of Environment* 112(3):1168–1180.

Hopkinson, C., Chasmer, L., Lim, K., Treitz, P., and Creed, I. 2006. Towards a universal lidar canopy height indicator. *Canadian Journal of Remote Sensing* 32(2):139–152.

Hopkinson, C., Chasmer, L., Young-Pow, C., and Treitz, P. 2004. Assessing forest metrics with a ground-based scanning lidar. *Canadian Journal of Forest Research* 34:573–583.

Hu, X., and Yuan, Y. 2016. Deep-learning-based classification for DTM extraction from ALS point cloud. *Remote Sensing* 8(9):730.

Huang, H., Li, Z., Gong, P., Cheng, X., Clinton, N., Cao, C., Ni, W., and Wang, L. 2011. Automated methods for measuring DBH and tree heights with a commercial scanning lidar. *Photogrammetric Engineering and Remote Sensing* 77:219–227.

Husch, B., Miller, C.I., and Beers, T.W. 1982. *Forest Mensuration*, 2nd ed. John Wiley & Sons.

Huuva, I. 2023. *Estimation of Change in Forest Variables Using Synthetic Aperture Radar*, Doctoral Thesis. Acta Universitatis Agriculturae Sueciae, Silvestria, No 2023:98. Swedish University of Agricultural Sciences, Umeå, Sweden.

Huuva, I., Wallerman, J., Fransson, J.E., and Persson, H.J. 2023. Prediction of site index and age using time series of TanDEM-X phase heights. *Remote Sensing* 15(17):4195.

Hyde, P., Nelson, R., Kimes, D., and Levine, E. 2007. Exploring LiDAR-RaDAR Synergy—Predicting Aboveground Biomass in a Southwestern ponderosa pine forest using LiDAR, SAR, and InSAR. *Remote Sensing of Environment* 106(1):28–38.

Hyyppä, E., Hyyppä, J., Hakala, T., Kukko, A., Wulder, M.A., White, J.C., Pyörälä, J., Yu, X., Wang, Y., Virtanen, J.P., and Pohjavirta, O. 2020a. Under-canopy UAV laser scanning for accurate forest field measurements. *ISPRS Journal of Photogrammetry and Remote Sensing* 164:41–60.

Hyyppä, E., Kukko, A., Kaartinen, H., Yu, X., Muhojoki, J., Hakala, T., and Hyyppä, J. 2022. Direct and automatic measurements of stem curve and volume using a high-resolution airborne laser scanning system. *Science of Remote Sensing* 5:100050.

Hyyppä, E., Kukko, A., Kaijaluoto, R., White, J.C., Wulder, M.A., Pyörälä, J., Liang, X., Yu, X., Wang, Y., Kaartinen, H., Virtanen, J.-P., and Hyyppä, J. 2020b. Accurate derivation of stem curve and volume using backpack mobile laser scanning. *ISPRS Journal of Photogrammetry and Remote Sensing* 161:246–262.

Hyyppä, E., Manninen, P., Maanpää, J., Taher, J., Litkey, P., Hyyti, H., Kukko, A., Kaartinen, H., Ahokas, E., Yu, X., and Muhojoki, J. 2023. Can the perception data of autonomous vehicles be used to replace mobile mapping surveys?—A case study surveying roadside city trees. *Remote Sensing* 15(7):1790.

Hyyppä, E., Yu, X., Kaartinen, H., Hakala, T., Kukko, A., Vastaranta, M., and Hyyppä, J. 2020c. Comparison of backpack, handheld, under-canopy UAV, and above-canopy UAV laser scanning for field reference data collection in boreal forests. *Remote Sensing* 12(20):3327.

Hyyppä, H., and Hyyppä, J. 1999. Comparing the accuracy of laser scanner with other optical remote sensing data sources for stand attribute retrieval. *Photogrammetric Journal of Finland* 16(2):5–15.

Hyyppä, J. 1993. *Development and Feasibility of Airborne Ranging Radar for Forest Assessment*, Doctor of Technology Thesis. Helsinki University of Technology, Laboratory of Space Technology, Report, December, 112 p.

Hyyppä, J., Hyyppä, H., Inkinen, M., Engdahl, M., Linko, S., and Zhu, Y.-H. 2000. Accuracy comparison of various remote sensing data sources in the retrieval of forest stand attributes. *Forest Ecology and Management* 128:109–120.

Hyyppä, J., Hyyppä, H., Leckie, D., Gougeon, F., Yu, X., and Maltamo, M. 2008. Review of methods of small-footprint airborne laser scanning for extracting forest inventory data in boreal forests. *International Journal of Remote Sensing* 29(5):1339–1366.

Hyyppä, J., and Inkinen, M. 1999. Detecting and estimating attributes for single trees using laser scanner. *Photogrammetric Journal of Finland* 16(2):27–42.

Hyyppä, J., Kelle, O., Lehikoinen, M., and Inkinen, M. 2001. A segmentation-based method to retrieve stem volume estimates from 3-dimensional tree height models produced by laser scanner. *IEEE Transactions of Geoscience and Remote Sensing* 39:969–975.

Hyyppä, J., Virtanen, J.P., Jaakkola, A., Yu, X., Hyyppä, H., and Liang, X. 2017. Feasibility of Google Tango and Kinect for crowdsourcing forestry information. *Forests* 9(1):6.

Hyyppä, J., Wagner, W., Hollaus, M., and Hyyppä, H. 2009. Airborne laser scanning. In: *The SAGE Handbook of Remote Sensing*. Edited by T.A. Warner, M.D. Nellis and G.M. Foody. SAGE Publications, 568 p.

Hyyppä, J., Yu, X., Hakala, T., Kaartinen, H., Kukko, A., Hyyti, H., Muhojoki, J., and Hyyppä, E. 2021. Under-canopy UAV laser scanning providing canopy height and stem volume accurately. *Forests* 12(7):856.

Hyyppä, J., Yu, X., Hyyppä, H., Vastaranta, M., Holopainen, M., Kukko, A., Kaartinen, H., Jaakkola, A., Vaaja, M., Koskinen, J., and Alho, P. 2012. Advances in forest inventory using airborne laser scanning. *Remote Sensing* 4:1190–1207.

Hyyppä, J., Yu, X., Rönnholm, P., Kaartinen, H., and Hyyppä, H. 2003. Factors affecting object-oriented forest growth estimates obtained using laser scanning. *Photogrammetric Journal of Finland* 18:16–31.

Hyyppä, M., Turppa, T., Yu, X., Kukko, A., Hyyti, H., Handolin, H., Virtanen, J.-P., and Hyyppä, J. 2024. Towards nation-wide single tree inventory and virtual forests. Manuscript Submitted.

Irwin, L., Coops, N.C., Queinnec, M., McCartney, G., and White, J.C. 2021. Single photon lidar signal attenuation under boreal forest conditions. *Remote Sensing Letters* 12(10):1049–1060.

Jaakkola, A., Hyyppä, J., Kukko, A., Yu, X., Kaartinen, M., Lehtomäki, M., and Lin, Y. 2010. A low-cost multi-sensoral mobile mapping system and its feasibility for tree measurements. *ISPRS Journal of Photogrammetry and Remote Sensing* 65(6):514–522.

Jaakkola, A., Hyyppä, J., Yu, X., Kukko, A., Kaartinen, H., Liang, X., Hyyppä, H., and Wang, Y. 2017. Autonomous collection of forest field reference—The outlook and a first step with UAV laser scanning. *RemoteSensing* 9(8):785.

Jakubowski, M.K., Guo, Q., and Kelly, M. 2013. Trade-offs between lidar pulse density and forest measurement accuracy. *Remote Sensing of Environment* 130:245–253.

Jensen, J.R. 2000. *Remote Sensing of the Environment—An Earth Resource Perspective*. Prentice Hall, 544 p.

Junttila, S., Campos, M., Hölttä, T., Lindfors, L., El Issaoui, A., Vastaranta, M., Hyyppä, H., and Puttonen, E. 2022. Tree water status affects tree branch position. *Forests* 13(5):728.

Kaartinen, H., and Hyyppä, J. 2008. EuroSDR/ISPRS project, commission II, "tree extraction". *Final Report, EuroSDR*. European Spatial Data Research, Official Publication No 53.

Kaartinen, H., Hyyppä, J., Yu, X., Vastaranta, M., Hyyppä, H., Kukko, A., Holopainen, M., Heipke, C., Hirschugl, M., Morsdorf, F., Næsset, E., Pitkänen, J., Popescu, S., Solberg, S., Bernd, M., and Wu, J. 2012. An international comparison of individual tree detection and extraction using airborne laser scanning. *Remote Sensing* 4(4):950–974.

Kangas, A., and Maltamo, M. (Editors). 2006. *Managing Forest Ecosystems: Forest Inventory: Methodology and Applications*. Springer.

Karila, K., Karjalainen, M., Yu, X., Vastaranta, M., Holopainen, M., and Hyyppä, J. 2015. Comparison of interferometric and stereo-radargrammetric 3D metrics in mapping of forest resources. *The International Archives of the Photogrammetry, Remote Sensing and Spatial Information Sciences* 40:425–431.

Karila, K., Matikainen, L., Karjalainen, M., Puttonen, E., Chen, Y., and Hyyppä, J. 2023. Automatic labelling for semantic segmentation of VHR satellite images: Application of airborne laser scanner data and object-based image analysis. *ISPRS Open Journal of Photogrammetry and Remote Sensing* 9:100046.

Karjalainen, M., Kankare, V., Vastaranta, M., Holopainen, M., and Hyyppä, J. 2012. Prediction of plot-level forest variables using TerraSAR-X stereo SAR data. *Remote Sensing of Environment* 17(2):338–347.

Katoh, M., and Gougeon, F.A. 2012. Improving the precision of tree counting by combining tree detection with crown delineation and classification on homogeneity guided smoothed high resolution (50 cm) multi-spectral airborne digital data. *Remote Sensing* 4(5):1411–1424.

Kellndorfer, J.M., Walker, W.S., LaPoint, E., Kirsch, K., Bishop, J., and Fiske, G. 2010. Statistical fusion of lidar, InSAR, and optical remote sensing data for forest stand height characterization: A regional-scale method based on LVIS, SRTM, Landsat ETM+, and ancillary data sets. *Journal of Geophysical Research* 115:G00E08.

Kellndorfer, J.M., Walker, W.S., Pierce, L., Dobson, C., Fites, J.A., Hunsaker, C., Vonad, J., and Clutter, M. 2004. Vegetation height estimation from shuttle radar topography mission and national elevation datasets. *Remote Sensing of Environment* 93(3):339–358.

Koch, B. 2010. Status and future of laser scanning, synthetic aperture radar and hyperspectral remote sensing data for forest biomass assessment. *ISPRS Journal of Photogrammetry and Remote Sensing* 65(6):581–590.

Koch, B., Heyder, U., and Weinacker, H. 2006. Detection of individual tree crowns in airborne lidar data. *Photogrammetric Engineering and Remote Sensing* 72:357–363.

Korhonen, L., Korhonen, K.T., Rautiainen, M., and Stenberg, P. 2006. Estimation of forest canopy cover: A comparison of field measurement techniques. *Silva Fennica* 40(4):577–588.

Korpela, I., Orka, H., Hyyppä, J., Heikkinen, V., and Tokola, T. 2010. Range and AGC normalization in airborne discrete-return LiDAR intensity data for forest canopies. *ISPRS Journal of Photogrammetry and Remote Sensing* 65(4):369–379.

Kraus, K., and Otepka, J. 2005. DTM modeling and visualization—The SCOP approach. In *Photogrammetric Week '05*. Edited by D. Fritsch. Herbert Wichmann Verlag, pp. 241–252.

Kraus, K., and Pfeifer, N. 1998. Determination of terrain models in wooded areas with airborne laser scanner data. *ISPRS Journal of Photogrammetry and Remote Sensing* 53:193–203.

Krieger, G., Moreira, A., Fiedler, H., Hajnsek, I., Werner, M., Younis, M., and Zink, M. 2007. TanDEM-X: A satellite formation for high-resolution SAR interferometry. *IEEE Transactions on Geoscience and Remote Sensing* 45(11):3317–3341.

Kwak, D., Lee, W., Lee, J., Biging, G.S., and Gong, P. 2007. Detection of individual trees and estimation of tree height using LiDAR data. *Journal of Forest Research* 12:425–434.

La Prade, G. 1963. An analytical and experimental study of stereo for radar. *Photogrammetric Engineering* 29(2):294–300.

Leberl, F. 1979. Accuracy analysis of stereo side-looking radar. *Photogrammetric Engineering and Remote Sensing* 45(8):1083–1096.

Lechner, A.M., Foody, G.M., and Boyd, D.S. 2020. Applications in remote sensing to forest ecology and management. *One Earth* 2(5):405–412.

Lefsky, M.A., Harding, D.J., Keller, M., Cohen, W.B., Carabajal, C.C., Del Bom Espirito-Santo, F., Hunter, M.O., and de Oliveira Jr., R. 2005. Estimates of forest canopy height and aboveground biomass using ICESat. *Geophysical Research Letters* 32:L22S02.

Lefsky, M.A., Keller, M., Pang, Y., De Camargo, P.B., and Hunter, M.O. 2007. Revised method for forest canopy height estimation from Geoscience Laser Altimeter System waveforms. *Journal of Applied Remote Sensing* 1(1).

Lei, Y., Treuhaft, R., and Gonçalves, F. 2021. Automated estimation of forest height and underlying topography over a Brazilian tropical forest with single-baseline single-polarization TanDEM-X SAR interferometry. *Remote Sensing of Environment* 252:112132.

Le Toan, T., Beaudoin, A., Riom, J., and Guyon, D. 1992. Relating forest biomass to SAR data. *IEEE Transactions on Geoscience and Remote Sensing* 30(2):403–411.

Li, Z.L., Liu, G.X., and Ding, X.L. 2006. Exploring the generation of digital elevation models from same-side ERS SAR images: Topographic and temporal effects. *Photogrammetric Record* 21(114):124–140.

Liang, X., and Hyyppä, J. 2013. Automatic stem mapping by merging several terrestrial laser scans at the feature and decision levels. *Sensors* 13(2):1614–1634.

Liang, X., Hyyppä, J., Kaartinen, H., Holopainen, M., and Melkas, T. 2012b. Detecting changes in forest structure over time with Bi-temporal terrestrial laser scanning data. *ISPRS International Journal of GeoInformation* 1(3):242–255.

Liang, X., Hyyppä, J., Kaartinen, H., Lehtomäki, M., Pyörälä, J., Pfeifer, N., Holopainen, M., Brolly, G., Francesco, P., Hackenberg, J., and Huang, H. 2018a. International benchmarking of terrestrial laser scanning approaches for forest inventories. *ISPRS Journal of Photogrammetry and Remote Sensing* 144:137–179.

Liang, X., Hyyppä, J., Kankare, V., and Holopainen, M. 2011. Stem curve measurement using terrestrial laser scanning. *Proceedings of Silvilaser 2011*. University of Tasmania, Australia, 6 p.

Liang, X., Hyyppä, J., Kukko, A., Kaartinen, H., Jaakkola, A., and Yu, X. 2014c. The use of a mobile laser scanning for mapping large forest plots. *IEEE Geoscience and Remote Sensing Letters* 11(9):1504–1508.

Liang, X., Kankare, V., Hyyppä, J., Wang, Y., Kukko, A., Haggrén, H., Yu, X., Kaartinen, H., Jaakkola, A., Guan, F., and Holopainen, M. 2016. Terrestrial laser scanning in forest inventories. *ISPRS Journal of Photogrammetry and Remote Sensing* 115:63–77.

Liang, X., Kankare, V., Yu, X., and Hyyppä, J. 2014a. Automated stem curve measurement using terrestrial laser scanning. *IEEE Transactions on Geoscience and Remote Sensing* 52(3):1739–1748.

Liang, X., Kukko, A., Balenović, I., Saarinen, N., Junttila, S., Kankare, V., Holopainen, M., Mokroš, M., Surový, P., Kaartinen, H., Jurjević, L., Honkavaara, E., Näsi, R., Liu, J., Hollaus, M., Tian, J., Yu, X., Pan, J., Cai, S., Virtanen, J-P., Wang, Y., and Hyyppä, J. 2022. Close-range remote sensing of forests: The state of the art, challenges, and opportunities for systems and data acquisitions. *IEEE Geoscience and Remote Sensing Magazine* 10(3):32–71.

Liang, X., Kukko, A., Hyyppä, J., Lehtomäki, M., Pyörälä, J., Yu, X., Kaartinen, H., Jaakkola, A., and Wang, Y. 2018b. In-situ measurements from mobile platforms: An emerging approach to address the old challenges associated with forest inventories. *ISPRS Journal of Photogrammetry and Remote Sensing* 143:97–107.

Liang, X., Kukko, A., Kaartinen, H., Hyyppä, J., Yu, X., Jaakkola, A., and Wang, Y. 2014b. Possibilities of a personal laser scanning system for forest mapping and ecosystem services. *Sensors* 14(1):1228–1248.

Liang, X., Litkey, P., Hyyppä, J., Kaartinen, H., Vastaranta, M., and Holopainen, M. 2012a. Automatic stem mapping using single-scan terrestrial laser scanning. *IEEE Transactions on Geoscience and Remote Sensing* 50(2):661–670.

Lin, Y., Jaakkola, A., Hyyppä, J., and Kaartinen, H. 2010. From TLS to VLS: Biomass estimation at individual tree level. *Remote Sensing* 2(8):1864–1879.

Lindberg, E. 2012. *Estimation of Canopy Structure and Individual Trees from Laser Scanning Data* [WWW Document]. Available online: http://pub.epsilon.slu.se/8888/ (accessed March 21, 2013).

Lindgren, N., Olsson, H., Nyström, K., Nyström, M., and Ståhl, G. 2022. Data assimilation of growing stock volume using a sequence of remote sensing data from different sensors. *Canadian Journal of Remote Sensing* 48(2):127–143.

Liu, A., Cheng, X., and Chen, Z. 2021. Performance evaluation of GEDI and ICESat-2 laser altimeter data for terrain and canopy height retrievals. *Remote Sensing of Environment* 264:112571.

Lovell, J.L., Jupp, D.L.B., Newnham, G.J., and Culvenor, D.S. 2011. Measuring tree stem diameters using intensity profiles from ground-based scanning lidar from a fixed viewpoint. *ISPRS Journal of Photogrammetry and Remote Sensing* 66:46–55.

Maas, H.G., Bienert, A., Scheller, S., and Keane, E. 2008. Automatic forest inventory parameter determination from terrestrial laser scanner data. *International Journal of Remote Sensing* 29(5):1579–1593.

Magnussen, S., and Wulder, M.A. 2012. Post-fire canopy height recovery in Canada's boreal forests using airborne laser scanner (ALS). *Remote Sensing* 4:1600–1616.

Mallet, C., and Bretar, F. 2009. Full-waveform topographic lidar: State-of-the-art. *ISPRS Journal of Photogrammetry and Remote Sensing* 64(1):1–16.

Maltamo, M., Malinen, J., Packalen, P., Suvanto, A., and Kangas, J. 2006. Nonparametric estimation of stem volume using airborne laser scanning, aerial photography, and stand-register data. *Canadian Journal of Forest Research* 36:426–436.

Maltamo, M., Næsset, E., and Vauhkonen, J. 2014. Forestry applications of airborne laser scanning. Concepts and case studies. *Managing Forest Ecosystems* 27:460.

Marselis, S.M., Yebra, M., Jovanovic, T., and van Dijk, A.I. 2016. Deriving comprehensive forest structure information from mobile laser scanning observations using automated point cloud classification. *Environmental Modelling & Software* 82:142–151.

Martone, M., Rizzoli, P., Wecklich, C., González, C., Bueso-Bello, J.L., Valdo, P., Schulze, D., Zink, M., Krieger, G., and Moreira, A. 2018. The global forest/non-forest map from TanDEM-X interferometric SAR data. *Remote Sensing of Environment* 205:352–373.

Massonnet, D., and Feigl, K.L. 1998. Radar interferometry and its application to changes in the earth's surface. *Reviews of Geophysics* 36(4):441–500.

Matasci, G., Hermosilla, T., Wulder, M.A., White, J.C., Coops, N.C., Hobart, G.W., Bolton, D.K., Tompalski, P., and Bater, C.W. 2018a. Three decades of forest structural dynamics over Canada's forested ecosystems using Landsat time-series and lidar plots. *Remote Sensing of Environment* 216:697–714.

Matasci, G., Hermosilla, T., Wulder, M.A., White, J.C., Coops, N.C., Hobart, G.W., and Zald, H.S.J. 2018b. Large-area mapping of Canadian boreal forest cover, height, biomass and other structural attributes using Landsat composites and lidar plots. *Remote Sensing of Environment* 209:90–106.

McGaughey, R.J. 2013. *FUSION/LDV: Software for LiDAR Data Analysis and Visualization.* Version 3.30. U.S. Department of Agriculture Forest Service, Pacific Northwest Research Station, University of Washington, Seattle, Wash. Available online: http://forsys.cfr.washington.edu/fusion/FUSION_manual.pdf (accessed May 2013).

Mokroš, M., Mikita, T., Singh, A., Tomaštík, J., Chudá, J., Wężyk, P., Kuželka, K., Surový, P., Klimánek, M., Zięba-Kulawik, K., Bobrowski, R., and Liang, X. 2021. Novel low-cost mobile mapping systems for forest inventories as terrestrial laser scanning alternatives. *International Journal of Applied Earth Observation and Geoinformation* 104:102512.

Mora, B., Wulder, M.A., Hobart, G.W., White, J.C., Bater, C.W., Gougeon, F.A., Varhola, A., and Coops, N.C. 2013a. Forest inventory stand height estimates from very high spatial resolution satellite imagery calibrated with lidar plots. *International Journal of Remote Sensing* 34(12):4406–4424.

Mora, B., Wulder, M.A., White, J.C., and Hobart, G. 2013b. Modeling stand height, volume, and biomass from very high spatial resolution satellite imagery and samples of airborne LiDAR. *Remote Sensing* 5:2308–2326.

Morsdorf, F., Meier, E., Allgöwer, B., and Nüesch, D. 2003. Clustering in airborne laser scanning raw data for segmentation of single trees. *International Archives of the Photogrammetry, Remote Sensing and Spatial Information Sciences* 34(3):W13.

Murphy, G.E., Acuna, M.A., and Dumbrell, I. 2010. Tree value and log product yield determination in radiata pine (Pinus radiata) plantations in Australia: Comparisons of terrestrial laser scanning with a forest inventory system and manual measurements. *Canadian Journal of Forest Research* 40:2223–2233.

Murray, B.A., Coops, N.C., Winiwarter, L., White, J.C., Dick, A., Barbeito, I., and Ragab, A. 2024. Estimating tree species composition from airborne laser scanning data using point-based deep learning models. *ISPRS Journal of Photogrammetry and Remote Sensing* 207:282–297.

Næsset, E. 1997a. Determination of mean tree height of forest stands using airborne laser scanner data. *ISPRS Journal of Photogrammetry and Remote Sensing* 52:49–56.

Næsset, E. 1997b. Estimating timber volume of forest stands using airborne laser scanner data. *Remote Sensing of Environment* 61(2):246–253.

Næsset, E. 2002. Predicting forest stand characteristics with airborne scanning laser using a practical two-stage procedure and field data. *Remote Sensing of Environment* 80: 88–99.

Næsset, E. 2007. Airborne laser scanning as a method in operational forest inventory: Status of accuracy assessments accomplished in Scandinavia. *Scandinavian Journal of Forest Research* 22:433–442.

Næsset, E., and Nelson, R. 2007. Using airborne laser scanning to monitor tree migration in the boreal–alpine transition zone. *Remote Sensing of Environment* 110(4):357–369.

NASA. 2014. *NASA Selects Instruments to Track Climate Impact on Vegetation.* Available online: https://www.nasa.gov/news-release/nasa-selects-instruments-to-track-climate-impact-on-vegetation/#.VAXwT_ldX1Z

Nelson, R. 2013. How did we get there? An early history of forestry lidar. *Canadian Journal of Remote Sensing* 39(1):6–17.

Nelson, R., Hyde, P., Johnson, P., Emessiene, B., Imhoff, M.L., Campbell, R., and Edwards, W. 2007. Investigating RaDAR-LiDAR synergy in a North Carolina pine forest. *Remote Sensing of Environment* 110(1):98–108.

Nelson, R., Krabill, W., and Tonelli, J. 1988. Estimating forest biomass and volume using airborne laser data. *Remote Sensing of Environment* 24:247–267.

Neuenschwander, A., Guenther, E., White, J.C., Duncanson, L., and Montesano, P. 2020. Validation of ICESat-2 terrain and canopy heights in boreal forests. *Remote Sensing of Environment* 251:112110.

Nurminen, K., Karjalainen, M., Yu, X., Hyyppä, J., and Honkavaara, E. 2013. Performance of dense digital surface models based on image matching in the estimation of plot-level forest variables. *ISPRS Journal of Photogrammetry and Remote Sensing* 83:104–115.

Nyström, M., Lindgren, N., Wallerman, J., Grafström, A., Muszta, A., Nyström, K., Bohlin, J., Willén, E., Fransson, J.E., Ehlers, S., and Olsson, H. 2015. Data assimilation in forest inventory: First empirical results. *Forests* 6(12):4540–4557.

Olesk, A., Voormansik, K., Vain, A., Noorma, M., and Praks, J. 2015. Seasonal differences in forest height estimation from interferometric TanDEM-X coherence data. *IEEE Journal of Selected Topics in Applied Earth Observations and Remote Sensing* 8(12):5565–5572.

Olofsson, K., Holmgren, J., and Olsson, H. 2014. Tree stem and height measurements using terrestrial laser scanning and the RANSAC algorithm. *Remote Sensing* 6(5):4323–4344.

Packalen, P., and Maltamo, M. 2007. The k-MSN method in the prediction of species specific stand attributes using airborne laser scanning and aerial photographs. *Remote Sensing of Environment* 109:328–341.

Päivinen, R., Nousiainen, M., and Korhonen, K. 1992. Puuntunnusten mittaamisen luotettavuus (Accuracy of certain tree measurements). *Folia Forestalia* 787:18 p.

Papathanassiou, K.P., and Cloude, S.R. 2001. Single-baseline polarimetric SAR interferometry. *IEEE Transactions on Geoscience and Remote Sensing* 39(11):2352–2363.

Parkan, M.J. 2019. *Combined Use of Airborne Laser Scanning and Hyperspectral Imaging for Forest Inventories*. Doctoral Thesis, No 9033. EPFL, Switzerland.

Perko, R., Raggam, H., Deutscher, J., Gutjahr, K., and Schardt, M. 2011. Forest assessment using high resolution SAR data in X-band. *Remote Sensing* 3(4):792–815.

Persson, Å., Holmgren, J., and Söderman, U. 2002. Detecting and measuring individual trees using an airborne laser scanner. *Photogrammetric Engineering and Remote Sensing* 68:925–932.

Persson, H.J., Olsson, H., Soja, M.J., Ulander, L.M.H., and Fransson, J.E.S. 2017. Experiences from large-scale forest mapping of Sweden using TanDEM-X data. *Remote Sensing* 9:1253.

Pierzchała, M., Giguère, P., and Astrup, R. 2018. Mapping forests using an unmanned ground vehicle with 3D LiDAR and graph-SLAM. *Computers and Electronics in Agriculture* 145:217–225.

Popescu, S.C. 2007. Estimating biomass of individual pine trees using airborne lidar. *Biomass & Bioenergy* 31(9):646–655.

Popescu, S.C., Wynne, R., and Nelson, R. 2003. Measuring individual tree crown diameter with lidar and assessing its influence on estimating forest volume and biomass. *Canadian Journal of Remote Sensing* 29(5):564–577.

Portillo-Quintero, C., Sanchez-Azofeifa, A., and Culvenor, D. 2014. Using VEGNET in-situ monitoring LiDAR (IML) to capture dynamics of plant area index, structure and phenology in aspen parkland forests in Alberta, Canada. *Forests* 5(5):1053–1068.

Puttonen, E., Lehtomäki, M., Litkey, P., Näsi, R., Feng, Z., Liang, X., Wittke, S., Pandžić, M., Hakala, T., Karjalainen, M., and Pfeifer, N. 2019. A clustering framework for monitoring circadian rhythm in structural dynamics in plants from terrestrial laser scanning time series. *Frontiers in Plant Science* 10:486.

Queinnec, M., Coops, N.C., White, J.C., Griess, V.C., Schwartz, N.B., and McCartney, G. 2022. Developing a forest inventory approach using airborne single photon lidar data: From ground plot selection to forest attribute prediction. *Forestry: An International Journal of Forest Research* 95(3):347–362.

Rahman, M.Z.A., and Gorte, B. 2008. Tree filtering for high density airborne LiDAR data. In *International Conference on LiDAR Applications in Forest Assessment and Inventory, SilviLaser 2008*, September 17–19, Edinburgh, UK.

Rauste, Y. 2005. Multi-temporal JERS SAR data in boreal forest biomass mapping. *Remote Sensing of Environment* 97(2):263–275.

Reitberger, J., Heurich, M., Krzystek, P., and Stilla, U. 2007. Single tree detection in forest areas with high-density LiDAR data. *International Archives of Photogrammetry, Remote Sensing and Spatial Information Sciences* 36(3):139–144.

Reitberger, J., Schnörr, C., Krzystek, P., and Stilla, U. 2009. 3D segmentation of single trees exploiting full waveform LIDAR data. *ISPRS Journal of Photogrammetry and Remote Sensing* 64(6):561–574.

Riofrío, J., White, J.C., Tompalski, P., Coops, N.C., and Wulder, M.A. 2022. Harmonizing multi-temporal airborne laser scanning point clouds to derive periodic annual height increments in temperate mixedwood forests. *Canadian Journal of Forest Research* 52(10):1334–1352.

Riofrío, J., White, J.C., Tompalski, P., Coops, N.C., and Wulder, M.A. 2023. Modelling height growth of temperate mixedwood forests using an age-independent approach and multi-temporal airborne laser scanning data. *Forest Ecology and Management* 543:121137.

Rönnholm, P., Hyyppä, J., Hyyppä, H., Haggrén, H., Yu, X., Pyysalo, U., Pöntinen, P., and Kaartinen, H. 2004. Calibration of laser-derived tree height estimates by means of photogrammetric techniques. *Scandinavian Journal of Forest Research* 19(6):524–528.

Rosen, P.A., Hensley, S., Joughin, I.R., Li, F.K., Madsen, S.N., Rodriguez, E., and Goldstein, R.M. 2000. Synthetic aperture radar interferometry. *Proceedings of the IEEE* 88(3):333–382.

Rosette, J.A.B., North, P.R.J., and Suarez, J.C. 2008. Vegetation height estimates for a mixed temperate forest using satellite laser altimetry. *International Journal of Remote Sensing* 29(5):1475–1493.

Sadeghi, Y., St-Onge, B., Leblon, B., Prieur, J.F., and Simard, M. 2018. Mapping boreal forest biomass from a SRTM and TanDEM-X based on canopy height model and Landsat spectral indices. *International Journal of Applied Earth Observation and Geoinformation* 68:202–213.

Santoro, M., Shvidenko, A., McCallum, I., Askne, J., and Schmullius, C. 2007. Properties of ERS-1/2 coherence in the Siberian boreal forest and implications for stem volume retrieval. *Remote Sensing of Environment* 106(2):154–172.

Schlund, M., Magdon, P., Eaton, B., Aumann, C., and Erasmi, S. 2019. Canopy height estimation with TanDEM-X in temperate and boreal forests. *International Journal of Applied Earth Observation and Geoinformation* 82:101904.

Schlund, M., Wenzel, A., Camarretta, N., Stiegler, C., and Erasmi, S. 2023. Vegetation canopy height estimation in dynamic tropical landscapes with TanDEM-X supported by GEDI data. *Methods in Ecology and Evolution* 14:1639–1656.

Schutz, B.E., Zwally, H.J., Shuman, C.A., Hancock, D., and DiMarzio, J.P. 2005. Overview of the ICESat mission. *Geophysical Research Letters* 32(21):L21S01.

Seely, H., Coops, N.C., White, J.C., Montwe, D., Winiwarter, L., and Ragab, A. 2023. Modelling tree biomass using direct and additive methods with point cloud deep learning in a temperate mixed forest. *Science of Remote Sensing* 8:100110.

Simonse, M., Aschoff, T., Spiecker, H., and Thies, M. 2003. Automatic determination of forest inventory parameters using terrestrial laserscanning. In *Proceedings of the ScandLaser Scientific Workshop on Airborne Laser Scanning of Forests*. University of Freiburg, pp. 252–258.

Sinha, S., Jeganathan, C., Sharma, L.K., and Nathawat, M.S. 2015. A review of radar remote sensing for biomass estimation. *International Journal of Environmental Science and Technology* 12:1779–1792.

Sithole, G., and Vosselman, G. 2004. Experimental comparison of filter algorithms for bare-Earth extraction from airborne laser scanning point clouds. *ISPRS Journal of Photogrammetry and Remote Sensing* 59:85–101.

Soininen, V., Kukko, A., Yu, X., Kaartinen, H., Luoma, V., Saikkonen, O., Holopainen, M., Matikainen, L., Lehtomäki, M., and Hyyppä, J. 2022. Predicting growth of individual trees directly and indirectly using 20-year bitemporal airborne laser scanning point cloud data. *Forests* 13(12):2040.

Solberg, S., Astrup, R., Bollandsas, O.M., Næsset, E., and Weydahl, D.J. 2010. Deriving forest monitoring variables from X-band InSAR SRTM height. *Canadian Journal of Remote Sensing* 36:68–79.

Solberg, S., Astrup, R., Breidenbach, J., Nilsen, B., and Weydahl, D. 2013. Monitoring spruce volume and biomass with InSAR data from TanDEM-X. Remote Sensing of Environment 139:60–67.

Solberg, S., Hansen, E.H., Gobakken, T., Næssset, E., and Zahabu, E. 2017. Biomass and InSAR height relationship in a dense tropical forest. *Remote Sensing of Environment* 192:166–175.

Spadavecchia, C., Campos, M.B., Piras, M., Puttonen, E., and Shcherbacheva, A. 2023. Wood-leaf unsupervised classification of silver birch trees for biomass assessment using oblique point clouds. *The International Archives of the Photogrammetry, Remote Sensing and Spatial Information Sciences* 48:1795–1802.

Sun, G., Ranson, K.J., Kimes, D.S., Blair, J.B., and Kovacs, K. 2008. Forest vertical structure from GLAS: An evaluation using LVIS and SRTM data. *Remote Sensing of Environment* 112(1):107–117.

Suvanto, A., Maltamo, M., Packalen, P., and Kangas, J. 2005. Kuviokohtaisten puustotunnustenennustaminen laserkeilauksella. *Metsätieteenaikakauskirja* 4(2005):413–428.

Tansey, K., Selmes, N., Anstee, A., Tate, N.J., and Denniss, A. 2009. Estimating tree and stand variables in a corsican pine woodland from terrestrial laser scanner data. *International Journal of Remote Sensing* 30(19):5195–5209.

Thiel, C., and Schmullius, C. 2014. The potential of ALOS PALSAR backscatter and InSAR coherence for forest growing stock volume estimation in Central Siberia. *Remote Sensing of Environment* 173:258–273.

Thies, M., Pfeifer, N., Winterhalder, D., and Gorte, B.G.H. 2004. Three-dimensional reconstruction of stems for assessment of taper, sweep and lean based on laser scanning of standing trees. *Scandinavian Journal of Forest Research* 19:571–581.

Tokola, T., Letoan, T., Poncet, F., V., Tuominen, S., and Holopainen, M. 2007. Forest reconnaissance surveys: Comparison of estimates based on simulated TerraSar, and optical data. *Photogrammetric Journal of Finland* 20:64–79.

Tompalski, P., Coops, N.C., White, J.C., Goodbody, T.R., Hennigar, C.R., Wulder, M.A., Socha, J., and Woods, M.E. 2021. Estimating changes in forest attributes and enhancing growth projections: A review of existing approaches and future directions using airborne 3D point cloud data. *Current Forestry Reports* 7:1–24.

Torres, R., Snoeij, P., Geudtner, D., Bibby, D., Davidson, M., Attema, E., Potin, P., Rommen, B., Floury, N., Brown, M., Traver, I., Deghaye, P., Duesmann, B., Rosich, B., Miranda, N., Bruno, C., L'Abbate, M., Croci, R., Pietropaolo, A., Huchler, M., and Rostan, F. 2012. GMES Sentinel-1 mission. *Remote Sensing of Environment* 120(S1):9–24.

Toutin, T., and Gray, L. 2000. State-of-the-art of elevation extraction from satellite SAR data. *ISPRS Journal of Photogrammetry and Remote Sensing* 55(1):13–33.

Tsitsi, B. 2016. Remote sensing of aboveground forest biomass: A review. *Tropical Ecology* 57: 125–132.

Tsui, O.W., Coops, N.C., Wulder, M.A., and Marshall, P.L. 2013. Integrating airborne LiDAR and spaceborne radar via multivariate kriging to estimate above-ground biomass. *Remote Sensing of Environment* 139:340–352.

van Aardt, J., Wynne, R., and Scrivani, J. 2008. Lidar-based mapping of forest volume and biomass by taxonomix group using structurally homogenous segments. *Photogrammetric Engineering & Remote Sensing* 74(8):1033–1044.

Vastaranta, M. 2012. Forest mapping and monitoring using active 3D remote sensing. *Dissertationes Forestales* 144:45 p.

Vastaranta, M., Holopainen, M., Karjalainen, M., Kankare, M., Hyyppä, J., and Kaasalainen, S. 2014. TerraSAR-X stereo radargrammetry and airborne scanning LiDAR height metrics in imputation of forest aboveground biomass and stem volume. *IEEE Transactions on Geoscience and Remote Sensing* 52:1197–1204.

Vastaranta, M., Kankare, V., Holopainen, M., Yu, X., Hyyppä, J., and Hyyppä, H. 2012. Combination of individual tree detection and area-based approach in imputation of forest variables using airborne laser data. *ISPRS Photogrammetry and Remote Sensing* 67:73–79.

Vastaranta, M., Wulder, M.A., White, J., Pekkarinen, A., Tuominen, S., Ginzler, C., Kankare, V., Holopainen, M., Hyyppä, J., and Hyyppä, H. 2013. Airborne laser scanning and digital stereo imagery measures of forest structure: Comparative results and implications to forest mapping and inventory update. *Canadian Journal of Remote Sensing* 39(5):382–395.

Vauhkonen, J., Ene, L., Gupta, S., Heinzel, J., Holmgren, J., Pitkänen, J., Solberg, S., Wang, Y., Weinacker, H., Hauglin, K.M., and Lien, V. 2012. Comparative testing of single-tree detection algorithms under different types of forest. *Forestry* 85(1):27–40.

Vosselman, G. 2000. Slope based filtering of laser altimetry data. *International Archives of Photogrammetry and Remote Sensing* 33(B3/2):935–942.

Wagner, W., Luckman, A., Vietmeier, J., Tansey, K., Balzter, H., Schmullius, C., Davidson, M., Gaveau, D., Gluck, M., Le Toan, T., Quegan, S., Shvidenko, A., Wiesmann, A., and Jiong Yu, J.J. 2003. Large-scale mapping of boreal forest in SIBERIA using ERS tandem coherence and JERS backscatter data. *Remote Sensing of Environment* 85(2):125–144.

Wang, Y., Hyyppä, J., Liang, X., Kaartinen, H., Yu, X., Lindberg, E., Holmgren, J., Qin, Y., Mallet, C., Ferraz, A., and Torabzadeh, H. 2016. International benchmarking of the individual tree detection methods for modeling 3-D canopy structure for silviculture and forest ecology using airborne laser scanning. *IEEE Transactions on Geoscience and Remote Sensing* 54(9):5011–5027.

Wang, Y., Lehtomäki, M., Liang, X., Pyörälä, J., Kukko, A., Jaakkola, A., Liu, J., Feng, Z., Chen, R., and Hyyppä, J. 2019. Is field-measured tree height as reliable as believed—A comparison study of tree height estimates from field measurement, airborne laser scanning and terrestrial laser scanning in a boreal forest. *ISPRS Journal of Photogrammetry and Remote Sensing* 147:132–145.

Wang, Y., Weinacker, H., and Koch, B. 2008. A lidar point cloud based procedure for vertical canopy structure analysis and 3D single tree modelling in forest. *Sensors* 8(6):3938–3951.

Wang, Y., Zhang, W., Gao, R., Jin, Z., and Wang, X. 2021. Recent advances in the application of deep learning methods to forestry. *Wood Science and Technology* 55(5):1171–1202.

Wästlund, A., Holmgren, J., Lindberg, E., and Olsson, H. 2018. Forest variable estimation using a high altitude single photon lidar system. *Remote Sensing* 10(9):1422.

Watt, P.J., and Donoghue, D.N.M. 2005. Measuring forest structure with terrestrial laser scanning. *International Journal of Remote Sensing* 26:1437–1446.

White, J.C., Coops, N.C., Wulder, M.A., Vastaranta, M., Hilker, T., and Tompalski, P. 2016. Remote sensing technologies for enhancing forest inventories: A review. *Canadian Journal of Remote Sensing* 42(5):619–641.

White, J.C., Penner, M., and Woods, M. 2021a. Assessing single photon lidar for operational implementation of an enhanced forest inventory in diverse mixedwood forests. *The Forestry Chronicle* 97(1):78–96.

White, J.C., Woods, M., Krahn, T., Papasodoro, C., Bélanger, D., Onafrychuk, C., and Sinclair, I. 2021b. Evaluating the capacity of single photon lidar for terrain characterization under a range of forest conditions. *Remote Sensing of Environment* 252:112169.

White, J.C., Wulder, M.A., Varhola, A., Vastaranta, M., Coops, N.C., Cook, B.D., Pitt, D., and Woods, M. 2013a. A best practices guide for generating forest inventory attributes from airborne laser scanning data using the area-based approach. In *Information Report FI-X-10*. Natural Resources Canada.

White, J.C., Wulder, M.A., Vastaranta, M., Coops, N.C., Pitt, D., and Woods, M. 2013b. The utility of image-based point clouds for forest inventory: A comparison with airborne laser scanning. *Forests* 4(3):518–536.

Wielgosz, M., Puliti, S., Xiang, B., Schindler, K., and Astrup, R. 2024. SegmentAnyTree: A sensor and platform agnostic deep learning model for tree segmentation using laser scanning data. *arXiv preprint arXiv:2401.15739*.

Winberg, O., Pyörälä, J., Yu, X., Kaartinen, H., Kukko, A., Holopainen, M., Holmgren, J., Lehtomäki, M., and Hyyppä, J. 2023. Branch information extraction from Norway spruce using handheld laser scanning point clouds in Nordic forests. *ISPRS Open Journal of Photogrammetry and Remote Sensing* 9:100040.

Wittke, S., Yu, X., Karjalainen, M., Hyyppä, J., and Puttonen, E. 2019. Comparison of two-dimensional multitemporal Sentinel-2 data with three-dimensional remote sensing data sources for forest inventory parameter estimation over a boreal forest. *International Journal of Applied Earth Observation and Geoinformation* 76:167–178.

Wu, B., Yu, B., Yue, W., Shu, S., Tan, W., Hu, C., Huang, Y., Wu, J., and Liu, H. 2013. A voxel-based method for automated identification and morphological parameters estimation of individual street trees from mobile laser scanning data. *Remote Sensing* 5(2):584–611.

Wulder, M.A., White, J.C., Bater, C.W., Coops, N.C., Hopkinson, C., and Chen, G. 2012b. Lidar plots a new large-area data collection option: Context, concepts, and case study. *Canadian Journal of Remote Sensing* 38(5):600–618.

Wulder, M.A., White, J.C., Nelson, R.F., Næsset, E., Ørka, H.O., Coops, N.C., Hilker, T., Bater, C.W., and Gobakken, T. 2012a. Lidar sampling for large-area forest characterization: A review. *Remote Sensing of Environment* 121:196–209.

Xiang, B., Wielgosz, M., Kontogianni, T., Peters, T., Puliti, S., Astrup, R., and Schindler, K. 2023. Automated forest inventory: Analysis of high-density airborne LiDAR point clouds with 3D deep learning. *arXiv preprint arXiv:2312.15084*.

Yrttimaa, T., Junttila, S., Luoma, V., Pyörälä, J., Puttonen, E., Campos, M., Hölttä, T., and Vastaranta, M. 2024. Tree height and stem growth dynamics in a Scots pine dominated boreal forest. *Trees, Forests and People* 15:100468.

Yu, X., Hyyppä, J., Kaartinen, H., and Maltamo, M. 2004. Automatic detection of harvested trees and determination of forest growth using airborne laser scanning. *Remote Sensing of Environment* 90:451–462.

Yu, X., Hyyppä, J., Kukko, A., Maltamo, M., and Kaartinen, H. 2006. Change detection techniques for canopy height growth measurements using airborne laser scanner data. *Photogrammetric Engineering & Remote Sensing* 72(12):1339–1348.

Yu, X., Hyyppä, J., Litkey, P., Kaartinen, H., Vastaranta, M., and Holopainen, M. 2017. Single-sensor solution to tree species classification using multispectral airborne laser scanning. *Remote Sensing* 9(2):108.

Yu, X., Hyyppä, J., Vastaranta, M., Holopainen, M., and Viitala, R. 2011. Predicting individual tree attributes from airborne laser point clouds based on the random forests technique. *ISPRS Journal of Photogrammetry and Remote Sensing* 66(1):28–37.

Yu, X., Kukko, A., Kaartinen, H., Wang, Y., Liang, X., Matikainen, L., and Hyyppä, J. 2020. Comparing features of single and multi-photon lidar in boreal forests. *ISPRS Journal of Photogrammetry and Remote Sensing* 168:268–276.

Zhao, K., Popescu, S., and Nelson, R. 2009. Lidar remote sensing of forest biomass: A scale-invariant estimation approach using airborne lasers. *Remote Sensing of Environment* 113(1):182–196.

Zhao, X., Liu, J., and Tan, M. 2006. A remote aerial robot for topographic survey. *Proc. International Conference on Intelligent Robots and Systems*, IEEE/RJS, Beijing, 9–15 October, pp. 3143–3148.

Zhao, Y., Hu, Q., Li, H., Wang, S., and Ai, M. 2018. Evaluating carbon sequestration and PM2.5 removal of urban street trees using mobile laser scanning data. *Remote Sensing* 10(11):1759.

Zwally, H.J., Schutz, B., Abdalati, W., Abshire, J., Bentley, C., Brenner, A., Bufton, J., Dezio, J., Hancock, D., Harding, D., Herring, T., Minster, B., Quinn, K., Palm, S., Spinhirne, J., and Thomas, R. 2002. ICESAT's laser measurements of polar ice, atmosphere, ocean, and land. *Journal of Geodynamics* 34:405–445.

3 Forest Biophysical and Biochemical Properties from Hyperspectral and LiDAR Remote Sensing

Gregory P. Asner, Susan L. Ustin, Philip Townsend, and Roberta E. Martin

ACRONYMS AND DEFINITIONS

ACD	Aboveground carbon density
ALI	Advanced Land Imager
APAR	Absorbed photosynthetically active radiation
CHRIS	Compact High Resolution Imaging Spectrometer
DLR	German Aerospace Center
EnMAP	Environmental Mapping and Analysis Program
ETM+	Enhanced Thematic Mapper+
GLAS	Geoscience Laser Altimeter System
HyspIRI	Hyperspectral Infrared Imager
ICESat	Instrument on board the Ice, Cloud, and land Elevation
IKONOS	A commercial earth observation satellite, typically, collecting sub-meter to 5 m data
LAI	Leaf area index
LiDAR	Light detection and ranging
LUE	Light use efficiency
NASA	National Aeronautics and Space Administration
NDVI	Normalized Difference Vegetation Index
NIR	Near-infrared
NPP	NPOESS Preparatory Project
PAR	Photosynthetically active radiation
PRI	Photochemical reflectance index
PROSPECT	Radiative transfer model to measure leaf optical properties spectra
SWIR	Shortwave infrared
UAV	Unmanned Aerial Vehicle

3.1 INTRODUCTION

Forests store about three-quarters of all carbon stocks in vegetation in the terrestrial biosphere and harbor an array of organisms that comprise most of this carbon (IPCC 2000). The distribution of carbon and biodiversity in forests is spatially and temporally heterogeneous. The complex, three-dimensional arrangement of plant species and their tissues has always challenged field-based studies of forests. Remote sensing has long endeavored to address these challenges

Forest Biophysical and Biochemical Properties

by mapping the cover, structure, composition, and functional attributes of forests, and new approaches are continually being developed to increase the breadth and accuracy of remote measurements.

Over the past several decades, two technologies—hyperspectral imaging (HSI) and light detection and ranging (LiDAR)—have rapidly advanced from their use in testbed-type research to operational applications ranging from ecology to land management. HSI, also known as imaging spectroscopy, involves the measurement of reflected solar radiance in narrow, contiguous spectral bands that form a spectrum for each image pixel. LiDAR uses emitted laser pulses, emitted in a scanning pattern to determine the distance between objects such as canopy foliage and ground surfaces. Individually, HSI and LiDAR are advancing the study of forests at landscape to global scales, uncovering new spatial and temporal patterns of forest biophysical and biochemical properties, as well as deeper understanding of physiological processes. When combined, HSI and LiDAR provide ecological detail at spatial scales unachievable in the field. This chapter discusses HSI and LiDAR data sources, techniques, applications, and challenges in the context of forest ecological research.

3.2 HSI AND LIDAR DATA

3.2.1 HSI Data Sources

The availability of hyperspectral imagery for ecological applications is growing as the utility of these data has increasingly been recognized. HSI can be collected either with airborne sensors that have limited spatial coverage but high spatial resolution or with spaceborne sensors capable of capturing data globally, but generally with coarser spatial resolution. There is an expanding number of government, private, and commercial airborne HSI sensors. One spaceborne HSI sensor—Earth Observing-1 Hyperion—was in operation as a technology demonstration from November 2000 until March 2017. Other orbital sensors include the NASA Earth Surface Mineral Dust Source Investigation (EMIT) and the DLR Earth Sensing Imaging Spectrometer (DESIS) on the International Space Station (ISS), a free-flying Indian Hyperspectral Imaging Satellite (HySIS) launched in 2018, the Italian Precursore Iperspettraie della Missione Applicativa (PRISMA) launched in 2019, the German Environmental Mapping and Analysis Program (EnMAP) launched in 2022, and Tanager-1 launched in 2024. Still others are in the planning or development stages to further extend the global coverage of imaging spectroscopy (Table 3.1).

Airborne HSI sensors have been operating since the 1980s. An early system was NASA's Airborne Imaging Spectrometer (AIS), followed later by the Airborne Visible/Infrared Imaging Spectrometer (AVIRIS) in 1987, which is still in operation as a third-generation sensor, providing data to NASA-supported investigators. Very similar instruments, including the Global Airborne Observatory (GAO; formerly the Carnegie Airborne Observatory) Visible to Shortwave Infrared (VSWIR) imaging spectrometer, in service since 2011; the US National Ecological Observatory Network (NEON) spectrometers; and the University of Zurich Compact Wide-swath Imaging Spectrometer (CWIS) provide a wide range of mission capabilities dotting the planet (Table 3.1).

Beyond government, university, and nonprofit instruments for research, a number of HSI sensors have been built for commercial applications. For example, the Compact Airborne Spectrographic Imager (CASI, CASI-2, CASI-1500) and the HyMap provide high-performance visible-to-near infrared (365–1052 nm) and visible-to-shortwave infrared (VSWIR; 440–2500 nm) measurements, respectively (Table 3.1). In addition, multiple commercial companies such as HySpex, SPECIM, and Headwall have developed consumer-grade imaging systems with VSWIR capability for both airplanes and drones, thus opening opportunities for the application of HSI to a wide range of potential users.

TABLE 3.1

Examples of Current and Planned Airborne and Spaceborne Hyperspectral Imagers (HSI)

Sensor	Spectral Range (nm)	Spectral Bands	Spectral Resolution (nm)	Spatial Resolution (m)	Reference
Airborne					
AVIRIS	400–2450	224	10	4.0 +	Green et al. (1998)
AVIS-2	400–900	64	9	2.0 +	Oppelt and Mauser (2007)
GAO VSWIR	380–2510	428	5	0.5 +	Asner et al. (2012a)
HYDICE	400–2500	206	8–15	1.0 +	Basedow et al. (1995)
NEON VSWIR	380–2500	212	10	0.5 +	www.neoninc.org
AISA	380–2500	275	3.5–12	1 +	www.specim.fi
CASI	365–1052	288	2–10	0.25 +	www.itres.com
HyMap	440–2500	100–200	10–20	2.0 +	Cocks et al. (1998)
Spaceborne					
EO-1 Hyperion	400–2500	220	10	30	Ungar et al. (2003)
Proba-1 CHRIS	415–1050	18–62	1.3–12	18, 36	Barnsley et al. (2004)
EMIT	380–2500	285	7.4	60	Green et al. (2020)
PRISMA	400–2505	229	6.5–12.5	30	Cogliati et al. (2021)
EnMAP	420–2450	98–130	6.5–10	30	Stuffler et al. (2007)
SBG VSWIR *(planned)*	380–2500	212	10	30	Cawse-Nicholson et al. (2021)

In comparison to airborne systems, there are fewer spaceborne sensors collecting hyperspectral data (Table 3.1). Prior to its mission conclusion in 2017, NASA's EO-1 Hyperion provided trailblazing hyperspectral analyses across a wide range of forest-related topics (Riebeek 2010). Thenkabail et al. (2004) showed that Hyperion data, when compared to data from even the most advanced broadband sensors (ETM+, IKONOS, and ALI) in orbit at that time, yielded models that explained 36–83% more of the variability in rainforest biomass and produced land use/land cover classifications with 45–52% higher accuracies. The European Space Agency (ESA) also had a hyperspectral sensor on board the Proba-1 satellite, which observed in the visible and near-infrared portion of the spectrum. Although at higher spatial resolutions than Hyperion, it was only able to capture 18 (selectable) bands at a time in this wavelength range (Barnsley et al. 2004). In addition, NASA is currently planning for the Surface Biology and Geology (SBG; Cawse-Nicholson et al. 2021) mission that replaced planning for the HyspIRI (Hyperspectral and Infrared Imager). Current plans expect it to be launched some time after 2027 and to be followed by the ESA's Copernicus Expansion mission, the Hyperspectral Imaging Mission for the Environment (CHIME) in the 2028–2030 period. The ongoing roll-out of spaceborne imaging spectrometers will continue to contribute to spatial and temporal hyperspectral coverage for forest research.

3.2.2 LiDAR Data Sources

LiDAR data sources are both numerous and variable, a reflection of the continued demand for LiDAR in a wide variety of scientific and engineering applications. Recent spaceborne LiDAR systems, described in this section, offer new data for forest monitoring. While the amount of LiDAR data being collected continues to increase, there is a great deal of variability in the quality, type (discrete return vs. waveform), and spatial resolution of the resulting data.

LiDAR datasets for the United States are publicly available from a variety of sources. The National Center for Airborne Laser Mapping (NCALM; www.ncalm.cive.uh.edu) uses commercially sourced LiDAR sensors to collect high-resolution data (> 2 laser spots m^{-2}) for NSF-funded projects, or for other select projects. These data are currently made available to the public within two years of collection through the NSF-supported OpenTopography program (www.opentopography.org), which provides a platform to access these data, along with other LiDAR datasets contributed by researchers. NASA's Land, Vegetation, and Ice Sensor (LVIS), which has been operating in North America since the late 1990s, provides waveform data at coarser resolution of 10–25 m diameter footprints in support of NASA studies (Blair et al. 1999). In addition, due to the increasing availability of commercial LiDAR acquisition services, many state and local governments have commissioned datasets. In the United States, the National Oceanic and Atmospheric Administration provides an inventory of these data (http://www.csc.noaa.gov/inventory/). There are no standard characteristics of these datasets, as they all vary with sensor parameters, elevation of data collection, and the density of returns collected. These heterogeneous data collection conditions hinder general assessments of the quality of these data.

In addition to airborne LiDAR data, NASA's GLAS Instrument (Geoscience Laser Altimeter System), on board the Ice, Cloud, and land Elevation Satellite (ICESat), was the first spaceborne LiDAR instrument (Abshire et al. 2005). GLAS collected waveform data with 70 m spot diameter and 170 m spot intervals. The GLAS instrument operated from 2003 to 2009, and the data are publicly available (icesat.gsfc.nasa.gov). The ICESat-2 launched in 2016 and, now extended to 2025, carries the Advanced Topographic Laser Altimeter System (ATLAS) with higher spatial resolution data. In addition, NASA carried the Global Ecosystem Dynamics Investigation (GEDI) on board the ISS and collected exceptional imagery from 2018 to 2023, when it was moved into storage to accommodate another mission. It is expected to be returned and operating from 2025 until the ISS is terminated, about 2031. There is no current plan for a dedicated space-based LiDAR that will monitor global forests, although this is of significant interest to the earth science community.

3.2.3 Data Quality

The vast majority of HSI and LiDAR instruments have been deployed on aircraft, so the geographic coverage, ground sampling distance (spatial resolution and/or laser spot spacing), flying altitudes and atmospheric conditions have varied enormously, making comparisons of instrument performances difficult to achieve. Nonetheless, comparative use of these instruments often reveals that sensor performance is paramount to achieving quality estimates of vegetation's biophysical and biochemical properties.

Three sensor qualities have proven particularly important in the effort to achieve high-fidelity data output. These include detector uniformity, instrument stability, and signal-to-noise (SNR) performance of the measurement (Green 1998). From the HSI perspective, each of these quality metrics is important. Uniformity refers to the collection of spectra in the cross-track and spectral directions on the instrument detector. Many HSI instruments fail to meet the often-cited 95–98% absolute uniformity standard. One of the most insidious errors in uniformity occurs in the spectral direction. Most area-array HSI sensors fail to keep the spectral measurement aligned "down spectrum" from the visible to near-infrared (e.g., 400–1100 nm) and throughout the shortwave-infrared (e.g., 1100–2500 nm), leading to a mismatch in different parts of the spectrum projected onto the Earth's surface. Another HSI performance issue is stability, which refers to the repeatability of the measurement across the imaging detector and/or over time. Much of the stability issue rests in the performance of the electronics and temperature stabilization subsystems. Finally, SNR is a quality that reports the strength and accuracy of the measurement signal relative to noise generated by the electronics and optics. SNR varies widely from instrument to instrument and also with environmental conditions such as temperature and humidity. Readers should be cautious when reviewing potential sources of HSI data, as providers may report SNR on either a bright target (e.g., white reference) or with enlarged camera apertures and/or inappropriately long integration times (equivalent to shutter speed). This will greatly inflate reported SNR values. For vegetation applications, SNR performances should be reported on dark targets in the 5–8% reflectance range, typical for plants in the visible spectrum (350–700 nm), and with integration times that are appropriate for airborne or spaceborne ground speeds (usually 10 ms).

LiDAR measurements also have signal-to-noise, uniformity, and stability challenges. The shape, noisiness, and strength of the outbound laser pulses largely affect LiDAR SNR. Commercial LiDAR systems come in a wide range of SNR performance levels. For forest science, strong pulse strength (e.g., high wattage laser diodes) is necessary to overcome absorption by the vegetation canopy. In addition, scientists tend to overlook uniformity prior to data source selection; it is highly advisable to select LiDAR instruments that deliver a uniform scan pattern across the swath of the dataset. Without strict control over this factor, the user will end up with high data density in the middle of the scan, and low data density at the edges of the scan. Finally, stability is a key issue with LiDAR instrumentation. Many commercial LiDAR systems exhibit instability as they change temperature, pressure, and humidity, resulting in variability in the quality of the laser data throughout the course of a mapping flight or research campaign.

3.3 HSI REMOTE SENSING OF FORESTS

Forests, as fundamental components of the Earth's biosphere, have been a major focus of study from the beginning of HSI data collection. HSI provides a quantitative measure of the wavelength-specific sunlight reflected from the forest canopy, and the properties therein. The extended range and high-fidelity narrowband resolution of HSI offers enhanced capability for mapping forest biochemical and biophysical constituents along with physiological processes that contribute to the shape of the reflectance spectrum (Table 3.2). HSI data are used in many ways to assess leaf and canopy properties, namely, semi-empirical methods utilizing narrowband spectral indices, regression modeling, and radiative transfer model inversion. As the HSI data quality improves, so do the results derived from these methods. Most recently, HSI combined with improved analytic methods has dramatically advanced species mapping and land cover classification (see reviews, Schneider et al. 2017; Cavender-Bares et al. 2020). At the leaf level, McManus et al. (2016) showed phylogenetic structure in multiple wavelength regions across the solar visible-infrared spectrum, and Miereles et al. (2020) further showed these spectral signals were present across lineages of seed plants. In addition, leaf-level spectroscopy has proven to be a valuable tool for tracking changes in properties of canopy leaves as they age (Chavana-Bryant et al. 2017; Wu et al. 2017).

3.3.1 Biophysical Properties

HSI data can uncover biophysical properties of ecological significance at both the leaf and canopy scales. Properties related to forest composition and leaf area index (LAI) are perhaps best retrieved from HSI data, whereas some properties, like canopy gap distribution and leaf angle distribution (LAD), are more readily determined from LiDAR. LAI (leaf area per unit of ground area, $m^2\ m^{-2}$) is one of the most important canopy properties because it is directly related to productivity and water use, but variation in LAI can also indicate stress resistance and competition for light (see Waring 1983; Asner et al. 2004a). Field data and models show that LAI and LAD are primary controls on canopy reflectance in dense vegetation (Asner 1998; Gong et al. 1992). While LAI is detectable from broadband sensors, studies show that HSI data and analysis methods optimized for HSI are more accurate (e.g., Gong et al. 1995; Spanner et al. 1994), although structural information from other sources such as LiDAR is generally required to best retrieve high LAI. Lee et al. (2004) examined four structurally different land-cover types and showed that HSI red-edge and SWIR

TABLE 3.2

Forest Biochemical and Physiological Properties Estimated from Hyperspectral Imaging, along with a Summary of Example Methods (Spectral Indices), Relevant Spectral Bands, Maturity, and References. Maturity Is a Metric of Relative Accuracy as Depicted in the Literature, with One Checkmark Indicating Low Maturity, and Three Checkmarks Indicating High Maturity. RT = Radiative Transfer; PLSR = Partial Least Squares Regression

Vegetation Property	Estimation Method(s)	Relevant Bands (nm)	Maturity Level	Example References
Foliar nitrogen	Normalized difference nitrogen index; Band depth analysis; PLSR; RT model inversion	1510, 1680; 400–2500	✓✓✓	Kokaly (2001), Serrano et al. (2002), Smith et al. (2003), Asner and Vitousek (2005), Dahlin et al. (2013), Asner et al. (2014, 2015), Chadwick and Asner (2016), Chavana-Bryant et al. (2017), Wang et al. (2018), Martin et al. (2020)
Light-use efficiency	Photochemical reflectance index	531, 570	✓✓	Gamon et al. (1992, 1997), Gamon and Surfus (1999), Stylinksi et al. (2000), Guo and Trotter (2004), Hilker et al. (2008), Filella et al. (2009), Garbulsky et al. (2011), Ripullone et al. (2011), Wu et al. (2019)
Foliar carotenoids	Various narrowband spectral indices	510, 550, 700; 445, 680, 800	✓✓✓	Gitelson et al. (2002), Peñuelas et al. (1995), Asner et al. (2014, 2015), Martin et al. (2020)
Foliar anthocyanin	Various narrowband spectral indices	400–700	✓	Gamon and Surfus (1999), Gitelson et al. (2001, 2006), Van den Berg and Perkins (2005)
APAR	Simple Ratio, NDVI	400–700	✓✓✓	Jordan (1969), Rouse et al. (1974)
LAI	Various narrowband spectral indices; RT model inversion	700–1300	✓✓✓	Rouse et al. (1974), Huete (1988), Rondeaux et al. (1996), Haboudane et al. (2002), Gitelson (2004), Lim et al. (2003)
LMA	PLSR; wavelet	400–2500	✓✓✓	Asner et al. (2011), Cheng et al. (2014), Chadwick and Asner (2016), Chavana-Bryant et al. (2017), Serbin et al. (2019), Martin et al. (2020), Wang et al. (2020), Yan et al. (2021)
Foliar chlorophylls	Various narrowband spectral indices; RT model inversion	550, 670, 700; 800–1300; 690–725	✓✓✓	Kim (1994), Zarco-Tejada et al. (2001), Gitelson et al. (2006), Zhang et al. (2008), Hoeppner et al. (2020), Martin et al. (2020), Wang et al. (2020), Yan et al. (2021)
Foliar water	Various narrowband spectral indices	820, 1600; 860, 1240; 900, 970	✓✓✓	Hunt and Rock (1989), Gao (1996), Peñuelas et al. (1997), Dahlin et al. (2013), Asner et al. (2014, 2015), Martin et al. (2020), Yan et al. (2021)
Canopy water	EWT and CWC; RT model inversion	800–2500	✓✓✓	Hunt and Rock (1989), Gao and Goetz (1990), Peñuelas et al. (1997), Roberts et al. (2004), Asner et al. (2016), Wang et al. (2020)
Foliar lignin and cellulose	Cellulose absorption index; Normalized difference lignin index	2015, 2106, 2195; 1680, 1754	✓✓	Daughtry (2001), Daughtry et al. (2005), Serrano et al. (2002), Asner et al. (2015), Wang et al. (2020)
Foliar carbon	PLSR	1500–2500	✓✓	Dahlin et al. (2013), Asner et al. (2014, 2015)

bands produced the best estimates of LAI. Equivalent Water Thickness (EWT, mm) produces better estimates of LAI than do pigment-based indices, such as the normalized difference vegetation index (NDVI) (Roberts et al. 1998). EWT has been shown to be sensitive to LAI values up to nine, far exceeding the sensitivity range of NDVI and other indices (Roberts et al. 2004). Water indices derived from HSI have also been used to quantify loss of LAI from pest-related defoliation and other factors (e.g., White et al. 2007).

At the leaf level, leaf mass per area (LMA: g m^{-2} and its reciprocal, specific leaf area or SLA: m^2 g^{-1}) is a key foliar property that is highly correlated with light harvesting and potential plant productivity (Niinemets 1999; Westoby et al. 2002). LMA can be defined for foliage throughout the canopy or in any given canopy layer, depending on the ecological question. While there is enormous range in LMA within a given plant functional type and among co-existing species, LMA is broadly correlated with temperature and precipitation at the global level (Wright et al. 2004). Higher temperatures, drier conditions, and higher irradiance are associated with higher values of LMA. Leaves with higher LMA are built for defense and longer lifespans, creating higher resource use efficiency per nutrient acquired (Poorter et al. 2009). Conversely, lower LMA values are found in fast-growing species, often with higher nutrient concentrations and photosynthetic rates (Wright et al. 2004). In addition, there is a strong degree of taxonomic organization to LMA within forest communities (Asner et al. 2014). Because LMA is a function of leaf thickness and is correlated with total carbon and nitrogen, it is uniquely detectable in HSI data and has been estimated from the inversion of radiative transfer models such as the PROSPECT model (Jacquemoud et al. 2009), chemometric analytical methods (Asner et al. 2011; Singh et al. 2015; Wang et al. 2020), and HSI optimized SWIR indices (le Maire et al. 2008). Results from these studies conform to field measurements, and synthesis studies using spectroscopy at the leaf level provide generalizable models to retrieve LMA from spectroscopic data in all vegetation types (Serbin et al. 2019). Almost all studies that retrieve LMA from HSI are restricted to top-of-canopy LMA, as LMA decreases with decreasing light levels in canopies (i.e., leaves become thinner) (Poorter et al. 2009). However, the integration of canopy structure information from LiDAR with HSI has enabled estimation of LMA variation within canopies (Chlus et al. 2020; Kamoske et al. 2020).

3.3.2 Biochemical Properties

The foremost motivation for biochemical detection is to better assess the spatio-temporal status and trends of forest canopy functioning, especially those related to fluxes of water, carbon, and nutrients. The list of plant biochemicals that have been identified and quantified using HSI data is extensive (Table 3.2) and has received several detailed reviews (Blackburn 2007; Kokaly et al. 2009; Ustin et al. 2009; Homolová et al. 2013). Wang et al. (2020) demonstrated the capacity to accurately map 26 foliar traits from airborne HIS across a broad range of ecosystems represented by the US National Ecological Observatory Network (NEON).

Many studies have found strong correlations between remotely sensed foliar nitrogen content and photosynthetic capacity or net primary production (Kokaly et al. 2009; Townsend et al. 2013; Serbin et al. 2015), despite the small fraction of biomass composed of nitrogen. Most of these studies have been based on partial least squares regression (PLSR) analysis (Ollinger et al. 2002, 2013; Smith et al. 2002; Martin et al. 2008) of the full spectrum or spectral matching and continuum removal techniques (Kokaly 2001). Additional data-driven methods, such as Gaussian Processes Regression (Wang et al. 2019) and hybrid empirical/physical methods (Berger et al. 2020), have emerged as plausible alternatives to PLSR. Verrelst et al. (2019) provide a detailed overview of the data-driven, physically based, and hybrid methods for retrieving vegetation biophysical variables from HSI. Feilhauer et al. (2011) and Homolová et al. (2013) show that multiple wavelengths throughout the 400–2500 nm range have enabled nitrogen detection, indicating that nitrogen-related spectral features may vary by site, species, or phenological state.

Vegetation indices (Zarco-Tejeda et al. 1999, 2001), semi-empirical indices (e.g., Gitelson et al. 2003, 2006), and radiative transfer models (Zarco-Tejeda et al. 2004, 2001; Féret et al. 2008, 2011) have been used to characterize growth-related foliar chemicals (e.g., nitrogen and chlorophyll pigments), yet other studies demonstrate that remote sensing of canopy structure also aids quantitative retrieval of biochemical properties (e.g., Zhang et al. 2008; Hernández-Clemente et al. 2012;

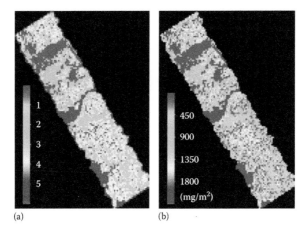

FIGURE 3.1 (a) LAI image of a black spruce forest (53.2% conifer, 16.1% deciduous species, and 21.1% grass) near Sudbury, Ontario, Canada. The image is derived based on a relationship between the simple ratio (NIR/R) and LAI ($r^2 = 0.88$). (b) Chlorophyll a + b content distribution per unit ground area. The image combines the retrieved leaf chlorophyll a + b content for the three cover types ($r^2 = 0.47$) times the LAI. The chlorophyll data was analyzed using the four-scale geometrical–optical model to characterize the effect of structure on above-canopy reflectance and inversion of the PROSPECT leaf model to estimate pigment concentration. Data from 72-band data Compact Airborne Spectrographic Imager (HSI) data averaged from 2 m pixel resolution to 20 m. Reprinted from Zhang et al. (2008).

Knyazikhin et al. 2013a, 2013b, 2013c; Ollinger et al. 2013; Townsend et al. 2013) (Figure 3.1). Asner and Warner (2003) conclude that quantitative information on gap fraction and tree structure is needed to validate or constrain remote sensing models to accurately estimate chemistry and energy exchange. Possible ways to account for structure in the retrieval of foliar chemistry include canopy radiative transfer models, LiDAR, and other methods that account for intra- and inter-canopy gaps, self-shading, and stand structure (see Section 3.5.1). Many proposed methods remain untested, including the Directional Area Scattering Factor (DASF), which is a function based on three wavelength invariant parameters: canopy interception, probability of recollision, and directional gap density (Knyazikhin et al. 2013a; Lewis and Disney 2007; Schull et al. 2007, 2011). Still other researchers have argued that the canopy architecture of a species is an integrated component of its strategy for resource capture and therefore should co-vary with chemistry (Ollinger et al. 2013; Townsend et al. 2013).

Foliar and canopy water content has also received a significant amount of attention due to its relationship with transpiration and plant water stress (Hunt et al. 2013; Ustin et al. 2012). The water absorption signal has a large effect on plant spectra, from small absorptions in the near infrared at 970 and 1240 nm, accessible through HSI data, to a large broad absorption across the entire SWIR (1300–2500 nm). Gao and Goetz (1995) developed one of the first narrowband indices for the quantification of equivalent water thickness (EWT) of vegetation. The values derived for EWT from AVIRIS data were tested against field data from the Harvard Forest, Massachusetts (Gao and Goetz 1995). HSI also offers the unique ability to differentiate between different phases of water (atmospheric water vapor and the moisture content of vegetation), for which the absorption maxima are offset by about 40–50 nm (Gao and Goetz 1990), a separation of four to ten bands in typical HSI data. This ability to quantify atmospheric water aids in the statistical modeling of the atmosphere such that water vapor signals can be removed, permitting proper estimation of the underlying liquid water stored in vegetation (Green et al. 1989). Cheng et al. (2013b) showed that even monitoring small diurnal changes in water content is possible to provide information on plant water status and whether root uptake can support full transpiration demand.

Non-pigment materials in the forest canopy range from foliar carbon constituents, such as lignin and cellulose, to dead leaves, stems, or remaining reproductive structures of flowers and fruits. The detection and quantification of these materials, sometimes referred to as dry matter or non-photosynthetic vegetation (NPV), is often used as an indicator of canopy stress and may be important for quantifying the contribution of plant litter to forest carbon pools. Particularly after foliage has lost pigments and water, the cellulose-lignin absorptions become easily detectable with HSI data through narrowband methods such as the Cellulose Absorption Index (Daughtry 2001; Daughtry et al. 2005), spectral mixture analysis (Asner and Lobell 2000; Roberts et al. 2003), chemometric approaches like PLSR (Asner et al. 2011), or radiative transfer models (Jacquemoud et al. 2009; Riaño et al. 2004a). Kokaly et al. (2007, 2009) used continuum removal combined with a spectral library to reveal a 2–3 nm shift in the cellulose-lignin absorption feature when the concentration of lignin increases, demonstrating the utility of HSI in quantifying subtle variations in canopy carbon. Numerous examples of forest NPV quantification also exist in the HSI literature (e.g., Ustin and Trabucco 2000; Roberts et al. 2004; Guerschman et al. 2009). Dry matter signatures in the HSI spectrum have been used to assess whether canopies were subjected to insect defoliation, drought stress (White et al. 2007; Fassnacht et al. 2014), or pathogen damage (Santos et al. 2010; Weingarten et al. 2022).

HSI data have significant potential for mapping forest composition at species and community levels, based largely on their biochemical attributes (Figure 3.2). Many examples have been published using various analytical approaches with airborne HSI images (e.g., Martin et al. 1998; Clark et al. 2005; Bunting and Lucas 2006; Bunting et al. 2010; Durán et al. 2019), EO-1 Hyperion satellite data (Townsend and Foster 2002), time series of Hyperion data (Kalacska et al. 2007; Somers and Asner 2013), and combinations of airborne HSI imagers and LiDAR (Dalponte et al. 2007; Jones et al. 2010; Colgan et al. 2012a; Naidoo et al. 2012; Baldeck et al. 2014). In recent years the ability to map species and detailed land cover has significantly improved (Asner 2013). It is likely that this is a consequence of improved instrument performance, especially for high-fidelity HSI data and for the adoption of a wide variety of new analytical methods, including radiative transfer models, segmentation, and object delineation, and numerous statistical methods such as ensemble classifiers,

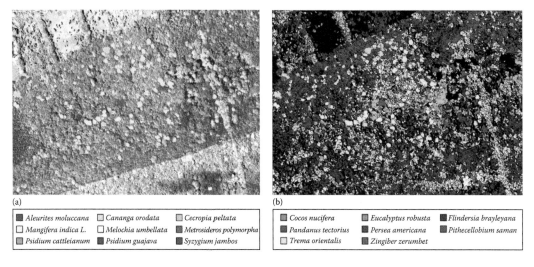

FIGURE 3.2 (Left) A false color composite image of Nanawale Forest Reserve, Hawaii Island ($R = 646$ nm; $G = 560.7$ nm; $B = 447$ nm), with colored polygons showing locations of species data from a field survey. (Right) Classification of 17 canopy species based on regularized discriminant analysis (n = 50 samples/species) using Carnegie Airborne Observatory (CAO) visible-to-near infrared imaging spectrometer data. Reprinted from Féret and Asner (2013).

discriminate analysis, support vector machines, and combined approaches. No one method has yet been shown to work universally across global land cover types with complex environment and terrain interactions. However, several general conclusions can be inferred from these and other studies: (1) the addition of SWIR bands along with VNIR bands often significantly increases the accuracy of mapping forest species; (2) species mapping is further enhanced if HSI data encompass multi-date periods that capture phenological patterns, as is consistent with improvements reported for multi-date multispectral data (e.g., Wolter et al. 2008); and (3) combining information on tree structure from LiDAR, such as canopy height, diameter, and volume, with HSI data improves results (Féret and Asner 2013).

3.3.3 Canopy Physiology

Imaging spectroscopy can be used to characterize three key physiological processes responsible for carbon uptake in forests: photochemistry, non-photochemical quenching (NPQ), and fluorescence. Solar radiation, and photosynthetically active radiation (PAR; 400–700 nm) in particular, supplies the energy that drives carbon uptake in forests. The first process, photochemistry, refers directly to the process by which the enzyme ribulose-1,5-bisphosphate carboxylase-oxygenase (RuBisCO) catalyzes RuBP to fix carbon from carbon dioxide. Within the Calvin Cycle of C_3 plants (which includes trees), photochemistry is driven by the energy supplied from light harvesting by pigment complexes. The second process, NPQ, relates directly to plant interactions with light. Plants down-regulate photosynthesis through a range of processes related to pigment concentrations to either make use of light energy or dissipate it (Demmig-Adams and Adams 2006). Photochemistry and NPQ processes can be characterized through estimation of pigment concentrations or through inference based on changes in leaf pigment pools associated with plant responses to excess light or stresses that prevent them from fully utilizing ambient light energy (Demmig-Adams and Adams 1996). Finally, all plants dissipate light energy through solar-induced fluorescence, which only occurs as a consequence of photosynthesis and has been found to scale directly to rates of photosynthetic activity (Baker 2008).

Quantifying foliar nitrogen, the key element in RuBisCO and a trait whose concentration within proteins in foliage scales directly with photosynthetic capacity (Field and Mooney 1986; Evans 1989; Reich et al. 1997), provides a measure of the functioning of forest canopies (as described earlier). This functioning includes the capacity for carbon uptake, but photosynthetic down-regulation limits carbon uptake under adverse environmental conditions. The most widely used models of photosynthesis employ the Farquhar model (Farquhar et al. 1980; Farquhar and von Caemmerer 1982), in which the potential photosynthetic performance of a leaf is characterized using two parameters, the maximum rate of carboxylation (V_{cmax}) governed by RuBisCO activity, and the maximum electron transport rate (ETR) (J_{max} is the maximum rate of ETR; Farqhuar and von Caemmerer 1982). Together, these limit the maximum rate of photosynthesis (A_{max}). V_{cmax} is strongly related to N concentration and LMA, i.e., the investment by a plant in light harvesting relative to construction and maintenance (Poorter et al. 2009). ETR and J_{max} are more closely related to the processes set in motion by light harvesting in PSI and PSII (PS = photosystem), necessary for the synthesis of adenosine triphosphate (ATP) to drive cellular reactions. Because the Calvin Cycle depends on ATP availability to sustain the regeneration of RuBP (which in turn permits carboxylation), photosynthetic capacity is limited by J_{max}. Therefore, optical properties of foliage related to light harvesting may also facilitate mapping J_{max} from HSI. It should be noted that all photosynthetic parameters of vegetation are sensitive to temperature and moisture, so any remotely sensed estimate of such parameters will be specific to the ambient conditions at the time of measurement (Serbin et al. 2012). HSI have also been used as part of multi-sensor approaches to characterize net ecosystem photosynthesis (e.g., Rahman et al. 2001; Asner et al. 2004a; Thomas et al. 2006, 2009).

Doughty et al. (2011) successfully related leaf-level spectroscopic measurements of sunlit leaves to A_{max} but had less success with the other parameters. Variations in V_{cmax} and J_{max} related to temperature were measured in cultivated aspen and cottonwood leaves and accurately predicted similar relationships in plantation trees (Serbin et al. 2012). More recently, the ability to remotely sense V_{cmax} across multiple species has been established (Serbin et al. 2015; Dechant et al. 2017; Wu et al. 2019), as well as the temperature sensitivity of V_{cmax}, the parameter usually referred to as E_v (Serbin et al. 2015). The ability to map V_{cmax} and J_{max} from imaging spectroscopy is most likely a consequence of the ability to infer these properties from traits that are directly detectable based on known or hypothesized absorption features (e.g., N, LMA, water; see Kattge et al. 2009; Cho et al. 2010) and the coordination of these traits with canopy structure (Ollinger et al. 2013). These studies show promise for developing remote sensing methods to map the properties used by modelers to characterize forest physiological function.

Efforts to map parameters directly associated with photochemistry are an area of continuing development in imaging spectroscopy. The discipline of physiological remote sensing using HSI has its roots in efforts to characterize NPQ, and how NPQ relates to photosynthetic rates and capacity. This work stems from the development of the Photochemical Reflectance Index (Gamon et al. 1992; Peñuelas et al. 1995). While typically associated with the de-epoxidation of xanthophylls for photosynthetic down-regulation during NPQ (Bilger and Björkman 1990; Demmig-Adams and Adams 1996), the PRI more generally correlates with total pigment pools and their variation with environmental context (Gamon and Bond 2013). As such, the PRI has been shown to be an indicator of photosynthetic rates and light use efficiency (Gamon and Surfus 1999). Accounting for species composition, environmental variability, and seasonal responses, the PRI is often correlated with the carotenoid to chlorophyll ratio (r^2 = 0.50–0.80), a property linked to photosynthesis and light harvesting (Garbulsky et al. 2011) (Figure 3.3). In addition, Stylinski et al. (2000) also showed close relationships between the PRI and xanthophyll cycle pigments and modeled electron transport capacity (J_{max}) in leaves of pubescent oak (*Quercus pubescens*). Kefauver et al. (2013) showed strong relationships between PRI and physiological damage to forests by ozone. A limitation of the PRI has been its species-level sensitivity, i.e., relationships between the PRI and photosynthesis are species-dependent (Guo and Trotter 2004; Filella et al. 2009; Ripullone et al. 2011) and also depend on illumination conditions (Gamon et al. 2023). However, Hilker et al. (2008) have shown that PRI data may facilitate retrieval of plant photosynthetic efficiency independent of species composition.

The key physiological processes responsible for productivity of forests can also be addressed by remote sensing through the estimation of light absorption by canopies and its presumed linkage to light harvesting and use in photosynthesis. Under non-stressed conditions, net primary production is linearly related to the absorbed photosynthetically active radiation (APAR; Montieth 1977). This relationship is modulated by light use efficiency (LUE). Traditionally, APAR has been successfully calculated from vegetation indices derived from spectral sensors of many varieties (e.g., Field et al. 1995; Sellers et al. 1996). The detection of forest light use efficiency using PRI, and thus potential carbon uptake, has been demonstrated in numerous systems, including boreal (Nichol et al. 2000) and conifer forests (Middleton et al. 2009; Atherton et al. 2013), but the utilization of remotely estimated APAR by the canopy for photosynthesis remains a more difficult task. The most common approach to assessing LUE using HSI has been through narrowband indices such as the PRI, which uses the reflectance at 570 and 531 nm (i.e., Gamon et al. 1997; Gamon et al. 2023), but future developments in retrieving Farquhar parameters (V_{cmax}, J_{max}) and solar-induced fluorescence are likely to provide more robust estimates of key drivers of physiological processes. Ultimately, linkages across methods, e.g., estimating LUE using derivations biochemistry (%N) and LAI, may provide a hybrid approach to best map factors important to NPP (Green et al. 2003).

Chlorophyll fluorescence provides another means of estimating photosynthetic performance and LUE from HSI data (Meroni et al. 2009; Mohammed et al. 2019). Numerous studies since the early 2000s have demonstrated the capacity of measurements of solar-induced fluorescence (SIF) to accurately characterize seasonal patterns of carbon uptake (Guanter et al. 2007; Frankenburg et al. 2011; Joiner et al. 2011; Magney et al. 2019; Campbell et al. 2019; Chang et al. 2020). Under natural conditions, fluorescence and photosynthesis are positively correlated. Energy absorbed in

Forest Biophysical and Biochemical Properties

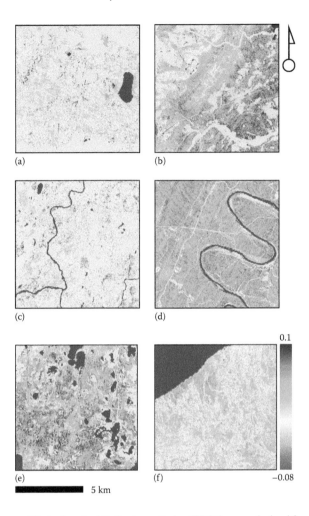

FIGURE 3.3   Midsummer Photochemical Reflectance Index (PRI) images derived from 2009 AVIRIS imagery of: (A) oak/pine forests in Baraboo/Devil's Lake, Wisconsin; (B) oak and tulip poplar forests in Fernow Experimental Forest, West Virginia; (C) northern hardwood and conifer forests in Flambeau River State Forest, Wisconsin; (D) xeric oak forests in Green Ridge State Forest, Maryland; (E) northern hardwood and sub-boreal conifers in Ottawa National Forest, Michigan; (F) hemlock, white pine, and deciduous hardwoods in the Porcupine Mountains, Michigan. Lower values indicate areas of greater vegetation stress. These images illustrate significant variability in forest physiological status across landscapes.

the photosystems is re-radiated at longer wavelengths than those absorbed, adding a subtle signal to reflected solar radiation, most notably with peaks around 685 and 740 nm. Measurements of SIF require narrowband data at specific wavelengths in the NIR in which the vegetation fluorescence signal in retrieved reflectance (about 2%) can be distinguished from NIR albedo (> 40%) (Berry et al. 2013). Most efforts to date have focused on retrievals of SIF in narrow wavebands (preferably < 0.3 nm) ± 20 nm around the solar Fraunhofer lines (wavelengths where there is no incoming solar energy, ~739 nm) or O_2–A band at 760 nm. Generally correlated with the PRI (Zarco-Tejeda et al. 2009; Cheng et al. 2013a; Campbell et al. 2019; Magney et al. 2019), fluorescence has also been measured at field sites differing in soil salinity and estimated spatially from airborne HSI data using the PRI index (531 and 570 nm) (Naumann et al. 2008). Zarco-Tejeda et al. (2009) estimated fluorescence from infilling of the O_2–A|bands at 757.5 nm and 760.5 nm measured in 1-nm wavelength bands, which minimized confounding effects from variance in chlorophyll and LAI. More recently

Zarco-Tejeda et al. (2013) used narrowband spectral indices and fluorescence infilling at 750, 762, and 780 nm, revealing that seasonal spectroscopic trends tracked changes in carbon fluxes. More recently Chang et al. (2020) and Magney et al. (2017, 2019) found strong slope-dependent correlations between reflectance at all wavelengths between 670 and 850 nm and fluorescence. HSI observations continue to pave additional avenues to insight on plant physiological processes.

3.4 LIDAR REMOTE SENSING OF FORESTS

Whereas HSI provides estimates of the chemical, physiological, and plant compositional properties of forests, LiDAR probes the structural and architectural traits of vegetation as well as the terrain below the canopy (Table 3.3). A large number of synthesis papers have been written on the use of LiDAR for studies of ecosystem structure (e.g., Dubayah and Drake 2000; Lefsky et al. 2002; Lim et al. 2003; Vierling et al. 2008; Wulder et al. 2012; Beland et al. 2019), including in other chapters of this book (e.g., Chapter 17 of Vol. II). Here we only briefly highlight the various uses of LiDAR in the context of forest structure, architecture, and biomass; the reader should also read Chapter 17 for further details.

3.4.1 Canopy Structure and Biomass

The height of a forest canopy is a fundamental characteristic that both discrete and waveform LiDAR sensors are capable of describing (Figure 3.4). Even discrete-return datasets that contain only the first and last return from the laser pulse will allow for the calculation of this parameter, after a ground elevation model has been generated from LiDAR data (Lim et al. 2003). While canopy height alone does not provide extensive information on forest structure, it is a parameter related to tree diameters (Feldpausch et al. 2012), and thus to aboveground biomass, as discussed next.

LiDAR can also be used to determine the vertical profile of canopy tissues, including foliar and some woody structures (Figure 3.5). Waveform LiDAR instruments collect the full shape of the

TABLE 3.3
Forest Structural Properties Estimated from Light Detection and Ranging (LiDAR), along with an Estimate of Scientific Maturity, and Example References. Maturity Is a Metric of Relative Accuracy as Depicted in the Literature, with One Checkmark Indicating Low Maturity, and Three Checkmarks Indicating High Maturity

Vegetation Property	Maturity Level	Example References
Total canopy height	✓✓✓	Dubayah and Drake (2000), Ni-Meister et al. (2001), Drake et al. (2002), Lim et al. (2003), Simard et al. (2011)
Mean canopy profile height	✓✓✓	Lefsky et al. (1999, 2002, 2005), Zhao et al. (2011), Ferraz et al. (2016), Shao et al. (2019), Lang et al. (2023)
Aboveground biomass	✓✓✓	Nelson (1988), Lefsky et al. (1999, 2002, 2005), Popescu et al. (2003), Næsset and Gobakken (2008), van Aardt et al. (2008), Asner et al. (2012c, 2014), Wulder et al. (2012), Ferraz et al. (2016), Asner et al. (2018), Shao et al. (2019)
Leaf area density	✓✓	Sun and Ranson (2000), Lovell et al. (2003), Riaño et al. (2004b), Morsdorf et al. (2006), Richardson et al. (2009), Soldberg et al. (2009), Vaughn et al. (2013), Oshio et al. (2015), Ferraz et al. (2016), Kamoske et al. (2019), Shao et al. (2019)
Understory presence	✓✓	Zimble et al. (2003), Asner et al. (2008), Wing et al. (2012), Campbell et al. (2018), Crespo-Peremarch et al. (2018)

Forest Biophysical and Biochemical Properties

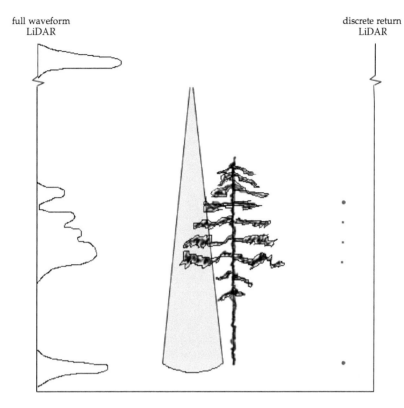

FIGURE 3.4 Illustration of waveform and discrete return measurements of a tree. While both provide information on the vertical structure of canopies, discrete return sampling records the returning laser pulse at specified peaks (e.g., first and last pulse) of the return wave, whereas waveform sampling collects the full shape of the returning pulse. Reprinted from Lim et al. (2003).

returning laser pulse, allowing for detailed information on the structure of the canopy (Blair and Hofton 1999; Dubayah and Drake 2000; Ni-Meister et al. 2001; Yao et al. 2012). If detailed canopy structure is of interest but only discrete-return LiDAR data are available, it is possible to use these data to generate a pseudo-waveform. This method aggregates discrete returns into bins over spatial extents that incorporate multiple laser spots in order to gain an aggregated understanding of the vertical vegetation profile in the absence of waveform data for each laser pulse (Muss et al. 2011). Vertical profiles are indicative of canopy density and volume, vertical distribution, and presence of undergrowth, all of which can provide information on the three-dimensional structure and habitat of forests (Parker 1995; Lefsky et al. 1999; Clark and Clark 2000; Weishampel et al. 2000; Drake et al. 2002; Asner et al. 2008; Vierling et al. 2008; Coops et al. 2021; Zhou and Li 2023).

One of the most widespread uses for LiDAR-derived canopy information is in the estimation of aboveground biomass, also known as aboveground carbon density (ACD). Such approaches have been applied in numerous studies of conifer, broadleaf temperate, and tropical forest ecosystems (Nelson 1988; Lefsky et al. 1999, 2002, 2005; Popescu et al. 2003; Næsset and Gobakken 2008; van Aardt et al. 2008; Asner et al. 2012c; Wulder et al. 2012). The mean canopy profile height (MCH) has been used as the canopy structural metric, which relates the LiDAR vertical structure data to ACD (Lefsky et al. 2002; Asner et al. 2009). However, recent studies have indicated that variations in sensor characteristics and settings can cause significant differences in the MCH metric between data acquisitions (Næsset 2009), strongly indicating that top-of-canopy height is a more reliable method for estimating ACD of tropical forests (Asner and Mascaro 2014). The use of LiDAR data to produce estimates of ACD that closely match plot-level estimates allows for the mapping and

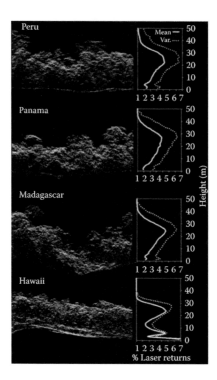

FIGURE 3.5 LiDAR cross-sectional views of four mature tropical forests in the Peruvian Amazon, Panamanian Neotropics, southeastern Madagascar, and Hawaii depict three-dimensional forest structures along a 100 m long x 20 m wide transect. Right-hand panels show mean and spatial variance of LiDAR vertical canopy profiles for all returns in a 1 km² area centered on each cross-section. Vertical canopy profiles are generally consistent across the four study sites, yet the Hawaiian forest contains the most pronounced ground-cover, understory, and canopy layers. Reprinted from Asner et al. (2012b).

monitoring of aboveground carbon stocks at landscape scales and, with the further development of spaceborne LiDAR, potentially regional/biome scales.

3.4.2 Light Penetration

Canopy gaps, or openings in forest canopies, influence population dynamics of forest trees by affecting forest structure, regeneration dynamics, and species composition (Brokaw 1985; Denslow 1987). Canopy gaps occur at scales ranging from single branches to multiple treefalls, and result from disturbances caused by natural tree life cycles (Asner 2013), human processes such as logging (e.g., Nepstad et al. 1999; Asner et al. 2004b; Curran et al. 2004), and environmental factors such large-scale blowdowns (Chambers et al. 2013). Recently, airborne LiDAR data from a number of tropical forests have enabled the measurement of millions of canopy gaps over large spatial scales, both as single measurements and with repeat collections, improving the understanding of static and dynamic gaps, respectively (e.g., Magnussen et al. 2002; Kellner et al. 2009; Udayalakshmi et al. 2011; Armston et al. 2013).

Static canopy-gap size-frequency distributions (known as λ) are strikingly similar across a wide range of tropical forest types on differing geologic substrates, and within differing disturbance regimes. This collective evidence suggests consistent turnover rates and similar mechanisms of gap-formation across tropical forests (Kellner and Asner 2009; Asner et al. 2013, 2014). Deviations from this stable range of observed λ values potentially provide another metric for detecting and mapping disturbance. In a recent study, repeat LiDAR collections permitted the quantification of positive height changes in a forest canopy in Hawaii and illustrated how size and the proximity to other canopies influenced the outcome of competition for space within this forest (Kellner and Asner 2014).

Forest Biophysical and Biochemical Properties

3.5 INTEGRATING HSI AND LIDAR

Since about 2005, HSI and LiDAR observations have been integrated using two approaches. One method involves the acquisition of HSI and LiDAR data from separate platforms, such as from different aircraft, followed by modeling and analysis steps to fuse the resulting datasets (e.g., Mundt et al. 2006; Anderson et al. 2008; Jones et al. 2010). This remains the most common approach, and following acquisition the data must be digitally co-aligned using techniques such as image pixel-based coregistration. These efforts usually yield an integrated data "cube" with an average misalignment of one pixel or so, although the scanning and/or array patterns of the HSI and LiDAR data may yield much higher coalignment errors.

A second approach to HSI and LiDAR data integration involves the co-mounting of instruments on the same platform, whether on board aircraft or an unmanned aerial vehicle (UAV) (Asner et al. 2007). Integration steps range from co-locating the instruments on the same mounting plate on board the aircraft or UAV, to precise time registration of each measurement, to final data fusion using ray tracing models for each instrument (Asner et al. 2012a). Each of these steps is key to producing a highly integrated dataset, reducing coalignment issues such that the data can be treated as one information vector per unit ground sample (e.g., one pixel). The onboard and post-flight fusion of HSI and LiDAR data developed and deployed by the Global Airborne Observatory (GAO; formerly the Carnegie Airborne Observatory) has been replicated and is currently being used by the US NEON's Airborne Operational Platform (AOP) program (http://www.neoninc.org/science/aop).

3.5.1 BENEFITS OF DATA FUSION

The benefits of HSI and LiDAR data fusion include increased data dimensionality, constraints on the interpretation of one portion of the dataset using another portion, and filtering of data to specific observation conditions or specifications. The dimensionality of, or degrees of freedom within, a fused dataset increases with the integration of complementary or orthogonal observations such as chemical or physiological metrics from HSI and structural or architectural measures from LiDAR. A highly demonstrative example can be taken from two integrated HSI-LiDAR datasets collected with the GAO Airborne Taxonomic Mapping System (Figure 3.6). One dataset was collected over a portion

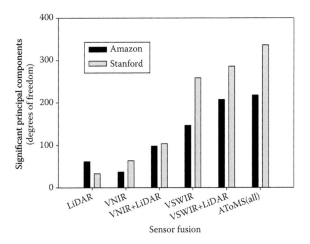

FIGURE 3.6 Integration of HSI and LiDAR sensor hardware, and data streams, provides a uniquely powerful way to greatly increase the inherent dimensionality of the data collected over forested and other ecosystems. For two example 200-ha areas (Stanford University and a lowland Amazonian forest), individual LiDAR and HSI sensors provide highly dimensional data as assessed with Principal Components Analysis. The dimensionality of the data increases when data are analyzed simultaneously. Here, VNIR is a visible-to-near infrared HSI and VSWIR is a visible-to-shortwave infrared HSI. All sensors combined are referred to as the Airborne Taxonomic Mapping Systems, or AToMS, on board the Global Airborne Observatory. Reprinted from Asner et al. (2012a).

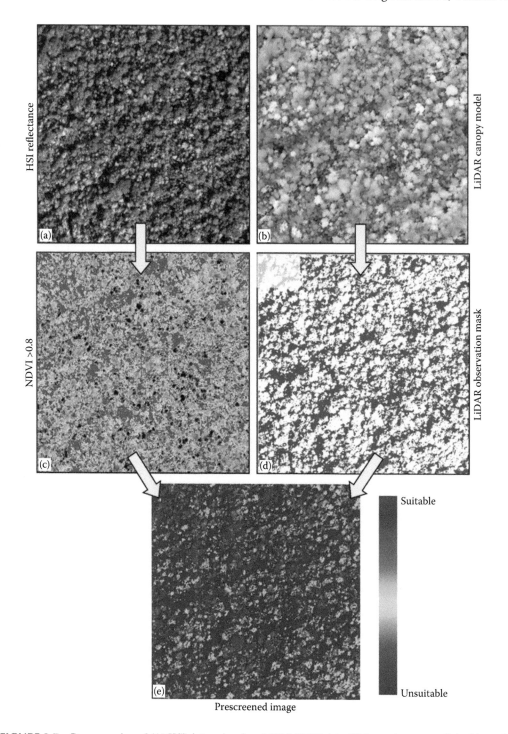

FIGURE 3.7 Pre-screening of (A) HSI data using fused (B) LiDAR data. This can be accomplished in various ways, and an example is shown here. (C) A minimum NDVI threshold of 0.8 ensures sufficient foliar cover in the analysis. (D) Combining LiDAR and solar-viewing geometry, a filtering mask is generated to remove pixels in shade or of ground and water surfaces. (E) The resulting suitability image provides an indication of pixels that can be used for biophysical, biochemical, and/or physiological analysis.

of Stanford University in 2011, and the other taken over a remote Amazonian rainforest in the same year. In the Stanford case, the LiDAR data alone contained about 25 degrees of freedom for a 200-ha area comprising buildings with varying architecture, vegetation ranging from grasses to trees, roads and pathways, and other built surfaces. Here, degrees of freedom are quantitatively assessed using Principal Component Analysis, so each degree is orthogonal to or unique from the others (Asner et al. 2012a). A 72-band visible to near-infrared (VNIR) image of the same Stanford scene, taken from the same aircraft, contains about 50 degrees of freedom. Combined, the VNIR HSI and LiDAR provide about 100 degrees of freedom. A visible to shortwave-infrared (VSWIR) imaging spectrometer on board the same aircraft provides about 260 degrees of freedom in the Stanford case. In conjunction, the LiDAR and VNIR and VSWIR HSI offer more than 330 orthogonally aligned sources of information. In the Amazon forest case, data fusion yields similar increases in data dimensionality, more than doubling the information content by sensor fusion over that which can be achieved by any one sensor.

A second powerful use of combined HSI and LiDAR data involves constraint of interpretation and/or filtering of one data stream relative to the other. Looking down upon a forest canopy, one observes strong variation in bright and dark portions of the canopy, as well as gaps and spectrally inconsistent observation conditions (Figure 3.7). As a result, reflectance analysis of forests is often an underdetermined problem involving variation in three-dimensional architecture, leaf layering (LAI), and foliar biochemical constituents. This variation in illumination conditions occurs between pixels in high-resolution HSI data, and within pixels in lower-resolution HSI data. One of many possible ways to constrain observation conditions for improved HSI-based analysis of forest canopy traits is to use the LiDAR (Asner and Martin 2008; Dalponte et al. 2008; Colgan et al. 2012b). For example, LiDAR maps of top-of-canopy structure can be used to precisely model sun and viewing geometry on the canopy surface in each pixel. Combined with simple filtering of the HSI data based on the normalized difference vegetation index (NDVI) or other narrowband indices, an HSI image can be partitioned into regions most suitable for a particular type of analysis. Biochemical analyses are particularly sensitive to this filtering process, and much higher performances in biochemical retrievals can be achieved based on combined HSI-LiDAR filtering (Asner and Martin 2008). The integration of HSI and LiDAR data has much promise for future research, such as in the full three-dimensional analysis and modeling of canopy structural and functional traits illustrated by the use of LiDAR to model within-canopy variation in LMA (Chlus et al. 2020). These approaches will become more common with the rise of integrated data fusion systems.

3.6 CONCLUSIONS

Hyperspectral imaging and LiDAR mapping provide independent and highly complementary data on forest canopies and whole ecosystems. Here we summarized sources of HSI and LiDAR data, their general uses in determining forest structural and functional properties, and the potential value of collecting and analyzing HSI and LiDAR together via hardware integration and data fusion. Much of the science of HSI and LiDAR analysis of forests will remain in the airborne domain until orbital instrumentation is deployed and made available to the scientific research and applications communities. In light of the myriad studies found throughout the remote sensing, forest science, and conservation research literature, it is clear that the time is right for a rapid expansion of HSI and LiDAR data collection and sharing efforts worldwide.

REFERENCES

Abshire, J. B., X. Sun, H. Riris, J. M. Sirota, J. F. McGarry, S. Palm, D. Yi, et al. 2005. Geoscience laser altimeter system (GLAS) on the ICESat Mission: On-orbit measurement performance. *Geophys Research Lett* 32:L21S02.

Anderson, J. E., L. C. Plourde, M. E. Martin, B. H. Braswell, M.-L. Smith, R. O. Dubayah, M. A. Hofton, et al. 2008. Integrating waveform lidar with hyperspectral imagery for inventory of a northern temperate forest. *Remote Sens Environ* 112:1856–1870.

Armston, J., M. Disney, P. Lewis, P. Scarth, S. Phinn, R. Lucas, P. Bunting, et al. 2013. Direct retrieval of canopy gap probability using airborne waveform lidar. *Remote Sens Environ* 134:24–38.

Asner, G. P. 1998. Biophysical and biochemical sources of variability in canopy reflectance. *Remote Sens Environ* 64:234–253.

Asner, G. P. 2013. Geography of forest disturbance. *Proc Nat Acad Sci* 110:3711–3712.

Asner, G. P., C. Anderson, R. E. Martin, D. E. Knapp, R. Tupayachi, F. Sinca, and Y. Malhi. 2014. Landscape-scale changes in forest structure and functional traits along an Andes-to-Amazon elevation gradient. *Biogeosci* 11:843–856.

Asner, G. P., P. G. Brodrick, C. B. Anderson, N. Vaughn, D. E. Knapp, and R. E. Martin. 2016. Progressive forest canopy water loss during the 2012–2015 California drought. *Proc Nat Acad Sci* 113 (2):E249–E255.

Asner, G. P., P. G. Brodrick, C. B. Philipson, N. R. Vaughn, R. E. Martin, D. E. Knapp, J. Heckler, L. J. Evans, T. Jucker, and B. Goossens. 2018. Mapped aboveground carbon stocks to advance forest conservation and recovery in Malaysian Borneo. *Biol Cons* 217:289–310.

Asner, G. P., J. K. Clark, J. Mascaro, G. A. Galindo García, K. Chadwick, D. A. Navarrete Encinales, G. Paez-Acosta, E. Cabrera Montenegro, T. Kennedy-Bowdoin, Á. Duque, A. Balaji, P. von Hildebrand, L. Maatoug, J. F. Phillips Bernal, A. P. Yepes Quintero, D. E. Knapp, M. C. García Dávila, J. Jacobson, and M. F. Ordóñez. 2012c. High-resolution mapping of forest carbon stocks in the Colombian Amazon. *Biogeosci* 9:2683–2696.

Asner, G. P., R. F. Hughes, T. A. Varga, D. E. Knapp, and T. Kennedy-Bowdoin. 2009. Environmental and biotic controls over aboveground biomass throughout a tropical rain forest. *Ecosystems* 12:261–278.

Asner, G. P., R. F. Hughes, P. M. Vitousek, D. E. Knapp, T. Kennedy-Bowdoin, J. Boardman, R. E. Martin, M. Eastwood, and R. O. Green. 2008. Invasive plants alter 3-D structure of rainforests. *Proc Nat Acad Sci* 105:4519–4523.

Asner, G. P., M. Keller, R. Pereira, J. C. Zweede, and J. N. Silva. 2004b. Canopy damage and recovery after selective logging in Amazonia: Field and satellite studies. *Ecol Appl* 14:S280–S298.

Asner, G. P., J. R. Kellner, T. Kennedy-Bowdoin, D. E. Knapp, C. Anderson, et al. 2013. Forest canopy gap distributions in the southern Peruvian amazon. *PLoS ONE* 8:e60875.

Asner, G. P., D. E. Knapp, J. Boardman, R. O. Green, T. Kennedy-Bowdoin, M. Eastwood, R. E. Martin, et al. 2012a. Carnegie airborne observatory-2: Increasing science data dimensionality via high-fidelity multi-sensor fusion. *Remote Sens Environ* 124:454–465.

Asner, G. P., D. E. Knapp, T. Kennedy-Bowdoin, M. O. Jones, R. E. Martin, J. Boardman and C. B. Field. 2007. Carnegie Airborne Observatory: In flight fusion of hyperspectral imaging and waveform light detection and ranging (LiDAR) for three-dimensional studies of ecosystems. *Journal of Applied Remote Sensing* 1:013536-36-21.

Asner, G. P., and D. B. Lobell. 2000. A biogeophysical approach for automated swir unmixing of soils and vegetation. *Remote Sens Environ* 74:99–112.

Asner, G. P., and R. E. Martin. 2008. Spectral and chemical analysis of tropical forests: Scaling from leaf to canopy levels. *Remote Sens Environ* 112:3958–3970.

Asner, G. P., R. E. Martin, C. B. Anderson, and D. E. Knapp. 2015a. Quantifying forest canopy traits: Imaging spectroscopy versus field survey. *Remote Sens Environ* 158:15–27.

Asner, G. P., R. E. Martin, R. Tupayachi, C. B. Anderson, F. Sinca, L. Carranza-Jimenez, and P. Martinez. 2014. Amazonian functional diversity from forest canopy chemical assembly. *Proc Nat Acad Sci* 111 (15):5604–5609.

Asner, G. P., R. E. Martin, R. Tupayachi, R. Emerson, P. Martinez, F. Sinca, G. V. N. Powell, et al. 2011. Taxonomy and remote sensing of leaf mass per area (LMA) in humid tropical forests. *Ecol Appl* 21:85–98.

Asner, G. P., and J. Mascaro. 2014. Mapping tropical forest carbon: Calibrating plot estimates to a simple lidar metric. *Remote Sens Environ* 140:614–624.

Asner, G. P., J. Mascaro, H. C. Muller-Landau, G. Vieilledent, R. Vaudry, M. Rasamoelina, J. S. Hall, et al. 2012b. A universal airborne lidar approach for tropical forest carbon mapping. *Oecologia* 168:1147–1160.

Asner, G. P., D. Nepstad, G. Cardinot, and D. Ray. 2004a. Drought stress and carbon uptake in an amazon forest measured with spaceborne imaging spectroscopy. *Proc Nat Acad Sci* 101:6039–6044.

Asner, G. P., and P. M. Vitousek. 2005. Remote analysis of biological invasion and biogeochemical change. *Proc Nat Acad Sci* 102:4383–4386.

Asner, G. P., and A. S. Warner. 2003. Canopy shadow in ikonos satellite observations of tropical forests and savannas. *Remote Sens Environ* 87:521–533.

Atherton, J. M., C. J. Nichol, M. Mencuccini, and K. Simpson. 2013. The utility of optical remote sensing for characterizing changes in the photosynthetic efficiency of Norway maple saplings following transplantation. *Int J Remote Sens* 34:655–667.

Baker, N. R. 2008. Chlorophyll fluorescence: A probe of photosynthesis in vivo. *Annual Rev Plant Biol* 59:89–113.

Baldeck, C. A., M. S. Colgan, J.-B. Féret, S. R. Levick, R. E. Martin, and G. P. Asner. 2014. Landscape-scale variation in plant community composition of an African savanna from airborne species mapping. *Ecol Appl* 24:84–93.

Barnsley, M. J., J. J. Settle, M. A. Cutter, D. R. Lobb, and F. Teston. 2004. The proba/chris mission: A low-cost smallsat for hyperspectral multiangle observations of the earth surface and atmosphere. *IEEE Trans Geosci Remote Sens* 42:1512–1520.

Basedow, R. W., D. C. Carmer, and M. E. Anderson. 1995, June. HYDICE system: Implementation and performance. Vol. 2480, Pp. 258–267 in *Imaging spectrometry*. SPIE.

Beland, M., G. Parker, B. Sparrow, D. Harding, L. Chasmer, S. Phinn, A. Antonarakis, and A. Strahler. 2019. On promoting the use of lidar systems in forest ecosystem research. *Forest Ecol Manage* 450:117484.

Berger, K., J. Verrelst, J.-B. Feret, T. Hank, M. Wocher, W. Mauser, and G. Camps-Valls. 2020. Retrieval of aboveground crop nitrogen content with a hybrid machine learning method. *Int J Appl Earth Obs Geoinfo* 92:102174.

Berry, J. A., C. Frankenberg, and P. Wennberg. 2013. New methods for measurements of photosynthesis from space. *Keck Institute for Space Studies* 1:72.

Bilger, W., and O. Björkman. 1990. Role of the xantophyll cycle in photoprotection elucidated by measurements of light-induced absorbance changes, fluorescence and photosynthesis in leaves of *Hedera canariensis*. *Photosyn Res* 25:173–185.

Blackburn, G. A. 2007. Hyperspectral remote sensing of plant pigments. *J Experim Botany* 58:855–867.

Blair, J. B., and M. A. Hofton. 1999. Modeling laser altimeter return waveforms over complex vegetation using high-resolution elevation data. *Geophys Res Lett* 26:2509–2512.

Blair, J. B., D. L. Rabine, and M. A. Hofton. 1999. The laser vegetation imaging sensor: A medium-altitude, digitisation-only, airborne laser altimeter for mapping vegetation and topography. *ISPRS J Photogramm Rem Sens* 54:115–122.

Brokaw, N. V. L. 1985. Treefalls, regrowth, and community structure in tropical forests. Pp. 53–69 in *The ecology of natural disturbance and patch dynamics*, edited by S. T. A. Pickett and P. S. White. New York: Academic Press.

Bunting, P., and R. Lucas. 2006. The delineation of tree crowns in Australian mixed species forests using hyperspectral compact airborne spectrographic imager (CASI) data. *Remote Sens Environ* 101:230–248.

Bunting, P., R. Lucas, K. Jones, and A. R. Bean. 2010. Characterisation and mapping of forest communities by clustering individual tree crowns. *Remote Sens Environ* 114:231–245.

Campbell, M. J., P. E. Dennison, A. T. Hudak, L. M. Parham, and B. W. Butler. 2018. Quantifying understory vegetation density using small-footprint airborne lidar. *Remote Sens Environ* 215:330–342.

Campbell, P. K., K. F. Huemmrich, E. M. Middleton, L. A. Ward, T. Julitta, C. S. Daughtry, A. Burkart, A. L. Russ, and W. P. Kustas. 2019. Diurnal and seasonal variations in chlorophyll fluorescence associated with photosynthesis at leaf and canopy scales. *Remote Sens* 11 (5):488.

Cavender-Bares, J., J. A. Gamon, and P. A. Townsend. 2020. *Remote sensing of plant biodiversity*. New York: Springer Nature.

Cawse-Nicholson, K., P. A. Townsend, D. Schimel, A. M. Assiri, P. L. Blake, M. F. Buongiorno, P. Campbell, N. Carmon, K. A. Casey, R. E. Correa-Pabón, et al. 2021. NASA's surface biology and geology designated observable: A perspective on surface imaging algorithms. *Remote Sens Environ* 257:112349.

Chadwick, K. D., and G. P. Asner. 2016. Organismic-scale remote sensing of canopy foliar traits in lowland tropical forests. *Remote Sens* 8 (2):87.

Chambers, J. Q., R. I. Negron-Juarez, D. M. Marra, A. Di Vittorio, J. Tews, et al. 2013. The steady-state mosaic of disturbance and succession across an old growth central amazon forest landscape. *Proc Nat Acad Sci* 110:3949–3954.

Chang, C. Y., L. Guanter, C. Frankenberg, P. Köhler, L. Gu, T. S. Magney, K. Grossmann, and Y. Sun. 2020. Systematic assessment of retrieval methods for canopy far-red solar-induced chlorophyll fluorescence using high-frequency automated field spectroscopy. *J Geophys Res Biogeosci* 125 (7):e2019JG005533.

Chavana-Bryant C., Y. Malhi, J. Wu, G. P. Asner, A. Anastasiou, B. J. Enquist, E. G. Cosio Caravasi, C. E. Doughty, S. R. Saleska, R. E. Martin, and F. F. Gerard. 2017. Leaf aging of Amazonian canopy trees as revealed by spectral and physiochemical measurements. *New Phytol* 214:1049–1063.

Cheng, T., D. Riaño, A. Koltunov, M. L. Whiting, S. L. Ustin, and J. Rodriguez. 2013b. Detection of diurnal variation in orchard canopy water content using modis/aster airborne simulator (master) data. *Remote Sens Environ* 132:1–12.

Cheng, T., B. Rivard, A. G. Sánchez-Azofeifa, J. B. Féret, S. Jacquemoud, and S. L. Ustin. 2014. Deriving leaf mass per area (LMA) from foliar reflectance across a variety of plant species using continuous wavelet analysis. *ISPRS J Photogramm Remote Sens* 87:28–38.

Cheng, Y.-B., E. M. Middleton, Q. Zhang, K. F. Huemmrich, P. K. E. Campbell, L. A. Corp, B. D. Cook, et al. 2013a. Integrating solar induced fluorescence and the photochemical reflectance index for estimating gross primary production in a cornfield. *Remote Sens* 5:6857–6879.

Chlus, A., E. L. Kruger, and P. A. Townsend. 2020. Mapping three-dimensional variation in leaf mass per area with imaging spectroscopy and lidar in a temperate broadleaf forest. *Remote Sens Environ* 250:112043.

Cho, M. A., J. A. van Aardt, R. Main, and B. Majeke. 2010. Evaluating variations of physiology-based hyperspectral features along a soil water gradient in a eucalyptus grandis plantation. *Int J Remote Sens* 31:3143–3159.

Clark, D. B., and D. A. Clark. 2000. Landscape-scale variation in forest structure and biomass in a tropical rain forest. *For Ecol Manage* 137:185–198.

Clark, M. L., D. A. Roberts, and D. B. Clark. 2005. Hyperspectral discrimination of tropical rain forest tree species at leaf to crown scales. *Remote Sens Environ* 96:375–398.

Cocks, T., R. Jenssen, A. Stewart, I. Wilson, and T. Shields. 1998, October. The HyMapTM airborne hyperspectral sensor: The system, calibration and performance. Pp. 37–42 in *Proceedings of the 1st EARSeL workshop on imaging spectroscopy*, edited by M. Schaepman, D. Schläpfer, and K. I. Itten, 6–8 October 1998, Paris, Zurich: EARSeL.

Cogliati, S., F. Sarti, L. Chiarantini, M. Cosi, R. Lorusso, et al. 2021. The PRISMA imaging spectroscopy mission: Overview and first performance. *Remote Sens Environ* 262:112499.

Colgan, M. S., G. P. Asner, S. R. Levick, R. E. Martin, and O. A. Chadwick. 2012b. Topo-edaphic controls over woody plant biomass in South African savannas. *Biogeosci* 9:1809–1821.

Colgan, M. S., C. Baldeck, J.-B. Féret, and G. Asner. 2012a. Mapping savanna tree species at ecosystem scales using support vector machine classification and BRDF correction on airborne hyperspectral and lidar data. *Remote Sens* 4:3462–3480.

Coops, N. C., P. Tompalski, T. R. Goodbody, M. Queinnec, J. E. Luther, D. K. Bolton, J. C. White, M. A. Wulder, O. R. van Lier, and T. Hermosilla. 2021. Modelling lidar-derived estimates of forest attributes over space and time: A review of approaches and future trends. *Remote Sens Environ* 260:112477.

Crespo-Peremarch, P., T. Piotyr, N. C. Coops, and R. L. Ángel. 2018. Characterizing understory vegetation in Mediterranean forests using full-waveform airborne laser scanning data. *Remote Sens Environ* 217:400–413.

Curran, L. M., S. N. Trigg, A. K. McDonald, D. Astiani, Y. M. Hardiono, P. Siregar, et al. 2004. Lowland forest loss in protected areas of Indonesian Borneo. *Science* 303:1000–1003.

Dahlin, K. M., G. P. Asner, and C. B. Field. 2013. Environmental and community controls on plant canopy chemistry in a Mediterranean-type ecosystem. *Proc Nat Acad Sci* 110:6895–6900.

Dalponte, M., L. Bruzzone, and D. Gianelle. 2007. Tree species classification in the southern alps based on the fusion of very high geometrical resolution multispectral/hyperspectral images and LiDAR data. *Remote Sens Environ* 123:258–270.

Dalponte, M., L. Bruzzone, and D. Gianelle. 2008. Fusion of hyperspectral and lidar remote sensing data for classification of complex forest areas. *IEEE Trans Geosci Remote Sens* 46:1416–1427.

Daughtry, C. S. T. 2001. Discriminating crop residues from soil by shortwave infrared reflectance. *Agronomy J* 93:125–131.

Daughtry, C. S. T., E. R. Hunt, P. C. Doraiswamy, and J. E. McMurtrey III. 2005. Remote sensing the spatial distribution of crop residues. *Agronomy J* 97:864–871.

Dechant, B., M. Cuntz, M. Vohland, E. Schulz, and D. Doktor. 2017. Estimation of photosynthesis traits from leaf reflectance spectra: Correlation to nitrogen content as the dominant mechanism. *Remote Sens Environ* 196:279–292.

Demmig-Adams, B., and W. W. Adams. 1996. The role of the xanthophyll cycle carotenoids in the protection of photosynthesis. *Trends Plant Sci* 1:21–26.

Demmig-Adams, B., and W. W. Adams. 2006. Photoprotection in an ecological context: The remarkable complexity of thermal energy dissipation. *New Phytol* 172:11–21.

Denslow, J. S. 1987. Tropical rainforest gaps and tree species diversity. *Annual Rev Ecol Evol Syst* 18:431–451.

Doughty, C. E., G. P. Asner, and R. E. Martin. 2011. Predicting tropical plant physiology from leaf and canopy spectroscopy. *Oecologia* 165:289–299.

Drake, J. B., R. O. Dubayah, D. B. Clark, R. G. Knox, J. B. Blair, M. A. Hofton, R. L. Chazdon, et al. 2002. Estimation of tropical forest structural characteristics using large-footprint lidar. *Remote Sens Environ* 79:305–319.

Dubayah, R., and J. Drake. 2000. Lidar remote sensing for forestry. *J Forestry* 98:44–46.

Durán, S. M., R. E. Martin, S. Díaz, B. S. Maitner, Y. Malhi, N. Salinas, A. Shenkin, M. R. Silman, D. J. Wieczynski, and G. P. Asner. 2019. Informing trait-based ecology by assessing remotely sensed functional diversity across a broad tropical temperature gradient. *Sci Adv* 5 (12):eaaw8114.

Evans, J. R. 1989. Photosynthesis and nitrogen relationships in leaves of c3 plants. *Oecologia* 78:9–19.

Farquhar, G. D., and S. von Caemmerer. 1982. Modelling of photosynthetic response to environmental conditions. Pp. 549–587 in *Physiological plant ecology ii. Water relations and carbon assimilation*, edited by O. L. Lange, P. S. Nobel, C. B. Osmond, and H. Ziegler. Berlin: Springer-Verlag.

Farquhar, G. D., S. von Caemmerer, and J. A. Berry. 1980. A biochemical model of photosynthetic co2 assimilation in leaves of c3 species. *Planta* 149:78–90.

Fassnacht, F. E., H. Latifi, A. Ghosh, P. K. Joshi, and B. Koch. 2014. Assessing the potential of hyperspectral imagery to map bark beetle-induced tree mortality. *Remote Sens Environ* 140:533–548.

Feilhauer, H., U. Faude, and S. Schmidtlein. 2011. Combining isomap ordination and imaging spectroscopy to map continuous floristic gradients in a heterogenous landscape. *Remote Sens Environ* 115:2513–2524.

Feldpausch, T. R., J. Lloyd, S. L. Lewis, R. J. W. Brienen, M. Gloor, A. Monteagudo Mendoza, G. Lopez-Gonzalez, et al. 2012. Tree height integrated into pantropical forest biomass estimates. *Biogeosci* 9:3381–3403.

Féret, J.-B., and G. P. Asner. 2013. Spectroscopic classification of tropical forest species using radiative transfer modeling. *Remote Sens Environ* 115:2415–2422.

Féret, J.-B., C. François, G. P. Asner, A. A. Gitelson, R. E. Martin, L. P. R. Bidel, S. L. Ustin, et al. 2008. Prospect-4 and -5: Advances in the leaf optical properties model separating photosynthetic pigments. *Remote Sens Environ* 112:3030–3043.

Féret, J.-B., C. François, A. Gitelson, G. P. Asner, K. M. Barry, C. Panigada, A. D. Richardson, et al. 2011. Optimizing spectral indices and chemometric analysis of leaf chemical properties using radiative transfer modeling. *Remote Sens Environ* 115:2742–2750.

Ferraz, A., S. Saatchi, C. Mallet, and V. Meyer. 2016. Lidar detection of individual tree size in tropical forests. *Remote Sens Environ* 183:318–333.

Field, C. B., and H. A. Mooney. 1986. The photosynthesis—nitrogen relationship in wild plants. Pp. 25–55 in *On the economy of plant form and function*, edited by T. J. Givnish. Cambridge: Cambridge University Press.

Field, C. B., J. T. Randerson, and C. M. Malmström. 1995. Global net primary production: Combining ecology and remote sensing. *Remote Sens Environ* 51:74–88.

Filella, I., A. Porcar-Castell, S. Munné-Bosch, J. Bäck, M. F. Garbulsky, and J. Peñuelas. 2009. Pri assessment of long-term changes in carotenoids/chlorophyll ratio and short-term changes in de-epoxidation state of the xanthophyll cycle. *Int J Remote Sens* 30:4443–4455.

Frankenburg, C., J. B. Fisher, J. Worden, G. Badgley, S. S. Saatchi, J-E. Lee, G. C. Toon, et al. 2011. New global observations of the terrestrial carbon cycle from gosat: Patterns of plant fluorescence with gross primary productivity. *Geophys Research Lett* 38:L17706.

Gamon, G. A., and B. Bond. 2013. Effects of irradiance and photosynthetic downregulation on the photochemical reflectance index in Douglas-fir and ponderosa pine. *Remote Sens Environ* 135:141–149.

Gamon, J. A., J. Peñuelas, and C. B. Field. 1992. A narrow-waveband spectral index that tracks diurnal changes in photosynthetic efficiency. *Remote Sens Environ* 41:35–44.

Gamon, J. A., L. Serrano, and J. S. Surfus. 1997. The photochemical reflectance index: An optical indicator of photosynthetic radiation use efficiency across species, functional types, and nutrient levels. *Oecologia* 112:492–501.

Gamon, J. A., and J. S. Surfus. 1999. Assessing leaf pigment content and activity with a reflectometer. *New Phytol* 143:105–117.

Gamon, J. A., R. Wang, and S. E. Russo. 2023. Contrasting photoprotective responses of forest trees revealed using PRI light responses sampled with airborne imaging spectrometry. *New Phytol* 238:1318–1332.

Gao, B. C. 1996. NDWI: A normalized difference water index for remote sensing of vegetation liquid water from space. *Remote Sens Environ* 58:257–266.

Gao, B. C., and A. F. H. Goetz. 1990. Column atmospheric water-vapor and vegetation liquid water retrievals from airborne imaging spectrometer data. *J Geophys Res Atmos* 95:3549–3564.

Gao, B. C., and A. F. H. Goetz. 1995. Retrieval of equivalent water thickness and information related to biochemical-components of vegetation canopies from AVIRIS data. *Remote Sens Environ* 52:155–162.

Garbulsky, M. F., J. Penuelas, J. Gamon, Y. Inoue, and I. Filella. 2011. The photochemical reflectance index (PRI) and the remote sensing of leaf, canopy and ecosystem radiation use efficiencies; a review and metaanalysis. *Remote Sens Environ* 115:281–297.

Gitelson, A. A. 2004. Wide dynamic range vegetation index for remote quantification of biophysical characteristics of vegetation. *J Plant Phys* 161:165–173.

Gitelson, A. A., U. Gritz, and M. N. Merzlyak. 2003. Relationships between leaf chlorophyll content and spectral reflectance and algorithms for non-destructive chlorophyll assessment in higher plant leaves. *J Plant Phys* 160:271–282.

Gitelson, A. A., Y. J. Kaufman, R. Stark, and D. Rundquist. 2002. Novel algorithms for remote estimation of vegetation fraction. *Remote Sens Environ* 80:76–87.

Gitelson, A. A., G. P. Keydan, and M. N. Merzlyak. 2006. Three-band model for noninvasive estimation of chlorophyll, carotenoids, and anthocyanin contents in higher plant leaves. *Geophys Res Lett* 33:L11402.

Gitelson, A. A., M. N. Merzlyak, and O. B. Chivkunova. 2001. Optical properties and non-destructive estimation of anthocyanin content in plant leaves. *J Photochem Photobiol* 74:38–45.

Gong, P., R. Pu, and I. R. Miller. 1992. Correlating leaf area index of ponderosa pine with hyperspectral CASI data. *Can J Remote Sens* 78:275–282.

Gong, P., R. Pu, and J. R. Miller. 1995. Coniferous forest leaf area index along the Oregon transect using compact airborne spectrographic imager data. *Photogramm Eng Remote Sensing* 61:107–117.

Green, D. S., J. E. Erickson, and E. L. Kruger. 2003. Foliar morphology and canopy nitrogen as predictors of light-use efficiency in terrestrial vegetation. *Agric For Meteorol* 115:163–171.

Green, R. O. 1998. Spectral calibration requirement for earth-looking imaging spectrometers in the solar-reflected spectrum. *Applied Optics* 37:683–690.

Green, R. O., V. Carrère, and J. E. Conel. 1989. Measurement of atmospheric water vapor using the airborne visible infrared imaging spectrometer. P. 6 in *Workshop imaging processing*. Sparks, Nevada: American Society of Photogrammetry and Remote Sensing.

Green, R. O., M. L. Eastwood, C. M. Sarture, T. G. Chrien, M. Aronsson, B. J. Chippendale, J. A. Faust, B. E. Pavri, C. J. Chovit, M. Solis, and M. R. Olah. 1998. Imaging spectroscopy and the airborne visible/infrared imaging spectrometer (AVIRIS). *Remote Sens Environ* 65 (3):227–248.

Green, R. O., N. Mahowald, C. Ung, D. R. Thompson, L. Bator, et al. 2020. The earth surface mineral dust source investigation: An earth science imaging spectroscopy mission. *IEEE Aerospace Conference*. Big Sky, MT. Pp. 1–15.

Guanter, L., L. Alonso, L. Gómez-Chova, J. Amorós-López, J. Vila, and J. Moreno. 2007. Estimation of solar-induced vegetation flourescence from space measurements. *Geophys Res Lett* 34:L08401.

Guerschman, J. P., M. J. Hill, L. J. Renzullo, D. J. Barrett, A. S. Marks, and E. J. Botha. 2009. Estimating fractional cover of photosynthetic vegetation, non-photosynthetic vegetation and bare soil in the Australian tropical savanna region upscaling the EO-1 Hyperion and MODIS sensors. *Remote Sens Environ* 113:928–945.

Guo, J., and C. M. Trotter. 2004. Estimating photosynthetic light-use efficiency using the photochemical reflectance index: Variations among species. *Funct Plant Biol* 31:255–265.

Haboudane, D., J. R. Miller, N. Tremblay, and P. J. Zarco-Tejada. 2002. Integrated narrow-band vegetation indices for prediction of crop chlorophyll content for application to precision agriculture. *Remote Sens Environ* 81:416–426.

Hernández-Clemente, R., R. M. Navarro-Cerrillo, and P. J. Zarco-Tejada. 2012. Carotenoid content estimation in a heterogeneous conifer forest using narrow-band indices and prospect+dart simulations. *Remote Sens Environ* 127:298–315.

Hilker, T., N. C. Coops, F. G. Hall, A. Black, M. A. Wulder, Z. Nesic, and P. Krishnan. 2008. Separating physiologically and directionally induced changes in PRI using BRDF models. *Remote Sens Env* 112:2777–2788.

Hoeppner, J. M., A. K. Skidmore, R. Darvishzadeh, M. Heurich, H. C. Chang, and T. W. Gara. 2020. Mapping canopy chlorophyll content in a temperate forest using airborne hyperspectral data. *Remote Sens* 12:3563.

Homolová, L., Z. Malenovský, J. Clevers, G. García-Santos, and M. E. Schaepman. 2013. Review of optical-based remote sensing for plant trait mapping. *Ecol Complexity* 15:1–16.

Huete, A. R. 1988. A soil-adjusted vegetation index (SAVI). *Remote Sens Environ* 25:295–309.

Hunt, E. R. Jr., and B. N. Rock. 1989. Detection of changes in leaf water content using near- and middle-infrared reflectances. *Remote Sens Environ* 30:43–54.

Hunt, E. R. Jr., S. L. Ustin, and D. Riaño. 2013. Remote sensing of leaf, canopy and vegetation water contents for satellite environmental data records. Pp. 335–358 in *Satellite-based applications of climate change*, edited by J. J. Qu, A. M. Powell, and M. V. K. Sivakumar. New York: Springer.

IPCC. 2000. *Special report on land use, land-use change, and forestry*. Edited by R. T. Watson, I. R. Noble, B. Bolin, N. H. Ravindranath, D. J. Verardo, and D. J. Dokeen. Cambridge: Cambridge University Press.

Jacquemoud, S., W. Verhoef, F. Baret, C. Bacour, P. J. Zarco-Tejada, G. P. Asner, C. François, and S. L. Ustin. 2009. Prospect + sail: A review of use for vegetation characterization. *Remote Sens Environ* 113:S56–S66.

Joiner, J., Y. Yoshida, A. P. Vasilkov, Y. Yoshida, L. A. Corp, and E. M. Middleton. 2011. First observations of global and seasonal terrestrial chlorophyll fluorescence from space. *Biogeosci* 7:8281:8318.

Jones, T. G., N. C. Coops, and T. Sharma. 2010. Assessing the utility of airborne hyperspectral and lidar data for species distribution mapping in the coastal pacific northwest, Canada. *Remote Sens Environ* 114:2841–2852.

Jordan, C. F. 1969. Leaf-area index from quality of light on the forest floor. *Ecology* 50:663–666.

Kalacska, M., S. Bohlman, G. A. Sanchez-Azofeifa, K. Castro-Esau, and T. Caelli. 2007. Hyperspectral discrimination of tropical dry forest lianas and trees: Comparative data reduction approaches at the leaf and canopy levels. *Remote Sens Environ* 109:406–415.

Kamoske, A. G., K. M. Dahlin, S. P. Serbin, and S. C. Stark. 2020. Leaf traits and canopy structure together explain canopy functional diversity: An airborne remote sensing approach. *Ecol App* 31:e02230.

Kamoske, A. G., K. M. Dahlin, S. C. Stark, and S. P. Serbin. 2019. Leaf area density from airborne LiDAR: Comparing sensors and resolutions in a temperate broadleaf forest ecosystem. *For Ecol Manage* 433:364–375.

Kattge, J., W. Knorr, T. Raddatz, and C. Wirth. 2009. Quantifying photosynthetic capacity and its relationship to leaf nitrogen content for global-scale terrestrial biosphere models. *Global Change Biol* 15:976–991.

Kefauver, S. C., J. Peñuelas, and S. Ustin. 2013. Using topographic and remotely sensed variables to assess ozone injury to conifers in the Sierra Nevada (USA) and Catalonia (Spain). *Remote Sens Environ* 139:138–148.

Kellner, J. R., and G. P. Asner. 2009. Convergent structural responses of tropical forests to diverse disturbance regimes. *Ecology Lett* 9:887–897.

Kellner, J. R., and G. P. Asner. 2014. Winners and losers in the competition for space in tropical forest canopies. *Ecology Lett*. http://doi.org/10.1111/ele.12256.

Kellner, J. R., D. B. Clark, and S. P. Hubbell. 2009. Pervasive canopy dynamics produce short-term stability in a tropical rain forest landscape. *Ecology Lett* 12:155–164.

Kim, M. S. 1994. *The use of narrow spectral bands for improving remote sensing estimation of fractionally absorbed photosynthetically active radiation(FAPAR)*. Master Thesis, Department of Geography, University of Maryland, College Park, MD.

Knyazikhin, Y., P. Lewis, M. I. Disney, P. Stenberg, M. Mõttus, M. Rautianinen, R. K. Kaufmann, et al. 2013c. Reply to Townsend et al.: Decoupling contributions from canopy structure and leaf optics is critical for remote sensing leaf biochemistry. *Proc Nat Acad Sci* 110:E1075–E1075.

Knyazikhin, Y., P. Lewis, M. I. Disney, P. Stenberg, M. Mõttus, M. Rautiainen, P. Stenberg, et al. 2013b. Reply to Ollinger et al.: Remote sensing of leaf nitrogen and emergent ecosystem properties. *Proc Nat Acad Sci* 110:E2438–E2438.

Knyazikhin, Y., M. A. Schull, P. Stenberg, M. Mõttus, M. Rautiainen, Y. Yang, A. Marshak, et al. 2013a. Hyperspectral remote sensing of foliar nitrogen content. *Proc Nat Acad Sci*. 110:E185–E192.

Kokaly, R. F. 2001. Investigating a physical basis for spectroscopic estimates of leaf nitrogen concentration. *Remote Sens Environ* 75:153–161.

Kokaly, R. F., G. P. Asner, S. V. Ollinger, M. E. Martin, and C. A. Wessman. 2009. Characterizing canopy biochemistry from imaging spectroscopy and its application to ecosystem studies. *Remote Sens Environ* 113:S78–S91.

Kokaly, R. F., B. W. Rockwell, S. L. Haire, and T. V. V. King. 2007. Characterization of post-fire surface cover, soils, and burn severity at the Cerro Grande fire, New Mexico, using hyperspectral and multispectral remote sensing. *Remote Sens Environ* 106:305–325.

Lang, N., W. Jetz, K. Schindler, and J. D. Wegner. 2023. A high-resolution canopy height model of the Earth. *Nature Ecol Evo*:1–12.

Lee, K.-S., W. B. Cohen, R. E. Kennedy, T. K. Maiersperger, and S. T. Gower. 2004. Hyperspectral versus multispectral data for estimating leaf area index in four different biomes. *Remote Sens Environ* 91:508–520.

Lefsky, M. A., W. B. Cohen, S. A. Acker, G. G. Parker, T. A. Spies, and D. Harding. 1999. Lidar remote sensing of the canopy structure and biophysical properties of Douglas-fir western hemlock forests. *Remote Sens Environ* 70:339–361.

Lefsky, M. A., W. B. Cohen, D. J. Harding, G. G. Parker, S. A. Acker, and S. T. Gower. 2002. Lidar remote sensing of above-ground biomass in three biomes. *Global Ecol Biogeog* 11:393–399.

Lefsky, M. A., D. J. Harding, M. Keller, W. B. Cohen, C. C. Carabajal, F. D. B. Espirito-Santo, M. O. Hunter, et al. 2005. Estimates of forest canopy height and aboveground biomass using ICESAT. *Geophys Res Lett* 35:L22S02.

le Maire, G., C. François, K. Soudani, D. Berveiller, J.-Y. Pontailler, N. Bréda, H. Genet, et al. 2008. Calibration and validation of hyperspectral indices for the estimation of broadleaved forest leaf chlorophyll content, leaf mass per area, leaf area index and leaf canopy biomass. *Remote Sens Environ* 112:3846–3864.

Lewis, P., and M. Disney. 2007. Spectral invariants and scattering across multiple scales from within-leaf to canopy. *Remote Sens Environ* 109:196–206.

Lim, K., P. Treitz, M. Wulder, B. St-Onge, and M. Flood. 2003. LiDAR remote sensing of forest structure. *Prog Phys Geogr* 27 (1):88–106.

Lovell, J. L., D. L. B. Jupp, D. S. Culvenor, and N. C. Coops. 2003. Using airborne and ground-based lidar to measure canopy structure in Australian forests. *Can J Remote Sens* 29:607–622.

Magney, T. S., D. R. Bowling, B. A. Logan, K. Grossmann, J. Stutz, P. D. Blanken, S. P. Burns, R. Cheng, M. A. Garcia, and P. Köhler. 2019. Mechanistic evidence for tracking the seasonality of photosynthesis with solar-induced fluorescence. *Proc Nat Acad Sci* 116 (24):11640–11645.

Magney, T. S., C. Frankenberg, J. B. Fisher, Y. Sun, G. B. North, T. S. Davis, A. Kornfeld, and K. Siebke. 2017. Connecting active to passive fluorescence with photosynthesis: A method for evaluating remote sensing measurements of Chl fluorescence. *New Phytol* 215 (4):1594–1608.

Magnussen, S., M. Wulder, and D. Seemann. 2002. Stand canopy closure estimated by line sampling. Pp. 1–12 in *Continuous cover forestry*, edited by K. von Gadow, J. Nagel, and J. Saborowski. Dordrecht, Netherlands: Springer.

Martin, M. E., S. D. Newman, J. D. Aber, and R. G. Congalton. 1998. Determining forest species composition using high spectral resolution remote sensing data. *Remote Sens Environ* 65:249–254.

Martin, M. E., L. C. Plourde, S. V. Ollinger, M-L. Smith, and B. E. McNeil. 2008. A generalizable method for remote sensing of canopy nitrogen across a wide range of forest ecosystems. *Remote Sens Environ* 112:3511–3519.

Martin, R. E., G. P. Asner, L. P. Bentley, A. Shenkin, N. Salinas, K. Q. Huaypar, M. M. Pillco, F. D. Ccori Álvarez, B. J. Enquist, and S. Diaz. 2020. Covariance of sun and shade leaf traits along a tropical forest elevation gradient. *Frontiers Plant Sci* 10:1810.

McManus, K. M., G. P. Asner, R. E. Martin, K. G. Dexter, W. J. Kress, and C. B. Field, 2016. Phylogenetic structure of foliar spectral traits in tropical forest canopies. *Remote Sens* 8 (3):196.

Meireles, J. E., J. Cavender-Bares, P. A. Townsend, S. Ustin, J. A. Gamon, A. K. Schweiger, M. E. Schaepman, G. P. Asner, R. E. Martin, A. Singh, and F. Schrodt. 2020. Leaf reflectance spectra capture the evolutionary history of seed plants. *New Phytol* 228 (2):485–493.

Meroni, M., M. Rossini, L. Guanter, L. Alonso, U. Rascher, R. Colombo, and J. Moreno. 2009. Remote sensing of solar-induced chlorophyll fluorescence: Review of methods and applications. *Remote Sens Environ* 113:2037–2051.

Middleton, E. M., Y.-B. Cheng, T. Hilker, T. A. Black, P. Krishnan, N. C. Coops, and K. F. Huemmrich. 2009. Linking foliage spectral responses to canopy-level ecosystem photosynthetic light-use efficiency at a Douglas-Fir forest in Canada. *Can J Remote Sens* 35:166–188.

Mohammed, G. H., R. Colombo, E. M. Middleton, U. Rascher, C. van der Tol, L. Nedbal, Y. Goulas, O. Pérez-Priego, A. Damm, and M. Meroni. 2019. Remote sensing of solar-induced chlorophyll fluorescence (SIF) in vegetation: 50 years of progress. *Remote Sens Environ* 231:111177.

Montieth, J. L. 1977. Climate and the efficiency of crop production in Britain. *Phil Trans Royal Soc B* 281:277–294.

Morsdorf, F., B. Kötz, K. I. Itten, and B. Allgöwer. 2006. Estimation of LAI and fractional cover from small footprint airborne laser scanning data based on gap fraction. *Remote Sens Environ* 29:607–622.

Mundt, J. T., D. R. Streutker, and N. F. Glenn. 2006. Mapping sagebrush distribution using fusion of hyperspectral and lidar classifications. *Photogramm Eng Remote Sensing* 72:47–54.

Muss, J. D., D. J. Mladenoff, and P. A. Townsend 2011. A pseudo-waveform technique to assess forest structure using discrete lidar data. *Remote Sens Environ* 115:824–835.

Næsset, E. 2009. Effects of different sensors, flying altitudes, and pulse repetition frequencies on forest canopy metrics and biophysical stand properties derived from small-footprint airborne laser data. *Remote Sens Environ* 113:148–159.

Næsset, E., and T. Gobakken. 2008. Estimation of above- and below-ground biomas across regions of the boreal forest zone using airborne laser. *Remote Sens Environ* 112:3079–3090.

Naidoo, L., M. A. Cho, R. Mathieu, and G. P. Asner. 2012. Classification of savanna tree species, in the Greater Kruger National Park region, by integrating hyperspectral and lidar data in a random forest data mining environment. *ISPRS J Photogramm Remote Sens* 69:167–179.

Naumann, J. C., J. E. Anderson, and D. R. Young. 2008. Linking physiological responses, chlorophyll fluorescence and hyperspectral imagery to detect salinity stress using the physiological reflectance index in the coastal shrub, Myrica cerifera. *Remote Sens Environ* 112:3865–3875.

Nelson, R. 1988. Estimating forest biomass and volume using airborne laser data. *Remote Sens Environ* 24:247–267.

Nepstad, D. C., A. Verissimo, A. Alencar, C. Nobre, E. Lima, P. Lefebvre, P. Schlesinger, et al.1999. Large-scale impoverishment of Amazonian forests by logging and fire. *Nature* 398:505–508.

Nichol, C. J., K. F. Huemmrich, T. A. Black, P. G. Jarvis, C. L. Walthall, J. Grace, and F. G. Hall. 2000. Remote sensing of photosynthetic-light-use efficiency of boreal forest. *Agric For Meteor* 101:131:142.

Niinemets, U. 1999. Research review. Components of leaf dry mass per area—thickness and density—alter leaf photosynthetic capacity in reverse directions in woody plants. *New Phytol* 144:35–47.

Ni-Meister, W., D. L. B. Jupp, and R. Dubayah. 2001. Modeling lidar waveforms in heterogenous and discrete canopies. *IEEE Trans Geosci Remote Sens* 39:1943–1958.

Ollinger, S. V., P. B. Reich, S. Frolking, L. C. Lepine, D. Y. Hollinger, and A. D. Richardson. 2013. Nitrogen cycling, forest canopy reflectance, and emergent properties of ecosystems. *Proc Nat Acad Sci* 110:E2437.

Ollinger, S. V., M. L. Smith, M. E. Martin, R. A. Hallett, C. L. Goodale, and J. D. Aber. 2002. Regional variation in foliar chemistry and n cycling among forests of diverse history and composition. *Ecology* 83:339–355.

Oppelt, N., and W. Mauser. 2007. Airborne visible/infrared imaging spectrometer AVIS: Design, characterization and calibration. *Sensors* 7 (9):1934–1953.

Oshio, H., T. Aswa, A. Hoyano, and S. Miyasaka. 2015. Estimation of the leaf area density distribution of individual trees using high-resolution and multi-return airborne LiDAR data. *Remote Sens Environ* 166:116–125.

Parker, G. G. 1995. Structure and microclimate of forest canopies. Pp. 73–106 in *Forest canopies*, edited by M. D. Lowman and N. M. Nadkarni. New York: Academic Press.

Peñuelas, J. F., F. Baret, and I. Filella. 1995. Semiempirical indexes to assess carotenoids chlorophyll-a ratio from leaf spectral reflectance. *Photosynthetica* 31:221–230.

Peñuelas, J. F., J. Pinol, R. Ogaya, and I. Lilella. 1997. Estimation of plant water content by the reflectance water index wi (r900/r970). *Int J Remote Sens* 18:2869:2875.

Poorter, H., U. Niinemets, L. Poorter, I. J. Wright, and R. Villar. 2009. Causes and consequences of variation in leaf mass per area (LMA): A meta-analysis. *New Phytol* 182:565–588.

Popescu, S. C., R. H. Wynne, and R. F. Nelson. 2003. Measuring individual tree crown diameter with lidar and assessing its influence on estimating forest volume and biomass. *Can J Remote Sens* 29:564–577.

Rahman, A. F., J. A. Gamon, D. A. Fuentes, D. A. Roberts, and D. Prentiss. 2001. Modeling spatially distributed ecosystem flux of boreal forest using hyperspectral indices from AVIRIS imagery. *J Geophys Res* 106:33579–33591.

Reich, P. B., M. B. Walters, and D. S. Ellsworth. 1997. From tropics to tundra: Global convergence in plant functioning. *Proc Nat Acad Sci* 94:13730–13734.

Riaño, D. P., F. Valladares, S. Condés, and E. Chuvieco. 2004b. Estimation of leaf area index and covered ground from airborne laser scanner (lidar) in two contrasting forests. *Agric For Meteor* 124:269–275.

Riaño, D. P., P. Vaughan, E. Chuvieco, P. J. Zarco-Tejada, and S. L. Ustin. 2004a. Estimation of fuel moisture content by inversion of radiative transfer models to simulate equivalent water thickness and dry matter content. Analysis at leaf and canopy level. *IEEE Trans Geosci Remote Sens* 43:819–826.

Richardson, J. J., L. M. Moskal, and S.-H. Kim. 2009. Modeling approaches to estimate effective leaf area index from aerial discrete-return lidar. *Agric For Meteor* 149:1152–1160.

Riebeek, H. 2010. Earth observing-1: Ten years of innovation. In *NASAearth observatory: Features*.

Ripullone, F., A. R. Rivelli, R. Baraldi, R. Guarini, R. Guerrieri, F. Magnani, J. Peñuelas, et al. 2011. Effectiveness of the photochemical reflectance index to track photosynthetic activity over a range of forest tree species and plant water statuses. *Funct Plant Biol* 38:177–186.

Roberts, D. A., P. Dennison, M. Gardner, Y. Hetzel, S. L. Ustin, and C. Lee. 2003. Evaluation of the potential of hyperion for fire danger assessment by comparison to the airborne visible infrared imaging spectrometer. *IEEE Trans Geosci Remote Sens* 41:1297–1310.

Roberts, D. A., M. Gardner, R. Church, S. L. Ustin, G. Scheer, and R. O. Green. 1998. Mapping chaparral in the Santa Monica mountains using multiple endmember spectral mixture models. *Remote Sens Environ* 65:267–279.

Roberts, D. A., S. L. Ustin, S. Ogunjemiyo, J. Greenberg, S. Z. Dobrowski, J. Chen, and T. M. Hinckley. 2004. Spectral and structural measures of northwest forest vegetation at leaf to landscape scale. *Ecosystems* 7:545–562.

Rondeaux, G., M. Steven, and F. Baret. 1996. Optimization of soil-adjusted vegetation indices. *Remote Sens Environ* 55:95–107.

Rouse, J. W., R. H. Haas, J. A. Schell, and D. W. Deering. 1974. Monitoring vegetation systems in the great plains with ERTS. Pp. 309–317 in *Third ERTS symposium NASA SP-351*, edited by S. C. Fraden, E. P. Marcanti, and M. A. Becker. Washington, DC: NASA.

Santos, M. J., J. A. Greenberg, and S. L. Ustin. 2010. Detecting and quantifying southeastern pine senescence effects to red-cockaded woodpecker (*Picoides borealis*) habitat using hyperspectral remote sensing. *Remote Sens Environ* 114:1242–1250.

Schneider, F. D., F. Morsdorf, B. Schmid, O. L. Petchey, A. Hueni, D. S. Schimel, and M. E. Schaepman. 2017. Mapping functional diversity from remotely sensed morphological and physiological forest traits. *Nat Comm* 8:1441.

Schull, M. A., S. Ganguly, A. Samanta, D. Huang, N. V. Shabanov, J. P. Jenkins, J. C. Chiu, et al. 2007. Physical interpretation of the correlation between multi-angle spectral data and canopy height. *Geophys Res Lett* 34:L18405.

Schull, M. A., Y. Knyazikhin, L. Xu, A. Samanta, P. L. Carmona, L. Lepine, J. P. Jenkins, et al. 2011. Canopy spectral invariants, part 2: Application to classification of forest types from hyperspectral data. *J Quant Spectrosc Radiative Trans* 112:736–750.

Sellers, P. J., S. O. Los, C. J. Tucker, C. O. Justice, D. A. Dazlich, G. J. Collatz, and D. A. Randall. 1996. A revised land surface parameterization (SiB2) for atmospheric GCMS. Part II: The generation of global fields of terrestrial biophysical parameters from satellite data. *J Climate* 9:706–737.

Serbin, S. P., D. N. Dillaway, E. L. Kruger, and P. A. Townsend. 2012. Leaf optical properties reflect variation in photosynthetic metabolism and its sensitivity to temperature. *J Experim Biol* 63:489–502.

Serbin, S. P., A. Singh, A. R. Desai, S. G. Dubois, A. D. Jablonski, C. C. Kingdon, E. L. Kruger, and P. A. Townsend. 2015. Remotely estimating photosynthetic capacity, and its response to temperature, in vegetation canopies using imaging spectroscopy. *Remote Sens Environ* 167:78–87.

Serbin, S. P., J. Wu, K. S. Ely, E. L. Kruger, P. A. Townsend, R. Meng, B. T. Wolfe, A. Chlus, Z. Wang, and A. Rogers. 2019. From the Arctic to the tropics: Multibiome prediction of leaf mass per area using leaf reflectance. *New Phyt* 224:1557–1568.

Serrano, L., J. F. Peñuelas, and S. L. Ustin. 2002. Remote sensing of nitrogen and lignin in Mediterranean vegetation from AVIRIS data: Decomposing biochemical from structural signals. *Remote Sens Environ* 81:355–364.

Shao, G., S. C. Stark, D. R. A. de Alemeida, and M. N. Smith. 2019. Towards high throughput assessment of canopy dynamics: The estimation of leaf area structure in Amazonian forests with multitemporal multisensor airborne lidar. *Remote Sens Environ* 221:1–13.

Simard, M., N. Pinto. J. B. Fisher, and A. Baccini. 2011. Mapping forest canopy height globally with spaceborne lidar. *J Geophys Res Biogeosci* 116:G4.

Singh, A., S. P. Serbin, B. E. McNeil, C. C. Kingdon, and P. A. Townsend. 2015. Imaging spectroscopy algorithms for mapping canopy foliar chemical and morphological traits and their uncertainties. *Ecol Appl* 25:2180–2197.

Smith, M. L., M. E. Martin, L. Plourde, and S. V. Ollinger. 2003. Analysis of hyperspectral data for estimation of temperate forest canopy nitrogen concentration: Comparison between an airborne (AVIRIS) and a spaceborne (Hyperion) sensor. *IEEE Trans Geosci Remote Sens* 41:1332–1337.

Smith, M. L., S. V. Ollinger, M. E. Martin, J. D. Aber, R. A. Hallett, and C. L. Goodale. 2002. Direct estimation of aboveground forest productivity through hyperspectral remote sensing of canopy nitrogen. *Ecol Appl* 12:1286–1302.

Soldberg, S., A. Brunner, K. H. Hanssen, H. Lange, E. Næsset, M. Rautiainen, and P. Stenberg. 2009. Mapping LAI in a Norway spruce forest using airborne laser scanning. *Remote Sens Environ* 113:2317–2327.

Somers, B., and G. P. Asner. 2013. Multi-temporal hyperspectral mixture analysis and feature selection for invasive species mapping in rainforests. *Remote Sens Environ* 136:14–27.

Spanner, M., L. Johnson, J. Miller, R. McCreight, J. Freemantle, J. Runyon, and P. Gong. 1994. Remote sensing of seasonal leaf area index across the Oregon transect. *Ecol Appl* 4:258–271.

Stuffler, T., C. Kaufmann, S. Hofer, K. P. Förster, G. Schreier, A. Mueller, A. Eckardt, H. Bach, B. Penné, U. Benz, and R. Haydn. 2007. The EnMAP hyperspectral imager—An advanced optical payload for future applications in Earth observation programmes. *Acta Astronaut* 61 (1–6):115–120.

Stylinski, C. D., W. C. Oechel, J. A. Gamon, D. T. Tissue, F. Miglietta, and A. Raschi. 2000. Effects of lifelong [co2] enrichment on carboxylation and light utilization of Quercus pubescens Willd. examined with gas exchange, biochemistry and optical techniques. *Plant Cell Environ* 23:1353–1362.

Sun, G., and K. J. Ranson. 2000. Modeling lidar returns from forest canopies. *IEEE Trans Geosci Remote Sens* 38:2617–2626.

Thenkabail, P. S., E. A. Enclona, M. S. Ashton, C. Legg, and M. J. De Dieu. 2004. Hyperion, IKONOS, ALI, and ETM+ sensors in the study of African rainforests. *Remote Sens Environ* 90:23–43.

Thomas, V., D. A. Flinch, J. H. McCaughey, T. Noland, L. Rich, and P. Treitz. 2006. Spatial modelling of the fraction of photosynthetically active radiation absorbed by a boreal mixedwood forest using a lidar—hyperspectral approach. *Agric For Meteorol* 140:287–307.

Thomas, V., J. H. M cCaughey, P. Treitz, D. A. Finch, T. Noland, and L. Rich. 2009. Spatial modelling of photosynthesis for a boreal mixedwood forest by integrating micrometeorological, lidar, and hyperspectral remote sensing data. *Agric For Meteorol* 149:639–654.

Townsend, P. A., and J. R. Foster. 2002. Comparison of EO-1 Hyperion to AVIRIS for mapping forest composition in the Appalachian mountains. *IGARSS* 2:793–795.

Townsend, P. A., S. P. Serbin, E. L. Kruger, and J. A. Gamon. 2013. Disentangling the contribution of biological and physical properties of leaves and canopies in imaging spectroscopy data. *Proc Nat Acad Sci* 110:E1074.

Udayalakshmi, V., B. St-Onge, and D. Kneeshaw. 2011. Response of a boreal forest to canopy opening: Assessing vertical and lateral tree growth with multi-temporal lidar data. *Ecol Appl* 21:99–121.

Ungar, S. G., J. S. Pearlman, J. A. Mendenhall, and D. Reuter. 2003. Overview of the earth observing one (EO-1) mission. *IEEE Transactions on Geoscience and Remote Sensing* 41 (6):1149–1159.

Ustin, S. L., A. A. Gitelson, S. Jacquemoud, M. Schaepman, G. P. Asner, J. A. Gamon, and P. Zarco-Tejada. 2009. Retrieval of foliar information about plant pigment systems from high resolution spectroscopy. *Remote Sens Environ* 113:S67–S77.

Ustin, S. L., D. Riaño, and E. R. Hunt Jr. 2012. Estimating canopy water content from spectroscopy. *Israel J Plant Sci* 60:9–23.

Ustin, S. L., and A. Trabucco. 2000. Analysis of AVIRIS hyperspectral data to assess forest structure and composition. *J Forestry* 98:47–49.

van Aardt, J. A. N., R. H. Wynne, and J. A. Scrivani. 2008. Lidar-based mapping of forest volume and biomass by taxonomic group using structurally homogenous segments. *Photogramm Eng Remote Sensing* 74:1033–1044.

Van den Berg, A. K., and T. D. Perkins. 2005. Nondestructive estimation of anthocyanin content in autumn sugar maple leaves. *HortScience* 40:685–686.

Vaughn, N. R., G. P. Asner, and C. P. Giardina. 2013. Polar grid fraction as an estimator of forest canopy structure using airborne lidar. *Int J Remote Sens* 34:7464–7473.

Verrelst, J., Z. Malenovsky, C. Van der Tol, G. Camps-Valls, J.-P. Gastellu-Etchegorry, P. Lewis, P. North, and J. Moreno. 2019. Quantifying vegetation biophysical variables from imaging spectroscopy data: A review on retrieval methods. *Surv Geophys* 40:589–629.

Vierling, K. T., L. A. Vierling, W. Gould, S. Martinuzzi, and R. Clawges. 2008. Lidar: Shedding new light on habitat characterization and modeling. *Frontiers Ecol Environ* 6:90–98.

Wang, Z., A. Chlus, R. Geygan, Z. Ye, A. Singh, J. J. Couture, J. Cavender-Bares, E. L. Kruger, and P. A. Townsend. 2020. Foliar functional traits from imaging spectroscopy across biomes in eastern North America. *New Phytol* 228:494–511.

Wang, Z., A. K. Skidmore, R. Darvishzadeh, and T. Wang. 2018. Mapping forest canopy nitrogen content by inversion of coupled leaf-canopy radiative transfer models from airborne hyperspectral imagery. *Agric For Meteorol* 253:247–260.

Wang, Z., P. A. Townsend, A. K. Schweiger, J. J. Couture, A. Singh, S. E. Hobbie, and J. Cavender-Bares. 2019. Mapping foliar functional traits and their uncertainties across three years in a grassland experiment. *Remote Sens Environ* 221:405–416.

Waring, R. H. 1983. Estimating forest growth and efficiency in relation to canopy leaf area. *Adv Ecol Res* 13:327–354.

Weingarten, E., R. E. Martin, R. F. Hughes, N. R. Vaughn, E. Shafron, and G. P. Asner. 2022. Early detection of a tree pathogen using airborne remote sensing. *Ecol Appl* 32:e2519.

Weishampel, J. F., J. B. Blair, R. G. Knox, R. Dubayah, and D. B. Clark. 2000. Volumetric lidar return patterns from an old-growth tropical rainforest canopy. *Intl J Remote Sens* 21:409–415.

Westoby, M., D. S. Falster, A. T. Moles, P. A. Vesk, and I. J. Wright. 2002. Plant ecological strategies: Some leading dimensions of variation between species. *Annu Rev Ecol Syst* 33:125–159.

White, J. C., N. C. Coops, T. Hilker, M. A. Wulder, and A. L. Carroll. 2007. Detecting mountain pine beetle red attack damage with EO-1 Hyperion moisture indices. *Intl J Remote Sens* 28:2111–2121.

Wing, B. M., M. W. Ritchie, K. Boston, W. B. Cohen, A. Gitelman, and M. J. Olsen 2012. Prediction of understory vegetation cover with airborne lidar in an interior ponderosa pine forest. *Remote Sens Environ* 124:730–741.

Wolter, P. T., P. A. Townsend, B. R. Sturtevant, and C. C. Kingdon. 2008. Remote sensing of the distribution and abundance of host species for spruce budworm in northern Minnesota and Ontario. *Remote Sens Environ* 112:3971–3982.

Wright, I. J., P. B. Reich, M. Westoby, D. D. Ackerly, Z. Baruch, F. Bongers, J. Cavender-Bares, et al. 2004. The worldwide leaf economics spectrum. *Nature* 428:821–827.

Wu, J., C. Chavana-Bryant, N. Prohaska, S. P. Serbin, K. Guan, L. P. Albert, X. Yang, W. J. van Leeuwen, A. J. Garnello, G. Martins, and Y. Malhi. 2017. Convergence in relationships between leaf traits, spectra and age across diverse canopy environments and two contrasting tropical forests. *New Phytologist* 214 (3):1033–1048.

Wu, J., A. Rogers, L. P. Albert, K. Ely, N. Prohaska, B. T. Wolfe, R. C. Oliveira Jr, S. R. Saleska, and S. P. Serbin. 2019. Leaf reflectance spectroscopy captures variation in carboxylation capacity across species, canopy environment and leaf age in lowland moist tropical forests. *New Phytol* 224:663–674.

Wulder, M. A., J. C. White, R. F. Nelson, E. Næsset, H. O. Ørka, N. C. Coops, T. Hilker, et al. 2012. Lidar sampling for large-area forest characterization: A review. *Remote Sens Environ* 121:196–209.

Yan, Z., Z. Guo, S. P. Serbin, G. Song, Y. Zhao, Y. Chen, S. Wu, J. Wang. X. Wang, J. Li, B. Wang, Y. Wu, Y. Su, H. Wang, A. Rogers, and L. Liu. 2021. Spectroscopy outperforms leaf trait relationships for predicting photosynthetic capacity across different forest types. *New Phytol* 232:134–147.

Yao, W., P. Krzystek, and M. Heurich. 2012. Tree species classification and estimation of stem volume and DBH based on single tree extraction by exploiting airborne full-waveform LiDAR data. *Remote Sens Environ* 123:368–380.

Zarco-Tejada, P. J., J. A. J. Berni, L. Suárez, G. Sepulcre-Cantó, F. Morales, and J. R. Miller. 2009. Imaging chlorophyll fluorescence from an airborne narrow-band multispectral camera for vegetation stress detection. *Remote Sens Environ* 113:1262–1275.

Zarco-Tejada, P. J., J. R. Miller, G. H. Mohammed, T. L. Noland, and P. H. Sampson. 1999. Optical indices as bioindicators of forest condition from hyperspectral CASI data. *Proc. 19th Symp European Assoc Remote Sensing Laboratories (EARSeL)*, Valladolid, Spain.

Zarco-Tejada, P. J., J. R. Miller, G. H. Mohammed, T. L. Noland, and P. H. Sampson. 2001. Scaling-up and model inversion methods with narrow-band optical indices for chlorophyll content estimation in closed forest canopies with hyperspectral data. *IEEE Trans Geosci Remote Sens* 39:1491–1507.

Zarco-Tejada, P. J., J. R. Miller, A. Morales, A. Berjón, and J. Agüera. 2004. Hyperspectral indices and model simulation for chlorophyll estimation in open-canopy tree crops. *Remote Sens Environ* 90:463–476.

Zarco-Tejada, P. J., A. Morales, L. Testi, and F. J. Villalobos. 2013. Spatio-temporal patterns of chlorophyll fluorescence and physiological and structural indices acquired from hyperspectral imagery as compared with carbon fluxes measured with eddy covariance. *Remote Sens Environ* 133:102–115.

Zhang, Y., J. M. Chen, J. R. Miller, and T. L. Noland. 2008. Leaf chlorophyll content retrieval from airborne hyperspectral remote sensing imagery. *Remote Sens Environ* 112:3234–3247.

Zhao, K., S. Popescu, X. Meng, Y. Pang, and M. Agca. 2011. Characterizing forest canopy structure with lidar composite metrics and machine learning. *Remote Sens Environ* 115 (8):1978–1996.

Zhou, X., and C. Li. 2023. Mapping the vertical forest structure in a large subtropical region using airborne LiDAR data. *Ecol Indic* 154:110731.

Zimble, D. A., D. L. Evans, G. C. Carlson, R. C. Parker, S. C. Grado, and P. D. Gerard. 2003. Characterizing vertical forest structure using small-footprint airborne lidar. *Remote Sens Environ* 87:171–182.

4 Optical Remote Sensing of Tree and Stand Heights

Sylvie Durrieu, Cédric Vega, Marc Bouvier, Frédéric Gosselin, Jean-Pierre Renaud, Laurent Saint-André, Anouk Schleich, and Maxime Soma

ACRONYMS AND DEFINITIONS

AGB	Aboveground biomass
ALOS	Advanced Land Observing Satellite
ALS	Airborne laser scanning
ASTER	Advanced spaceborne thermal emission and reflection radiometer
ATLAS	Advanced Topographic Laser Altimeter System
CHM	Canopy height model
CHRIS	Compact High Resolution Imaging Spectrometer
CNES	The Centre national d'études spatiales or the National Center of Space Studies of France
DEM	Digital Elevation Model
DESDynI	Deformation, Ecosystem Structure and Dynamics of Ice
DTM	Digital terrain model
EnMAP	Environmental Mapping and Analysis Program
ENVISAT	Environmental satellite
ERS	European remote sensing satellites
GEDI	Global Ecosystem Dynamics Investigation
GLAS	Geoscience Laser Altimeter System
GPS	Global Positioning System
ICESat	Ice, Cloud, and land Elevation Satellite
IKONOS	A commercial earth observation satellite, typically, collecting sub-meter to 5 m data
IRS	Indian Remote Sensing Satellites
LAI	Leaf area index
LiDAR	Light detection and ranging
MODIS	Moderate-resolution imaging spectroradiometer
NASA	National Aeronautics and Space Administration
NIR	Near-infrared
RADARSAT	Radar satellite
REDD	Reducing Emissions from Deforestation in Developing Countries
SAR	Synthetic aperture radar
SPOT	Satellite Pour l'Observation de la Terre, French Earth Observing Satellites
SRTM	Shuttle Radar Topographic Mission
TerraSAR-X	A radar Earth observation satellite with its phased array synthetic aperture radar
TIN	Triangulated irregular network
UAV	Unmanned Aerial Vehicle
UNFCCC	United Nations Framework Convention on Climate Change

4.1 INTRODUCTION

Forests cover 30% of continental surfaces and play a key role in climate change regulation, in raw material and renewable energy supply, and in biodiversity conservation. Successfully maintaining all the functions of forest ecosystems through their sustainable management is thus crucial for the future of mankind. However, developing appropriate policies and management practices requires an in-depth knowledge of forest ecosystems within a fast-evolving context, as well as appropriate models to forecast how they will respond to management practices and global change.

Remote sensing data supported by ground observations are considered as the key to obtaining quantitative and timely information on forest ecosystems at a variety of scales in space and time, thereby allowing effective monitoring (Fassnacht et al., 2024; Coops et al., 2023; Stahl et al., 2023; Sagar et al., 2022; Senf et al., 2017; White et al., 2016; DeFries et al., 2007; Fuller, 2006; Kleinn, 2002; Simonett, 1969), enhanced ecosystem modeling (Blanco et al. 2023; Senf, 2022; Masek et al., 2015; Marsden et al., 2013; Cabello et al., 2012; le Maire et al., 2011; Wang et al., 2010), and appropriate management of forest resources (Pandey et al., 2022; Goodbody et al., 2021a; Camarretta et al., 2020; Fardusi et al., 2017; Le Goff et al., 2010; Thürig and Kaufmann, 2010; Liu and Han, 2009).

In addition to forest composition, i.e., abundance and distribution of species, forest structure is a key descriptor of forest ecosystems (Fischer et al., 2019a; Wynne, 2006) that can, to some extent, be retrieved from remote sensing data. Structure refers to the 3D arrangement and characteristics of vegetation compartments, including trunks, branches, twigs, and leaves. A given structure is both a result and a driver of ecosystem functions (Shugart et al., 2010). Forest structure is also directly related to the main biogeochemical (water, nutrient, and carbon) cycles, thus determining stocks and driving fluxes (Zhou et al., 2021; Ali, 2019; Ellison et al., 2012). It can affect local abiotic factors and is also essential in providing habitats, therefore possibly impacting biodiversity (Toivonen et al., 2023; Gril et al., 2023; Lenoir et al., 2022; Bohn and Huth, 2017; Bouvier et al., 2017; Götmark, 2013; Couteron et al., 2005). Accurate measurements of forest structure based on remote sensing data would thus represent a major step toward an in-depth knowledge of forest ecosystems. Several remote sensing technologies can provide highly valuable information for sustainable forest management. Some of them are particularly promising for providing information on forest structure. The aim of this chapter is to present two remote sensing approaches: airborne laser scanning (ALS) and digital photogrammetry (DP) that have both proven their efficiency in characterizing forest structure based on vegetation height measurements, either at the individual tree level or at the stand level. The first section explains why the knowledge of height structure is of major importance for both forest management and ecosystem modeling. In this section, our focus on the two aforementioned optical remote sensing approaches as promising solutions to extend height measurements in space and time is also justified. In the second section, the concepts and history of both technologies are briefly described, and the resulting 3D data are compared in the framework of forest applications. Sections 3 and 4 detail how forest height structures can be measured from these 3D data at the tree and stand levels, respectively. After a brief conclusion, the last section presents some promising prospects offered by the possibility to monitor changes in vegetation height using ALS or DP 3D data. In this last section we will also discuss some issues related to spaceborne systems. These systems offer the opportunity to overcome coverage limitations encountered with systems operated from air platforms, thus enabling to extend forest ecosystems monitoring at a global scale. Existing spaceborne imaging systems can be used to provide DP 3D data from space. For LiDAR, the ICESat experiment (2003–2009), which was the first spaceborne LiDAR designed for Earth surface monitoring, has demonstrated the possibility to derive global height (Simard et al., 2011) and biomass maps (Baccini et al., 2012; Saatchi et al., 2011) by combining LiDAR sampling measurements with other satellite data and global products. It paved the way for the era of Earth observation spaceborne LiDAR systems, as evidenced by NASA's launch of two new space LiDAR missions in 2018: ICESat-2 (Markus et al., 2017; Neumann et al., 2019) and GEDI (Dubayah et al., 2020).

4.2 WHY MEASURE TREE HEIGHTS?

4.2.1 THE DETERMINANTS OF HEIGHTS

Height growth, also called primary tree growth, is a complex process that combines the production of the internodes by the apical bud and their elongation. Every year, one or several growth units can be produced. Apical growth is one of the three main processes that govern tree architecture along with branching and reiteration (Martin-Ducup et al., 2020; Barthelemy and Caraglio, 2007; Guédon et al., 2007). Tree height growth and tree architecture results from an equilibrium between endogenous features (cell arrangements, cell properties) and exogenous factors (competition for light, water, and nutrients). Growth phase patterns and drivers (e.g., climate, between tree competition, forest management) can be disentangled using mathematical or statistical models (e.g., Matrix models [Liang and Picard, 2013]; Markov chains [Chaubert-Pereira et al., 2009], non-linear models [Saint-André et al., 2008], or process-based models such as the MAESPA, 3-PG or the iLand models [Rammer et al., 2024; Gupta and Sharma, 2019; Duursma and Medlyn, 2012]). In natural forest ecosystems, stand height structure is mainly dependent on the tree species and their traits (e.g., shade tolerant, light demander, among others), while it is greatly determined by forester strategies in managed forests leading to stand structures such as even-aged high forest with either a single or mixed species, coppices, coppices with standards, or selection forests. The secondary growth, which results from the cambium activity, increases tree size and biomass. Primary growth and secondary growth are correlated, and the proportions between height and diameter, or between biomass and height and diameter, follow rules that are the same for all trees of a given species growing under the same conditions (climate, forest management). These allometries are widely used to calculate volume and biomass at tree and plot scale (Latifah et al., 2021, Picard et al., 2012b) and the research domain is very active (Pretzsch, 2020; Fischer et al., 2019b; Henry et al., 2013). Tree height at a given time is then integrating all previous growth phases and can be used as an indicator of the current forest status, e.g., biomass or carbon stocks, or future growth.

4.2.2 IMPORTANCE OF TREE HEIGHT DISTRIBUTION FOR FOREST MANAGEMENT AND ECOLOGY

Among forest attributes, tree height plays a central role in forest inventories, where it is critical for calculating volumes, site quality indexes, and assessing needs for silvicultural treatments (Pretzsch et al., 2021; Bontemps and Bouriaud, 2014; Pardé, 1956). An evocative example is given by thinning operations in young stands, which are normally planned based on threshold crossings in canopy height. In more mature stands, height indicators such as top, dominant, or Lorey's heights are frequently used for volume or growth estimations. This underlines the importance of height indicators for forest managers.

Regarding ecological modeling, growth and yield models are mainly based on height growth. These models generally apply to even-aged forests, which are characterized by a relative tree population homogeneity: same age (or age range in the case of natural regeneration) and a dominant tree species. The growth of these populations has been widely studied (De Perthuis, 1788 in Batho and Garcia, 2006), which gave rise to other generic principles (García, 2011; Pretzsch, 2009; Skovsgaard and Vanclay, 2008; Dhôte, 1991; Assmann, 1970). It is customary to distinguish the population as a whole, then the tree in the stand. This distinction is useful to separate the different factors involved in tree growth into three main components: (1) site fertility in a broad sense, including the ability of the soil to feed the trees (nutrients and water availability) and the climate characteristics of the area; (2) the overall pressure within the population that is appreciated by different indices of density; and, finally, (3) the social status of each individual tree, which will define its ability to mobilize resources in its immediate environment.

For these three components, height is of primary importance since dominant height is widely used to define site fertility (or site index, for which preliminary principles were given by De Perthuis in 1788 (Batho and Garcia, 2006) and concepts formalized by Eichhorn (1904) and later discussed by Assmann (1970), and more recently by Bontemps and Bouriaud (2014) or by Tarmu et al. (2020). It can also be used to define some stand density/global pressure indicators (e.g., Hart-Becking factor), even if to a less significant extent than stand density or density indicators based on tree diameters. Finally, height can be used to define the social status of the trees, e.g., tree height related to the stand dominant height.

Owing to the importance of forest in climate mitigation, through its capacity to accumulate carbon, improved knowledge of carbon stocks and fluxes in forest ecosystems is needed to better understand the carbon cycle and to develop improved climate models (Kumar and Sharma, 2015; Malhi et al., 2002). This knowledge is also needed for carbon accounting. After the Kyoto protocol (1997), the Cancún Agreements (2010) provides strong backing for a REDD+ (Reducing Emissions from Deforestation and Forest Degradation) mechanism under the United Nations Framework Convention on Climate Change (UNFCCC) whereby developed countries would provide positive benefits to developing ones for reducing deforestation, forest degradation, enhancement of forest carbon stocks, and forest conservation. It was followed by the Paris Agreement (2015) that includes commitments from all countries, both developed and developing, to address climate change. Countries are required to set their own targets (NDCs, i.e., nationally determined contributions) for reducing emissions and to report on their progress (Krug, 2018). However, in order to implement these mechanisms, forest services in most countries must make more accurate assessment of the forest carbon stocks and carbon stock changes (Herold et al., 2019). Changes in forest carbon stocks through time are best appraised by a combination of remote sensing and field-based measurement where height is an essential variable. Most of the current methods used to assess carbon emissions from deforestation and forest degradation are indeed based on the measurement of changes in surface area of the main forest types and on the assessment of a mean biomass value for each type (De Sy et al., 2012; ESA, 2008). Therefore, improving accuracy of both surface areas of forest types and biomass estimations would lead to improved carbon flux predictions. For biomass estimation, a consensus exists stating that, for a biomass map with a 1 ha resolution, estimation accuracy should not exceed 20 t.ha-1 or 20% of field estimations without exceeding 50 t.ha-1 (Hall et al., 2011b; Le Toan et al., 2011; Zolkos et al., 2013). However, reaching such accuracy all over the world is highly challenging (Réjou-Méchain et al., 2019; Pelletier et al., 2011; Angelsen, 2008; Chave et al., 2003). Another approach to study carbon stocks and fluxes relies on the use of forest ecosystems functioning models. Up to now, two kinds of models have coexisted (Maréchaux et al., 2021; Bellassen et al., 2011). The first one consists of models adapted to stand scale, like growth models previously discussed, which are process-based models that can include information on silvicultural practices. Provided an important local calibration is performed, these models can output reliable simulations of local carbon stock evolutions. The second type consists of Global Vegetation Models (GVM), which can provide carbon stocks and fluxes at regional scales, but with lower accuracy levels. There are two main causes for inaccuracy. The first one lies in pedoclimatic data, which are required to drive the models but are too coarse to reflect local variations. The second cause for inaccuracy is the non-integration of management impacts, which hampers a reliable modeling of age-dependant variables like biomass (Maréchaux et al., 2021; Bellassen et al., 2011). This is why GVM are currently evolving to better manage intra-cell variability within coarse grid models and to take into account management impacts. ORCHIDEE-FM adapted from the ORCHIDEE model (Bellassen et al., 2010), the second version of the Ecosystem Demography model, ED2 (Medvigy and Moorcroft, 2012; Medvigy et al., 2009), or the FORMIND model (Fischer et al., 2019a) are examples of this new model generation that can assimilate height or biomass information from NFI field plots or remote sensing data. This assimilation led to significant improvement in aboveground biomass dynamic characterization and productivity assessment (Fisher et al., 2019a; Antonarakis et al., 2011; Bellassen et al., 2011), e.g., a decrease in error rate of 30% and 50% for total ecosystem respiration (RECO) and net productivity (NEP), respectively, in the study by Bellassen et al. (2011).

Therefore, whatever the approach used and due to its tie link with biomass, vegetation height is an essential parameter to develop scientific knowledge of the carbon cycle and to address carbon accounting issues.

Another main aspect of forest structure is its major role in ecology (Mori et al., 2017; Jaskierniak, 2011; Vepakomma et al., 2008). Stand structure affects micro-climate, habitat quality, and therefore biodiversity potential. In particular the gradient in gap sizes is known to influence many parts of biodiversity. Gaps are either part of the natural forest cycle or result from silvicultural treatments or are caused by accidental disturbances such as a fire, storm, or plant health problems. They have a short-term impact on biodiversity through different mechanisms. The most obvious one is an increased irradiance that can benefit heliophilous species, but it also has an impact on the microclimate in the patch, with a higher temperature variance in gaps than in forests. Through the removal of trees, gaps also release some soil resources, such as water and nutrients, which can positively impact vascular plant development, for example. Soil disturbances associated with gaps (e.g., pits and mounds), were also found to impact biodiversity. Bouget (2005) reported that the gap effect was, on the whole, favorable to biodiversity, with the notable exception of shade-preferring groups. And, while small gaps generally have a weaker effect on biodiversity (see Bouget, 2005), surprisingly, benefits were observed for forest vascular plants, for which small openings might be very positive (Duguid and Ashton, 2013). Gaps not only have an effect in the area they cover, as some species shun closed forests in the vicinity of the gaps. These species are generally called forest-interior species and have particularly been found among birds (Germaine et al., 1997). However, similar behaviors were also noted for plant species in Bouvier et al. (2017). For all these reasons, biodiversity indicators were shown to be strongly correlated to the three-dimensional spatial patterns of vegetation (Toivonen et al., 2023; Williams et al., 1994), and wildlife richness was shown to be related to the three-dimensional features of the canopy (Davison et al., 2023; Zellweger et al., 2016; Magnussen et al., 2012). Temporal effects have also been reported. Through the succession that occurs after gap formation, gap impacts can evolve and even be reversed. For vascular plants in particular, silvicultures based on large cuttings have been found to have an adverse effect on floristic biodiversity, which occurs decades after gap formation (Lenoir et al., 2022; Duguid and Ashton, 2013). This may be why some authors promote the imitation of the small-scale natural gap dynamics in managed stands (Næsset, 2002b; Angelstam, 1998).

4.2.3 Limitations of Field Measurements and How Remote Sensing Can Help Meet Information Requirements

From the previous sub-sections we can see how important it is to measure tree heights and to monitor the dynamic of both canopy height and gaps at several scales in time and space in order to improve both forest ecosystem modeling and to contribute to their sustainable management.

Total height of a tree is either defined as "the distance between the top and the base of the tree, measured along a perpendicular, dropped from the top" (van Laar and Akça, 2007) or as the stem length. Several methods are used to measure tree height in the field (Larjavaara and Muller-Landau, 2013; Williams et al., 1994). Graduated poles can be used, but without climbing the tree, their use is limited to relatively small trees (Larjavaara and Muller-Landau, 2013), usually under 15 m. For bigger trees, height measurements are usually derived from angle and distance measurements (Figure 4.1). Angles are measured using a clinometer. Distances are measured with either a measuring tape, an ultrasonic measuring system, or a laser rangefinder (Larjavaara and Muller-Landau, 2013). With the latter, which can be used to measure the distance between the operator and the treetop, the sine method can be applied (Figure 4.1). This method is very effective in dense forests, as it can be used to carry out measurements from a distance close from the trunk, even by shooting directly up the tree, whereas the tangent method is highly inaccurate in these situations.

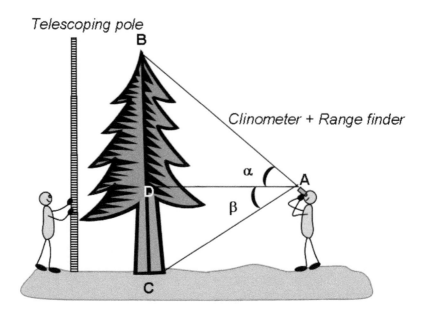

FIGURE 4.1 Height measurement of a vertical tree using a graduated pole (left side of the figure) or a hypsometer measuring angles and distances (right side of the figure). If distance AD is measured, using, for example, an ultrasonic system, tree height can be retrieved using the tangent method (H_{tree} = BD + DC = AD.tgα + AD.tgβ); if distances AB and AC are measured, using, for example, a laser system, sine method can be used (H_{tree} = BD + DC = AB.sinα + AC.sinβ). Several variants inspired by this basic principle exist depending on the measurement system and the targeted accuracy (e.g., DC is sometimes approximated from the height of the surveyor) and also to account for slope or leaning trees.

Mean absolute errors were found to be in the order of 1 m in temperate forests (e.g., from 0.9 to 1.3 m, according to the measurement method used in Williams et al. [1994]) and higher for tall trees. For example, in the study by Williams et al. (1994), for one of the two classes with the tallest trees, i.e., trees higher than 24 m, errors ranged from 1.65 to 3.45 m according to the measurement method used. Measurements are more difficult to achieve and more error prone for broadleaved trees than for coniferous species, especially during the leaf-on period, and also for leaning trees. Stereńczak et al. (2019) also found that absolute errors in measurements were mostly influenced by tree length and species, followed by slope of the terrain. At the tree level, they found absolute and relative errors ranging from −7.77 m to 7.1 m and from −30% to up to 45%, respectively. In a moist tropical semi-deciduous forest and during the leaf-on period, which can be considered among one of the most complex measurement conditions, the residual mean squared error was found to be on average between 5.05 and 6.85 m, using sine and tangent methods, respectively (Larjavaara and Muller-Landau, 2013). Part of the error can be attributed to a bias, and errors depend not only on the measurement method and the stand characteristics but also on the surveyor's expertise (Larjavaara and Muller-Landau, 2013; Kitahara et al., 2010).

Within the inventory process, height is certainly among the most costly data to collect. This is why different height indicators have been developed to minimize the measurement effort (Van Laar and Akça, 2007). Such estimators rely on the relationship between tree size and stand density and depend on other indicators used to assess the biophysical characteristics of the stand (Vanclay, 2009). For example, when using the quadratic mean diameter, which is favored over the arithmetic-mean diameter due to its higher correlation with stand volume and biomass, surveyors will only measure and average the heights of a sample of trees whose diameters at breast height (DBH) are closest to the plot quadratic mean diameter. In even-aged stands, the dominant height or top height is most commonly used. It can be defined as the arithmetic mean of the 100 largest trees per hectare. It is favored over mean height when assessing site quality and modeling growth because it is less affected by stand density and silvicultural practices (Tran-Ha et al., 2011). However, dominant height definitions may vary among users (Jaskierniak, 2011), and more importantly, this measurement can be biased, as it depends upon the sampling area (García, 1998). Another measured height indicator is the Lorey's mean height. Here, the contribution of each tree to the stand height is weighted by its basal area. The Lorey's mean height is thus calculated by multiplying the tree height (h) by the tree basal area (g) and by dividing the sum of these weighted heights by the total stand basal area. The Lorey's mean height is less impacted than the aforementioned height indicators by both the mortality and harvesting of smaller trees and is also well correlated with stand volume (Tran-Ha et al., 2011).

In parallel, efforts have been made to develop technologies for automatic height measurements in the field. Instruments such as terrestrial, backpack, or handheld LiDARs are playing a central role in these aims and are seen as promising tools for precision forestry (Hyyppä et al., 2020; Su et al., 2020; Beland et al., 2019; Cabo et al., 2018; Newnham et al., 2015). However, there are still challenges in clearly describing the top of the canopy with such instruments and in automating the processing of collected data.

Field measurements are essential but remain labor-intensive and costly, and their scope must therefore be limited in space and time. Therefore, a technology accounting for extensive areas and for tree heights at individual or stand scale, as well as providing gap measurements at regular time scale and at lower costs would represent a real quantum leap in the development of forest ecosystem modeling. It would also help managers who are calling for more precise inventories with fine spatial and temporal resolutions in order to deal with the growing societal and economic pressure exerted on forests.

The potential of remote sensing–based technologies to extend forest structure measurements in space and time is widely acknowledged (see, e.g., Fassnacht et al., 2024; White et al., 2016; De Sy et al., 2012; Ostendorf, 2011; De Leeuw et al., 2010; Wang et al., 2010; DeFries et al., 2007). Table 4.1 summarizes forest information that can be derived from the main remote sensing data types. Forest information was classified into three main categories: forest maps, forest structural and biophysical properties, and information on forest changes. The remote sensing technologies that can be used to characterize forest structure can be identified from Table 4.1 along with the level of data availability.

Forest structure can be indirectly assessed by analyzing, for example, texture in very-high-resolution imagery (Aquino et al., 2024; Ploton et al., 2017; Proisy et al., 2007; Couteron et al., 2005). But several remote sensing techniques, based on LiDAR, optical imagery, or radar technologies, allow direct measurements of the forest 3D structure.

The utility of LiDAR, which is an active optical remote-sensing technology, has been widely demonstrated with respect to forest structure measurements and biomass estimation (Coops et al., 2021; Beland et al., 2019; Wulder et al., 2012; Hyyppä et al., 2008; Næsset, 2007; Lim et al., 2003). LiDAR remains efficient in closed-canopy tropical areas supporting high biomass forests greater than 200 t.ha^{-1} (Rodda et al., 2024; Leitold et al., 2015; Kellner et al., 2009; Lefsky et al., 2005), where optical vegetation indices and volumetric radar measurements typically saturate (Joshi et al., 2017; Castro et al., 2003). Photogrammetry has long been used for forest applications, particularly forest inventories, since accurate 3D measurements were rendered possible thanks to analytic stereo plotters and comparators that became commonly available from the mid-1970s (Vastaranta et al., 2013). This technology has been neglected for many years, probably partly due to the great potential of LiDAR to measure the 3D structure of vegetation and the fast development of airborne laser scanning (ALS) systems, systems combining a LiDAR with a scanning device. But photogrammetry has recently received renewed interest for forest applications thanks to developments in digital photogrammetry (DP) that enables automatic reconstruction of 3D canopy models based on very high-resolution optical images (e.g., Goodbody et al., 2019; Pearse et al., 2018; Fassnacht et al., 2017).

TABLE 4.1
Overview of the main remote sensing data and of their use for forest ecosystem monitoring. Remote sensing technology capabilities are evaluated through the quality level of forest information or parameters derived from each data source. Capabilities are classified into four classes: *Contribution but poor product quality, **Average product quality, ***Good Product Quality—Limited Contribution to No Technical Capabilities or No relevant published study identified. Change monitoring capability is a consequence of land cover type discrimination capability. The operationally effective level results from a combination of data availability, scale coverage, and information quality level. The potential synergy between technologies or between remote sensing data and ancillary information is not presented in this table.

	Remote Sensing data				
	Multispectral optical imagery Reflected radiations measured for several discrete wavebands in the spectral range 400–3000 nm Luminance and reflectance can be computed from measurements [1,2,3,36]		Hyperspectral optical imagery Reflected radiations measured for many narrow wavebands in the spectral range 400–3000 nm Higher spectral resolution and lower geometric resolution than for multispectral imagery [6,7,37,38,39]		Synthetic Aperture Radar (SAR) Backscattered microwave radiation at wavelengths between 1 cm and few meters Retrodiffusion coefficient can be computed from measurements. Several polarizations available (HH, HV, VH, VV) [1,2,8,40]
	Very High resolution data Airborne (airborne): Scale: ~1:5000 to < 1:80000 Digital: Resolution : sub-metric to few meters	High to low resolution data Resolution : tens of meters to few hundreds of meters	High resolution Resolution : 1 to around 30 m, depending on flight height and system	Low resolution Resolution: about 100 m to 1200m	
		2D imagery			
				Airborne Cloud penetration capability	Spaceborne Cloud penetration capability
Data availability (from a technological perspective)	Airborne and spaceborne (IKONOS, QuickBird, Pléiades, WorldView, SPOT, Planet-Scope-Dove,...) High but expensive Restricted by cloud cover Operational	Spaceborne - High (Sentinel2, Landsat, Aster, RapidEye, IRS, MODIS...) Restricted by cloud cover Several free sources	Low for very high resolution (few experimental airborne systems) Low for high resolution with spaceborn systems (CHRIS/Proba-1, PRISMA, EnMAP, DESIS, HISUI) but likely to increase in the future with several missions under study Restricted by cloud cover	Low Spaceborne (Hyperion, HICO, PACE) Restricted by cloud cover	Low (few commercial systems)	High (Sentinel-1, RADARSAT, TerraSAR-X, TanDEM-X, ALOS...)
Scale & Coverage	From local to national scale Mapping scale greater than 1:50000	From regional to global scale according to spatial resolution	From local to national scale	From regional to global scale according to system	Local to regional	Regional to global
Forest maps						
Forest/non forest	***	***	***	***	*** Using multitemporal datasets	
Coniferous/Deciduous/Mixed	***	***	***	***	** Using polarisation information	
Additional information on forest types	** Structural information by texture analysis [4]	* Information on species by time series analysis	*** Higher species classification accuracy than with multispectral data (e.g. + 27% => 88% final accuracy) [7]	* Hampered by the coarse resolution	** Using multifrequency/multipolarization data	
Tree level	* Generalization & computing issues [5]	-	** If sufficient resolution Technically difficult	-	-	
Forest structural and biophysical properties						
Cover rate (%)	**	*	***	*	** Using multifrequency/multipolarization data	
Density (nb of trees/ha)	*	-	** If sufficient resolution	** If sufficient resolution	-	
Height (m)	*	-	*	*	*	
Basal area (m²)	*	-	*	*	-	
LAI / PAI (m²/m²)	* Saturation at low to medium LAI level	*	** Using appropriate vegetation indices; Possibility to also retrieve chlorophyll or other pigments content, as well as water content	** Using appropriate vegetation indices	** [9,10]	
Wood volume (m3/ha) AGB (Mg/Ha)	Saturation at low biomass level	*	** Using appropriate vegetation indices	** Using appropriate vegetation indices	** Using polarization information Saturation for AGB > 200 t.ha-1 (for L or P band, lower saturation levels for X and C bands) [8,11,12]	
Forest change monitoring						
Land cover change (km²)	**	**	**	**	**	
Afforestation/Reforestation/Deforestation (km²)	**	**	**	**	**	
Degradation	-	-	**	** If sufficient resolution	**	
Disaster (km²)	*** For fire ** For storm and phytosanitary problems	**	***	**	*	
Growth	-	-	-	-	-	
Gap dynamics	**	* Depends on Resolution /gap size	** Depends on data availability	*	-	

(Continued)

TABLE 4.1 (Continued)

Overview of the main remote sensing data and of their use for forest ecosystem monitoring. Remote sensing technology capabilities are evaluated through the quality level of forest information or parameters derived from each data source. Capabilities are classified into four classes: *Contribution but poor product quality, **Average product quality, ***Good Product Quality—Limited Contribution to No Technical Capabilities or No relevant published study identified. Change monitoring capability is a consequence of land cover type discrimination capability. The operationally effective level results from a combination of data availability, scale coverage, and information quality level. The potential synergy between technologies or between remote sensing data and ancillary information is not presented in this table.

	Remote Sensing data			
	2.5 D imagery Digital surface or terrain model		3 D information	
	Digital Photogrammetry	Radar interferometry	Radar tomography	Lidar
Surface geometry retrieved from two or multi point of view images [13,14,15,16,17,41]		Difference in the phase information between two SAR images provides an ambiguous measurement of the relative terrain altitude due to the periodic nature of the signal. After the phase unwrapping step, aiming at solving the ambiguity, an accurate DSM is obtained. [14,19,20,40]	Raw vegetation profiles (several tens of centimeters from airborne data to several meters expected from BIOMASS mission) are computed using multiple-baseline images Long wavelengths sensitive to the whole vegetation layer must be used (P and L bands) [20,22,23,42,43,44,45]	Part of the 3D information can be used to provide DSM, DTM and CHM Additional information on vegetation vertical structure [30,31,32,33,34,35,46,47,48,49,50]
	Very High resolution DSM Accuracy: 3 cm to 14 cm Acquisition parameters affecting elevation accuracy: flying height, image resolution, B/H ratio Aerial photographs or VHR images (< 4 m)	Spaceborne X, C or L bands (e.g. ERS, J-ERS, ENVISAT, ALOS, TerraSAR, TanDEM-X, SRTM) Z accuracy: better than 2m up to 15m (for SRTM)	Spaceborne BIOMASS (P-band) and NISAR (L and S bands) planned to be launched end of 2024	Spaceborne ICESat (2003–2009) not optimized for forest applications GEDI and ICESat-2 launched in 2018
	High Low resolution DSM with depointing imagers Accuracy: 5 m to 40 m	Airborne (e.g. commercial intermap products) Z accuracy: 0.5 m	Airborne	Airborne commercial systems (ALS)
	Airborne and spaceborne High but expensive Restricted by cloud cover Operational	Low No cloud cover restriction	Low	High But expensive
	High Available globally (e.g. from ASTER) but with low accuracy for global products	High No cloud cover restriction No P based InSAR data yet available from space; but one paned mission: the ESA BIOMASS mission	Low – Comming soon	Low- No certainty of continuity
From local to regional or national	Regional to global	Local to regional	Local to regional	Regional to Global
		Regional (high accurate z data) to global (medium accurate z data)	Regional to global	Not wall to wall coverage
*** Using image spectral information when available	** Using image spectral information when available	** Using the associated amplitude images	** Using the associated amplitude images	** based on vegetation height
*** Using image spectral information when available	** Using image spectral information when available	** Using the associated amplitude images	** Using the associated amplitude images	*
** Direct using the associated images Indirect information on topographic environment can help to identify species ** Improved results when using both DSM and images	** Indirect: information on topographic environment can help to identify species	** Using the associated amplitude images	** Using the associated amplitude images	** Information on forest structure * Map quality depends on other RS data used for structural information extrapolation
	-	-	-	** Generalization & computing issues
** Using image spectral information when available	* Using external DTM or associated images	** Using the associated amplitude images	Using the associated amplitude images	*** GEDI (**) and ICESat2 (*) are more suitable than ICESat but still limitations for dense covers and on slopes [25,26,49,50]
** Using image spectral information when available	-	-	-	**
*** for dominant strata & using external DTM Highly accurate DSM possible (few cm)	*** Using external DTM Height information quality depends on elevation accuracy	*** Using external DTM Modeled surface depends on the used frequency ==> Direct or indirect height assessement	** Raw vertical profiles	*** for dominant strata ** for understorey *** Improved height measurements with GEDI and ICESat2 compared to ICESat but limitations remain for dense covers and slopes ** Height maps obtained by fusing spaceborne Lidar data with well-to-wall optical or radar data: issues with saturation, high local uncertainty and significant discrepancies at the local level between maps [25,26,27,49,50,51,52]

(Continued)

TABLE 4.1 (Continued)

Overview of the main remote sensing data and of their use for forest ecosystem monitoring. Remote sensing technology capabilities are evaluated through the quality level of forest information or parameters derived from each data source. Capabilities are classified into four classes: *Contribution but poor product quality, **Average product quality, ***Good Product Quality—Limited Contribution to No Technical Capabilities or No relevant published study identified. Change monitoring capability is a consequence of land cover type discrimination capability. The operationally effective level results from a combination of data availability, scale coverage, and information quality level. The potential synergy between technologies or between remote sensing data and ancillary information is not presented in this table.

Remote Sensing data

	2.5 D imagery Digital surface or terrain model			Radar interferometry		Radar tomography		Lidar	
	Digital Photogrammetry Surface geometry retrieved from two or multi point of view images [13,14,15,16,17,41]			Difference in the phase information between two SAR images provides an ambiguous measurement of the relative terrain altitude due to the periodic nature of the signal. After the phase unwrapping step, solving the ambiguity, an accurate DSM is obtained. [14,19,20,40]		Raw vegetation profiles (several tens of centimeters from airborne data to several meters expected from BIOMASS mission) are computed using multiple-baseline images Long wavelengths sensitive to the whole vegetation layer must be used (P and L bands) [20,22,23,42,43,44,45]		Part of the 3D information can be used to provide DSM, DTM and CHM Additional information on vegetation vertical structure [30,31,32,33,34,35,46,47,48,49,50]	
	Very High resolution DSM Accuracy: 3 cm to 14 m Acquisition parameters affecting elevation accuracy: flying height, image resolution, B/H ratio Aerial photographs or VHR images (<4 m)	High to Low resolution DSM with depointing imagers Accuracy: 5 m to 40 m		Spaceborne X,C or L bands (e.g. ERS, J-ERS, ENVISAT, ALOS, TerraSAR, TanDEM-X, SRTM) Z accuracy: better for AGB > 300 t.ha-1 (P band) [20,21]	Airborne (e.g. commercial Intermap products) Z accuracy: 0.5 m	Spaceborne BIOMASS (P-band) and NISAR (L and S bands) planned to be launched end of 2024	Airborne	Spaceborne ICESat (2003–2009) not optimized for forest applications GEDI and ICESat-2 launched in 2018	Airborne commercial systems (ALS)
3 D information									
** Using external DTM Highly accurate DSM possible (few cm)	* Using external DTM	-		** Using the associated amplitude images	** Using the associated amplitude images	** Using vertical profiles		**	*** Using fullwaveform data (i.e. ICESat and GEDI); ICESat-2 not adapted [28,49] ** Biomass maps obtained by fusing spaceborne Lidar data with wall-to-wall optical or radar: issues with saturation, high local uncertainty and significant discrepancies at the local level between maps. [29,53,54]
	-	** Using external DTM Height information quality depends on elevation accuracy		** Using polarization information and height information saturation for AGB > 300 t.ha-1 (P band) [20,21]				***	
	** Using image spectral information when available	** Using image spectral information when available		*** Using the associated amplitude images	** Using the associated amplitude images & height information from interferometry	*** Using the associated amplitude images & height information from tomography		-	
	-	** Using image spectral information when available		** Using the associated amplitude images, height & biomass information from interferometry		** Using the associated amplitude images, height & biomass information from tomography		** based on vegetation height	
	** Using image spectral information when available	** Using image spectral information or external DTM		**				* the most promising technology but yet little studied topic e.g. [24]	
	** [18]	-		-		-		* Through changes in structure Kind of disaster difficult to identify	
	*	-		-		-		*** Depends on data availability (expensive)	
								*** Depends on data availability (expensive)	

References quoted in the table: [1] Boyd and Danson (2005); [2] Ke and Quackenbush (2011); [3] Goodenough et al. (2004); [4] Li et al. (2009); [5] Manninen et al. (2005); [6] Castro et al. (2003); [7] Kasischke et al. (1997); [8] Mitchard et al. (2009); [9] Véga and St-Onge (2008); [10] Garestier et al. (2009); [11] Le Toan et al. (2011); [12] Weishampel et al. (2012); [13] Durrieu and Nelson (2013); [14] Rosette et al. (2013); [15] Bolton et al. (2013); [16] Luo et al. (2013); [17] Mitchard et al. (2013); [18] Véga and St-Onge (2013); [19] Ferretti et al. (2009); [20] Le Toan et al. (2011); [21] Garestier et al. (2009); [22] Ho Tong Minh et al. (2014); [23] Tebaldini and Rocca (2012); [24] Weishampel et al. (2012); [25] Durrieu and Nelson (2013); [26] Rosette et al. (2013); [27] Bolton et al. (2013); [28] Luo et al. (2013); [29] Mitchard et al. (2013); [30] Lim et al. (2003); [31] Næsset (2007); [32] Zolkos et al. (2013); [33] Holmgren et al. (2003); [34] Evans et al. (2009); [35] Ustin and Middleton (2021); [36] Coops-Nichelson et al. (2021); [37] Hill et al. (2019); [38] Qian (2021); [39] Papathanassiou et al. (2021); [40] Goodbody et al. (2019); [42] Fefo et al. (2018); [43] Tebaldini et al. (2019); [44] Ramachandran et al. (2023); [45] Fatoyinbo et al. (2021); [46] Beland et al. (2019); [47] Coops et al. (2021); [48] Goodbody et al. (2021a); [49] Dubayah et al. (2020); [50] Neuenschwander et al. (2020); [51] Potapov et al. (2021); [52] Lang et al. (2023); [53] Shendryk (2022); [54] Silva et al. (2021).

Using small wavelength polarimetric radar systems (e.g., band X [~2.5 to 3.75 cm] radar), accurate digital surface models of the top of the canopy can also be produced using interferometry, also referred as the PolInSAR technique (Lei et al., 2021; Chen et al., 2018; Karila et al., 2015). Indeed, the signal barely penetrates the vegetation at these wavelengths and is then mainly backscattered by the elements of the upper canopy (Soja and Ulander, 2013). Polarimetric radar tomography applied to data acquired with large wavelength systems with a signal that penetrates deeper into the vegetation, e.g., band P (> 1m), has also been used recently to decompose the signal into a few vertical strata producing low-resolution vertical signal profiles in forests (Liu et al., 2024; Pardini et al., 2024; Ho Tong Minh et al., 2014).

However, unlike for radar, the high technological readiness of ALS systems and of both airborne and spaceborne optical imagery systems, as well as the great versatility of the produced data, means that both these technologies are particularly suited to providing information for operational applications or for research in the field of ecosystem modeling. Indeed the high availability of data has favored the development of processing approaches, some of which are currently used on an operational level.

4.3 TWO PROMISING OPTICAL REMOTE SENSING TECHNIQUES FOR TREE HEIGHT MEASUREMENTS: LIDAR AND DIGITAL PHOTOGRAMMETRY

4.3.1 LiDAR

4.3.1.1 Principle and Brief History

LiDAR (Light Detection And Ranging) is an active remote-sensing technology based on emission-reception of a laser beam. Several kinds of LiDAR systems exist, e.g., differential absorption LiDAR, Doppler, range finder, but most of the systems dedicated to continental surface observation belong to the range finder class. They assess the distance between the sensor and a target by measuring the roundtrip time for a short laser pulse (in general, NIR or green wavelengths) to travel between the sensor and a target. By combining these range measurements with information on both sensor position and attitude, obtained thanks to a differential GPS and an inertial measurement unit on board the platform, the position of the target on the Earth's surface can be accurately computed (Li et al., 2020a; Wagner, 2010; Mallet and Bretar, 2009; Baltsavias, 1999a, 1999b; Wehr and Lohr, 1999). For semi-transmitting mediums, such as forests, the incident pulse might be partly backscattered by the top of the canopy, the understory vegetation, and the ground. The backscattered signal (waveform) thus embeds information on the 3D structure of vegetation covers. It is collected by a telescope and recorded using photodiodes (Figure 4.2).

The idea of measuring distance using light can be traced back to the late 1930s, i.e., before the development of lasers. Barthélémy (1946) reported the development, in 1938, of a device that could be used to measure cloud height using high-power flashes of light produced by a spark gap and lasting no more than a few microseconds. In favorable situations, cloud heights up to 700 m could be measured with this system (Barthélémy, 1946).

Attempts to build the first operational laser (Light Amplification by Stimulated Emission of Radiation) started in the late 1950s and were successful in 1960 (Nelson, 2013).

In his paper tracing the history of the use of LiDAR in forest applications, Nelson (2013) reported that range finder lasers were initially used over continental surfaces for topographic purposes thanks to their ability to penetrate forest canopies. Tree measurements became the primary objective rather than just a source of noise in 1976 (Nelson, 2013). The development of digital recording devices and of positioning and scanning systems were key steps leading to the current systems, which can be classified depending on (Stoker et al., 2016; Wulder et al., 2007; Dubayah and Drake, 2001; Baltsavias, 1999b):

(1) whether they fully digitize the return signal (full waveform systems), record multiple echoes (multi-echo systems)—with current commercial systems capturing from one (first or last) up to eight returns—or detect individual photons with high sensitivity, such as Geiger-mode (GML) or Single Photon LiDAR (SPL) systems, which are suitable for high-altitude and long-range measurements while requiring lower-power lasers;

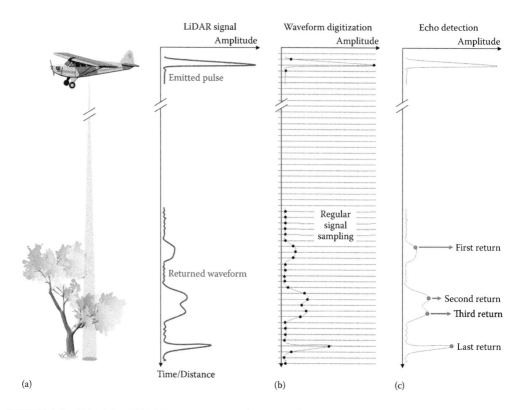

FIGURE 4.2 Principle of LiDAR measurement: photons are backscattered toward the sensor every time the laser beam is partially or totally intercepted by an obstacle. The resulting signal is a waveform that is either (a) digitized at a high frequency, e.g., every 1 ns, with fullwaveform systems, or (b) processed in real time by multi-echo systems that only record a few returns (c). Depending on the system used, time information might be complemented with additional information, such as signal intensity and echo width.

(2) whether they are small footprint (typically with footprints at the ground level in the order of one to a few decimeters) or large footprint systems (tens of meters); and
(3) their sampling rate and scanning pattern.

Another major feature of LiDAR systems is the laser wavelength. Excluding X-ray and free electron lasers, lasers exist in a wide spectrum, i.e., 50–30,000 nm (Baltsavias, 1999c). However, most of the range finder LiDAR developed for Earth surface studies operate in the near infrared (NIR) domain (900, 1040–1064, or 1550 nm) (Baltsavias, 1999c). NIR is particularly well fitted because of high atmospheric transmission in this spectral range (Baltsavias, 1999a) and because most targets at the Earth's surface, except water, have a high reflectivity in the NIR. In particular, their reflectivity is higher than in visible wavelengths. In addition, higher powers are allowed with NIR wavelengths than with visible ones while respecting eye safety. This is especially the case for 1550 nm lasers that stay eye safe at higher power than 1064 nm ones. Signal-to-noise ratio is thus increased by using NIR lasers. NIR LiDAR instruments are also currently capable of significantly higher pulse rates than green LiDAR (Faux et al., 2009). The latter are mainly used for bathymetry due to the capacity of green wavelengths to penetrate the water (Pricope and Bashit, 2023; Allouis et al., 2010). However, after ICESat, which acquired NIR wavelength data from 2003 to 2009, the ATLAS instrument on board ICESat-2, the second space LiDAR mission dedicated to Earth surface monitoring and launched in 2018, is using a green laser (Magruder et al., 2021). ICESat-2 includes among its secondary scientific objectives measurement of vegetation canopy height. The choice to use a low-energy, high-frequency—i.e., photon counting—instrument, constrained the choice of the wavelength. Indeed, this technology was not mature enough and not yet space qualified in the NIR wavelength. Airborne UV (Allouis et al., 2011) and dual wavelength LiDAR (Axelsson et al., 2023; Hancock et al., 2012) have also been successfully used for forest applications. The later systems were found well suited for separating the vegetation from the ground in complex topography (Hancock et al., 2012) as well as improving land cover classification and species composition in forests (Axelsson et al., 2023; Wang et al., 2014).

The fast development of airborne LiDAR systems (most of which are multi-echo sensors [Figure 4.2]), has been driven by commercial opportunities presented by environmental issues. Advances in LiDAR technology (e.g., increased pulse rates, increased number of echo digitization or full waveform recording, scanning devices) and advances in data geolocation have rapidly turned LiDAR into a fully functional and operational technology with a steady increase in the availability of ALS systems operated by data providers. Full waveform LiDAR systems were first designed as experimental systems (e.g., LVIS, Laser Vegetation Imaging Sensor), and developed by NASA as an improved version of a former experimental system called SLICER, developed in 1994 (Blair et al., 1999). GLAS, a full waveform LiDAR system on ICESat, an ice-centric designed mission that collected data from 2003 to late 2009, was the first spaceborne LiDAR system to measure terrestrial surfaces (Zwally et al., 2002). However, the first commercial full waveform airborne LiDAR system only became available in 2004 (LiteMapper-5600 LiDAR system based on the Riegl LMS-Q560 laser scanner) (Hug et al., 2004). The full waveform spaceborne LiDAR GEDI is the first spaceborne system primarily designed for vegetation monitoring. Embedded on the International Space Station at the end of 2018, it measures the 3D structure of forest canopies to better understand carbon storage, biodiversity, and the impacts of climate change on global ecosystems (Dubayah et al., 2020). Even though it focuses on ice surface monitoring and is based on a completely different technology, i.e., a photon counting system, ICESat-2 (Markus et al., 2017; Neumann et al., 2019), launched in 2018, also provides valuable height information over forests that complement GEDI's data (Liu et al., 2021). While spaceborne LiDAR has not yet made the transition to systematic monitoring like what is done with optical imagery (e.g., Landsat or Sentinel missions), these two NASA missions showcase its potential and drive the development of future systems.

4.3.1.2 Measuring Vegetation Height from 3D ALS Data

The increasing use of ALS data for forest applications demonstrates the suitability of ALS technology for forest survey (Beland et al., 2019; White et al., 2016; Wulder et al., 2007) (see also Table 4.1). However, the assessment of forest parameters derived from ALS data is faced with a number of issues related to both the horizontal and vertical sampling characteristics of the measurements and to the nature of the interactions between the signal and the vegetation. The main factors that influence vegetation height accuracy at the footprint, tree, and stand levels are reported in Table 4.2.

By modifying the point distributional properties, the acquisition setup, which includes the laser instrument characteristics (wavelength, energy, pulse frequency, beam divergence, scanning pattern) and the flight parameters (flying altitude and speed, overlap of flight lines), was found to impact the quality of height estimates (Dayal et al., 2020; Keränen et al., 2016; Hopkinson, 2007; Næsset, 2009b). Additionally, because the laser signal interacts with vegetation components on its path to the ground, both the structure and the optical properties of the vegetation also influence the vertical sampling of the vegetation layer (Disney et al., 2010). Consequently, the quality of the information on tree height distribution derived from ALS data depends on the capacity of the data to describe the top of the canopy and the underlying ground, but also on vegetation layering.

When a laser pulse interacts with a tree apex, the total tree height is accurate on condition that the amount of vegetation material in a narrow elevation range is sufficient to backscatter, in a very short lapse of time, more energy than required to trigger a return (see, e.g., Wagner et al. [2006] for the theoretical background). According to the nature of the target—i.e., its shape, leaf density, and reflectance properties—canopy height is underestimated to varying degrees (Disney et al., 2006;

TABLE 4.2
Main factors influencing accuracy of vegetation height measurements from LiDAR data at three levels of analysis: individual footprint (pulse), tree, and stand

Level of Analysis	Factors Influencing Height Accuracy	Comments	References (Non-exhaustive)
Footprint	Signal triggering method in multi-echo systems or echo detection method in full waveform data	Impacts both number and position of detected points	Disney et al. (2010) Holmgren et al. (2003)
	Vegetation structure and spectral properties	Height underestimation ranges from ~4% to 7% and is more important within conifers	Wagner et al. (2006)
	Emitted energy	The higher the energy, the lower the time for triggering a return	Wagner et al. (2004)
	Footprint size	Underestimation increases with footprint size due to a decrease in irradiance (power/unit area)	
		But the probability to sample a tree apex increases with footprint size	
	Scan angle	No significant impact on measurements for angles < 10°	
Individual tree	Pulse density (function of ALS system and flight parameters)	The probability to sample a tree apex increases with pulse density	Hirata (2004)
		The minimum required density depends on crown size: 2–10 pulses/m² for mature stands and saplings	Kaartinen et al. (2012)
		Tree structure and terrain are better characterized when density increases	Véga et al. (2014) Véga and Durrieu (2011)
	Scan angle	Impact of scan angle on point distribution is greater for elongated crowns, e.g., conifer crowns	Alexander et al. (2018)
		Flight line overlapping reduces occlusions, thus improving both crown shape and height descriptors	
	Vegetation structure and composition	Quality of height estimates depends on crown shape and radiometric properties	
		Tree detection algorithms perform better in homogenous stands	
		Dominated trees are more difficult to detect and measure	
	Topography	Accuracy of height assessment decreases with slope	
		Height of slanting trees might be biased	
		Crown structure can be distorted by slope normalization	
Plot/Stand	Pulse density (function of ALS system and flight parameters)	Height parameters estimated through models; results are less sensitive to pulse density than for tree-level analysis; multipulse mode provides more accurate results than the single-pulse mode	Disney et al. (2010) Evans et al. (2009) Hodgson and Bresnahan (2004)
	Scan angle	Scan angles < 15° off-nadir to be preferred; or multiple viewing angles for the same plot	Hopkinson (2007) Næsset (2009b)
	Vegetation structure and composition	Accuracy is higher in simple structure (e.g., even-age, single-layer stands)	Véga et al. (2014)
		Accuracy is higher in coniferous stands	Keränen et al. (2016)
		Local calibration is required.	Dayal et al. (2020)
	Topography	DTM quality impacts the accuracy of height parameters	

Nelson, 1997). Disney et al. (2010), using a simulation approach, reported underestimations of canopy height of approximately 4% and 16% for broadleaves and conifers, respectively.

Also, point positions, and hence height estimates, may depend on the analog detection method used to identify an echo in the backscattered waveform (Disney et al., 2010; Wagner et al., 2004).

The impacts of the ALS system and flight setting on height retrieval have been investigated in numerous studies. Increasing the flight altitude leads to an increase in the footprint size at the Earth surface level and a simultaneous reduction in the pulse energy per unit area. Therefore, the emitted pulse has to penetrate deeper within the canopy before sufficient energy is backscattered to trigger a return, thus leading to higher underestimations of canopy height (Hopkinson, 2007; Lovell et al., 2005; Persson et al., 2002). For instance, Anderson et al. (2006b) reported mean errors of -0.76 m (± 0.43 m) and -1.12 m (± 0.56) with footprint sizes of 0.33 m and 0.8 m, respectively. However, the increase in footprint size that occurs with higher flight altitude increases the probability of sampling tree tops (Hirata, 2004; Hopkinson, 2007; Næsset, 2009b). Hirata (2004) showed that when increasing the footprint size from 0.3 m to 1.2 m by increasing flying altitude from 300 to 1200 m a canopy height increase of 0.9 m was obtained for a mountainous stand of Japanese cedar (*Cryptomeria japonica* L.f.). As a result, some authors have concluded that flying altitude has a minimal effect on height accuracy (e.g., Keränen et al., 2016).

Another important factor is the point density (Borgogno et al., 2020; Wilkes et al., 2015; Véga et al., 2012; Reutebuch et al., 2003; Nelson, 1997). Point density is closely linked to flight altitude and speed, pulse and scan frequencies, as well as scan angle. It also depends on the overlap between flight lines. Because the likelihood of sampling tree apices increases with point density, lower densities were found to generate higher height underestimations (Disney et al., 2010) and prevent accurate estimation of parameters at the tree level (Borgogno et al., 2020). However, Hopkinson (2007) stressed that it remains difficult to distinguish the impact of each component. When working at the tree level, height is directly assessed using treetop elevation (see Section 4.5), and some authors recommend a point density above 5 points/m^2 to maximize both tree crown detection and the probability of sampling tree apices (Falkowski et al., 2009; Hirata, 2004). In their study based on UAV-LiDAR data, Peng et al. (2021) found that tree height accuracy improved with point densities from 12 to 108 points/m^2, with the most significant gains occurring between 12 and 17 points/m^2. On the contrary, assessments of height at the plot level, which are mainly performed using models and no longer by direct measurements (see Section 4.4), were found to be unaffected by point density (Borgogno et al., 2020; Montealegre et al., 2016; Jakubowski et al., 2013; Lim et al., 2008; Treitz et al., 2012) (Table 4.2).

Besides flight parameters, and because vegetation height is typically computed by subtracting a digital terrain model (DTM) from the elevation of either the non-ground points or the digital surface model (DSM) of the outer canopy layer, the quality of vegetation height estimates is closely correlated with DTM quality. Thanks to the ability of a light signal to penetrate through vegetation openings, ALS acquisitions can sample the ground. Dedicated algorithms have been developed to classify points into ground and non-ground categories and to produce a digital terrain model (DTM) (Li et al., 2020b; Cățeanu and Arcadie, 2017; Guo et al., 2015; Meng et al., 2010; Sithole and Vosselman, 2004). Overall, ground elevation errors in LiDAR DTMs are usually less than 30 cm under forest covers (Chen, 2010; Chauve et al., 2008; Hodgson and Bresnahan, 2004; Reutebuch et al., 2003). Lower sampling densities result in coarser terrain modeling, but ground elevation accuracy also depends on the chosen interpolation method (Adedapo and Zurqani, 2024; Bui and Glennie, 2023; Cățeanu and Ciubotaru, 2020). In addition, significant variations in errors were found depending on both vegetation structure and density, which impacts the way the ground is sampled (Adedapo and Zurqani, 2024; Simpson et al., 2017; Cățeanu and Ciubotaru, 2020; Hodgson and Bresnahan, 2004). Ackermann (1999) reported penetration rates around 20–40% for coniferous and deciduous forest types in Europe, but this rate can be locally inferior to a few percent, i.e., 2–3%, in very dense tropical forests. Under a conifer forest, Reutebuch et al. (2003) reported mean DTM errors of 0.22 m (± 0.24 m SD) with errors increasing with canopy density and ranging from 0.16 m (± 0.23 m) within clearcuts to 0.31 m (± 0.29 m) within uncut areas. In their study, Clark et al. (2004) obtained the highest RMSE (0.95 m) within dense, multi-layered evergreen canopy in old-growth forests, for which ground elevation was overestimated. In addition, Leckie et al. (2003) noticed that variation of the ground surface at the base of the tree could easily reach ±50 cm. For these authors, the local micro-topography, which is hard to model, partly explained the observed 1.3 m (± 1.0 SD) tree height underestimation.

Slope can also affect the quality of DTM under vegetation, and elevation errors were found to increase with slope (Næsset, 2015; Aguilar et al., 2010; Hodgson et al., 2005; Hodgson and Bresnahan, 2004). For example, under a multi-layered tropical forest, Clark et al. (2004) reported a 0.67 m elevation RMSE increase within steep slopes. This can be explained by the fact that point classification algorithms, which are used to identify ground points before producing the DTM, are mainly based on geometric properties. Due to more similar geometric characteristics between ground and low vegetation point clouds, in the presence of slope they perform less efficiently. The difficulties involved in classifying ground points in relief and forested environments led to the development of many algorithms (Qin et al., 2023; Zhao et al., 2016; Meng et al., 2010). Among the several classification algorithms, the TIN iterative approach (Dong et al., 2018; Axelsson, 2000) is widely used and is acknowledged to be quite robust and to produce good results in a wide range of environments (Chen et al., 2021; Véga et al., 2012). Methods developed to process full waveform data acquired by commercial small footprint systems led to improved geometric information (more echoes extracted and higher target localization accuracy) and also provide additional features, such as echo intensity and width, that are linked to target properties (Chauve et al., 2009; Reitberger et al., 2006; Wagner et al., 2004). Despite the difficulty involved in decorrelating the influence of geometric and radiometric characteristics of the targets on these features (Ducic et al., 2006), they proved to be very useful in some studies (Chehata et al., 2009) when attempting to improve ground point classification when used in addition to echo locations.

DTM errors in sloping areas clearly impact height assessments. Over a complex terrain in a mountainous area, Véga and Durrieu (2011) found that tree height errors increased as a function of slope. Heights were underestimated for low slopes (i.e., below 25%), while an overestimation trend was found for steeper slopes (i.e., above 25%). However, errors in height were not attributed solely to DTM inaccuracy but were also due to the way heights are derived from ALS data (Alexander et al., 2018; Véga and Durrieu, 2011). Indeed, when a treetop is identified, height is assessed as the difference between the treetop elevation and the ground elevation and does not always represent the actual tree height. When trees slant, the crown and its associated local maxima move toward the slope and thus lead to an overestimated height (Figure 4.3). Overall, ALS was found to underestimate vegetation height with a magnitude of underestimation that varied depending on the sensor used, the flight parameters, and the characteristics of the vegetation (Table 4.2).

4.3.1.3 Measuring Vegetation Height with Spaceborne LiDAR

The development of spaceborne LiDAR systems for Earth observation is still in its early stages, and the continuity of these missions is not yet assured. This uncertainty limits the use of space LiDAR data for long-term operational forest monitoring. Therefore, this chapter focuses on airborne LiDAR. Nonetheless, spaceborne LiDAR missions offer new opportunities for forest monitoring and have been the subject of many research studies, particularly in the last five years following the launch of NASA's GEDI and ICESat-2 missions. While GEDI is dedicated to vegetation monitoring, ICESat-2 includes vegetation height monitoring as a secondary objective. Processing data from spaceborne LiDAR requires specific methods that are not covered here. However, NASA provides several levels of products, especially vegetation height data. We hereafter briefly summarize the quality of height measurements available from current space LiDAR missions.

The GEDI instrument operates at a wavelength of 1064 nm, and each LiDAR pulse covers 25 m diameter circular footprints on the ground (Dubayah et al., 2020). For each footprint, the raw first-level data is a full waveform, a digitized representation of the backscattered signal, which provides insights into the vertical distribution of vegetation within the footprint. Higher-level GEDI products estimate forest characteristics, such as vegetation heights, from the analysis of these waveforms.

Optical Remote Sensing of Tree and Stand Heights

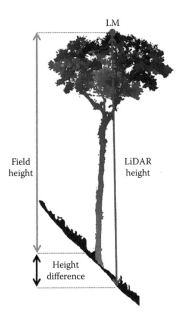

FIGURE 4.3 Error in tree height measurement associated with terrain slope (adapted from Véga and Durrieu, 2011).

The last significant return in the full waveform is considered the pulse hitting the ground, i.e., ground peak, and allows for estimating the ground elevation. The beginning of the backscattered signal corresponds to the part of the vegetation first intercepted by the laser beam and translates the elevation of the top of the canopy within the footprint. Using the ground elevation and the top-of-canopy elevation for each 25-m diameter footprint, the canopy height is estimated. Additionally, relative heights (RH) within the canopy are calculated, indicating the height at which a specific quantile of the returned energy is reached, with these heights given relative to the ground. The resulting RH profiles reflect the vertical distribution of vegetation within the footprint, with measurements provided in 1% increments ranging from RH0 to RH100.

Different studies have considered various relative heights as a measurement for either the canopy top height or the stand dominant height, e.g., RH100, RH99, RH98, or RH95 (Schleich, 202;, Dorado-Roda et al., 2021; Duncanson et al., 2021). GEDI vegetation height errors can reach up to several meters (Dorado-Roda et al., 2021; Guerra-Hernández and Pascual, 2021; Urbazaev et al., 2021; Adam et al., 2020), e.g., RMSE of ~6 m for a complex study area in northeastern France with relief and highly diverse forest stands (Schleich et al., 2023) and of ~4 m for 40 sites located in the U.S. mainland (Liu et al., 2021).

A significant part of the errors can be attributed to GEDI's low georeferencing accuracy that can impair the ability to compare with reference measurements (Lang et al., 2023; Schleich et al., 2023; Potapov et al., 2021; Roy et al., 2021; Urbazaev et al., 2021; Adam et al., 2020). Schleich (2024) estimated that 84% of GEDI height estimation errors could be attributed to georeferencing errors. Season (leaf-on vs. leaf-off for broadleaf forests) and stand type (broadleaf, coniferous, and mixed forest) were also found to be factors affecting height accuracy. GEDI vegetation height quality, like that obtained from ALS and DP, is also impacted by the quality of GEDI's ground elevation estimation. Slope and canopy cover density, which may degrade the accuracy of ground elevation retrieval, have also been shown to influence GEDI height estimations (Oliveira et al., 2023; Yun et al., 2023; Wang et al., 2022; Dorado-Roda et al., 2021; Liu et al., 2021; Adam et al., 2020). Moreover, GEDI height quality assessment is very sensitive to the quality filters applied to the dataset. Putting more effort into poor-quality data filtering and developing an approach to correct GEDI heights from the slope effect would largely improve GEDI height estimates (Schleich, 2024).

The second spaceborne LiDAR mission currently operating and acquiring data is NASA's ICESAT-2 (Ice, Cloud, and land Elevation Satellite 2). While the primary goal of ICESat-2 is to monitor Earth's ice sheets, it also provides valuable products for land surfaces, oceans, and the atmosphere. Since 2018, the Advanced Topographic Laser Altimeter System (ATLAS) instrument on board ICESat-2 has been routinely measuring forest structural characteristics across the Earth (Neuenschwander and Pitts, 2019). Unlike GEDI, ATLAS is a photon-counting LiDAR operating at 532 nm, and the laser is split into six beams—three weak and three strong—sampling Earth's surface simultaneously along ground tracks. Ground track orbits are spaced by 28.8 km, with each orbit being revisited every 90 days. Within a given orbit, the beams are spaced 3.3 km apart. The elevations and coordinates of backscattered photons are collected and registered. Within an 11-m beam footprint, the LiDAR system records backscattered photons with a spacing of 70 cm between consecutive detections (Magruder et al., 2021). Photon profiles are then post-processed to: (1) classify photons as noise, ground, canopy, or top canopy and (2) compute ground elevation and canopy heights (Malambo and Popescu, 2021; Neuenschwander and Pitts, 2019). Photons are processed in 100-m and 20-m geolocated segments for ground elevation and height estimates.

The geolocation accuracy of ICESat-2 segments has been demonstrated to be very high, with only a few meters offset compared to ground references (Magruder et al., 2021) and less than 1 m offset when compared to ALS datasets (Soma et al., 2022). Additionnaly, ICESat-2 is the only spaceborne LiDAR mission that covers the Earth's poles, including boreal forests (Feng et al., 2023; Mulverhill et al., 2022). In boreal forests, ICESat-2 ground elevation estimates show strong agreement with ALS references for ground elevation (< 0.3 m bias and 3 m RMSE) and for canopy heights (< 1 m bias and ~4m RMSE) (Feng et al., 2023). However, while a 30% RMSE was observed for trees between 10 and 30 m in height, significant bias (> 30%) and errors (> 60%) were found for shorter vegetation (overestimation) and taller trees (underestimation). Additionally, a strong slope effect was noted (+10% RMSE at 30 m resolution), exacerbated by the absence of snow and during daytime conditions.

Validation of ICESat-2 in temperate forests remains scarce. In heterogenous forests, a strong effect of day versus night conditions was also observed for 100-m segments, with ~ −2 m and 0.2 m bias (6 m and 3 m RMSE) for ground elevation during day and night measurements, respectively (Soma et al., 2022). As a result, 100 m canopy height measurements were significantly affected, with RMSE close to 7 m in the daytime. In contrast, 20-m segments showed much less sensitivity to daytime sampling, with a bias lower than 0.2 m and ~2 m RMSE. As a result, canopy height estimations were consistently underestimated by 2.5 m for both day and night conditions, although the RMSE was higher at night compared to daytime (8 m versus 5.5 m, respectively).

In both boreal and temperate contexts, filtering of ICESat-2 products was a critical step to remove noisy estimates. Several variables embedded in ATLAS products provide information on the quality and reliability of the estimates (e.g., multiple scattering warning flag, cloud contamination, DEM outliers, canopy RH confidence). Therefore, it is recommended to apply a set of filters during pre-processing before any estimations. Research on the ICESat-2 photon processing algorithm is still ongoing, and new product releases are processed with regularly improved methods, enhancing the filtering of outliers and the accuracy of canopy height estimates. Hence, although the sparse density of ICESat-2 estimates might not be suitable for generating high-resolution height and biomass maps on its own, its reliable georeferencing and accuracy make it a valuable candidate for combining height estimates with 2D remote sensing data to produce such maps.

4.3.2 3D Modeling of the Canopy by Digital Photogrammetry

4.3.2.1 Principle and Brief History

Since the post–World War I period, large-scale aerial photographs have been extensively used in both forest inventory and monitoring (Spurr, 1960) for many purposes: locating forest areas; mapping forest types; inventorying forest conditions; assessing wood production; monitoring damage due to insects, diseases, and fires, etc. Information on stand structure and composition has been intensively used as the basis of stratification to improve the efficiency of field data collection and the accuracy of results in multi-stages sampling designed forest inventory (Korpela, 2004). Stereo-photogrammetry, introduced during the same period (Andrews, 1936), was used to assess qualitative and quantitative forest stand structural characteristics and was further developed in the 1940s (Korpela, 2004). The stereo-photogrammetry process is analogous to our own perception of depth with normal binocular vision. It is based upon the principle of parallax, which is the apparent displacement of a stationary object due to changes in the observer's position (White et al., 2013). It thus requires two images, which are taken from two different viewpoints, and the 3D measurements are deduced from the analysis of the parallaxes that change according to the distance between the objects and the sensor (Figure 4.4).

Stereoscopes and parallax bars have been long used as low-cost viewing and measurement instruments (Korpela, 2004). Analytic stereo plotters and comparators, which have been commonly available since the mid-1970s, enable accurate 3D measurements. The emergence of digital photogrammetric (DP) dates back to the 1990s (Maltamo et al., 2009) and was made possible by the development in computer technology that provided significant computing power required by 3D image analysis algorithms (Morgan and Gergel, 2013). DP has benefited from the research works conducted in the graphic, vision, and photogrammetry communities (Iglhaut et al., 2019; Remondino and El-hakim, 2006), the automation in photogrammetric workflows (Qin and Gruen, 2021), and deep learning approaches for both geometric and semantic problems (Hu et al., 2023; Heipke and Rottensteiner, 2020).

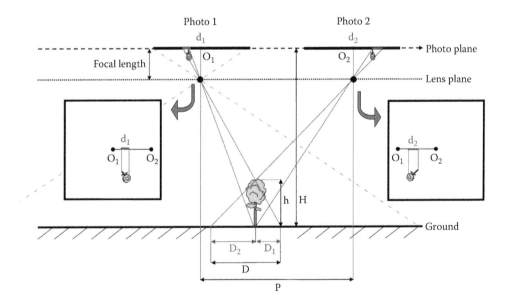

FIGURE 4.4 Principle of stereoscopic measurement, with O1 and O2: the centers of the photographs; H: height of the camera above the ground level; h: height of the tree; P: absolute parallax of the tree base; and D: parallax difference of the treetop with reference to the base plane (D1–D2). Tree height (h) is computed as the ratio (H-D)/(P-D)

To retrieve the three dimensions of surfaces from airborne or spaceborne images one must fully characterize the geometry of the acquisition system in order to trace the geometric path of the sunrays, from the target to its images, reflected by the observed target. The workflow consists in solving the internal orientation, which defines the geometry of the imaging system, and the exterior orientation, which gives the position and orientation of the sensor at the acquisition time and is often divided into two steps, i.e., solving the relative and absolute orientations (Heipke, 1997). The relative orientation is solved by matching points from one image to the other to generate tie points, and the absolute orientation by using control points whose coordinates are known in both the image and a mapping reference frame. Once the orientation of the photogrammetric model is known, parallax differences can be used to compute the elevation for each pixel of one image, provided a conjugate point can be identified in another image. This latter stage is also referred to as dense matching. Point matching, also known as the correspondence problem, is thus crucial in the photogrammetric workflow. And many algorithms have been developed to solve it, including surface-based or object-based approaches (Brown et al., 2003; Barnard and Fischler, 1982) or a combination of both (Lisein et al., 2013;). Remondino et al. (2014) provided an extensive review of dense matching algorithms. Albanwan and Qin (2022) reviewed and compared dense matching algorithms involving deep learning. Finally, Han et al. (2020a) provided a review of commercial and open-source image matching softwares, with an emphasis on digital surface models rather than point clouds.

With the current state-of-the-art computers and computational methods, digital photogrammetry (DP) does indeed offer several advantages. The complex calculation process named aerial triangulation can be used to process image blocks and not only stereo pairs, thus leading to a reduced number of ground control points required to process an area covered by several images. Furthermore, a multi-ray matching strategy, made possible by multi-view acquisitions, can significantly improve results of both the aero-triangulation and the object reconstruction steps (Thurgood et al., 2004).

The development of DP was also favored by the emergence of novel large-format digital aerial cameras producing significantly improved image quality, as they combine very high spatial resolution and high radiometric sensitivity, e.g., radiometric information coded on more than 8 bits (Leberl et al., 2010). The sensors thus overcame some issues linked to the complex interdependency between the radiometric range and the pixel size encountered in analog film imagery, i.e., grain noise (Leberl et al., 2010). Image texture is also enhanced, which is an important advantage for image-matching algorithms (White et al., 2013). The processing workflow is also simplified by the elimination of some tricky and time-consuming tasks, such as the scanning of analog photographs. State-of-the-art digital aerial cameras, e.g., the Z/I Imaging DMC (digital mapping camera) series from Intergraph, the ADS (airborne digital sensor) series from Leica, or the UltraCam series from Microsoft (formerly Vexcel), can capture large-format multispectral images typically in the red, green, blue, and near-infrared wavebands (Petrie and Walker, 2007). Some may also have the capability to record simultaneously very large-format panchromatic images (Z/I Imaging DMC [Hinz and Heier, 2000] and UltraCam [http://www.microsoft.com/en-us/ultracam/]).

Digital images can nowadays be acquired with increased within and between flight-line overlaps. While traditional acquisition mainly used around 60% and 70% overlap within and between flight-lines, respectively, state-of-the-art technology currently allows for up to 90% and 60% overlap with an add-on cost only linked to the additional air time when between flight-line overlap is increased (Leberl et al., 2010; Thurgood et al., 2004). With an emphasis on forest ecosystems, Goodbody et al. (2021) indicated that metrics from DP and ALS were more similar, with along- and across-track overlap of at least 60%.

The more recently developed algorithms can also be used to process data acquired by non-metric cameras, thereby extending the processing capacity beyond standard photogrammetric geometry and products. Data acquired with low-cost light acquisition systems, like consumer-grade cameras embedded on unmanned aerial vehicles (UAV), can now be processed. These cameras have high distortion levels and low geometric stability. Therefore, images that are also characterized by high rotational and angular variations between successive images and significant perspective distortions (Lisein et al., 2013) cannot be processed using conventional photogrammetric software. To address such image constraints, new solutions have been developed. For example, Lisein et al. (2013) used MICMAC (Multi Image Matches for Auto Correlation Methods), an open-source software combining a photogrammetric approach and newly developed computer vision algorithms referred to as Structure from Motion, to retrieve canopy structure from very high-resolution images acquired using an UAV. A review of Structure from Motion approaches with a focus on forestry applications is provided in Iglhaut et al. (2019).

The combined and recent developments in sensor technology, positioning systems (triangulation), and processing algorithms offer numerous benefits, including reduced occlusion, higher levels of automation, limited manual editing, and increased geometry accuracy (Goodbody et al., 2019; Leberl et al., 2010).

4.3.2.2 Measuring Tree Height Using Digital Photogrammetry and Resulting 3D Models

Even if some authors consider that image-based methods are now better at creating 3D point clouds than ALS surveys (Leberl et al., 2010), height information retrieval remains challenging in forest environments, and several issues must be addressed (Dietmaier et al., 2019; Iizuka et al., 2018; Fassnacht et al., 2017).

Firstly, a major drawback of photogrammetry is its inability to provide ground elevation under dense forests, thus preventing vegetation height assessment. Secondly, only the trees whose crowns are at least partially in direct sunlight are detectable on aerial photographs. And despite progress in sensor technology, suppressed or small shaded trees are still undetectable (Korpela, 2004). Finally, image matching in forest environments remains challenging and an error-prone task due to occlusion, repetitive texture, multilayered objects, or moving objects, i.e., change in position of treetops in windy conditions (Dietmaier et al., 2019; Lisein et al., 2013).

With visual or semi-automatic stereo interpretation, the accuracy of tree height measurement has been widely studied (Gyawali et al., 2022; Mielcarek et al., 2020; Ganz et al. 2019; Korpela, 2004; Gong et al., 2002). Accuracy depends on factors related to both images, e.g., scale and quality, and targets, e.g., crown geometry, radiometric properties, stand structure and topography, and leaf-on versus leaf-off conditions (Gyawali et al., 2022). Some authors announced 10 cm accuracy rates (Gagnon et al., 1993) and others mean absolute errors of 1.8 m (Gong et al., 2002). Korpela (2004), who compiled results from several studies, reported that heights are underestimated due to the inability to measure the very highest top shoots, and this bias tends to increase as photograph scale decreases. Despite this, Dempwolf et al. (2017) showed that short time series of DP models could be used to assess height growth for various species in a mixed temperate forest in Germany.

But tree height measurement accuracy is also highly dependent on tree base elevation assessment capacities. Ground elevation has been measured within open areas in the neighborhood of the trees (Gong et al., 2002) or, in closed-canopies, on targets positioned on the ground (Kovats, 1997). It was then assimilated to the tree base elevation. More conveniently, a range of digital terrain models (DTMs) have been used in combination with photogrammetric measurements or DSMs to derive forest parameters at the tree or stand level. DTMs extracted from field surveys (Fujita et al., 2003) or from a triangulation of manual photogrammetric measurement in open areas (Næsset, 2002a) were sometimes used. However, such solutions are limited to small areas. St-Onge et al. (2004) first tested the co-registration of a photogrammetric model with an ALS DTM to estimate individual tree height based on manual measurements. Using scanned photographs with an 11.3 cm pixel size, they reported an average underestimation in height of 0.59 m in a white cedar (*Thuja occidentalis*) stand. Korpela (2007) developed a semi-automatic, single-scale template-matching approach, to position treetops on stereo-images and then to estimate tree height and crown size using an ALS DTM. The spatial registration of both photo-derived DSM and ALS-derived DTM proved to be a critical step toward canopy height estimation (St-Onge et al., 2004), and various approaches have been proposed to co-register these models, at both point and pixel level (Balestra et al., 2024; Xu et al., 2023; St-Onge et al., 2008).

Despite the potential of DP to accurately measure elevations, most of the image-matching algorithms, originally developed for DTM extraction; proved incapable of consistently generating accurate DSMs over forested areas. St-Onge et al. (2008) reported that, when using precise ALS DTMs, the photo-ALS Canopy Height Model (CHM=DSM-DTM) could reconstruct general height patterns, but was unable to provide details about both individual tree crowns and small gaps. Similarly, using images at a 1:15,000 scale digitized at 2000 dpi and processed using an automatic image-matching algorithm (Match-T), Næsset (2002a, 2002b) reported significant underestimation of mean plot height. The author also indicated that the algorithm was not flexible enough to reconstruct abrupt changes in elevation, such as those that characterize canopies. Occlusions, due to the shape of the trees and to the complex 3D structure of the forest canopy, hinder image matching (Lisein et al., 2013). Compared to single-stereo models, image block approaches allowing multi-view processing were found to improve 3D canopy modeling and tree height estimation due to the reduction of occluded areas (Goodbody et al., 2021b; Hirschmugl et al., 2007; Magnani et al., 2000). The quality of the results also depends on the parameters defining the matching strategy (Figure 4.5). For example, Lisein et al. (2013), who tested several matching strategies, explained that some omitted isolated trees and that those that were optimized for deciduous canopy reconstruction did not perform very well when used for coniferous crown reconstruction.

Due to the high sensitivity of the canopy model quality to the matching strategy, and whose optimization depends on the cover type, the production of accurate CHM over large areas is a tricky task. To tackle this problem, Baltsavias et al. (2008) proposed a complex method combining both area and feature-based matching, the self-tuning of matching parameters, the generation of redundant matches, an automatic blunder detection, and a coarse-to-fine hierarchical matching strategy. The method was able to achieve better results than when using an ALS dataset for canopy height assessment, thus suggesting that methods developed for processing ALS CHM could also be efficient to process high-quality photo-ALS CHM over forests.

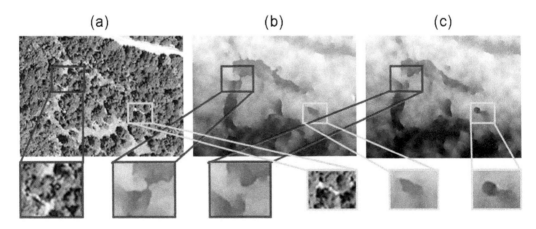

FIGURE 4.5 Illustration of the impact of changes in matching strategy on forest DSM using the MICMAC software. (a) Orthorectified image of the area. The two DSM (b and c) were both computed using a 0.28 m spatial resolution and a 9 × 9 pixel correlation window. The regularization coefficients (RS) and the correlation thresholds (CT) were different: RS = 0.010 and CT = 0.2 for (b) and RS = 0.005 and CT = 0.0 for (c). Changes in matching parameters impact crown shape (red area) and gap shape (yellow area) that were better reconstructed with the first set of parameters (b).

4.3.3 Comparison of ALS and Photogrammetric Products

Few studies have compared photogrammetric and ALS products (Ullah et al., 2019; Kangas et al., 2018; White et al., 2013; Leberl et al., 2010; Baltsavias, 1999c). As a complement to Table 4.1, Table 4.3 compares the main strengths and limitations of both photogrammetric and ALS technologies and derived products.

For acquisition purposes, imagery is considered the most advantageous (Table 4.3). Imaging systems, which have a greater field of view, can be operated at both a higher flying speed and altitude than ALS systems for which flight height is dependent on available laser power. As a result, survey planning is easier and less costly with an imager (White et al., 2013). For example, assuming an equal flying altitude and speed, and an equal sidelap, the same area is covered in about one-third of the time with an imager, considering a typical 75° field of view, than with an ALS with a 30° scan angle (Baltsavias, 1999c). Comparing typical acquisitions under optimal flying height and speed, Leberl et al. (2010) found that the flying time would be 13 times longer with an ALS system when compared to a digital camera system to obtain comparable point clouds and elevation results. In addition, imaging continuously benefits from the development of satellite stereo-imagery. Despite a lower geometric resolution leading to a decrease in CHM quality (St-Onge et al., 2008), satellite solutions can be used to cover very large areas at low cost (Neigh et al., 2014) and is a viable alternative to ALS for small-scale forest monitoring when high-quality DTMs already exist. Besides these considerations, another advantage of ALS is that, as an active sensor, it can be operated at any time of the day while the quality of aerial photographs is highly influenced by solar illumination. Indeed, as shadows hamper image matching, flying hours must be carefully chosen to minimize shadowing in the forest canopy (White et al., 2013). Furthermore, new ALS technologies such as GEIGER mode imagers and photon-counting systems can be operated at higher altitudes and generate larger swath, thus reducing the gap with DP surveys (Liu et al., 2022; Mandlburger et al., 2019; Degnan, 2016; Stoker et al., 2016).

Georeferenced point clouds or a 3D point cloud can be obtained more quickly when using ALS systems, as coordinates can be, under ideal conditions, automatically computed (Baltsavias, 1999c). However, as digital photogrammetry workflows become both increasingly efficient and automated, this advantage is gradually diminishing (White et al., 2013). Baltsavias (1999c) gives detailed information on the comparative geometric quality of point clouds generated from photogrammetry and ALS. At the same flying height, the geometric resolution of a laser measurement, given by the footprint size and depending on the laser beam divergence—typically 1 mrad—is coarser than the pixel resolution obtained by a digital camera (e.g., with a 15 µm pixel). Concerning relative planimetric/altimetric accuracies, planimetry is typically one-third more accurate than elevation with photogrammetry, while it is two to six times less accurate with ALS data. These higher planimetric errors will also significantly influence elevation accuracy on sloped terrain (Baltsavias, 1999c). Comparing the accuracy values for identical flying height in the 400–1000 m range, shared by both technologies, the photogrammetric accuracy is, on average, slightly better than with ALS. However, in practice, airborne LiDAR and imagers are not operated in the same conditions, thus making it difficult to compare their real performance levels.

Height point clouds, obtained by subtracting ground elevation from the elevation of ALS and photo-derived 3D points, were also compared in terms of height distribution by comparing percentiles on 400 m^2 forest plots (Lisein et al., 2013; Vastaranta et al., 2013). Low correlations between lower height percentile values of both point clouds (Lisein et al., 2013; Vastaranta et al., 2013) may be explained by the presence of ALS points within the canopy when DP CHM only describes the top of the canopy (Iqbal et al., 2019). However the very high correlations observed for higher percentiles—e.g., Vastaranta et al. (2013) reported mean differences below 0.2 m for the 70th height percentile values and beyond for 500 circular plots—reveal that most of the points in the ALS dataset were located at the same level as photo-derived points. They mainly describe the outer canopy shape, and there is very little information remaining to describe both the understory and the ground due to occlusion effects.

TABLE 4.3
Comparative summary of main strengths and limitations of both ALS and DP technologies and of resulting 3D point clouds and forest products (Baltsavias, 1999c; Leberl et al., 2010; Lisein et al., 2013; St-Onge et al., 2008a; Vastaranta et al., 2013; White et al., 2013)

		Best Rated	ALS (Airborne Laser Scanning)	Digital Photogrammetry (DP)
Data acquisition	System lifetime	DP	Determined by the number of pulses that can be emitted by the laser; equivalent to ~10,000 operating hours Rapid deterioration may occur with a drastic decrease in output power	Decades for robust aerial cameras
	Mission planning	DP	Flight height limited by eye safety (min. height) and laser power (max. height); typically 500–1000 m Higher flight height possible (up to 4000 m) but with reduced pulse frequency to avoid signal mixture between successive backscattered signals Typical scan angle: 20° to 40° ➔ Mission planning difficult in mountainous areas	Flight height: from 500 to 3500 m according to plane type (up to 12 km with high-altitude aircraft). Typical effective FOV: 75° ➔ Large areas covered with fewer flight-lines and flying time up to 13 times shorter compared to ALS ➔ Multi-view stereo and optimized B/H might provide higher-quality forest canopy reconstruction
	Flying conditions	ALS	Few illumination constraints (LiDAR can be operated day and night, winter and summer) System can be operated in both leaf-on and leaf-off conditions	Acquisitions constrained by solar illumination (impacts radiometric quality), and view angles due to sensitivity of image matching to occlusions and shadows Leaf-on conditions only to provide forest height products
Data processing	Production of geolocated 3D point clouds	ALS	3D coordinates automatically computed by combining information recorded by the LiDAR, the IMU, and the DGPS ➔ reduced processing time	With increased efficiency and automation level of DP workflows, advantages of ALS have gradually diminished Final quality of geolocation is less dependent on DGPS and IMU measurements quality than for ALS

Product quality	3D products	ALS	Planimetry two to six times less accurate than altimetry Sampling pattern and pulse density depend on flight parameters Information on both the vegetation surface and the ground, allowing to extract a DTM and to further assess vegetation height and structure
			Planimetry 1/3 more accurate than altimetry Sampling: in theory, regular (once to twice the image resolution); in practice, depends on image-matching results At a given cost, higher point density achievable than with ALS Information only for object surfaces visible in at least two images
	Forest height products	ALS	Height product accuracy depends on sensor specifications, flight parameters, and resulting point density Small gaps and treetops better described than on DP products
			External DTM needed to provide vegetation height Height product accuracy depends on image scale, B/H, image quality (radiometry and texture), shadow patterns, image-matching algorithm, number of images used during matching, and external DTM quality
			Accuracy of both ALS and DP height products is function of vegetation type and structure. In both cases, it is higher within coniferous and even-aged stands. In-depth evaluations of forest height products are still required for accuracy comparisons.
	Forest inventory and monitoring	ALS and DP	ALS is the leading technology for acquiring DTMs over forested areas, but the potential of ALS for updating forest information is limited due to cost considerations
			Higher potential than DP for the assessment of biophysical parameters linked to structural parameters
			Large area wall-to-wall coverage at low cost Multiple use of data: stand delineation, species identification, and monitoring of deforestation and disasters
			Large archive datasets exist in some countries for retrospective mapping
			ALS and DP are complementary for forest mapping and inventory → necessity to optimize alternated acquisitions under cost constraints

When further assessing heights at the tree level or assessing dominant heights at the plot level using height distribution metrics (see Section 4.4.2), estimation accuracy was found to be slightly better with ALS data compared to photo-derived data (Lisein et al., 2013; Vastaranta et al., 2013; White et al., 2013). For instance, Lisein et al. (2013) obtained an adjusted R^2 and a RMSE (%) of 0.94 and 3.7% and, respectively, 0.91 and 4.7% for height estimation at the tree level and a R^2 and RMSE (%) of 0.86 and 7.4% and 0.82 and 8.4% for dominant height estimation at the stand level (Lisein et al., 2013). Vastaranta (2013) also found a lower RMSE (%) for mean height prediction at the plot level with ALS data (7.8%) compared to photo-derived products (11.2%). However, White et al. (2013) also pointed out that there is no rigorous comparison of the relative accuracy of canopy heights derived from ALS and image-based point clouds over a range of forest types and terrain complexities.

The product that is common to both ALS and DP technologies is the 3D model of the Earth's surface. Focusing on the structure of the outer canopy surface approximated by computing a raster CHM, several studies reported that tree crowns were wider and less defined in photo-derived CHM (Lisein et al., 2013; Vastaranta et al., 2013; Barbier et al., 2010). Moreover, small gaps and treetops as well as fine-scale peaks and gaps in the outer canopy were not perfectly modeled on the photo-derived CHM that tend to behave as a smoothed version of ALS CHM (Lisein et al., 2013; St-Onge et al., 2008).

The ability of ALS to provide information on sub-canopy forest structure as well as on ground topography, even below closed canopies, means this technology is very suitable for forest inventory. Despite its sensitivity to solar illumination, optical imagery can provide consistent spectral information that can be used to report, for example, on species composition and to assist forest inventory in a way that cannot be done by ALS (Baltsavias, 1999c).

Finally, due to their respective advantages and to their equivalent potential for top vegetation height measurements, ALS and photogrammetry are more complementary than mutually exclusive, and the issue should be how to best optimize alternated acquisitions under cost constraints. For example, Vastaranta et al. (2013) suggest that, for forest mapping and monitoring purposes, ALS data could be acquired every 10 or 20 years, depending on forest and management considerations, and digital stereo imagery could be used to update forest information in the intervening period. However, for an initial inventory, both ALS and imagery—but not necessarily stereo imagery in this case—are acknowledged to be very useful.

4.4 ASSESSING HEIGHT CHARACTERISTICS AT THE STAND LEVEL

Early LiDAR studies showed an underestimation bias when measuring tree height from LiDAR data, which is due to several factors (see Section 4.3.1.2). Models based on empirical relationships between LiDAR data and forest attributes measured in the field at the plot level (Figure 4.6) were thus used to correct this bias and produce stand-level estimations and maps of stand height characteristics (Coops et al., 2021; White et al., 2016; Laurin et al., 2019). These approaches, which have been extended to predict other stand characteristics, are widely used in forest applications and are often referred to as area-based approaches (Coops et al., 2021; Bouvier et al., 2015; Næsset, 2002b).

4.4.1 General Presentation of Area-Based Approaches

The development of models at the plot level is usually achieved in two or three stages. The first two stages, which are the establishment of a predictive model to assess a forest parameter from 3D data and then its application to the area covered by the 3D dataset, are described in many studies (e.g., Bouvier et al., 2015; White et al., 2013; Wulder et al., 2012; Naesset, 2002b). When wall-to-wall 3D data are available these stages lead directly to maps. When only 3D measurement samples are available, local forest parameter assessments have to be further extrapolated as part of a third stage to produce maps over large areas using other remote sensing data, e.g., optical or radar imageries

Optical Remote Sensing of Tree and Stand Heights

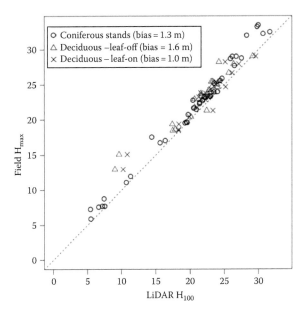

FIGURE 4.6 Maximum LiDAR height (H_{100}) is compared to the height of the tallest tree measured in the field for 93 forest plots. Regardless of the stand type (here coniferous [39 plots] and deciduous [twice 27 plots]) and of acquisition conditions (leaf-on and leaf-off for the deciduous stand) maximum height was underestimated by LiDAR by at least 1 m on average (biases ranging from 1 to 1.6 m). The coniferous stands are pine plantations, and trees have a relatively flat crown compared to other coniferous species. In addition, LiDAR datasets used for the comparison of maximum heights have point densities high enough to limit the probability to miss treetops (8, 20, and 18 pulses/m² for coniferous, deciduous leaf-on, and deciduous leaf-off, respectively, and with a footprint size of about 27 cm in all cases).

and other map products available at a national or global scale (e.g., SRTM, vegetation maps). The two first stages are briefly presented next and are illustrated in Figure 4.7 for 3D ALS point clouds. However, the principle remains the same when using either DP point clouds or raster CHM. It can also be extended to large footprint LiDAR data processing, even if in this case the metrics used are different from those derived from point clouds.

First a model is developed at the plot level to infer forest parameters from metrics derived from either ALS or photogrammetric 3D data (Figure 4.7a,b,c). The model is calibrated and validated using reference plots measured in the field for which stand characteristics such as dominant height, Lorey's height, basal area, volume, and aerial biomass are inferred from the field measurements (Figure 4.7a). This step requires allometric equations to assess some biophysical parameters, e.g., volume and biomass, from structural characteristics such as diameter at breast height and height.

For each plot the corresponding 3D ALS data subset is clipped and several metrics are derived from either the sub–point clouds or the CHM (Figure 4.7b). Next, using a subset of the available plots as a training set or a cross-validation procedure (e.g., leave-one-out), a model is built that predicts stand characteristics measured in the field from the most explanative 3D metrics. Multiple approaches can be used to establish the model. They might be parametric or non-parametric and include maximum likelihood, discriminant analysis, nearest neighbors, random forests, and various forms of regressions (Coops et al., 2021; McRoberts et al., 2010). The selection of the most explanatory variables is either made based on a preliminary statistical analysis or is part of the model construction, e.g., when stepwise regression approaches or random forest analysis are used. When enough field reference data are available, more recent deep-learning approaches can also be implemented (Dayal et al., 2023; Lahssini et al., 2022; Nguyen et al., 2022). The model is then

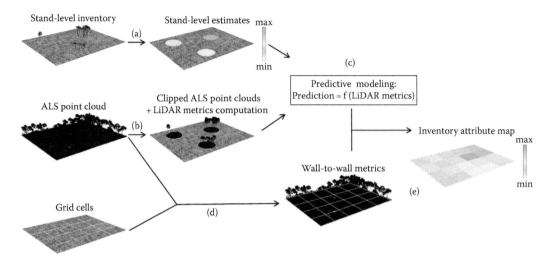

FIGURE 4.7 Principle and main steps of area-based approaches used to predict forest parameters from 3D data. (a) Field inventory at plot level; (b) extraction of sub–point clouds for all the inventoried plots; (c) establishment of a predictive model linking forest characteristics to 3D ALS data characteristics and validation of the model; (d) segmentation of the whole area into grid cells with a size similar to the one of the inventory plots; and (e) computation of the explanative ALS metrics for each grid cell and application of the model, cell by cell, to obtain a map of forest parameters.

validated using either the set of remaining reference ground plots or the plots that were left aside at each repetition when applying a bootstrap approach. This latter approach is to be favored especially when the total number of reference plots is limited.

Once built, the model can be extrapolated to the whole area covered by the ALS 3D dataset. To achieve this, the area is first subdivided into grid cells, the size of which are similar to the size of the reference ground plots (Figure 4.7d), the explanative 3D metrics are then computed for each cell, and the model is used to predict the value of the forest parameter at the cell level for the whole area (Figure 4.7e).

If 3D data do not encompass the whole area, a third stage is required to obtain a map. This consists of building another model that in turn uses the forest parameters assessed using the 3D dataset and links them to new variables extracted from another ancillary dataset that covers the whole area and that is usually made up of optical images. Once built, the model is applied to assess the parameter of interest on the areas that were not covered by the 3D data. Nearest neighbors techniques have been widely used to predict continuous variables based on satellite image data (McRoberts et al., 2010). Other non-parametric approaches, like random forests or neural networks, or parametric approaches based on regression analysis can also be used to achieve this aim (McRoberts et al., 2010). In the next subsection we will focus on the first step, i.e., the construction of the model linking 3D metrics to forest parameters.

4.4.2 Area-Based Model Implementation

The objective of this subsection is threefold. First, it aims to illustrate and compare two families of approaches that can be used to build a model to predict stand height characteristics derived from 3D ALS data. Second, it seeks to compare ALS and DP products to assess stand height characteristics through the comparison of results obtained with a model using only the information from the top of the canopy to the models using the whole 3D ALS information. This comparison should illustrate on the added value of having information from the understory, even if it is attenuated by occlusion effects. Third, it presents some issues regarding model accuracy. To achieve the first two objectives, two sites with contrasted stand types were chosen in order to compare the behavior of the three model types by considering several key elements involved in the model construction process. All the models were tested to predict both dominant and Lorey's heights. The first site is a coniferous forest located in the Landes region in the southwest of France (44.40° N, 0.50° W). The site is dominated by monospecific stands of maritime pine (*Pinus pinaster*) in even-aged plantations. The second site is a deciduous forest located in northeastern France (48.53° N, 5.37° E). The site comprises multi-layered broadleaved stands, dominated by European beech (*Fagus sylvatica*), hornbeams (*Carpinus betulus*), and sycamore maple (*Acer pseudoplatanus*). ALS data were collected over the coniferous and deciduous study sites with a point density of 10 pt/m² and 30 pt/m², respectively. Field data were collected on 39 15-m radius circular plots within the two months that followed the ALS survey for the coniferous site and on 42 circular plots within the year prior to the ALS survey for the broadleaved forest. Lorey's height (H_L) and dominant height (H_{dom}) were estimated from field measurements for each field plot. H_L was computed as the mean height weighted by basal area. H_{dom} was computed as the mean height of the six largest trees according to their diameter at breast height. H_L and H_{dom} were estimated at the stand level from ALS metrics using two different approaches that are keeping with the general methodology for area-based approaches described in the previous subsection.

On the one hand, we applied a practical process to predict H_L and H_{dom} from ALS data proposed by Næsset (2002b). This approach is typical and widely used to build predictive models of stand characteristics from ALS data. Numerous metrics were derived from the height distributions of first or last LiDAR returns: maximum values (H_{maxf}, H_{maxl}), mean values (H_{meanf}, H_{meanl}), coefficients of variation (H_{cvf}, H_{cvl}), percentiles of the distributions (H_{0f}, H_{10f}, ..., H_{90f} and H_{0l}, ..., H_{90l}), and canopy densities (d_{0f}, d_{10f}, ..., d_{90f} and d_{0l}, ..., d_{90l}) computed as proportions of ALS hits above a given percentile of the distribution. Stepwise regression was performed in order to select the most explanative ALS metrics that would remain in the final models. No metric with a partial Fisher statistic greater than 0.05 was selected (Næsset, 2002b). A linear relationship among log-transformed variables was applied. Log-transformation of stand attributes and ALS metrics was used to accommodate non-linearity. H_L and H_{dom} were estimated following this methodology for both sites. We used adjusted R² (R^2_{adj}) to account for the number of ALS metrics in the final models. RMSE was also calculated to assess the accuracy of the predictions. This approach is referred to hereafter as the point distribution approach.

On the other hand, we proposed to use a conceptual model to predict H_L and H_{dom} from only four ALS metrics (Bouvier et al., 2015). This model skips the step aiming at selecting the best metrics amongst a large set of potential ALS metrics. Metrics have been defined to characterize the natural variability of stand structures. In area-based approaches, individual tree heights are not determined. Instead, an average canopy height (μ_{CH}) is easily measured from the 3D LiDAR point cloud and is an important predictive variable (Lefsky et al., 2002). Thereby μ_{CH} metric was chosen as the first variable and calculated by averaging first return elevations. An indicator of tree height heterogeneity at plot level should be used in addition to μ_{CH}. We used the variance in canopy height (σ^2_{CH}) to characterize tree height heterogeneity as suggested by Magnussen et al. (2012). These first two variables were calculated without taking into account returns that were below a 2 m height threshold, so as to describe the parts of the plots that were actually covered by trees. However, the fact that tree attributes measured in the field are related to the whole plot area means that the rate of open areas in each plot must also be evaluated. Gap fraction (P) has been calculated from ALS data as the ratio between the number of first returns below a specified height threshold and the total number of first returns. When calculated in this way, P is related to the penetration of light through the canopy but was found to be well correlated with fractional cover (Hopkinson and Chasmer, 2009). P was thus the third selected metric. The previously defined metrics only refer to the structural properties of the top of the canopy and can be calculated for either ALS or photo-derived point clouds. To take advantage of the capacity of LiDAR to penetrate the vegetation and provide information on the vertical crown size and on overtopped trees, we defined an additional metric, H_{crown}, as the mean of the crown heights. Crown heights were estimated for each 1 m x 1 m area included in the plot based on the vertical distribution of ALS points. H_L and H_{dom} were estimated using the four metrics in a log-transformed model. This approach is referred to as the mechanistic model.

A third type of model was built in a way similar to the mechanistic model but using only the three metrics that can be computed when using photo-derived point clouds or raster CHMs, i.e., μ_{CH}, σ^2_{CH}, and P. This last approach is referred to as the top of canopy mechanistic model.

H_L and H_{dom} have been predicted using the three approaches. A summary of the results is displayed in Table 4.4.

The three approaches satisfactorily estimated height attributes in the coniferous forest. Models provided high R^2_{adj}, all equal to 0.99 despite a higher number of variables for the mechanistic model and low RMSE. Only one metric remained in the final point distribution model, H_{90f} and H_{maxl}, for H_L and H_{dom}, respectively. RMSE was slightly reduced using the mechanistic model, i.e., −0.16 m for both H_L and H_{dom} when compared to the point distribution model. The majority of forest studies have focused on coniferous forests characterized by a quite simple structure (Lim et al., 2003). In such stand types, both stand homogeneity and absence of understory may explain the very good and similar performances obtained with the three approaches. A single metric describing the top of the canopy (e.g., H_{90f} or H_{maxl}) is sufficient to summarize stand height characteristics. Results were different for the more complex deciduous forest. The first approach provided H_L and H_{dom} predictions with an R^2_{adj} of 0.89 and 0.95, respectively. Three metrics have been selected to predict H_L, while two metrics have been selected to predict H_{dom}. The mechanistic model provided more accurate estimates with an R^2_{adj} of 0.97 and RMSEs reduced by 0.86 m and 0.4 m compared to the point distribution model for H_L and H_{dom}, respectively. A more significant improvement was observed for the prediction of H_L, which is a parameter that takes into account overtopped trees and not only the tallest ones. Figure 4.8 illustrates the improvement obtained using the mechanistic model by showing the observed values of H_L against the ones predicted by the three models for the deciduous forest. Applying the mechanistic model across diverse forest area types only required a calibration of the parameters, as both the metrics and the model shape were kept from one area to the other one. From these examples we can see that the mechanistic model proved robust and more efficient than the point distribution model, which is currently the most widely used approach. In this approach, point metrics other than the height distributional and density metrics used in the previous example can be included, such as canopy cover, kurtosis, and skewness (Borsah et al., 2023; Coops et al., 2021; White et al., 2017). Véga et al. (2016) also demonstrated the successful use of less conventional

TABLE 4.4

Goodness-of-fit statistics for H_L and H_{dom} predictions using the point distribution, the mechanistic model, and the top of canopy mechanistic model

		Coniferous Forest			Deciduous Forest		
Approach	Predicted Attribute	ALS Metrics	R^2_{adj}	RMSE (m)	ALS Metrics	R^2_{adj}	RMSE (m)
Point distribution model	H_L	H_{90f}	0.99	0.84	H_{100f}, H_{60l}, d_{90f}	0.89	1.97
	H_{dom}	H_{maxl}	0.99	0.88	H_{maxf}, H_{80l}	0.95	1.53
Mechanistic model	H_L	μ_{CH}, σ^2_{CH}, P, H_{crown}	0.99	0.68	μ_{CH}, σ^2_{CH}, P, H_{crown}	0.97	1.11
	H_{dom}	μ_{CH}, σ^2_{CH}, P, H_{crown}	0.99	0.72	μ_{CH}, σ^2_{CH}, P, H_{crown}	0.97	1.08
Top of canopy mechanistic model	H_L	μ_{CH}, σ^2_{CH}, P	0.99	0.68	μ_{CH}, σ^2_{CH}, P	0.96	1.22
	H_{dom}	μ_{CH}, σ^2_{CH}, P	0.99	0.73	μ_{CH}, σ^2_{CH}, P	0.97	1.18

Optical Remote Sensing of Tree and Stand Heights

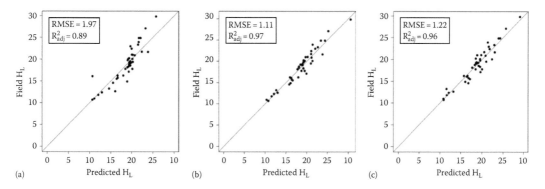

FIGURE 4.8 Observed values of HL against predicted values for the deciduous forest dataset using (a) the point distribution model, (b) the mechanistic model, and (c) the mechanistic top of canopy model.

metrics, such as LiDAR penetration depths and canopy volume metrics, which can be more easily related to forest characteristics.

Except when considering dominant heights, for which the top-of-canopy information is sufficient, this example also underlines the value of the information coming from the understory to improve H_L assessments. First, in the point distribution model, one of the selected variables was H_{60l}, which is a parameter that is decorrelated from the variability in the height of the canopy surface as seen in Section 4.3.3. Second, when comparing the mechanistic and the top of canopy mechanistic models, the latter performed slightly less well, with a similar R^2_{adj} (0.96 against 0.97) despite using one fewer parameter than the complete mechanistic model and obtained a slightly increased RMSE (+ 0.11 m).

Even if not as widely used as point distribution approaches, a mechanistic approach appeared to provide a viable alternative solution regarding both robustness and accuracy in order to develop models predicting stand height characteristics from 3D data metrics. Furthermore, even if they performed slightly less well, the models using only the top-of-canopy information might provide interesting results in a range of stand types, including quite complex ones such as multilayered broadleaved stands. However, their performance is likely to be lower if the area or the strata on which they are calibrated and applied is characterized by changes in stand structure. In that particular case, a more complete mechanistic model that partly includes the diversity of structures thanks to an additional parameter, such as H_{crown} used in our example, is expected to be more robust.

To better cope with the impact of stand characteristics on ALS metrics, a preliminary stratification of forest stands is recommended and would be expected to improve ALS-derived predictions (Næsset, 2002b). Aerial images or very high-resolution imagery from space (sub-meter to few meters resolution) can be used to that aim (see Table 4.1) (Maltamo et al., 2021; Kukkonen et al., 2018). Height distributions derived from ALS could also give some precious information for stratification purposes. But the fact that some metrics derived from 3D data are influenced by acquisitions characteristics (such as laser footprint, ALS sensors, pulse repetition frequency or scan angle; see Section 4.3.1.2) also hampers the development of generic models. In general, field calibration plots associated to each new ALS acquisition are required (Næsset, 2009a; Gobakken and Næsset, 2008; Thomas et al., 2006). Trying to identify metrics that are least impacted by the changes in acquisition characteristics would represent an additional step required to provide more robust models. Computing plant area density profiles from point clouds could be one of the paths to partially normalize LiDAR information, as shown by Dayal et al. (2022). Apart from both stand and acquisition characteristics, other factors such as plot size and co-registration precision were shown to influence the models' quality and their derived maps (e.g., Bouvier et al., 2019; Strunk et al., 2012; Frazer et al., 2011). For example, using large plots reduces the probability of edge effects, which may occur when parts of the crowns belonging to trees located outside but close enough to the boundary of a plot are included in the ALS or DP 3D data subsets corresponding to the plot or, conversely, when part of the crowns of trees located inside a plot are excluded when using the plot area to extract the plot-related 3D data subset.

It is also important to bear in mind that LiDAR "sees" all the vegetation within the plot, irrespective of tree social status, while in calibration plots, several trees can be ignored since field measurements may start only at a given minimum diameter. On the contrary, photogrammetric point clouds "see" only the dominant stratum while part of the overtopped trees are measured in the field (Table 4.3). Field measurement protocols, either designed for NFI or for the purpose of a specific study, can be different from one study to another. Sampling designs can vary, not only in terms of sampling scheme and density, but also in terms of plot design (e.g., fixed area, concentric plots with various diameters, fixed number of trees, threshold defining the limit size for measurable trees). Furthermore, tree measurements can also change, e.g., measurement of diameter at breast height only or both diameters and heights, or alternatively all diameters but heights for only a subsample of trees. All these elements are likely to change the field reference estimations and therefore also change the models (Bouvier et al., 2019). Therefore, reflexions and efforts made to harmonize forest inventory procedures (Ferretti, 2010; McRoberts et al., 2009) are likely to increase the generalization level of the models and should contribute to a more widespread use of remote sensing data in forest applications.

When a model is developed, errors propagate through the entire process (Bouvier et al., 2019). Five main sources of error should be considered in area-based approaches. First, the field measurement errors; they may range from several decimeters to a few meters for height measurements (see Section 4.2.3). Second, for some parameters that cannot be directly measured in the field, such as biomass, another source of error resides in the allometric equations available to estimate the parameters from field measurements. But, unless heights are assessed in the field from diameters at breast height measurements, this error will not affect height predictions. A third source of error is the geolocation inaccuracy of both field plots and 3D datasets. The latter was identified as the most significant source of error in Bouvier et al. (2019). A fourth one is the sampling error, which is often neglected and mainly linked to the sampling design and the extent of sampling effort. If the whole population is inadequately sampled, the sampling error might be significant. Finally, the models produced using metrics derived from point clouds or CHMs have their own inherent uncertainty. All these sources of error will combine and impact the quality of the model results.

In the previous part of this section, we focused on models developed to predict the structural and biophysical characteristics of stands. The use of LiDAR in landscape ecology and biodiversity studies is a more recent field of research (Bouvier et al., 2019; Lenoir et al., 2022). However, metrics extracted from LiDAR data or elevation models have already been proposed for the characterization of landscape patterns and structure at several scales (Goodbody et al., 2021a; Karasov et al., 2021; Vauhkonen and Ruosalainen, 2017; Mücke and Hollaus, 2010; Uuemaa et al., 2009). In addition, the relationships between metrics describing the 3D distribution of the vegetation, and the presence or abundance of a given species, or the assemblage of species, have been increasingly investigated (Toivonon et al., 2023). For example, in earlier studies, Nelson et al. (2005) used LiDAR data to identify forested sites that might support populations of Delmarva fox squirrels, while Müller and Brandl (2009) and Müller et al. (2014) predicted arthropod diversity and assemblages. Bird species abundance and assemblages have also been at the core of several studies (Melin et al., 2019; Zellweger et al., 2013; Lesak et al., 2011; Müller et al., 2010; Goetz et al., 2007). All the aforementioned studies deal with fauna biodiversity. And despite the existence of relationships between floristic biodiversity and forest structure (Zilliox and Gosselin, 2013), only a few attempts have been made to study these relationships using either ALS or photo-derived 3D data (e.g., Bouvier et al., 2017; Mao et al., 2018; Vehmas et al., 2009; Simonson et al., 2012). Modeling the link between biodiversity and environmental conditions remains challenging. Indeed, biodiversity is driven by many processes, and taking into account vegetation or landscape structural characteristics exclusively is not sufficient. The ecological context, with regards to abiotic variables, is also of great importance when attempting to explain biodiversity indicators. While several studies do not take into account the ecological context, complementing LiDAR metrics with abiotic variables was shown to improve the predictive power of models (e.g., Kemppinen et al., 2024; Haesen et al., 2023; Bouvier et al., 2017; Zellweger et al., 2014). The time-lag dynamics in the response of biodiversity to contemporary environmental changes should also be taken into account for plant diversity (Lenoir et al., 2022). Furthermore, to tackle the complexity of ecological modeling, multiple regressions, which are widely used to predict forest biophysical parameters, are no longer the most appropriate approach (He et al., 2015). They were sometimes replaced by statistical approaches that are more suited to the modeling of various parametric distributions and deal with discrete variables while allowing the utilization of non-parametric functions for variable weighting. To this aim, generalized additive models (GAM) (Goetz et al., 2007), boosted regression trees (Zellweger et al., 2013), and Bayesian models (Bouvier et al., 2017; Zilliox and Gosselin, 2013) have been occasionally used.

4.4.3 Model Extrapolation and Inferences for Large-Area Inventories

For a number of operational inventory applications, many authors (see, e.g., Maltamo et al., 2021; Næsset, 2007, 2002b) have demonstrated how wall-to-wall ALS coverage could efficiently be used to improve the accuracy and spatial resolution of field surveys. In Scandinavian countries, LiDAR-derived forest inventories have been made since 1995 (Næsset, 2004). Statistical inferences, sampling design, and statistical properties of LiDAR-derived estimations have been receiving increasing attention over the past two decades because of the often complex structure of LiDAR surveys (Maltamo et al., 2021; Di Biase et al., 2018; Véga et al., 2014; Li et al., 2012; Ben-Arie et al., 2009; Morsdorf et al., 2004).

Some studies rely directly on existing NFI plots (e.g., in Maltamo et al., 2009), whereas, in other studies, specific field campaigns have to be carried out. Sampling design associated with LiDAR surveys are frequently established based on a model-based design (Smits et al., 2012), as opposed to design-based sampling, where systematic or random locations of field calibration plots are being performed (Henry et al., 2013; Næsset, 2007). This aspect is important since model-based estimators are not design-unbiased and may reveal potential bias depending on model correctness (McRoberts et al., 2022; Sagar et al., 2022; Magnussen and Nord-Larsen, 2021; Picard et al., 2012a).

Wall-to-wall ALS coverages over extended areas of forests have become more frequent. However, such coverage efforts are still complex and costly to undertake and generate a considerable amount of data to be processed (Wulder et al., 2012; Uuemaa et al., 2009). Furthermore, some systems, such as early LiDAR systems, e.g., the PAL profiling system (Nelson et al., 2003), or spaceborne systems cannot provide full coverage. ICESat, the first LiDAR mission aimed at measuring terrestrial surfaces, acquired data from 2003 to 2009 and was a profiling system. Due to technological limitations, future space LiDAR missions are likely to embed, at best, multi-beam systems. This is already the case for both the GEDI and ICESat-2 systems, which acquire data at each orbit along eight tracks (four with full-power beams and four with half-power beams) and six tracks (three pairs of strong and weak beam tracks), respectively (Dubayah et al., 2020; Markus et al., 2017). Designed to either reduce survey costs or process data acquired by multi-beams or profiling systems, some methods using LiDAR measurement samples were developed to perform extensive forest inventories. These methods also incorporate remote sensing sources other than LiDAR into existing large-area sample-based forest inventory frameworks (Persson et al., 2022; Wulder et al., 2012; Ståhl et al., 2011). These approaches still require local calibration of models to link remote sensing data to forest parameters. It is worth noting that all applications do not require map production and part of the required information can be obtained by analyzing sets of characteristics assessed on samples. Using simulation, Ene et al. (2013) demonstrated that ALS data enhanced forest inventory results and that ALS-aided surveys can be a cost-efficient alternative to traditional field inventories. The latter also provided more accurate results if sampling intensity was optimized (i.e., by optimizing the distance between regular flight-lines covering only part of the NFI plots) (Ene et al., 2013).

When models are extrapolated, two points are worth recalling. First it is important to check if the predictions have been made within the calibration domain. For example, models developed for even-aged stands, or for one specific species, could yield erroneous predictions in multilayer stands or in stands composed of mixed species. Model robustness therefore represents a major issue and should be evaluated across several stand types. It is therefore crucial to be able to characterize the stand types consistently with the calibration domains of the models. This was, for example, done in Renaud et al. (2022) using a convex hull computed on the variable space of the training sample to assess the validity domain of the model. However, further difficulties may arise when predictions must be made in edge areas covering several stand types. In some countries edges between stand types or between forest and non-forest areas can represent a significant share of the NFI plots. In France, for instance, at least 20% of NFI plots are located on an edge. Therefore, forest heterogeneity must not be underestimated when building, validating, and extrapolating models (Nguyen et al., 2023; Frazer et al., 2011). Finally, when maps are considered as an operational outcome of the models, one must remember, as McRoberts et al. (2010) explain, "the utility of maps is greatly increased when they form an appropriate basis for inferring values of maps parameters describing the populations represented by the maps." Inference means being able to calculate estimates of the population mean for a given attribute, and its variance, in order to be able to define a 1-α confidence interval for this parameter (McRoberts et al., 2010). Both extrapolation of forest structural or biophysical parameters and inference issues are at the core of active research work (Ben-Arie et al., 2009; Ene et al., 2013; Li et al., 2012; McRoberts, 2010; McRoberts et al., 2010). And various methods were recently proposed to provide uncertainty assessment at the pixel level (Saarela et al., 2020), including geostatical (Kangas et al., 2023) and geometrical (Renaud et al., 2022) approaches, among others.

Access to data acquired by the recent spaceborne LiDAR systems has led to extensive research efforts aimed at developing methods to produce height maps from regional to global scales. LiDAR height measurements, whether at non-contiguous footprint positions (ICESat or GEDI) or along transects (ICESat-2), serve as reference data and are combined with wall-to-wall optical or radar remote sensing data to spatialize heights and produce these maps. Deep learning approaches are often promoted for developing height prediction models (e.g., Lang et al., 2023; Schwartz et al., 2023; Potapov et al., 2021). However, they can only spatialize a single height measurement and are not well-suited to quantify uncertainty. Besic et al. (2024) and Schleich (2024) have proposed methods based on a simple multilayer perceptron and K-Nearest Neighbors (KNN)-bagging, respectively, to attempt to spatialize the entire vegetation profile. These more traditional models remain interesting from this perspective. Additionally, the KNN-bagging approach allows for the quantification of uncertainties.

4.5 APPROACHES FOR INDIVIDUAL TREE HEIGHT ASSESSMENT

Unlike area-based approaches that are poorly sensitive to point density (see Section 4.3.1.2), methods for individual tree height assessment require several height measurements per tree crowns, hence at least a few points/m²; and the optimal point density is likely to change according to the stand type and age (Lindberg and Holmgren, 2017; Zhen et al., 2016; Eysn et al., 2015). Whereas 2 pts./m^{-2} was found to be sufficient in mature stands, at least 10 pts./m^{-2} might be required for saplings (Kaartinen et al., 2012). Early methods used for detecting and characterizing individual trees from ALS data were based on techniques developed to process very high-resolution optical images (Leckie et al., 2005). To locate individual trees, brightness or color gradients were used in optical imagery (Leckie et al., 2005; Wulder et al., 2000), while ALS CHM-based methods can make use of the geometrical properties of the CHMs, including height, slope, and curvature

(Bongers, 2001) for the same purpose. Now that accurate photo-DSMs can be produced with digital photogrammetry, raster-based tree detection approaches can be applied either to ALS-derived or photo-derived CHMs. These approaches are described in Section 4.5.1. Section 4.5.2 presents more recent approaches that directly use 3D point clouds, to improve both tree crown characterization and overtopped tree detection. Finally, section 4.5.3 presents hybrid approaches based on algorithms that either use both point clouds and raster information or exploit the complementarity between structural and radiometric information.

4.5.1 Raster-based Approaches

Standard methods for extracting individual trees from a CHM are based on three steps: CHM modeling and optimization, detection of tree apices using local maxima, and development of crown segments around each tree apex (Yun et al. 2021; Solberg et al., 2006; Persson et al., 2002; Popescu et al., 2002). When solely focusing on tree height assessment, this last step is not always required in the processing workflow.

4.5.1.1 CHM Modeling and Optimization

The efficiency of raster-based approaches is intimately linked to the quality of the CHM derived from the point clouds. Grid cell size is the first critical parameter (Lindberg and Holmgren, 2017). Various studies reported that the optimal size should be of the same order of magnitude as the original point spacing (Vepakomma et al., 2008). Besides pixel size, the point-to-grid transformation has been widely investigated. Two main approaches are commonly used to compute the initial CHM, from either the first returns or the points classified as non-ground points. One popular method consists in assigning the maximum Z-value of the points of each grid cell and to then estimate a value for the empty cells. This can be achieved, for example, by averaging the values of the connected filled cells (Brandtberg et al., 2003; Hyyppä et al., 2001) or by using an inverse distance weighted (IDW) method applied to a given number of neighboring points among those belonging to the canopy surface (Véga and Durrieu, 2011; Vepakomma et al., 2008). Alternative methods involve interpolating a value at the center of each grid cell from the Z-values of the neighboring points using kriging (Popescu and Wynne, 2004), IDW (Vepakomma et al., 2008), active contour (Persson et al., 2002), or minimum curvature (Solberg et al., 2006) algorithms. When the task is measuring trees, exact interpolation methods might be favored (Kato et al., 2009), but overall, simple interpolation techniques like IDW were found to be sufficiently accurate (Anderson et al., 2006a; Vepakomma et al., 2008). To enhance both treetop detection and crown segmentation algorithms, different procedures can be used to improve the surface described by the CHM prior to further processing. These include point cloud thinning, hole filling, as well as CHM filtering (Yun et al., 2021). Thinning procedures are implemented upstream from CHM generation to filter out points within the canopy. For example, an initial outer canopy surface can be defined by using an active contour algorithm to trace the outer part of the crowns (Persson et al., 2002), and points that are too far below this surface are discarded (Solberg et al., 2006: Persson et al., 2002). Hole-filling algorithms are applied once an initial CHM has been calculated and is aimed at removing irregularities within the canopy surface partly due to crown porosity. Ben-Arie et al. (2009) introduced a six-step semi-automated pit-filling algorithm based on a Laplacian edge detector to remove pits while preserving edges. Véga and Durrieu (2011) developed an iterative method based on four to eight connectivity kernels to automatically detect and recalculate local minima (Figure 4.9).

In general, the workflows include a last smoothing step to reduce commission errors induced by noise within the CHM prior to tree parameter extraction. Gaussian filtering is commonly used (Solberg et al., 2006; Morsdorf et al., 2004; Persson et al., 2002; Hyyppä et al., 2001). But median filtering was also used, as it can preserve the original values in the CHM (Popescu et al., 2003). A single smoothing filter can be applied (Popescu et al., 2003) with smoothing intensity, which may be driven by local vegetation height (Koch et al., 2006). Nevertheless, iterative Gaussian filtering

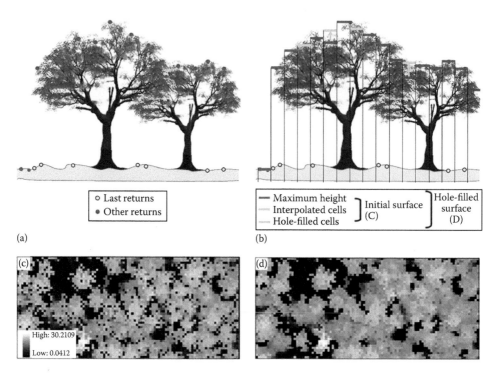

FIGURE 4.9 CHM modeling and optimization: from the point cloud (A) to the CHM (B). C and D, respectively, represent an initial CHM (0.5 m resolution) and its hole-filled version over a coniferous forest (Draix, France).

is the most widely used approach (Solberg et al., 2006; Maltamo et al., 2004; Hyyppä et al., 2001). In some studies the results obtained at several filtering levels were combined, as optimal filtering intensity level depends on crown size (Véga and Durrieu, 2011; Persson et al., 2002).

4.5.1.2 Detecting Tree Apices

Treetops are usually considered as local maxima (LM) in the smoothed CHM. A LM can be defined as a pixel containing the highest value in a given neighboring defined by a kernel.

Along with the impact of CHM quality, the efficiency of treetop detection using LM identification mainly depends on the optimization of both kernel size and shape. As explained in various studies, the selection of either an overly small or overly large kernel might lead to commission (i.e., false detection) or omission errors, respectively (Figure 4.10) (Yun et al., 2021; Reitberger et al., 2009; Popescu et al., 2002). In addition, while a single kernel size might be sufficient to process CHM acquired over forest plantations with homogenous crowns, approaches using variable kernel sizes had to be developed to improve treetop detection over complex forest structures. Popescu et al. (2002) fixed the window size according to a relationship between tree height and crown diameter that he established from field measurements. The method was further improved by introducing a circular window and by using specific relationships for both conifer and hardwood stands (Popescu and Wynne, 2004) (Figure 4.10). Chen et al. (2006) extended the concept of window size dependence on tree height by using a non-linear power model linking window size to tree height and by then estimating a prediction interval for the optimal window size around the value predicted by the model. This helped to reduce the omission of trees with crowns smaller than those predicted by the model.

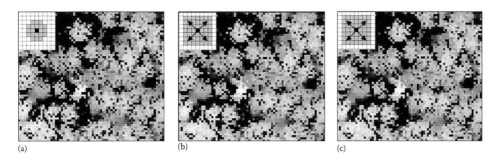

FIGURE 4.10 Treetop detection in a coniferous area using either a fixed 20 pixel kernel (left), or variable window sizes from using a generic model (center) or a model derived for conifers (right) (Popescu et al., 2002).

Many other approaches have been proposed to identify trees, such as the 2nd derivative of blob signature (Brandtberg et al., 2003), a multiple morphological opening method to tackle the problem of heterogeneity in crown dimensions (Hu et al., 2014), marked point process models optimized using a multiple births and deaths approach (Zhou et al., 2010), Gaussian smoothing (Yun et al., 2021) or h-minima transform using distance-transformed images (Chen et al., 2006). The latter was found to be particularly suited for the detection of trees with flat or aggregated crowns.

4.5.1.3 Measuring Tree Crowns

Several methods have also been put forward for reconstructing crown segments and subsequently estimating tree characteristics, such as tree height, crown diameter or area, or crown base height (Dong et al., 2020; Zhen et al., 2016). Most common methods include region growing approaches (Persson et al., 2002; Hyyppä et al., 2001), watershed analysis (Kwak et al., 2007; Chen et al., 2006; Mei and Durrieu, 2004), morphological analysis (Wang and Glenn, 2008), valley following approaches (Leckie et al., 2003), fitting functions (Popescu and Wynne, 2004), ellipse fitting (Véga and Durrieu, 2011), wavelet analysis (Falkowski et al., 2006), or a combination of methods, as in Koch et al. (2006), Hu et al. (2014), and Holmgren et al. (2022).

Unlike approaches such as region growing or watershed approaches, some can be implemented without a treetop detection step, such as the valley following approach (Leckie et al., 2003) or the spatial wavelet analysis (SWA) proposed by Falkowski et al. (2006). In a few cases, treetop identification and tree delineation are interdependent. For example, in Véga and Durrieu (2011), the initial treetop set evolves as crown contours are identified and refined in a multi-scale iterative process. Several methods—most of the region growing or watershed approaches—tend to produce irregularly shaped crown segments and thus necessitate a shape control process step. Sets of rules have been developed to qualify and constrain the segmentation process based on either height values, areas, distance from the gravity center, or on shapes (Hu et al., 2014; Koch et al., 2006; Solberg et al., 2006; Weinacker et al., 2004; Hyyppä et al., 2001). Other methods directly model tree crowns using circular or elliptic shapes (Véga and Durrieu, 2011; Popescu et al., 2003). Here, results might be improved by allowing the treetop to be different from the center of symmetry of the shape used by the model. This is, for example, the case for the approach proposed by Véga and Durrieu (2011) where LM were used to estimate the crown radius in various directions, and crown elliptic shapes were further adjusted while only using the set of the radius endpoints that were assumed to characterize the crown edge.

Kaartinen et al. (2012) compared several individual tree detection approaches on a given study area and for three ALS datasets of different point densities (2, 4, and 8 pts/m²). The comparison was made with regard to the number of correctly detected trees, i.e., the number of correct matches with field reference, commissions, and omissions. The results were analyzed according to both height classes and tree status. For well-detected trees, tree location accuracy and the accuracy of the assessment of both height and crown dimensions were analyzed. The percentage of detected trees ranged from 25% to 102%. Surprisingly, several automated methods performed better, even in the case of co-dominant or suppressed trees, than manual detection, which identified 70% of the trees (Kaartinen et al., 2012). This point deserves particular attention because, unlike in many other fields, manual detection cannot be used as a reliable reference to assess the performance of automatic tree detection approaches derived from imagery or ALS data (Kaartinen et al., 2012). As expected, higher commission errors were mainly found for the smallest trees. Tree detection was not significantly improved by the increase in point density from 2 to 8 pts/m². On the contrary, height assessment accuracy was affected by point density even if to a lesser extent than by the method chosen. While 75% of the raster-based approaches showed a RMSE below 2 m, the best models provided RMSEs ranging from 60 to 80 cm (Kaartinen et al., 2012). In an operational context, the authors emphasized the efficiency of quite simple methods based on local maxima findings, which achieved: a tree detection rate of over 70%, a percentage of matched trees of 95%, and of 95% when considering only dominant trees, a commission error of 18%, and a RMSE for height estimates around 80 cm.

4.5.2 Point-based Approaches

In addition to raster-based approaches, point-based approaches have been developed in order to fully take advantage of the 3D information of ALS point clouds and were expected to detect not only dominant but also overtopped trees (Yun et al., 2021). If there is no theoretical reason to not apply point-based approaches to photo-derived point clouds the utility is limited, as these point clouds mainly describe canopy surface, like CHMs. Morsdorf et al. (2004) introduced a voxel-based approach with k-means clustering to detect individual trees and modeled crowns using a paraboloid model. However, as the k-means seeds were based on CHM local maxima, the method suffers from the same limitations as raster-based approaches. Ferraz et al. (2012) proposed another clustering approach based on a mean-shift algorithm that does not require seed points. The method is used to identify several vegetation layers and then define an adaptive kernel bandwidth parameter optimized for each layer. The method gave very promising results and could detect 98.6% of the dominant trees and approximately 13% of the suppressed ones within stands dominated by eucalyptus and pines. However, the authors stressed that a more sophisticated approach would be required to process more complex forest structures (Ferraz et al., 2010). The normalized cut segmentation introduced by Reitberger et al. (2009) is also not dependent on seed identification. Tested in a mixed mountain forest, the method was used to detect 77%, 32%, and 18% of the trees in the dominant, intermediate, and lower canopy layer, respectively. Li et al. (2012) proposed a distance-based algorithm to sequentially detect trees. The method assumes relative spacing between trees and overcomes the issue of finding local maxima by using "global maxima," defined during the segmentation process as the highest point not yet associated with a crown. The method was not evaluated in hardwood forests but achieved 86% global detection rate in a mixed conifer forest with a 94% correct detection rate (Li et al., 2012). Recently, Véga et al. (2014) proposed a multi-scale dynamic segmentation approach. As in Li et al. (2012), the algorithm is based on global maxima. Each point is considered a new tree apex or assigned to an existing crown. In the latter case, the algorithm allocates the point to the tree segment, i.e., the projection of a given crown in the 2D plane, which is the least changed by the inclusion of this new point considering the change in the surface area of the convex hull of each candidate segment. With this approach, 86%, 32%, and 17% of the trees were correctly detected on average when considering three different forest types (Figure 4.11) (Véga et al., 2014). In the comparative analysis performed by Kaartinen et al. (2012), only one point-based approach was selected to be compared to raster-based approaches. As expected, it enabled better detection of under-layered vegetation than that offered by the raster-based approaches.

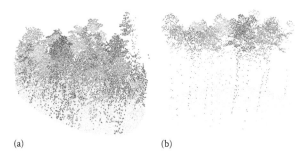

FIGURE 4.11 Example of point-based segmentation over a broadleaved forest (566 stem/ha) (a) and a pine plantation (206 stem/ha) (b).

In addition to pure 3D point-cloud approaches, other approaches combining raster and point cloud data recently emerged and have proven effective at improving individual tree extraction performance (Zhou et al., 2021; Ganz et al., 2019; Balenović et al., 2017). Using a high-density dataset Reitberger et al. (2009) obtained improved results by combining their point-based approach with a watershed segmentation of the CHM and a stem detection applied to the point cloud (7–8% of correct detection for the dominant, intermediate, and lower canopy layer, respectively, instead of 77%, 32%, and 18% (see Section 4.5.2). Hu et al. (2014) first segmented a CHM using a multi-scale morphological opening approach and flagged uncertain shapes based on shape and height information. Points within the flagged segments were then reprocessed by multi-Gaussian fitting applied to point distribution. The number of fitted Gaussian functions corresponds to the number of trees inside a segment. Splitting and merging operations were then applied to refine the segmentation and derive the final crowns (Hu et al., 2014). In order to tackle issues involved in CHM-based crown extraction, Wang and Glenn (2008) adopted a voxel structure to generate 2D horizontal projection images of the point cloud. For each horizontal layer, a hierarchical morphological algorithm is used to generate crown segments. Tree crowns were then reconstructed using a pre-order forest transversal approach to connect the 2D crown segments of the successive layers. The method was able to identify both canopy and overtopped trees but was found to be highly sensitive to the parameters (Wang and Glenn, 2008).

Other studies aimed at exploiting the complementarities between structural information provided by canopy models and radiometric information contained in optical images. This helped to improve individual tree detection and subsequent height measurement accuracy. Popescu et al. (2004) used optical data to identify and locate forest types that were then used to adapt the segmentation parameters of their adaptive window size algorithm, thus leading to improved results. Leckie et al. (2003) applied a valley following approach to both multispectral images and ALS CHM in order to extract and characterize individual trees. They reported that crown segmentation was more efficient in dense forests when using spectral images (80–90% detection) as poor crown outlines were obtained using the ALS CHM. On the contrary, better results were obtained by processing the CHM within open areas where sunlit ground vegetation can cause false detections within images. These results show that combining ALS data with multispectral data might improve crown segmentation, which also holds true for treetop detection (Smits et al., 2012). Suárez et al. (2005) applied an object-based segmentation using both spectral and height data to identify individual trees implemented in e-Cognition® software by Definiens. Tested on 345 trees, results showed that ALS underpredicted individual tree heights by 7–8%. The author also claimed that an ALS and imagery combination could be used to work with lower ALS point densities, thus reducing costs. But, as reported by Kaartinen et al. (2012), methods developed to process ALS data alone are more mature and several approaches based on raster CHM gave better results than hybrid approaches.

A qualitative comparison of the three types of approaches, namely, raster based, point based, and hybrid, is provided in Table 4.5. Current trends in individual tree detection and assessment relies on deep learning approaches, which requires large sets of training datasets. Examples of such approaches can be found in Sun et al. (2022) and Li et al. (2023).

Compared to area-based approaches, the first major drawback of individual tree approaches is practical in nature. The latter require higher point densities and thus induce an increase in data acquisition costs and in processing time. A second disadvantage is their inconsistent behavior depending on the type of forest on which they are applied. Currently, their performance is insufficient to extract understory trees, whatever the individual tree approach method used. However, the main advantage of individual tree approaches over area-based ones is that they might one day provide height distributions despite synthetic height indicators. Such distributions would enable

TABLE 4.5
Summary of main strengths and limitations of raster-based, point-based, and hybrid approaches developed for tree detection and tree height assessment

Approach	Advantages	Limitations
Raster-based	• Computational efficiency • Well-established methods • Widely available image-processing tools can be used • Several methods perform better than manual detection • Simple methods can achieve tree detection rates > 95% for dominated trees, > 70% for all trees • RMSE ranging from 0.6 to 0.8 m for height assessment with the best-performing methods	• High sensitivity to point density • Point to raster interpolation can impact height estimation • CHM smoothing impacts detection of local maxima (LM) and crown boundaries • Approaches based on LM identification are not well adapted to "flat" crowns • Crown segments: quality is function of the method and the chosen parameters (e.g., region growing, watershed, model fitting) • Detection rates highly dependent on the method (from 25% to 102%) • Dominated trees cannot be detected • Crown parameters such as crown-based height cannot be assessed
Point-based	• No need for height interpolation nor for a smoothing step of a raster CHM • Improved description of crown shape and attributes; extraction of additional tree parameters such as crown base height • More adapted to identify small and dominated trees ➔ detection rates > 77% considering all trees even in complex stands	• Sensitive to both pulse sampling pattern and to point density • Lower computational efficiency • Tree detection and crown segmentation are impacted by the point clustering approach used • Might be sensitive to sampling patterns
Hybrid	• Improved efficiency by combining advantages of both raster-based and point-based approaches ➔ optimization of computational efficiency by processing raster data and improvement of results by using additional information on understory provided by 3D point clouds • When radiometric information is used in addition to 3D point clouds, results are enhanced due to information complementarity (e.g., 9% increase in detection rate of dominant trees in (Reitberger et al., 2009)	• Lower computational efficiency compared to pure raster-based approaches • Might be affected by the limitations of each source of information depending on the processing workflow

improved predictions of other forest parameters, such as timber volume or over-ground biomass (Kaartinen et al., 2012). This is why there is still a lot of commitment to developing powerful and operational approaches in this research field despite the significant challenges involved.

4.6 CONCLUSION AND PERSPECTIVES

Tree height and vegetation height distribution are structural characteristics of major importance when attempting to understand and monitor biological and ecological processes at the tree, stand, and landscape scales.

Tree height plays a central role in forest inventories, and height growth is a main driver for growth and yield models. Forest structure features, and in particular gaps, also play a major role in forest ecology because of their influence on microclimate and habitat quality and thus on biodiversity potential.

Accurate measurement of tree heights and careful monitoring of the dynamics of both canopy height and gaps at several scales in space and time is known to be crucial to improve both forest ecosystem monitoring and modeling, and thus contribute to their sustainable management. But, as field evaluations of tree height and forest structure are labor-intensive and costly, they cannot fully meet data requirements. In this chapter we focused on two remote sensing technologies: ALS and digital photogrammetry, which are deemed to be mature enough to provide practical and accurate measurements of 3D forest structure. Developing their use to measure and monitor tree height and vegetation structure at several scales, and at regular time intervals, is expected to represent a major breakthrough regarding the development of forest ecosystem modeling and to provide precious information to help managers deal with the increasing societal and economic pressures weighing on forests.

Each technology, although hampered by some major drawbacks, possesses major assets. Today, LiDAR is probably the most promising and mature technology capable of providing direct measurements of 3D forest structure from airborne systems. Despite continuous system improvements, data acquisition from airplanes remains costly, which partly explains the development of inventory methods based on samples of ALS measurements (Ene et al., 2013; Wulder et al., 2012). These high costs likely also explain the growing interest in low-cost, lightweight LiDAR systems mounted on UAVs (Hu et al., 2020). However, the areas covered by such systems remain limited and are not suitable for forest monitoring at broad scales. In contrast, ongoing developments of spaceborne missions are paving the way for large-scale monitoring of forest height. The first ICESat mission was designed for ice monitoring. ICESat-2, which NASA launched in 2018, was the second LiDAR mission dedicated to Earth surface monitoring. Although primarily designed to characterize and monitor polar ice, vegetation height and biomass measurement is one of its secondary scientific objectives. However, with its current design, ICESat-2 cannot adequately replace the shelved DESDynI vegetation LiDAR mission (Hall et al., 2011a). The instrument originally planned for the latter gave birth to the GEDI mission on board the ISS, which is the first spaceborne mission optimized for forest monitoring (Dubayah et al., 2020) and demonstrates the interest of such kinds of systems for forest structure characterization. Looking ahead, two mission projects are taking shape. The multi-wavelength LiDAR for vegetation (MOLI) instrument from JAXA, with two beams and footprints of 25 m in diameter, is scheduled to be installed aboard the ISS in 2026 (CEOS, 2024). Additionally, earth dynamics geodetic explorer (EDGE), which includes a swath-mapping laser altimeter, is one of the four missions recently shortlisted as part of NASA's Earth System Explorers Program. It could be launched in the early 2030s if chosen as one of the two selected missions following preliminary studies (NASA, 2024).

The coincident development of deep learning approaches allowed wall-to-wall mapping of GEDI height metrics using optical and radar images (Potapov et al., 2021; Lang et al., 2023; Schwartz et al., 2023), supporting information needs for monitoring purposes.

Due to recent developments in digital photogrammetry that have provided improved canopy surface reconstruction, this technology has recently benefited from renewed interest for forest applications. Unlike with LiDAR, information is limited to the top of the canopy. However, the ease of access to data offers several advantages over LiDAR. Firstly, due to the relatively long history of both airborne photography and photogrammetry, large databases of aerial photographs, dating back over 60 years and even beyond, exist in several countries (Véga and St-Onge, 2008). This enables researchers to monitor forest structure over long-past time periods. Secondly, regarding the acquisitions, imagery is considered to offer more advantages than ALS due to its easier survey planning and lower costs. Moreover, the radiometric information provided by imagers is complementary to 3D information and is of high value for forest type characterization. Finally, digital imagers have long been deemed fit for space applications, and several spaceborne systems today provide very high-resolution multispectral stereoscopic images from which accurate DSM, with resolutions of 1 m or less, can be obtained (e.g., Pearse et al., 2018; Aguilar et al., 2014). Such DSMs are becoming increasingly available and are expected to complement airborne DP products. For instance, regional-scale DSMs can be obtained from satellites such as Pleiades, Pleiades Neo, GeoEye, and WorldView (e.g., Piermattei et al., 2019; Han et al., 2020b). On a global scale, PlanetScope images from a large constellation of small satellites (Huang et al., 2022) or the CO3D mission, which is expected to provide the first global 1-m accuracy DSM by 2025 (Lebègue et al., 2020), offer additional sources of high-resolution DP products.

Overall, due to their respective advantages and to similar potential for top vegetation height measurements, ALS and photogrammetry are more complementary than mutually exclusive, and the objective should be to optimize alternated acquisitions under cost constraints. In particular, combining satellite stereo-imagery and ALS DTM may allow the monitoring of vegetation height over large areas at a lower cost than with airborne-based solutions.

Concerning methodological issues, producing accurate surface model in forest environments based on digital photogrammetry remains challenging and, despite considerable improvements made over the last years, further efforts are needed to develop methods that can be used to process different forest types with increased self-tuning functionalities in order to provide user-friendly and powerful tools to foresters. Currently, the methods used to retrieve information on vegetation heights from either ALS or photo-derived 3D data can be classified within two families: area-based approaches that assess height characteristics at the stand level, and approaches for individual tree height assessment. A major advantage of the latter is their capacity to provide height distributions instead of the synthetic height indicators provided by area-based approaches. Such distributions could help improve predictions of other forest parameters such as timber volume or aboveground biomass. However, developing efficient approaches remains very challenging due to the difficulties involved in accurately segmenting trees, in particular those belonging to the understory. In addition, these approaches require high-density point clouds, which increases both acquisition and data management costs, as well as processing time. Therefore, despite their interesting features, their operational value for forest monitoring at broad scales is limited. On the contrary, area-based approaches are very efficient and have already been incorporated in the NFI workflow of several countries. One way to make the most of both approaches could be the development of methods aimed at assessing height distributions, in a way similar to the one developped for stem diameter distribution (Gobakken and Næsset, 2004; Thomas et al., 2008). The development of databases for training deep learning algorithms is challenging established principles and will probably revolutionize methods and approaches in the forthcoming years.

Finally, another major perspective concerns 3D data time series analysis used to monitor changes in forest structural characteristics (Coops et al., 2021). As the development of ALS is quite recent compared to airborne photography, digital photogrammetry should be used whenever long time series are needed. In the future, combining spaceborne LiDAR data with time series of optical and radar data is likely to become a widely used means for generating high-resolution time-series height maps (e.g., with resolutions in the range of ten to a few tens of meters). However, time series analysis raises specific methodological challenges, such as data calibration, or managing data of varying quality in change detection and characterization processes. Monitoring changes in forest structure over time can provide key information on forest growth and disturbances, as well as on forest functioning through the estimation of biomass changes or carbon fluxes. Forest modelers and forest managers also need accurate information on height changes to develop models of forest dynamics, estimate annual allowable cuts based on prediction of future yields, or monitor fluctuations of forest carbon stocks under changing climatic and disturbance regimes (see, e.g., Coops and Waring, 2001). Véga and St-Onge (2009) demonstrated the potential of height growth monitoring to assess forest site productivity (Figure 4.12). In addition to canopy growth, time series of elevation models also provide insights into disturbance regimes such as gaps and gap dynamics (e.g., Vepakomma and Fortin, 2010; Tanaka and Nakashizuka, 1997). While recent studies have focused on the analysis of structural forest changes and gap dynamics, little attention has been paid to the potential of such information to characterize biodiversity. Indeed, regarding the role of canopy gaps in diversity richness, recent knowledge has mostly been acquired using synchronic approaches. Less work has been devoted to monitoring biodiversity according to gap characteristics or linking past gap dynamics to present-day biodiversity (Jucker et al., 2023; Lenoir et al., 2022). Yet, this issue warrants further investigation, as this dynamic approach is linked to theoretical constructs about disturbance theory and their forestry counterparts (McCarthy and Burgman, 1995) and, as an integral part of forest biodiversity, could be linked to disturbance dynamics. Documenting gap dynamics could therefore lead to the identification of new and interesting biodiversity indicators. This type of indicators could also be valuable for forest management, serving as promising candidates for sustainable management indicators of forest biodiversity. It could provide useful information to identify effective sustainable management practices. Indeed, it has been shown that structurally complex canopy enhanced both biodiversity and productivity (Ishii et al., 2004), and natural disturbance emulation has been proposed as a general approach for ecologically sustainable forests (see, e.g., Kuuluvainen and Grenfell, 2012). To reach these economic and ecological goals will require increased vegetation height monitoring in order to develop and implement silvicultural prescriptions that aim at maintaining forest ecosystem functions and biodiversity (Ishii et al., 2004). Thanks to their potential for measuring vegetation height, both optical remote sensing technologies presented in this chapter are likely to play a major role in reaching these goals.

4.7 ACKNOWLEDGMENTS

This work is partly based on data and results obtained in the frame of two projects, the FORESEE project (ANR- 2010-BIOE-008) granted by the French National Research Agency (ANR) and the GNB project granted by the French ministry in charge of ecology through the Biodiversité, Gestion

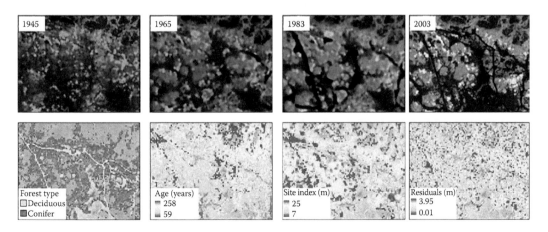

FIGURE 4.12 Time series of DP and ALS CHM (upper line) and derived maps of age and site index along with model residuals at 20 m resolution (lower line). Forest types were computed from a Quickbird image. Adapted from Véga and St-Onge (2009).

Forestière et Politiques Publiques (BGF) program (convention RESINE CVOJ 000 150, convention MBGD-BGF-1-CVS-092, n°CHORUS 2100 214 651). This work also benefited from data acquired by the Scientific Interest Group Bassins de Draix, étude de l'érosion en montagne, led by Irstea. The R&D department of ONF and the INRA BEF research unit are supported by the ANR as part of the Investissements d'Avenir program (ANR-11-LABX-0002–01, Lab of Excellence ARBRE). This work also benefited from the support of the SLIM project, funded by the TOSCA Continental Surface Program of the Centre National d'Études Spatiales (CNES) under Grant 4500066524. Additionally, it received funding from CNES for Maxime Soma's postdoctoral research on ICESat-2, as well as co-funding from INRAE and IGN for Anouk Schleich's PhD thesis on GEDI and multisource inventories. The work was further supported by Labex ARBRE under Grant ANR-11-LABX-0002-01, by the French Agency for Ecological Transition (ADEME) (Grant 1703C0069 for the PROTEST project, GRAINE program) and by the FRISBEE project (TOSCA Continental Surface program of the Centre National d'Etudes Spatiales (CNES) (order N° 4500070632)). The authors also greatly thank all the people involved in the field measurements from ANDRA, INRA, IRSTEA, and ONF. Special thanks to Laurent Albrech for his highly valuable work during field campaigns and for field data georeferencement and to Xavier Lucie, who provided images of digital photogrammetric DSMs.

REFERENCES

Ackermann, F., 1999, Airborne laser scanning—present status and future expectations. *ISPRS Journal of Photogrammetry and Remote Sensing*, 54(2):64–67.

Adam, M., Urbazaev, M., Dubois, C. & Schmullius, C., 2020, Accuracy assessment of GEDI terrain elevation and canopy height estimates in European temperate forests: Influence of environmental and acquisition parameters. *Remote Sensing*, 12:3948. https://doi.org/10/gpfhdv.

Adedapo, S. M. & Zurqani, H. A., 2024, Evaluating the performance of various interpolation techniques on digital elevation models in highly dense forest vegetation environment. *Ecological Informatics*, 81:102646.

Aguilar, F. J., Mills, J. P., Delgado, J., Aguilar, M. A., Negreiros, J. G. & Pérez, J. L., 2010, Modelling vertical error in LiDAR-derived digital elevation models. *ISPRS Journal of Photogrammetry and Remote Sensing*, 65(1).

Aguilar, M. A., Del Mar Saldana, M. & Aguilar, F. J., 2014, Generation and quality assessment of stereo-extracted DSM from geoeye-1 and worldview-2 imagery. *IEEE Transactions on Geoscience and Remote Sensing*, 52(2):1259–1271.

Albanwan, H. & Qin, R., 2022, A comparative study on deep-learning methods for dense image matching of multi-angle and multi-date remote sensing stereo-images. *The Photogrammetric Record*, 37:385–409.

Alexander, C., Korstjens, A. H. & Hill, R. A., 2018, Influence of micro-topography and crown characteristics on tree height estimations in tropical forests based on LiDAR canopy height models. *International Journal of Applied Earth Observation and Geoinformation*, 65:105–113.

Ali, A. 2019, Forest stand structure and functioning: Current knowledge and future challenges. *Ecological Indicators*, 98:665–677.

Allouis, T., Bailly, J. S., Pastol, Y. & Le Roux, C. 2010, Comparison of LiDAR waveform processing methods for very shallow water bathymetry using Raman, near-infrared and green signals. *Earth Surface Processes and Landforms: The Journal of the British Geomorphological Research Group*, 35(6):640–650.

Allouis, T., Durrieu, S., Chazette, P., Bailly, J. S., Cuesta, J., Véga, C., Flamant, P. & Couteron, P., 2011, Potential of an ultraviolet, medium-footprint lidar prototype for retrieving forest structure. *ISPRS Journal of Photogrammetry and Remote Sensing*, 66(6 Suppl.):S92–S102.

Anderson, E. S., Thompson, J. A., Crouse, D. A. & Austin, R. E., 2006a, Horizontal resolution and data density effects on remotely sensed LIDAR-based DEM. *Geoderma*, 132(3–4):406–415.

Anderson, J., Martin, M. E., Smith, M. L., Dubayah, R. O., Hofton, M. A., Hyde, P., Peterson, B. E., Blair, J. B. & Knox, R. G., 2006b, The use of waveform lidar to measure northern temperate mixed conifer and deciduous forest structure in New Hampshire. *Remote Sensing of Environment*, 105(3):248–261.

Andrews, G., 1936, Tree-heights from air photographs by simple parallax measurements. *The Forestry Chronicle*, 12(2):152–197.

Angelsen, A., 2008, *Moving Ahead with REDD: Issues, Options and Implications*, CIFOR.

Angelstam, P. K., 1998, Maintaining and restoring biodiversity in European boreal forests by developing natural disturbance regimes. *Journal of Vegetation Science*, 9(4):593–602.

Antonarakis, A. S., Saatchi, S. S., Chazdon, R. L. & Moorcroft, P. R., 2011, Using Lidar and Radar measurements to constrain predictions of forest ecosystem structure and function. *Ecological Applications*, 21(4):1120–1137.

Aquino, C., Mitchard, E. T., McNicol, I. M., Carstairs, H., Burt, A., Vilca, B. L., Mayta, S. & Disney, M. 2024, Detecting selective logging in tropical forests with optical satellite data: An experiment in Peru shows texture at 3 m gives the best results. *Remote Sensing in Ecology and Conservation*.

Assmann, E., 1970, *Principles of Forest Yield Study. Studies in the Organic Production, Structure, Increment and Yield of Forest Stands*, Pergamon Press, 504 p.

Axelsson, C. R., Lindberg, E., Persson, H. J. & Holmgren, J., 2023, The use of dual-wavelength airborne laser scanning for estimating tree species composition and species-specific stem volumes in a boreal forest. *International Journal of Applied Earth Observation and Geoinformation*, 118:103251.

Axelsson, P., 2000, DEM generation from laser scanner data using adaptive TIN models. *International Archives of Photogrammetry and Remote Sensing*, 33(B4/1; PART 4):111–118.

Baccini, A., Goetz, S. J., Walker, W. S., Laporte, N. T., Sun, M., Sulla-Menashe, D., Hackler, J., Beck, P. S. A., Dubayah, R., Friedl, M. A., Samanta, S. & Houghton, R. A., 2012, Estimated carbon dioxide emissions from tropical deforestation improved by carbon-density maps. *Nature Climate Change*, 2(3):182–185.

Balenović, I., Milas, A. S. & Marjanović, H., 2017, A comparison of stand-level volume estimates from image-based canopy height models of different spatial resolutions. *Remote Sensing*, 9(3):205. https://doi.org/10.3390/rs9030205

Balestra, M., Marselis, S., Sankey, T. T., Cabo, C., Liang, X., Mokroš, M., Peng, X., Singh, A., Stereńczak, K., Vega, C., Vincent, G. & Hollaus, M., 2024, LiDAR data fusion to improve forest attribute estimates: A review. *Current Forestry Reports*, 10:281–297.

Baltsavias, E. P., 1999a, Airborne laser scanning: Basic relations and formulas. *ISPRS Journal of Photogrammetry and Remote Sensing*, 54(2):199–214.

Baltsavias, E. P., 1999b, Airborne laser scanning: Existing systems and firms and other resources. *ISPRS Journal of Photogrammetry and Remote Sensing*, 54(2–3):164–198.

Baltsavias, E. P., 1999c, A comparison between photogrammetry and laser scanning. *ISPRS Journal of Photogrammetry and Remote Sensing*, 54(2–3):83–94.

Baltsavias, E. P., Gruen, A., Eisenbeiss, H., Zhang, L. & Waser, L., 2008, High-quality image matching and automated generation of 3D tree models. *International Journal of Remote Sensing*, 29(5):1243–1259.

Barbier, N., Couteron, P., Proisy, C., Malhi, Y. & Gastellu-Etchegorry, J., 2010, The variation of apparent crown size and canopy heterogeneity across lowland Amazonian forests. *Global Ecology and Biogeography*, 19(1):72–84.

Barnard, S. T. & Fischler, M. A., 1982, Computational stereo. *ACM Computing Surveys (CSUR)*, 14(4):553–572.

Barthelemy, D. & Caraglio, Y., 2007, Plant architecture: A dynamic, multilevel and comprehensive approach to plant form, structure and ontogeny. *Annals of Botany*, 99(3):375–407.

Barthélémy, R., 1946, Atmospheric physics—About air soundings—Note from Mr. René Barthélémy, presented by Mr. Camille Guttoni, Session on 11 February 1946. *Comptes rendus hebdomadaires des séances de l'Académie de Sciences- Janvier–Juin 1946, Weekly Reports of the Sessions of the French Academy of Sciences- January–June 1946*. Paris:220, 450–451.

Batho, A. & Garcia, O., 2006, De Perthuis and the origins of site index: A historical note. *FBMIS*, 1:1–10.

Beland, M., Parker, G., Sparrow, B., Harding, D., Chasmer, L., Phinn, S., Antonarakis, A. & Strahler, A. 2019, On promoting the use of lidar systems in forest ecosystem research. *Forest Ecology and Management*, 450:117484.

Bellassen, V., Delbart, N., Le Maire, G., Luyssaert, S., Ciais, P. & Viovy, N., 2011, Potential knowledge gain in large-scale simulations of forest carbon fluxes from remotely sensed biomass and height. *Forest Ecology and Management*, 261(3):515–530.

Bellassen, V., Le Maire, G., Dhôte, J., Ciais, P. & Viovy, N., 2010, Modelling forest management within a global vegetation model—Part 1: Model structure and general behaviour. *Ecological Modelling*, 221(20):2458–2474.

Ben-Arie, J. R., Hay, G. J., Powers, R. P., Castilla, G. & St-Onge, B., 2009, Development of a pit filling algorithm for LiDAR canopy height models. *Computers & Geosciences*, 35(9):1940–1949.

Besic, N., Durrieu, S., Schleich, A. & Vega, C. 2024, Using structural class pairing to address the spatial mismatch between GEDI measurements and NFI plots. *IEEE Journal of Selected Topics in Applied Earth Observations and Remote Sensing*.

Blair, J. B., Rabine, D. L. & Hofton, M. A., 1999, The laser vegetation imaging sensor: A medium-altitude, digitisation-only, airborne laser altimeter for mapping vegetation and topography. *ISPRS Journal of Photogrammetry and Remote Sensing*, 54(2–3):115–122.

Blanco, J. A., & Lo, Y. H., 2023, Latest trends in modelling forest ecosystems: New approaches or just new methods? *Current Forestry Reports*, 9:219–229. https://doi.org/10.1007/s40725-023-00189-y

Bohn, F. J. & Huth, A., 2017, The importance of forest structure to biodiversity–productivity relationships. *Royal Society Open Science*, 4(1):160521.

Bolton, D. K., Coops, N. C. & Wulder, M. A., 2013, Investigating the agreement between global canopy height maps and airborne Lidar derived height estimates over Canada. *Canadian Journal of Remote Sensing*, 39(Suppl.1):S139–S151.

Bongers, F., 2001, Methods to assess tropical rain forest canopy structure: An overview. *Plant Ecology*, 153(1–2):263–277.

Bontemps, J.-D. & Bouriaud, O., 2014, Predictive approaches to forest site productivity: Recent trends, challenges and future perspectives. *Forestry*, 87(1):109–128.

Borgogno Mondino, E., Fissore, V., Falkowski, M. J. & Palik, B., 2020, How far can we trust forestry estimates from low-density LiDAR acquisitions? The Cutfoot Sioux experimental forest (MN, USA) case study. *International Journal of Remote Sensing*, 41(12).

Borsah, A. A., Nazeer, M. & Wong, M. S., 2023, LIDAR-based forest biomass remote sensing: A review of metrics, methods, and assessment criteria for the selection of allometric equations. *Forests*, 14(10):2095.

Bouget, C., 2005, Short-term effect of windstorm disturbance on saproxylic beetles in broadleaved temperate forests. Part I: Do windthrow changes induce a gap effect? *Forest Ecology and Management*, 216(1–3):1–14.

Bouvier, M., Durrieu, S., Fournier, R. A., Saint-Geours, N., Guyon, D., Grau, E. & De Boissieu, F., 2019, Influence of sampling design parameters on biomass predictions derived from airborne lidar data. *Canadian Journal of Remote Sensing*, 45(5):650–672.

Bouvier, M., Durrieu, S., Fournier, R. A. & Renaud, J. P., 2015, Generalizing predictive models of forest inventory attributes using an area-based approach with airborne LiDAR data. *Remote Sensing of Environment*, 156:322–334.

Bouvier, M., Durrieu, S., Gosselin, F. & Herpigny, B., 2017, Use of airborne lidar data to improve plant species richness and diversity monitoring in lowland and mountain forests. *PLoS One*, 12(9):e0184524.

Boyd, D. & Danson, F., 2005, Satellite remote sensing of forest resources: Three decades of research development. *Progress in Physical Geography*, 29(1):1–26.

Brandtberg, T., Warner, T. A., Landenberger, R. E. & McGraw, J. B., 2003, Detection and analysis of individual leaf-off tree crowns in small footprint, high sampling density lidar data from the eastern deciduous forest in North America. *Remote Sensing of Environment*, 85(3):290–303.

Brown, M. Z., Burschka, D. & Hager, G. D., 2003, Advances in computational stereo. *Pattern Analysis and Machine Intelligence, IEEE Transactions on*, 25(8):993–1008.

Buckingham, R. & Staenz, K., 2008, Review of current and planned civilian space hyperspectral sensors for EO. *Canadian Journal of Remote Sensing*, 34(sup1):S187–S197.

Bui, L. K. & Glennie, C. L., 2023, Estimation of lidar-based gridded DEM uncertainty with varying terrain roughness and point density. *ISPRS Open Journal of Photogrammetry and Remote Sensing*, 7:100028.

Cabo, C., Ordóñez, C., López-Sánchez, C. A. & Armesto, J., 2018, Automatic dendrometry: Tree detection, tree height and diameter estimation using terrestrial laser scanning. *International Journal of Applied Earth Observation and Geoinformation*, 69:164–174.

Cabello, J., Fernández, N., Alcaraz-Segura, D., Oyonarte, C., Piñeiro, G., Altesor, A., Delibes, M. & Paruelo, J. M., 2012, The ecosystem functioning dimension in conservation: Insights from remote sensing. *Biodiversity and Conservation*, 21(13):3287–3305.

Camarretta, N., Harrison, P. A., Bailey, T., Potts, B., Lucieer, A., Davidson, N. & Hunt, M., 2020, Monitoring forest structure to guide adaptive management of forest restoration: A review of remote sensing approaches. *New Forests*, 51(4):573–596.

Castro, K. L., Sanchez-Azofeifa, G. A. & Rivard, B., 2003, Monitoring secondary tropical forests using spaceborne data: Implications for Central America. *International Journal of Remote Sensing*, 24(9):1853–1894.

Cățeanu, M. & Arcadie, C., 2017, ALS for terrain mapping in forest environments: An analysis of LiDAR filtering algorithms. *EARSeL eProceedings*, 16(1):9–20.

Cățeanu, M. & Ciubotaru, A., 2020, Accuracy of ground surface interpolation from airborne laser scanning (ALS) data in dense forest cover. *ISPRS International Journal of Geo-Information*, 9(4):224.

Cawse-Nicholson, K., Townsend, P. A., Schimel, D., Assiri, A. M., Blake, P. L., Buongiorno, M. F., ... & SBG Algorithms Working Group, 2021, NASA's surface biology and geology designated observable: A perspective on surface imaging algorithms. *Remote Sensing of Environment*, 257:112349.

CEOS, 2024, *CEOS Database*. https://database.eohandbook.com/database/missionsummary.aspx?missionID=1384. Accessed 14 August 2024.

Chaubert-Pereira, F., Caraglio, Y., Lavergne, C. & Guédon, Y., 2009, Identifying ontogenetic, environmental and individual components of forest tree growth. *Annals of Botany*, 104(5):883–896.

Chauve, A., Véga, C., Bretar, F., Allouis, T., Pierrot Deseilligny, M. & Puech, W., 2008, Processing full-waveform lidar data in alpine coniferous forest: Assessing terrain and tree height quality. *International Journal of Remote Sensing*, 30(19):5211–5228.

Chauve, A., Véga, C., Durrieu, S., Bretar, F., Allouis, T., Deseilligny, M. P. & Puech, W., 2009, Advanced full-waveform lidar data echo detection: Assessing quality of derived terrain and tree height models in an alpine coniferous forest. *International Journal of Remote Sensing*, 30(19):5211–5228.

Chave, J., Condit, R., Lao, S., Caspersen, J. P., Foster, R. B. & Hubbell, S. P., 2003, Spatial and temporal variation of biomass in a tropical forest: Results from a large census plot in Panama. *Journal of Ecology*, 91(2):240–252.

Chehata, N., Guo, L. & Mallet, C., 2009, Airborne lidar feature selection for urban classification using random forests. *International Archives of the Photogrammetry, Remote Sensing and Spatial Information Sciences*, 39(Part 3/W8):207–212.

Chen, C., Guo, J., Wu, H., Li, Y. & Shi, B., 2021, Performance comparison of filtering algorithms for high-density airborne LiDAR point clouds over complex LandScapes. *Remote Sensing*, 13(14):2663.

Chen, H., Cloude, S. R., Goodenough, D. G., Hill, D. A. & Nesdoly, A., 2018, Radar forest height estimation in mountainous terrain using Tandem-X coherence data. *IEEE Journal of Selected Topics in Applied Earth Observations and Remote Sensing*, 11(10):3443–3452.

Chen, Q., 2010, Assessment of terrain elevation derived from satellite laser altimetry over mountainous forest areas using airborne lidar data. *ISPRS Journal of Photogrammetry and Remote Sensing*, 65(1):111–122.

Chen, Q., Baldocchi, D., Gong, P. & Kelly, M., 2006, Isolating individual trees in a savanna woodland using small footprint lidar data. *Photogrammetric Engineering & Remote Sensing*, 72(8):923–932.

Clark, M. L., Clark, D. B. & Roberts, D. A., 2004, Small-footprint lidar estimation of sub-canopy elevation and tree height in a tropical rain forest landscape. *Remote Sensing of Environment*, 91(1):68–89.

Coops, N. C., Tompalski, P., Goodbody, T. R., Achim, A. & Mulverhill, C., 2023, Framework for near real-time forest inventory using multi source remote sensing data. *Forestry*, 96(1):1–19.

Coops, N. C., Tompalski, P., Goodbody, T. R., Queinnec, M., Luther, J. E., Bolton, D. K., White, J. C., Wulder, M. A., van Lier, O. R. & Hermosilla, T., 2021, Modelling lidar-derived estimates of forest attributes over space and time: A review of approaches and future trends. *Remote Sensing of Environment*, 260:112477.

Coops, N. & Waring, R., 2001, The use of multiscale remote sensing imagery to derive regional estimates of forest growth capacity using 3-PGS. *Remote Sensing of Environment*, 75(3):324–334.

Couteron, P., Pelissier, R., Nicolini, E. A. & Paget, D., 2005, Predicting tropical forest stand structure parameters from Fourier transform of very high-resolution remotely sensed canopy images. *Journal of Applied Ecology*, 42(6):1121–1128.

Davison, C. W., Assmann, J. J., Normand, S., Rahbek, C. & Morueta-Holme, N., 2023, Vegetation structure from LiDAR explains the local richness of birds across Denmark. *Journal of Animal Ecology*, 92(7):1332–1344.

Dayal, K. R., Durrieu, S., Alleaume, S., Revers, F., Larmanou, E., Renaud, J. P. & Bouvier, M., 2020, Scan angle impact on LiDAR-derived metrics used in ABA models for prediction of forest stand characteristics: A grid based analysis. *The International Archives of the Photogrammetry, Remote Sensing and Spatial Information Sciences*, 43:975–982.

Dayal, K. R., Durrieu, S., Lahssini, K., Alleaume, S., Bouvier, M., Monnet, J. M., Renaud, J-P. & Revers, F., 2022, An investigation into lidar scan angle impacts on stand attribute predictions in different forest environments. *ISPRS Journal of Photogrammetry and Remote Sensing*, 193:314–338.

Dayal, K. R., Durrieu, S., Lahssini, K., Ienco, D. & Monnet, J. M., 2023, Enhancing forest attribute prediction by considering terrain and scan angles from lidar point clouds: A neural network approach. *IEEE Journal of Selected Topics in Applied Earth Observations and Remote Sensing*, 16:3531–3544.

DeFries, R., Achard, F., Brown, S., Herold, M., Murdiyarso, D., Schlamadinger, B. & de Souza Jr, C., 2007, Earth observations for estimating greenhouse gas emissions from deforestation in developing countries. *Environmental Science and Policy*, 10(4):385–394.

Degnan, J. J., 2016, Scanning, multibeam, single photon lidars for rapid, large scale, high resolution, topographic and bathymetric mapping. *Remote Sensing*, 8:958.

De Leeuw, J., Georgiadou, Y., Kerle, N., de Gier, A., Inoue, Y., Ferwerda, J., Smies, M. & Narantuya, D., 2010, The function of remote sensing in support of environmental policy. *Remote Sensing*, 2(7):1731–1750.

De Sy, V., Herold, M., Achard, F., Asner, G. P., Held, A., Kellndorfer, J. & Verbesselt, J., 2012, Synergies of multiple remote sensing data sources for REDD+ monitoring. *Current Opinion in Environmental Sustainability*, 4(6):696–706.

Di Biase, R. M., Fattorini, L. & Marchi, M., 2018, Statistical inferential techniques for approaching forest mapping. A review of methods. *Annals of Silvicultural Research*, 42(2):46–58.

Dietmaier, A., McDermid, G. J., Rahman, M. M., Linke, J. & Ludwig, R., 2019, Comparison of LiDAR and digital aerial photogrammetry for characterizing canopy openings in the boreal Forest of Northern Alberta. *Remote Sensing*, 11:1919.

Dhôte, J., 1991, Modélisation de la croissance des peuplements réguliers de hêtre: Dynamique des hiérarchies sociales et facteurs de production. *Annales des Sciences forestières*, 48(4):389–416.

Disney, M. I., Kalogirou, V., Lewis, P., Prieto-Blanco, A., Hancock, S. & Pfeifer, M., 2010, Simulating the impact of discrete-return lidar system and survey characteristics over young conifer and broadleaf forests. *Remote Sensing of Environment*, 114(7):1546–1560.

Disney, M. I., Lewis, P. & Saich, P., 2006, 3D modelling of forest canopy structure for remote sensing simulations in the optical and microwave domains. *Remote Sensing of Environment*, 100(1):114–132.

Dong, T., Zhang, X., Ding, Z. & Fan, J., 2020, Multi-layered tree crown extraction from LiDAR data using graph-based segmentation. *Computers and Electronics in Agriculture*, 170:105213.

Dong, Y., Cui, X., Zhang, L. & Ai, H., 2018, An improved progressive TIN densification filtering method considering the density and standard variance of point clouds. ISPRS *International Journal of Geo-Information*, 7(10):409.

Dorado-Roda, I., Pascual, A., Godinho, S., Silva, C., Botequim, B., Rodríguez-Gonzálvez, P., González-Ferreiro, E., Guerra, J., 2021, Assessing the accuracy of GEDI data for canopy height and aboveground biomass estimates in mediterranean forests. *Remote Sensing*, 13:2279. https://doi.org/10/gkjf2c.

Dubayah, R., Blair, J. B., Goetz, S., Fatoyinbo, L., Hansen, M., Healey, S., Hofton, M., Hurtt, G., Kellner, J., Luthcke, S., Armston, J., Tang, H., Duncanson, L., Hancock, S., Jantz, P., Marselis, S., Patterson, P. L., Qi, W. & Silva, C., 2020, The Global Ecosystem Dynamics Investigation: High-resolution laser ranging of the Earth's forests and topography. *Science of Remote Sensing*, 1:100002.

Dubayah, R. O. & Drake, J. B., 2001, *Lidar Remote Sensing for Forestry Applications*. www.geog.umd.edu/vcl/pubs/jof.pdf-2001-01-25

Ducic, V., Hollaus, M., Ullrich, A., Wagner, W. & Melzer, T., 2006, 3D Vegetation mapping and classification using full-waveform laser scanning. *Workshop on 3D Remote Sensing in Forestry, Vienna, Austria, International Archives of Photogrammetry, Remote Sensing and Spatial Information Sciences*, 211–217.

Duguid, M. C. & Ashton, M. S., 2013, A meta-analysis of the effect of forest management for timber on understory plant species diversity in temperate forests. *Forest Ecology and Management*, 303:81–90.

Duncanson, L., Kellner, J., Armston, J., Dubayah, R., Minor, D., Hancock, S., Healey, S., Patterson, P., Saarela, S., Marselis, S., Silva, C., Bruening, J., Goetz, S., Tang, H., Hofton, M., Blair, B., Luthcke, S., Fatoyinbo, L., Abernethy, K. & Zgraggen, C., 2021. Aboveground biomass density models for NASA's global ecosystem dynamics investigation (GEDI) lidar mission. *Remote Sensing of Environment*. https://doi.org/10/gn3jrm.

Durrieu, S. & Nelson, R. F., 2013, Earth observation from space—The issue of environmental sustainability. *Space Policy*, 29(4):238–250.

Duursma, R. & Medlyn, B., 2012, MAESPA: A model to study interactions between water limitation, environmental drivers and vegetation function at tree and stand levels, with an example application to [CO 2]× drought interactions. *Geoscientific Model Development Discussions*, 5(1):459–513.

Eichhorn, F., 1904, Beziehungen zwischen Bestandshöhe und Bestandsmasse. *Allgemeine Forst-und Jagdzeitung*, 80:45–49.

Ellison, D. N. Futter, M. & Bishop, K., 2012, On the forest cover–water yield debate: From demand-to supply-side thinking. *Global Change Biology*, 18(3):806–820.

Ene, L. T., Næsset, E., Gobakken, T., Gregoire, T. G., Ståhl, G. & Holm, S., 2013, A simulation approach for accuracy assessment of two-phase post-stratified estimation in large-area LiDAR biomass surveys. *Remote Sensing of Environment*, 133;210–224.

ESA, 2008, *Biomass, Mission Assessment Report*, ESA:132 p. http://esamultimedia.esa.int/docs/SP1313-2_BIOMASS.pdf

Evans, J. S., Hudak, A. T., Faux, R. & Smith, A., 2009, Discrete return lidar in natural resources: Recommendations for project planning, data processing, and deliverables. *Remote Sensing*, 1(4):776–794.

Eysn, L., Hollaus, M., Lindberg, E., Berger, F., Monnet, J.-M., Dalponte, M., Kobal, M., Pellegrini, M., Lingua, E., Mongus, D. & Pfeifer, N., 2015, A benchmark of lidar-based single tree detection methods using heterogeneous forest data from the alpine *Space. Forests*, 6:1721–1747.

Falkowski, M. J., Smith, A. M. S., Hudak, A. T., Gessler, P. E., Vierling, L. A. & Crookston, N. L., 2006, Automated estimation of individual conifer tree height and crown diameter via two-dimensional spatial wavelet analysis of lidar data. *Canadian Journal of Remote Sensing*, 32(2):153–161.

Falkowski, M. J., Wulder, M. A., White, J. C. & Gillis, M. D., 2009, Supporting large-area, sample-based forest inventories with very high spatial resolution satellite imagery. *Progress in Physical Geography*, 33(3):403–423.

Fardusi, M. J., Chianucci, F. & Barbati, A., 2017, Concept to practice of geospatial-information tools to assist forest management and planning under precision forestry framework: A review. *Annals of Silvicultural Research*, 41(1):3–14.

Fassnacht, F. E., Mangold, D., Schäfer, J., Immitzer, M., Kattenborn, T., Koch, B. & Latifi, H., 2017, Estimating stand density, biomass and tree species from very high resolution stereo-imagery—towards an all-in-one sensor for forestry applications? *Forestry: An International Journal of Forest Research*, 90(5):613–631. https://doi.org/10.1093/forestry/cpx014

Fassnacht, F. E., White, J. C., Wulder, M. A. & Næsset, E., 2024, Remote sensing in forestry: Current challenges, considerations and directions. *Forestry: An International Journal of Forest Research*, 97(1):11–37.

Fatoyinbo, T., Armston, J., Simard, M., Saatchi, S., Denbina, M., Lavalle, M., . . . & Hibbard, K., 2021, The NASA AfriSAR campaign: Airborne SAR and lidar measurements of tropical forest structure and biomass in support of current and future space missions. *Remote Sensing of Environment*, 264:112533.

Faux, R., Buffington, J. M., Whitley, G., Lanigan, S. & Roper, B., 2009, Use of airborne near-infrared LiDAR for determining channel cross-section characteristics and monitoring aquatic habitat in Pacific Northwest rivers: A preliminary analysis. *Proceedings of the American Society for Photogrammetry and Remote Sensing. Pacific Northwest Aquatic Monitoring Partnership*, Cook, Washington:43–60.

Feng, T., Duncanson, L., Montesano, P., Hancock, S., Minor, D., Guenther, E. & Neuenschwander, A., 2023, A systematic evaluation of multi-resolution ICESat-2 ATL08 terrain and canopy heights in boreal forests. *Remote Sensing of Environment*, 291:113570. https://doi.org/10.1016/j.rse.2023.113570.

Ferraz, A., Bretar, F., Jacquemoud, S., Gonçalves, G., Pereira, L., Tomé, M. & Soares, P., 2012, 3-D mapping of a multi-layered Mediterranean forest using ALS data. *Remote Sensing of Environment*, 121:210–223.

Ferraz, A., Bretar, F., Jacquemoud, S., Gonçalves, G. & Pereira, L., 2010, 3D segmentation of forest structure using an adaptive mean shift based procedure. *SilviLaser 2010, Conference on Lidar Applications for Assessing Forest Ecosytems Freiburg*, Germany:13 p.

Ferretti, A., Monti-Guar, A., Prati, C., Rocca, F. & Massonnet, D., 2007, *InSAR Principles: Guidelines for SAR Interferometry Processing and Interpretation—Part A*, ESA Publications-TM 19:40 p. http://www.esa.int/esapub/tm/tm19/TM-19_ptA.pdf

Ferretti, M., 2010, Harmonizing forest inventories and forest condition monitoring—the rise or the fall of harmonized forest condition monitoring in Europe? *IForest*, 3(January):1–4.

Fischer, F. J., Maréchaux, I. & Chave, J., 2019b, Improving plant allometry by fusing forest models and remote sensing. *New Phytologist*, 223(3):1159–1165.

Fischer, R., Knapp, N., Bohn, F., Shugart, H. H. & Huth, A., 2019a, The relevance of forest structure for biomass and productivity in temperate forests: New perspectives for remote sensing. *Surveys in Geophysics*, 40:709–734.

Frazer, G., Magnussen, S., Wulder, M. & Niemann, K., 2011, Simulated impact of sample plot size and co-registration error on the accuracy and uncertainty of LiDAR-derived estimates of forest stand biomass. *Remote Sensing of Environment*, 115(2):636–649.

Fujita, T., Itaya, A., Miura, M., Manabe, T. & Yamamoto, S.-I., 2003, Long-term canopy dynamics analysed by aerial photographs in a temperate old-growth evergreen broad-leaved forest. *Journal of Ecology*, 91(4): 686–693.

Fuller, D. O., 2006, Tropical forest monitoring and remote sensing: A new era of transparency in forest governance? *Singapore Journal of Tropical Geography*, 27(1):15–29.

Gagnon, P., Agnard, J. & Nolette, C., 1993, Evaluation of a soft-copy photogrammetry system for tree-plot measurements. *Canadian Journal of Forest Research*, 23(9):1781–1785.

Ganz, S., Käber, Y. & Adler, P., 2019, Measuring tree height with remote sensing—A comparison of photogrammetric and LiDAR data with different field measurements. *Forests*, 10:694.

Gao, J., 2007, Towards accurate determination of surface height using modern geoinformatic methods: Possibilities and limitations. *Progress in Physical Geography*, 31(6):591–605.

García, O., 1998, Estimating top height with variable plot sizes. *Canadian Journal of Forest Research*, 28(10):1509–1517.

García, O., 2011, Dynamical implications of the variability representation in site-index modelling. *European Journal of Forest Research*, 130(4):671–675.

Garestier, F., Dubois-Fernandez, P. C., Guyon, D. & Le Toan, T., 2009, Forest biophysical parameter estimation using L- and P-band polarimetric SAR data. *IEEE Transactions on Geoscience and Remote Sensing*, 47(10):3379–3388.

Germaine, S. S., Vessey, S. H. & Capen, D. E., 1997, Effects of small forest openings on the breeding bird community in a Vermont hardwood forest. *Condor*, 99(3):708–718.

Gobakken, T. & Næsset, E., 2004, Estimation of diameter and basal area distributions in coniferous forest by means of airborne laser scanner data. *Scandinavian Journal of Forest Research*, 19(6):529–542.

Gobakken, T. & Næsset, E., 2008, Assessing effects of laser point density, ground sampling intensity, and field sample plot size on biophysical stand properties derived from airborne laser scanner data. *Canadian Journal of Forest Research*, 38(5):1095–1109.

Goetz, S., Steinberg, D., Dubayah, R. & Blair, B., 2007, Laser remote sensing of canopy habitat heterogeneity as a predictor of bird species richness in an eastern temperate forest, USA. *Remote Sensing of Environment*, 108(3):254–263.

Gong, P., Sheng, Y. & Biging, G., 2002, 3D model-based tree measurement from high-resolution aerial imagery. *Photogrammetric Engineering and Remote Sensing*, 68(11):1203–1212.

Goodbody, T. R., Coops, N. C., Luther, J. E., Tompalski, P., Mulverhill, C., Frizzle, C., Fournier, R., Furze, S. & Herniman, S., 2021a, Airborne laser scanning for quantifying criteria and indicators of sustainable forest management in Canada. *Canadian Journal of Forest Research*, 51(7):972–985.

Goodbody, T. R., Coops, N. C. & White, J. C., 2019, Digital aerial photogrammetry for updating area-based forest inventories: A review of opportunities, challenges, and future directions. *Current Forestry Reports*, 5: 55–75.

Goodbody, T. R., White, J. C., Coops, N. C. & LeBoeuf, A., 2021b, Benchmarking acquisition parameters for digital aerial photogrammetric data for forest inventory applications: Impacts of image overlap and resolution. *Remote Sensing of Environment*, 265:112677.

Goodenough, D. G., Pearlman, J., Chen, H., Dyk, A., Han, T., Li, J., Miller, J. & Niemann, K. O., 2004, Forest information from hyperspectral sensing. *IGARSS 2004. 2004 IEEE International Geoscience and Remote Sensing Symposium*, IEEE, Vol. 4:2585–2589.

Götmark, F., 2013, Habitat management alternatives for conservation forests in the temperate zone: Review, synthesis, and implications. *Forest Ecology and Management*, 306:292–307.

Gril, E., Laslier, M., Gallet-Moron, E., Durrieu, S., Spicher, F., Le Roux, V., Brasseur, B., Haesen, S., Van Meerbeek, K., Decocq, G., Marrec, R. & Lenoir, J., 2023, Using airborne LiDAR to map forest microclimate temperature buffering or amplification. *Remote Sensing of Environment*, 298:113820.

Guédon, Y., Caraglio, Y., Heuret, P., Lebarbier, E. & Meredieu, C., 2007, Analyzing growth components in trees. *Journal of Theoretical Biology*, 248(3):418–447.

Guerra-Hernández, J., Pascual, A., 2021, Using GEDI lidar data and airborne laser scanning to assess height growth dynamics in fast-growing species: A showcase in Spain. *Forest Ecosystems*, 8:14. https://doi.org/10/gpfhdx.

Guo, B., Huang, X., Zhang, F. & Sohn, G., 2015, Classification of airborne laser scanning data using JointBoost. *ISPRS Journal of Photogrammetry and Remote Sensing*, 100:71–83.

Gupta, R. & Sharma, L. K., 2019, The process-based forest growth model 3-PG for use in forest management: A review. *Ecological Modelling*, 397:55–73.

Gyawali, A., Aalto, M., Peuhkurinen, J., Villikka, M. & Ranta, T., 2022, Comparison of individual tree height estimated from LiDAR and digital aerial photogrammetry in young forests. *Sustainability*, 14:3720.

Haesen, S., Lenoir, J., Gril, E., De Frenne, P., Lembrechts, J. J., Kopecký, M., Macek, M., Man, M., Wild, J. & Van Meerbeek, K., 2023, Microclimate reveals the true thermal niche of forest plant species. *Ecology Letters*, 26(12):2043–2055.

Hall, F. G., Bergen, K., Blair, J. B., Dubayah, R., Houghton, R., Hurtt, G., Kellndorfer, J., Lefsky, M., Ranson, J., Saatchi, S., Shugart, H. H. & Wickland, D., 2011b, Characterizing 3D vegetation structure from space: Mission requirements. *Remote Sensing of Environment*, 115(11):2753–2775.

Hall, F. G., Saatchi, S. & Dubayah, R., 2011a, Preface: DESDynI VEG-3D special issue. *Remote Sensing of Environment*, 115(11):2752.

Han, Y., Qin, R. & Huang, X., 2020a, Assessment of dense image matchers for digital surface model generation using airborne and spaceborne images—An update. *The Photogrammetric Record*, 35:58–80.

Han, Y., Wang, S., Gong, D., Wang, Y. & Ma, X., 2020b, State of the art in digital surface modelling from multi-view high-resolution satellite images. *ISPRS Annals of the Photogrammetry, Remote Sensing and Spatial Information Sciences*, 2:351–356.

Hancock, S., Lewis, P., Foster, M., Disney, M. & Muller, J.-P., 2012, Measuring forests with dual wavelength lidar: A simulation study over topography. *Agricultural and Forest Meteorology*, 161:123–133.

Harding, D. J., Lefsky, M. A., Parker, G. G. & Blair, J. B., 2001, Laser altimeter canopy height profiles methods and validation for closed-canopy, broadleaf forests. *Remote Sensing of Environment*, 76(3):283–297.

He, K. S., Bradley, B. A., Cord, A. F., Rocchini, D., Tuanmu, M. N., Schmidtlein, S., Turner, W., Wegmann, M. & Pettorelli, N., 2015, Will remote sensing shape the next generation of species distribution models? *Remote Sensing in Ecology and Conservation*, 1(1):4–18.

Heipke, C., 1997, Automation of interior, relative, and absolute orientation. *ISPRS Journal of Photogrammetry and Remote Sensing*, 52(1):1–19.

Heipke, C. & Rottensteiner, F., 2020, Deep learning for geometric and semantic tasks in photogrammetry and remote sensing. *Geo-Spatial Information Science*, 23:10–19.

Henry, M., Bombelli, A., Trotta, C., Alessandrini, A., Birigazzi, L., Sola, G., Vieilledent, G., Santenoise, P., Longuetaud, F. & Valentini, R., 2013, GlobAllomeTree: International platform for tree allometric equations to support volume, biomass and carbon assessment. *iForest-Biogeosciences & Forestry*, 6(6):326.

Herold, M., Carter, S., Avitabile, V., Espejo, A. B., Jonckheere, I., Lucas, R., . . . & De Sy, V., 2019, The role and need for space-based forest biomass-related measurements in environmental management and policy. *Surveys in Geophysics*, 40:757–778.

Hill, J., Buddenbaum, H. & Townsend, P. A., 2019, Imaging spectroscopy of forest ecosystems: Perspectives for the use of space-borne hyperspectral earth observation systems. *Surveys in Geophysics*, 40(3):553–588.

Hinz, A. & Heier, H., 2000, The Z/I imaging digital camera system. *The Photogrammetric Record*, 16(96):929–936.

Hirata, Y., 2004, The effects of footprint size and sampling density in airborne laser scanning to extract individual trees in mountainous terrain. *International Archives of Photogrammetry, Remote Sensing and Spatial Information Sciences*, 36(Part 8):102–107.

Hirschmugl, M., Ofner, M., Raggam, J. & Schardt, M., 2007, Single tree detection in very high resolution remote sensing data. *Remote Sensing of Environment*, 110(4):533–544.

Hodgson, M. E. & Bresnahan, P., 2004, Accuracy of airborne lidar-derived elevation: Empirical assessment and error budget. *Photogrammetric Engineering and Remote Sensing*, 70(3):331–339.

Hodgson, M. E., Jensen, J., Raber, G., Tullis, J., Davis, B. A., Thompson, G. & Schuckman, K., 2005, An evaluation of lidar-derived elevation and terrain slope in leaf-off conditions. *Photogrammetric Engineering and Remote Sensing*, 71(7):817–823.

Holmgren, J., Lindberg, E., Olofsson, K. & Persson, H. J., 2022, Tree crown segmentation in three dimensions using density models derived from airborne laser scanning. *International Journal of Remote Sensing*, 43:299–329.

Holmgren, J., Nilsson, M. & Olsson, H., 2003, Estimation of tree height and stem volume on plots using airborne laser scanning. *Forest Science*, 49(3):419–428.

Hopkinson, C., 2007, The influence of flying altitude, beam divergence, and pulse repetition frequency on laser pulse return intensity and canopy frequency distribution. *Canadian Journal of Remote Sensing*, 33(4):312–324.

Hopkinson, C. & Chasmer, L., 2009, Testing LiDAR models of fractional cover across multiple forest ecozones. *Remote Sensing of Environment*, 113(1):275–288.

Ho Tong Minh, D., Le Toan, T., Rocca, F., Tebaldini, S., D'Alessandro, M. M. & Villard, L., 2014, Relating P-band synthetic aperture radar tomography to tropical forest biomass. *IEEE Transactions on Geoscience and Remote Sensing*, 52(2):967–979.

Hu, B., Li, J., Jing, L. & Judah, A., 2014, Improving the efficiency and accuracy of individual tree crown delineation from high-density LiDAR data. *International Journal of Applied Earth Observation and Geoinformation*, 26:145–155.

Hu, H., Su, L., Mao, S., Chen, M., Pan, G., Xu, B. & Zhu, Q., 2023, Adaptive region aggregation for multi-view stereo matching using deformable convolutional networks. *The Photogrammetric Record*, 38:430–449.

Hu, T., Sun, X., Su, Y., Guan, H., Sun, Q., Kelly, M. & Guo, Q., 2020, Development and performance evaluation of a very low-cost UAV-LiDAR system for forestry applications. *Remote Sensing*, 13(1):77.

Huang, D., Tang, Y. & Qin, R., 2022, An evaluation of PlanetScope images for 3D reconstruction and change detection–experimental validations with case studies. *GIScience & Remote Sensing*, 59(1):744–761.

Huang, S., Hager, S. A., Halligan, K. Q., Fairweather, I. S., Swanson, A. K. & Crabtree, R. L., 2009, A comparison of individual tree and forest plot height derived from lidar and InSAR. *Photogrammetric Engineering and Remote Sensing*, 75(2):159–167.

Hug, C., Ullrich, A. & Grimm, A., 2004, LiteMapper-5600—A waveform digitizing lidar terrain and vegetation mapping system. *International Conference "Laser-scanners for Forest and Landscape Assessment"*, Frieburg, Germany, International archives of photogrammetry, remote sensing and spatial information sciences, XXXVI, part 8/W2:24–29.

Hyyppä, E., Yu, X., Kaartinen, H., Hakala, T., Kukko, A., Vastaranta, M. & Hyyppä, J., 2020, Comparison of backpack, handheld, under-canopy UAV, and above-canopy UAV laser scanning for field reference data collection in boreal forests. *Remote Sensing*, 12(20):3327.

Hyyppä, J., Hyyppä, H., Leckie, D., Gougeon, F., Yu, X. & Maltamo, M., 2008, Review of methods of small-footprint airborne laser scanning for extracting forest inventory data in boreal forests. *International Journal of Remote Sensing*, 29(5):1339–1366.

Hyyppä, J., Kelle, O., Lehikoinen, M. & Inkinen, M., 2001, A segmentation-based method to retrieve stem volume estimates from 3-D tree height models produced by laser scanners. IEEE Transactions on Geoscience and Remote Sensing, 39(5):969–975.

Iglhaut, J., Cabo, C., Puliti, S., Piermattei, L., O'Connor, J. & Rosette, J., 2019, Structure from motion photogrammetry in forestry: A review. *Current Forestry Reports*, 5:155–168.

Iqbal, I. A., Musk, R. A., Osborn, J., Stone, C. & Lucieer, A., 2019, A comparison of area-based forest attributes derived from airborne laser scanner, small-format and medium-format digital aerial photography. *International Journal of Applied Earth Observation and Geoinformation*, 76:231–241.

Ishii, H. T., Tanabe, S. I. & Hiura, T., 2004, Exploring the relationships among canopy structure, stand productivity, and biodiversity of temperate forest ecosystems. *Forest Science*, 50(3):342–355.

Jakubowski, M. K., Guo, Q. & Kelly, M., 2013, Tradeoffs between lidar pulse density and forest measurement accuracy. *Remote Sensing of Environment*, 130:245–253.

Jaskierniak, D., 2011, *Modelling the Effects of Forest Regeneration on Streamflow Using Forest Growth Models*. PhD dissertation, School of Geography and Environmental Studies, University of Tasmania:240 p.

Joshi, N., Mitchard, E. T., Brolly, M., Schumacher, J., Fernández-Landa, A., Johannsen, V. K., Marchamalo, M. & Fensholt, R., 2017, Understanding 'saturation' of radar signals over forests. *Scientific Reports*, 7(1):3505.

Jucker, T., Gosper, C. R., Wiehl, G., Yeoh, P. B., Raisbeck-Brown, N., Fischer, F. J., Graham, J., Langley, H., Newchurch, W., O'Donnell, A. J., Page, G. F. M., Zdunic, K. & Prober, S. M., 2023, Using multi-platform LiDAR to guide the conservation of the world's largest temperate woodland. *Remote Sensing of Environment*, 296:113745.

Kaartinen, H., Hyyppä, J., Yu, X., Vastaranta, M., Hyyppä, H., Kukko, A., Holopainen, M., Heipke, C., Hirschmugl, M., Morsdorf, F., Næsset, E., Pitkänen, J., Popescu, S., Solberg, S., Wolf, B. M. & Wu, J. C., 2012, An international comparison of individual tree detection and extraction using airborne laser scanning. *Remote Sensing*, 4(4):950–974.

Kangas, A., Myllymäki, M. & Mehtätalo, L., 2023, Understanding uncertainty in forest resources maps. *Silva Fennica*, 57(2):22026.

Kangas, A., Gobakken, T., Puliti, S., Hauglin, M. & Naesset, E., 2018, Value of airborne laser scanning and digital aerial photogrammetry data in forest decision making. *Silva Fennica*, 52:9923.

Karasov, O., Külvik, M. & Burdun, I., 2021, Deconstructing landscape pattern: Applications of remote sensing to physiognomic landscape mapping. *GeoJournal*, 86(1):529–555.

Karila, K., Vastaranta, M., Karjalainen, M. & Kaasalainen, S., 2015, Tandem-X interferometry in the prediction of forest inventory attributes in managed boreal forests. *Remote Sensing of Environment*, 159:259–268.

Kasischke, E. S., Melack, J. M. & Craig Dobson, M., 1997, The use of imaging radars for ecological applications—a review. *Remote Sensing of Environment*, 59(2):141–156.

Kato, A., Moskal, L. M., Schiess, P., Swanson, M. E., Calhoun, D. & Stuetzle, W., 2009, Capturing tree crown formation through implicit surface reconstruction using airborne lidar data. *Remote Sensing of Environment*, 113(6):1148–1162.

Ke, Y. & Quackenbush, L. J., 2011, A review of methods for automatic individual tree-crown detection and delineation from passive remote sensing. *International Journal of Remote Sensing*, 32(17):4725–4747.

Kellner, J. R., Clark, D. B. & Hofton, M. A., 2009, Canopy height and ground elevation in a mixed-land-use lowland Neotropical rain forest landscape. *Ecology*, 90(11):3274–3274.

Kemppinen, J., Lembrechts, J. J., Van Meerbeek, K., Carnicer, J., Chardon, N. I., Kardol, P., . . . & De Frenne, P., 2024, Microclimate, an important part of ecology and biogeography. *Global Ecology and Biogeography*, 33(6):e13834.

Keränen, J., Maltamo, M. & Packalen, P., 2016, Effect of flying altitude, scanning angle and scanning mode on the accuracy of ALS based forest inventory. *International Journal of Applied Earth Observation and Geoinformation*, 52:349–360.

Kitahara, F., Mizoue, N. & Yoshida, S., 2010, Effects of training for inexperienced surveyors on data quality of tree diameter and height measurements. *Silva Fennica*, 44(4):657–667.

Kleinn, C., 2002, New technologies and methodologies for national forest inventories. *Unasylva*, 53(210):10–15.

Koch, B., Heyder, U. & Welnacker, H., 2006, Detection of individual tree crowns in airborne lidar data. *Photogrammetric Engineering and Remote Sensing*, 72(4):357–363.

Korpela, I., 2004, *Individual Tree Measurements by Means of Digital Aerial Photogrammetry*. Vol. 3, Helsinki, Finland, Finnish Society of Forest Science:1–93. http://www.scopus.com/inward/record.url?eid=2-s2.0-33746484995&partnerID=40&md5=5384dddf6344157bbc40b5a7a5d4c283

Korpela, I., Dahlin, B., Schäfer, H., Bruun, E., Haapaniemi, F., Honkasalo, J., ... & Virtanen, H. (2007, September). Single-tree forest inventory using lidar and aerial images for 3D treetop positioning, species recognition, height and crown width estimation. In *Proceedings of ISPRS workshop on laser scanning* (pp. 12–14).

Kovats, M., 1997, A large-scale aerial photographic technique for measuring tree heights on long-term forest installations. *Photogrammetric Engineering and Remote Sensing*, 63(6):741–747.

Kraus, K. & Pfeifer, N., 1998, Determination of terrain models in wooded areas with airborne laser scanner data. *Journal of Photogrammetry and Remote Sensing*, 53:193–203.

Krug, J. H. A., 2018, Accounting of GHG emissions and removals from forest management: A long road from Kyoto to Paris. *Carbon Balance and Management*, 13:1. https://doi.org/10.1186/s13021-017-0089-6

Kukkonen, M., Korhonen, L., Maltamo, M., Suvanto, A. & Packalen, P., 2018, How much can airborne laser scanning based forest inventory by tree species benefit from auxiliary optical data? *International Journal of Applied Earth Observation and Geoinformation*, 72:91–98.

Kumar, A. & Sharma, M. P., 2015, Assessment of carbon stocks in forest and its implications on global climate changes. *Journal of Materials and Environmental Science*, 6(12):3548–3564.

Kuuluvainen, T. & Grenfell, R., 2012, Natural disturbance emulation in boreal forest ecosystem management—Theories, strategies, and a comparison with conventional even-aged management. *Canadian Journal of Forest Research*, 42(7):1185–1203.

Kwak, D.-A., Lee, W.-K., Lee, J.-H., Biging, G. & Gong, P., 2007, Detection of individual trees and estimation of tree height using LiDAR data. *Journal of Forest Research*, 12(6):425–434.

Lahssini, K., Dayal, K. R., Durrieu, S. & Monnet, J. M., 2022, June, Joint use of airborne LiDAR metrics and topography information to estimate forest parameters via neural networks. *2022 IEEE 21st Mediterranean Electrotechnical Conference (MELECON)*, IEEE:442–447.

Lang, N., Jetz, W., Schindler, K. & Wegner, J. D., 2023, A high-resolution canopy height model of the Earth. *Nature Ecology & Evolution*, 7:1778–1789.

Larjavaara, M. & Muller-Landau, H. C., 2013, Measuring tree height: A quantitative comparison of two common field methods in a moist tropical forest. *Methods in Ecology and Evolution*, 4(9):793–801.

Latifah, S., Purwoko, A., Hartini, K. S. & Fachrudin, K. A., 2021, March, Allometric models to estimate the aboveground biomass of forest: A literature review. *IOP Conference Series: Materials Science and Engineering,* IOP Publishing, Vol. 1122, No. 1:012047.

Lebègue, L., Cazala-Hourcade, E., Languille, F., Artigues, S. & Melet, O., 2020, CO3D, a worldwide one one-meter accuracy dem for 2025. *The International Archives of the Photogrammetry, Remote Sensing and Spatial Information Sciences*, 43:299–304.

Leberl, F., Irschara, A., Pock, T., Meixner, P., Gruber, M., Scholz, S. & Wiechert, A., 2010, Point clouds: Lidar versus 3D vision. *Photogrammetric Engineering and Remote Sensing*, 76(10):1123–1134.

Leckie, D., Gougeon, F., Hill, D., Quinn, R., Armstrong, L. & Shreenan, R., 2003, Combined high-density lidar and multispectral imagery for individual tree crown analysis. *Canadian Journal of Remote Sensing*, 29(5):633–649.

Leckie, D. G., Gougeon, F. A., Tinis, S., Nelson, T., Burnett, C. N. & Paradine, D., 2005, Automated tree recognition in old growth conifer stands with high resolution digital imagery. *Remote Sensing of Environment*, 94(3):311–326.

Lefsky, M. A., Cohen, W. B., Harding, D. J., Parker, G. G., Acker, S. A. & Gower, S. T., 2002, Lidar remote sensing of above-ground biomass in three biomes. *Global Ecology and Biogeography*, 11(5):393–399.

Lefsky, M. A., Harding, D. J., Keller, M., Cohen, W. B., Carabajal, C. C., Del Bom Espirito-Santo, F., Hunter, M. O. & de Oliveira Jr, R., 2005, Estimates of forest canopy height and aboveground biomass using ICESat. *Geophysical Research Letters*, 32(22):1–4.

Le Goff, H., De Grandpré, L., Kneeshaw, D. & Bernier, P., 2010, Sustainable management of old-growth boreal forests: Myths, possible solutions and challenges. *The Forestry Chronicle*, 86(1):63–76.

Lei, Y., Treuhaft, R. & Gonçalves, F., 2021, Automated estimation of forest height and underlying topography over a Brazilian tropical forest with single-baseline single-polarization TanDEM-X SAR interferometry. *Remote Sensing of Environment*, 252:112132.

Leitold, V., Keller, M., Morton, D. C., Cook, B. D. & Shimabukuro, Y. E., 2015, Airborne lidar-based estimates of tropical forest structure in complex terrain: Opportunities and trade-offs for REDD+. *Carbon Balance and Management*, 10:1–12.

le Maire, G., Marsden, C., Nouvellon, Y., Grinand, C., Hakamada, R., Stape, J.-L. & Laclau, J.-P., 2011, MODIS NDVI time-series allow the monitoring of Eucalyptus plantation biomass. *Remote Sensing of Environment*, 115(10):2613–2625.

Lenoir, J., Gril, E., Durrieu, S., Horen, H., Laslier, M., Lembrechts, J. J., . . . & Decocq, G., 2022, Unveil the unseen: Using LiDAR to capture time-lag dynamics in the herbaceous layer of European temperate forests. *Journal of Ecology*, 110(2):282–300.

Lesak, A. A., Radeloff, V. C., Hawbaker, T. J., Pidgeon, A. M., Gobakken, T. & Contrucci, K., 2011, Modeling forest songbird species richness using LiDAR-derived measures of forest structure. *Remote Sensing of Environment*, 115(11):2823–2835.

Le Toan, T., Quegan, S., Davidson, M. W. J., Balzter, H., Paillou, P., Papathanassiou, K., Plummer, S., Rocca, F., Saatchi, S., Shugart, H. & Ulander, L., 2011, The BIOMASS mission: Mapping global forest biomass to better understand the terrestrial carbon cycle. *Remote Sensing of Environment*, 115(11):2850–2860.

Li, K., Shao, Y., Zhang, F., Xu, M., Li, X. & Xia, Z., 2009, Forest parameters estimation using polarimetric SAR data. *Sixth International Symposium on Multispectral Image Processing and Pattern Recognition, International Society for Optics and Photonics*, 7498, Article ID. 749857:1–8.

Li, S., Brandt, M., Fensholt, R., Kariryaa, A., Igel, C., Gieseke, F., Nord-Larsen, T., Oehmcke, S., Carlsen, A. H., Junttila, S., Tong, X., d'Aspremont, A. & Ciais, P., 2023, Deep learning enables image-based tree counting, crown segmentation, and height prediction at national scale. *PNAS Nexus 2*, pgad076.

Li, W., Wang, F. D. & Xia, G. S., 2020b, A geometry-attentional network for ALS point cloud classification. *ISPRS Journal of Photogrammetry and Remote Sensing*, 164:26–40.

Li, W., Guo, Q., Jakubowski, M. K. & Kelly, M., 2012, A new method for segmenting individual trees from the lidar point cloud. *Photogrammetric Engineering & Remote Sensing*, 78(1):75–84.

Li, X., Liu, C., Wang, Z., Xie, X., Li, D. & Xu, L., 2020a, Airborne LiDAR: State-of-the-art of system design, technology and application. *Measurement Science and Technology*, 32(3):032002.

Liang, J. & Picard, N., 2013, Matrix model of forest dynamics: An overview and outlook. *Forest Science*, 59(3):359–378.

Lim, K., Hopkinson, C. & Treitz, P., 2008, Examining the effects of sampling point densities on laser canopy height and density metrics. *The Forestry Chronicle*, 84(6):876–885.

Lim, K., Treitz, P., Wulder, M., St-Onge, B. & Flood, M., 2003, LiDAR remote sensing of forest structure. *Progress in Physical Geography*, 27(1):88–106.

Lindberg, E. & Holmgren, J., 2017, Individual tree crown methods for 3D data from remote sensing. *Current Forestry Reports*, 3:19–31.

Lisein, J., Pierrot-Deseilligny, M., Bonnet, S. & Lejeune, P., 2013, A photogrammetric workflow for the creation of a forest canopy height model from small unmanned aerial system imagery. *Forests*, 4(4): 922–944.

Liu, A., Cheng, X., Chen, Z., 2021, Performance evaluation of GEDI and ICESat-2 laser altimeter data for terrain and canopy height retrievals. *Remote Sensing of Environment*, 264:112571. https://doi.org/10/gkzw4v.

Liu, F., He, Y., Chen, W., Luo, Y., Yu, J., Chen, Y., Jiao, C. & Liu, M., 2022, Simulation and design of circular scanning airborne Geiger mode lidar for high-resolution topographic mapping. *Sensors*, 22:3656.

Liu, G. & Han, S., 2009, Long-term forest management and timely transfer of carbon into wood products help reduce atmospheric carbon. *Ecological Modelling*, 220(13–14):1719–1723. https://doi.org/10.3390/f9070398.

Liu, X., Neigh, C. S., Pardini, M. & Forkel, M., 2024, Estimating forest height and above-ground biomass in tropical forests using P-band TomoSAR and GEDI observations. International *Journal of Remote Sensing*, 45(9):3129–3148.

Lovell, J., Jupp, D., Newnham, G., Coops, N. & Culvenor, D., 2005, Simulation study for finding optimal lidar acquisition parameters for forest height retrieval. *Forest Ecology and Management*, 214(1):398–412.

Luo, S., Wang, C., Li, G. & Xi, X., 2013, Retrieving leaf area index using ICESat/GLAS full-waveform data. *Remote Sensing Letters*, 4(8):745–753. https://doi.org/10.3390/f14030454

Magnani, F., Mencuccini, M. & Grace, J., 2000, Age-related decline in stand productivity: The role of structural acclimation under hydraulic constraints. *Plant, Cell and Environment*, 23(3):251–263.

Magnussen, S., Næsset, E., Gobakken, T. & Frazer, G., 2012, A fine-scale model for area-based predictions of tree-size-related attributes derived from LiDAR canopy heights. *Scandinavian Journal of Forest Research*, 27(3):312–322.

Magnussen, S. & Nord-Larsen, T., 2021, Forest inventory inference with spatial model strata. *Scandinavian Journal of Forest Research*, 36(1):43–54.

Magruder, L., Neumann, T. & Kurtz, N., 2021, ICESat-2 early mission synopsis and observatory performance. *Earth and Space Science*, 8(5):e2020EA001555.

Magruder, L., Brunt, K., Alonzo, M., 2020, Early ICESat-2 on-orbit geolocation validation using ground-based corner cube retro-reflectors. *Remote Sensing*, 12:3653.

Malambo, L. & Popescu, S. C., 2021, Assessing the agreement of ICESat-2 terrain and canopy height with airborne lidar over US ecozones. *Remote Sensing of Environment*, 266, 112711.

Malhi, Y., Meir, P. & Brown, S., 2002, Forests, carbon and global climate. *Philosophical Transactions of the Royal Society of London. Series A: Mathematical, Physical and Engineering Sciences*, 360(1797):1567–1591.

Mallet, C. & Bretar, F., 2009, Full-waveform topographic lidar: State-of-the-art. *ISPRS Journal of Photogrammetry and Remote Sensing*, 64(1):1–16.

Maltamo, M., Mustonen, K., Hyyppä, J., Pitkänen, J. & Yu, X., 2004, The accuracy of estimating individual tree variables with airborne laser scanning in a boreal nature reserve. *Canadian Journal of Forest Research*, 34:1791–1801.

Maltamo, M., Packalen, P. & Kangas, A., 2021, From comprehensive field inventories to remotely sensed wall-to-wall stand attribute data—A brief history of management inventories in the Nordic countries. *Canadian Journal of Forest Research*, 51(2):257–266.

Maltamo, M., Packalén, P., Suvanto, A., Korhonen, K., Mehtätalo, L. & Hyvönen, P., 2009, Combining ALS and NFI training data for forest management planning: A case study in Kuortane, Western Finland. *European Journal of Forest Research*, 128(3):305–317.

Mandlburger, G., Lehner, H. & Pfeifer, N., 2019, A comparison of single photon and full waveform lidar. *ISPRS Annals of the Photogrammetry, Remote Sensing and Spatial Information Sciences*, IV-2-W5:397–404.

Manninen, T., Stenberg, P., Rautiainen, M., Voipio, P. & Smolander, H., 2005, Leaf area index estimation of boreal forest using ENVISAT ASAR. *Geoscience and Remote Sensing, IEEE Transactions on*, 43(11):2627–2635.

Mao, L., Dennett, J., Bater, C. W., Tompalski, P., Coops, N. C., Farr, D., . . . & Nielsen, S. E., 2018, Using airborne laser scanning to predict plant species richness and assess conservation threats in the oil sands region of Alberta's boreal forest. *Forest Ecology and Management*, 409:29–37.

Maréchaux, I., Langerwisch, F., Huth, A., Bugmann, H., Morin, X., Reyer, C. P., . . . & Bohn, F. J., 2021, Tackling unresolved questions in forest ecology: The past and future role of simulation models. *Ecology and Evolution*, 11(9):3746–3770.

Markus, T., Neumann, T., Martino, A., Abdalati, W., Brunt, K., Csatho, B., . . . & Zwally, J., 2017, The Ice, Cloud, and land Elevation Satellite-2 (ICESat-2): Science requirements, concept, and implementation. *Remote Sensing of Environment*, 190:260–273.

Marsden, C., Nouvellon, Y., Laclau, J.-P., Corbeels, M., McMurtrie, R. E., Stape, J. L., Epron, D. & le Maire, G., 2013, Modifying the G'DAY process-based model to simulate the spatial variability of Eucalyptus plantation growth on deep tropical soils. *Forest Ecology and Management*, 301:112–128.

Martin-Ducup, O., Ploton, P., Barbier, N., Momo Takoudjou, S., Mofack, G., Kamdem, N. G., Fourcaud, T., Sonké, B., Couteron, P., & Pélissier, R., 2020, Terrestrial laser scanning reveals convergence of tree architecture with increasingly dominant crown canopy position. *Functional Ecology*, 34(12):2442–2452.

Masek, J. G., Hayes, D. J., Hughes, M. J., Healey, S. P. & Turner, D. P., 2015, The role of remote sensing in process-scaling studies of managed forest ecosystems. *Forest Ecology and Management*, 355:109–123.

McCarthy, M. A. & Burgman, M. A., 1995, Coping with uncertainty in forest wildlife planning. *Forest Ecology and Management*, 74(1):23–36.

McRoberts, R. E., 2010, Probability-and model-based approaches to inference for proportion forest using satellite imagery as ancillary data. *Remote Sensing of Environment*, 114(5):1017–1025.

McRoberts, R. E., Cohen, W. B., Erik, N., Stehman, S. V. & Tomppo, E. O., 2010, Using remotely sensed data to construct and assess forest attribute maps and related spatial products. *Scandinavian Journal of Forest Research*, 25(4):340–367.

McRoberts, R. E., Næsset, E., Saatchi, S. & Quegan, S., 2022, Statistically rigorous, model-based inferences from maps. *Remote Sensing of Environment*, 279:113028.

McRoberts, R. E., Tomppo, E., Schadauer, K., Vidal, C., Ståhl, G., Chirici, G., Lanz, A., Cienciala, E., Winter, S. & Smith, W. B., 2009, Harmonizing national forest inventories. *Journal of Forestry*, 107(4):179–187.

Medvigy, D. & Moorcroft, P. R., 2012, Predicting ecosystem dynamics at regional scales: An evaluation of a terrestrial biosphere model for the forests of northeastern North America. *Philosophical Transactions of the Royal Society B: Biological Sciences*, 367(1586):222–235.

Medvigy, D., Wofsy, S., Munger, J., Hollinger, D. & Moorcroft, P., 2009, Mechanistic scaling of ecosystem function and dynamics in space and time: Ecosystem Demography model version 2. *Journal of Geophysical Research:Biogeosciences (2005–2012)*:114(G1).

Mei, C. & Durrieu, S., 2004, Tree crown delineation from digital elevation models and high resolution imagery. *Proceedings of the ISPRS Working Group Part*, 8(2):3–6.

Melin, M., Hill, R. A., Bellamy, P. E. & Hinsley, S. A., 2019, On bird species diversity and remote sensing—utilizing lidar and hyperspectral data to assess the role of vegetation structure and foliage characteristics as drivers of avian diversity. *IEEE Journal of Selected Topics in Applied Earth Observations and Remote Sensing*, 12(7):2270–2278.

Meng, X., Currit, N. & Zhao, K., 2010, Ground filtering algorithms for airborne LiDAR data: A review of critical issues. *Remote Sensing*, 2(3):833–860.

Mielcarek, M., Kamińska, A. & Stereńczak, K., 2020, Digital aerial photogrammetry (DAP) and airborne laser scanning (ALS) as sources of information about tree height: Comparisons of the accuracy of remote sensing methods for tree height estimation. *Remote Sensing*, 12:1808.

Mitchard, E. T., Saatchi, S. S., Baccini, A., Asner, G. P., Goetz, S. J., Harris, N. L. & Brown, S., 2013, Uncertainty in the spatial distribution of tropical forest biomass: A comparison of pan-tropical maps. *Carbon Balance and Management*, 8:1–13.

Mitchard, E. T. A., Saatchi, S. S., Woodhouse, I. H., Nangendo, G., Ribeiro, N. S., Williams, M., Ryan, C. M., Lewis, S. L., Feldpausch, T. R. & Meir, P., 2009, Using satellite radar backscatter to predict aboveground woody biomass: A consistent relationship across four different African landscapes. *Geophysical Research Letters*, 36(23):L23401.

Montealegre, A. L., Lamelas, M. T., De La Riva, J., García-Martín, A. & Escribano, F., 2016, Use of low point density ALS data to estimate stand-level structural variables in Mediterranean Aleppo pine forest. *Forestry: An International Journal of Forest Research*, 89(4):373–382.

Morgan, J. L. & Gergel, S. E., 2013, Automated analysis of aerial photographs and potential for historic forest mapping. *Canadian Journal of Forest Research*, 43(8):699–710.

Mori, A. S., Lertzman, K. P. & Gustafsson, L., 2017, Biodiversity and ecosystem services in forest ecosystems: A research agenda for applied forest ecology. *Journal of Applied Ecology*, 54(1):12–27.

Morsdorf, F., Meier, E., Kötz, B., Itten, K. I., Dobbertin, M. & Allgöwer, B., 2004, LIDAR-based geometric reconstruction of boreal type forest stands at single tree level for forest and wildland fire management. *Remote Sensing of Environment*, 92(3):353–362.

Mücke, W. & Hollaus, M., 2010, Derivation of 3D landscape metrics from airborne laser scanning data. *SilviLaser 2010, Conference on Lidar Applications for Assessing Forest Ecosytems Freiburg*, Germany:11 p.

Müller, J., Bae, S., Röder, J., Chao, A. & Didham, R. K., 2014, Airborne LiDAR reveals context dependence in the effects of canopy architecture on arthropod diversity. *Forest Ecology and Management*, 312:129–137.

Müller, J. & Brandl, R., 2009, Assessing biodiversity by remote sensing in mountainous terrain: The potential of LiDAR to predict forest beetle assemblages. *Journal of Applied Ecology*, 46(4):897–905.

Müller, J., Stadler, J. & Brandl, R., 2010, Composition versus physiognomy of vegetation as predictors of bird assemblages: The role of lidar. *Remote Sensing of Environment*, 114(3):490–495.

Mulverhill, C., Coops, N. C., Hermosilla, T., White, J. C. & Wulder, M. A., 2022, Evaluating ICESat-2 for monitoring, modeling, and update of large area forest canopy height products. *Remote Sensing of Environment*, 271:112919. https://doi.org/10.1016/j.rse.2022.112919.

Næsset, E., 2002a, Determination of mean tree height of forest stands by digital photogrammetry. *Scandinavian Journal of Forest Research*, 17(5):446–459.

Næsset, E., 2002b, Predicting forest stand characteristics with airborne scanning laser using a practical two-stage procedure and field data. *Remote Sensing of Environment*, 80(1):88–99.

Næsset, E., 2004, Accuracy of forest inventory using airborne laser-scanning: Evaluating the first Nordic full-scale operational project. *Scandinavian Journal of Forest Research*, 19:554–557.

Næsset, E., 2007, Airborne laser scanning as a method in operational forest inventory: Status of accuracy assessments accomplished in Scandinavia. *Scandinavian Journal of Forest Research*, 22(5):433–442.

Næsset, E., 2009a, Effects of different sensors, flying altitudes, and pulse repetition frequencies on forest canopy metrics and biophysical stand properties derived from small-footprint airborne laser data. *Remote Sensing of Environment*, 113(1):148–159.

Næsset, E., 2009b, Influence of terrain model smoothing and flight and sensor configurations on detection of small pioneer trees in the boreal–alpine transition zone utilizing height metrics derived from airborne scanning lasers. *Remote Sensing of Environment*, 113(10):2210–2223.

Næsset, E., 2015, Vertical height errors in digital terrain models derived from airborne laser scanner data in a boreal-alpine ecotone in Norway. *Remote Sensing*, 7(4):4702–4725.

NASA, 2024, *New Proposals to Help NASA Advance Knowledge of Our Changing Climate*. https://www.nasa.gov/news-release/new-proposals-to-help-nasa-advance-knowledge-of-our-changing-climate/. Accessed 14 August 2024.

Neigh, C. S., Masek, J. G., Bourget, P., Cook, B., Huang, C., Rishmawi, K. & Zhao, F., 2014, Deciphering the precision of stereo IKONOS canopy height models for US forests with G-LiHT airborne LiDAR. *Remote Sensing*, 6(3):1762–1782.

Nelson, R., 1997, Modeling forest canopy heights: The effects of canopy shape. *Remote Sensing of Environment*, 60(3):327–334.

Nelson, R., 2013, How did we get here? An early history of forestry lidar. *Canadian Journal of Remote Sensing*, 39(Suppl.1):S6–S17.

Nelson, R., Keller, C. & Ratnaswamy, M., 2005, Locating and estimating the extent of Delmarva fox squirrel habitat using an airborne LiDAR profiler. *Remote Sensing of Environment*, 96(3):292–301.

Nelson, R., Parker, G. & Hom, M., 2003, A portable airborne laser system for forest inventory. *Photogrammetric Engineering and Remote Sensing*, 69(3):267–273.

Neumann, T. A., Martino, A. J., Markus, T., Bae, S., Bock, M. R., Brenner, A. C., . . . & Thomas, T. C., 2019, The Ice, Cloud, and Land Elevation Satellite-2 mission: A global geolocated photon product derived from the advanced topographic laser altimeter system. *Remote Sensing of Environment*, 233:111325.

Neuenschwander, A. & Pitts, K., 2019, The ATL08 land and vegetation product for the ICESat-2 mission. *Remote Sensing of Environment*, 221:247–259. https://doi.org/10.1016/j.rse.2018.11.005.

Neuenschwander, A., Guenther, E., White, J. C., Duncanson, L. & Montesano, P., 2020, Validation of ICESat-2 terrain and canopy heights in boreal forests. *Remote Sensing of Environment*, 251:112110.

Newnham, G. J., Armston, J. D., Calders, K., Disney, M. I., Lovell, J. L., Schaaf, C. B., . . . & Danson, F. M., 2015, Terrestrial laser scanning for plot-scale forest measurement. *Current Forestry Reports*, 1:239–251.

Nguyen, T. A., Ehbrecht, M. & Camarretta, N., 2023, Application of point cloud data to assess edge effects on rainforest structural characteristics in tropical Sumatra, Indonesia. *Landscape Ecology*, 38(5):1191–1208.

Nguyen, T. A., Kellenberger, B. & Tuia, D., 2022, Mapping forest in the Swiss Alps treeline ecotone with explainable deep learning. *Remote Sensing of Environment*, 281:113217.

Ostendorf, B., 2011, Overview: Spatial information and indicators for sustainable management of natural resources. *Ecological Indicators*, 11(1):97–102.

Oliveira, V. C. P., Zhang, X., Peterson, B. & Ometto, J. P., 2023. Using simulated Gedi waveforms to evaluate the effects of beam sensitivity and terrain slope on Gedi l2a relative height metrics over the Brazilian amazon forest. *Science of Remote Sensing*, 7:100083. https://www.sciencedirect.com/science/article/pii/S2666017223000081; https://doi.org/10.1016/j.srs.2023.100083.

Pandey, S., Kumari, N., Dash, S. K. & Al Nawajish, S., 2022, Challenges and monitoring methods of forest management through geospatial application: A review. *Advances in Remote Sensing for Forest Monitoring*, 289–328.

Papathanassiou, K. P., Cloude, S. R., Pardini, M., Quiñones, M. J., Hoekman, D., Ferro-Famil, L., . . . & Soja, M. J., 2021, Forest applications. *Polarimetric Synthetic Aperture Radar: Principles and Application*, 59–117.

Pardé, J., 1956, Une notion pleine d'intérêt: La hauteur dominante des peuplements forestiers. *Revue forestière française*, 12:850–856.

Pardini, M., Guliaev, R., Romero Puig, N., Papathanassiou, K. & Hajnsek, I., 2024, Addressing Tropical forest structure changes with SAR tomography: The AfriSAR 2016-GabonX 2023 case. *International Geoscience and Remote Sensing Symposium (IGARSS)*. IEEE:1–4.

Pearse, G. D., Dash, J. P., Persson, H. J. & Watt, M. S., 2018, Comparison of high-density LiDAR and satellite photogrammetry for forest inventory. *ISPRS Journal of Photogrammetry and Remote Sensing*, 142:257–267.

Pelletier, J., Ramankutty, N. & Potvin, C., 2011, Diagnosing the uncertainty and detectability of emission reductions for REDD+ under current capabilities: An example for Panama. *Environmental Research Letters*, 6(024005):12.

Peng, X., Zhao, A., Chen, Y., Chen, Q. & Liu, H., 2021, Tree height measurements in degraded tropical forests based on UAV-LiDAR data of different point cloud densities: A case study on Dacrydium pierrei in China. *Forests*, 12(3):328.

Persson, A., Holmgren, J. & Söderman, U., 2002, Detecting and measuring individual trees using an airborne laser scanner. *Photogrammetric Engineering & Remote Sensing*, 68(9):925–932.

Persson, H. J., Olofsson, K. & Holmgren, J., 2022, Two-phase forest inventory using very-high-resolution laser scanning. *Remote Sensing of Environment*, 271:112909.

Petrie, G., 2005, Les Caméras Numériques Aéroportées. *Geomatique Expert* (No. 41/42):50–61.

Petrie, G. & Walker, A. S., 2007, Airborne digital imaging technology: A new overview. *Photogrammetric Record*, 22(119):203–225.

Picard, N., Saint-André, L. & Henry, M., 2012a, Manual for building tree volume and biomass allometric equations: From field measurement to prediction. *Food and Agricultural Organization of the United Nations, Rome, and Centre de Coopération Internationale en Recherche Agronomique pour le Développement*, Montpellier:215 p. http://www.fao.org/docrep/018/i3058e/i3058e.pdf

Picard, N., Saint-André, L. & Henry, M., 2012b, *Manual for Building Tree Volume and Biomass Allometric Equations: From Field Measurement to Prediction*.

Piermattei, L., Marty, M., Ginzler, C., Pöchtrager, M., Karel, W., Ressl, C., Pfeifer, N., & Hollaus, M., 2019, Pléiades satellite images for deriving forest metrics in the Alpine region. *International Journal of Applied Earth Observation and Geoinformation*, 80:240–256.

Ploton, P., Barbier, N., Couteron, P., Antin, C. M., Ayyappan, N., Balachandran, N., . . . & Pélissier, R., 2017, Toward a general tropical forest biomass prediction model from very high resolution optical satellite images. *Remote Sensing of Environment*, 200:140–153.

Popescu, S. C. & Wynne, R. H., 2004, Seeing the trees in the forest: Using lidar and multispectral data fusion with local filtering and variable window size for estimating tree height. *Photogrammetric Engineering & Remote Sensing*, 70(5):589–604.

Popescu, S. C., Wynne, R. H. & Nelson, R. F., 2002, Estimating plot-level tree heights with lidar: Local filtering with a canopy-height based variable window size. *Computers and Electronics in Agriculture*, 37:71–95.

Popescu, S. C., Wynne, R. H. & Nelson, R. F., 2003, Measuring individual tree crown diameter with lidar and assessing its influence on estimating forest volume and biomass. *Canadian Journal of Remote Sensing*, 29(5):564–577.

Popescu, S. C., Wynne, R. H. & Scrivani, J. A., 2004, Fusion of small-footprint lidar and multispectral data to estimate plot-level volume and biomass in deciduous and pine forests in Virginia, USA. *Forest Science*, 50(4):551–565.

Potapov, P., Li, X., Hernandez-Serna, A., Tyukavina, A., Hansen, M. C., Kommareddy, A., Pickens, A., Turubanova, S., Tang, H., Silva, C. E., Armston, J., Dubayah, R., Blair, J. B. & Hofton, M., 2021, Mapping global forest canopy height through integration of GEDI and Landsat data. *Remote Sensing of Environment*. https://doi.org/10.1016/j.rse.2020.112165.

Pretzsch, H., 2009, Forest dynamics, growth, and yield. *From Measurement to Model*, Springer-Verlag:664 p.

Pretzsch, H., 2020, The course of tree growth. Theory and reality. *Forest Ecology and Management*, 478:118508.

Pretzsch, H., Poschenrieder, W., Uhl, E., Brazaitis, G., Makrickiene, E. & Calama, R., 2021, Silvicultural prescriptions for mixed-species forest stands. A European review and perspective. *European Journal of Forest Research*, 140(5):1267–1294.

Pricope, N. G. & Bashit, M. S., 2023, Emerging trends in topobathymetric LiDAR technology and mapping. *International Journal of Remote Sensing*, 44(24):7706–7731.

Proisy, C., Couteron, P. & Fromard, F., 2007, Predicting and mapping mangrove biomass from canopy grain analysis using Fourier-based textural ordination of IKONOS images. *Remote Sensing of Environment*, 109(3):379–392.

Qian, S. E., 2021, Hyperspectral satellites, evolution, and development history. *IEEE Journal of Selected Topics in Applied Earth Observations and Remote Sensing*, 14:7032–7056.

Qin, N., Tan, W., Guan, H., Wang, L., Ma, L., Tao, P., Fatholahi, S., Hu, X. & Li, J., 2023, Towards intelligent ground filtering of large-scale topographic point clouds: A comprehensive survey. *International Journal of Applied Earth Observation and Geoinformation*, 125:103566.

Qin, R. & Gruen, A., 2021, The role of machine intelligence in photogrammetric 3D modeling—An overview and perspectives. *International Journal of Digital Earth*, 14:15–31.

Ramachandran, N., Saatchi, S., Tebaldini, S., d'Alessandro, M. M. & Dikshit, O., 2023, Mapping tropical forest aboveground biomass using airborne SAR tomography. *Scientific Reports*, 13(1):6233.

Rammer, W., Thom, D., Baumann, M., Braziunas, K., Dollinger, C., Kerber, J., Mohr, J. & Seidl, R., 2024, The individual-based forest landscape and disturbance model iLand: Overview, progress, and outlook. *Ecological Modelling*, 110785.

Reitberger, J., Krzystek, P. & Stilla, U., 2006, Analysis of full waveform lidar data for tree species classification. *Symposium of ISPRS Commission III- Photogrammetric Computer Vision PCV'06*, Bonn, Germany, ISPRS Commission III, XXXVI- part 3:228–233.

Reitberger, J., Schnörr, C., Krzystek, P. & Stilla, U., 2009, 3D segmentation of single trees exploiting full waveform LIDAR data. *ISPRS Journal of Photogrammetry and Remote Sensing*, 64:561–574.

Réjou-Méchain, M., Barbier, N., Couteron, P., Ploton, P., Vincent, G., Herold, M., Mermoz, S., Saatchi, S., Chave, J., de Boissieu, F., Féret, J-B., Takoudjou, S. M. & Pélissier, R., 2019, Upscaling forest biomass from field to satellite measurements: Sources of errors and ways to reduce them. *Surveys in Geophysics*, 40: 881–911.

Remondino, F. & El-hakim, S., 2006, Image-based 3D modelling: A review. *Photogrammetric Record*, 21(115):269–291.

Renaud, J. P., Sagar, A., Barbillon, P., Bouriaud, O., Deleuze, C. & Vega, C., 2022, Characterizing the calibration domain of remote sensing models using convex hulls. *International Journal of Applied Earth Observation and Geoinformation*, 112:102939.

Reutebuch, S. E., McGaughey, R. J., Andersen, H.-E. & Carson, W. W., 2003, Accuracy of a high-resolution lidar terrain model under a conifer forest canopy. *Canadian Journal of Remote Sensing*, 29(5):527–535.

Rodda, S. R., Fararoda, R., Gopalakrishnan, R., Jha, N., Réjou-Méchain, M., Couteron, P., . . . & Ploton, P., 2024, LiDAR-based reference aboveground biomass maps for tropical forests of South Asia and Central Africa. *Scientific Data*, 11(1):334.

Rosette, J., North, P. R., Rubio-Gil, J., Cook, B., Los, S., Suarez, J., Sun, G., Ranson, J. & Blair, J., 2013, Evaluating prospects for improved forest parameter retrieval from satellite LiDAR using a physically-based radiative transfer model. *Selected Topics in Applied Earth Observations and Remote Sensing, IEEE Journal of*, 6(1):45–53.

Roy, D. P., Kashongwe, H. B., Armston, J., 2021, The impact of geolocation uncertainty on GEDI tropical forest canopy height estimation and change monitoring. *Science of Remote Sensing*, 4:100024. https://doi.org/10/gktzv5.

Saarela, S., Wästlund, A., Holmström, E., Mensah, A. A., Holm, S., Nilsson, M., Fridman, J. & Ståhl, G., 2020, Mapping aboveground biomass and its prediction uncertainty using LiDAR and field data, accounting for tree-level allometric and LiDAR model errors. *Forest Ecosystems*, 7:43.

Saatchi, S. S., Harris, N. L., Brown, S., Lefsky, M., Mitchard, E. T. A., Salas, W., Zutta, B. R., Buermann, W., Lewis, S. L., Hagen, S., Petrova, S., White, L., Silman, M. & Morel, A., 2011, Benchmark map of forest carbon stocks in tropical regions across three continents. *Proceedings of the National Academy of Sciences of the United States of America*, 108(24):9899–9904.

Sagar, A., Vega, C., Bouriaud, O., Piedallu, C. & Renaud, J. P., 2022, Multisource forest inventories: A model-based approach using k-NN to reconcile forest attributes statistics and map products. *ISPRS Journal of Photogrammetry and Remote Sensing*, 192:175–188.

Saint-André, L., Laclau, J., Deleporte, P., Gava, J., Gonçalves, J., Mendham, D., Nzila, J., Smith, C., Du Toit, B. & Xu, D., 2008, Slash and litter management effects on Eucalyptus productivity: A synthesis using a growth and yield modelling approach. *Site Management and Productivity in Tropical Plantation Forests: Proceedings of Workshops in Piracicaba (Brazil), 22–26 November 2004 and Bogor (Indonesia), 6–9 November 2006*, Center for International Forestry Research (CIFOR):173–189.

Schleich, A., 2024, *Contribution of Spaceborne Lidar to the Development of Multisource Forest Inventory Methods Adapted to Sustainable Forest Management in a Context of Global Change*. Doctoral thesis, AgroParisTech - University of Montpellier.

Schleich, A., Durrieu, S., Soma, M. & Vega, C., 2023, Improving GEDI footprint geolocation using a high resolution digital elevation model. *IEEE Journal of Selected Topics in Applied Earth Observations and Remote Sensing*, 16:7718–7732. https://doi.org/10.1109/JSTARS.2023.3298991.

Schwartz, M., Ciais, P., De Truchis, A., Chave, J., Ottlé, C., Vega, C., . . . & Fayad, I., 2023, FORMS: Forest multiple source height, wood volume, and biomass maps in France at 10 to 30 m resolution based on Sentinel-1, Sentinel-2, and Global Ecosystem Dynamics Investigation (GEDI) data with a deep learning approach. *Earth System Science Data*, 15(11):4927–4945.

Senf, C., Seidl, R. & Hostert, P., 2017, Remote sensing of forest insect disturbances: Current state and future directions. *International Journal of Applied Earth Observation and Geoinformation*, 60:49–60.

Senf, C., 2022, Seeing the system from above: The use and potential of remote sensing for studying ecosystem dynamics. *Ecosystems*, 25:1719–1737. https://doi.org/10.1007/s10021-022-00777-2.

Shendryk, Y., 2022, Fusing GEDI with earth observation data for large area aboveground biomass mapping. *International Journal of Applied Earth Observation and Geoinformation*, 115:103108.

Shugart, H. H., Saatchi, S. & Hall, F. G., 2010, Importance of structure and its measurement in quantifying function of forest ecosystems. *Journal of Geophysical Research: Biogeosciences (2005–2012)*, 115(D24):1–16.

Silva, C. A., Duncanson, L., Hancock, S., Neuenschwander, A., Thomas, N., Hofton, M., Fatoyinbo, L., Simard, M., Marshak, C. Z., Armston, J., Lutchke, S. & Dubayah, R., 2021, Fusing simulated GEDI, ICESat-2 and NISAR data for regional aboveground biomass mapping. *Remote Sensing of Environment*, 253:112234.

Simard, M., Pinto, N., Fisher, J. B. & Baccini, A., 2011, Mapping forest canopy height globally with spaceborne lidar. *Journal of Geophysical Research: Biogeosciences (2005–2012)*, 116(G4):1–12.

Simonett, D., 1969, Editor's preface. *Remote Sensing of Environment*, 1(1):v.

Simonson, W. D., Allen, H. D. & Coomes, D. A., 2012, Use of an airborne lidar system to model plant species composition and diversity of Mediterranean oak forests. *Conservation Biology*, 26(5):840–850.

Simpson, J. E., Smith, T. E. & Wooster, M. J., 2017, Assessment of errors caused by forest vegetation structure in airborne LiDAR-derived DTMs. *Remote Sensing*, 9(11):1101.

Sithole, G. & Vosselman, G., 2004, Experimental comparison of filter algorithms for bare-Earth extraction from airborne laser scanning point clouds. *ISPRS Journal of Photogrammetry and Remote Sensing*, 59(1–2):85–101.

Skovsgaard, J. P. & Vanclay, J. K., 2008, Forest site productivity: A review of the evolution of dendrometric concepts for even-aged stands. *Forestry*, 81(1):13–31.

Smits, I., Prieditis, G., Dagis, S. & Dubrovskis, D., 2012, Individual tree identification using different LIDAR and optical imagery data processing methods. *Biosystems and Information Technology*, 1(1):19–24.

Soja, M. J. & Ulander, L. M. H., 2013, Digital canopy model estimation from TanDEM-X interferometry using high-resolution lidar DEM:165–168.

Solberg, S., Næsset, E. & Bollandsas, O. M., 2006, Single tree segmentation using airborne laser scanner data in a structurally heterogeneous spruce forest. *Photogrammetric Engineering and Remote Sensing*, 72(12):1369–1378.

Soma, M. et al., 2022, August, Qualification of ICESat2 ground elevation and canopy height products: Evaluating the potential of 100 m and 20 m resolution products for forest inventory in temperate heterogeneous forests. *ForestSAT*, Forestsat 2022, 29 August–3 September 2022, Berlin, Germany.

Spurr, S. H., 1960, *Photogrammetry and Photo-Interpretation with a Section on Applications to Forestry*, Ronald Press Comp.:472 p.

Stahl, A. T., Andrus, R., Hicke, J. A., Hudak, A. T., Bright, B. C. & Meddens, A. J., 2023, Automated attribution of forest disturbance types from remote sensing data: A synthesis. *Remote Sensing of Environment*, 285:113416.

Ståhl, G., Holm, S., Gregoire, T. G., Gobakken, T., Næsset, E. & Nelson, R., 2011, Model-based inference for biomass estimation in a LiDAR sample survey in Hedmark County, Norway. *Canadian Journal of Forest Research*, 41(1):96–107.

Stereńczak, K., Mielcarek, M., Wertz, B., Bronisz, K., Zajączkowski, G., Jagodziński, A. M., . . . & Skorupski, M., 2019, Factors influencing the accuracy of ground-based tree-height measurements for major European tree species. *Journal of Environmental Management*, 231:1284–1292.

Stoker, J. M., Abdullah, Q. A., Nayegandhi, A. & Winehouse, J., 2016, Evaluation of single photon and Geiger mode LiDAR for the 3D elevation program. *Remote Sensing*, 8(9):767.

St-Onge, B., Jumelet, J., Cobello, M. & Véga, C., 2004, Measuring individual tree height using a combination of stereophotogrammetry and lidar. *Canadian Journal of Forest Research*, 34(10):2122–2130.

St-Onge, B., Véga, C., Fournier, R. & Hu, Y., 2008, Mapping canopy height using a combination of digital stereo-photogrammetry and lidar. *International Journal of Remote Sensing*, 29(11):3343–3364.

Strunk, J., Temesgen, H., Andersen, H.-E., Flewelling, J. P. & Madsen, L., 2012, Effects of lidar pulse density and sample size on a model-assisted approach to estimate forest inventory variables. *Canadian Journal of Remote Sensing*, 38(5):644–654.

Su, Y., Guo, Q., Jin, S., Guan, H., Sun, X., Ma, Q., . . . & Li, Y., 2020, The development and evaluation of a backpack LiDAR system for accurate and efficient forest inventory. *IEEE Geoscience and Remote Sensing Letters*, 18(9):1660–1664.

Suárez, J. C., Ontiveros, C., Smith, S. & Snape, S., 2005, Use of airborne LiDAR and aerial photography in the estimation of individual tree heights in forestry. *Computers & Geosciences*, 31(2):253–262.

Sun, C., Huang, C., Zhang, H., Chen, B., An, F., Wang, L. & Yun, T., 2022, Individual tree crown segmentation and crown width extraction from a heightmap derived from aerial laser scanning data using a deep learning framework. *Frontiers in Plant Science*, 13:914974.

Tanaka, H. & Nakashizuka, T., 1997, Fifteen years of canopy dynamics analyzed by aerial photographs in a temperate deciduous forest. *Japan. Ecology*, 78(2):612–620.

Tarmu, T., Laarmann, D. & Kiviste, A., 2020, Mean height or dominant height–what to prefer for modelling the site index of Estonian forests? *Forestry Studies*, 72(1):121–138.

Tebaldini, S., Ho Tong Minh, D., Mariotti d'Alessandro, M., Villard, L., Le Toan, T. & Chave, J., 2019, The status of technologies to measure forest biomass and structural properties: State of the art in SAR tomography of tropical forests. *Surveys in Geophysics*, 40:779–801.

Tebaldini, S. & Rocca, F., 2012, Multibaseline polarimetric SAR tomography of a boreal forest at P-and L-bands. *Geoscience and Remote Sensing, IEEE Transactions on*, 50(1):232–246.

Tello, M., Cazcarra-Bes, V., Pardini, M. & Papathanassiou, K., 2018, Forest structure characterization from SAR tomography at L-band. *IEEE Journal of Selected Topics in Applied Earth Observations and Remote Sensing*, 11(10):3402–3414.

Thomas, V., Oliver, R., Lim, K. & Woods, M., 2008, LiDAR and Weibull modeling of diameter and basal area. *The Forestry Chronicle*, 84(6):866–875.

Thomas, V., Treitz, P., McCaughey, J. H. & Morrison, I., 2006, Mapping stand-level forest biophysical variables for a mixedwood boreal forest using lidar: An examination of scanning density. *Canadian Journal of Forest Research*, 36:34–47.

Thurgood, J., Gruber, M. & Karner, K., 2004, Multi-ray matching for automated 3D object modeling. *The International Archives of the Photogrammetry, Remote Sensing and Spatial Information Sciences*, 35:1682–1777.

Thürig, E. & Kaufmann, E., 2010, Increasing carbon sinks through forest management: A model-based comparison for Switzerland with its Eastern Plateau and Eastern Alps. *European Journal of Forest Research*:1–10.

Toivonen, J., Kangas, A., Maltamo, M., Kukkonen, M. & Packalen, P., 2023, Assessing biodiversity using forest structure indicators based on airborne laser scanning data. *Forest Ecology and Management*, 546:121376.

Tran-Ha, M., Cordonnier, T., Vallet, P. & Lombart, T., 2011, Estimation du volume total aérien des peuplements forestiers à partir de la surface terrière et de la hauteur de Lorey. *Revue forestière française*, 63(3):361–378.

Treitz, P., Lim, K., Woods, M., Pitt, D., Nesbitt, D. & Etheridge, D., 2012, LiDAR sampling density for forest resource inventories in Ontario, Canada. *Remote Sensing*, 4(4):830–848.

Turner, W., Spector, S., Gardiner, N., Fladeland, M., Sterling, E. & Steininger, M., 2003, Remote sensing for biodiversity science and conservation. *Trends in Ecology and Evolution*, 18(6):306–314.

Ullah, S., Dees, M., Datta, P., Adler, P., Schardt, M. & Koch, B., 2019, Potential of modern photogrammetry versus airborne laser scanning for estimating forest variables in a mountain environment. *Remote Sensing*, 11:661.

Urbazaev, M., Hess, L., Hancock, S., Ometto, J., Thiel, C., Dubois, C., Adam, M., Schmullius, C., 2021, Accuracy assessment of terrain and canopy height estimates from ICESat-2 and GEDI LiDAR missions in temperate and tropical forests: First results.

Ustin, S. L. & Middleton, E. M., 2021, Current and near-term advances in Earth observation for ecological applications. *Ecological Processes*, 10(1):1.

Uuemaa, E., Antrop, M., Roosaare, J., Marja, R. & Mander, Ü., 2009, Landscape metrics and indices: An overview of their use in landscape research. *Living Reviews in Landscape Research*, 3(1):1–28.

Vanclay, J. K., 2009, Tree diameter, height and stocking in even-aged forests. *Annals of Forest Science*, 66(7):1–7.

Van Laar, A. & Akça, A., 2007, *Forest Mensuration*, Springer.

Vastaranta, M., Wulder, M. A., White, J. C., Pekkarinen, A., Tuominen, S., Ginzler, C., Kankare, V., Holopainen, M., Hyyppä, J. & Hyyppä, H., 2013, Airborne laser scanning and digital stereo imagery measures of forest structure: Comparative results and implications to forest mapping and inventory update. *Canadian Journal of Remote Sensing*, 39(5):382–395.

Vauhkonen, J. & Ruotsalainen, R., 2017, Reconstructing forest canopy from the 3D triangulations of airborne laser scanning point data for the visualization and planning of forested landscapes. *Annals of Forest Science*, 74:1–13.

Véga, C. & Durrieu, S., 2011, Multi-level filtering segmentation to measure individual tree parameters based on Lidar data: Application to a mountainous forest with heterogeneous stands. *International Journal of Applied Earth Observation and Geoinformation*, 13(4):646–656.

Véga, C., Durrieu, S., Morel, J. & Allouis, T., 2012, A sequential iterative dual-filter for Lidar terrain modeling optimized for complex forested environments. *Computers & Geosciences*, 44:31–41.

Véga, C., Hamrouni, A., El Mokhtari, S., Morel, J., Bock, J., Renaud, J.-P., Bouvier, M. & Durrieu, S., 2014, PTrees: A point-based approach to forest tree extraction from lidar data. *International Journal of Applied Earth Observation and Geoinformation*, 33:98–108.

Véga, C., Renaud, J. P., Durrieu, S. & Bouvier, M., 2016, On the interest of penetration depth, canopy area and volume metrics to improve Lidar-based models of forest parameters. *Remote Sensing of Environment*, 175:32–42.

Véga, C. & St-Onge, B., 2008, Height growth reconstruction of a boreal forest canopy over a period of 58 years using a combination of photogrammetric and lidar models. *Remote Sensing of Environment*, 112(4):1784–1794.

Véga, C. & St-Onge, B., 2009, Mapping site index and age by linking a time series of canopy height models with growth curves. *Forest Ecology and Management*, 257(3):951–959.

Vehmas, M., Eerikäinen, K., Peuhkurinen, J., Packalén, P. & Maltamo, M., 2009, Identification of boreal forest stands with high herbaceous plant diversity using airborne laser scanning. *Forest Ecology and Management*, 257(1):46–53.

Vepakomma, U. & Fortin, M.-J., 2010, Scale-specific effects of disturbance and environment on vegetation patterns of a boreal landscape. *SilviLaser 2010, Conference on Lidar Applications for assessing Forest Ecosytems Freiburg*, Germany:18 p.

Vepakomma, U., St-Onge, B. & Kneeshaw, D., 2008, Spatially explicit characterization of boreal forest gap dynamics using multi-temporal lidar data. *Remote Sensing of Environment*, 112(5):2326–2340.

Wagner, W., 2010, Radiometric calibration of small-footprint full-waveform airborne laser scanner measurements: Basic physical concepts. *ISPRS Journal of Photogrammetry and Remote Sensing*, 65(6):505–513.

Wagner, W., Ullrich, A., Ducic, V., Melzer, T. & Studnicka, N., 2006, Gaussian decomposition and calibration of a novel small-footprint full-waveform digitising airborne laser scanner. *ISPRS Journal of Photogrammetry and Remote Sensing*, 60(2):100–112.

Wagner, W., Ullrich, A., Melzer, T., Briese, C. & Kraus, K., 2004, From single-pulse to full-waveform airborne laser scanners: Potential and practical challenges. *International Archives of Photogrammetry and Remote Sensing*, 35(B3):201–206.

Wang, C., Elmore, A. J., Numata, I., Cochrane, M. A., Shaogang, L., Huang, J., Zhao, Y., Li, Y., 2022, Factors affecting relative height and ground elevation estimations of GEDI among forest types across the conterminous USA. *GIScience & Remote Sensing*, 59:975–999. https://doi.org/10.1080/15481603.2022.2085354.

Wang, C.-K. & Glenn, N. F., 2008, A linear regression method for tree canopy height estimation using airborne lidar data. *Canadian Journal of Remote Sensing*, 34(2):S217–S227.

Wang, C.-K., Tseng, Y.-H. & Chu, H.-J., 2014, Airborne dual-wavelength LiDAR data for classifying land cover. *Remote Sensing*, 6(1):700–715.

Wang, K., Franklin, S. E., Guo, X. & Cattet, M., 2010, Remote sensing of ecology, biodiversity and conservation: A review from the perspective of remote sensing specialists. *Sensors*, 10(11):9647–9667.

Wehr, A. & Lohr, U., 1999, Airborne laser scanning—an introduction and overview. *ISPRS Journal of Photogrammetry and Remote Sensing*, 54(2–3):68–82.

Weinacker, H., Koch, B., Heyder, U. & Weinacker, R., 2004, Development of filtering, segmentation and modelling modules for lidar and multispectral data as a fundament of an automatic forest inventory system. *International Archives of Photogrammetry, Remote Sensing and Spatial Information Sciences*, 36(Part 8):50–55.

Weishampel, J. F., Hightower, J. N., Chase, A. F. & Chase, D. Z., 2012, Use of airborne LiDAR to delineate canopy degradation and encroachment along the Guatemala-Belize border. *Tropical Conservation Science*, 5(1):12–24.

White, J., Tompalski, P., Vastaranta, M., Wulder, M. A., Saarinen, N., Stepper, C. & Coops, N. C., 2017, *A Model Development and Application Guide for Generating an Enhanced Forest Inventory Using Airborne Laser Scanning Data and an Area-Based Approach*, Canadian Wood Fibre Centre. Information report, no. FI-X-018, Natural Resources Canada, Victoria, BC. http://cfs.nrcan.gc.ca/publications?id=38945

White, J. C., Coops, N. C., Wulder, M. A., Vastaranta, M., Hilker, T. & Tompalski, P., 2016, Remote sensing technologies for enhancing forest inventories: A review. *Canadian Journal of Remote Sensing*, 42(5):619–641.

White, J. C., Wulder, M. A., Vastaranta, M., Coops, N. C., Pitt, D. & Woods, M., 2013, The utility of image-based point clouds for forest inventory: A comparison with airborne laser scanning. *Forests*, 4(3):518–536.

Wilkes, P., Jones, S. D., Suarez, L., Haywood, A., Woodgate, W., Soto-Berelov, M., ... & Skidmore, A. K., 2015, Understanding the effects of ALS pulse density for metric retrieval across diverse forest types. *Photogrammetric Engineering & Remote Sensing*, 81(8):625–635.

Williams, M. S., Bechtold, W. A. & LaBau, V., 1994, Five instruments for measuring tree height: An evaluation. *Southern Journal of Applied Forestry*, 18(2):76–82.

Wulder, M. A., Bater, C. W., Coops, N. C., Hirata, Y. & Sweda, T., 2007, Advances in laser remote sensing of forests. *Sustainable Development Research Advances*. Ed. B. A. Larson, Nova Science Publishers, Inc. chap. 8: 223–234.

Wulder, M. A., Niemann, K. O. & Goodenough, D. G., 2000, Local maximum filtering for the extraction of tree locations and basal area from high spatial resolution imagery. *Remote Sensing of Environment*, 73(1):103–114.

Wulder, M. A., White, J. C., Nelson, R. F., Næsset, E., Ørka, H. O., Coops, N. C., Hilker, T., Bater, C. W. & Gobakken, T., 2012, Lidar sampling for large-area forest characterization: A review. *Remote Sensing of Environment*, 121:196–209.

Wynne, R. H., 2006, Lidar remote sensing of forest resources at the scale of management. *Photogrammetric Engineering and Remote Sensing*, 72(12):1310–1314.

Xu, N., Qin, R. & Song, S., 2023, Point cloud registration for LiDAR and photogrammetric data: A critical synthesis and performance analysis on classic and deep learning algorithms. *ISPRS Open Journal of Photogrammetry and Remote Sensing*, 8:100032.

Yun, T., Jiang, K., Li, G., Eichhorn, M. P., Fan, J., Liu, F., Chen, B., An, F. & Cao, L., 2021, Individual tree crown segmentation from airborne LiDAR data using a novel Gaussian filter and energy function minimization-based approach. *Remote Sensing of Environment*, 256:112307.

Yun, Z., Zheng, G., Monika Moskal, L., Li, J. & Gong, P., 2023, Stratifying forest overstory and understory using the global ecosystem dynamic investigation laser scanning data. *International Journal of Applied Earth Observation and Geoinformation*, 124:103538. https://www.sciencedirect.com/science/article/pii/S156984322300362X; https://doi.org/10.1016/j.jag.2023.103538.

Zhao, X., Guo, Q., Su, Y. & Xue, B., 2016, Improved progressive TIN densification filtering algorithm for airborne LiDAR data in forested areas. *ISPRS Journal of Photogrammetry and Remote Sensing*, 117:79–91.

Zellweger, F., Baltensweiler, A., Ginzler, C., Roth, T., Braunisch, V., Bugmann, H. & Bollmann, K., 2016, Environmental predictors of species richness in forest landscapes: Abiotic factors versus vegetation structure. *Journal of Biogeography*, 43(6):1080–1090.

Zellweger, F., Braunisch, V., Baltensweiler, A. & Bollmann, K., 2013, Remotely sensed forest structural complexity predicts multi species occurrence at the landscape scale. *Forest Ecology and Management*, 307: 303–312.

Zellweger, F., Morsdorf, F., Purves, R. S., Braunisch, V. & Bollmann, K., 2014, Improved methods for measuring forest landscape structure: LiDAR complements field-based habitat assessment. *Biodiversity and Conservation*, 23(2):289–307.

Zhen, Z., Quackenbush, L. J. & Zhang, L., 2016, Trends in automatic individual tree crown detection and delineation—Evolution of LiDAR data. *Remote Sensing*, 8:333.

Zhou, J., Proisy, C., Descombes, X., Hedhli, I., Barbier, N., Zerubia, J., Gastellu-Etchegorry, J.-P. & Couteron, P., 2010, Tree crown detection in high resolution optical and LiDAR images of tropical forest. *Remote Sensing, International Society for Optics and Photonics*, 78240Q-78240Q-78246.

Zhou, T., Wang, C. & Zhou, Z., 2021, Thinning promotes the nitrogen and phosphorous cycling in forest soils. *Agricultural and Forest Meteorology*, 311:108665.

Zilliox, C. & Gosselin, F., 2013, Tree species diversity and abundance as indicators of understory diversity in French mountain forests: Variations of the relationship in geographical and ecological space. *Forest Ecology and Management*, 321:105–116.

Zolkos, S. G., Goetz, S. J. & Dubayah, R., 2013, A meta-analysis of terrestrial aboveground biomass estimation using lidar remote sensing. *Remote Sensing of Environment*, 128:289–298.

Zwally, H. J., Schutz, B., Abdalati, W., Abshire, J., Bentley, C., Brenner, A., Bufton, J., Dezio, J., Hancock, D., Harding, D., Herring, T., Minster, B., Quinn, K., Palm, S., Spinhirne, J. & Thomas, R., 2002, ICESat's laser measurements of polar ice, atmosphere, ocean, and land. *Journal of Geodynamics*, 34(3–4):405–445.

Part II

Biodiversity

5 Biodiversity of the World
A Study from Space

Thomas W. Gillespie, Morgan Rogers, Chelsea Robinson, and Duccio Rocchini

ACRONYMS AND DEFINITIONS

ASAR	Advanced synthetic aperture radar on board ENVISAT
ASTER	Advanced spaceborne thermal emission and reflection radiometer
AVHRR	Advanced very high resolution radiometer
DMSP	Defense Meteorological Satellite Program
EOS	Earth Observing System on board Aqua satellite
ETM+	Enhanced Thematic Mapper+
GIS	Geographic Information System
IKONOS	A commercial earth observation satellite, typically, collecting sub-meter to 5 m data
JERS	Japanese Earth Resources Satellite
LiDAR	Light detection and ranging
MERIS	Medium-resolution imaging spectrometer
MODIS	Moderate-resolution imaging spectroradiometer
MSS	Multi-spectral scanner
NASA	National Aeronautics and Space Administration
NDVI	Normalized Difference Vegetation Index
NOAA	National Oceanic and Atmospheric Administration
OLI	Operational land imager
RADARSAT	Radar satellite
SPOT	Satellite Pour l'Observation de la Terre, French Earth Observing Satellites
SRTM	Shuttle Radar Topographic Mission
TerraSAR-X	A radar Earth observation satellite with its phased array synthetic aperture radar
TRMM	Tropical Rainfall Measuring Mission
UAV	Unmanned Aerial Vehicle
VIIRS	Visible Infrared Imaging Radiometer Suite

5.1 INTRODUCTION

The Earth is undergoing an accelerated rate of native ecosystem conversion and degradation (Myers et al. 2000; Sage 2020), and there is increased interest in measuring, modeling, and monitoring biodiversity using remote sensing from spaceborne sensors (Nagendra 2001; Secades et al. 2014; Skidmore et al. 2021). Biodiversity can be defined as the variation of life forms (genetic, species) within a given ecosystem, region, or the entire Earth. Terrestrial biodiversity, rare, and threatened species tend to be highest near the equator and generally decrease toward the poles because of decreases in temperature and precipitation (Jenkins et al. 2013) (Figure 5.1). However, the distribution of biodiversity is complex and based on a number of environmental and anthropogenic factors over different spatial scales (Whittaker et al. 2001; Field et al. 2009; Rahbek et al. 2019).

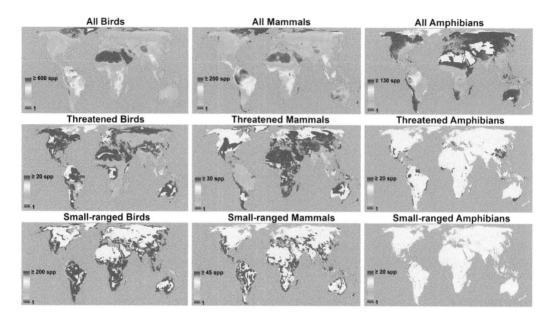

FIGURE 5.1 Global species richness, threatened species, and small-ranged species patterns for mammals, birds, and amphibians at 10 × 10 km (Jenkins et al. 2013).

Remote sensing has considerable potential as a source of information on biodiversity at site, landscape, continental, and global spatial scales (Turner et al. 2003; Cavender-Bares et al. 2022). The main attractions of remote sensing as a source of information on biodiversity are that it offers an inexpensive means of deriving complete spatial coverage of environmental information for large areas in a consistent manner that may be updated regularly (Duro et al. 2007; Gillespie et al. 2008; Wang and Gamon 2019). There has been an increase in studies and reviews of biodiversity and remote sensing taking advantage of advances in sensor technology or focusing on broad patterns of variables related to biodiversity (Rocchini 2007a; Pfeifer et al. 2012; Cavender-Bares et al. 2022).

These advances in remote sensing are generally divided into measuring, modeling, and monitoring biodiversity (Gillespie et al. 2008). Measuring uses spaceborne sensors to identify either species or individuals, such as the identification of tree species and density, or land cover types associated with species assemblages (such as redwood forest). Modeling uses spaceborne sensors to create probability models of species distributions and the distributions of biodiversity and associated metrics such as species richness. Monitoring is the use of time series spaceborne data of measured or modeled biodiversity to study dynamics over time, and this has significant applications for endangered species and ecosystem conservation.

This chapter reviews recent advances in remote sensing that can be used to study biodiversity from space. In particular, this chapter examines ways to measure, model, and monitor biodiversity patterns and processes using spaceborne imagery. First, we examine advances currently being used to measure biodiversity from space. Second, we examine advances in modeling patterns of species and biodiversity. Third, we examine monitoring applications of remote sensing for the conservation of biodiversity. Finally, we identify spaceborne sensors that can be used to study biodiversity from space.

5.2 MEASURING BIODIVERSITY FROM SPACE

5.2.1 Mapping Species and Vegetation Types

There is an increasing desire to identify species and map vegetation types associated with biodiversity and native and non-native species within landscapes from high-resolution spaceborne sensors

Biodiversity of the World

that have been launched in recent years (Goodwin et al. 2005; He et al. 2011; Pu 2021). High spatial resolution imagery has been used to accurately identify some plant species and plant assemblages (Martin et al. 1998; Carleer and Wolff 2004; Foody et al. 2005).

5.2.2 Mapping Individual Trees

Much progress has been made in identifying single species of plants, such as trees or non-native invasive species, that are of particular interest in natural resource management (He et al. 2011; Gholizadeh et al. 2022). For instance, QuickBird was used to map the invasive non-native giant reed (*Arundo donax*) in southern Texas with 86–100% accuracy (Everitt et al. 2006). There has also been significant progress in identifying tree canopies within forest ecosystems. For instance, high-resolution data has been used to identify non-native invasive species (Fuller 2005; He et al. 2011), native trees, and urban trees (Christian and Krishnayya 2009; Pu 2021). High spatial resolution imagery (GeoEye, QuickBird, IKONOS, WorldView) from space has also allowed researchers to undertake research at the scale of individual tree crowns over large areas (Pu and Landry 2012; Ma et al. 2023) (Figure 5.2). Moreover, it may sometimes be possible to achieve high levels of accuracy for some species from satellite as well as airborne sensor data (Carleer and Wolff 2004; Pu 2021). There is great potential to manually or digitally identify tree species and canopy attributes from high-resolution imagery. High-resolution imagery is collected primarily from commercial satellites that are still expensive to acquire ($1000 to $4000 for a 10 km^2). While the increased pricing of such imagery has put it out of the reach of many ecologists, especially those located in developing countries where the need is perhaps greatest, such cost has decreased with the competition, and an increasing number of archived images in the visible spectrum are readily available on Google Earth. Indeed, it is possible to identify species from moderate to high degrees of accuracy within many landscapes and forests already, and the accuracies will only increase with increased resolutions (spatial, spectral, and temporal) in the near future.

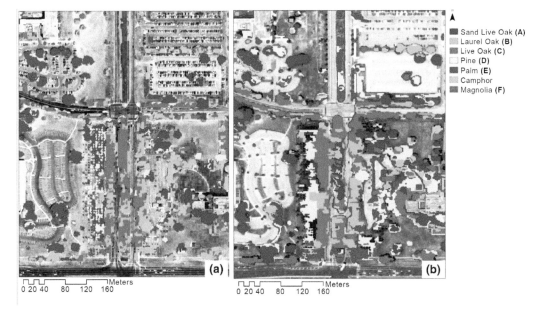

FIGURE 5.2 Mapping results of seven tree species in Tampa, Florida, classified using (a) IKONOS imagery and (b) WorldView-2 (Pu and Landry 2012).

5.2.3 Mapping Animals from Space

The identification of animals from space is currently difficult because most of the Earth's species are smaller than the largest pixel of current public access satellites (0.5 m) and revisit times are too infrequent for meaningful comparisons. However, high-resolution spaceborne imagery has been used to map large species. Indeed, some groups like whales can be monitored from space with high-resolution WorldView-2 imagery at a 50 cm pixel resolution in the panchromatic (Fretwell et al. 2014; Figure 5.3), and spaceborne remote sensing has been used to survey and discover large colonies of animals like penguin colonies in Antarctica (Schwaller et al. 2013; Fretwell and Trathan 2021). However, most remote sensing studies on monitoring fauna have focused on mapping vegetation types or habitat associated with endangered fauna. For instance, giant panda habitat has been monitored over time in China (Jian et al. 2011). Potential great progress in monitoring animals from remote sensing will take off when unmanned aerial vehicles (UAVs) start gathering data from a wide array of sensors. UAVs, technology wise, are already quite mature and have been used to monitor elephants in Africa (Schiffman 2014). However, UAVs have several limitations, especially with regard to covering large areas, potential costs, and security concerns. Thus, high-resolution satellites such as WorldView-3 have recently been used to count African elephants from space (Duporge et al. 2021).

5.2.4 Mapping Species Assemblages

The production of thematic maps of species assemblages is one of the most common applications of spaceborne remote sensing (Foody 2002). In particular, plant species assemblages or ecosystem distributional patterns within the landscapes, regions, and continents have important applications to natural resource management. In countries with strong and well-funded institutions dedicated to natural resource management, such as the US National Park Service, there is a need to update and standardize landscape dynamic protocols related to vegetation, landcover, and unique resource management needs by region using remote sensing (Fancy et al. 2009). Natural resource agencies in the tropics, where some of the largest decreed protected areas exist (Brooks et al. 2006), do not always have access to the remote sensing technology or trained individuals to develop and maintain a landscape dynamics change database (Laurance et al. 2012). Numerous large-area, multi-image-based, multiple-sensor land cover mapping programs exist that have resulted in robust and repeatable large-area land cover classifications (Duro et al. 2007; Gillespie et al. 2008; Hansen et al. 2013). Franklin and Wulder (2002) undertook an excellent review of large-scale land cover classifications, such as GlobCover, CORINE, and GAP, that generally seek to attain 85% accuracy across all mapping classes using a variety of passive sensors (TM, SPOT, AVHRR, MODIS, ENVI) and to a lesser extent active sensors (RADARSAT, JERS). For instance, mangrove ecosystems have been mapped using remote sensing at a global spatial scale (Wang et al. 2004; Giri et al. 2011; Heumann 2011). These land cover classifications provide measurements on the distribution of species assemblages and ecosystems. Recently, there have been a number of advances in methods that can improve the resolution and accuracy of land cover classification. Increased integration of LiDAR and radar data may significantly improve classification accuracy (Boyd and Danson 2005; Li and Chen 2005; Huang et al. 2010; Marselis et al. 2022).

5.3 MODELING BIODIVERSITY FROM SPACE

There are a number of spaceborne metrics that are associated with species and ecosystem distributions that can be used to create probability maps of the distribution of biodiversity (Rocchini et al. 2013) (Table 5.1).

5.3.1 Species Distribution Modeling

Species distribution modeling, also known as ecological niche modeling or spatial modeling, has been growing at a striking rate in the last 20 years, providing both estimates of species distributions

Biodiversity of the World

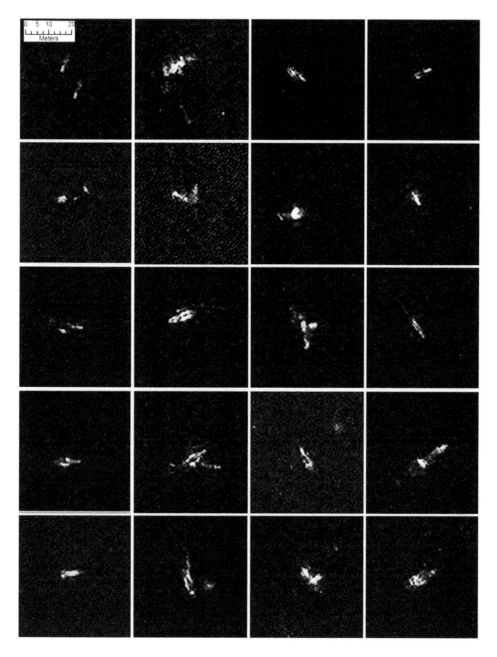

FIGURE 5.3 Counting of whales at Golfo Nuevo, Argentina, with WorldView-2 imagery (Fretwell et al. 2014).

over space together with estimates of bias in the models (Swanson et al. 2012). Species distribution models are based on presence, absence, or abundance data from museum vouchers or field surveys and environmental predictors to create probability models of species distributions within landscapes, regions, and continents. Most environmental predictors used in these species distribution models have been based on geographic information system data over different scales (Pearson and Dawson 2003) (Figure 5.4).

There has been an increase in the incorporation of spaceborne remote sensing data on climate, topography, and land cover that has great potential to improve models of species over different spatial scales (Turner et al. 2003; Randin et al. 2020). Remote sensing data on precipitation at 0.1

TABLE 5.1
Remote Sensing Variables Used for Modeling Biodiversity from Space

Remote Sensing Variable	Satellite or Sensor	Pixel Size (m)	Reference
Climate			
Rainfall	GPM	10,000	Hou et al. 2014
Temperature	MODIS	1000	Albright et al. 2011
Topography			
Elevation and topography	GEDI, Sentinel	10	Lang et al. 2023
Land cover			
Vegetation type	Landsat TM, MSS	30, 80	Gottschalk et al. 2005
NDVI	SPOT, Landsat	20, 30	Leyequien et al. 2007
Heterogeneity	SPOT, Landsat	20, 30	Rocchini et al. 2010
Vegetation structure	GEDI	25	Marselis et al. 2022
Forest cover/change	Landsat, MODIS	30, 1000	Hansen et al. 2013
Old growth	GEDI	25	Marselis et al. 2022
Fire	MODIS/VIIRS	1000, 375	https://firms.modaps.eosdis.nasa.gov/
Burned areas	MODIS	500	http://modis-fire.umd.edu/
Light pollution	DMSP	1100	http://ngdc.noaa.gov/eog/

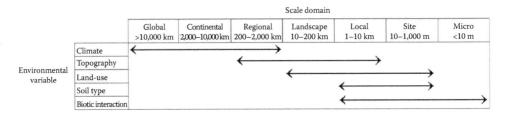

FIGURE 5.4 Environmental variables and their impact of species distributions across spatial domain (Pearson and Dawson 2003).

degree from NOAA satellites (Pearson et al. 2007) and from the Global Precipitation Measurement Mission (Hou et al. 2014) have been used in conjunction with ground-based measurements. This may be superior to traditional GIS estimates of precipitation based on interpolation among widely dispersed climate stations in isolated regions. Topography data has also been a fundamental component of species distribution models. Topography data is usually collected from digitized elevation maps, but 90 m and 30 m elevation and topography data are available at a near global extent due to the Shuttle Radar Topography Mission and ASTER. These data are increasingly being used in species distribution models (Chaves et al. 2007) (Figure 5.5).

5.3.2 Land Cover and Diversity

Land cover classifications collected from spaceborne sensors has long been used to link species distributions based on vegetation types and associated habitat preference (Nagendra 2001; Gottschalk et al. 2005; Leyequien et al. 2007). The greatest accuracy was found with non-mobile species such as plants (Pearson et al. 2004; Wang and Gamon 2019). However, vegetation maps as a surrogate for habitat preference have provided insights into the distributions of birds (Peterson et al. 2006; Regos et al. 2020), herpetofauna (Raxworthy et al. 2003), and insects (Luoto et al. 2002; Rhodes et al. 2021).

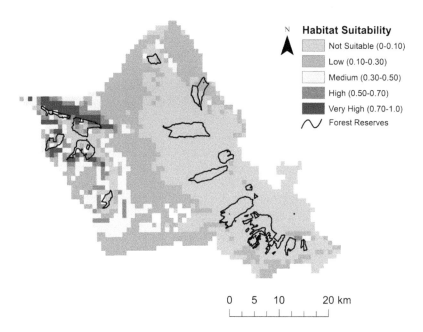

FIGURE 5.5 Species distribution model that incorporated remote sensing variable on topography and NDVI for the endangered *Hibiscus brackenridgei* on Oahu, Hawaii.

There have been a number of advances in modeling or predicting species richness, alpha diversity, and beta diversity using multi-sensors that examine relationships over different temporal and spatial scales with increasingly sophisticated methods to improve accuracy. The simplest measure of biodiversity is species richness or the number of species per unit area (i.e., trees per hectare, reptiles per km^2). The term "diversity" is more complex and technically refers to a combination of species richness and weighted abundance or evenness data and is generally quantified as an index (Simpson index, Shannon index, or Fisher alpha). These indices are used to define alpha diversity, which is the species diversity in one area, community, or ecosystem. Beta diversity refers to the amount of turnover in species composition from one site to another or identifies taxa unique to each area, community, or ecosystem. Beta diversity is more closely related to changes in species similarity or turnover with space. Typically, studies have focused on assessments of species richness with limited attention to other aspects such as species abundance and composition that are difficult to detect from spaceborne sensors (Foody and Cutler 2003; Wang and Gamon 2019). Information on species richness or diversity may be extracted from remotely sensed data in a variety of ways, such as land cover classifications, measures of productivity, and measures of heterogeneity (Nagendra 2001; Kerr and Ostrovsky 2003; Leyequien et al. 2007).

Many studies have related species richness or diversity to information on the land cover mosaic derived from satellite imagery (Gould 2000; Griffiths et al. 2000; Kerr et al. 2001; Oindo et al. 2003; Gottschalk et al. 2005; Leyequien et al. 2007; Gillespie et al. 2008). Through relationships with land cover and habitat suitability, it is possible to assess the diversity of species and assess impacts associated with changes in the habitat mosaic such as fragmentation based on landscape metrics (i.e., area and connectivity) (Kerr et al. 2001; Luoto et al. 2002, 2004; Cohen and Goward 2004; Fuller et al. 2007).

5.3.3 Spectral Indices and Diversity

Most attention has focused on the use of the popular normalized difference vegetation index (NDVI) from passive sensors because it is easy to calculate using the red and near infrared bands common to almost all passive spaceborne sensors (Oindo and Skidmore 2002; Seto et al. 2004; Gillespie 2005; Pettorelli 2013). NDVI has been associated with photosynthetic activity, primary productivity, and has been hypothesized to quantify species richness and diversity based on the species-energy theory (Currie 1991; Evans et al. 2005). There have been an increasing number of studies and reviews that have found significant associations between NDVI and diversity (Nagendra 2001; Kerr and Ostrovsky 2003; Leyequien et al. 2007; Wang and Gamon 2019). For plants, many studies have reported significant positive correlations between plant species richness or diversity from plot or regions data and NDVI in both temperate (Fairbanks and McGwire 2004; Levin et al. 2007; Rocchini 2007b) and tropical ecosystems (Bawa et al. 2002; Gillespie 2005; Feeley et al. 2005; Cayuela et al. 2006; Wang and Gamon 2019). NDVI can explain between 30% and 87% of the variation in species richness or diversity within a vegetation type, landscape, or region. Results for terrestrial fauna are more complicated given the mobility of faunal species and because NDVI does not directly quantify animal species but species habitats (Leyequien et al. 2007). Similar relationships between NDVI and diversity have been noted for animal taxa such as birds and butterflies within landscapes (Seto et al. 2004; Goetz et al. 2007) and regions (Hurlbert and Haskell 2003; Foody 2004b; Ding et al. 2006; Bino et al. 2008). Over the last decade, the NDVI has also proven extremely useful in predicting guild distributions, abundance, and life history traits in space and time (Hurlbert and Haskell 2003; Pettorelli et al. 2011). However, NDVI does not always have a positive relationship with animal species richness, and there is no consensus as to which scale results in the greatest accuracy.

Heterogeneity in land cover types, spectral indices, and spectral variability derived from satellite imagery has also been be correlated with species richness (Gould 2000; Rocchini 2007b; Rocchini et al. 2010). This is largely based on the hypothesis that heterogeneity in land cover, spectral indices, or spectral variability within an area or landscape is an indicator of habitat heterogeneity, which allows more species to coexist and hence greater species richness (Simpson 1949; Palmer et al. 2002; Carlson et al. 2007; Rocchini et al. 2007, Rocchini et al. 2010). The variation in land cover types within an area has been associated with species richness for a number of taxa (Gould 2000; Kerr et al. 2001; Leyequien et al. 2007). Variation in spectral indices has been shown to be positively associated with species richness and diversity for a number of taxa in different regions (Gould 2000; Oindo and Skidmore 2002; Fairbanks and McGwire 2004; Levin et al. 2007).

5.3.4 Multiple Sensors and Diversity

Recently there has been a move toward the use of multiple remote sensing sensors over different time periods and increasingly sophisticated approaches to modeling diversity over different spatial scales. There is an increasing number of diversity studies that are undertaken using multiple passive sensors (i.e., Landsat, ASTER, QuickBird) (Levin et al. 2007; Rocchini 2007b) or that examine relationships with diversity over different time periods (Fairbanks and McGwire 2004; Foody 2005; Levin et al. 2007; Leyequien et al. 2007). These studies are important in the assessment of individual sensors and the effects of seasonality. There has also been an increasing interest in the combination of passive and active sensors to improve species diversity models (Gillespie et al. 2008; Zarnetske et al. 2019). Active spaceborne sensors can provide data on the vegetation structure that has been associated with diversity, especially avian diversity, across a number of spatial scales (Imhoff et al. 1997; Bergen et al. 2007; Goetz et al. 2007; Leyequien et al. 2007; Torresani et al. 2023).

5.4 MONITORING BIODIVERSITY FROM SPACE

It is well-established that biodiversity is greatly threatened by human activity (Myers et al. 2000; Gaston 2005). In particular, land cover changes such as those linked to human-induced habitat loss, fragmentation, and degradation represent the largest current threat to biodiversity (Gaston 2005; Gillespie et al. 2008). Remote sensing can be used to derive metrics on fragmentation, often in the

TABLE 5.2
Spaceborne Sensors Commonly Used in Monitoring Biodiversity

Platform (Sensor)	Monitoring Use	Strengths	Limitations
High resolution			
Google Earth	Validation, communication	High resolution, free	Temporal gaps, no IR
PlanetScope	High resolution	High repeat time	Cross calibration
QuickBird, IKONOS	Area, degradation	High resolution	Cost, coverage
WorldView-3	Area, degradation	Multi-bands	Cost, coverage
Moderate resolution			
Landsat (TM, ETM+)	Land cover, fragmentation	Long time series, free	Clouds
Landsat (OLI, TIRS)	Temperature, water quality	High resolution	Since 2013[1]
EOS (ASTER)	Land cover, fragmentation	High spectral resolution	On-demand system
SPOT	Land cover, fragmentation,	Spectral and spatial resolution	Cost
EOS (Hyperion)	Ecosystem chemistry	Spectral resolution	Signal-to-noise ratio
Low resolution			
NOAA (AVHRR)	Vegetation indices, thermal	Time series from the 1980s	1.1 km pixels
EOS (MODIS)	Vegetation indices, thermal, fire	Time series from 2001	Under utilized
ESA (MERIS)	Vegetation indices, land cover	Highest global land cover map	Not for change detection
Active sensors			
SRTM	Elevation	Bench of canopy height 2001	One off
QSCAT	Canopy structure and moisture	Canopy moisture	2.25 km pixel
TRMM (TMI)	Rainfall	Global comparisons	Under utilized
Envisat (ASAR)	Heterogeneity	High-resolution comparisons	Side angle
ISS (GEDI)	Canopy structure, biomass	Vegetation height, biomass	Extent of Earth (4%)

[1] This is from Landsat 8 launched 2013.

form of landscape pattern and connectivity indices calculated from a thematic map produced with an image classification analysis (Foody 2001; Gillespie 2005; Lung and Schaab 2006; Kupfer 2012). These metrics can be monitored over time (Table 5.2). Remote sensing may be used to monitor a habitat of interest with a one-class classification approach adopted to focus effort and resources on the class of interest (Foody et al. 2006; Sanchez-Hernandez et al. 2007). This can also reduce problems associated with not satisfying the assumptions of an exhaustively defined set of classes that is commonly made in a standard classification analysis (Foody 2004a). For instance, Hansen et al. (2013) mapped the spatial extent of all forests at a global spatial scale to 30 m pixel resolution using Landsat imagery (Figure 5.6).

5.4.1 Remote Sensing of Protected Areas

Protected areas are one of the best ways to conserve biodiversity. Remote sensing has had a major role to play in helping to monitor changes in and around protected areas (Gross et al. 2013; Secades et al. 2014; Gillespie et al. 2015). Remote sensing offers a repeatable, systematic, and spatially exhaustive source of information on key variables such as productivity, disturbance, and land cover that impact biodiversity (Duro et al. 2007; Wright et al. 2007; Gillespie et al. 2008). The provision of data for monitoring large areas is especially attractive in remote and often inaccessible regions (Cayuela et al. 2006; Conchedda et al. 2011). Remote sensing is also often a cost-effective data source (Luoto et al. 2004) and enables rapid biodiversity assessments (Foody 2003).

The spatial coverage provided by remote sensing offers the potential to monitor the effectiveness of protected areas, allowing comparisons of changes inside and outside of reserves to be evaluated (Southworth et al. 2006; Wright et al. 2007). For example, even relatively severely logged forest

FIGURE 5.6 Example of forest cover loss (red) from 2000 to 2013 in Northern Sumatra based on Landsat imagery (Hansen et al. 2013).

outside of a reserve may represent a significant resource for biodiversity conservation (Tang et al. 2010) (Figure 5.7). Thus, actions inside and outside of the protected areas are important, supporting the view that biodiversity conservation activities should be undertaken at the level or scale of the landscape (Nagendra et al. 2013). This activity may benefit from remote sensing, as its synoptic overview provides information on the entire landscape. Indeed, Crabtree et al. (2009) provide a modeling and spatio-temporal framework for monitoring environmental change using net primary productivity as an ecosystem health indicator.

Remote sensing may be a useful component to general biodiversity assessments, especially in providing data at appropriate spatial and temporal scales. For example, the biodiversity intactness index was proposed recently as a general indicator of the overall state of biodiversity to aid monitoring and decision-making (Scholes and Biggs 2005; Schipper et al. 2020). Although there are concerns for its use, notably with the impacts of land degradation, remote sensing may be an important source of data for its derivation (Rouget et al. 2006).

5.4.2 Remote Sensing of Urban Areas

Cities are developing urban biodiversity conservation plans and looking for cost-effective methods for monitoring urban biodiversity (Nilon et al. 2017). Remote sensing has been an invaluable tool in global biodiversity monitoring (Wang and Gamon 2019); however, its application in urban landscapes has been limited due to resolution constraints and the structural complexities of the urban environment. Fortunately, recent advancements in spaceborne technology now offer measures of both spectral and structural attributes of vegetation at resolutions suited for highly heterogeneous urban landscapes (Velasquez-Camacho et al. 2021). Scholars have utilized multispectral, hyperspectral, and LiDAR to monitor vegetation diversity and phenology in cities and even identify urban tree species richness, providing a low-cost and efficient method that offers vital urban biodiversity metrics (Gillespie et al. 2017; Velasquez-Camacho et al. 2021).

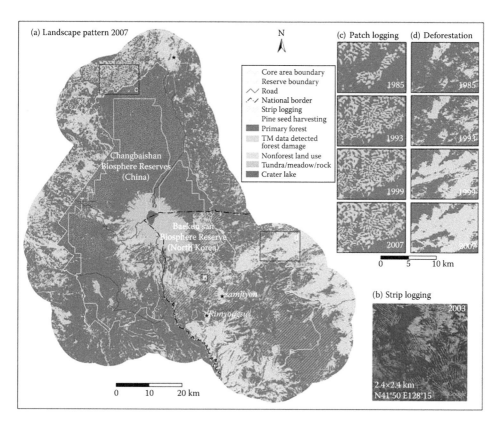

FIGURE 5.7 Two adjacent biosphere reserves across the border of China and North Korea. Remote sensing monitoring with Landsat shows over one-half of primary forest landscapes have been deteriorated by exploitive uses, including systematic logging (Tang et al. 2010).

In an urban biodiversity context, LiDAR excels, spanning its utility from pinpointing tree crowns to delving into the broader composition and structure of forests (Velasquez-Camacho et al. 2021). Numerous studies highlight the synergy between LiDAR and other spaceborne remote sensing methods, such as multispectral sensors like WorldView, and LiDAR for urban tree species identification (Pu 2021; Velasquez-Camacho et al. 2021). Using a combined approach of integrating data from multiple sensors, including LiDAR, has been validated for its accuracy in detailing vegetation diversity and structure (Parent et al. 2015). Despite its efficacy, the limited global availability and cost of obtaining LiDAR data remains a significant hurdle. One promising source of globally available LiDAR data is GEDI, a spaceborne LiDAR on the International Space Station, which could provide potential comparative data on tree height metrics in urban areas across the globe and correlations with tree and bird species richness (Marselis et al. 2022; Torresani et al. 2023).

5.5 SPACEBORNE SENSORS AND BIODIVERSITY

There has been a dramatic increase in Earth observation satellites and sensors over the last decade, which have been used to measure, model, and monitor biodiversity from space (Table 5.3).

5.5.1 Spectral Sensors and Biodiversity

Passive sensors, which record reflected (visible and infrared wavelengths) and emitted energy (thermal wavelengths), are most frequently used in biodiversity studies. The highest spatial resolution data

TABLE 5.3
Satellites That Can Be Used to Measure, Model, and Monitor Biodiversity from Space

Satellite (Sensor)	Pixel Size (m)	Bands	Revisit Time	Year Operation
Spectral sensors		Spectral bands		
High resolution				
QuickBird	0.65, 2.6	5	1–3.5 days	2001
IKONOS	0.8, 3.2	5	3 days	1999
GeoEye 1	0.46, 1.8	5	3 days	2008
WorldView-3	0.3, 1.24, 3.7, 30	29	1 day	2014
Moderate resolution				
Landsat 7 (ETM+)	15, 30, 60, 120	8	16 days	1999
Landsat 8 (OLI, TIRS)	15, 30, 100	11	16 days	2013
Landsat 9 (OLI2, TIRS2)	15, 30, 100	11	16 days	2021
Sentinel 2 A (MSI)	10, 20, 60	13	5 days	2015
SPOT 5	2.5, 20, 1150	5	26 days	2002
Low resolution				
NOAA (AVHRR)	1100	5	1 day	1979
EOS (MODIS)	250, 500, 1000	36	1 day	1999
Envisat (MERIS)	300	15	3 days	2002–2012
Active sensors		Bands		
SRTM	30, 90	X, C	none	2000
QSCAT	2500	Ku	4 days	1999–2009
Radarsat 2	9–100	C	24 days	2007
TRMM (TMI)	5000	X, K, Ka, W	0.5 days	1997
Envisat (ASAR)	30–150	C	35 days	2002–2012
Sentinel-1A	5–100	C	12 days	2014
ISS (GEDI)	25	1064 nm	none	2019

comes from commercial satellites, such as Planet, GeoEye, WorldView, QuickBird, and IKONOS series, which contain visible and infrared bands used in species and species assemblage mapping. The NASA Landsat series is the most widely used sensor for biodiversity studies due to the ease in which the data can be obtained, long time series, and low cost (Leimgruber et al. 2005). The Landsat series has been used extensively in land cover classifications, diversity models, and conservation studies. The recently launched Landsat 9 satellite will be useful for continuing this time series and biodiversity-related research. Other satellites and sensors, such as the European Space Agency's Sentinel missions, are free, high spatial resolution, and can be scaled up to all regions of the globe. The MODIS sensors on EOS satellites and VIIRS sensors have provided extremely useful data for sites, landscape, regional, continental, and global studies of land cover classification and diversity models. These sensors also provide data on temperature, cloud cover, and fire that have been incorporated into biodiversity studies.

5.5.2 Radar Sensors and Biodiversity

Radar is the most common active spaceborne sensor used in biodiversity studies (Bae et al. 2019; Zarnetske et al. 2019). Radar sensors send and receive a microwave pulse in different wavelengths (i.e., X-, C-, and L-bands) to create an image based on radar backscatter, or interferometric radar can be used to provide high-resolution data on elevation, topography, and tree canopy height estimates. Unlike passive sensors, radar can penetrate cloud cover, providing imagery both day and night regardless of weather conditions. The Shuttle Radar Topography Mission (SRTM) provides

30–90 m resolution data on elevation and topography that has been used in species and diversity models. Recently, Sentinel-1 has been used to map the biodiversity of 12 taxa across five temperate forest regions in Central Europe and further demonstrate the predictive ability of radar-derived data on the species composition of birds and insects (Bae et al. 2019).

5.5.3 LiDAR Sensors and Biodiversity

Airborne and spaceborne LiDAR has revolutionized the way that we study biodiversity. Airborne LiDAR has now been collected for a number of regions, and large-scale maps of individual tree canopies and associated metrics (e.g., height, crown area) are becoming common (Ma et al. 2023). Indeed, Ma et al. (2023) mapped the tree locations, height, and crown area for over six million individual urban trees in New York City boroughs. Recently, spaceborne data from the Global Ecosystem Dynamics Investigation (GEDI) on the International Space Station offers an unprecedented opportunity for studying biodiversity. GEDI has three lasers that produce full waveforms within a 25 m footprint that can measure the three-dimensional structure of forests and terrain in temperate and tropical forests (Marselis et al. 2022; Torresani et al. 2023). Each footprint is separated by 60 m along a track, with an across-track distance of approximately 600 m between each of the eight tracks. GEDI covers areas between 51°North and South latitude and will gather data for approximately 4% of the Earth's surface. GEDI canopy structure metrics can explain up to 66% of the variation in tree species richness in natural forests without a history of recent disturbance across the globe (Marselis et al. 2022; Torresani et al. 2023). GEDI has improved the spatial resolution and point density necessary to study fine-scale processes associated with biodiversity. Researchers have been able to fuse sparse GEDI data with Sentinel-2 optical images, which has global coverage at 10–60 m resolution, to create a high-resolution canopy height model of the Earth with 10 m resolution (Lang et al. 2023) (Figure 5.8). While the challenges associated with LiDAR's accessibility and cost persist, the integration of LiDAR with other remote sensing products offers a promising avenue for detailed and comprehensive biodiversity studies.

5.5.4 Ideal Biodiversity Satellites and Sensors

Is it possible to design a satellite specifically for measuring, modeling, and monitoring biodiversity? Remote sensing of biodiversity in theory could benefit from a satellite that includes both a high-resolution spectral sensor and a high-resolution LiDAR sensor as well as a rapid repeat time. Ideally 0.5 m or less pixel resolution is needed for the spectral sensors, and it should capture the visible, infrared, and thermal bands. The actual wavelengths should correspond to those of Landsat 8 with two bands in the blue wavelength to study water, infrared bands similar to WorldView-3 to study plant species and vegetation, and two bands in the thermal to identify large animals and disturbances such as fires. The waveform LiDAR sensors should have a 1-m footprint because this clearly improves estimates of vegetation height and subcanopy topography, both of which are important for modeling and monitoring vegetation structure, species distribution models, and estimates of species richness and biomass.

However, the future of biodiversity and remote sensing most likely lies in using a constellation of satellites, sensors, and field data that already exist. Indeed, to really study biodiversity you need historic and real-time data on climate, topography, land cover, and soils. The increase in high-resolution Planetscope and WorldView archived imagery should increase our ability to identify plants to species and improve vegetation classifications. This can also be used to monitor changes over individual endangered plant species and habitats of endangered animal species. Publicly available Landsat and Sentinel images are central to near real-time change detection, and Landsat provides important information on long-term vegetation change since the 1980s. Combining full coverage of WorldView-3, Sentinel-2, and GEDI may come closest to the ideal combination of satellites and sensors needed to monitor biodiversity.

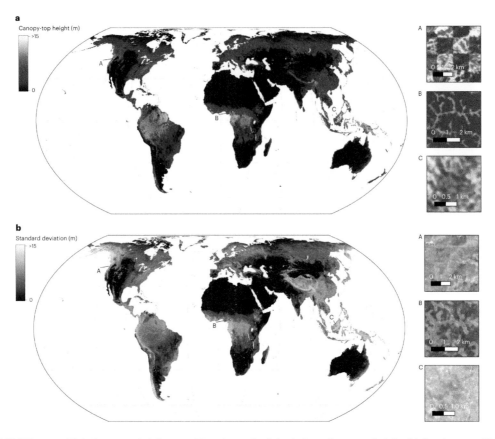

FIGURE 5.8 Global canopy height map (a) and standard deviation of canopy height (b) for the year 2020 estimated from GEDI and Sentinel-2 imagery at a 10 m spatial resolution (Lang et al. 2023).

5.6 CONCLUSIONS

Spaceborne imagery has made significant contributions to measuring, modeling, and monitoring biodiversity patterns and processes from space. Future research should focus on incorporating historic and new spaceborne sensors; more extensive integration of available field, GIS, and passive and active imagery that can be used across spatial scales; and the collection and dissemination of high-quality field data (Cavender-Bares et al. 2022). The recent developments in satellite and sensor technology will further improve our abilities to measure and model patterns of biodiversity from space. The increase in high-resolution spectral satellites will make it possible to acquire data at enhanced spatial (0.5 m), spectral (visible, infrared, thermal), and radiometric resolutions (11 bit) that can be used to map individual species. Indeed, Google Earth has led the way by providing high-resolution airborne and spaceborne imagery, and Google Earth Engine can now process comparative Landsat, Sentinel, and MODIS datasets in an easy-to-use interface. Current radar satellites may be ideal for studying species distributions and diversity patterns, especially in regions with high cloud cover like the tropics. There are multiple satellites (SAR-Lupe, COSMO- SkyMed, TerraSAR-X) that provide elevation and radar backscatter data to 1 m pixel resolution. This will provide valuable multidimensional datasets (vegetation structure, biomass, landcover classifications) that should result in a richer characterization of the environment than conventional passive image datasets. Remote sensing scientists interested in biodiversity have taken and can take advantage of the different satellite datasets that integrate climate, topography, spectral, LiDAR, and radar data over landscape, regional, continental, and global spatial scales. This has increased our understanding of biodiversity, and remote sensing is a useful tool for measuring, modeling, and monitoring biodiversity in near real time and across multi-spatial scales.

REFERENCES

Albright, T. P., A. M. Pidgeon, C. D. Rittenhouse, M. K. Clayton, C. H. Flather, P. D. Culbert, and V. C. Radeloff. 2011. Heat waves measured with MODIS land surface temperature data predict changes in avian community structure. *Remote Sensing of Environment* 115:245–254.

Bae, S., S. R. Levick, L. Heidrich, P. Magdon, B. F. Leutner, S. Wöllauer, A. Serebryanyk, T. Nauss, P. Krzystek, M. M. Gossner, and P. Schall. 2019. Radar vision in the mapping of forest biodiversity from space. *Nature Communications* 10:4757.

Bawa, K., J. Rose, K. N. Ganeshaiah, N. Barve, M. C. Kiran, and R. Umashaanker. 2002. Assessing biodiversity from space: An example from the Western Ghats, India. *Conservation Ecology* 6(2):7.

Bergen, K. M., A. M. Gilboy, and D. G. Brown. 2007. Multi-dimensional vegetation structure in modeling avian habitat. *Ecological Informatics* 2:9–22.

Bino, G., N. Levin, S. Darawshi, N. Van der Hal, A. Reich-Solomon, and S. Kark. 2008. Landsat derived NDVI and spectral unmixing accurately predict bird species richness patterns in an urban landscape. *International Journal of Remote Sensing* 29(13):3675–3700.

Boyd, D. S. and F. M. Danson. 2005. Satellite remote sensing of forest resources: Three decades of research development. *Progress in Physical Geography* 29:1–26.

Brooks, T. M., R. A. Mittermeier, G. A. da Fonseca, J. Gerlach, M. Hoffmann, J. F. Lamoreux, C. G. Mittermeier, J. D. Pilgrim, and A. S. Rodrigues. 2006. Global biodiversity conservation priorities. *Science* 313:58–61.

Carleer, A. and E. Wolff. 2004. Exploitation of very high resolution satellite data for tree species identification. *Photogrammetric Engineering and Remote Sensing* 70:135–140.

Carlson, K. M., G. P. Asner, R. F. Hughes, R. Ostertag, and R. E. Martin. 2007. Hyperspectral remote sensing of canopy biodiversity in Hawaiian lowland rainforests. *Ecosystems* 10:536–549.

Cavender-Bares, J., F. D. Schneider, M. J. Santos, A. Armstrong, A. Carnaval, K. M. Dahlin, L. Fatoyinbo, G. C. Hurtt, D. Schimel, P. A. Townsend, and S. L. Ustin. 2022. Integrating remote sensing with ecology and evolution to advance biodiversity conservation. *Nature Ecology & Evolution* 6:506–519.

Cayuela, L., J. M. Benayas, A. Justel, and J. Salas-Rey. 2006. Modelling tree diversity in a highly fragmented tropical montane landscape. *Global Ecology and Biogeography* 15:602–613.

Chaves, J. A., J. P. Pollinger, T. B. Smith, and G. LeBuhn. 2007. The role of geography and ecology in shaping the phylogeography of the speckled hummingbird (*Adelomyia melanogenys*) in Ecuador. *Molecular Phylogenetics and Evolution* 43:795–807.

Christian, B. and N. S. R. Krishnayya. 2009. Classification of tropical trees growing in a sanctuary using Hyperion (EO-1) and SAM algorithm. *Current Science* 96:1601–1607.

Cohen, W. B. and S. N. Goward. 2004. Landsat's role in ecological applications of remote sensing. *Bioscience* 54:535–545.

Conchedda, G., E. F. Lambin, and P. Mayaux. 2011. Between land and sea: Livelihoods and environmental changes in mangrove ecosystems of Senegal. *Annals of the Association of American Geographers* 101:1259–1284.

Crabtree, R., C. Potter, R. Mullen, J. Sheldon, S. Huang, J. Harmsen, A. Rodman, and C. Jean. 2009. A modeling and spatio-temporal analysis framework for monitoring environmental change using NPP as an ecosystem indicator. *Remote Sensing of Environment* 113:1486–1496.

Currie, D. J. 1991. Energy and large-scale patterns of animal-and plant-species richness. *American Naturalist* 137:27–49.

Ding, T., H. Yuan, S. Geng, C. Koh, and P. Lee. 2006. Macro-scale bird species richness patterns of East Asian mainland and islands: Energy, area, and isolation. *Journal of Biogeography* 33:683–693.

Duporge, I., O. Isupova, S. Reece, D. W. Macdonald, and T. Wang. 2021. Using very-high-resolution satellite imagery and deep learning to detect and count African elephants in heterogeneous landscapes. *Remote Sensing in Ecology and Conservation* 7:369–381.

Duro, D., N. C. Coops, M. A. Wulder, and T. Han. 2007. Development of a large area biodiversity monitoring system driven by remote sensing. *Progress in Physical Geography* 31:235–260.

Evans, K. L., J. J. D. Greenwood, and K. J. Gaston. 2005. Dissecting the species-energy relationship. *Proceedings of the Royal Society B-Biological Sciences* 272:2155–2163.

Everitt, J. H., C. Yang, and C. J. Deloach Jr. 2006. Remote sensing of giant reed with QuickBird satellite imagery. *Journal of Aquatic Plant Management* 43:81–85.

Fairbanks, D. H. K. and K. C. McGwire. 2004. Patterns of floristic richness in vegetation communities of California: Regional scale analysis with multi-temporal NDVI. *Global Ecology and Biogeography* 13:221–235.

Fancy, S. G., J. E. Gross, and S. L. Carter. 2009. Monitoring the condition of natural resources in US national parks. *Environmental Monitoring and Assessment* 151:161–174.

Feeley, K. J., T. W. Gillespie, and J. W. Terborgh. 2005. The utility of spectral indices from Landsat ETM+ for measuring the structure and composition of tropical dry forests. *Biotropica* 37:508–519.

Field, R., B. A. Hawkins, H. V. Cornell, D. J. Currie, J. A. F Diniz-Filho, J. F. Guégan, and J. R. Turner. 2009. Spatial species-richness gradients across scales: A meta-analysis. *Journal of Biogeography* 36:132–147.

Foody, G. M. 2001. Monitoring the magnitude of land-cover change around the southern limits of the Sahara. *Photogrammetric Engineering and Remote Sensing* 67:841–847.

Foody, G. M. 2002. Status of land cover classification accuracy assessment. *Remote Sensing of Environment* 80:185–201.

Foody, G. M. 2003. Remote sensing of tropical forest environments: Towards the monitoring of environmental resources for sustainable development. *International Journal of Remote Sensing* 24:4035–4046.

Foody, G. M. 2004a. Supervised image classification by MLP and RBF neural networks with and without an exhaustively defined set of classes. *International Journal of Remote Sensing* 25:3091–3104.

Foody, G. M. 2004b. Spatial nonstationarity and scale-dependency in the relationship between species richness and environmental determinants for the sub-Saharan endemic avifauna. *Global Ecology and Biogeography* 13:315–320.

Foody, G. M. 2005. Mapping the richness and composition of British breeding birds from coarse spatial resolution satellite sensor imagery. *International Journal of Remote Sensing* 26:3943–3956.

Foody, G. M., P. M. Atkinson, P. W. Gething, N. A. Ravenhill, and C. K. Kelly. 2005. Identification of specific tree species in ancient semi-natural woodland from digital aerial sensor imagery. *Ecological Applications* 15:1233–1244.

Foody, G. M. and M. E. J. Cutler. 2003. Tree biodiversity in protected and logged Bornean tropical rain forests and its measurement by satellite remote sensing. *Journal of Biogeography* 30:1053–1066.

Foody, G. M., A. Mathur, C. Sanchez-Hernandez, and D. S. Boyd. 2006. Training set size requirements for the classification of a specific class. *Remote Sensing of Environment* 104:1–14.

Franklin, S. E. and M. A. Wulder. 2002. Remote sensing methods in medium spatial resolution satellite data land cover classification of large areas. *Progress in Physical Geography* 26:173–205.

Fretwell, P. T., I. J. Staniland, and J. Forcada. 2014. Whales from space: Counting southern right whales by satellite. *PLoS ONE* 9(2):e88655.

Fretwell, P. T. and P. N. Trathan. 2021. Discovery of new colonies by Sentinel 2 reveals good and bad news for emperor penguins. *Remote Sensing in Ecology and Conservation* 7(2):139–153.

Fuller, D. O. 2005. Remote detection of invasive Melaleuca trees (*Melaleuca quinquenervia*) in South Florida with multispectral IKONOS imagery. *International Journal of Remote Sensing* 26:1057–1063.

Fuller, R. M., B. J. Devereux, S. Gillings, R. A. Hill, and G. S. Amable. 2007. Bird distributions relative to remotely sensed habitats in Great Britain: Towards a framework for national modeling. *Journal of Environmental Management* 84:586–605.

Gaston, K. J. 2005. Biodiversity and extinction: Species and people. *Progress in Physical Geography* 29:239–247.

Gholizadeh, H., M. S. Friedman, N. A. McMillan, W. M. Hammond, K., Hassani, A. V. Sams, M. D., Charles, D. R. Garrett, O. Joshi, R. G. Hamilton, and S. D. Fuhlendorf. 2022. Mapping invasive alien species in grassland ecosystems using airborne imaging spectroscopy and remotely observable vegetation functional traits. *Remote Sensing of Environment* 271:112887.

Gillespie, T. W. 2005. Predicting woody-plant species richness in tropical dry forests: A case study from south Florida, USA. *Ecological Applications* 15:27–37.

Gillespie, T. W., J. de Goede, L. Aguilar, G. D. Jenerette, G. A. Fricker, M. L. Avolio, S. Pincetl, T. Johnston, L. W. Clarke, and D. E. Pataki. 2017. Predicting tree species richness in urban forests. *Urban Ecosystems* 20:839–849.

Gillespie, T. W., G. M. Foody, D. Rocchini, A. P. Giorgi, and S. Saatchi. 2008. Measuring and modelling biodiversity from space. *Progress in Physical Geography* 32:203–221.

Gillespie, T. W., K. S. Willis, and S. Ostermann-Kelm. 2015. Spaceborne remote sensing of the world's protected areas. *Progress in Physical Geography* 39:388–404.

Giri, C., E. Ochieng, L. L. Tieszen, Z. Zhu, A. Singh, T. Loveland, J. Masek, and N. Duke. 2011. Status and distribution of mangrove forests of the world using earth observation satellite data. *Global Ecology and Biogeography* 20:154–159.

Goetz, S., D. Steinberg, R. Dubayah, and B. Blair. 2007. Laser remote sensing of canopy habitat heterogeneity as a predictor of bird species richness in an eastern temperate forest, USA. *Remote Sensing of Environment* 108:254–263.

Goodwin, N., R. Turner, and R. Merton. 2005. Classifying Eucalyptus forests with high spatial and spectral resolution imagery: An investigation of individual species and vegetation communities. *Australian Journal of Botany* 53:337–345.

Gottschalk, T. K., F. Huettmann, and M. Ehler. 2005. Thirty years of analyzing and modeling avian habitat relationships using satellite imagery: A review. *International Journal of Remote Sensing* 26:2631–2656.

Gould, W. 2000. Remote Sensing of vegetation, plant species richness, and regional biodiversity hot spots. *Ecological Applications* 10:1861–1870.

Griffiths, G. H., J. Lee, and B. C. Eversham. 2000. Landscape pattern and species Richness: Regional scale analysis from remote sensing. *International Journal of Remote Sensing* 21:2685–2704.

Gross, D., G. Dubois, J. F. Pekel, P. Mayaux, M. Holmgren, H. H. T. Prins, C. Rondinini, and L. Boitani. 2013. Monitoring land cover changes in African protected areas in the 21st century. *Ecological Informatics* 14:31–37.

Hansen, M. C., P. V. Potapov, R. Moore, M. Hancher, S. A. Turubanova, A. Tyukavina, D. Thau, S. V. Stehman, S. J. Goetz, T. R. Loveland, A. Kommareddy, A. Egorov, L. Chini, C. O. Justice, and J. R. G. Townshend. 2013. High-resolution global maps of 21st-century forest cover change. *Science* 342:850–853.

He, K. S., D. Rocchini, M. Neteler, and H. Nagendra. 2011. Benefits of hyperspectral remote sensing for tracking plant invasions. *Diversity and Distributions* 17:381–392.

Heumann, B. W. 2011. Satellite remote sensing of mangrove forests: Recent advances and future opportunities. *Progress in Physical Geography* 35:87–108.

Hou, A. Y., R. K. Kakar, S. Neeck, A. A. Azarbarzin, C. D. Kummerow, M. Kojima, R. Oki, K. Nakamura, and T. Iguchi. 2014. The global precipitation measurement mission. *Bulletin of the American Meteorological Society* 95:701–722.

Huang, S., C. Potter, R. L. Crabtree, S. Hager, and P. Gross, 2010. Fusing optical and radar data to estimate sagebrush, herbaceous, and bare ground cover in Yellowstone. *Remote Sensing of Environment* 114:251–264.

Hurlbert, A. H. and J. P. Haskell. 2003. The effect of energy and seasonality on avian species richness and community composition. *American Naturalist* 161:83–97.

Imhoff, M. L., T. D. Sisk, A. Milne, G. Morgan, and T. Orr. 1997. Remotely sensed indicators of habitat heterogeneity: Use of synthetic aperture radar in mapping vegetation structure and bird habitat. *Remote Sensing of Environment* 60:217–227.

Jenkins, C. N., S. L. Pimm, and L. N. Joppa. 2013. Global patterns of terrestrial vertebrate diversity and conservation. *Proceedings of the National Academy of Sciences* 110:2602–2610.

Jian J., H. Jiang, G. Zhou, Z. Jiang, S. Yu, S. Peng, S. Liu, and J. Wang. 2011. Mapping the vegetation changes in giant panda habitat using Landsat remotely sensed data. *International Journal of Remote Sensing* 32:1339–1356.

Kerr, J. T. and M. Ostrovsky. 2003. From space to species: Ecological applications for remote sensing. *Trends in Ecology and Evolution* 18:299–305.

Kerr, J. T., T. R. E. Southwood, and J. Cihlar. 2001. Remotely sensed habitat diversity predicts butterfly species richness and community similarity in Canada. *Proceedings of the National Academy of Sciences of the United States of America* 98:11365–11370.

Kupfer, J. A. 2012. Landscape ecology and biogeography: Rethinking landscape metrics in a post-FRAGSTATS landscape. *Progress in Physical Geography* 36:400–420.

Lang, N., W. Jetz, K. Schindler, and J. D. Wegner. 2023. A high-resolution canopy height model of the Earth. *Nature Ecology & Evolution*:1–12.

Laurance, W. F., et al. 2012. Averting biodiversity collapse in tropical forest protected areas. *Nature* 489:290–294.

Leimgruber, P., C. A. Christen, and A. Laborderie. 2005. The impact of Landsat satellite monitoring on conservation biology. *Environmental Monitoring and Assessment* 106:81–101.

Levin, N., A. Shimida, O. Levanoni, H. Tamari, and S. Kark. 2007. Predicting mountain plant richness and rarity from space using satellite-derived vegetation indices. *Diversity and Distribution* 13:1–12.

Leyequien, E., J. Verrelst, M. Slot, G. Schaepman-Strub, I. M. A. Heitkonig, and A. Skidmore. 2007. Capturing the fugitive: Applying remote sensing to terrestrial animal distribution and diversity. *International Journal of Applied Earth Observation and Geoinformation* 9:1–20.

Li, J. and W. Chen. 2005. A rule-based method for mapping Canada's wetlands using optical, radar, and DEM data. *International Journal of Remote Sensing* 26:5051–5069.

Lung, T. and G. Schaab. 2006. Assessing fragmentation and disturbance of west Kenyan rainforests by means of remotely sensed time series data and landscape metrics. *African Journal of Ecology* 44:491–506.

Luoto, M., M. Kuussaari, and T. Toivonen. 2002. Modelling butterfly distribution based on remote sensing data. *Journal of Biogeography* 29:1027–1037.

Luoto, M., R. Virkkala, R. K. Heikkinen, and K. Rainio. 2004. Predicting bird species richness using remote sensing in boreal agricultural-forest mosaics. *Ecological Applications* 14:1946–1962.

Ma, Q., J. Lin, Y. Ju, W. Li, L. Liang, and Q. Guo. 2023. Individual structure mapping over six million trees for New York City USA. *Scientific Data* 10:102.

Marselis, S. M., P. Keil, M. J. Chase, and R. Dubayah. 2022. The use of GEDI canopy structure for explaining variation in tree species richness in natural forests. *Environmental Research Letters* 17:045003.

Martin, M. E., S. D. Newman, J. D. Aber, and R. G. Congalton. 1998. Determining forest species composition using high spectral resolution remote sensing data. *Remote Sensing of Environment* 65:249–254.

Myers, N., R. A. Mittermeier, C. G. Mittermeier, G. A. Da Fonseca, and J. Kent. 2000. Biodiversity hotspots for conservation priorities. *Nature* 403:853–858.

Nagendra, H. 2001. Using remote sensing to assess biodiversity. *International Journal of Remote Sensing* 22:2377–2400.

Nagendra, H., R. Lucas, J. P. Honrado, R. H. Jongman, C. Tarantino, M. Adamo, and P. Mairota. 2013. Remote sensing for conservation monitoring: Assessing protected areas, habitat extent, habitat condition, species diversity, and threats. *Ecological Indicators* 33:45–59.

Nilon, C. H., M. F. J. Aronson, S. S. Cilliers, C. Dobbs, L. J. Frazee, M. A. Goddard, K. M. O'Neill, D. Roberts, E. K. Stander, P. Werner, M. Winter, and K. P. Yocom. 2017. Planning for the future of urban biodiversity: A global review of city-scale initiatives. *BioScience* 67:332–342.

Oindo, B. O. and A. K. Skidmore. 2002. Interannual variability of NDVI and species richness in Kenya. *International Journal of Remote Sensing* 23:285–298.

Oindo, B. O., A. K. Skidmore, and P. De Salvo. 2003. Mapping habitat and biological diversity in the Maasai Mara ecosystem. *International Journal of Remote Sensing* 24:1053–1069.

Palmer, M. W., P. Earls, B. W. Hoagland, P. S. White, and T. Wohlgemuth. 2002. Quantitative tools for perfecting species lists. *Environmetrics* 13:121–137.

Parent, J. R., J. C. Volin, and D. L. Civco. 2015. A fully-automated approach to land cover mapping with airborne LiDAR and high resolution multispectral imagery in a forested suburban landscape. *ISPRS Journal of Photogrammetry and Remote Sensing*, 104:18–29.

Pearson, R. G. and T. P. Dawson. 2003. Predicting the impacts of climate change on the distribution of species: Are bioclimate envelope models useful. *Global Ecology and Biogeography* 12:361–371.

Pearson, R. G., T. P. Dawson, and C. Liu. 2004. Modelling species distribution in Britain: A hierarchical integration of climate and land-cover data. *Ecography* 27:285–298.

Pearson, R. G., C. J. Raxworthy, M. Nakamura, and A. Townsend Peterson. 2007. Predicting species distributions from small numbers of occurrence records: A test case using cryptic geckos in Madagascar. *Journal of Biogeography* 34:102–117.

Peterson, A. T., V. Sánchez-Cordero, E. Martínez-Meyer, and A. G. Navarro-Sigüenza. 2006. Tracking population extirpations via melding ecological niche modeling with land-cover information. *Ecological Modelling* 195:229–236.

Pettorelli, N. 2013. *The Normalized Difference Vegetation Index*. Oxford University Press.

Pettorelli, N., S. Ryan, T. Mueller, N. Bunnefeld, B. Jedrzejewska, B, M. Lima, and K. Kausrud. 2011. The Normalized Difference Vegetation Index (NDVI): Unforeseen successes in animal ecology. *Climate research* 46(1):15–27.

Pfeifer, M., M. Disney, T. Quaife, and R. Marchant. 2012. Terrestrial ecosystems from space: A review of earth observation products for macroecology applications. *Global Ecology and Biogeography* 21:603–624.

Pu, R. 2021. Mapping tree species using advanced remote sensing technologies: A state-of-the-art review and perspective. *Journal of Remote Sensing* 2021:26.

Pu, R. and S. Landry. 2012. A comparative analysis of high spatial resolution IKONOS and WorldView-2 imagery for mapping urban tree species. *Remote Sensing of Environment* 124:516–533.

Rahbek, C., M. K. Borregaard, R. K. Colwell, B. O. Dalsgaard, B. G. Holt, N. Morueta-Holme, D. Nogues-Bravo, R. J. Whittaker, and J. Fjeldså. 2019. Humboldt's enigma: What causes global patterns of mountain biodiversity? *Science* 365:1108–1113.

Randin, C. F., M. D. Ashcroft, J. Bolliger, J. Cavender-Bares, N. C. Coops, S. Dullinger, T. Dirnböck, S. Eckert, E. Ellis, N. Fernández, and G. Giuliani. 2020. Monitoring biodiversity in the Anthropocene using remote sensing in species distribution models. *Remote Sensing of Environment* 239:111626.

Raxworthy, C. J., E. Martinez-Meyer, N. Horning, R. A. Nussbaum, G. E. Schneider, M. A. Ortega-Huerta, and A. T. Peterson. 2003. Predicting distributions of known and unknown reptile species in Madagascar. *Science* 426:837 841.

Regos, A., P. Gómez-Rodríguez, S. Arenas-Castro, L. Tapia, M. Vidal, and J. Domínguez. 2020. Model-assisted bird monitoring based on remotely sensed ecosystem functioning and atlas data. *Remote Sensing* 12:2549.

Rhodes, M. W., J. J. Bennie, A. Spalding, R. H. ffrench-Constant, and I. M. Maclean. 2021. Recent advances in the remote sensing of insects. *Biological Reviews* 97:343–360.

Rocchini, D. 2007a. Distance decay in spectral space in analyzing ecosystem-diversity. *International Journal of Remote Sensing* 28:2635–2644.

Rocchini, D. 2007b. Effects of spatial and spectral resolution in estimating ecosystem α-diversity by satellite imagery. *Remote Sensing of Environment* 111:423–434.

Rocchini, D., N. Balkenhol, G. A. Carter, G. M. Foody, T. W. Gillespie, K. S. He, S. Kark, N. Levin, K. Lucas, M. Luoto, H. Nagendra, J. Oldelnad, C. Ricotta, J. Southworth, and M. Neteler. 2010. Remotely sensed spectral heterogeneity as a proxy of species diversity: Recent advances and open challenges. *Ecological Informatics* 5(5):318–329.

Rocchini, D., G. M. Foody, H. Nagendra, C. Ricotta, M. Anand, K. S. He, V. Amici, B. Kleinschmit, M. Förster, S. Schmidtlein, H. Feilhauer, A. Ghisla, M. Metz, and M. Neteler. 2013. Uncertainty in ecosystem mapping by remote sensing. *Computers & Geosciences* 50:128–135.

Rocchini, D., C. Ricotta, and A. Chiarucci. 2007. Using satellite imagery to assess plant species richness: The role of multispectral systems. *Applied Vegetation Science* 10:25–331.

Rouget, M., R. M. Cowling, J. Vlok, M. Thompson, and A. Balmford. 2006. Getting the biodiversity intactness index right: The importance of habitat degradation data. *Global Change Biology* 12:2032–2036.

Sage, R. F. 2020. Global change biology: A primer. *Global Change Biology* 26(1):3–30.

Sanchez-Hernandez, C., D. S. Boyd, and G. M. Foody. 2007. One-class classification for mapping a specific land-cover class: SVDD classification of fenland. *IEEE Transactions on Geoscience and Remote Sensing* 45:1061–1073.

Schiffman, R. 2014. Drones flying high as new tool for field biologists. *Science* 344:459

Schipper, A. M., J. P. Hilbers, J. R. Meijer, L. H. Antão, A. Benítez-López, M. M. de Jonge, L. H. Leemans, E. Scheper, R. Alkemade, J. C. Doelman, and S. Mylius. 2020. Projecting terrestrial biodiversity intactness with GLOBIO 4. *Global Change Biology* 26(2):760–771.

Scholes, R. J. and R. Biggs. 2005. A biodiversity intactness index. *Nature* 434:45–49.

Schwaller, M. R., C. J. Southwell, and L. M. Emmerson. 2013. Continental-scale mapping of Adelie penguin colonies from Landsat imagery. *Remote Sensing of Environment* 139:353–364.

Secades, C., B. O'Connor, C. Brown, and M. Walpole. 2014. Earth observation for biodiversity monitoring: A review of current approaches and future opportunities for tracking progress towards the Aichi Biodiversity Targets. *Secretariat of the Convention on Biological Diversity, Montréal, Canada. Technical Series No. 72*, 183 pages. https://geobon.org/downloads/biodiversity-monitoring/technical-reports/other/2014/cbd-ts-72-en.pdf

Seto, K. C., E. Fleishman, J. P. Fay, and C. J. Betrus. 2004. Linking spatial patterns of bird and butterfly species richness with Landsat TM derived NDVI. *International Journal of Remote Sensing* 25:4309–4324.

Simpson, E. H. 1949. Measurement of diversity. *Nature* 163(4148):688.

Skidmore, A. K., N. C. Coops, E. Neinavaz, A. Ali, M. E. Schaepman, M. Paganini, W. D. Kissling, P. Vihervaara, R. Darvishzadeh, H. Feilhauer, and M. Fernandez. 2021. Priority list of biodiversity metrics to observe from space. *Nature Ecology & Evolution* 5(7):896–906.

Southworth, J., H. Nasendra, and D. K. Munroe. 2006. Are parks working? Exploring human-environment tradeoffs in protected area conservation. *Applied Geography* 26:87–95.

Swanson, A. K., S. Z. Dobrowski, A. O. Finley, J. H. Thorne, and M. K. Schwartz. 2012. Spatial regression methods capture prediction uncertainty in species distribution model projections through time. *Global Ecology and Biogeography* 22:242–251.

Tang, L., G. Shao, Z. Piao, L. Dai, M. A. Jenkins, S. Wang, G. Wu, J. Wu, and J. Zhao. 2010. Forest degradation deepens around and within protected areas in East Asia. *Biological Conservation* 143:1295–1298.

Torresani, M., D. Rocchini, A. Alberti, V. Moudrý, M. Heym, E. Thouverai, P. Kacic, and E. Tomelleri. 2023. LiDAR GEDI derived tree canopy height heterogeneity reveals patterns of biodiversity in forest ecosystems. *Ecological Informatics* 76:102082.

Turner, W., S. Spector, N. Gardiner, M. Fladeland, E. Sterling, and M. Steininger. 2003. Remote sensing for biodiversity science and conservation. *Trends in Ecology and Evolution* 18:306–314.

Velasquez-Camacho, L., A. Cardil, M. Mohan, M. Etxegarai, G. Anzaldi, and S. de-Miguel. 2021. Remotely sensed tree characterization in Urban areas: A review. *Remote Sensing* 13:4889.

Wang, L., W. P. Sousab, P. Gong, and G. S. Biging. 2004. Comparison of IKONOS and QuickBird images for mapping mangrove species on the Caribbean coast of Panama. *Remote Sensing of the Environment* 91:432–440.

Wang, R. and J. A. Gamon. 2019. Remote sensing of terrestrial plant biodiversity. *Remote Sensing of Environment* 231:111218.

Whittaker, R. J., K. J. Willis, and R. Field. 2001. Scale and species richness: Towards a general, hierarchical theory of species diversity. *Journal of Biogeography* 28:453–470.

Wright, S. J., G. A. Sanchez-Azofeifa, C. Portillo-Quintero, and D. Davies. 2007. Poverty and corruption compromise tropical forest reserves. *Ecological Applications* 17:1259–1266.

Zarnetske, P. L., Q. D. Read, S. Record, K. D. Gaddis, S. Pau, M. L. Hobi, S. L. Malone, J. Costanza, J., K. Dahlin, A. M. Latimer, and A. M. Wilson. 2019. Towards connecting biodiversity and geodiversity across scales with satellite remote sensing. *Global Ecology and Biogeography* 28(5):548–556.

6 Multi-Scale Habitat Mapping and Monitoring Using Satellite Data and Advanced Image Analysis Techniques

Stefan Lang, Christina Corbane, Palma Blonda, Kyle Pipkins, and Michael Forster

ACRONYMS AND DEFINITIONS

ALOS	Advanced Land Observing Satellite
ASTER	Advanced spaceborne thermal emission and reflection radiometer
AVHRR	Advanced very high resolution radiometer
CHRIS	Compact High Resolution Imaging Spectrometer
EO	Earth Observation
ETM+	Enhanced Thematic Mapper+
FAO	Food and Agriculture Organization of the United Nations
GEO	Group on Earth Observation
GEOSS	Global Earth Observation System of Systems
GIS	Geographic Information System
GLAS	Geoscience Laser Altimeter System
ICESat	Instrument on board the Ice, Cloud, and land Elevation
IKONOS	A commercial earth observation satellite, typically, collecting sub-meter to 5 m data
LiDAR	Light detection and ranging
MERIS	Medium-resolution imaging spectrometer
MIR	Mid-infrared
MODIS	Moderate-resolution imaging spectroradiometer
NIR	Near-infrared
OBIA	Object-Oriented Image Analysis
PALSAR	Phased Array type L-band Synthetic Aperture Radar
PROBA	Project for On Board Autonomy
RADARSAT	Radar satellite
SAR	Synthetic aperture radar
SPOT	Satellite Pour l'Observation de la Terre, French Earth Observing Satellites
SWIR	Shortwave infrared
SVM	Support vector machines
TerraSAR-X	A radar Earth observation satellite with its phased array synthetic aperture radar
UAV	Unmanned Aerial Vehicle
VHRI	Very high resolution imagery

6.1 INTRODUCTION—THE POLICY FRAMEWORK

6.1.1 Monitoring Global Change

"Global change"—a short formula for a multitude of anticipated shifts in societal and environmental domains due to global drivers—calls for spatial monitoring and modeling techniques to understand the implications and potential dynamics of such changes (Alkhuzaei and Brolly, 2024; Amani et al., 2023; An et al., 2023; Iglseder et al., 2023; Larsen et al., 2023; Price et al., 2023; Zhang et al., 2023; van der Reijden et al., 2021; Wan and Ma, 2020; Wicaksono et al., 2019; Roelfsema et al., 2018; Setyawidati et al., 2018; Onojeghuo and Onojeghuo, 2017; Lang et al., 2013a). International initiatives, programs, and visions strive for unified frameworks and quality standards for data, products, and services to establish optimized observation capacity for global monitoring. This includes land surfaces and oceans, climate and atmosphere, as well as sustainable social systems such as public health, human security, and energy consumption. The 2005 founded intergovernmental Group on Earth Observations (GEO[1]) envisions "a future where decisions and actions for the benefit of humankind are informed by coordinated, comprehensive and sustained Earth observations" (EEA, 1995). GEO distinguishes nine societal benefit areas (SBAs) of civilian observations systems, among which biodiversity and ecosystems are two. The term Earth observation (EO) comprises all observation systems that use sensor technologies to capture various kinds of biophysical parameters. This includes space- or airborne measurement devices (sensors mounted on satellites, aircrafts, or unmanned aerial vehicles [UAVs]), as well as mobile ground devices (e.g., unmanned ground vehicles [UGVs]) and fixed measurement instruments (e.g., ground sensors, buoys, terrestrial laser scanners). All these observation systems taken together form a GEO System of Systems, the GEOSS, which was generated from 2005 to and beyond 2015 (GEO, 2005). For example, the GEO Biodiversity Observation Network (GEO-BON[2]), evaluated the adequacy of existing biodiversity observation systems to support the Convention on Biological Diversity (CBD) 2020 targets. A list of Essential Biodiversity Variables (EBV) (Pereira et al., 2013) has been established to be monitored worldwide in order to follow up the state of biodiversity adequately. Many of these global indicators are relying on remotely sensed imagery (Price et al., 2023; Wicaksono et al., 2019; Roelfsema et al., 2018).

This GEOSS implementation plan adheres to certain quality criteria, to be assessed by quality indices (QI) referring to data provision, data pre-processing, and the information supplied. According to the Quality Assurance Framework for Earth Observation (QA4EO), it is ensured that GEOSS is "implemented in a harmonious and consistent manner throughout all EO communities to the benefit of all stakeholders" (GEO, 2010). The QA4EO highlights the role of the user, who should be the "driver for any specific quality requirements" and assess "if any supplied information . . . is fit for purpose."

In the context of this chapter, we mainly refer to satellite-borne EO techniques, whose general assets as compared to conventional terrestrial field mapping are summarized in Table 6.1.

The growing need for a civilian use of satellite remote sensing and other EO technologies has born the European initiative Copernicus,[3] originally called as GMES (Global Monitoring for the Environment and Security), as a conjoint endeavor of the European Commission (EC) and the European Space Agency (ESA). The space programme Copernicus is considered the European contribution to GEO fostering the provision of geospatial information and monitoring services in six principal domains, i.e., land, water, atmosphere, climate change, emergency response, and human security. It builds on advanced space infrastructure and the technological capability to turn data into information services. The Sentinel family of EO satellites, developed by ESA, provides global coverage, with radar and optical data ranging from 10 m to 20 m and 60 m *spatial resolution* (in the VNIR to SWIR spectral range). Additional EO data from satellites of contributing missions increase both the variety of available data types and the temporal coverage. Next to the provision of frequently updated satellite data we need the adequate means for an intelligent usage of such data and an efficient analysis of these (Iglseder et al., 2023; Wan and Ma, 2020; Onojeghuo

TABLE 6.1
How General Strengths of Satellite EO-Based Mapping Translates to Habitat Delineation and Characterization

Wide area coverage	• Depending on the spatial resolution, area extents per scene range from a few to several hundreds of square kilometers.
	• Within the instantaneous view-field of a shot, similar atmospheric and illumination conditions apply so that image statistics are homogenous over a scene and variations within the image can be—by and large—related to changing habitat types or conditions.
Multi-scale option	• The technical trade-off between spatial resolution (grain) and area covered (extent) allows for purpose-driven usage.
	• Habitat mapping can be performed in a multi-scale approach, assuming fine-scaled observations are required for limited areas ("hot spot") only, and otherwise national or continental investigations cope with lesser detail.
From a distance, no direct contact	• A bird's-eye view and the lack of physical contact with the object of interest takes the complex structure of natural habitats, terrain inaccessibility, or disturbance of protected areas.
Multi-temporal/ multi-seasonal coverage	• A (semi-)permanent installation of space infrastructure enables a repetitive, standardized observation pattern within a given timeframe, whether over several years or several seasons.
	• Recursive observation under standardized conditions is a crucial prerequisite for monitoring activities prescribed in national and international environmental policies.

and Onojeghuo, 2017; Lang, 2008). The Copernicus initiative has opened new fields of activity to industry (including SMEs) and research organizations. The financing of dedicated core services and GMES initial operations (GIOs) as fundamental information services in all Copernicus domains has had initially launched, the stimulation of downstream services in new emerging areas. A key prerequisite for the creation of versatile applications across domains and related business cases is the provision of the EO data at low or no cost. The Sentinel family has been designed in such a way that satellite data are distributed for free and with no limitation for whatsoever (civilian) use. The next generation of Sentinel satellites is currently in its design phase to ensure continuity of the missions.

Biodiversity and habitat monitoring represent an emerging societal benefit area. Biodiversity, the variety of life forms, has become a key word for shaping and bundling political will, and—if thought of as the information content of life—requires adequate technology to observe and monitor it. Satellite Earth observation has turned into a ubiquitous means, a "democratic tool" to observe what is going on at the different levels of political implementation (Price et al., 2023; van der Reijden et al., 2021; Lang et al., 2013a).

Two collaborative EU-funded projects MS.MONINA[4] and BIO_SOS,[5] both started in 2010, have explored EO data combined with data from ground surveys (Larsen et al., 2023; Wicaksono et al., 2019; Blonda et al., 2012b; Lang et al., 2014). The idea was to set up EO-based (pre-)operational, yet economically priced solutions to provide timely information on pressures and impacts, to establish spatial priority for conservation and to evaluate its effectiveness. MS.MONINA developed advanced data-driven, EO-based analysis and modeling tools specifically tailored to user requirements on all levels of policy implementation. Three (sub-)services were designed, the so-called .EU, .State, and .Site level services, addressing agencies on the EU level (e.g., ETC Biodiversity, EEA, and DG Environment), national and federal agencies, and local management authorities using advanced mapping methods for status assessment and change maps. BIO_SOS provided cost-effective, knowledge-driven, EO-based analysis and modeling tools for meeting regulation obligations and for the definition (and effectiveness assessment) of related management strategies and actions. The project developed a pre-operational open-source processing system that combined multi-seasonal EO satellite data and *in situ* measurements on the basis of prior expert rules to map land cover (LC) and habitats, their changes and modifications over time, and to quantify anthropogenic pressures.

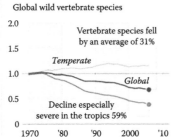
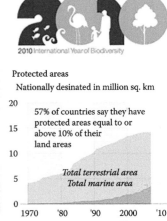

FIGURE 6.1 In 2010, the United Nations celebrated the International Year of Biodiversity. In that year the international community realized that the 1992 target to "halt the loss of biodiversity" has not been reached and more restrictive and better observable measures need to be taken. Source: savingspecies.org.

Expert rules include spectral, spatial, and temporal features characterizing LC classes and habitat classes. The system is cost-effective for mapping large or non-accessible areas as in-field reference data are not required for training the system but only for validating the output products.

In this chapter we distill the projects' technical outcomes and scientific achievements described in a Special Issue on "Earth observation for habitat mapping and biodiversity monitoring," edited by S. Lang and others in 2014. We highlight the great potential of EO data and the achievements of recent technologies, but also their challenges and limitations, in support of biodiversity and ecosystem monitoring. In Europe, nature conservation rests upon a strong and ambitious policy framework with legally binding directives. As in other parts of the world, the environmental legislation follows ambitious goals that often have to compete with other societal premises, such as growth, production, and expansion. Thus, geospatial information products are required at all levels of implementation. With recent advances in EO data availability and forthcoming capable analysis tools, we enter a new dimension of satellite-based information services. Recent achievements and challenges are brought in this article, using spearheading examples from inside and outside Europe.

6.1.2 Biodiversity and Related Policies

Biodiversity—our "natural capital and life-insurance" (EuropeanCommission, 2011)—is on the decline (Iglseder et al., 2023; van der Reijden et al., 2021; Onojeghuo and Onojeghuo, 2017; Isbell, 2010; Trochet and Schmeller, 2013). This is expected to directly influence the integrity of ecosystem functioning and stability and thus, ultimately, to human well-being (Naeem et al., 2009). In 1992, the United Nations joined forces in the International Convention on Biological Diversity (CBD) to halt or at least lower the accelerated loss of biodiversity. Next to the challenging nature of this aim as a key global challenge, it remains demanding to monitor and evaluate its success, which requires a concerted, effective use of the latest technology (Lang et al., 2014). As by the end of 2010 (the International Year of Biodiversity) the global society became aware that the ambitious goal of "halting biodiversity" had not been reached, the importance of observation techniques became even more important (see Figure 6.1).

Multi-Scale Habitat Mapping and Monitoring

The integrity of species and ecosystems is a global phenomenon with continental, regional, and ultimately local implications. This makes biodiversity a *glocalized* phenomenon (Price et al., 2023; Zhang et al., 2023; Setyawidati et al., 2018; Lang et al., 2014). Geographically this manifests in a hierarchy of scales, from biomes to (systems of) ecosystems down to communities, populations, and species. Observing and monitoring aspects of biodiversity, at any level and scale, can be approximated by analyzing the composition, variability, and changes of tangible entities (i.e., habitats) and their spatial patterns (Bock et al., 2005). Remote sensing information complements data obtained through standardized, *in situ* surveys related to local aspects of biodiversity, by representing integrated higher-level characteristics such as those of ecological neighborhoods (Addicot et al., 1987), defined by the upper (extent, object/scene size) and lower (grain, spatial resolution) limits of data information content and perception (Wiens, 1989). The matching of various resolution levels of satellite sensor families with the organizational levels of biological systems and organism perception is one aspect—the correspondence with spatial and temporal domains of environmental policies another. Today, satellite EO is a key technology to observe what is happening at the different levels of political implementation (Amani et al., 2023; Zhang et al., 2023; Lang et al., 2014).

The EU responded to the recognition that the biodiversity target 2010 would not be met, despite some major successes and the adoption of a global Strategic Plan for Biodiversity 2011–2020 at the tenth Conference of the Parties (CoP10) to the CBD, with the *EU biodiversity strategy to 2020*.[6] The EU Biodiversity Strategy complements the (general) EU Strategy 2020, the EU's growth strategy for the current decade where five main targets are established: Employment, R&D and Innovation, Climate Change and Energy, Education, and Poverty and Social Exclusion. The strategy's main target was—again—to halt biodiversity loss and the degradation of ecosystem services in the EU by 2020. To meet this target several sub-targets and actions had been framed. Projects such as MS.MONINA and BIO_SOS[7] and subsequent activities comply with the actions under Target 1 "to fully implement the Birds and Habitats Directives." Additionally, they support action 5 (mapping and assessment of ecosystems and their services) and action 6a (set priorities for ecosystem restoration) and 6b (Development of a Green Infrastructure Strategy 2012) (see Figure 6.2) of the Strategy's Target 2.

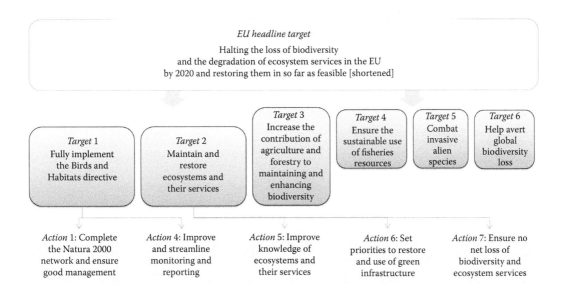

FIGURE 6.2 General and specific objectives of the EU Biodiversity Strategy 2020 as a strategic policy framework to MS.MONINA and BIO_SOS (taken from MS.MONINA user requirement dossier, modified).

6.1.3 Mapping the State of Ecosystems

A key action of the EU Biodiversity Strategy is the Mapping and Assessment of Ecosystems and their Services in Europe (An et al., 2023; van der Reijden et al., 2021; Setyawidati et al., 2018; Maes et al., 2013). Action 5, which aims to improve knowledge on ecosystems and their services, entails EU Member States to map and assess the state of ecosystems and their services in their national territory by 2014, to assess the economic value of such services, and to promote the integration of these values into accounting and reporting systems at the EU and national level by 2020.

The supply of ecosystem services (ES) relates to functions and characteristics of biodiversity aspects, including genes, species, and habitats. Next to the collection of biological data such as functional traits of plants, the mapping focuses mainly on ecosystem structure and habitat data. Ideally all ecosystem types that act as functional units in delivering services should be mapped and evaluated separately. Despite the complexity of the matter, it would be desirable to directly map the "total ecosystem service" for a certain area as cumulative effect of the (sub-)services. Strategies from multidimensional mapping exist and could be adapted, in order to integrate a larger set of single information layers or indicators (cf. Lang et al., Vol. 2., Chapter 9). The main challenges with mapping ES, however, are the lack of coverage and resolution in the available habitat maps, a general shortage in time and efforts required, and most importantly, the need for coherence and standardization.

Thus, in the two projects mentioned earlier—which are considered "downstream" to the Copernicus land monitoring service, we followed the EU Biodiversity 2010 Baseline approach for ecosystem mapping. This implies that land cover classes as monitored in Copernicus are aggregated into habitats and ecosystem types, in the most meaningful way possible to represent broad-scale ecosystems and combined with ecosystem-relevant information. This aggregation is based on detailed expert analysis of relationships between land cover classes (as derived from the Food and Agricultural Organization [FAO] Land Cover Classification System [LCCS] taxonomy, and habitat classification systems (i.e., EUNIS) to ensure consistency between these approaches (Larsen et al., 2023; Wicaksono et al., 2019; Onojeghuo and Onojeghuo, 2017; Tomaselli et al., 2013). Harmonization of the assessment activities of the EU Member States (MS) is an important ongoing activity but must leave some degrees of freedom to reflect the specific ecological, social, and historical context of each MS. Accordingly, MS are encouraged to use a more detailed habitat typology if available, with the only restriction that the more detailed classes are linkable to the EU-level typology. Habitat maps produced at the local scales usually include detailed data on the associated biodiversity that enable links to ecosystem services. Mapping programs have incorporated qualitative descriptors of the mapped habitats under various names (ecosystem state, ecosystem health, ecological integrity, naturalness, vegetation condition, degradation level, etc.). Action 5 of the Biodiversity strategy emphasizes the "need to map . . . the state of ecosystems and not simply to map the ecosystems" (see Figures 5.5 and 5.7). In order for maps of habitat conservation status to become a useful input to policies at the transnational level, the classification schemes used to evaluate the degradation levels, the habitat categories, and the methods used to assess them should be harmonized (Ichter et al., 2014).

6.1.4 The EU Habitats Directive

The EU Habitats Directive (short HabDir, Council Directive 92/43/EEC), an essential part of the European endeavor toward the CBD (Amani et al., 2023; An et al., 2023; van der Reijden et al., 2021; Trochet and Schmeller, 2013), is an ambitious legal instrument to safeguard biodiversity and set aside a network of protected areas (Gruber et al., 2012), called *Natura 2000*,[8] constantly progressing towards completion (Evans, 2012). HabDir entails standardized and frequent (every six years) monitoring and reporting activities with specific responsibilities on all political levels of implementation: (1) the local management authorities for the monitoring of individual protected sites, (2) the EU

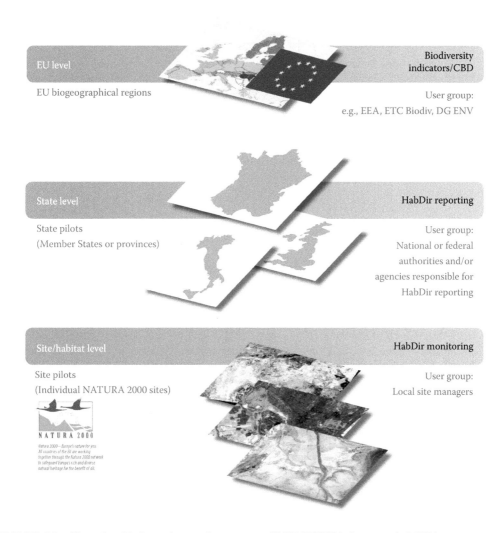

FIGURE 6.3 Three-level information service concept of MS.MONINA (Lang et al., 2013b).

Member States for reporting on the status of the network of protected sites and habitat distribution over the entire territory, and (3) the European Union for aggregating this information and the reporting toward the CBD. Updated geospatial information products are required at all three levels, not only by upscaling ground-level information but also to provide additional independent information on each level. In this framework, EO data and related techniques offer objective, yet economically priced solutions to (1) provide timely information on pressures and impacts, (2) establish spatial priority for conservation, and (3) collect long-term, multi-scale baseline information for evaluating the effectiveness of conservation strategies.

6.2 SATELLITE SENSOR CAPABILITIES

Earlier we claimed that habitats, as natural systemic areal features, are mappable through satellite remote sensing. With advances in EO imagery, we had entered a new dimension in habitat and biodiversity mapping around the year 2010. Still challenges are ahead in terms of data integration, advanced pre-processing and calibration, automated information extraction, ground verification and product validation, as well as semantic interoperability and exchange (Alkhuzaei and Brolly, 2024; Price et al., 2023; Roelfsema et al., 2018; Onojeghuo and Onojeghuo, 2017; Lang et al., 2014). In the following sections, we take a closer look at the specific requirements habitat mapping poses toward satellite imagery; in other words, what can be achieved in terms of habitat categorization and change quantification. In this section, we shall first look at the capabilities space-based observation technology hold to support such demand. Note that the "demand view" and the "supply view" are not distinct realms; in fact, there are interaction and interdependencies among them, a mutually fertilizing process that research and development activities, as the projects mentioned earlier, try to catalyze and stimulate.

The key technical characteristics of satellite technology playing a considerable role in habitat mapping are the following:

1. **Spatial resolution**. Habitats as areal features have a certain extent ranging from a few square meters to hundreds of square kilometers. In order to find a commensurate scale of observation, a range of sensor families are at disposal, whereby very high-resolution (VHR) and high-resolution (HR) sensors play the most important roles, while lower-resolution sensors are more used for differentiating large ecosystems, biomes, etc. on a continental scale. Spatial resolution is also a key factor in characterizing within-habitat conditions, such as structure or composition, whereby the resolution needs to be significantly smaller than the extent of the habitat of concern. The number of quantization levels (radiometric resolution) shall match the spatial resolution. There is no need for having a lot of neighboring pixels with all the same digital number (DN). There are, for example, 256 levels (8 bit) for Landsat 8 (30/15 m resolution) and 2048 (11 bit) for WorldView-2 (2/0.5 m resolution).

2. **Spectral resolution**. The main feature to be looked at when categorizing and delineating habitats is plant composition. The latter can only be characterized in a satisfying manner if spectral resolution suffices. The differentiation of vegetation types and plant species require specific wavelengths, such as the so-called red-edge, or—by and large—the infrared bands. Multi- and hyperspectral data provide the range of wavelength that can be used for plant species discrimination. When spectral resolution is not adequate for vegetation discrimination, multi-temporal (seasonal) data can be used.

3. **Revisiting time**. The revisiting time of nonflexible sensor units is bound to the orbiting time of the satellite. This is about 16 days for Landsat and ten days for one Sentinel-2. Rotating and side-looking sensors increase the revisiting time. Another factor of repeatability can be gained through constellations, reducing Sentinel-2A/B to five days. Identical satellites orbit in turn to cover each spot on Earth in multiples compared to a standard revisiting time. Multi-temporal, in particular multi-seasonal, observations are critical to many, if not most, habitat assessments.

According to Corbane et al. (2014) an obstacle is that, despite the tremendous progress in the applications of remote sensing to habitat mapping, many data types discussed here (hyperspectral, LiDAR, and radar) might go be beyond the actual capabilities of the community of practice (Larsen et al., 2023; Wicaksono et al., 2019; Rannow et al., 2014). Furthermore, the costs for imagery and other geospatial data products are still fairly high, though there has been an overall declining trend due to market competition. The trend is supported by the recent release of free VHRI (e.g., USGS released Orbview-3 data in January 2012) or the Sentinel HR family in general (Berger et al., 2012). The availability of free geospatial data (e.g., the Open Street Map initiative and open-source image processing and GIS software and tools (e.g., Orfeo Toolbox, QGIS, GRASS, R, etc.) is also contributing to the democratization of remote sensing and to the decline in the costs of the image processing packages. Still, a challenge for ecologists and conservation

Multi-Scale Habitat Mapping and Monitoring

biologists is the technical expertise required to handle imagery and other data products (Turner et al., 2003) and thus not recognizing the full potential provided by such data sets and related techniques.

6.2.1 SPATIAL RESOLUTION—WHAT DETAIL CAN BE MAPPED?

Satellite-based EO systems are categorized according to their spatial resolution, while the grouping schema is relative to the highest technical resolution and thereby subject to change. Over the last years, the term "very high resolution" (VHR) has been used for images with a resolution (i.e., pixel size) around or smaller than 1 m. We speak of high-resolution (HR) images at resolution levels of up to 5 m (see Figure 6.4 and Table 6.2). The highest resolution operationally available is around 0.5 m, increasing to 0.3 m at the latest generation of VHR satellites, such as WorldView-3.

Habitats as areal features with a certain extent, cover a wide range of scales, from a few square meters, like specific springs or lichen patches, to several square kilometers in the case of the savanna. While the average size of habitats clearly depends on the hierarchical level of habitat categorization (broad habitat categories vs. more specific ones, see later in this chapter), there are even great differences within one scale domain. The extent of habitats considered internally homogenous (i.e., residing on the same hierarchical level of organization from an ecological point of view), varies depending on the physical conditions and the species living there. Therefore, it is important to discriminate individual plants, such as single trees, or distinct features, such as shrubs, in a specific habitat composition. Table 6.2 provides an overview of the suitability of different sensor groups for distinguishing within broad, physiognomic habitat types.

With the highest resolution available, pan-sharpening routines are frequently used to optimize both spatial and spectral resolution. Still, the fidelity of the pixel information may be too high as well. Aiming, for example, at the extraction of sunlit tree crowns, a resolution of 1–2 m may just be the right resolution to tackle this (Strasser and Lang, 2014b). In other words, sensor resolution needs to be commensurate to the observed features, from both directions so to say, not too coarse and not too fine. LiDAR systems, are capable of representing in great detail the physical attributes of vegetation canopy structure that are highly correlated with the basic plant community measurements of interest to ecologists (Mücher et al., 2014).

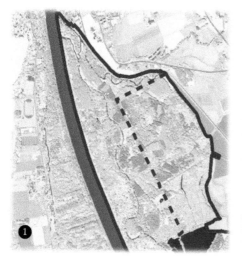 

FIGURE 6.4 How WorldView-2 data (left, red outlines indicate boundaries of designated protected areas) translate into habitat classes in a semi-automated image classification process (here EUNIS-3 categories).

TABLE 6.2

Sensor's Suitability for the Characterization of Natural Habitats Based on Corbane et al. (2014) and Ichter et al. (2014), modified and completed. Various Remote Sensing Techniques (Sensors and Resolution) Are Compared for Distinguishing between Broad Physiognomic Habitat Types. The Degree of Sensor Suitability Is Indicated as Follows: – = Unsuitable, –/+ = Partly Suitable, + = Suitable, ++ = Recommended. See Detailed References in Corbane et al. (2014)

Low Spatial Resolution and High Temporal Resolution[9]	Medium to High Spatial/ Temporal Resolution[10]	Very High Spatial Resolution[11]	Hyperspectral[12]	Laser Scanning (LiDAR)[13]	Active Microwave Sensors (e.g., SAR)
Forests					
–/+ Deciduous/coniferous/mixed forest, evergreen/deciduous, dense/fragmented	+ Broad types, dominant species using multitemporal imagery	++ Tree species classification, differentiation of structure and age classes, multitemporal	++	–/+ Assessment of forest parameters (stand density, height, crown width, crown length) species distributions	–/+ Often complementary to the information provided by multi-spectral imaging
Grasslands					
–	++ (multi-seasonal imagery): distinction between marshy grasslands (*Molinia*- or *Juncus*-dominated), unimproved (*Festuca*-dominated), semi-improved and improved	+ (multi-seasonal imagery): grassland types with different levels of agricultural improvement, levels of mowing intensity	++ Detection of floristic gradients, determination of homogenous cover types	–/+	+ Distinction between natural grasslands and improved pastures, mowing intensity via swath detection

	++	++	−/+	
		Heathlands		
−	(multi-seasonal imagery): Distinction between heath types (e.g., Genista, Erica), four heath types, including ancillary data	Seasonal phenological variation can discriminate evergreen Calluna vulgaris from deciduous Vaccinium myrtillus	Distinction between dry and wet heathland, heathland types (Calluna, Molinia Deschampsia, Erica, etc.) and heather age classes	−
	+	++	−/+	
		Wetlands[14]		
−	Seasonal imagery: mapping extent of seasonally submerged wetlands and some vegetation species, freshwater swamp vegetation, functional wetland types	Detection of riparian vegetation species, shallow, submerged vegetation	Distinction between aquatic macrophyte species (Typha, Phragmites, Scirpus)	Surface and terrain models used to better understand characteristics of wetland vegetation species

6.2.2 Spectral Resolution—Plant and Plant Feature Discrimination

6.2.2.1 Sun-Source Systems—Optical Sensors

Plant species respond characteristically to light emitted by the sun or an artificial energy source, with specific reflection behavior in the electromagnetic spectrum (EMS). Ideally, remotely sensed data of adequate spectral and spatial resolution can be used to distinguish different species, but to identify the appropriate sensor and the appropriate spectral bands can be challenging. Sensors can be grouped into passive and active sensors, depending on the source of energy involved. Passive sensors record the reflectance of sunlight on surfaces, while the spectral resolution corresponds to the number of bands that a sensor is able to acquire from a distinct part of the EMS. Panchromatic sensors, as the name indicates, cover a broad range of the EMS, usually including the visible and the (short) infrared ranges. Multispectral sensors are sensitive to certain, well-defined portions of the EMS, which are categorized according to their relative and absolute position, such as visible light (VIS, 400–700 nm), near infrared (NIR, 750–900 nm), mid-infrared (MIR, 1.55–1.75 μm), etc.

Multi-spectral sensors have been used for several decades, collecting data from fairly broad wavebands (typically four to, more recently, eight). Multi-spectral data recorded by spaceborne sensors (e.g., Landsat 9, ASTER, SPOT-5, WorldView-2 & 3, RapidEye Planet) are useful for land cover assessments of from a local to regional scale (depending on the spatial resolution). Such data are being used to assess habitat conditions for monitoring purposes (Price et al., 2023; Zhang et al., 2023; Roelfsema et al., 2018; Setyawidati et al., 2018; Onojeghuo and Onojeghuo, 2017; Förster et al., 2008; Franke et al., 2012; Spanhove et al., 2012). Advanced studies investigate spectral characteristics of specific sensors for such tasks and address the suitability of multi-spectral data for an assessment of detailed floristic variation, using multi-spectral sensors (Nagendra and Rocchini, 2008) to assess the spectral information of multi-spectral data in terms of variable floristic composition within a certain natural habitat (mainly grassland and wet heath and floodplain meadows). Tree species differentiation has been accomplished using 8-band WorldView-2 data for forest management in general (Immitzer et al., 2012) and riparian forest assessment in particular (Strasser and Lang, 2014b). Table 6.3 shows a panel of WorldView-2 scenes as used for fine-scaled habitat delineation in MS.MONINA. WorldView-3 sensor acquires data with 30 cm maximum spatial resolution.

Imaging spectroscopy (i.e., hyperspectral) data usually contain several hundreds of narrow spectral bands in the range of 400–2500 nm. Some sensors cover only parts of this range (e.g., CHRIS/PROBA focuses on VIS/NIR). Due to the large number of wavebands, digital image processing can discriminate based on biochemical and structural properties of vegetation. Underwood et al., 2003 demonstrate the potential of hyperspectral data to extract information regarding plant properties (such as leaf pigment, water content, and chemical composition), thus discriminating tree species in landscapes and identifying different species. Applications of Hyperion hyperspectral imagery include forest biodiversity (Peng et al., 2003), grasslands (Psomas et al., 2011), and invasive species monitoring (Walsh et al., 2008).

6.2.3 Active Systems—Radar and Lidar

Active sensors send an electromagnetic signal and record the travel time of the signal and its reflection by a given surface. Active sensors are differentiated by wavelength into microwave (radar) and laser scanning (LiDAR) systems. Active systems are increasingly used for vegetation mapping with a number of radar-based satellites (e.g., TerraSAR-X, Sentinel-1). Gillespie (2005) discusses opportunities for landscape monitoring at finer spatial resolution. The returned signal supplies information about the height and structure of vegetation, especially of woody vegetation, with these relating to forest conditions and disturbance regimes (Huang et al., 2013). In particular, X-band radar backscattering has been recommended to differentiate plant species on the basis of canopy architecture (Bouman and van Kasteren, 1990). Schuster et al.

TABLE 6.3
Details (Location, Country, Date, Band Combination) and Subsets in Equal Size (2.5 × 2.2 Kilometers) and Scale (Original: 1:10,000) of WorldView-2 Satellite Scenes Used for Test and Service Cases in MS.MONINA

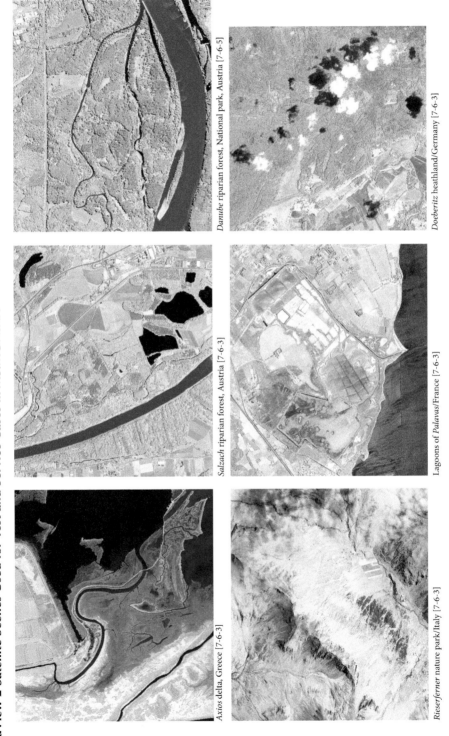

Danube riparian forest, National park, Austria [7-6-5]

Doeberitz heathland/Germany [7-6-3]

Salzach riparian forest, Austria [7-6-3]

Lagoons of *Palavas*/France [7-6-3]

Axios delta, Greece [7-6-3]

Rieserferner nature park/Italy [7-6-3]

(*Continued*)

TABLE 6.3 (Continued)
Details (Location, Country, Date, Band Combination) and Subsets in Equal Size (2.5 × 2.2 Kilometers) and Scale (Original: 1:10,000) of WorldView-2 Satellite Scenes Used for Test and Service Cases in MS.MONINA

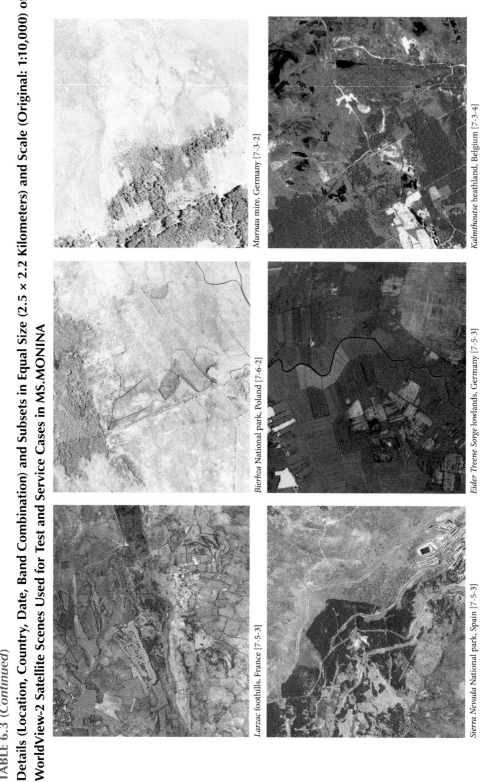

(2011) proved that distinguishing different grassland swath types for NATURA 2000 monitoring is possible with the TerraSAR-X sensor. The SAR sensor of the Sentinel family, Sentinel-1 is increasingly used for volumetric estimation. ALOS PALSAR and RADARSAT-2 SAR have shown great potential for mapping wildlife habitats, particularly when combined with optical remote sensing through data fusion (Wang et al., 2009). Moreover, there are attempts to monitor different shrubland, grassland, and forest habitats with COSMO SkyMed (Ali et al., 2013). While still under research the ability to penetrate the canopies makes microwave instruments a potential tool for measuring biomass and determining vegetation structure (Price et al., 2023; Wicaksono et al., 2019; Roelfsema et al., 2018; Setyawidati et al., 2018; Onojeghuo and Onojeghuo, 2017).

Light detection and ranging (LiDAR) provides highly accurate information on the three-dimensional vegetation structure, derived from pulse characteristics, over a limited area (Puech et al., 2012). For example, data from the airborne Laser Vegetation Imaging Sensor (LVIS) enable the mapping of sub-canopy topography and canopy heights to within 1 m. More generally, LiDAR is used for the extraction of information on forest structure (vertical information), e.g., canopy and tree height, biomass, and volume. According to Turner et al. (2003), the recording of numerous LiDAR return signals (pulses) enables the estimation of vegetation density at different heights throughout the canopy and enables three-dimensional profiles of vegetation structure. Besides airborne LiDAR with the limitations of large data volumes, footprint size, and high costs, spaceborne laser technology has been launched by ICESat/GLAS (Ice, Cloud, and Land Elevation Satellite/Geoscience Laser Altimeter System), the first laser-ranging instrument for continuous global observations. While LiDAR provides structural attributes of vegetation, little can be derived on the actual species composition, suggesting that laser scanning is more a complementary technology (Mücher et al., 2014). In general, optical and LiDAR data acquired simultaneously increase the differentiation of vegetation species. It has also been demonstrated that species distribution models can be improved through airborne LiDAR quantifying vegetation structure within a landscape (Goetz et al., 2007).

6.2.4 REVISITING TIME—PHENOLOGY

A major asset, and a quality-related one in using satellite remote sensing for habitat monitoring is that data can be acquired on a regular and repetitive basis, therefore allowing consistent comparisons between image scenes (Larsen et al., 2023; Wan and Ma, 2020; Wicaksono et al., 2019; Roelfsema et al., 2018). The frequency of observation by optical spaceborne sensors ranges from several times daily (for coarse spatial resolution sensors, e.g., NOAA AVHRR, MODIS, and MERIS) to every 16–18 days (e.g., Landsat) although cloud cover, haze, and smoke often limit the number of usable scenes. The Copernicus satellites with a revisit time of five days have been operational since 2015. With a five-day revisit capacity, the Sentinel-2 satellites will acquire intra-annual data, thereby allowing the temporal variations in reflectance to be exploited for mapping and monitoring natural habitats. The changes in the seasonality (i.e., phenology) of plants are significant for the differentiation into classes. So far, this information has been widely neglected due to limited access of spatial high-resolution imagery with a high temporal domain (Förster et al., 2012). To include an increasing number of intra-annual images increases the classification accuracy until a certain threshold is reached (Schuster et al., 2015). Images taken over two phenological stages help in discriminating species and habitat classification (Lucas et al., 2011), although some studies conclude that monotemporal data suffice when the SWIR is included (Feilhauer et al., 2013). Moreover, it is possible to analyze the optimal season for acquiring a data set (Schmidt et al., 2014). Multi-annual coverage also helps in assessing changes in habitat (such as loss, degradation, and fragmentation) through change detection approaches. Monitoring experts appreciate this because it directs fieldwork on these areas, possibly yielding a significant increase in cost-efficiency (Vanden Borre et al., 2011).

6.2.5 ADVANCED IMAGE ANALYSIS TECHNIQUES

The advancement of imaging technology is a crucial, but not the only prerequisite to cope with the challenges of habitat mapping. As important as high-quality, high-suitable imagery is the utilization of appropriate image analysis techniques (Alkhuzaei and Brolly, 2024; Strasser and Lang, 2015; Larsen et al., 2023; Roelfsema et al., 2018; Setyawidati et al., 2018; Lang et al., 2014). Figure 6.6 illustrates the importance of analysis steps that follow the actual provision of imagery. Various information types can be extracted from imagery, ranging from biophysical parameters, including vegetation indices (Adamczyk and Osberger, 2014), and structural parameters (Mairota et al., 2014) to ultimately nominal land cover, or more specifically, habitat categories (Adamo et al., 2014) (see Figure 6.5). Beyond that, existing or image-derived habitat delineations can be combined with the parameters in order to assess habitat quality, e.g., Riedler et al. (2014) (Figure 6.5). Kuenzer et al. (2014) provide a detailed overview of existing sensor types and related vegetation indices being useful for biodiversity mapping.

Reproducibility, objectivity, transferability, and quantification have been reported as the main advantages of mapping approaches based on EO data. Semi-automated classification methodologies for EO data provide a more objective, i.e., reproducible and transparent, outcome compared to visual interpretation (Lang and Langanke, 2006). Over the last years, great advantages have been reported in the use of remote sensing technology for the mapping and assessment of habitats in Europe: for an overview see Vanden Borre et al. (2011). This likewise applies to different broad habitat types (forests, grasslands, wetlands, etc.) and different scales of observations as fine as the sub-habitat level (Lucas et al., 2011; Strasser and Lang, 2015). Object models can be stored and explicitly called for semi-automated mapping routines (Lucas et al., 2014), as can advanced habitat or biotope class models (Tiede et al., 2010).

One strategy to tackle this variability is the use of object-based image analysis (OBIA) (Price et al., 2023; Roelfsema et al., 2018; Onojeghuo and Onojeghuo, 2017; Lang, 2008). This is typically done through the combination of spectral behavior and spatial variability, either from the details available through high spatial resolution imagery (Johansen et al., 2007; Strasser and Lang,

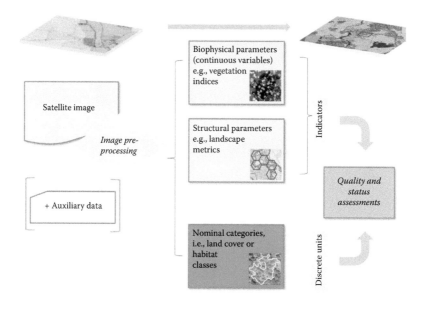

FIGURE 6.5 From imagery via habitat categories to habitat quality assessment.

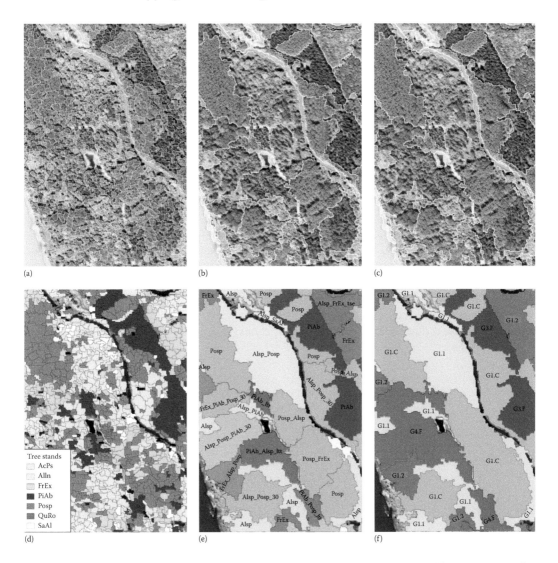

FIGURE 6.6 Hierarchical representation of riparian forest habitats, achieved by multi-level segmentation and class modeling, from Strasser and Lang (2015).

2015) or through the inclusion of data from active sensors such as LiDAR (Mücher et al., 2014). Object-based class modeling allows for mapping complex, hierarchical habitat systems, such as forest habitats. Forest composition, including intermixture of non-native tree species, was modeled in a six-level hierarchical representation in a riparian semi-natural forest by Strasser and Lang (2014b). VHRI from WorldView-2 provided the required spatial and spectral details for a multi-scale image segmentation and rule-based composition. An image object hierarchy was established to delineate forest stands, stands of homogenous tree species, and single trees represented by sunlit tree crowns.

6.3 EO-BASED BIODIVERSITY AND HABITAT MAPPING

According to Turner et al. (2003), there are two general approaches to remotely "sense" biodiversity: (1) direct mapping of individual organisms, species assemblages, or ecological communities using airborne or satellite sensors, and (2) indirect sensing of biodiversity-related aspects using environmental parameters as proxies. Many species are confined in their distribution to specific habitats such as woodland, grassland, or sea grass beds that can be directly identified with remote sensing data. Habitats (see Figure 6.3) as the spatial expression of ecosystems do have a certain extent to be mapped and observed, they function as living space for specific species (both animals and plants) and bear a certain constancy in the 4D space-time physical world.

6.3.1 Land Cover, Habitats, and Indicators

First, we would like to distinguish between land cover and habitats, two concepts that frequently cause confusion in the remote sensing literature (Alkhuzaei and Brolly, 2024; Amani et al., 2023; An et al., 2023; Wan and Ma, 2020; Lang et al., 2013a): According to EUROSTAT,[15] land cover "corresponds to a physical description of space, the observed (bio) physical cover of the earth's surface. It is that which overlays or currently covers the ground." Land cover classes representing biophysical categories, such as grassland, woodland, water bodies, etc., are usually derived from multi-spectral remote sensing data by multivariate clustering methods in the feature space. Land cover data are used at different scales (local, regional, and global) as input variables for biosphere-atmosphere models and terrestrial ecosystem models and respective change assessments as well as proxies of biodiversity distribution (Grillo and Venora, 2011). The FAO-LCCS (Di Gregorio and Jansen, 2005), as a universally applicable classification system, enables a comparison of land cover classes regardless of data source, economic sector, or country.

A habitat is a

> three-dimensional spatial entity that comprises at least one interface between air, water and ground spaces. It includes both the physical environment and the communities of plants and animals that occupy it. It is a fractal entity in that its definition depends on the scale at which it is considered.
>
> *(Blondel, 1979)*

Habitats are often distinguished into two (or more) stages of naturalness (hemeroby) according to the level of human modification: Natural habitats are considered the land and water areas where the ecosystem's biological communities are formed largely by native plant and animal species and human activity has not essentially modified the area's primary ecological functions (European Environmental Agency, EEA). Semi-natural habitats are considered to be managed or altered by humans but still "natural" in terms of species diversity and species interrelation complexity (i.e., diversity).

In general, a perfect correspondence of conventional biotope types and spectrally derived vegetation cover is hardly to be achieved (yet from today's point of view not impossible), due to the practice of manually delineating biotope types from aerial photos and field surveys (Weiers et al., 2004). The EUNIS (European Nature Information System[16]) habitat classification scheme is a hierarchical scheme with six levels (Davies et al., 2004). Alternative classification systems such as the General Habitat Categories (GHC) (Bunce et al., 2008) or the Terrestrial Ecosystem Mapping system (Johansen et al., 2007), using vegetation attributes such as height and leaf phenology, have been proposed in order to more successfully employ EO data in the classification of habitats. A comprehensive mapping system starting from LC classes expressed in FAO-LCCS taxonomy, translated to GHC (Tomaselli et al., 2013), and finally addressing habitat classes according to HabDir has been proposed by Lucas et al. (2014). Such systems require robust and reliable remote sensing–based methods from the start. In this context, Corbane et al. (2014) reviewed the ability of remote sensing to physiognomically distinguish between habitat types at different scales. They report about advances in the use of remote sensing technology for the mapping and the assessment of habitats in Europe (Vanden Borre et al., 2011). This applies to different broad habitat types (forests, grasslands, wetlands, etc.) and scales of observations as fine as sub-habitat level. Mapping broad habitat types using remote sensing is a common

Multi-Scale Habitat Mapping and Monitoring

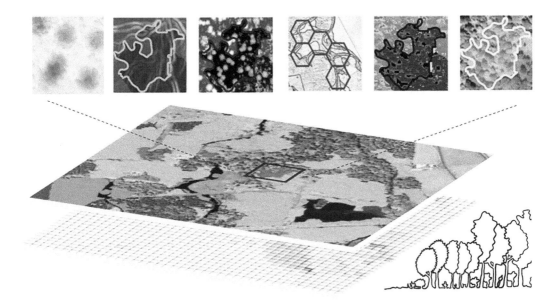

FIGURE 6.7 Habitats are the physical and mappable expression of ecosystems. Satellite EO enables the mapping and monitoring of a variety of habitat distribution, quality, and change aspects in different spatial and temporal scales, from Lang et al. (2014).

practice from the perspective of land cover mapping and is generally done at a relatively coarse scale of analysis (Wulder et al., 2004).

6.3.2 Distinguishing between and within Broad Habitat Categories

Mapping broad habitat types using remote sensing converges with land cover mapping done at a relatively coarse scale of analysis (Iglseder et al., 2023; Wan and Ma, 2020; Wulder et al., 2004). Global land cover mapping has been accomplished using the MODIS satellite, at 500 m resolution (Friedl et al., 2010), while country- and regional-level land cover classifications have been accomplished using medium-resolution sensors such as Landsat or SPOT (Fuller et al., 1994; Tiede et al., 2010). More detailed land cover boundaries can be obtained using a higher spatial resolution, multi-temporal coverage (Förster et al., 2010), or by including ancillary data (Tiede et al., 2010) or active sensors. In the following we will discuss how vegetation categories can be distinguished within several broad physiognomic types: forest, grassland, heathland, and wetland (Corbane et al., 2014). Generally, satellite-based habitat mapping can be supported, in terms of plausibility and reliability, through advanced GIS modeling techniques to derive probabilities for the presence of habitats in different biogeographical regions (Förster et al., 2005) and potential habitat ranges under specific assumptions or even changing conditions. The advantages of a high spectral, temporal, or spatial resolution as well as active sensors are summarized with references in Table 6.4.

6.3.2.1 Forest Habitats

Low spatial resolution data allow rough differentiation of the main forest cover types (deciduous, coniferous, mixed) (Alkhuzaei and Brolly, 2024; Iglseder et al., 2023; Larsen et al., 2023; Wicaksono et al., 2019; Corbane et al., 2014). The number of differentiated forest classes can be improved when ancillary data, such as terrain (Woodcock et al., 1994), additional geodata (Förster and Kleinschmit, 2014), or time series are used. More detailed analyses can be performed using

TABLE 6.4
Influence of Increasing Resolution and the Utilization of Active Sensors on the Possible Detection and Differentiation within Different Habitat Categories

Habitat Type	Influence of Increasing Resolution			Use of Active Sensors
	Spatial	Spectral	Temporal	
Temperate Forest	Allows for object-based classification of tree crowns (Immitzer et al., 2012)	Reduces spectral overlap of tree species with similar spectral characteristics (Dalponte et al., 2012)	Enables phenological information (leaf unfolding, coloring, leaf fall) (Wolter et al., 1995)	Improves accuracy via incorporation of canopy height, canopy architecture, and forest structure (Ghosh et al., 2014)
Tropical Forest	Reduces the number of mixed pixels between tree species (Nagendra and Rocchini, 2008)	Allows for species identification using unique biochemical signatures (Asner, Gregory et al., 2008b)	Allows a spectral endmember analysis and a detection of tree types (Somers and Asner, 2014)	Incorporates forest structural properties to mask non-tree gaps (Asner, Gregory et al., 2008a)
Riparian Forest	Allows for object-based classification of tree types (Strasser and Lang, 2014a; Suchenwirth et al., 2012)	Allows the estimation of different levels of biomass and subsequent distinction of different types of riparian forest (Filippi et al., 2014)	Possible detection of changes in crown extend and derived health status (Gaertner et al., 2014)	Improves accuracy by means of the structural properties as additional information (Akay et al., 2012)
Grasslands	Allows for object-based classification of homogenous grassland patches (Corbane et al., 2013)	Allows for the detection of floristic gradients (Schmidtlein and Sassin, 2004)	Increased accuracy if images are timed to differentiate warm/cool season grasses (Price et al., 2002; Price et al., 2002)	Incorporates information on grassland management practices, such as mowing intensity (Schuster, Christian et al., 2011; Schuster et al., 2011)
Heathland	Allows for the use of object shape complexity in differentiating successional stages (Mac Arthur and Malthus, 2008; Mac Arthur and Malthus, 2008)	Allows for the discrimination of heather age classes (Thoonen et al., 2013; Thoonen et al., 2013)	Multi-seasonal imagery allows for distinguishing evergreen and deciduous heath species (Mac Arthur and Malthus, 2008)	Can be used to separate shrubs from trees and grassland (Hellesen and Matikainen, 2013)
Coastal Wetlands	Reduced spectral mixing in heterogeneous species patches (Belluco et al., 2006)	Accuracy is marginally increased, feature selection is necessary (Belluco et al., 2006); spectral libraries may improve classification in properly calibrated images (Schmidt and Skidmore, 2003)	Allows for the incorporation of seasonal differences in vegetation communities (Gilmore et al., 2008)	Can be used to separate vegetation height of plant communities (Prisloe et al., 2006)
Inland Wetlands	Allows for the distinguishing of small vegetation patches (Everitt et al., 2004)	Hyperspectral imagery has consistently higher accuracy than multi-spectral (Jollineau and Howarth, 2008)	Wetland types may be differentiated by seasonal water regimes and differences in growing season (Davranche et al., 2010)	Can be used to separate vegetation structure to the genus level (Zlinszky et al., 2012)

high spatial resolution sensors, including image texture analysis, which is indicative for tree species and age classes at the canopy level (Immitzer et al., 2012; Johansen et al., 2007). The use of time series with high spatial resolution data helps distinguish between individual trees through the use of phenological characteristics, such as leaf development and senescence (Key et al., 2001).

The use of hyperspectral imagery allows for an even greater level of detail (Corbane et al., 2014), enabling the distinction of tree types based on reflectance in response to pigment, nutrient, and structural differences between species (Asner et al., 2008). Still, within-class variability of VHRI (through features such as branches, shadow, and undergrowth) and the between-class spectral mixing of low-resolution imagery (Nagendra and Rocchini, 2008) needs to be taken into account. Spectral unmixing is possible for lower-resolution sensors but is limited by the respective sensitivity of the sensor. Forest habitat categories that differ by their understory vegetation, such as *Stellario-Carpinetum* and *Galio-Carpinetum*, which are both oak-hornbeam forest types, can be approached by using additional geodata, such as soil data, etc.

6.3.2.2 Grassland Habitats

In contrast to forests, grassland species are detectable primarily as assemblages and may occur in complex mixtures within habitats (Larsen et al., 2023; Price et al., 2023; Wicaksono et al., 2019; Roelfsema et al., 2018; Setyawidati et al., 2018; Onojeghuo and Onojeghuo, 2017; Corbane et al., 2014). Thus, direct remote sensing approaches are generally limited to the detection of relatively homogenous grassland habitat types, while indirect approaches were found to be successful, such as those that use environmental gradients (Fuller et al., 1994) or usage intensity for mowed semi-natural grasslands (Buck et al., 2014; Schuster et al., 2011) (see Figure 6.8). Moderate spatial resolution sensors such as Landsat Thematic Mapper (TM) and Landsat-7 Enhanced Thematic Mapper (ETM+) were used by Lucas et al. (2007) for the classification of grasslands, allowing only for a broad level of class distinction with respect to grassland improvement levels. Higher spatial resolution

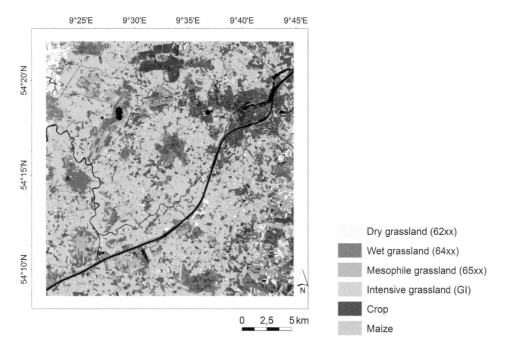

FIGURE 6.8 Grassland classification using SVM classifier of information layers and RapidEye imagery, from Buck et al. (2014), modified.

imagery, using an object-based approach with ancillary data, such as elevation and soil type, has proven successful in differentiating between a few dominant grassland species (Laliberte et al., 2007), although other grassland habitats are more difficult to distinguish (Corbane et al., 2013). For relatively homogeneous grasslands, the use of a spectral-temporal library (instead of training areas) with multitemporal HR data has been shown to accurately differentiate grassland types; still, object-based classification methods have been shown to perform better for more heterogeneous grassland types (Förster et al., 2012).

Hyperspectral imagery has been applied to determine floristic gradients, proven to be more useful in habitat identification than single-species classifications (Price et al., 2023; Zhang et al., 2023; Setyawidati et al., 2018; Schmidtlein and Sassin, 2004). Given this, the necessity of hyperspectral imagery has been debated, as recent research has shown that the spectrum from visible to short-wave infrared is the most significant for detecting wet and dry grassland floristic gradients (Feilhauer et al., 2013).

6.3.2.3 Heathland Habitats

Generally, heathland habitats are characterized by a mixture of ericaceous dwarf shrub species (e.g., *Calluna vulgaris*), grassland species, and open soil. Since all three components can be reliably spectrally distinguished, these habitats are rather straightforward to detect and to monitor (Corbane et al., 2014). Wet and dry heathland types are more difficult to distinguish with limited spectral separability on moderate-resolution imagery (Diaz Varela et al., 2008). OBIA has successfully been applied to high spatial resolution imagery in identifying dominant heather areas based on indicator species (Förster et al., 2008) or structural parameters (Langanke et al., 2007). Similar, more detailed approaches have been done using hyperspectral imagery, where kernel-based reclassification was used to transform the resultant land cover classes into heathland habitats (Thoonen et al., 2013). Few specific studies used active sensors in detecting heathland habitats, e.g., using LiDAR (Hellesen and Matikainen, 2013) data for vertical differentiation of shrub and grassland forms and using high-resolution, multi-channel SAR data (Bargiel, 2013).

6.3.2.4 Wetland Habitats

Wetland vegetation is characterized by high spatial and spectral variability and is influenced by soil moisture, atmospheric moisture, and the respective hydrological properties of the wetland type (Larsen et al., 2023; Wicaksono et al., 2019; Onojeghuo and Onojeghuo, 2017; Corbane et al., 2014). This makes traditional vegetation mapping approaches based on the mid-to-near infrared range difficult, due to the relatively higher absorption of this wavelength by water (Adam et al., 2009). Nevertheless, medium resolution imagery such as Landsat has been used to classify broad wetland habitat types (Mac Alister and Mahaxy, 2009). Additionally, the use of ancillary data such as soil type, combined with multispectral imagery, can be used to help differentiate spectrally similar classes (Bock, 2003). In terms of mapping dominant species, high spatial resolution imagery was successful (Everitt et al., 2004), including submerged vegetation types (Dogan et al., 2009). WorldView-2 data were used by Keramitsoglou et al. (2014) to perform kernel-based classification of river delta vegetation and habitats in Greece. These habitats form a rich yet fine-scaled mosaic of brackish lagoons, saline soils, extensive mudflat, saltwater and freshwater, sand dunes, but also rich vegetation. Mapping remnants of these delta habitats, which are exposed to anthropogenic, mainly agricultural pressure, has proven to be successful to be transferred between similar delta situations.

LiDAR has been used alone to perform genus-level wetland, as well as in combination with VHRI for object-based classification. Using hyperspectral imagery, Schmidt and Skidmore (2003) indicated that it is possible to distinguish between 27 types of salt marsh vegetation using spectral signatures. It remains challenging to distinguish between submerged vegetation types, due to factors such as water turbidity, depth, and bottom reflectance (Jollineau and Howarth, 2008).

6.4 OBSERVING QUALITY, PRESSURES, AND CHANGES

As the Millennium Ecosystem Assessment stated in 2005,[17] the Earth's ecosystems have been altered rapidly by human pressure in a short time frame, about half a century. In addition to directly reducing existing habitat through land-use conversions, the human impact highly affects the quality of the remaining habitats (Mairota et al., 2014). Indicators derived from remotely sensed data—for an overview see, e.g., Strand et al. (2007)—can help assess and monitor habitat states and conditions. This section describes means and methods of how habitat quality can be assessed, pressures on habitat can be characterized, and changes can be quantified.

6.4.1 MEASURING HABITAT QUALITY

Even more challenging than detecting species and delineating habitat, is to obtain robust information about the quality and conservation status of habitat types using remote sensing data. Biodiversity surrogates (BS) include parameters such as species presence, abundance, probability of site occupancy, and aggregate measures such as species richness, diversity, or carrying capacity, and they can be used to observe the degradation of habitat quality characteristics related to resource availability (e.g., nutrients, refugia), phytomass, vegetation structure, and microclimate (Mairota et al., 2014). Habitat "quality," according to Lindenmayer et al. (2002), is inherently taxon-specific and scale-dependent with respect to extent and grain, *sensu* Kotliar and Wiens (1990). Since it can be prohibitively expensive to obtain fine-grained habitat-quality data at large spatial extents through field surveys alone, remote sensing is used to estimate environmental heterogeneity at differing grains across differing spatial extents (Mairota et al., 2014) and relate it to the variation in species diversity and distribution (Nagendra et al., 2014). VHR EO data are of particular power (Nagendra et al., 2013), as they enable multiple scale levels to be extracted from one single image (Strasser and Lang, 2015). BS exhibit different relevance to taxa and functional groups with varying spectral and textural diversity measurements at different spatial scales and can be predicted with reasonable accuracy using habitat modeling (based on remotely sensed measures of environmental attributes [Mairota et al., 2014]).

Approaches to infer habitat quality from remote sensing data are abundant (Costanza et al., 2011; Rocchini et al., 2010; Townsend et al., 2009). In addition, spatial analysis techniques can be applied in order to quantitatively assess and compare structural indicators related to the actual state (Riedler et al., 2014; Vaz et al., 2015) and conditions of habitats. Related to this is the influence of the complexity of landscape structure (Corbane et al., 2014). Overall, mapping becomes more challenging when landscapes are heterogeneous and fine-grained and the variation between habitats is more continuous (Diaz Varela et al., 2008). Also, the complexity of landscape structure differs between protected areas and their surroundings, and thus different approaches to mapping need to be considered. As landscapes become more heterogeneous and the numbers of potential habitat types increase, modeling approaches of the relationship between species distribution patterns and remotely sensed data gain importance (Schmidtlein and Sassin, 2004).

National quality parameters for monitoring the conservation status according to HabDir are defined by different European states, such as Austria (Ellmauer, 2005) and Germany (Balzer et al., 2008), as well as other European Member States, including Belgium, France, and Denmark. In most cases, the conservation status is assessed by habitat structures (e.g., horizontal and vertical variation, age structure), presence of typical species (mostly flora) in the habitat, abiotic factors (e.g., flooding), and pressures or disturbances of the habitat type (e.g., eutrophication indicators, invasive species). Indicators are usually framed for broad habitat types, as discussed earlier. Within forest habitats, many indicators, such as "percentage of characteristic tree species," can be derived by means of remote sensing (Corbane et al., 2014). Others (e.g., "habitat trees," very old and degraded living microhabitat-bearing trees with hollows for nesting) remain tricky and require very detailed vertical representations by means of LiDAR data. This also applies to other broad habitat types:

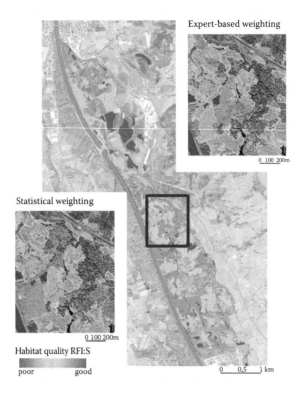

FIGURE 6.9 Composite indicator for riparian forest habitat quality assessment, from Riedler et al. (2014).

Heathlands have been studied extensively in this respect (Delalieux et al., 2012; Spanhove et al., 2012), including degradation stages in bog areas (Langanke et al., 2007). For grasslands, recent advances have been made in monitoring grassland use intensity (Schuster et al., 2011) and shrub encroachment (Lang and Langanke, 2006). In addition, "negative" quality criteria can be used, as can the share of invasive species. Many such neophyte species can be detected on remotely sensed imagery based on spectral, phonological, or structural characteristics. This can be supported by modeling potential distribution and susceptibility of specific areas to invasion.

Few approaches exist (as yet), where several indicators are integrated in a quantitative and spatially explicit way in order to receive an aggregated view on habitat quality. Riedler et al. (2014) propose such a strategy for habitat quality assessment in riparian forests. They use a composite indicator (RFI_S) for integrated assessment of habitat quality on the patch level and the identification of "hot-spots" where management action may be focused. RFI_S is composed of seven indicators (derived from VHRI and LiDAR data) addressing four important attributes of riparian forest quality: (1) tree species composition, (2) vertical forest structure, (3) horizontal forest structure, and (4) water regime. For the aggregation of the RFI_S, expert-based and statistical weighting were applied. Measures of improvement or conservation can be specifically designed through the decomposition of the overall indicator into its underlying components (see Figure 6.9). An advancement of the methodology toward habitat quality on patch level by analyzing the significance of contributing indicators is demonstrated by Riedler and Lang (2018).

6.4.2 Identifying Pressures and Changes

Both loss of habitat and decline of habitat quality can be linked to anthropogenic pressures that affect the provisioning of essential ecosystem services (Nagendra et al., 2014) for human well-being.

International bodies such as the Intergovernmental Panel on Biodiversity and Ecosystem Services (IPBES[18]) have stressed the need to assess human pressures on biodiversity and ecosystem services across all scales. Even protected areas, through direct or indirect anthropogenic impact, continue to experience anthropogenic pressure, thus for effective management response, spatial knowledge of the type and location of pressure is required (Nagendra et al., 2013).

EO data and associated techniques, coupled with landscape pattern analysis and habitat modeling as well as BS estimates, can provide critical information on changes in state and condition of habitats, which in turn can be used to infer evidence of pressures. Nagendra et al. (2014) propose a unified approach to facilitate the provision of value-added products from EO sources for biodiversity conservation purposes. The proposed approach builds on the DPSIR (driving forces, pressure, state, impact and response) framework (EEA, 1995) and is based on the definition of four broad categories of changes in state, which can be mapped and monitored through EO analyses.

Within the BIO_SOS EODHaM system (Lucas et al., 2014), six types of change assessments with respect to LC/habitat classes or geometry are distinguished, namely, changes in (1) LCCS classes (or GHCs), (2) specific LCCS component codes, (3) the number of extracted objects belonging to the same category, (4) object size and geometry (splitting or merging), (5) EO-derived measurements (e.g., LiDAR-derived height or vegetation indices), and (6) calculated landscape indicators useful for subsequent biodiversity indicator quantification, e.g., species distribution in Ficetola et al. (2014).

In the following[19] we highlight some specific (Jaeger, 2000) types of pressure and/or change indicators with respect to broad habitat categories, which often have a dual function in terms of their actual species-hosting function and any form of natural resource for anthropogenic use. Among the most generic indicators are habitat extent and habitat fragmentation. Habitat can change in area due to a multitude of driving forces that can each pose different pressures and threats (Nagendra et al., 2014). Changing habitat areas may pinpoint to expanding or shrinking areas of competing land cover types like agricultural, industrial, and urban land, or might reveal the impact of pollution, climate change, or catastrophes. Forest habitats, for example, degrade as a result of forest plantation that affects their species composition, or artificial fires, a loss of carbon stock and biomass by (over)exploitation, and a decrease of old-growth forest. Species composition of grasslands and their layered structures are influenced by human management regime (cutting, grazing, burning) and intensity (fertilization, management frequency) influencing stages of growth and regrowth. Pressures on grasslands most often relate to agricultural practices with potential impact on biodiversity, especially through the use of fertilization, irrigation, and pesticides, with associated pressures and threats related to water, air, and soil pollution, as well as drainage. The presence of shrubs and trees might indicate a decrease or even a total abandonment of traditional management practices like hay-making or grazing, including active afforestation. Changing indicator values through time may reveal recurrent burning practices, the loss of habitat for species that depend on open grassland on the one hand or grasslands with interspersed trees on the other hand, or even climate change affecting high-elevation grasslands. The presence of open water implies a multitude of potential pressures to biodiversity present in wetland or riverine habitats. Areas with open water can be subject to the extraction of sand, gravel, clay, or other minerals, destructing the habitat of species that entirely or partly rely on open water. The use of open water as a source of gravitational energy or drinking water production can be a threat too, especially when subjected to a management that impedes the development of habitats for particular species (e.g., by recurrent cleaning of basins and water bodies or large and irregular water table fluctuations). The same applies to water bodies used for transport or recreation, or marine and freshwater aquaculture. Fishing and harvesting of aquatic resources can be detrimental too when applied in a non-sustainable way. Non-native aquatic species grown for food by aquaculture can invade neighboring water bodies and even entire catchments when they escape from nurseries. In areas with a high human population density, pollution of open water by all kinds of activities may be ubiquitous, as are changes in the hydrological conditions of the open water or the neighboring areas (e.g., by dredging).

Habitat fragmentation can be evaluated by structural indices (Jaeger, 2000) depending on scale and level of spatial explicitness. Fragmentation indices not only consider the absolute shrinking of habitat area but also, and with particular emphasis, the decrease or even loss of functional connectivity. The dynamic of such indices provides information not only on one habitat type, but usually habitat fragmentation is coupled with an interplay of changing land-use patterns with mutual influences.

6.5 TOWARD A BIODIVERSITY MONITORING SERVICE

Remote sensing has long been used as a tool for environmental monitoring, especially for vegetation. But while at the global and continental scales, applications using broad land cover/habitat categories have been quite successful, success has been harder to achieve at detailed local scales (Alkhuzaei and Brolly, 2024; Amani et al., 2023; An et al., 2023; Iglseder et al., 2023; Larsen et al., 2023; Price et al., 2023; Zhang et al., 2023; van der Reijden et al., 2021; Wan and Ma, 2020; Wicaksono et al., 2019; Roelfsema et al., 2018; Setyawidati et al., 2018; Onojeghuo and Onojeghuo, 2017). Indeed, applications in detailed vegetation mapping and monitoring are often demanding in terms of data (requiring both high spectral and high spatial resolution), placing them at the forefront of technological development, with many new approaches still being in a research phase to tackle the widely divergent user needs. In contrast to global mapping initiatives, which have received wide attention and critical evaluations (Bartholome and Belward, 2005), local mapping exercises rarely receive any evaluation and validation other than by the user for which they were intended. Hence, it remains unknown whether the chosen method was most appropriate for the situation and problem at hand, and whether the method would yield comparable results in a different setting (i.e., the robustness of the method). This impairs the wider application of such methods and adoption by other users with similar problems and their further development toward operational use.

The above mentioned projects, have addressed these needs by exploring the potential of EO data in combination with data from ground surveys for supporting management options and reporting of obligations.[20] Both projects have prepared the ground for establishing services to support a successful implementation of European environmental legislation on all levels. Services, developed in a pre-operational mode, underlay four suitability criteria as identified by Vanden Borre et al. (2011): (1) multi-scale, i.e., addressing multiple scales on all levels of implementation; (2) versatile, with algorithms tailored to the habitat type of interest and different image types; (3) user-friendly, allowing integration of the products into existing workflows; and (4) cost-efficient, providing reliable and reproducible products at an affordable cost compared to traditional field methods.

The three (sub-)services .EU, .State, and .Site reflecting the different levels of operation (see Figure 6.10). This requires a concordant multi-user approach. Each of the service developments was tailored to the user and technical requirements that are specific for each level of implementation. User requirement surveys collected all details on existing workflows, data usages, and the responsibilities imposed by HabDir. Based on these requirements, the testing, comparison, and integration of state-of-the-art methodologies was performed. Demonstrators, accompanied by a full-fledged user validation exercise, complete the service evolution plan and the final scoping toward the market. MS.MONINA thereby addressed: (1) agencies on the EU level, i.e., ETC Biodiversity, the EEA, and DG Environment; (2) national and federal agencies in their reporting on sensitive sites and habitats within biogeographical regions on the entire territory; (3) local management authorities by advanced mapping methods for status assessment and change maps of sensitive sites; and (4) all three groups by providing transferable and interoperable monitoring results for improved information flow between all levels.

Within BIO_SOS, a pre-operational, knowledge-driven, open-source, three-stage processing system was developed capable of combining multi-seasonal EO data (HR and VHR) and *in situ* data (including ancillary information and ground measurements) and for subsequent translation of LC to

Multi-Scale Habitat Mapping and Monitoring

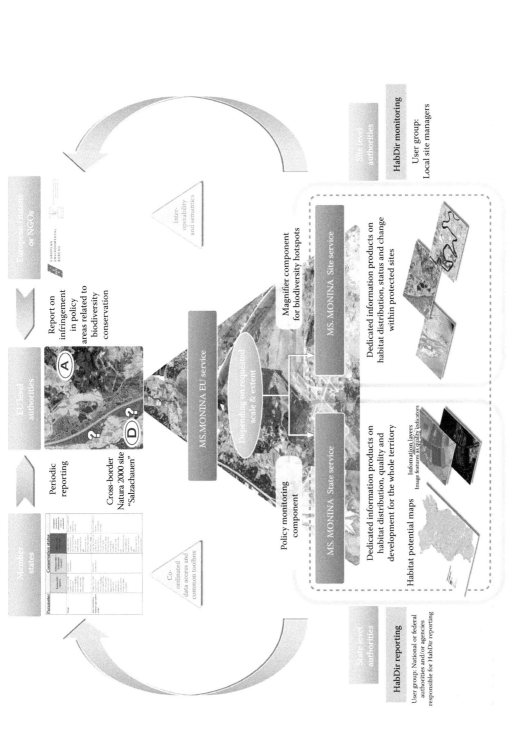

FIGURE 6.10 MS.MONINA three-level concept of operational service implementation for HabDir-related monitoring requirements (Lang and Pernkopf, 2013). The EU service builds on two main components: (1) Magnifier component: on-demand provision of habitat distribution and quality indicator maps for biodiversity hotspot sites (e.g., riparian areas, coastal areas) and (2) Policy monitoring component: on-demand provision of maps in "rush mode" as a means for external/independent validation of the congruence of national biodiversity reports in the case of trans-boundary protected sites.

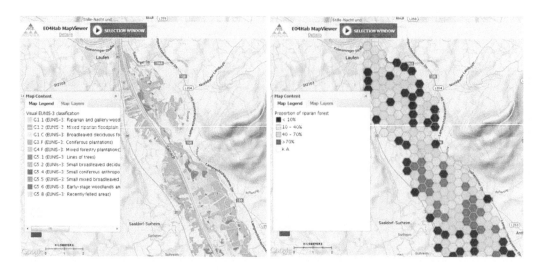

FIGURE 6.11 Transboundary riparian forest site 'Salzachauen' (Germany/Austria): Screenshots taken from MS.MONINA map viewer.

habitat maps. This system, named EO Data for Habitat Monitoring (EODHaM) (Lucas et al., 2014), is based on expert knowledge elicited from botanists, ecologists, remote sensing experts, and management authorities in order to monitor large and not accessible areas without any ground reference data. Ontologies are used to formally represent the expert knowledge (Arvor et al., 2013). The FAO LCCS and the GHC taxonomies, to which HabDir Annex I habitats can be linked, were used for describing LC/LU and habitat categories (Tomaselli et al., 2013) and for subsequent translation to habitats. In addition, BIO_SOS focused on the development of a modeling framework for (1) filling the gap between LC/LU and habitat domains (Blonda et al., 2012a) by coupling FAO LCCS taxonomy with GHCs and EUNIS classification schemes and providing a reliable, cost-effective, knowledge-driven, long-term biodiversity monitoring scheme of protected areas and their surroundings (Adamo et al., 2014; Kosmidou et al., 2014; Lucas et al., 2014; Tomaselli et al., 2013); (2) analyzing appropriate spatial and temporal scales of EO data sources for pressure assessment in the context of existing reference frameworks for pressure assessment and trend extraction (Lang et al., 2014); and (3) handling uncertainty in habitat mapping (Petrou et al., 2014).

Public-access web platforms were built to integrate and showcase these achievements. These mainly include an online portfolio with specific details on the services to be offered and further information on the respective service cases. Web-GIS and OGC-conform metadata geoportals, including an external quality evaluation module (fitness-for-use and fitness-to-purpose), contains relevant geospatial information (see Figure 6.11). A tool repository allows for searching, both thematically and spatially, the available data and inputs for the the methodological components and algorithms to perform the respective image analysis and geospatial analysis tasks.

NOTES

1. www.earthobservations.org/
2. http://www.earthobservations.org/geobon.shtml
3. www.copernicus.eu/
4. www.ms-monina.eu
5. www.biosos.eu

6 Our life insurance, our natural capital: an EU biodiversity strategy to 2020—COM(2011) 244 final—http://ec.europa.eu/environment/nature/biodiversity/comm2006/pdf/2020/1_EN_ACT_part1_v7%5B1%5D.pdf
7 See White Paper "Copernicus Biodiversity Monitoring Services," available at http://www.biosos.eu/publ/White_Paper_Biodiversity_Monitoring_BIOSOS_MSMONINA.pdf.
8 https://www.eea.europa.eu/themes/biodiversity/natura-2000
9 e.g., NOAA-AVHRR, MODIS
10 e.g., Landsat, IRS, SPOT
11 e.g., IKONOS, QuickBird, GeoEye, WorldView-2, Pléiades
12 e.g., HyMap, CASI, Hyperion
13 To be combined with multi-spectral/hyperspectral imagery
14 Wetlands are not a physiognomic type per se but are various physiognomic types that have adapted to the continuous or temporary presence of water.
15 http://ec.europa.eu/eurostat/ramon/nomenclatures
16 http://eunis.eea.europa.eu/
17 http://www.millenniumassessmnet.org
18 http://www.ipbes.net/
19 Taken from MS.MONINA Deliverable 5.6 "Framework for identifying threats and pressures on sensitive sites, from remote sensing derived (change) indicators in the site surroundings," available at www.ms-monina.eu.
20 See White Paper on "Copernicus Biodiversity Monitoring Services," available at http://www.biosos.eu/publ/White_Paper_Biodiversity_Monitoring_BIOSOS_MSMONINA.pdf.

REFERENCES

Adam, E., O. Mutanga and D. Rugege 2009. Multispectral and hyperspectral remote sensing for identification and mapping of wetland vegetation: A review. *Wetl Ecol Manag*.

Adamczyk, J. and A. Osberger 2015. Red-edge vegetation indices for detecting and assessing disturbances in Norway spruce dominated mountain forests. *Int J Appl Earth Obs Geoinf* 37, pp. 90–99.

Adamo, M., et al. 2014. Expert knowledge for translating land cover/use maps to general habitat categories (GHC). *Landsc. Ecol.* 29, 1045–1067.

Addicot, J. F., J. M. Aho, M. F. Antolini, D. K. Padilla, J. S. Richardson and D. A. Soluk 1987. Ecological neighborhoods: Scaling environmental patterns. *Oikos*, pp. 340–346.

Akay, A. E., M. G. Wing and J. Sessions 2012. Estimating structural properties of riparian forests with airborne lidar data. *Int J Remote Sens*, pp. 7010–7023.

Ali, I., C. Schuster, M. Zebisch, M. Förster, B. Kleinschmit and C. Notarnicola 2013. First results of monitoring nature conservation sites in alpine region by using very high resolution (VHR) X-band SAR data. *IEEE J Sel Top Appl Earth Obs Remote Sens*, pp. 2265–2274.

Alkhuzaei, M. and M. Brolly 2024. Advancing coastal habitat mapping in bahrain: A comparative study of remote sensing classifiers. *Model Earth Syst Environ*. https://doi.org/10.1007/s40808-024-01957-w

Amani, M., F. Foroughnia, A. Moghimi, S. Mahdavi and S. Jin 2023. Three-dimensional mapping of habitats using remote-sensing data and machine-learning algorithms. *Remote Sens* 15, no. 17, pp. 4135. https://doi.org/10.3390/rs15174135

An, N. V., N. H. Quang, T. P. Hoang Son and T. T. An 2023. High-resolution benthic habitat mapping from machine learning on PlanetScope imagery and ICESat-2 data. *Geocarto Int* 38, no. 1. http://doi.org/10.1080/10106049.2023.2184875

Arvor, D., L. Durieux, S. Andrés and L. Marie-Angélique 2013. Advances in geographic object-based image analysis with ontologies: A review of main contributions and limitations from a remote sensing perspective. *ISPRS J Photogramm*, pp. 125–137.

Asner, G. P., M. O. Jones, R. E. Martin, D. E. Knapp and R. F. Hughes 2008a. Remote sensing of native and invasive species in Hawaiian forests. *Remote Sens Environ*, pp. 1912–1926.

Asner, G. P., D. E. Knapp, T. Kennedy-Bowdoin, M. O. Jones, R. E. Martin, J. Boardman and R. F. Hughes 2008b. Invasive species detection in Hawaiian rainforests using airborne imaging spectroscopy and LiDAR. *Remote Sens Environ*, pp. 1942–1955.

Balzer, S., G. Ellwanger, U. Raths, E. Schröder and A. Ssymank 2008. Verfahren und erste Ergebnisse des nationalen Berichts nach Artikel 17 der FFH-Richtlinie (in German). *Natur und Landschaft*, pp. 111–117.

Bargiel, D. 2013. Capabilities of high resolution satellite radar for the detection of semi-natural habitat structures and grasslands in agricultural landscapes. *Ecol Inform*, pp. 9–16.

Bartholome, E. and A. Belward 2005. GLC2000: A new approach to global land cover mapping from Earth observation data. *Int J Remote Sens*, pp. 1959–1977.

Belluco, E., M. Camuffo, S. Ferrari, L. Modenese, S. Silvestri, A. Marani and M. Marani 2006. Mapping salt-marsh vegetation by multispectral and hyperspectral remote sensing. *Remote Sens Environ*, pp. 54–67.

Berger, M., J. Moreno, J. A. Johannssen, P. F. Levelt and R. F. Hansen 2012. ESA's sentinel missions in support of Earth system science. *Remote Sens Environ*, pp. 84–90.

Blonda, P., K. B. Jones, J. Stutte and P. Dimipoulos 2012a. From space to species. Safeguarding biodiversity in Europe. *Int Innovat Environ*, pp. 86–88.

Blonda, P., R. Lucas and J. P. Honrado 2012b. From Space to species: Solutions for biodiversity monitoring. In: S. Ourevitch (ed.) *Window on Copernicus: Discover What GMES Can Do for European Regions and Cities*, Brussels: SpaceTec Partners, pp. 66–73.

Blondel, J. 1979. *Biogéographie et écologie: Synthèse sur la structure, la dynamique et l'évolution des peuplements de vertébrés terrestres*, Paris and New York: Masson.

Bock, M. 2003. Remote sensing and GIS-based techniques for the classification and monitoring of biotopes: Case examples for a wet grass- and moor land area in Northern Germany. *J Nat Conserv*, pp. 145–155.

Bock, M., G. Rossner, M. Wissen, K. Remm, T. Langanke, S. Lang and H. Klug 2005. Spatial indicators for nature conservation from European to local scale. *Ecol Indic*, pp. 322–338.

Bouman, B. A. M. and H. W. J. van Kasteren 1990. Ground-based X-band (3-cm wave) radar backscattering of agricultural crops. I. Sugar beet and potato; backscattering and crop growth. *Remote Sens Environ*, pp. 93–105.

Buck, O., V. E. G. Millán, A. Klink and K. Pakzad 2015. Using information layers for mapping grassland habitat distribution at local to regional scales. *Int J Appl Earth Obs Geoinf* 37, pp. 83–89.

Bunce, R. H. G., et al. 2008. A standardized procedure for surveillance and monitoring European habitats and provision of spatial data. *Landsc Ecol* 23, 11–25.

Corbane, C., S. Alleaume and M. Deshayes 2013. Mapping natural habitats using remote sensing and sparse partial least square discriminant analysis. *Int J Remote Sens*, pp. 7625–7647.

Corbane, C., S. Lang, K. Pipkins, S. Alleaume, M. Deshayes, V. E. García Millán, T. Strasser, J. Vanden Borre, S. Toon and F. Michael 2015. Remote sensing for mapping natural habitats and their conservation status – New opportunities and challenges. *Int J Appl Earth Obs Geoinf* 37, pp. 7–16.

Costanza, J. K., A. Moody and R. K. Peet 2011. Multi-scale environmental heterogeneity as a predictor of plant species richness. *Landsc Ecol*, pp. 851–864.

Dalponte, M., L. Bruzzone and D. Gianelle 2012. Tree species classification in the Southern Alps based on the fusion of very high geometrical resolution multispectral/hyperspectral images and LiDAR data. *Remote Sens Environ* 123, pp. 258–270.

Davies, C. E., M. D. and M. O. Hill 2004. *EUNIS Habitat Classification*. https://www.eea.europa.eu/data-and-maps/data/eunis-habitat-classification-1/documentation/eunis-2004-report.pdf

Davranche, A., G. Lefebvre and B. Poulin 2010. Wetland monitoring using classification trees and SPOT-5 seasonal time series. *Remote Sens Environ*, pp. 552–562.

Delalieux, S., B. Somers, B. Haest, T. Spanhove, J. Vanden Borre and C. A. Mücher 2012. Heathland conservation status mapping through integration of hyperspectral mixture analysis and decision tree classifiers. *Remote Sens Environ* 126, pp. 222–231.

Diaz Varela, R., P. Ramil Rego, S. Calvo Iglesias and C. Muñoz Sabrino 2008. Automatic habitat classification methods based on satellite images: A practical assessment in the NW Iberia coastal mountains. *Environ Monit Assess*, pp. 229–250.

Di Gregorio, A. and L. J. M. Jansen 2005. *Land Cover Classification System (LCCS): Classification Concepts and User Manual*, Rome: Food and Agriculture Organization of the United Nations.

Dogan, O. K., Z. Akyurek and M. Beklioglu 2009. Identification and mapping of submerged plants in a shallow lake using quickbird satellite data. *J Environ Manage*, pp. 2138–2143.

EEA 1995. *Europe's Environment: The Dobris Assessment*, Copenhagen: EEA.

Ellmauer, T. 2005. *Entwicklung von Kriterien, Indikatoren und Schwellenwerten zur Beurteilung des Erhaltungszustandes der Natura 2000-Schutzgüter (in German)*. https://www.researchgate.net/publication/301547944

EuropeanCommission 2011. *COM(2011) 244 Final. Our Life Insurance, Our Natural Capital: An EU Biodiversity Strategy to 2020.*

Evans, D. 2012. Building the European Union's Natura 2000 network. *Nat Conserv*, pp. 11–26.

Everitt, J. H., C. S. Yang, R. S. Fletcher, M. R. Davis and D. L. Drawe 2004. Using aerial color-infrared photography and QuickBird satellite imagery for mapping wetland vegetation. *Geocarto Int*, pp. 15–22.

Feilhauer, H., F. Thonfeld, U. Faude, K. S. He, D. Rocchini and S. Schmidtlein 2013. Assessing floristic composition with multispectral sensors—A comparison based on monotemporal and multiseasonal field spectra. *Int J Appl Earth Obs Geoinf*, pp. 218–229.

Ficetola, G. F., M. Adamo, A. Bonardi, V. De Pasquale, C. Liuzzi, F. Lovergine, F. Marcone, F. Mastropasqua, C. Tarantino, P. Blonda and E. Padoa-Schioppa 2015. Importance of landscape features and Earth observation derived habitat maps for modelling amphibian distribution in the Alta Murgia National Park. *Int J Appl Earth Obs Geoinf* 37, pp. 152–159.

Filippi, A. M., I. Gueneralp and J. Randall 2014. Hyperspectral remote sensing of aboveground biomass on a river meander bend using multivariate adaptive regression splines and stochastic gradient boosting. *Remote Sens Lett*, pp. 432–441.

Förster, M., A. Frick, H. Walentowski and B. Kleinschmit 2008. Approaches to utilising QuickBird data for the monitoring of NATURA 2000 habitats. *Community Ecol*, pp. 155–168.

Förster, M. and B. Kleinschmit 2014. Significance analysis of different types of ancillary geodata utilized in a multisource classification process for forest identification in Germany. *IEEE Trans Geosci Remote Sens*, pp. 3453–3463.

Förster, M., B. Kleinschmit and H. Waltentowski 2005. Comparison of three modelling approaches of potential natural forest habitats in Bavaria, Germany. *Waldökologie Online*, pp. 126–135.

Förster, M., T. Schmidt, C. Schuster and B. Kleinschmit 2012. Multi-temporal detection of grassland vegetation with RapidEye imagery and a spectral-temporal library. *IEEE Trans Geosci Remote Sens*, pp. 4930–4933.

Förster, M., C. Schuster and B. Kleinschmit 2010. Significance analysis of multi-temporal RapidEye satellite images in a land cover classification. In: N. J. Tate and P. F. Fisher (eds.) *Accuracy 2010*, Leicester, pp. 273–276.

Franke, J., V. Keuck and F. Siegert 2012. Assessment of grassland use intensity by remote sensing to support conservation schemes. *J Nat Conserv*, pp. 125–134.

Friedl, M. A., D. Sulla-Menashe, B. Tan, A. Schneider, N. Ramankutty, A. Sibley and X. Huang 2010. MODIS Collection 5 global land cover: Algorithm refinements and characterization of new datasets. *Remote Sens Environ*, pp. 168–182.

Fuller, R., G. B. Groom and A. Jones 1994. The land cover map of Great Britain: An automated classification of Landsat Thematic Mapper data. *Photogramm Eng Remote Sens*, pp. 553–562.

Gaertner, P., M. Foerster, A. Kurban and B. Kleinschmit 2014. Object based change detection of Central Asian Tugai vegetation with very high spatial resolution satellite imagery. *Int J Appl Earth Obs Geoinf*, pp. 110–121.

GEO 2005. *The Global Earth Observation System of Systems (GEOSS) 10-Year Implementation Plan, adopted 16 February 2005.* http://www.earthobservations.org/docs/10-Year%20Implementation%20Plan.pdf

GEO 2010. *A Quality Assurance Framework for Earth Observation, version 4.0.* http://qa4eo.org/docs/QA4EO_Principles_v4.0.pdf

Ghosh, A., F. E. Fassnacht, P. K. Joshi and B. Koch 2014. A framework for mapping tree species combining hyperspectral and LiDAR data: Role of selected classifiers and sensor across three spatial scales. *Int J Appl Earth Obs Geoinf*, pp. 49–63.

Gillespie, T. W. 2005. Predicting woody-plant species richness in tropical dry forests: A case study from south Florida, USA. *Ecol Appl*, pp. 27–37.

Gilmore, M. S., E. H. Wilson, N. Barrett, D. L. Civco, S. Prisloe, J. D. Hurd and C. Chadwick 2008. Integrating multi-temporal spectral and structural information to map wetland vegetation in a lower Connecticut River tidal marsh. *Remote Sens Environ*, pp. 4048–4060.

Goetz, S., D. Steinberg, R. Dubayah and B. Blair 2007. Laser remote sensing of canopy habitat heterogeneity as a predictor of bird species richness in an eastern temperate forest, USA. *Remote Sens Environ*, pp. 254–263.

Grillo, O. and G. Venora 2011. *Biodiversity Loss in a Changing Planet*, Rijeka, Croatia: InTech.

Gruber, B. et al. 2012. "Mind the gap!"—How well does Natura 2000 cover species of European interest? *Nat Conserv*, pp. 45–62.

Hellesen, T. and L. Matikainen 2013. An object-based approach for mapping shrub and tree cover on grassland habitats by use of LiDAR and CIR orthoimages. *Remote Sens*, pp. 558–583.

Huang, W., G. Sun, R. Dubayah, B. Cook, P. Montesano, W. Ni and Z. Zhang 2013. Mapping biomass change after forest disturbance: Applying LiDAR footprint-derived models at key map scales. *Remote Sens Environ*, pp. 319–332.

Ichter, J., D. Evans and D. Richard 2014. *Habitat and Vegetation Mapping in Europe—An Overview*, Copenhagen: EEA-MNHN.

Iglseder, A., M. Immitzer, A. Dostálová, A. Kasper, N. Pfeifer, C. Bauerhansl, S. Schöttl and M. Hollaus 2023. The potential of combining satellite and airborne remote sensing data for habitat classification and monitoring in forest landscapes. *Int J Appl Earth Obs Geoinf* 117, p. 103131. ISSN 1569-8432. https://doi.org/10.1016/j.jag.2022.103131. https://www.sciencedirect.com/science/article/pii/S1569843222003193

Immitzer, M., C. Atzberger and T. Koukal 2012. Tree species classification with random forest using very high spatial resolution 8-Band WorldView-2 satellite data. *Remote Sens* 4, no. 9, pp. 2661–2693.

Isbell, F. 2010. Causes and consequences of biodiversity declines. *Nat Educ Knowledg*, p. 54.

Jaeger, J. 2000. Landscape division, splitting index, and effective mesh size: New measures of landscape fragmentation. *Landsc Ecol*, pp. 115–130.

Johansen, K., N. C. Coops, S. E. Gergel and Y. Stange 2007. Application of high spatial resolution satellite imagery for riparian and forest ecosystem classification. *Remote Sens Environ*, pp. 29–44.

Jollineau, M. Y. and P. J. Howarth 2008. Mapping an inland wetland complex using hyperspectral imagery. *Int J Remote Sens*, pp. 3609–3631.

Keramitsoglou, I., D. Stratoulias, E. Fitoka, C. Kontoes and N. Sifakis 2015. A transferability study of the kernel-based reclassification algorithm for habitat delineation. *IntJ Appl Earth Obs Geoinf* 37, pp. 38–47.

Key, T., T. A. Warner, J. B. McGraw and M. A. Fajvan 2001. A comparison of multispectral and multitemporal information in high spatial resolution imagery for classification of individual tree species in a temperate hardwood forest. *Remote Sens Environ*, pp. 100–112.

Kosmidou, V. et al. 2014. Harmonization of the Land Cover Classification System (LCCS) with the General Habitat Categories (GHC) classification system: Linkage between remote sensing and ecology. *Ecol Indic*, pp. 290–300.

Kotliar, N. B. and J. Wiens 1990. Multiple scales of patchiness and patch structure: A hierarchical framework for the study of heterogeneity. *Oikos*. 59, no. 2.

Kuenzer, C., et al. 2014. Earth observation satellite sensors for biodiversity monitoring: Potentials and bottlenecks. *Int J Remote Sens*, pp. 6599–6647.

Laliberte, A. S., E. L. Frederickson and A. Rango 2007. Combining decision trees with hierarchical object-oriented image analysis for mapping arid rangelands. *Photogramm Eng Remote Sens*, pp. 197–207.

Lang, S. 2008. Object-based image analysis for remote sensing applications: Modeling reality—dealing with complexity. In: T. Blaschke, S. Lang and G. J. Hay (eds.) *Object-Based Image Analysis—Spatial Concepts for Knowledge-Driven Remote Sensing Applications*, Berlin: Springer, pp. 3–28.

Lang, S., C. Corbane and L. Pernkopf 2013a. Earth observation for habitat and biodiversity monitoring. In: S. Lang and L. Pernkopf (eds.) *Ecosystem and Biodiversity Monitoring: Best Practice in Europe and Globally [Section Title]*, Heidelberg: Wichmann, pp. 478–486.

Lang, S. and T. Langanke 2006. Object-based mapping and object-relationship modelling for land-use classes and habitats. *Photogramm Fernerkun Geoinform*, pp. 5–18.

Lang, S. and L. Pernkopf 2013. EO-based monitoring of Europe's most precious habitats inside and outside protected areas. In: K. Bauch (ed.) *5th Symposium for Research in Protected Areas*, Mittersill: Salzburger Nationalparkfonds, pp. 443–448.

Lang, S., L. Pernkopf, P. Mairota and E. P. Schioppa 2014. Earth observation for habitat mapping and biodiversity monitoring. *Int J Appl Earth Obs Geoinf* 37, pp. 1–6.

Lang, S., G. Smith and J. Vanden Borre 2013b. Monitoring Natura2000 habitats at local, regional and European scale. In: S. Ourevitch (ed.) *Discover What GMES Can Do for European Regions and Cities*, Brussels: SpaceTec Partners, pp. 58–65.

Langanke, T., C. Burnett and S. Lang 2007. Assessing the mire conservation status of a raised bog site in Salzburg using object-based monitoring and structural analysis. *Landsc Urban Plan*, pp. 160–169.

Larsen, S., J. M. Alvarez-Martinez, J. Barquin, M. C. Bruno, L. C. Zubiri, L. Gallitelli, M. Jonsson, M. Laux, G. Pace, M. Scalici and R. Schulz. 2023. RIPARIANET—Prioritising riparian ecotones to sustain and connect multiple biodiversity and functional components in river networks. *Res Ideas Outcomes* 9. http://doi.org/10.3897/rio.9.e108807

Lindenmayer, D. B., R. B. Cunnigham, C. F. Donnelly and H. A. Nix 2002. The distribution of birds in a novel landscape context. *Ecol Monogr*, pp. 1–18.

Lucas, R., et al. 2011. Updating the phase 1 habitat map of Wales, UK, using satellite sensor data. *ISPRS J Photogramm*, pp. 81–102.

Lucas, R., P. Blonda, P. Bunting, G. Jones, J. Inglada, M. Arias, V. Kosmidou, Z. I. Petrou, I. Manakos, M. Adamo, R. Charnock, C. Tarantino, C. A. Mücher, R. H. G. Jongman, H. Kramer, D. Arvor, J. P. Honrado and P. Mairota 2015. The Earth Observation Data for Habitat Monitoring (EODHaM) system. *Int J Appl Earth Obs Geoinf* 37, pp. 17–28.

Lucas, R., A. Rowlands, A. Brown, S. Keyworth and P. Bunting 2007. Rule-based classification of multi-temporal satellite imagery for habitat and agricultural land cover mapping. *ISPRS J Photogramm Remote Sens* 62, no. 3, pp. 165–185.

Mac Alister, C. and M. Mahaxy 2009. Mapping wetlands in the Lower Mekong Basin for wetland resource and conservation management using Landsat ETM images and field survey data. *J Environ Manage*, pp. 2130–2137.

Mac Arthur, A. A. and J. T. Malthus 2008. An object-based image analysis approach to the classification and mapping of calluna vulgaris canopies. In *Remote Sensing and Photogrammetry Society Annual Conference*.

Maes, J., et al. 2013. *Mapping and Assessment of Ecosystems and their Services. An Analytical Framework for Ecosystem Assessments Under Action 5 of the EU Biodiversity Strategy to 2020*. In P. o. o. t. E. Union (ed.). Discussion paper, Luxembourg.

Mairota, P., B. Cafarelli, R. Labadessa, F. P. Lovergine, C. Tarantino, H. Nagendra and R. K. Didham 2015. Very high resolution Earth observation features for testing the direct and indirect effects of landscape structure on local habitat quality. *Int J Appl Earth Obs Geoinf* 37, pp. 96–102.

Mücher, C. A., L. Roupioz, H. Kramer, M. M. B. Bogers, R. H. G. Jongman, R. M. Lucas, V. E. Kosmidou, Z. Petrou, I. Manakos, E. Padoa-Schioppa, M. Adamo and P. Blonda 2015. Synergy of airborne LiDAR and Worldview-2 satellite imagery for land cover and habitat mapping: A BIO_SOS-EODHaM case study for the Netherlands. *Int J Appl Earth Obs Geoinf* 37, pp. 48–55.

Naeem, S., D. E. Bunker, A. Hector, M. Loreau and C. Perrings 2009. *Biodiversity, Ecosystem Functioning, and Human Wellbeing: An Ecological and Economic Perspective*, Oxford: Oxford University Press.

Nagendra, H., R. Lucas, J. P. Honrado, R. H. G. Jongman, C. Tarantino and M. Adamo 2013. Remote sensing for conservation monitoring: Assessing protected areas, habitat extent, habitat condition, species diversity and threat. *Ecol Indic.* 33, 45–59.

Nagendra, H., P. Mairota, C. Marangi, R. Lucas, P. Dimopoulos, J. P. Honrado, M. Niphadkar, C. A. Mücher, V. Tomaselli, M. Panitsa, C. Tarantino, I. Manakos, P. and P. Blonda 2015. Satellite Earth observation data to identify anthropogenic pressures in selected protected areas. *Int J Appl Earth Obs Geoinf* 37, pp. 124–132.

Nagendra, H. and D. Rocchini 2008. High resolution satellite imagery for tropical biodiversity studies: The devil is in the detail. *Biodivers Conserv*, pp. 3431–3442.

Onojeghuo, A. O. and A. R. Onojeghuo 2017. Object-based habitat mapping using very high spatial resolution multispectral and hyperspectral imagery with LiDAR data. *Int J Appl Earth Obs Geoinf* 59, pp. 79–91. ISSN 1569-8432. https://doi.org/10.1016/j.jag.2017.03.007. https://www.sciencedirect.com/science/article/pii/S0303243417300703

Peng, G., R. Pu, G. S. Biging and M. R. Larrieu 2003. Estimation of forest leaf area index using vegetation indices derived from Hyperion hyperspectral data. *IEEE Trans Geosci Remote Sens*, pp. 1355–1362.

Pereira, H. M., et al. 2013. Essential biodiversity variables. *Science*, pp. 277–278.

Petrou, Z., et al. 2014. A rule-based classification methodology to handle uncertainty in habita mapping employing evidential reasoning and fuzzy logic. *Pattern Recognit Lett*, pp. 24–33.

Price, B., N. Huber, A. Nussbaumer and C. Ginzler. 2023. The habitat map of Switzerland: A remote sensing, composite approach for a high spatial and thematic resolution product. *Remote Sens* 15, no. 3, p. 643. https://doi.org/10.3390/rs15030643

Price, K. P., X. Guo and J. M. Stiles 2002. Optimal Landsat TM band combinations and vegetation indices for discrimination of six grassland types in eastern Kansas. *Int J Remote Sens*, pp. 5031–5042.

Prisloe, S., M. Wilson, D. Civco, J. Hurd and M. Gilmore 2006. Use of Lidar data to aid in discriminating and mapping plant communities in tidal marshes of the lower Connecticut River—preliminary results. In: *Annual Conference of the American Society for Photogrammetry and Remote Sensing*, Nevada: American Society for Photogrammetry and Remote Sensing.

Psomas, A., M. Kneubühler, S. Huber, K. Itten and N. E. Zimmermann 2011. Hyperspectral remote sensing for estimating aboveground biomass and for exploring species richness patterns of grassland habitats. *Int J Remote Sens*, pp. 9007–9031.

Puech, C., S. Durrieu and J. S. Bailly 2012. Airborne lidar for natural environments research and applications in France. *Rev Fr Photogramm Télédétect*, pp. 54–68.

Rannow, S., et al. 2014. Managing protected areas under climate change: Challenges and priorities. *Environ Manag*, pp. 732–743.

Riedler, B. and S. Lang 2018. A spatially explicit patch model of habitat quality, integrating spatio-structural indicators. *Ecol Indic* 94, pp. 128–141.

Riedler, B., L. Pernkopf, T. Strasser, G. Smith and S. Lang 2015. A composite indicator for assessing habitat quality of riparian forests derived from Earth observation data. *Int J Appl Earth Obs Geoinf* 37, pp. 114–123.

Rocchini, D., et al. 2010. Remotely sensed spectral heterogeneity as a proxy of species diversity: Recent advances and open challenges. *Ecol Inform*, pp. 318–329.

Roelfsema, C., E. Kovacs, J. C. Ortiz, N. H. Wolff, D. Callaghan, M. Wettle, M. Ronan, S. M. Hamylton, P. J. Mumby and S. Phinn 2018. Coral reef habitat mapping: A combination of object-based image analysis and ecological modelling. *Remote Sens Environ* 208, pp. 27–41. ISSN 0034-4257. https://doi.org/10.1016/j.rse.2018.02.005. https://www.sciencedirect.com/science/article/pii/S0034425718300117

Schmidt, K. S. and A. K. Skidmore 2003. Spectral discrimination of vegetation types in a coastal wetland. *Remote Sens Environ*, pp. 92–108.

Schmidt, T., C. Schuster, B. Kleinschmit and M. Förster 2014. Evaluating an intra-annual time series for grassland classification—How many acquisitions and what seasonal origin are optimal? *IEEE J Sel Top Appl Earth Obs Remote Sens*, pp. 3428–3439.

Schmidtlein, S. and J. Sassin 2004. Mapping of continuous floristic gradients in grasslands using hyperspectral imagery. *Remote Sens Environ*, pp. 126–138.

Schuster, C., I. Ali, P. Lohmann, A. Frick, M. Förster and B. Kleinschmit 2011. Towards detecting swath events in TerraSAR-X time series to establish Natura 2000 grassland habitat swath management as monitoring parameter. *Remote Sens*, pp. 1308–1322.

Schuster, C., T. Schmidt, C. Conrad, B. Kleinschmit and M. Förster 2015. Grassland habitat mapping by intra-annual time series analysis—Comparison of RapidEye and TerraSAR-X satellite data. *Int J Appl Earth Obs Geoinf*, pp. 25–34.

Setyawidati, N., A. H. Kaimuddin, I. P. Wati, et al. 2018. Percentage cover, biomass, distribution, and potential habitat mapping of natural macroalgae, based on high-resolution satellite data and in situ monitoring, at Libukang Island, Malasoro Bay, Indonesia. *J Appl Phycol* 30, pp. 159–171. https://doi.org/10.1007/s10811-017-1208-1

Somers, B. and G. P. Asner 2014. Tree species mapping in tropical forests using multi-temporal imaging spectroscopy: Wavelength adaptive spectral mixture analysis. *Int J Appl Earth Obs Geoinf*, pp. 57–66.

Spanhove, T., J. Vanden Borre, S. Delalieux, B. Haest and D. Paelinckx 2012. Can remote sensing estimate fine-scale quality indicators of natural habitats? *Ecol Indic*, pp. 403–412.

Strand, H., R. Höft, J. Strittholt, L. Miles, N. Horning, E. Fosnight and W. Turner 2007. *Sourcebook on Remote Sensing and Biodiversity Indicators*, Montreal: Secretariat of the Convention on Biological Diversity.

Strasser, T. and S. Lang 2014. Class modelling of complex riparian forest habitats. *South-Eastern Eur J Earth Observ Geomat*, pp. 219–222.

Strasser, T. and S. Lang, S. 2015. Object-based class modelling for multi-scale Riparian forest habitat mapping. *Int J Appl Earth Obs Geoinf* 37, pp. 29–37.

Suchenwirth, L., M. Foerster, A. Cierjacks, F. Lang and B. Kleinschmit 2012. Knowledge-based classification of remote sensing data for the estimation of below- and above-ground organic carbon stocks in riparian forests. *Wetl Ecol Manag*, pp. 151–163.

Thoonen, G., T. Spanhove, J. Vanden Borre and P. Scheunders 2013. Classification of heathland vegetation in a hierarchical contextual framework. *Int J Remote Sens*, pp. 96–111.

Tiede, D., S. Lang, F. Albrecht and D. Hölbling 2010. Object-based class modeling for cadastre constrained delineation of geo-objects. *Photogramm Eng Remote Sens*. 76, 193–202.

Tomaselli, V., et al. 2013. Translating land cover/land use classifications to habitat taxonomies for landscape monitoring: A Mediterranean assessment. *Landsc Ecol*, pp. 905–930.

Townsend, P. A., T. R. Lookingbill, C. C. Kingdon and R. H. Gardner 2009. Spatial pattern analysis for monitoring protected areas. *Remote Sens Environ*, pp. 1410–1420.

Trochet, A. and D. Schmeller 2013. Effectiveness of the Natura 2000 network to cover threatened species. *Nat Conserv*, pp. 35–53.

Turner, W., S. Spector, N. Gardiner, M. Fladeland, E. Sterling and M. Steininger 2003. Remote sensing for biodiversity science and conservation. *Trends Ecol Evolu*, pp. 306–314.

Underwood, E. C., S. L. Ustin and C. M. Ramirez 2003. A comparison of spatial and spectral image resolution for mapping invasive plants in coastal California. *Environ Manag*, pp. 63–83.

Vanden Borre, J., D. Paelinckx, C. A. Mücher, A. Kooistra, B. Haest, G. DeBlust and A. M. Schmitdt 2011. Integrating remote sensing in Natura 2000 habitat monitoring: Prospects on the way forward. *J Nat Conserv*, pp. 116–125.

van der Reijden, K. J., L. L. Govers, L. Koop, J. H. Damveld, P. M. J. Herman, S. Mestdagh, G. Piet, A. D. Rijnsdorp, G. E. Dinesen, M. Snellen and H. Olff 2021. Beyond connecting the dots: A multi-scale, multi-resolution approach to marine habitat mapping. *Ecol Indic* 128, p. 107849. ISSN 1470-160X. https://doi.org/10.1016/j.ecolind.2021.107849. https://www.sciencedirect.com/science/article/pii/S1470160X21005148

Vaz, A. S., B. Marcos, J. Gonçalves, A. Monteiro, P. Alves, E. Civantos, R. Lucas, P. Mairota, J. Garcia-Robles, J. Alonso, P. Blonda, A. Lomba and J. P. Honrado 2015. Can we predict habitat quality from space? A multi-indicator assessment based on an automated knowledge-driven system. *Int J Appl Earth Obs Geoinf* 37, pp. 106–113.

Walsh, S. J., A. L. McCleary, C. F. Mena, Y. Shao, J. P. Tuttle and A. A. Gonzales 2008. QuickBird and Hyperion data analysis of an invasive plant species in the Galapagos Islands of Ecuador: Implications for control and land use management. *Remote Sens Environ*, pp. 1927–1941.

Wan, J. X. and Y. Ma 2020. Multi-scale spectral-spatial remote sensing classification of coral reef habitats using CNN-SVM. In: H.-S. Jung, S. Lee, J.-H. Ryu and T. Cui (eds.) Advances in Geospatial Research of Coastal Environments. *J Coast Res*, Special Issue, no. 102, pp. 11–20. Coconut Creek (Florida). ISSN 0749-0208.

Wang, K., S. E. Franklin, X. Guo, Y. He and G. J. McDermid 2009. Problems in remote sensing of landscapes and habitats. *Prog Phys Geogr*, pp. 747–768.

Weiers, S., M. Bock, M. Wissen and G. Rossner 2004. Mapping and indicator approaches for the assessment of habitats at different scales using remote sensing and GIS methods. *Landsc Urban Plan*, pp. 43–65.

Wicaksono, P., P.A. Aryaguna and W. Lazuardi 2019. Benthic habitat mapping model and cross validation using machine-learning classification algorithms. *Remote Sens.* 11, no. 11, p. 1279. https://doi.org/10.3390/rs11111279

Wiens, J. 1989. Spatial scaling in ecology. *Funct Ecol*, pp. 385–397.

Wolter, P. T., D. J. Mladenoff, G. E. Host and T. R. Crow 1995. Improved forest classification in the northern lake states using multi-temporal Landsat imagery. *Photogramm Eng Remote Sens*, pp. 1129–1143.

Woodcock, C. E., et al. 1994. Mapping forest vegetation using Landsat TM imagery and a canopy reflectance model. *Remote Sens Environ*, pp. 240–254.

Wulder, M. A., R. J. Hall, N. C. Coops and S. E. Franklin 2004. High spatial resolution remotely sensed data for ecosystem characterization. *BioScience*, pp. 511–521.

Zhang, X., W. Huang, H. Ye and L. Lu. 2023. Study on the identification of habitat suitability areas for the dominant locust species dasyhippus barbipes in inner mongolia. *Remote Sens* 15, no. 6, p. 1718. https://doi.org/10.3390/rs15061718

Zlinszky, A., W. Mücke, H. Lehner, C. Briese and N. Pfeifer 2012. Categorizing wetland vegetation by airborne laser scanning on Lake Balaton and Kis-Balaton, Hungary. *Remote Sens*, pp. 1617–1650.

Part III

Ecology

7 Ecological Characterization of Vegetation Using Multi-Sensor Remote Sensing in the Solar Reflective Spectrum

Conghe Song, Jing Ming Chen, Taehee Hwang, Alemu Gonsamo, Holly Croft, Quanfa Zhang, Matthew Dannenberg, Yulong Zhang, Christopher Hakkenberg, and Junxiang Li

ACRONYMS AND DEFINITIONS

ALOS	Advanced Land Observing Satellite
APAR	Absorbed photosynthetically active radiation
AVHRR	Advanced very high resolution radiometer
EOS	Earth Observing System on board Aqua satellite
ETM+	Enhanced Thematic Mapper+
EVI	Enhanced Vegetation Index
GIMMS	Global Inventory Modeling and Mapping Studies
GLAS	Geoscience Laser Altimeter System
GPP	Gross primary productivity
HyspIRI	Hyperspectral Infrared Imager
ICESat	Instrument on board the Ice, Cloud, and land Elevation
IKONOS	A commercial earth observation satellite, typically, collecting sub-meter to 5 m data
IRS	Indian Remote Sensing Satellites
LAI	Leaf area index
LiDAR	Light detection and ranging
LSP	Land surface phenology
LUE	Light use efficiency
MERIS	Medium-resolution imaging spectrometer
MIR	Mid-infrared
MODIS	Moderate-resolution imaging spectroradiometer
MSS	Multi-spectral scanner
NASA	National Aeronautics and Space Administration
NDVI	Normalized Difference Vegetation Index
NIR	Near-infrared
NOAA	National Oceanic and Atmospheric Administration
NPP	NPOESS Preparatory Project
OLI	Operational land imager
PALSAR	Phased Array type L-band Synthetic Aperture Radar
PAR	Photosynthetically active radiation

PolSAR	RADARSAT-2 polarimetric SAR
PRI	Photochemical reflectance index
PROSPECT	Radiative transfer model to measure leaf optical properties spectra
SAIL	Scattering by arbitrary inclined leaves (SAIL)—a physically based model to measure and model canopy bidirectional reflectance
SAVI	Soil-adjusted vegetation index
SPOT	Satellite Pour l'Observation de la Terre, French Earth Observing Satellites
SWIR	Shortwave infrared
VIIRS	Visible Infrared Imaging Radiometer Suite

7.1 INTRODUCTION

Vegetation is the primary producer in the terrestrial ecosystem. Vegetation absorbs the energy of electromagnetic radiation from the Sun and converts it to the energy that consumers in the ecosystem can use. As a result, vegetation is the foundation for nearly all the goods and services that terrestrial ecosystems provide to humanity. The advent of optical remote sensing revolutionized our ability to map the characteristics of vegetation wall-to-wall in space and to do so repeatedly, in a cost-efficient manner. Many of these vegetation parameters serve as key inputs to ecological models aiming to understand terrestrial ecosystem functions, at regional to global scales. This chapter summarizes the progress made in characterizing vegetation structure and its ecological functions with optical remote sensing. We first provide a brief review of the development of optical sensors designed primarily for vegetation monitoring. Second, we synthesize the progress made in mapping the physical structure of vegetation with optical sensors, including vegetation cover, vegetation successional stages, biomass, and leaf area index (LAI) and its spatial organization, i.e., leaf clumping. Third, we review the achievements made in understanding vegetation function with optical remote sensing, particularly vegetation primary productivity and related ecologically important functions. Primary production provides the energy that drives all subsequent ecosystem processes. Optical remote sensing has made it possible to estimate the primary productivity of vegetation over the entire Earth's land surface (Running et al. 1994; Zhao et al. 2005).

7.2 A BRIEF HISTORY OF KEY OPTICAL SENSORS FOR VEGETATION MAPPING

Optical remote sensing is a technique that detects the properties of the Earth's surface from space, using sensors that capture reflected radiation spanning the visible, near infrared, and short-wave infrared wavelengths (~0.4–2.5 μm [Richards 2013]). Different materials absorb and reflect light differently at various wavelengths. Thus, targets can be distinguished by their unique spectral reflectance signatures. Compared to water and bare soil, healthy vegetation generally absorbs more blue and red light in the visible spectrum for photosynthesis but reflects more near-infrared light (0.7–1.1μm) to prevent tissue damage (Jones and Vaughan 2010). This unique spectral signature of vegetation is the key to monitoring vegetation structure and function with optical sensors. Since the first man-made satellite (Sputnik 1) was launched in 1957, the development of artificial satellites, which provide the platform for optical sensors, has significantly enhanced the collection of remotely sensed data and offers an efficient platform to obtain vegetation information over large areas (Campbell 2002). Here we briefly review the history of major optical sensors for the remote sensing of vegetation launched since the first International Symposium on Remote Sensing of Environment (ISRSE) held at the University of Michigan in 1962. These programs for optical remote sensing, which follow in order of spatial resolution in this chapter, include NOAA/AVHRR, MODIS/MISR, Suomi NPP, Landsat, SPOT, and a series of commercial high-resolution satellites since 1999 (Figure 7.1).

Ecological Characterization of Vegetation

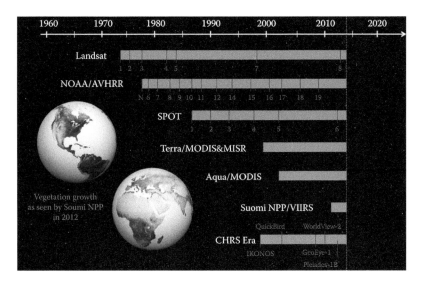

FIGURE 7.1 History of major optical sensors for remote sensing of vegetation reviewed in this chapter. The red line and font indicate the commission date and related satellite, respectively. CHRS is the abbreviation of Commercial High-Resolution Satellite. The two hemisphere images in the lower left corner show the global vegetation growth in terms of Normalized Difference Vegetation Index derived from the Visible-Infrared Imager/Radiometer Suite, or VIIRS, instrument aboard the Suomi NPP satellite (Source: http://www.nasa.gov/mission_pages/NPP/news/vegetation.html#.Ut3_LRAo7IU).

7.2.1 THE NOAA/AVHRR PROGRAM

The Advanced Very High Resolution Radiometer (AVHRR) is a multi-channel radiometer carried on the US National Oceanic and Atmospheric Administration (NOAA) family of polar orbiting platforms (POES) (Table 7.1; http://www.ospo.noaa.gov/Operations/POES/index.html). The AVHRR sensor is active on two POES satellites in opposite orbits (ascending and descending), ensuring that every place on Earth can be observed every six hours. The first AVHRR carried sensors in four spectral channels on TIROS-N (launched in October 1978). This was subsequently improved to a five-channel sensor (AVHRR2) that was initially carried on NOAA-7 (launched in June 1981). The latest sensor is AVHRR3, with six channels, first carried on NOAA-15 (launched May 1998). All AVHRR sensors have the same spatial resolution of 1.09 km at nadir (Table 7.2). The primary purpose of AVHRR is to monitor clouds and to measure the thermal emission of the Earth. However, the first two bands of AVHRR are sensitive to visible/near-infrared radiation, which can be used to detect changes of terrestrial vegetation (Tucker, Townshend, and Goff 1985; Gutman and Ignatov 1998; Tucker et al. 2005). Based on NOAA/AVHRR, several long-term global vegetation index datasets have been established, including NOAA/NASA Pathfinder Normalized Difference Vegetation Index (NDVI) (1981–2000) (http://iridl.ldeo.columbia.edu/SOURCES/.NASA/.GES-DAAC/.PAL/.vegetation/.pal_ndvi.html), GIMMS NDVI (1981–2011) (http://cliveg.bu.edu/modismisr/lai3g-fpar3g.html), and NOAA's Global Vegetation Index (GVI) data (1981–2014) (http://www.ospo.noaa.gov/Products/land/gvi/NDVI.html).

7.2.2 MODIS AND MISR

As the centerpiece of NASA's Earth Science Enterprise (ESE), the Earth Observing System (EOS) consists of a coordinated series of polar-orbiting satellites for continuous observations of the

TABLE 7.1
Summary of NOAA/POES Satellite Family

Satellite	Launch Date	Decommission Date	Sensor
TIROS-N	13-Oct-78	30-Jan-80	AVHRR1
NOAA-6	27-Jun-79	16-Nov-86	AVHRR1
NOAA-7	23-Jun-81	7-Jun-86	AVHRR2
NOAA-8	28-Mar-83	31-Oct-85	AVHRR1
NOAA-9	12-Dec-84	11-May-94	AVHRR2
NOAA-10	17-Sep-86	17-Sep-91	AVHRR1
NOAA-11	24-Sep-88	13-Sep-94	AVHRR2
NOAA-12	13-May-91	15-Dec-94	AVHRR2
NOAA-14	30-Dec-94	23-May-07	AVHRR2
NOAA-15	13-May-98	Present	AVHRR3
NOAA-16	21-Sep-00	Present	AVHRR3
NOAA-17	24-Jun-02	10-Apr-13	AVHRR3
NOAA-18	20-May-05	Present	AVHRR3
NOAA-19	6-Feb-09	Present	AVHRR3

TABLE 7.2
Spectral Specifications of NOAA/AVHRR Sensors

Channel Number	Ground Resolution (km)	Spectral Range (μm)	AVHRR1	AVHRR2	AVHRR3
1	1.09	0.58–0.68	√	√	√
2	1.09	0.725–1.00	√	√	√
3A	1.09	1.58–1.64			√
3B	1.09	3.55–3.93	√	√	√
4	1.09	10.30–11.30	√	√	√
5	1.09	11.50–12.50		√	√

Earth's land, atmosphere, and ocean that offers us a detailed understanding of the biosphere and the dynamics of global change (Justice et al. 2002; Xiong and Barnes 2006). The Terra and Aqua satellites, launched in December 1999 and May 2002, respectively, are two flagships of the EOS. The Moderate Resolution Imaging Spectroradiometer (MODIS) is a key scientific instrument operating on both the Terra and Aqua satellites and is considered to be a major advance over the spectral, spatial, and temporal characteristics of previous sensors (Xiong and Barnes 2006). It has 36 discrete spectral bands ranging from visible through thermal emission bands (wavelengths from 0.4 μm to 14.4 μm) and three ground spatial resolutions (250 m for bands 1 and 2, 500 m for bands 3–7, and 1 km for bands 8–36) (Table 7.3). With complementary morning (local time 10:30 AM for Terra) and afternoon (local time 1:30 PM for Aqua) observations, the Terra and Aqua sensors can image the entire Earth within two days with a swath of 2330 km. The MODIS Characterization Support Team (MCST) from NASA (http://mcst.gsfc.nasa.gov/) is responsible for converting instrument responses (digital numbers) to the primary calibrated products (radiance and reflectance) (Xiong and Barnes 2006), from which over 50 geophysical science products have been developed by the MODIS Science Team (https://lpdaac.usgs.gov/products/modis_products_table). The MODIS Ecosystem Products include vegetation index (Huete et al. 2002), leaf area index (Myneni et al. 2002), vegetation continuous fields (Hansen et al. 2003), gross and net primary productivity (Zhao and Running 2010), and global evapotranspiration (Mu, Zhao, and Running 2011), among others. These products offer unprecedented perspectives of ecosystem structure and function of the biosphere.

The Multi-angle Imaging SpectroRadiometer (MISR) is another innovative sensor on board the Terra satellite (http://www-misr.jpl.nasa.gov/). It is designed to measure the reflected solar radiation of the Earth system from nine discrete viewing angles and four visible/near-infrared spectral bands (Diner et al. 1998). The MISR instrument has nine digital cameras, with one pointing toward nadir and others pointing at forward and backward view angles of 26.1°, 45.6°, 60°, and 70.5°. For each direction, the cameras record reflected radiation in four spectral bands (blue, green, red, and near-infrared). During each orbit, MISR obtains a swath of imagery that is 360 km wide by about 20,000 km long with spatial resolutions of 250 m at nadir and 275 m at other angles. The multi-angle viewing strategy of MISR provides a unique opportunity to characterize the structure and dynamics of the atmosphere and land surface (Diner et al. 1998). Among other applications, Terra/MISR has been used to retrieve aerosol distribution (Martonchik et al. 2002), measure cloud height (Davies and Molloy 2012), estimate leaf area index (Hu et al. 2003), extract canopy structure (Chen, Menges, and Leblanc 2005a), and improve the classification of land cover (Liesenberg, Galvão, and Ponzoni 2007).

7.2.3 Suomi NPP/VIIRS

The Suomi National Polar-orbiting Partnership (Suomi NPP), as a major component of NOAA's Joint Polar-orbiting Satellite System (JPSS), was designed to provide continuity with NASA's Earth Observing System (EOS) (Justice et al. 2013). The satellite was launched on October 28, 2011, and was named after Verner E. Suomi, a meteorologist at the University of Wisconsin-Madison who is widely recognized as "the father of satellite meteorology" (http://www.nasa.gov/mission_pages/NPP/news/suomi.html). The Visible Infrared Imaging Radiometer Suite (VIIRS) is a key scanning radiometer on board Suomi NPP, which signifies a new era of moderate-resolution imaging capabilities following the legacy of AVHRR and MODIS (Cao et al. 2013). VIIRS is designed to collect imagery of radiometric measurements for the Earth in wavelengths ranging from 0.4 to 12.5 µm (Oudrari et al. 2012). It has 22 spectral bands, including five imagery bands (I-bands) with 375 m spatial resolution, one day-night band (DNB) with 375 m spatial resolution, and 16 moderate resolution bands (M bands) with 750 m spatial resolution (Table 7.4). VIIRS has a large ground swath of about 3040 km and provides daily coverage of the entire globe. After about two years of calibration and validation, the VIIRS data have achieved provisional maturity (Cao et al. 2013) and are now being used to produce more than 20 land and cryosphere products by NOAA and NASA (Justice et al. 2013), including vegetation index (Vargas et al. 2013), active fire (Csiszar et al. 2014), surface albedo (Wang et al. 2013), and nighttime light distribution (Miller et al. 2012).

7.2.4 The Landsat Program

The US Landsat program (http://landsat.usgs.gov) has been collecting images of Earth's land surface for over four decades, providing the longest continuous archive of the Earth's surface conditions (Markham and Helder 2012). The first Landsat satellite, originally named "Earth Resources Technology Satellite," was launched in 1972. To date, eight Landsat satellites have been launched.

TABLE 7.3
Spectral and Spatial Resolutions of MODIS Sensors On Board Terra/Aqua

Channel Number	Spectral Range (μm)	Usage	Ground Resolution (m)
1	0.620–0.670	Land cover transformation, vegetation chlorophyll	250
2	0.841–0.876	Cloud amount, vegetation land cover transformation	250
3	0.459–0.479	Soil/vegetation differences	500
4	0.545–0.565	Green vegetation	500
5	1.230–1.250	Leaf/canopy differences	500
6	1.628–1.652	Snow/cloud differences	500
7	2.105–2.155	Cloud properties, land properties	500
8	0.405–0.420	Chlorophyll	1000
9	0.438–0.448	Chlorophyll	1000
10	0.483–0.493	Chlorophyll	1000
11	0.526–0.536	Chlorophyll	1000
12	0.546–0.556	Sediments	1000
13h	0.662–0.672	Atmosphere, sediments	1000
13l	0.662–0.672	Atmosphere, sediments	1000
14h	0.673–0.683	Chlorophyll fluorescence	1000
14l	0.673–0.683	Chlorophyll fluorescence	1000
15	0.743–0.753	Aerosol properties	1000
16	0.862–0.877	Aerosol properties, atmospheric properties	1000
17	0.890–0.920	Atmospheric properties, cloud properties	1000
18	0.931–0.941	Atmospheric properties, cloud properties	1000
19	0.915–0.965	Atmospheric properties, cloud properties	1000
20	3.660–3.840	Sea surface temperature	1000
21	3.929–3.989	Forest fires and volcanoes	1000
22	3.929–3.989	Cloud temperature, surface temperature	1000
23	4.020–4.080	Cloud temperature, surface temperature	1000
24	4.433–4.498	Cloud fraction, troposphere temperature	1000
25	4.482–4.549	Cloud fraction, troposphere temperature	1000
26	1.360–1.390	Cloud fraction (thin cirrus), troposphere temperature	1000
27	6.535–6.895	Mid-troposphere humidity	1000
28	7.175–7.475	Upper troposphere humidity	1000
29	8.400–8.700	Surface temperature	1000
30	9.580–9.880	Total ozone	1000
31	10.78–11.28	Cloud and surface temperature, forest fires and volcanoes	1000
32	11.77–12.27	Cloud height, forest fires and volcanoes, surface temperature	1000
33	13.19–13.49	Cloud fraction, cloud height	1000
34	13.49–13.79	Cloud fraction, cloud height	1000
35	13.79–14.09	Cloud fraction, cloud height	1000
36	14.09–14.39	Cloud fraction, cloud height	1000

All but Landsat 6 successfully reached orbit (Table 7.5). The most recent Landsat satellite, the eighth in the series, was launched in February 2013. The Return Beam Vidicon (RBV), a television camera carried on board Landsat 1 through 3, obtained visible and near infrared photographic images, while the Multispectral Scanner (MSS) sensors, which were carried on board Landsat 1

TABLE 7.4
Spectral and Spatial Resolutions of Suomi NPP/VIIRS

Channel Number	Spectral Range (μm)	Description	Ground Resolution (m)
I1	0.6–0.68	Visible/reflective	375
I2	0.85–0.88	Near IR	375
I3	1.58–1.64	Shortwave IR	375
I4	3.55–3.93	Medium-wave IR	375
I5	10.5–12.4	Long-wave IR	375
DNB	0.5–0.9	Visible/reflective	750
M1	0.402–0.422	Visible/reflective	750
M2	0.436–0.454		750
M3	0.478–0.488		750
M4	0.545–0.565		750
M5	0.662–0.682		750
M6	0.739–0.754	Near IR	750
M7	0.846–0.885		750
M8	1.23–1.25	Shortwave IR	750
M9	1.371–1.386		750
M10	1.58–1.64		750
M11	2.23–2.28		750
M12	3.61–3.79	Medium-wave IR	750
M13	3.97–4.13		750
M14	8.4–8.7	Long-wave IR	750
M15	10.26–11.26		750
M16	11.54–12.49		750

through 5, acquired digital images around the globe nearly continuously from July 1972 to October 1992. Compared with MSS, RBV was rarely used scientifically but considered only for engineering evaluation purposes. It was replaced by the Thematic Mapper (TM) sensor on board Landsat 4 and 5, which consisted of seven spectral bands with a 16-day repeat cycle and a spatial resolution of 30 m (the thermal infrared band 6 was collected at 120 m spatial resolution). By the time of its decommission at the end of 2012, Landsat 5 had orbited the Earth for 29 years—an extraordinary success for NASA—far exceeding its original three-year design life (https://landsat.usgs.gov/Landsat5Tribute.php).

On Landsat 7, the TM sensor was replaced by the Enhanced Thematic Mapper Plus (ETM+), which included the addition of a panchromatic band 8 at 15 m spatial resolution (Table 7.6) that can be used to "sharpen" the other bands. However, the Scan Line Corrector (SLC) on the satellite failed in May 2003, causing a permanent loss of about 25% of data toward the scanning edges in all subsequent Landsat 7 images. Fortunately, the successful launch of Landsat 8 in 2013 ensured the continuity of Landsat data. The Operational Land Imager (OLI) sensors on board Landsat 8 include refined versions of the seven TM and ETM+ heritage bands, along with two new bands: a deep blue band for coastal/aerosol studies and a shortwave infrared band for cirrus cloud detection (Table 7.6). Landsat 8 Thermal InfraRed Sensors (TIRS) are composed of two thermal bands with a spatial resolution of 100 m (Table 7.6). Both OLI and TIRS sensors provide improved signal-to-noise radiometric performance quantized over a 12-bit dynamic range compared with the 8-bit instruments for TM and ETM+ sensors.

Conceived in the 1960s, the Landsat program has kept improving its imaging capability and quality while ensuring continuity over the full instrument record (Loveland and Dwyer 2012). To date, it has provided the longest and most geographically comprehensive record of the Earth's surface. Thanks to a data policy change in 2008, all new and archived Landsat images have been made freely available to the public by the US Geological Survey (Woodcock et al. 2008), which has spurred a dramatic increase in scientific applications using Landsat imagery (Wulder et al. 2012).

TABLE 7.5
Landsat Satellites Launched

Satellite	Launch Date	Decommission Date	Orbit Height (km)	Revisit Time (days)	Sensors
Landsat 1	Jul. 1972	Jan. 1978	917	18	RBV/MSS
Landsat 2	Jan. 1975	Feb. 1982	917	18	RBV/MSS
Landsat 3	Mar. 1978	Mar. 1983	917	18	RBV/MSS
Landsat 4	Jul. 1982	Dec. 1993	705	16	MSS/TM
Landsat 5	Mar. 1984	Dec. 2012	705	16	MSS/TM
Landsat 6	Oct. 1993	Failed	—	—	—
Landsat 7	Apr. 1999	Present	705	16	ETM+
Landsat 8	Feb. 2013	Present	705	16	OLI/TIRS

RBV: Return Beam Vidicon; MSS: Multispectral Scanner; TM: Thematic Mapper; ETM+: Enhanced Thematic Mapper Plus; OLI: Operational Land Imager; TIRS: Thermal InfraRed Sensor

TABLE 7.6
Spectral and Spatial Resolutions of Landsat Sensors (See Table 7.5 for Sensor Abbreviations)

Sensor	Channel Number	Spectral Range (µm)	Description	Ground Resolution (m)
RBV[a]	1	4.75–5.75	Blue	80
	2	5.80–6.80	Orange-Red	80
	3	6.90–8.30	Red-NIR	80
MSS	4	0.5–0.6	Green	57×79
	5	0.6–0.7	Red	57×79
	6	0.7–0.8	NIR	57×79
	7	0.8–1.1	NIR	57×79
	8[b]	10.4–12.6	Thermal	57×79
TM	1	0.45–0.52	Blue	30
	2	0.52–0.60	Green	30
	3	0.63–0.69	Red	30
	4	0.76–0.90	NIR	30
	5	1.55–1.75	SWIR	30
	6	10.40–12.50	Thermal	120
	7	2.09–2.35	SWIR	30

TABLE 7.6 (Continued)
Spectral and Spatial Resolutions of Landsat Sensors (See Table 7.5 for Sensor Abbreviations)

Sensor	Channel Number	Spectral Range (µm)	Description	Ground Resolution (m)
ETM+	1	0.45–0.52	Blue	30
	2	0.52–0.60	Green	30
	3	0.63–0.69	Red	30
	4	0.77–0.90	NIR	30
	5	1.55–1.75	SWIR	30
	6	10.40–12.50	Thermal	60
	7	2.08–1.35	SWIR	30
	8	0.52–0.90	Pan	15
OLI	1	0.43–0.45	Deep Blue	30
	2	0.45–0.51	Blue	30
	3	0.53–0.59	Green	30
	4	0.64–0.67	Red	30
	5	0.85–0.88	NIR	30
	6	1.57–1.65	SWIR1	30
	7	2.11–2.29	SWIR2	30
	8	0.50–0.68	Panchromatic	15
	9	1.36–1.38	Cirrus clouds	30
TIRS	10	10.60–11.19	Thermal	100
	11	11.50–12.51	Thermal	100

[a] Lansat 3 had two RBV cameras with 40 m ground resolution.
[b] Only Landsat 3 had this thermal channel.

NIR: Near Infrared; SWIR: Shortwave Infrared; Pan: Panchromatic

7.2.5 THE SPOT PROGRAM

The SPOT (Satellite Pour l'Observation de la Terre) program is a joint Earth observing satellite family initiated by France in partnership with Belgium and Sweden (http://www.vgt.vito.be/). Since 1986, six SPOT satellites have been successfully launched (Table 7.7). Currently, SPOT 5 and 6 are operational. The High Resolution Visible (HRV) sensor with one panchromatic (10 m spatial resolution) and three multispectral bands (20 m spatial resolution; green, red, near infrared) were carried on board SPOT 1 through 3 (Table 7.8). They have a scene size of 60×60 km² and a revisit interval of one to four days, depending on the latitude. SPOT 4 featured the High Resolution Visible Infrared (HRVIR) instrument, which was similar to the HRV but with the addition of a short-wave infrared (SWIR) band and a narrower panchromatic band (Table 7.8). SPOT 5 carries the High Resolution Geometrical (HRG) sensor (derived from HRVIR), offering a finer resolution of 2.5–5 m in panchromatic mode and 10 m in multispectral mode (20 m for SWIR) (Table 7.8). SPOT 6 was launched in September 2012, carrying the New Astrosat Optical Modular Instrument (NAOMI). NAOMI is capable of imaging the Earth with a resolution of 1.5 m panchromatic and 6 m multispectral (blue, green, red, NIR) with daily revisit capability, providing the finest level of spatial detail in the history of the SPOT family of satellites (Table 7.8). It is worth noting that the Vegetation sensor was carried on board SPOT 4 and 5 (launched in 1998 and 2002, respectively). SPOT/Vegetation was designed to provide daily coverage of the entire globe with a spatial resolution of 1.15 km. Unlike many other commercial high-resolution images, some SPOT/Vegetation products are publicly available. The ten-day 1 km global NDVI, for example, is available from May 1998 to the present (http://www.vgt.vito.be/) and has been valuable for studying agriculture, deforestation, and other vegetation changes on a broad scale (Kamthonkiat et al. 2005; Liu et al. 2010).

TABLE 7.7
Summary of SPOT Satellite Family

Satellite	Launch Date	Decommission Date	Orbit Height (km)	Revisit Time (day)	Sensors
SPOT 1	Feb. 1986	Dec. 1990	832	1–4	HRV
SPOT 2	Jan. 1990	July 2009	832	1–4	HRV
SPOT 3	Sep. 1993	Nov. 1997	832	1–4	HRV
SPOT 4	Mar. 1998	July 2013	832	1–4	Vegetation/HRVIR
SPOT 5	May 2002	Present	832	1–4	Vegetation/HGR
SPOT 6	Sep. 2012	Present	694	1–4	NAOMI

HRV: High Resolution Visible; HRVIR: High Resolution Visible InfraRed; NAOMI: New Astrosat Optical Modular Instrument

TABLE 7.8
Spectral and Spatial Resolutions of Optical Sensors On Board SPOT Satellites (See Table 7.7 for Sensor Abbreviations)

Sensor	Mode	Description	Spectral Range (μm)	Ground Resolution (m)
HRV	Multispectral	Green	0.50–0.59	20
		Red	0.61–0.68	20
		NIR	0.78–0.89	20
	Panchromatic	PAN	0.50–0.73	10
HRVIR	Multispectral	Green	0.50–0.59	20
		Red	0.61–0.68	20
		NIR	0.79–0.89	20
		MIR	1.58–1.75	20
	Panchromatic	PAN	0.61–0.68	10
HGR	Multispectral	Green	0.50–0.59	10
		Red	0.61–0.68	10
		NIR	0.79–0.89	10
		SWIR	1.58–1.75	20
	Panchromatic	PAN	0.51–0.73	5/2.5
NAOMI	Multispectral	Blue	0.45–0.53	6
		Green	0.53–0.59	6
		Red	0.63–0.70	6
		NIR	0.76–0.89	6
	Panchromatic	PAN	0.45–0.75	1.5
Vegetation	Multispectral	Blue	0.43–0.47	1150
		Red	0.61–0.68	1150
		NIR	0.78–0.89	1150
		SWIR	1.58–1.75	1150

7.2.6 COMMERCIAL HIGH-RESOLUTION SATELLITE ERA

IKONOS, which was launched in 1999, is the first high-resolution commercial Earth observation satellite that collects imagery at sub-meter (0.82 m for panchromatic band) spatial resolution. This marked the start of a new era of high-resolution Earth observation by commercial satellites, which may revolutionize the future of the entire photogrammetric and remote sensing community (Dial et al. 2003). After IKONOS, a series of commercial civilian satellites with optical sensors were launched to produce panchromatic images with spatial resolutions ranging from less than 0.5 m to 3 m and multispectral images with spatial resolution ranging from 2 to 10 m (Table 7.9). The finer-resolution panchromatic bands can be used to sharpen the coarser-resolution multispectral bands, increasing the spatial detail of multispectral images (Zhang and Mishra 2012). Based on the previously launched satellites IKONOS and OrbView-3, the US commercial company GeoEye Inc. (merged with DigitalGlobe since January 2013) launched by far the finest spatial resolution civilian Earth observation satellite (GeoEye-1) in September 2008. GeoEye-1 provides 0.41 m panchromatic and 1.65 m multispectral (blue, green, red, NIR) imagery and features a revisit time of less than three days with a swath of 22.2 km. Based on QuickBird and WorldView-1, the US company DigitalGlobe Inc. launched the first high-resolution commercial satellite with eight multispectral imaging bands in October 2009. This satellite, known as WorldView-2, is capable of collecting panchromatic imagery at 0.46 m spatial resolution and multispectral (coastal, blue, green, yellow, red, red edge, near infrared 1 and 2) imagery at 1.84 m spatial resolution with an average revisit time of 1.1 days. Compared to the four standard multispectral bands (blue, green, red, NIR), the additional bands increase the spectral information used for vegetation analysis at high spatial resolutions. Depending on budget and usage purposes, other high resolution commercial satellite images that can be employed include France's Pleiades-1A/B (0.5 m pan, 2 m multispectral), Korea's KOMPSAT-2 (1 m pan, 4 m multispectral), China-Taiwan's FORMOSAT-2 (2 m pan, 4 m multispectral), Japan's ALOS (2.5 m pan, 10 m multispectral), among others.

TABLE 7.9
Major Commercial High-Resolution Satellites since 1999

Satellite	Year Launched	Country	Pan Band(μm)/ Ground Resolution (m)	Multispectral Bands[b] Ground Resolution (m)	Swath (km)	Revisit Time (day)
IKONOS	1999	US	(0.45–0.90) 0.82	(Blue, green, red, NIR) 4	11.3 × 11.3	3–4
QuickBird	2001	US	(0.405–1.053) 0.61	(Blue, green, red, NIR) 2.44	16.5 × 16.5	1–3.5
OrbView-3	2003	US	(0.45–0.90) 1	(Blue, green, red, NIR) 4	8 × 8	1–3
FORMOSAT-2	2004	China-Taiwan	(0.45–0.90) 2	(Blue, green, red, NIR) 4	24 × 24	1
CartoSat-1	2005	India	(0.5–0.85) 2.5	—	25 × 25	5
ALOS	2005	Japan	(0.52–0.77) 2.5	10 (Blue, green, red, NIR)	70 × 70	2
EROS-B	2006	Israel	(0.5–0.9) 0.7	—	7 × 7	5–6
KOMPSAT-2	2006	Korea	(0.50–0.90) 1	(Blue, green, red, NIR) 4	15 × 15	1–3

(*Continued*)

TABLE 7.9 (Continued)
Major Commercial High-Resolution Satellites since 1999

Satellite	Year Launched	Country	Pan Band(µm)/ Ground Resolution (m)	Multispectral Bands[b] Ground Resolution (m)	Swath (km)	Revisit Time (day)
WorldView-1	2007	US	(0.40–0.90) 0.46[a]	—	17.7 × 17.7	1–5
GeoEye-1	2008	US	(0.45–0.90) 0.41[a]	(Blue, green, red, NIR) 1.65	22.2 × 22.2	1–3
RapidEye	2008	German	—	(Blue, green, red, NIR) 5	77 × 77	1
WorldView-2	2009	US	(0.45–0.80) 0.46[a]	(Coastal, blue, green, yellow, red, red edge, NIR 1 and 2)[c] 1.85	16.4 × 16.4	1–5
Pleiades-1A	2011	France	(0.48–0.83) 0.5	(Blue, green, red, NIR) 2	20 × 20	1
SPOT 6	2012	France	(0.450–0.745) 1.5	(Blue, green, red, NIR) 5	60 × 60	1–3
ZY-3	2012	China	(0.50–0.80) 2.1–3.5	(Blue, green, red, NIR) 6	52 × 52	3–5
Pleiades-1B	2012	France	(0.48–0.83) 0.5	(Blue, green, red, NIR) 2	20 × 20	1
GF-1	2013	China	(0.45–0.90) 2	(Blue, green, red, NIR) 8	60 × 60	4

[a] Due to US Government Licensing, the imagery will be made available commercially at ground resolution of 0.5 m.
[b] Different satellites may have slightly different spectral ranges for each of their multispectral bands.
[c] The spectral ranges for WorldView-2 are: 0.40–45µm (Coastal), 0.45–0.51µm (Blue), 0.51–0.58µm (Green), 0.585–0.625µm (Yellow), 0.63–0.69µm (Red), 0.705–0.745µm (Red Edge), 0.77–0.895µm (NIR1), and 0.86-1.04µm (NIR2), respectively.

7.2.7 Future Direction of Optical Remote Sensing

Optical sensors are poised to acquire increasingly high-quality data across a wide range of spatial, temporal, and spectral resolutions. Following the merger of Digital Global and MacDonald, Dettwiler and Associates merged to become Maxar Technologies in 2017. Maxar currently operates the WorldView Legion satellites with 34 cm spatial resolution for panchromatic band and 1.36 m spatial resolution for eight multispectral bands. Meanwhile, the European Space Agency (ESA) is carrying out one of the most ambitious Earth observation programs to date, called Copernicus. To satisfy the operational needs of Copernicus, up to 30 Sentinel satellites with various sensors will be developed (http://www.esa.int/ESA). The first Sentinel satellite (S1) was successfully put in orbit in April 2014. Undoubtedly, integrating multiple sources of optical remote sensing will offer a valuable opportunity for the scientific community to investigate and understand the structures and functions of terrestrial ecosystems at different spatial and temporal resolutions (Richards 2013; Weng 2011).

7.3 OPTICAL REMOTE SENSING OF VEGETATION STRUCTURE

Optical remotely sensed signals originate from the photons in the solar spectrum after interactions with the land surface. Remote sensing signals over vegetated areas are determined by the abundance and spatial organization of vegetation (Li and Strahler 1985; Asrar, Myneni, and Choudhury 1992).

Ecological Characterization of Vegetation

Therefore, information about vegetation structure can be derived from optical remotely sensed data. In this section, we review the capabilities of optical remote sensing in deriving information about vegetation cover, forest successional stage, leaf area index, and biomass, all of which are essential biophysical information to understand terrestrial ecosystem functions.

7.3.1 Vegetation Cover

Vegetation cover is perhaps the simplest measure of vegetation structure that can be derived from remote sensing. The most common approach for mapping vegetation cover from remotely sensed imagery is to assign a single class to each pixel. Vegetation cover can then be estimated as the percentage of pixels classified as vegetation. This approach makes an implicit assumption that each pixel represents a homogenous cover type. This assumption may be a reasonable one when the pixel size is significantly smaller than the average vegetation patch size. However, this assumption is rarely valid for coarse resolution remotely sensed imagery because coarse-resolution pixels generally comprise a mixture of several cover types. Assuming homogeneous land cover composition at the pixel level can lead to substantial errors in estimates of areal abundance (Foody and Cox 1994; Moody and Woodcock 1994).

More accurate estimation of vegetation cover from remotely sensed imagery is usually based on sub-pixel land cover composition, i.e., the fraction of a pixel that is covered by vegetation. The fractional vegetation cover (fc) concept, introduced by Deardorff (1978), is a key component of the current generation of climate models (Zeng et al. 2000). Many methods have been proposed to derive fc from remotely sensed imagery.

7.3.1.1 Regression

A common approach to estimate vegetation fraction cover or percent tree cover is to develop an empirical relationship between ground-based measurements with remotely sensed signals, such as spectral vegetation indices (e.g., NDVI, EVI) or suites of other remotely sensed measurements. A variety of model types have been used for this purpose, including ordinary least squares regression (Jiapaer, Chen, and Bao 2011), generalized linear models (Schwarz and Zimmermann 2005), stepwise multiple regression (Cohen et al. 2001), reduced major axis regression (Hayes et al. 2008), and a variety of machine learning methods, such as decision trees and neural networks (Colditz et al. 2011; Verrelst et al. 2012). At the global scale, the MODIS Vegetation Continuous Fields product (MOD44B) estimates sub-pixel percentages of tree cover, non-tree vegetation cover, and bare ground at 250 m spatial resolution using regression trees and a large suite of metrics calculated from MODIS reflectance data. The algorithm estimates a mean vegetation cover for each node in the regression tree and then uses a linear model fit to the independent variable to fine tune the tree cover estimation for each node (Hansen et al. 2002; Hansen et al. 2003; Hansen et al. 2005).

7.3.1.2 Fuzzy Classification

In a typical application of supervised classification of remotely sensed imagery, a single land-cover/land-use class is assigned to each pixel based on its spectral similarity to training classes (so-called hard classifiers). Some of these classifiers can also be modified to predict gradients of class membership—"fuzzy" or "soft" classifications—that provide a relative measure of the similarity of the pixel spectral signature to the class signature. The posterior probabilities from maximum likelihood classifiers, for example, have been used to estimate sub-pixel land cover fractions, though with limited success (Bastin 1997). Artificial neural networks, frequently used in hard classifications, have also been used for deriving sub-pixel membership functions (Foody 1996; Atkinson, Cutler, and Lewis 1997).

Some classifiers, such as the fuzzy c-means algorithm, are specifically designed to provide fuzzy membership functions (Foody and Cox 1994). In these approaches, each pixel generally receives a membership value (ranging from 0 to 1) for each class, with the membership values summing to 1. Relatively pure pixels are likely to receive large membership values for a single class, while mixed pixels are more likely to receive intermediate values for multiple classes. The relationship between fuzzy membership values and sub-pixel land cover fractions can be further improved through a simple regression model based on reference data (Foody and Cox 1994). Fractional land cover obtained using these fuzzy classifiers generally compares favorably with other methods and provides considerable improvement in areal estimates of forest cover over those obtained from hard classifications (Foody and Cox 1994; Atkinson, Cutler, and Lewis 1997; Bastin 1997).

7.3.1.3 Mixture Models with Spectral Vegetation Indices

Simple two-class mixture models typically assume that pixels in the natural environment are composed of vegetation and soil background. The radiance received at the satellite sensor is therefore assumed to be a mixture of the spectral signatures of vegetation and soil, weighted by their respective fractions. Gutman and Ignatov (1998) proposed a simple linear mixture model based on NDVI to estimate the proportions of vegetation and soil within a pixel:

$$f_c = \frac{NDVI - NDVI_S}{NDVI_V - NDVI_S}, \tag{7.1}$$

where NDVI is the vegetation index for a given pixel and $NDVI_V$ and $NDVI_S$ are the vegetation indices corresponding to pixels completely covered with dense vegetation and soil, respectively. Other studies have suggested that multiple scattering in vegetation canopies can result in non-linear relationships between fc and NDVI and have therefore proposed similar alternative models (Carlson and Ripley 1997):

$$f_c = \left(\frac{NDVI - NDVI_S}{NDVI_V - NDVI_S} \right), \tag{7.2}$$

Equation (7.1) has been applied for global scale estimation of fc (Zeng et al. 2000), with NDVI derived from the maximum 12-month NDVI of each pixel in AVHRR imagery, $NDVI_V$ computed separately for each vegetated land cover class in the IGBP database based on NDVI histograms, and a globally uniform $NDVI_S$ of 0.05 (corresponding to the fifth percentile of the NDVI histogram for the barren or sparsely vegetated category). Results from this model were comparable with, but systematically less than, fc calculated from a more complex global linear mixture model (DeFries, Townshend, and Hansen 1999; DeFries et al. 2000).

7.3.1.4 Spectral Mixture Analysis

The procedure described in the previous section is a special case of a more general technique called spectral mixture analysis (SMA). The generalized formulation of SMA techniques can be represented in matrix notation as:

$$x = Mf + e, \tag{7.3}$$

where $\mathbf{x}$ is a column vector of the observed reflectance (with one element per spectral band), $\mathbf{M}$ is a matrix of spectral endmembers (with each column representing the spectral signature of pure pixels for each endmember), $\mathbf{f}$ is a column vector of sub-pixel proportions for each endmember, and $\mathbf{e}$ is a column vector of error residuals. Once $\mathbf{x}$ and $\mathbf{M}$ are known, equation (7.3) can be inverted and solved for the unknown $\mathbf{f}$ using a variety of techniques (including ordinary least squares), typically with the constraint that the sum of elements in $\mathbf{f}$ equals unity and each element takes a value within (0,1). (Somers et al. 2011).

Ecological Characterization of Vegetation

The major challenge in the use of SMA is the selection of appropriate endmembers and their spectral signatures. Endmembers can be derived either directly from remotely sensed imagery (image endmembers) (e.g., DeFries, Townshend, and Hansen 1999; Song 2004) or from field or laboratory measurements (e.g., Adams et al. 1995; Roberts et al. 1998). The number of endmembers that can be used is limited by the dimensionality of the remotely sensed image data. In the case of Landsat imagery, for example, SMA techniques are generally limited to three to four endmembers. In many complex landscapes, three to five endmembers may be insufficient to represent the spectral and spatial variability within an image. A variety of techniques exist to account for endmember variability (reviewed in Somers et al. 2011), including the multiple endmember spectral mixture analysis (MESMA) technique (Roberts et al. 1998), in which endmember models are selected separately for each pixel in the image from a large library of spectral endmembers to construct numerous candidate models, from which the "best" candidate model is selected for each pixel to perform SMA. Somers et al. (2011) suggest that these types of iterative endmember selection approaches can provide a more effective representation of endmember variability than simple SMA approaches (in which endmember signatures are assumed constant across the entire image). Song (2005) developed the Bayesian Spectral Mixture Analysis (BSMA) to account for endmember signature variation when estimating fractional vegetation cover in a pixel. The endmember spectral signature in BSMA is represented by a probability mass function instead of a constant. Deng and Wu (2013) further developed an algorithm that adaptively generates endmember spectral signatures over space to account for endmember spectral signature variations.

7.3.2 Forest Successional Stages

Forest succession can be defined as the change in the three-dimensional architecture and species composition of forest communities through time (Pickett, Cadenasso, and Meiners 2013). Successional stage serves as a useful proxy for forest age, as well as competition-mediated demographic and structural development (Peet and Christensen 1988). Successional processes have a profound impact on the provision of ecological goods and services, including productivity (Gower, McMurtrie, and Murty 1996), nutrient cycling (Law et al. 2001), and biodiversity (Denslow 1980). Though succession is a continuous process, most models characterize succession as a four-stage process including (1) stand initiation/establishment, (2) stem exclusion/thinning, (3) understory re-initiation/transition, and (4) old growth/steady state (Oliver and Larson 1996; Peet and Christensen 1987). Remote sensing technologies, including physical and empirical-based models, offer an efficient method for monitoring forest succession over large spatial extents.

7.3.2.1 Physical-Based Models

Physical-based models simulate vegetation canopy reflectance based on the physical principles of interaction among incoming solar radiation and canopy structural elements. Forward models like the Li-Strahler model (Li and Strahler 1985) have proven useful for understanding the relationship between vegetation structure and canopy reflectance. The Li-Strahler model is a geometric-optical model that simulates canopy reflectance as viewed by the sensor based on the weighted average of individual scene components within a pixel created by the sun—tree crown geometry. This model can be inverted to estimate key canopy structure parameters that manifest successional stage, including mean crown size and canopy cover (Franklin and Strahler 1988; Wu and Strahler 1994). Li, Strahler, and Woodcock (1995) improved the Li-Strahler model by representing tree crowns as ellipsoids rather than cones and incorporating multiple scatterings of photons with a turbid medium radiative transfer model (GORT).

Song, Woodcock, and Li (2002) coupled the GORT model with a forest succession model ZELIG (Urban 1990), which simulates stand growth and development, to understand how forest succession changes in the spectral/temporal domain. Using this hybrid model to simulate Landsat TM reflectance of stand succession from open conditions to young, mature, and old-growth stages, they found forest succession produces highly nonlinear temporal trajectories in the Tasseled Cap Brightness/Greenness space. The nonlinear spectral/temporal trajectory pattern produced by the GORT-ZELIG simulation compared well with that derived from a time series of Landsat TM images and stand age information from Forest Inventory and Analysis (FIA) stand data collected by the US Forest Service (Song, Schroeder, and Cohen 2007).

7.3.2.2 Empirical-Based Approaches

Empirical-based approaches to the remote sensing of forest succession include (1) indirect space for time substitutions using single or multi-date imagery and (2) direct monitoring of successional change using multi-temporal change detection and time-series analysis. While the former is more effective at distinguishing successional stands over large landscapes, the latter is better adapted to capture ongoing successional change in individual stands. Numerous studies show both approaches to produce robust results, though factors of uncertainty remain, such as atmospheric and ground conditions as well as the confounding effects of topography, sun and view angles, and phenology (Song and Woodcock 2003).

7.3.2.2.1 Space-for-Time Substitution

Given that the short historical record of remotely sensed imagery is often insufficient to capture temporal processes of forest succession that could stretch over centuries, space-for-time substitution uses stands in different successional stages at different locations in space to construct a proxy successional trajectory for a single stand through time. This approach is particularly useful for distinguishing mature and old-growth forests. For example, Fiorella and Ripple found most raw Landsat TM bands to be inversely correlated with forest age in the Pacific Northwest (PNW), with mean TM spectral values tending to be lower for old-growth stands compared with those for mature stands (Fiorella and Ripple 1993a; Fiorella and Ripple 1993b). Jakubauskas (1996) found a nonlinear trend in TM spectral reflectance from early to late successional forests in Wyoming resulting from the combined effects of overstory canopy development, increasing canopy shadow, and understory conditions. Spectral indices have likewise proven effective at distinguishing successional stages. For instance, TM 4/5 ratio (Fiorella and Ripple 1993a) and Tasseled Cap wetness have been used to distinguish successional stage in the conifer forests of the PNW (Cohen and Spies 1992; Fiorella and Ripple 1993b; Cohen, Spies, and Fiorella 1995), while the NDVI/ETM+ band 5 ratio successfully distinguished four successional stages of tropical secondary forests in Brazil (Vieira et al. 2003).

Sabol et al. (2002) mapped structural development and stand age in Washington State, USA, using a spectral mixture analysis (SMA) approach consisting of four spectral endmembers: green vegetation (GV), nonphotosynthetic vegetation (NPV), soil, and shade (topographic shading and canopy shadows). They found successional stage to follow a non-linear trajectory, characterized by high NPV from slash after clearcut, to high GV during canopy closure, and finally to higher shade fractions as forests mature and gaps develop. Other techniques using the space-for-time substitution approach include those utilizing spatial predictors. Cohen, Spies, and Bradshaw (1990) identified the crown-gap patterning characteristics of different successional stages in Douglas fir forests by interpreting semivariograms of DN values at different spatial resolutions. They found a pronounced periodicity for the more spatially clumped old-growth crowns compared with the more texturally homogeneous early successional stands. Cohen and Spies (1992) compared spatial versus spectral variables to predict stand structural attributes in Douglas fir forests, finding the most robust results using textural measures from a 10 m panchromatic SPOT HRV image.

Liu et al. (2008) demonstrated the advantages of multi-temporal vs. single-date Landsat TM images to distinguish successional groups in the PNW, a result corroborated by others (Song, Schroeder, and Cohen 2007). Multi-temporal Landsat imagery has been used to evaluate tropical secondary forest regrowth in Brazil (Steininger 1996) and distinguish secondary forests from agricultural lands and old-growth forests in southern Costa Rica (Helmer, Brown, and Cohen 2000). Jiang et al. (2004) used a dense stack of Landsat ETM+ images for successional classification in the PNW, achieving high accuracy for late seral forests. Bergen and Dronova (2007) used multi-temporal Landsat ETM+ data to demonstrate the relationship between ecological land units and the successional pathways of hardwood forests in northern Michigan.

7.3.2.2.2 Multi-Temporal Change Detection and Time-Series Analysis

Multi-temporal imagery can be used to capture successional processes by assessing change at the stand/pixel level between two or more dates. While this approach circumvents errors from spatial extrapolation in the space-for-time substitution approach, observation of successional development is limited by the temporal extent of the satellite record. In addition, the success of time-series analysis in monitoring subtle successional change over time ultimately hinges on the successful calibration of the image series (Song and Woodcock 2003; Schroeder et al. 2006).

In a classic paper on the subject, Hall et al. (1991) used Landsat MSS images from 1973 and 1983 to infer transition rates in ecological states associated with forest succession in the boreal forests of Minnesota, USA. McDonald, Halpin, and Urban (2007) used change vector analysis to validate the prediction of successional models (e.g., Oosting 1942) that predict the transition from pine to hardwood forests in North Carolina using a Landsat time series from 1986 and 2000. Brandt et al. (2012) employed MSS and TM/ETM+ images from 1974 and 2009 to distinguish successional pathways differentially affected by anthropogenic pressure in Yunnan, China.

Provided forest stands were initiated within the satellite record for the area in question, one approach to infer the approximate age for primary (Lawrence and Ripple 1999) and secondary forest (Lucas et al. 2002; Cohen et al. 2002; Schroeder, Cohen, and Yang 2007) is to estimate time since the last stand-replacing disturbance. More recently, a number of time-series methods have been developed to automate forest disturbance and recovery monitoring in early successional forests by exploiting the relatively long, and growing archive available from the Landsat and Landsat-like family of sensors. The central premise of this approach is that changes in vegetation cover, such as disturbance and early successional regrowth, leave a distinct temporal signal in spectral space that can be identified to derive metrics such as disturbance date and intensity, as well as regeneration rate (Healey et al. 2005; Kennedy, Cohen, and Schroeder 2007). Prominent examples of such automated approaches include the vegetation change tracker (VCT) (Huang et al. 2010), the Landsat-based detection of Trends in Disturbance and Recovery (LandTrendr) algorithm (Kennedy, Yang, and Cohen 2010) and TimeSync (Cohen, Yang, and Kennedy 2010), a software tool used to aid in image interpretation and validation of time-series products. LandTrendr was used to predict current forest structure attributes based on disturbance history, with Landsat-derived predictors performing comparably with, and in some cases better than, LiDAR-based models (Pflugmacher, Cohen, and Kennedy 2012). More recently, forest disturbance detection algorithms have employed multi-sensor fusion to provide near real-time vegetation change monitoring (Zhu, Woodcock, and Olofsson 2012a; Xin et al. 2013).

7.3.3 Remote Sensing of Leaf Area Index and Clumping Index

Since a leaf surface is a substrate on which major physical and biological processes of plants occur, leaf area index (LAI) is arguably the most important vegetation structural parameter and is indispensable for all process-based models for estimating terrestrial fluxes of energy, water, carbon, and other masses. It is therefore of interest not only to the remote sensing community that produces LAI maps but also to ecological, hydrological, and meteorological communities that use LAI products for various modeling purposes (Sellers et al. 1997; Chen et al. 2005b; Dai et al. 2003).

7.3.3.1 Definitions and Ground Measurement Techniques

LAI is defined as one-half the total (all-sided) leaf area per unit ground surface area (Chen and Black 1992, see also the review by Jonckheere et al. 2004). It is often indirectly measured using optical instruments that acquire transmitted radiation through a plant canopy, from which the canopy gap fraction is derived. The canopy gap fraction, $P(\theta)$, at zenith angle θ, is related to the plant area index, denoted as L_t, which includes both green leaves and non-green materials, such as stems and branches, that intercept radiation. This relation is given by the following equation:

$$p(\theta) = e^{-G(\theta)\Omega L_t / \cos\theta}, \tag{7.4}$$

where $G(\theta)$ is the projection coefficient, which is determined by the leaf angular distribution (Monsi and Saeki 1953; Campbell 1990), and Ω is the clumping index, which is related to the leaf spatial distribution pattern (Nilson 1971). If $P(\theta)$ is measured at one angle and $G(\theta)$ and Ω are known, L_t can be inversely calculated using equation (7.4). However, both $G(\theta)$ and Ω are generally unknown, therefore different optical instruments have been developed to measure these unknown parameters.

The Li-Cor LAI 2000 Plant Canopy Analyzer is an optical instrument developed to address the issue of unknown $G(\theta)$ due to non-spherical leaf angle distribution. It measures the diffuse radiation transmission simultaneously in five concentric rings covering the zenith angle range from 0° to 75°, i.e., $P(\theta)$ at five angles. These measurements are used to calculate the LAI based on Miller's theorem (Miller 1967):

$$L_e = 2 \int_0^{\pi/2} \ln\frac{1}{p(\theta)} \cos\theta \sin\theta \, d\theta, \tag{7.5}$$

The original Miller's equation was developed for canopies with random leaf spatial distributions, i.e., $\Omega = 1$, and allows the calculation of LAI without the knowledge of $G(\theta)$ when $P(\theta)$ is measured over the full zenith angle range and its azimuthal variation is ignored. LAI and $G(\theta)$ can also be derived simultaneously using multiple angle measurements (Norman and Campbell 1989). For spatially non-random canopies, Miller's theorem actually calculates the effective LAI (Chen, Black, and Adams 1991), expressed as:

$$L_e = \Omega L, \tag{7.6}$$

Equation (7.5) can be discretized to calculate L_e using the $P(\theta)$ measurements at five zenith angles by LAI–2000. L_e calculated this way includes all green and non-green materials above the instrument. With measured L_e, the following equation is proposed to calculate LAI (Chen 1996a):

$$L = (1-\alpha) L_e / \Omega, \tag{7.7}$$

where α is the woody-to-total area ratio. The total area includes both green leaves and non-green materials such as stems, branches, and attachments (e.g., moss) on branches. The α value is generally in the range of 0.05–0.3 depending mostly on forest age (Chen et al. 2006).

There are also optical techniques for indirect measurement of the clumping index (Chen and Cihlar 1995). These techniques are based on the canopy gap size distribution theory of Miller and Norman (1971). An optical instrument named Tracing Radiation and Architecture of Canopies (TRAC, Chen and Cihlar 1995) was developed to measure the canopy gap size distribution using the solar beam as the probe. In conifer canopies, the gaps between needles within a shoot (a basic collection of needles around the smallest twig) are obscured due to the penumbra effect, and the clumping index derived from TRAC measurements represents the clumping effects at scales larger

Ecological Characterization of Vegetation

than the shoot (treated as the foliage element), denoted as Ω_E. According to a random gap size distribution curve based on Miller and Norman's theory, large gaps caused by the non-random foliage element distribution, i.e., those caused by tree crowns and branches, are identified and removed to reconstruct a random gap size distribution. With this gap removal technique, Ω_E is calculated from the following equation (Chen and Cihlar 1995; Leblanc 2002):

$$\Omega_E(\theta) = \frac{\ln[F_m(0,\theta)][1-F_{mr}(0,\theta)]}{\ln[F_m(0,\theta)][1-F_{mr}(0,\theta)]}, \quad (7.8)$$

where $F_m(0,\theta)$ is the total canopy gap fraction at zenith angle, i.e., the accumulated gap fraction from the largest to the smallest gaps, and $F_{mr}(0,\theta)$ is the total canopy gap fraction after removing large gaps resulting from the non-random foliage element distribution due to canopy structures such as tree crowns and branches.

Clumping within individual shoots depends on the density of needles on a shoot. This level of foliage clumping was recognized and estimated in various ways by Oker-Blom (1986), Gower and Norman (1991), Stenberg et al. (1994), Fassnacht et al. (1994), etc. Based on a theoretical development by Chen (1996a), this clumping is quantified using the needle-to-shoot area ratio (γ_E) as follows:

$$\gamma_E = A_n / A_s, \quad (7.9)$$

where A_n is half the total needle area (including all sides) in a shoot, and A_s is half the shoot area (for a shoot that can be approximated by an ellipsoid, the total shoot area is the ellipsoid surface area, not the projected elliptical area). To obtain γ_E, shoots need to be sampled from trees of different sizes at different heights, and A_n and A_s need to be measured using laboratory equipment (Chen et al. 1997; Kucharik, Norman, and Gower 1999). For broadleaf forests, the individual leaves are the foliage elements, and therefore $\gamma_E = 1$.

The total clumping of a stand can therefore be written as:

$$\Omega = \Omega_E / \gamma_E, \quad (7.10)$$

and the final equation for deriving LAI from indirect measurements is

$$L = (1-\alpha)L_e \gamma_E / \Omega_E, \quad (7.11)$$

Different instruments can be used to measure the different variables in this equation in order to determine LAI.

7.3.3.2 LAI Retrieval Using Remote Sensing Data

Plant leaves intercept solar radiation and selectively absorb part of it for conversion into stored chemical energy by photosynthesis. The unabsorbed radiation is either reflected by the leaf surface or transmitted through the leaves. Healthy plant leaves have distinct reflectance and transmittance spectra relative to soil and other non-living materials. Optical remote sensing makes use of the contrast between leaf and soil spectral characteristics for retrieving LAI of vegetation. However, vegetation stands have complex three-dimensional canopy architecture, such as tree crowns, branches and shoots in forests, plantation rows in crops, and foliage clumps in shrubs. Remote sensing signals acquired over vegetated area are not only influenced by the amount of leaf area in the canopy but also by the canopy architecture. Seasonal variations of the vegetation background, such as moss/

grass cover and snow cover on the forest floor, also greatly influence the total reflectance from a vegetated surface. It has therefore been a challenge to produce consistent and accurate LAI products using satellite measurements. Many remote sensing algorithms have been developed to retrieve LAI with full or partial consideration of the aforementioned factors influencing remote sensing measurements from vegetation. These algorithms are described in the following sections.

7.3.3.2.1 LAI Algorithms Based on Spectral Vegetation Indices

Reflectance spectra of healthy leaves show distinct low values in the red (620–750 nm) wavelengths and high values in NIR (800–1300 nm) wavelengths, and therefore many vegetation indices (VIs) have been developed using remote sensing measurements in red and NIR bands for estimating LAI and other vegetation parameters (Table 7.10). Liquid water in aboveground living biomass absorbs MIR (1300–2500 nm) radiation, lowering the reflectance in the MIR band. Since foliage biomass interacts most with solar radiation, the MIR reflectance is expected to correlate well with LAI, and some two-band and three-band vegetation indices utilizing the additional information from MIR have been developed for LAI retrieval (Table 7.10).

TABLE 7.10
Vegetation Indices Useful for LAI Retrieval (see also Chen 1996b)

Vegetation Index	Definition	Reference
Normalized Difference Vegetation Index (NDVI)	$NDVI = \dfrac{(\rho_n - \rho_r)}{(\rho_n + \rho_r)}$	Rouse et al. (1974)
Simple Ratio (SR)	$SR = \dfrac{\rho_n}{\rho_r}$	Jordan (1969)
Modified Simple Ratio (MSR)	$MSR = \dfrac{\dfrac{\rho_n}{\rho_r} - 1}{\sqrt{\dfrac{\rho_n}{\rho_r} + 1}}$	Chen (1996b)
Renormalized Difference Vegetation Index (RDVI)	$RDVI = \dfrac{\rho_n - \rho_r}{\sqrt{\rho_n + \rho_r}}$	Roujean and Bren (1995)
Weighted Difference Vegetation Index (WDVI)	$WDVI = \rho_n - \alpha \cdot \rho_r$ where $\alpha = \dfrac{\rho_{n,\,soil}}{\rho_{r,\,soil}}$	Clevers (1989)
Soil Adjusted Vegetation Index (SAVI)	$SAVI = \dfrac{(\rho_n - \rho_r)(1+L)}{(\rho_n + \rho_r + L)}$ $L = 0.5$	Huete (1988)
Soil Adjusted Vegetation Index 1 (SAVI1)	$SAVI1 = \dfrac{(\rho_n - \rho_r)(1+L)}{(\rho_n + \rho_r + L)}$ $L = 1 - 2.12 \cdot NDVI \cdot WDVI$	Qi et al. (1994)
Global Environmental Monitoring Index (GEMI)	$GEMI = \dfrac{\eta(1 - 0.25 \cdot \eta) - (\rho_r - 0.125)}{(1 - \rho_r)}$ $\eta = \dfrac{2(\rho_n^2 - \rho_r^2) + 1.5\rho_n + 0.5\rho_r}{\rho_n + \rho_r + 0.5}$	Pinty and Verstraete (1992)
Non-Linear Index (NLI)	$NLI = \dfrac{(\rho_n^2 - \rho_r)}{(\rho_n^2 + \rho_r)}$	Goel and Qin (1994)

TABLE 7.10 (*Continued*)
Vegetation Indices Useful for LAI Retrieval (see also Chen 1996b)

Vegetation Index	Definition	Reference
Atmospherically Resistant Vegetation Index (ARVI)	$ARVI = \dfrac{(\rho_n - \rho_{rb})}{(\rho_n + \rho_{rb})}$ $\rho_{rb} = \rho_r - \gamma(\rho_b - \rho_r)$	Kaufman and Tanre (1992)
Soil and Atmospheric Resistant Vegetation Index (SARVI)	$SARVI = \dfrac{(\rho_n - \rho_{rb})(1+L)}{(\rho_n + \rho_{rb} + L)}$ $L = 0.5$	Huete and Liu (1994)
Soil and Atmospheric Resistant Vegetation Index 2 (SARVI2)	$SARVI2 = \dfrac{2.5(\rho_n - \rho_r)}{(1 + \rho_n + 6\rho_r - 7.5\rho_b)}$	Huete and Liu (1994)
Modified NDVI (MNDVI)	$MNDVI = \dfrac{(\rho_n - \rho_r)}{(\rho_n + \rho_r)}\left(1 - \dfrac{\rho_s - \rho_{s\,min}}{\rho_{s\,max} - \rho_{s\,min}}\right)$	Nemani et al. (1993)
Reduced SR (RSR)	$RSR = \dfrac{\rho_n}{\rho_r}\left(1 - \dfrac{\rho_s - \rho_{s\,min}}{\rho_{s\,max} - \rho_{s\,min}}\right)$	Brown, Chen, and Leblanc (2000)

Not all two-band and three-band VIs are well correlated to LAI. The significance level of the correlation of two-band VIs with LAI varies greatly even though they are constructed using the same two-band reflectance data because these two data are combined in different ways, under different assumptions. An ideal VI for LAI retrieval should preferably have the following properties: (1) it is more or less linearly related to LAI and (2) it can minimize the impacts of both random and systematic biases that remote sensing errors have on its value. A linear relationship between a VI and LAI is preferred because it is insensitive to the surface heterogeneity within a pixel and induces less error in spatial scaling (Chen 1999). No VIs have so far been found to be linearly related to LAI for all plant functional types. However, some are more linearly related to LAI than others. SR, for example, is more linearly related to LAI than NDVI and SAVI (Chen and Cihlar 1996; Chen et al. 2002). Ideally, VIs would vary with LAI only, or the effects of surface variations other than LAI can be considered by adjusting coefficients or constants in the algorithm. Measured reflectance in different spectral bands are affected by environmental noise, such as subpixel clouds and their shadows, that are not identified in image processing, mixtures of non-vegetative surface features (small water bodies, rock, etc.), fog, smoke, etc. This unwanted noise frequently exists in remote sensing imagery and can dramatically alter the values of VIs. However, the impacts of these types of noise on the reflectances in different spectral bands are often correlated. For example, subpixel clouds would cause red and NIR reflectance to increase simultaneously, while cloud shadows would decrease them in about the same proportion. The same is true for other types of noise mentioned earlier. Variations in solar and view angle also cause variations of reflectance in various spectral bands in the same direction and in about the same proportions. VIs that are based on ratios of these two bands, such as NDVI and SR, can greatly reduce the impacts of various sources of noise. However, some VIs with sophisticated manipulations of two-band data, such as GEMI, may amplify noise. VIs that cannot be expressed as a function of the ratio of these two-band reflectances, such as SAVI, NLI, and RDVI will retain the noise. MSR, for example, is developed with the same purpose as RDVI to increase its linearity with LAI, but it is better correlated to LAI than RDVI because it can be expressed as a function of the ratio of NIR and red reflectances while RDVI cannot. The ability of a VI to minimize unwanted measurement noise is of paramount importance in LAI retrieval because noise in the reflectance measurements can come from many sources and is unavoidable.

Three-band VIs have been developed for various purposes. ARVI and SARVI modify NDVI and SAVI, respectively, with the reflectance in the blue band to reduce the atmospheric effect. They are useful when there are insufficient simultaneous atmospheric data to conduct atmospheric correction. MNDVI and RSR introduce a multiplier to NDVI and SR, respectively, based on the reflectance in a MIR band (1600–1800 nm *or* 2100–2300 nm). RSR has several advantages over SR for LAI retrieval (Brown, Chen, and Leblanc 2000): (1) it is more significantly correlated with LAI for different forest types because it is more sensitive to LAI variation; (2) the differences in the LAI-RSR relationship among different forest types are greatly reduced from those in the LAI-SR relationship, and therefore RSR is particularly useful for mixed cover types; and (3) the influence of the variable background optical properties is much smaller on RSR than on SR because MIR reflectance is highly sensitive to the greenness of the background due to the strong absorption of MIR radiation by grass, moss, and understory. These advantages of RSR over SR for forest LAI retrieval are confirmed by several independent studies (Eklundh et al. 2003; Stenberg et al. 2004; Wang et al. 2004; Chen et al. 2005c; Tian et al. 2007; Heiskanen et al. 2011). However, RSR is sensitive to soil and vegetation wetness and can increase greatly immediately after rainfall or irrigation, and therefore it is only suitable for forests in LAI retrieval algorithms (Deng et al. 2006).

7.3.3.2.2 LAI Algorithms Based on Radiative Transfer Models

The relationships between LAI and reflectances in individual spectral bands can be simulated using plant canopy radiative transfer models, and LAI algorithms can be developed based on these modeled relationships. Models are useful alternatives to empirical relationships established through correlating VIs or reflectances with LAI measurements because the empirical data are often limited in spatial and temporal coverage and are often location-specific. These empirical relationships are also dependent on the quality of ground LAI data, the spectral response functions of remote sensing sensors, the angle of measurements, atmospheric effects, etc. The quality of LAI data can be influenced by the method of LAI measurements, the definition of LAI, and the measurement protocol. Some reported LAI values are actually the effective LAI without considering the clumping effect, and some optical measurements do not include the correction for non-green materials (see equation 7.11). Radiative transfer models can theoretically avoid these shortcomings of empirical data, but they need to be calibrated with ground data. In the calibration process, misconceptions and errors in empirical data can also bias the model outcome. For example, some destructive LAI values used for model validation are incorrectly based on the projected area rather than half the total leaf area.

There have been many LAI algorithms developed using radiative transfer models (Table 7.11) for regional and global LAI retrieval. These algorithms are characterized by the radiative transfer modeling method, the ways to consider foliage clumping and background optical properties, and the ways to combine the individual bands. A radiative transfer model, however sophisticated, is an abstract representation of the complex reality, and therefore the modeled relationship between LAI and remote sensing data depends not only on the aforementioned factors but also on how radiative transfer is simulated, such as the ways to consider multiple scattering in the canopy, the assumed leaf angle distributions, the treatments of diffuse sky radiation, etc. As radiative transfer methods are diverse, it is expected that the simulated relationships between remote sensing data and LAI are quite different among the existing model-based global LAI algorithms. There is a need to calibrate radiative transfer models and LAI algorithms against an accurate ground and remote sensing dataset covering the diverse plant structural types around the globe. The Radiation Transfer Model Intercomparison efforts (Pinty et al. 2004; Widlowski et al. 2007) have laid a foundation for further activities to satisfy this need.

7.3.4 BIOMASS

Biomass is the accumulated net primary production (NPP) in living plants, including both the above- and belowground components. Because of litterfall and mortality, biomass is always less

TABLE 7.11

Global LAI Products and Their Main Characteristics (modified after Garrigues et al. 2008). They Include CYCLOPES (Carbon Cycle and Change in Land Observational Products from an Ensemble of Satellites), ECOCLIMAP, GLOBCARBON (GLOBal Biophysical Products for Terrestrial CARBON Studies), and MODIS (MODerate Resolution Imaging Spectroradiometer)

	CYCLOPES	ECOCLIMAP	GLOBCARBON	MODIS
Algorithm Development	One-dimensional turbid media radiative transfer model	Empirical LAI-NDVI relationships	Geometric-optical model	Lookup tables produced using a three-dimensional radiative transfer model
Clumping Consideration	No clumping consideration except consideration of the differences among cover types at the landscape level	Clumping within shoot and canopy is considered, but clumping at the landscape level is not considered	Clumping is fully considered based on TRAC-measured, cover type–specific values	Clumping is considered through a parameter related to the 3D canopy structure
Background Optical Property	Assigned constant values	Assigned constant values	Assigned constant values	Assigned constant values
Seasonal Smoothing	No	No	Yes	No
Reference	Baret et al. (2007)	Masson et al. (2003)	Deng et al. (2006)	Knyazikhin et al. (1998), Yang et al. (2006)

than the sum of annual NPP over the plant's lifetime. It is relatively straightforward to measure the biomass for perennials, but measuring biomass for forests in the real world is extremely laborious. Due to the fact that the majority of the terrestrial biomass is stored in forest ecosystems (Dixon et al. 1994), measuring forest ecosystem biomass has become a major task in global carbon budget studies. In fact, measuring forest biomass defines the state of the art of biomass measurement. Since clearing an extensive area just for measuring biomass would represent an undue disturbance to the ecosystem, measuring forest biomass in the real world usually involves several steps. First, a species-specific allometric relationship is established between biomass and some easy-to-measure structural parameter(s) (typically diameter at breast height (DBH) and/or occasionally height). This step involves destructive sampling of a number of individuals for each species and is quite expensive. However, these allometric relationships can be reused once they are developed. Such relationships have been documented for the majority of tree species in North America (Gholz et al. 1979; Grier and Logan 1977; Jenkins et al. 2003; Smith, Heath, and Jenkins 2003; Ter-Mikaelian and Korzukhin 1997). Second, a series of sample plots are made in a region either systematically or randomly (Zhang and Song 2006). Each individual tree species within a sample plot is tallied, and its biomass is calculated using the allometric equations developed. The total biomass for a plot is calculated as the sum of biomass for all individuals. Lastly, the total biomass of a geographic region is estimated based on these sampling plots.

Depending on the rate of growth and length of time accumulating NPP, biomass density is strongly dependent on vegetation successional stage (Song and Woodcock 2002; Pregitzer and Euskirchen 2004). Therefore, an accurate estimation of biomass over a geographic region requires a large number of sampling plots, which are often not practical to make. Optical remote sensing offers a significant advantage over the traditional fieldwork approach in mapping biomass over large areas, as remote sensing–based approaches provide wall-to-wall coverage in space and are much more cost-effective. Numerous approaches using optical remotely sensed data have been developed in the literature and can be summarized into a few categories: (1) nearest neighbor imputation, (2) regression, (3) machine learning algorithms, and (4) biophysical approaches.

7.3.4.1 K-Nearest Neighbor Imputation

K-nearest neighbor imputation (kNN) takes advantage of the spatial autocorrelation of biomass in space. The approach estimates the biomass for a particular location or pixels in a remotely sensed image from the spatial interpolation of biomass in k nearby sampling plots based on distance weighted average (Fazakas and Olsson 1999; Franco-Lopez, Ek, and Bauer 2001). Using the field plots from the Swedish National Forest Inventory, Tomppo et al. (2002) used the kNN approach to produce a biomass map with Landsat imagery, then rescaled the biomass map to match that of IRS-1C WiFS imagery, and produced biomass maps over large areas using non-linear multiple regression. For kNN to be effective, a large number of sampling plots are needed to represent the spatial pattern and the whole range of biomass variation in space. The approach was used to develop national biomass maps for Sweden and Finland since 1990 using the National Forest Inventory sampling plots (Tomppo et al. 2008). Recently, kNN was used with various imputation approaches and remotely sensed data from multiple sensors (Latifi and Koch 2012) and achieved encouraging results for mapping biomass.

7.3.4.2 Regression

Estimating biomass via regression with remotely sensed imagery involves two steps. The first step is the development of an empirical regression model between biomass measured on the ground and remotely sensed data, which can be surface reflectance or a transformation of surface reflectance, such as spectral vegetation indices. Once a robust regression model is established, the second step is to apply the model to the rest of the valid pixels in the image. Many successful applications of this approach have been reported in the literature (Anderson, Hanson, and Haas 1993; Fazakas and Olsson 1999; Heiskanen 2006; Muukkonen and Heiskanen 2007; Roy and Ravan 1996; Steininger 2000; Tomppo et al. 2002). Careful examination found that these successful applications were conducted in areas with low biomass density.

Although significant challenges have been encountered using relatively high spatial resolution optical imagery to estimate biomass, some success has been achieved using coarse resolution imagery over large areas. Myneni et al. (2001) and Dong et al. (2003) developed an empirical relationship between cumulative NDVI over the growing season from AVHRR over the forested areas and the woody biomass derived from forest inventory for six countries (Canada, Finland, Norway, Russia, Sweden, and the USA) during 1981–1999 and found a large carbon sink in Eurasian boreal and North American temperate forests. Similarly, Piao et al. (2005) used the GIMMS NDVI (Tucker et al. 2001) and China's forest inventory data to estimate aboveground forest biomass via a nonlinear regression model. The model predicted the aboveground biomass well for the majority of the provinces except for a few outliers, which might be due to errors from the inventory data. Zhang and Kondragunta (2006) used MODIS LAI, land cover types, and vegetation continuous fields to estimate the aboveground biomass for the conterminous USA with a RMSE of 12 t/ha at the state level compared with estimates from US Forest Service Forest Inventory and Analysis data. Le Maire et al. (2011) successfully ($R^2 \approx 0.9$) estimated the forest biomass for young eucalyptus plantations using MODIS time series NDVI images and simple bioclimatic variables. It is counter-intuitive that spectral vegetation indices at high spatial resolution performed more poorly in predicting biomass at the plot level than those at coarse spatial resolution that estimate biomass at the continental scale. However, these empirical models provide little insight on the biophysical basis for the strong performance in estimating biomass using coarse spatial resolution remotely sensed data over large areas.

7.3.4.3 Machine Learning Algorithms

Machine learning algorithms have several advantages over conventional regression approaches in mapping biomass from remotely sensed data. First, the algorithms do not require normally distributed data. Second, the algorithms do not require the input data layers to be independent from each other. Data layers from multispectral remotely sensed imagery are often correlated. Third, machine learning algorithms are capable of handling nonlinear relationships between biomass and remotely sensed signals. In a Bornean tropical rainforest, Foody et al. (2001) found that artificial neural networks (NN) using the reflectance of the six optical bands from Landsat TM sensors mapped biomass better than a regression approach using spectral vegetation indices only or the kNN imputation. Foody, Boyd, and Cutler (2003) further confirmed that NN outperformed the conventional regression approach in mapping biomass in three tropical forest sites in Brazil, Malaysia, and Thailand. Powell et al. (2010) compared reduced major axis regression, kNN, and random forest (RF) algorithms in mapping aboveground biomass in Arizona and Minnesota, USA, using biomass derived from Forest Inventory and Analysis plots and Landsat imagery, and they found all three approaches predict biomass at the pixel level with RMSE well above 50% of the mean biomass. Mutanga, Adam, and Cho (2012) compared RF and multiple regression to estimate biomass for a densely vegetated wetland using narrowband vegetation indices computed from WorldView-2 imagery and found that RF performed better in biomass estimation with RMSE of 0.441 kg/m^2 (12.9% of observed mean biomass). Because it is easy to use data layers from multiple sources with machine learning algorithms, they proved to be effective in integrating remotely sensed data from multiple sensors for mapping aboveground biomass, particularly combining multispectral and LiDAR data. Latifi and Koch (2012) compared kNN imputation with RF to map aboveground biomass using data from airborne scanning LiDAR data and color infrared optical imagery and found that RF produced more accurate results.

In addition to high-resolution imagery, machine learning algorithms have also been used to map biomass with low-resolution imagery, particularly data from MODIS. Combining surface reflectance of the first seven MODIS bands with climate and topographic data, Baccini et al. (2004) mapped aboveground forest biomass for 18 national forests in California using tree-based regression with reasonable accuracy, but the approach tends to underestimate biomass for stands with high biomass density (> 250 *t/ha*). Houghton et al. (2007) used RF to map aboveground forest biomass for the Russian Federation using MODIS Nadir BRDF Adjusted Reflectance (NBAR) and forest inventory data. They produced biomass estimates comparable with previous independent estimates. Blackard et al. (2008) mapped the aboveground biomass of the conterminous USA, Alaska, and Puerto Rico with multiple sources of data, including MODIS imagery, FIA plots, climatic and topographic variables, and other ancillary data. They first divided the conterminous USA into 65 ecological zones and separated the FIA plots into forest and non-forest categories. A separate regression tree model was developed for each ecological zone, Alaska, and Puerto Rico. The models tended to overestimate areas with low biomass and underestimate areas with high biomass. Baccini et al. (2008) developed a regression tree model for aboveground biomass using seven-band 1×1 km MODIS NBAR data and biomass data derived from NFI for the Republic of the Congo, Cameroon, and Uganda. The model was then applied to the entire tropical Africa. The RMSE was 50.5 *t/ha* for a biomass range up to 454 *t/ha*. The estimated aboveground biomass was also highly correlated (R^2 = 0.90) with height metrics from ICESat Geospatial Laser Altimeter System (GLAS). However, the predicted biomass had a positive bias for low biomass and negative bias for high biomass.

A promising recent development in mapping biomass with optical imagery is to combine it with remotely sensed data from LiDAR and/or radar sensors, and produce improved biomass maps.

Andersen et al. (2011) integrated Landsat imagery with airborne LiDAR and dual-polarization synthetic aperture radar (PolSAR) from ALOS PALSAR to map forest biomass in the interior of Alaska. The ICESat GLAS data were successfully used to map biomass with Landsat (Duncanson, Niemann, and Wulder 2010; Helmer, Lefsky, and Roberts 2009) and MODIS (Baccini et al. 2012; Nelson et al. 2009) imagery.

7.3.4.4 Biophysical Approaches

Biophysical approaches rely on the physical principles that govern the relationship between vegetation structure and remotely sensed signals. One of the earliest attempts to estimate standing total biomass via optical remote sensing was by Wu and Strahler (1994). The physical principle is the geometric optical theory for remotely sensed imagery over vegetated landscapes (Li and Strahler 1985). They inverted the remotely sensed data from Landsat TM sensors for tree density and mean tree crown size on a stand basis with a GO model, and then used allometry to estimate total standing biomass. However, Wu and Strahler (1994) only tested the model with a limited number of stands. More comprehensive studies by Woodcock et al. (1994, 1997) found that the Li-Strahler model could be used to estimate tree cover effectively, but separation of tree size and stem count was poor. Hall, Townshend, and Engman (1995) proved that remotely sensed spectral signals over black spruce forests can be calculated as a linear mixture of sunlit crown, sunlit background, and shadow. Based on the geometric optical theory, they derived these fractions from stand-level reflectance obtained at nadir by the helicopter-mounted Modular Multiband Radiometer (MMR) and found that the fraction of shadows was highly correlated to aboveground biomass. Hall, Peddle, and Ledrew (1996) and Peddle, Hall, and Ledrew (1999) further confirmed the usefulness of fraction of shadows in estimating aboveground biomass. Peddle, Hall, and Ledrew (1999) found that the fraction of canopy shadows performed 20% better than numerous vegetation indices for estimating aboveground biomass. More recently, Soenen et al. (2010) demonstrated promising results using a geometric optical canopy reflectance model to estimate tree crown size and stem density and further estimate aboveground biomass with SPOT imagery. Chopping et al. (2008, 2011) developed a similar approach based on the Simple Geometric Optical model to estimate biomass by taking bidirectional reflectance functions from MODIS and MISR. The approach produced biomass estimation that is comparable with independent data in a low biomass region, but the approach is computationally intensive and requires some detailed canopy structural parameters that may not be available as a priori knowledge. Hall et al. (2006) developed the BioSTRUCT algorithm to map aboveground biomass and volume in two steps: (1) estimate canopy height and crown closure with Landsat ETM+ imagery; and (2) estimate aboveground biomass and stand volume using canopy height and crown closure based on allometric relationships.

7.3.5 UNCERTAINTIES, ERRORS, AND ACCURACY

Despite the many successes of optical remote sensing in extracting information about vegetation structure, there are varying degrees of uncertainty associated with them. Even for the most simple vegetation estimate, vegetation cover, there is about 10% uncertainty (Hansen et al. 2002; Hayes et al. 2008). Given the same vegetation cover, ecological functions could differ tremendously depending on its successional stages (Law et al. 2001). For example, Liu et al. (2008) found that using Landsat imagery, forest succession in the PNW can only be reliably separated into three broad successional stages (young, mature, and old growth).

LAI is perhaps the most sought-after measure of vegetation structure due to the important role it plays in energy and matter exchange between the land surface and the atmosphere. Several factors prevent the accurate estimate of LAI with optical remote sensing. First, remote sensing signals saturate in high LAI (Baret and Guyot 1991; Turner et al. 1999). Second, remnant cloud contamination remains a problem. Lastly, LAI cannot be derived from remotely sensed data analytically. It is an ill-posed mathematical problem because there are too many other factors influencing the remotely sensed signal in addition to LAI (Eklundh, Harrie, and Kuuse 2001; Gobron, Pinty, and Verstraete 1997). As a result, significant uncertainties remain in most of the current LAI products (Song 2013). Numerous studies found that current MODIS LAI products tend to overestimate LAI (Cohen et al. 2003, 2006; Wang et al. 2004; Aragao et al. 2005; Pisek and Chen 2007; Sprintsin et al. 2009).

Numerous studies have found that remotely sensed signals saturate in high biomass density areas, and spectral vegetation indices poorly predict biomass for forests (Sader et al. 1989; Hall, Townshend, and Engman 1995; Peddle, Hall, and Ledrew 1999). Steininger (2000) found that Landsat TM surface reflectance correlates well with stand structure only when biomass is under 150 t/ha and age is under 15 years in Brazil, and even this is not the case for another study site in Bolivia. Nelson et al. (2000) found a single Landsat TM imagery could not be used to reliably differentiate tropical forest age classes. Lu (2005) found that the spectral signals from Landsat TM imagery can only be used to estimate aboveground biomass for forests with simple structure. Although optical remote sensing has been used to produce numerous key vegetation structure parameters wall-to-wall, improving the accuracy of the estimation remains the major challenge for the foreseeable future.

7.4 OPTICAL REMOTE SENSING OF VEGETATION FUNCTIONS

Vegetation plays a key role in the terrestrial ecosystem that provides vital goods and services upon which the welfare of the humanity depends, such goods as food, fiber, and medicine, and services as soil and water conservation and the preservation of biodiversity (Dobson et al. 1997; Salim and Ullsten 1999). Photosynthesis is the entry point of inorganic materials, e.g., CO_2, water, and nutrients, into organic forms, such as carbohydrates. The product of photosynthesis provides the matter and energy that drive all subsequent ecosystem processes. Therefore, vegetation primary production is at the core of almost all terrestrial ecosystem goods and services. In this section, we review remote sensing of vegetation functions that are tied to plant photosynthesis, including its seasonal cycles (phenology), the amount of energy used in the process (the fraction of absorbed photosynthetically active radiation), the abundance of photosynthesis apparatus (chlorophyll concentration), and the efficiency of converting the absorbed photosynthetically active radiation to carbohydrate (light use efficiency).

7.4.1 VEGETATION PHENOLOGY

Vegetation phenology is the natural rhythm of plant life cycle events, and the timing of these events is largely dependent on climate signals (Körner and Basler 2010). In many temperate forests, winter dormancy must be broken by extended exposure to cold temperatures, and after this chilling requirement has been met, increases in temperature and photoperiod can trigger leaf emergence in spring (Archibold 1995; Zhang, Tarpley, and Sullivan 2007). Leaf expansion and shoot growth in some arid and seasonally moist ecosystems can be triggered by the start of the rainy season, while soil moisture depletion may trigger senescence and leaf abscission, though low temperatures may also limit photosynthesis in cooler deserts (Archibold 1995; Jolly and Running 2004; Jolly, Nemani, and Running 2005).

Optical remote sensing offers unprecedented opportunities to observe the synoptic patterns of the timing of plant life cycle events, known as land surface phenology (LSP), (Gonsamo et al. 2012a). While the seasonal patterns of LSP variability are related to plant biological traits, LSP derived from spaceborne optical sensors is distinct from the traditional definition of plant phenology, which aims to understand the timing of recurring biological events, the causes of their timing with regard to biotic and abiotic forces, and the interrelation among phases of the same or different species (Lieth 1974). LSP has strong effects and feedback both on climate (Keeling, Chin, and

Whorf 1996; Peñuelas, Rutishauser, and Filella 2009) and on terrestrial ecosystem functions. The carbon balance of terrestrial ecosystems is highly sensitive to climatic changes in early and late growing seasons (Piao et al. 2007; Piao et al. 2008; Richardson et al. 2010; Wu et al. 2012a; Wu et al. 2012b). Vegetation phenology has been known to be a key and first element in ecosystem response to climate change (Menzel et al. 2006), as well as a major determinant of species distributions (Chuine and Beaubien 2001).

Therefore changes in LSP events have the potential to broadly impact global carbon fixation, nitrogen cycles, evapotranspiration, and ecosystem respiration (Morisette et al. 2009; Richardson et al. 2010), surface meteorology (Schwartz 1992; Bonan 2008a, 2008b; Richardson et al. 2013), interspecific interactions both among plants and between plants and insects, vegetation community structure, and success of invasive species (Willis et al. 2008, 2010; Wolkovich and Cleland 2011; Cleland et al. 2012; Fridley 2012), crop production, frost damage, pollination (Brown and De Beurs 2008), and spreading of diseases (Morisette et al. 2009).

The 4th Assessment Report ("AR4," Parry et al. 2007) of the Intergovernmental Panel on Climate Change (IPCC)—which found that spring onset has been advancing at a rate of between 2.3 and 5.2 days per decade since the 1970s—emphatically concluded that phenology "is perhaps the simplest process in which to track changes in the ecology of species in response to climate change" (Rosenzweig et al. 2007). The spatially integrated nature of LSP—as derived from optical satellite observations of land surface reflectance and their combination in the form of vegetation indices (VIs) which are associated with the biophysical and biochemical properties of vegetation—has thus received much attention due to its role as a surrogate in detecting the impact of climate change

7.4.1.1 Vegetation Index for Land Surface Phenology Study

Remote sensing LSP studies use data gathered by satellite sensors that measure wavelengths of visible light as absorbed by leaf pigments, near-infrared light as reflected by leaf internal structure, and shortwave-infrared light as absorbed by leaf *in vivo* water content. As a plant canopy changes from early spring growth to late-season maturity and senescence, these reflectance and absorptance properties also change. Vegetation indices rather than land surface reflectance are especially useful for continental- to global-scale LSP monitoring because they can compensate for changing illumination conditions, surface slope, and viewing angle. Although there are several vegetation indices, the following four are common:

$$NDVI = (NIR - RED)/(NIR + RED), \quad (7.12)$$

$$NDMI = (NIR - SWIR)/(NIR + SWIR), \quad (7.13)$$

$$PI = \begin{cases} 0, \text{if } NDVI \text{ or } NDMI < 0 \\ (NDVI + NDMI)(NDVI - NDMI) = NDVI^2 - NDMI^2, \\ 0, \text{if } PI < 0 \end{cases} \quad (7.14)$$

$$EVI = G \times (NIR - RED)/(NIR + C1 \times RED - C2 \times BLUE + L), \quad (7.15)$$

where the BLUE, RED, NIR, and SWIR are surface reflectances in blue, red, near-infrared, and shortwave-infrared spectral bands, respectively. L is a canopy background adjustment that addresses non-linear, differential NIR and red radiant transfer through a canopy, and C1 and C2 are the coefficients of the aerosol resistance term, which uses the blue band to correct for aerosol influences in the red band. G is a gain factor that limits the EVI value to the −1 to +1 range.

One of the most widely used vegetation indices is the normalized difference vegetation index (NDVI). NDVI values range from −1 to + 1. Areas of barren rock, sand, water, ice, and snow usually show very low NDVI values (<0.1), sparse vegetation such as shrubs and grasslands or senescing crops may result in moderate NDVI values, and high NDVI values (>0.6) correspond to dense vegetation such as that found in temperate and tropical forests or crops in their peak growth stage. Numerous studies have shown that NDVI is closely correlated with LAI and fraction of photosynthetically active radiation absorbed by vegetation canopy. The NOAA AVHRR archive of NDVI data (Tucker et al. 2005) is generated in the framework of the Global Inventory Monitoring and Modeling System (GIMMS) project by careful assembly from different AVHRR sensors, accounting for various deleterious effects, such as calibration loss, orbital drift, volcanic eruptions, etc. The AVHRR archive is the longest time series for LSP studies, and the results from the analysis of AVHRR-based NDVI revealed significant changes in spring phenology of vegetation during the 1980s and 1990s (Eastman et al. 2013; Myneni et al. 1997). The latest version of the GIMMS NDVI dataset spans the period from July 1981 to December 2011 and is termed NDVI3g (third-generation GIMMS NDVI) from AVHRR sensors (Zhu et al. 2013).

To help discriminate the seasonal dynamics of vegetation phenology from background phenomena such as accumulation and melting of snow, alternative vegetation indices, such as normalized difference moisture index (NDMI), have been used (Delbart et al. 2005; Gonsamo et al. 2012b). NDMI is comparable with normalized difference infrared index (NDII) and normalized difference water index (NDWI). NDMI first decreases with snowmelt and then increases during vegetation greening. NDWI is related to the quantity of water per unit area in the canopy and soil and therefore increases during leaf development and increase in soil moisture content. NDMI time series show that greening-up may start before or after complete snowmelt. If snow did not totally melt before leaf appearance, NDMI first decreases and then increases, displaying a trough at its minimum. If snowmelt is complete before leaf appearance, then NDMI remains stable during a period that may last between a few days and a few weeks before increasing. If greening-up occurs during snowmelt, the NDMI decrease due to snowmelt may mask the NDMI increase due to greening up, and NDMI may start increasing later than the actual onset of greening-up. If snowmelt and greening-up overlap during a long period, NDMI variations with snowmelt and greening-up compensate for each other, making the NDWI increase start later than the actual onset of greening-up. NDWI time series is sensitive to water intercepted by leaves and to abrupt increases in soil moisture.

Vegetation indices are not intrinsic physical quantities, and several attempts have been made to remove the confounding effect of brightness (mainly soil) and wetness (mainly from land surface moisture) from greenness that responds to the development of photosynthetic biomass. For example, NDMI alone cannot capture LSP, since it responds to land surface moisture both from the landscape and vegetation components. NDVI is also affected significantly by both soil moisture and brightness. Therefore, both NDVI and NDMI respond not only to the development of photosynthetic biomass (greenness) but also to soil exposure (brightness) and snow, soil, and land surface moisture (wetness), suggesting that NDVI or NDMI alone cannot remove the confounding effect of brightness and wetness on a LSP time series. NDVI and NDMI exhibit opposing trends with increasing brightness and wetness, and similar trends with increasing greenness. Given these premises, the phenology index (PI) was constructed (Gonsamo et al. 2012b) by combining the merits of NDVI and NDMI (Figure 7.2). PI takes the difference of squared greenness and wetness to remove the soil and snow cover dynamics from key vegetation LSP cycles. The following rationale explains the formulation of PI: (1) NIR reflectance is less than red reflectance for ice, snow, and water, resulting in NDVI < 0 for which PI becomes 0; (2) NIR reflectance is less than SWIR for soil and for non-photosynthetic vegetation, resulting in NDMI < 0 and NDVI > 0 for which PI becomes 0; (3) if NDMI > NDVI, the green vegetation or land surface is covered by snow for which PI becomes 0; (4) the use of PI instead

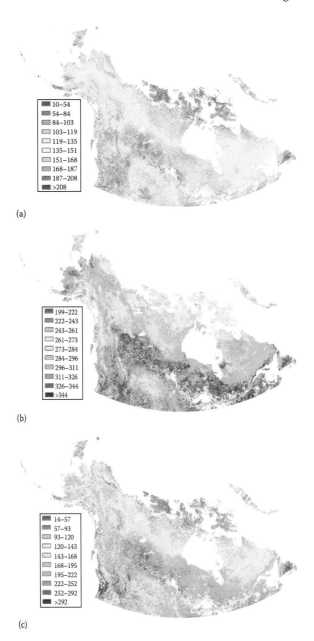

FIGURE 7.2 Start, end, and length of growing season for circumpolar North America (> 45°) derived from SPOT VGT sensors using the Phenology Index (PI) for the year 1999.

of NDVI or NDMI masks out the time series of permanently non-vegetated landscape for which NDVI or NDMI may result in a spurious time series due to moisture variations resembling vegetation LSP; and (5) the product of the sum and the difference of NDVI and NDMI gives a pronounced and smooth curve, removes the effect of wetness from the greenness, and avoids the local solution if we simply consider the use of NDVI once the previous criteria (1–4) are met, which may particularly occur in boreal forests due to intermittent loading and unloading of snow. PI is actually the squared greenness minus squared wetness in the growing season and follows the seasonal dynamics of gross (GPP) and net ecosystem productivity (NEP) better than NDVI or NDMI (Gonsamo et al. 2012a; Gonsamo et al. 2012b).

The enhanced vegetation index (EVI) is developed for use with the MODIS Land Cover Dynamics product (informally called the MODIS Global Vegetation Phenology product). The EVI is a modified NDVI with a soil adjustment factor (L), gain factor (G), and two coefficients (C1 and C2) that describe the use of the blue band in correction of the red band for atmospheric aerosol scattering. The coefficients, G, C1, C2, and L, are empirically determined 2.5, 6.0, 7.5, and 1.0, respectively. This algorithm has improved sensitivity to high biomass regions and improved vegetation monitoring through a de-coupling of the canopy background signal and a reduction in atmospheric influences (Huete et al. 2002). While NDVI is chlorophyll sensitive, the EVI is more responsive to canopy structural variations, including LAI, canopy type, plant physiognomy, and canopy architecture. EVI is used as a standard vegetation index for LSP study for the NASA MODIS project and has shown not to saturate at high photosynthetic biomass vegetation areas, a great improvement compared to NDVI.

7.4.1.2 Land Surface Phenology Metrics Derivation and Validation

LSP studies pay more attention to critical annual events, such as start (SOS), end (EOS), and length of season (LOS). Earlier studies that used AVHRR NDVI time series focused on a phenologically important threshold value from multi-temporal NDVI time series, which values can be global (like 0.3 NDVI) or locally driven (mid-values of max and min NDVI; White et al., 2009). Since there is no fixed threshold value for SOS and EOS that can be applicable globally, recent LSP studies have focused on curve geometry fitting to extract important dates (e.g., Delbart et al. 2005; Gonsamo et al. 2012a; Gonsamo et al. 2012b; Zhang et al. 2003). Most of these methods involve one or several forms of logistic functions to fit sinusoid models. Jönsson and Eklundh (2004) developed open-source LSP extraction software (TIMESAT) that includes double logistic and asymmetric Gaussian functions. These methods to retrieve SOS and EOS estimates from time series of remote sensing data are well-summarized in White et al. (2009).

Traditional validation is usually carried out by comparing remote sensing SOS and EOS with specific life cycle events, such as budbreak, flowering, or leaf senescence using *in situ* observations of individual plants or species by network volunteer observations (e.g., PlantWatch Canada Network, USA National Phenology Network) or experts at intensive study sites (e.g., Harvard Forest). Given the lack of *in situ* data that are comparable with LSP in spatial coverage and landscape representativeness, currently more attention is given to synoptic sensors that have comparable footprints with remote sensing pixels. One of these includes the increasing availability of networks of web cameras (Richardson et al. 2009) that match LSP validation from remote sensing footprints rather than the traditional individual plant phenology networks. However, much work is needed to derive a robust method to extract SOS and EOS from near-surface remote sensing obtained from networked webcams that record repeat LSP in association with the existing eddy covariance flux towers (e.g., the PhenoCam program). Another more objective validation method is the use of GPP estimated based on eddy covariance flux tower measurements (Gonsamo et al. 2012a; Gonsamo et al. 2012b). Several LSP metrics are developed from GPP using curve geometry fitting (Gonsamo, Chen, and D'Odorico 2013), which can be used to validate one or more LSP estimates from remote sensing. The temporal dynamics of GPP as a true photosynthesis phenology provides an objective measure of SOS and EOS. There are also evolving developments in LSP reference measures, such as ground-based spectral and photosynthetic radiation measurements (Richardson et al. 2012), which are expected to help extract the subtle LSP inter-annual and spatial variability across plant functional types. Remote sensing LSP measures, compared to individual plant or plant organ phenology cycles, integrate the collective effects of atmospheric, environmental, and edaphic conditions as well as interspecific responses to the changing climate.

7.4.2 Fraction of Absorbed Photosynthetically Active Radiation

FPAR (Fraction of absorbed Photosynthetically Active Radiation) is defined as the fraction of incoming photosynthetically active radiation (400–700 nm) absorbed by vegetation. It is a key input variable to models that estimate terrestrial ecosystem primary production based on light-use-efficiency theory driven by remotely sensed data (Potter et al. 1993; Running et al. 1994; Landsberg and Waring 1997). As a result, FPAR has become a much sought-after biophysical product from remote sensing.

7.4.2.1 Estimating FPAR with Empirical Models

FPAR depends on various leaf optical properties (e.g., clumping, leaf angle distribution, and spatial heterogeneity) and is non-linearly correlated with LAI (e.g., Asrar et al. 1984). FPAR has been empirically related to several vegetation indices from remote sensing data, such as SR, NDVI, greenness, and PVI (Perpendicular Vegetation Index) since the 1980s (Table 7.12). These indices are estimated from various combinations of remotely sensed data from different spectral bands. The

TABLE 7.12

Empirical Models for the Fraction of Absorbed Photosynthetically Active Radiation (FPAR) with Spectral Vegetation Indices (SR: Simple Ratio, NDVI: Normalized Difference Vegetation Index, Greenness, PVI: Perpendicular Vegetation Index)

Index	FPAR Model	Notes	Reference
SR	$0.34SR - 0.63$	Various crop	Kumar and Monteith (1981)
SR	$0.369\ln(SR) - 0.0353$	Field spectro-photometer/sugar beet	Steven, Biscore, and Jaggard (1983)
SR	$0.0026SR^2 + 0.102SR - 0.006$	Landsat/corn	Gallo, Daughtry, and Bauer (1985)
SR	$0.0294SR + 0.3669$	SPOT/wheat	Steinmetz et al. (1990)
SR	Not specified	Modeling study	Sellers (1987)
NDVI	$1.253NDVI - 0.109$	Landsat/spring wheat	Asrar et al. (1984)
NDVI	$1.200NDVI - 0.184$ (growing) $0.257NDVI + 0.684$ (senescence)	Landsat/wheat	Hatfield, Asrar, and Kanemasu (1984)
NDVI	$2.9NDVI^2 - 2.2NDVI + 0.6$ (growing)	Landsat/corn	Gallo, Daughtry, and Bauer (1985)
NDVI	$1.00NDVI - 0.2$	Landsat/coniferous	Peterson et al. (1987)
NDVI	$1.23NDVI - 0.06$	Wheat	Baret and Olioso (1989)
NDVI	$1.33NDVI - 0.31$	Modeling study	Baret and Olioso (1989)
NDVI	$1.240NDVI - 0.228$	Modeling study	Baret, Guyot, and Major (1989)
NDVI	$0.229\exp(1.95NDVI) - 0.344$ (growing) $1.653NDVI - 0.450$ (senescence)	SPOT/cotton and corn	Wiegand et al. (1991)
NDVI	$1.222NDVI - 0.191$*	Modeling study	Asrar, Myneni, and Choudhury (1992)
NDVI	$1.254NDVI - 0.205$	Field spectro-radiometer/corn and soybean	Daughtry et al. (1992)
NDVI	$1.075NDVI - 0.08$	Modeling study	Goward and Huemmrich (1992)
NDVI	$1.386NDVI - 0.125$	Cereal crop/modeling study	Begue (1993)
NDVI	$1.164NDVI - 0.143$	Modeling study	Myneni and Williams (1994)
NDVI	$1.21NDVI - 0.04$	AVHRR/mixed forests	Goward et al. (1994)
NDVI	$0.95NDVI - 0.02$	Landsat/modeling study	Friedl et al. (1995)
Greenness	Not specified	Landsat/corn	Daughtry, Gallo, and Bauer (1983)
PVI	$0.036PVI - 0.015$ (growing) $0.037PVI + 0.114$ (senescence)	SPOT and videography/cotton and corn	Wiegand et al. (1991)

* intercept was converted to a negative value based on a scatter plot (Figure 11 in Asrar, Myneni, and Choudhury [1992]).

Ecological Characterization of Vegetation

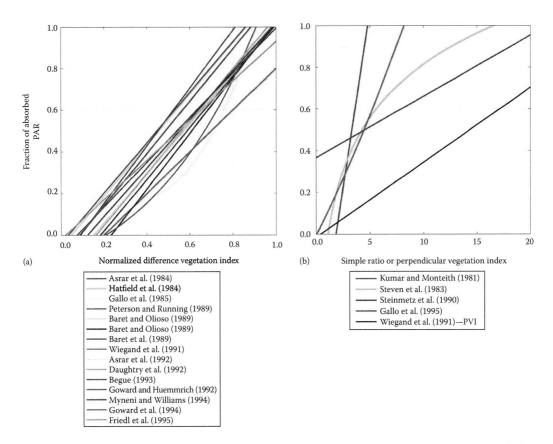

FIGURE 7.3 Graphical representations of the relations between (a) normalized difference vegetation index (NDVI) and fraction of absorbed photosynthetically active radiation (FPAR) and between (b) simple ratio (SR) (or perpendicular vegetation index, PVI) and FPAR.

SR-FPAR relationship developed in the earlier research was usually done for crops in field using spectroradiometers and was often used to estimate crop production. Kumar and Monteith (1981) developed a linear relationship between SR and FPAR for various crops, and the relationship was further examined by Steven, Biscoe, and Jaggard (1983). Asrar et al. (1984) provided the theoretical basis for the NDVI-FPAR relationship, which was further validated with field measurements.

Many studies have found that the intercept (b) of this linear NDVI-FPAR relationship is not zero for non-black background but slightly negative (Table 7.13; Figure 7.3). These relations are often dependent on vegetation phenological phases, such as green-up and senescence (e.g., Hatfield, Asrar, and Kanemasu 1984; Wiegand et al. 1991). The slope values (a) in the NDVI-FPAR relations range from 0.95 to 1.386, except for senescence periods. NDVI seems to provide more consistent linear relations with FPAR than SR across different biomes and sensors (Figure 7.3). Although both NDVI and FPAR are not scale-invariant, the NDVI-FPAR relationship has been proven to be scale invariant due to its linearity (Myneni et al. 1995; Friedl et al. 1995; Myneni and Williams 1994). For this reason, FPAR was often incorporated as a key scaling measure when fusing multi-temporal remote sensing datasets at different scales (Hwang et al. 2011).

7.4.2.2 Estimating FPAR with Biophysical Models

Since Asrar et al. (1984), the NDVI-FPAR relations have been explored for different sensors (Landsat MSS or TM, SPOT, AVHRR etc.) and biome types at different scales. This linear relationship between FPAR and vegetation index has also been reproduced with several radiative transfer models (Sellers 1985, 1987; Asrar, Myneni, and Choudhury 1992; Myneni and Williams 1994; Knyazikhin et al. 1998). These studies usually found that the NDVI-FPAR relationship is mostly linear regardless of vegetation spatial heterogeneity and leaf optical properties. However, this relationship is sensitive to soil background reflectance. Knyazikhin et al. (1998) mathematically proved the NDVI-FPAR proportionality (FPAR = a·NDVI) if background soil is ideally black.

One of the best-known models estimating FPAR is perhaps the MODIS/FPAR algorithm, which simulates Bidirectional Reflectance Factor (BRF) using a 3D canopy radiative transfer model by biome types (Myneni et al. 2002; Ganguly et al. 2008a). This algorithm assumes biome-specific canopy structure (e.g., leaf orientation distribution; Knyazikhin et al. 1998) and leaf/soil optical properties (Myneni et al. 2002). The algorithm also assumes spectral invariance in canopy transmittance and absorptance at a reference wavelength to those in other wavelengths, which also provides the theoretical basis of the linear NDVI-FPAR relationship (Knyazikhin et al. 1998). In this algorithm, FPAR is calculated by integrating the weighted spectral absorptance over the PAR spectral region (Knyazikhin et al. 1998; Ganguly et al. 2008a). The algorithm was also successfully applied to retrieve FPAR values from AVHRR and Landsat imagery with different spatial and spectral resolutions (Ganguly et al. 2008b; Ganguly et al. 2012).

7.4.3 Leaf Chlorophyll Content

Leaf pigments such as chlorophyll a and b play a crucial role in plant photosynthesis through the conversion of solar radiation into stored chemical energy, via a series of electron transfers that occur on the thylakoid membranes in chloroplasts. As the amount of solar radiation absorbed by a leaf is primarily a function of the foliar photosynthetic pigments, low concentrations of chlorophyll can limit photosynthetic potential and primary production (Richardson, Duigan, and Berlyn 2002). With a large proportion of nitrogen contained within chlorophyll molecules, leaf chlorophyll is intrinsically linked to carbon and nitrogen cycles. Further, its role in photosynthesis and net primary productivity is important within regional and global carbon models. Decreases in foliar chlorophyll can indicate plant disturbance and stress, for example, from disease, limited water availability, extreme temperature or pests, and thus acts as a bio-indicator of plant physiological condition. The importance of chlorophyll to a range of ecological processes has led to a considerable body of research dedicated to deriving chlorophyll content from leaf and canopy reflectance, from laboratory- and field-based studies to airborne and satellite platforms.

Leaf reflectance is controlled by a range of biochemical and physical variables, including chlorophyll, nitrogen, carotenoids, anthocyanins, water, and internal leaf structure, with chlorophyll dominating in the visible wavelengths (400–700 nm). Chlorophyll absorbs strongly in red and blue spectral regions, with maximum absorbance in red wavelengths between 660 and 680 nm and maximum reflectance in green wavelengths (~560 nm) within the visible spectrum. Overlapping absorption from the presence of carotenoids in blue wavelengths often prevents this region from being useful in chlorophyll estimation (Sims and Gamon 2002). Research has also identified that the absorption feature in red wavelengths readily saturates at relatively low chlorophyll contents, leading to a reduced sensitivity to higher chlorophyll content. This has led to the use of "off-center" wavelengths, with reflectance in wavelengths along the red-edge region (690–750 nm) showing greater sensitivity to subtler changes in chlorophyll content (Curran, Dungan, and Gholz 1990). This improved sensitivity along the red-edge is because increasing chlorophyll content causes a broadening of the absorption feature centered around 680 nm; shifting the position of the red-edge to longer wavelengths. Whilst the relationship between leaf reflectance and chlorophyll content is reasonably well established, particularly for broadleaf species, reflectance sampled from remote platforms is also governed by additional canopy contributions. These include leaf architecture, leaf area index (LAI), clumping, tree density, non-photosynthetic canopy elements, along with solar/

TABLE 7.13
Application of Empirical Models for the Fraction of Photosynthetically Active Radiation (FPAR) with the Spectral Vegetation Indices into Global Production Efficiency or Process-Based Biogeochemical Models

FPAR	Model	Reference
$\min((SR - SR_{min})/(SR_{max} - SR_{min}), 0.95)^*$	CASA	Potter et al. (1993)
$\{(SR - SR_{min})/(SR_{max} - SR_{min})\}(FPAR_{max} - FPAR_{min})$	SiB2	Sellers et al. (1994)
$1.25 NDVI - 0.025$	TURC	Ruimy, Saugier, and Dedieu (1994)
$0.11 SR - 0.12$	TURC	Ruimy, Saugier, and Dedieu (1994)
$1.67 NDVI - 0.08$	3-PGS	Coops, Waring, and Landsberg (1998)
$1.67 NDVI - 0.08$	Glo-PEM2	Goetz et al. (1999)
$1.21 NDVI - 0.04$	3-PGS	Coops (1999)
NDVI	Biome-BGC	Running et al. (2004)
$1.24 NDVI - 0.168$	EC-LUE	Rahman et al. (2004)
$0.279 SR - 0.294$	NASA-CASA	Potter et al. (2003)
EVI	VPM	Xiao et al. (2004)

* SR_{max} and SR_{min} are biome type dependent

viewing geometry, ground cover, and understory vegetation, making the relationship between leaf chlorophyll and canopy reflectance complex.

7.4.3.1 Laboratory Extraction of Leaf Chlorophyll Content

Many studies calibrate or validate chlorophyll estimates derived from handheld chlorophyll meters and remotely sensed platforms using laboratory-derived *in vitro* chlorophyll content. Chlorophyll content is determined by extraction from leaf samples and subsequent spectrophotometric measurements, and expressed by weight, or in most cases by area. A range of organic solvents are typically used to extract chlorophylls and carotenoids from plant tissues, including acetone, methanol, ethanol, dimethyl sulfoxide (DMSO), and N,N-dimethylformamide (DMF), which range in optimal extraction time and performance. In a comprehensive study, Minocha et al. (2009) compared the efficiencies of acetone, ethanol, DMSO, and DMF for chlorophyll and carotenoid extraction for 11 species, finding that extraction efficiencies of ethanol and DMF were comparable for analyzing chlorophyll concentrations. DMF was the most efficient solvent for the extraction of carotenoids; however, the toxicity of DMF requires care when using this solvent.

7.4.3.2 Measuring Leaf Chlorophyll Content with SPAD Meter

For rapid, non-destructive leaf chlorophyll measurements taken in the field, a handheld chlorophyll SPAD meter, developed by Minolta Corporation, Ltd., is often used, particularly in agricultural applications. The current model (SPAD-502) measures leaf transmittance through a leaf clamped within the meter at two wavelengths: 650 and 940 nm. The 650-nm is selected to coincide with the chlorophyll maximum absorbance feature, and 940 nm is used as a reference to compensate for factors such as leaf moisture content and internal structure (Zhu, Tremblay, and Liang 2012b). The measured SPAD unit value is converted to chlorophyll content using a calibration equation. However, relatively few studies perform their own calibrations with *in vitro* chlorophyll content, which are likely to vary according to plant species, leaf thickness, and leaf age. Uddling et al. (2007) tested the relationships between chlorophyll content and SPAD values for birch, wheat, and potato. For all three species, the relationships were non-linear, although for birch and wheat it was strong (~R^2 = 0.9), while the potato relationship was weaker (~R^2 = 0.5). It may therefore be appropriate to develop species-specific calibrations for robust chlorophyll estimation, with consideration for leaf developmental stage.

7.4.3.3 Estimating Leaf Chlorophyll Content with Spectral Vegetation Indices

Empirical spectral vegetation indices (VIs) are perhaps the most popular and straightforward means of retrieving chlorophyll content from remotely sensed data. There has been a wealth of research devoted to deriving statistical relationships between VIs and biochemical constituents, in order to retrieve chlorophyll content. Spectral indices are usually formulated using ratios of wavelengths that are sensitive to a particular leaf pigment to spectral regions where scattering is mainly driven by leaf internal structure or canopy structure (i.e., the near infrared). Indices including "off-center" wavelengths (690–740 nm) have been shown to be strong indicators of chlorophyll content (Croft, Chen, and Zhang 2014) compared to indices containing chlorophyll absorption wavelengths (660–680 nm) due to ready saturation even at low chlorophyll content (Daughtry et al. 2000). Many VIs used in chlorophyll studies are focused along the red-edge, including the MERIS Terrestrial Chlorophyll Index (Dash and Curran 2004). Recent research has focused on improving the generality and applicability of spectral indices through testing and modification over a range of species and physiological conditions, using empirical and simulated data. However, many indices have been developed and tested using a few closely related species, at the leaf scale and under controlled laboratory conditions. At the leaf level, surface scattering, internal structural characteristics, and leaf water content affect the relationship between VI and chlorophyll content estimation. Scaling up to a branch or canopy, other factors such as LAI, solar/viewing geometry and canopy architecture also affect the VI. Background contributions have also been shown to perturb the relationship between chlorophyll and VIs, particularly in sparse or clumped canopies, with low LAI values (Croft et al. 2013). However, these confounding influences are less of a concern in closed broadleaf canopies, which essentially behave as a big leaf (Gamon et al. 2010).

7.4.3.4 Estimating Leaf Chlorophyll Content with Radiative Transfer Models

Physically based modeling approaches use radiative transfer models to account for variations in canopy architecture, image acquisition conditions, and background vegetation that may vary in space and time. As radiative transfer models are underpinned by physical laws governing the interaction of radiation at the canopy surface, and within the canopy, they provide a direct physical relationship between canopy reflectance and canopy biophysical properties. The most recognized approach for modeling leaf chlorophyll content from remotely sensed data in this manner is through the coupling of a canopy radiative transfer model and a leaf optical model, to firstly retrieve leaf-level reflectance and then derive leaf biochemical constituents from the modeled leaf reflectance (Croft et al. 2013). Several canopy models have been used for this purpose, with the most popular including SAIL (Verhoef 1984), DART (Gastellu-Etchegorry, Martin, and Gascon 2004), 4-SCALE (Chen and Leblanc 1997), GeoSAIL (Huemmrich 2001), and FLIGHT (North 1996). The models range from turbid medium models (SAIL) to hybrid geometric optical and radiative transfer models (4-SCALE, GeoSAIL, DART) in which the turbid media are constrained into a geometric form (i.e., a leaf, shoot, branch, and/or crown) to ray-tracing techniques (FLIGHT). At the leaf level, the PROSPECT model (Jacquemoud and Baret 1990) has had widespread validation across a wide range of vegetation species and functional types. In the inverse mode, PROSPECT models chlorophyll and carotenoids from leaf reflectance, along with dry matter, a structural parameter, and equivalent water thickness. The smaller number of input parameters compared to other leaf-level models, such as LIBERTY (Dawson, Curran, and Plummer 1998), means that it is readily inverted. A range of different techniques have been employed to invert the leaf and canopy radiative transfer models, including iterative numerical optimization methods, artificial neural networks, vector machine regression, and lookup-tables (LUT). However, the "ill-posed" problem means that different combinations of the same structural and image acquisition parameters can result in the same canopy reflectance, indicating that some *a priori* scene information is required to constrain the inversion (Kimes et al. 2000).

7.4.4 Light Use Efficiency

7.4.4.1 Biophysical Basis of Light Use Efficiency

Light provides the necessary energy for plant photosynthesis. Light use efficiency (LUE) measures carbohydrate produced by plants absorbing a unit amount of photosynthetically active radiation (PAR). Monteith (1972, 1977) initially proposed the theory in estimating crop production. Kumar and Monteith (1981) first applied the LUE theory to estimate crop growth with remote sensing. They successfully estimated crop growth with a constant LUE. This triggered tremendous interest in estimating LUE for different plant communities. Prince (1991) and Ruimy, Saugier, and Dedieu (1994) reviewed LUE values published in the literature and found it varied greatly. In addition to the inherent difference in LUE among different plant communities, numerous external factors also contributed to the apparent variation in LUE values reported in the literature (Gower, Kucharik, and Norman 1999; Prince 1991), including:

- Realized LUE vs. maximum LUE
- LUE based on aboveground growth vs. total plant growth
- LUE based on PAR absorbed vs. intercepted
- LUE based on PAR vs. total global radiation
- LUE based on NPP vs. GPP
- LUE based on PAR absorbed by green vegetation vs. total foliage

In order to reduce confusion, Prince (1991); Gower, Kucharik, and Norman (1999); and Song, Dannenberg, and Hwang (2013) advocated that LUE should be based on absorbed PAR by total foliage in plant canopies, treating the proportion of non-photosynthetic vegetation component as an inherent part of the plant communities.

Use of light for photosynthesis by a single leaf can be characterized by the photosynthesis-light (P-L) response curve (Figure 7.4). The net photosynthesis rate and PAR density is a straight line only when PAR is low. After reaching the light saturation point, the net photosynthesis rate is nonlinearly related to photon flux density. Therefore, leaves in a plant canopy may have very different LUE at any given time of day because leaves could have very different PAR density due to mutual shading and variation in leaf orientation. Plant production of a given plant community is the sum of photosynthesis from all leaves, and the LUE of such production varies with time of day due to changes in incident solar radiation angle and cloudiness. Song, Dannenberg, and Hwang (2013) showed that the shorter the time span, the bigger the variation of LUE based on simulations with a Farquhar photosynthesis model coupled with a sophisticated canopy radiation transfer model (Song et al. 2009). It takes about a month for LUE to converge to a stable value (Figure 7.5). Despite the highly nonlinear relationship between net photosynthesis rate and PAR density at the instantaneous time scale, the combination of mutual shading of leaves, variation of leaf orientation, and change in solar angle tends to linearize the relationship between absorbed PAR and plant growth over time (Sellers 1985). It is important to note that the LUE estimated by Monteith (1972, 1977) was done over a whole growing season. Goetz and Prince (1999) argued based on plant functional convergence hypothesis (Field 1991) that LUE should converge to a narrow range for gross primary productivity among a wide range of plant functional types.

Incident radiation arrives at the top of plant canopies as direct and diffuse radiation. Their relative composition is a function of sun angle, cloudiness of the atmosphere, and the characteristics of canopy structure (Ni et al. 1997), thus varying with location and canopy structure. Because diffuse radiation can penetrate plant canopies deeper than direct light, plants have a higher LUE for

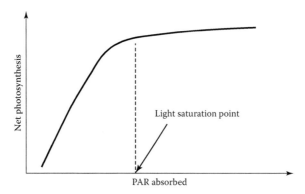

FIGURE 7.4 Photosynthesis-light response curve.

diffuse light. Therefore, LUE should converge to different values for different vegetation biomes, a conclusion that is now generally accepted (Ruimy, Saugier, and Dedieu 1994; Running et al. 2004; Turner et al. 2003). Due to landscape heterogeneity in vegetation biome composition, the biome-dependent LUE creates spatial variation in LUE at the pixel level, particularly with coarse spatial resolution imagery, such as that from MODIS (Turner et al. 2002), making direct mapping of LUE from remotely sensed data attractive.

7.4.4.2 Remote Sensing LUE

PAR absorbed by leaf pigments has three pathways within the chloroplast: photochemical quenching (i.e., used for photosynthesis), non-radiative quenching, or photoprotection through which excess energy is dissipated through the xanthophyll cycle and chlorophyll fluorescence (ChF) through which radiation is emitted at longer wavelength than the absorbed light (Coops et al. 2010). Energy directed toward non-radiative quenching and chlorophyll fluorescence reduces LUE in photosynthesis (Coops et al. 2010; Meroni et al. 2009). Remote sensing of LUE is based on the detection of energy that is not used in the photochemical quenching.

One such approach was initially developed by Gamon, Peñuelas, and Field (1992) based on their discovery that leaf reflectance at 531 nm is related to the xanthophyll cycle (Gamon et al. 1990). When absorbed PAR exceeds photosynthetic capacity, the "excess light" is dissipated through deepoxidation of violaxanthin to zeaxanthin via the intermediate antheraxanthin pigment. The process is reversed when there is insufficient supply of energy (Demmig-Adams 1990). Because zeaxanthin and antheraxanthin have higher absorption coefficients for radiation near 531 nm than violaxanthin, leaf reflectance at 531 nm changes with the xanthophyll cycle. Gamon, Peñuelas, and Field (1992) developed the photochemical reflectance index (PRI) to measure the leaf reflectance changes at 531 nm

$$PRI = \frac{R_{531} - R_{570}}{R_{531} + R_{570}}, \tag{7.16}$$

where R_{531} and R_{570} are leaf reflectance at 531 nm and 570 nm, respectively. Here R_{570} is used as reference reflectance as leaf reflectance changes little at 570 nm with the xanthophyll cycle. Numerous studies confirm that PRI is highly correlated with photosynthetic radiation use efficiency (Gamon and Surfus 1999; Nichol et al. 2000; Peñuelas, Filella, and Gamon 1995). Great interest has been generated to map LUE remotely with PRI, particularly with remotely sensed data from MODIS since band 11 is centered at 531 nm. However, MODIS does not have a band centered at 570 nm that can be used as the reference band. Drolet et al. (2005) used bands 11 and 13 to calculate PRI and found that PRI from backscatter images (i.e., near hotspot view) is significantly correlated with LUE estimated from flux tower measurements. Drolet et al. (2008) further demonstrated that

Ecological Characterization of Vegetation

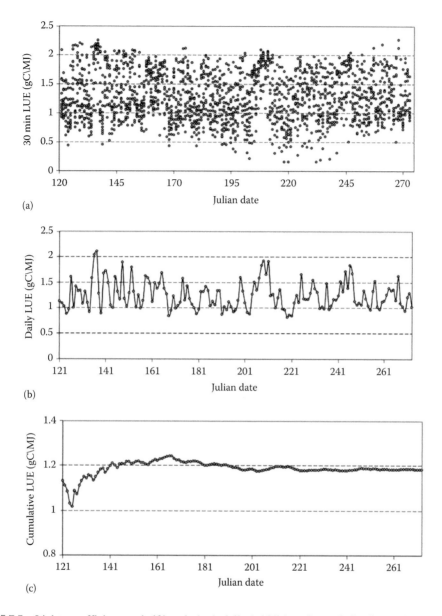

FIGURE 7.5 Light use efficiency at half hourly (top), daily (middle), and cumulative (bottom) temporal scale based on simulation from Song et al. (2009).

PRI calculated with band 14 as the reference band also had a strong relationship with 90-minute LUE for a boreal forest in Saskatchewan, Canada. Garbulsky et al. (2008) found that band 12 can also be used as the reference band for characterizing the relationship between PRI and LUE over a Mediterranean forest.

Testing PRI from MODIS with flux tower measurements offers insight on the potential of mapping LUE in space. However, the PRI-LUE relationship can be compromised by the mismatch in scale between the footprint of a flux tower and the corresponding MODIS pixel. Hall et al. (2008) and Hilker et al. (2008, 2010) conducted a series of fieldwork studies that measure PRI on a flux tower with a spectroradiometer in two forests, with one on Vancouver Island, BC, in a Douglas fir stand, and the other in a mature aspen stand in central Saskatchewan. They found that the down regulation of photosynthesis at the forest scale governs the relationship between PRI and LUE. They also found a stronger relationship between PRI and LUE for sunlit canopies than shaded canopies, providing evidence to explain why only PRI from backscatter MODIS imagery correlated with LUE (Drolet et al. 2005). Hall, Hilker, and Coops (2011) proposed a satellite mission, PHOTOSYNSAT, a multiangle along-track satellite, to map LUE based on their findings that: (1) the first derivative of PRI with respect to the fraction of shadows in the sensor field of view is proportional to LUE, and (2) PRI response is independent of vegetation structure and optical properties.

Another approach to estimating LUE remotely is through detecting the energy dissipated in chlorophyll fluorescence at wavelengths ranging from 650 to 800 nm (Meroni et al. 2009). Remote sensing of ChF is based on three Fraunhofer lines where incident solar radiation is low due to absorption by hydrogen in the solar atmosphere (656.4 nm) and oxygen in the Earth's atmosphere (760.5 and 787.5 nm) (Guanter et al. 2010; Liu and Cheng 2010; Meroni et al. 2009). The European Space Agency's Fluorescence Explorer (FLEX) mission is aimed at measuring solar-induced ChF in the Fraunhofer lines created by oxygen (Guanter et al. 2010; Mohammed et al. 2012). The feasibility has been demonstrated on the ground with high spectral resolution spectroradiometers and fluorimeters (Damm et al. 2010; Liu and Cheng 2010; Liu et al. 2013), and in the air and space (Frankenberg et al. 2011; Guanter et al. 2007; Zarco-Tejada et al. 2009; Zarco-Tejada et al. 2013). Due to the weak ChF signal, it is mandatory for FLEX to be successful that the remotely sensed data be collected with high spectral resolution and data have to be corrected with a precise modeling of atmospheric effect in the visible and near-infrared spectra (Guanter et al. 2010; Malenovsky et al. 2009; Grace et al. 2007).

7.4.5 Gross Primary Productivity (GPP)/Net Primary Productivity (NPP)

Gross primary productivity (GPP) measures the net carbohydrate produced during photosynthesis after dark respiration over a unit area in a given time. Net primary productivity (NPP) is the balance of GPP after plant autotrophic respiration. Both GPP and NPP are key carbon fluxes between the terrestrial ecosystem and the atmosphere in the global carbon cycle. However, current eddy-covariance technology cannot measure GPP directly, but only measures net ecosystem exchange (NEE), i.e., the net difference between GPP and ecosystem respiration, which includes both autotrophic and heterotrophic respiration. Estimation of GPP is accomplished by resorting to models to estimate autotrophic respiration during the daytime. Through careful fieldwork, NPP can be estimated as the total organic matter produced on an annual basis (Clark et al. 2001a; Clark et al. 2001b). However, remote sensing is the only viable option to provide wall-to-wall estimation of primary production.

7.4.5.1 Remote Sensing of Primary Production Based on LUE

Optical remote sensing of terrestrial ecosystem primary production is predominantly based on the LUE theory (Monteith 1972; Monteith 1977) because of its sound biophysical basis and simplicity. The general form of LUE models for primary productivity is:

$$P = \varepsilon_{APRA} \times IPAR \times f(E), \quad (7.17)$$

where P is the primary productivity. The maximum LUE, ε_{max}, is the conversion factor of APAR to NPP or GPP under optimal conditions. APAR is generated as the product of the fraction of APAR (f_{APAR}) and incident PAR (IPAR). The environmental scalar function, $f\varepsilon$, produces a scalar that is between 0 and 1 to reduce the maximum LUE to actual LUE due to environmental stress.

The product of a LUE model can either be NPP or GPP, as listed in Table 7.14. For models with their initial product being GPP, autotrophic respiration is usually simulated to produce NPP, such as the Biome-BGC model that produces the MODIS NPP and GPP products (Zhao et al. 2005).

TABLE 7.14
Typical LUE Models for Terrestrial Ecosystem Primary Productivity

Model	Productivity measure	LUE (ε_{max}) (gC/MJ)	Environmental Scalars — Temperature (T), vapor pressure deficit (D), and soil water (θ)	References
Kumar and Monteith	NPP	1.3	N/A	Kumar and Monteith 1981
BIOMASS	GPP	A	$f(T_{min}) = \begin{cases} 0 & T_{min} < T_{min1} \\ \dfrac{T_{min} - T_1}{T_2 - T_1} & T_{min1} \leq T_{min} < T_{min2} \\ 1 & T_{min} \geq T_{min2} \end{cases}$ $f(\theta) = \begin{cases} 0 & \theta < \theta_{min} \\ \dfrac{\theta - \theta_{min}}{\theta_{max} - \theta_{min}} & \theta_{min} \leq \theta < \theta_{max} \\ 1 & \theta \geq \theta_{max} \end{cases}$ $f(VPD) = \begin{cases} 0 & D \geq D_2 \\ \dfrac{D_2 - D}{D_2 - D_1} & D_1 \leq D < D_2 \\ 1 & D < D_1 \end{cases}$	McMurtrie, Rook, and Kelliher 1990
Ruimy et al.	NPP	0.37–2.07 (biome-dependent)	N/A	Ruimy et al. 1994
GLO-PEM	GPP	ε_{max}=55.2a	$f(T_{min}) = 0$, when $T_{min} < 0$, $f(D) = 1.2\exp(-0.35 \cdot D) - 0.2$ $f(\theta) = \dfrac{\alpha_0 - \alpha_d}{\alpha_w - \alpha_d}$	Prince and Goward 1995
TURC	GPP	1.1	N/A	Ruimy et al. 1996
3-PGS	GPP	1.8	$f(D) = \exp(-2.5 \cdot D)$, $f(\theta) = \dfrac{1}{1 + \{(1-\theta)/c\}^d}$	Coops et al. 1998
NASA-CASA	NPP	0.389/0.506	$f_1(T) = 0.8 + 0.02 T_{opt} - 0.0005 T_{opt}^2$ $f_2(T) = \dfrac{1.18/\left(1 + \exp\{0.2(T_{opt1} - 10 - T_a)\}\right)}{1 + \exp\{0.3(-T_{opt1} - 10 - T_a)\}}$ $f(\theta) = 0.5 + \dfrac{AET}{PET}$	Potter et al. 1993; Frield et al. 1995; Potter et al. 2003
MODIS (MOD17)	GPP	0.604–1.259 (biome-dependent)	$f(T_{min}) = \begin{cases} 0 & T_{min} < T_1 \\ \dfrac{T_{min} - T_1}{T_2 - T_1} & T_1 \leq T_{min} < T_2 \\ 1 & T_{min} \geq T_2 \end{cases}$ $f(D) = \begin{cases} 0 & D \geq D_2 \\ \dfrac{D_2 - D}{D_2 - D_1} & D_1 \leq D < D_2 \\ 1 & D < D_1 \end{cases}$	Running et al. 2000; Zhao et al. 2005

(Continued)

TABLE 7.14 *(Continued)*
Typical LUE Models for Terrestrial Ecosystem Primary Productivity

Model	Productivity measure	LUE (ε_{max}) (gC/MJ)	Environmental Scalars — Temperature (T), vapor pressure deficit (D), and soil water (θ)	References
VPM	GPP	Quantum yield	$f(T) = \dfrac{(T_a - T_3)(T_a - T_4)}{\{(T_a - T_3)(T_a - T_4)\} - (T_a - T_{opt})^2}$ $f(\theta) = \dfrac{1 + LSWI}{1 + LSWI_{max}}$ $LSWI = \dfrac{\rho_{NIR} - \rho_{SWIR}}{\rho_{NIR} + \rho_{SWIR}}$	Xiao et al. 2004a, 2004b
TOPS (diagnostic version)	GPP	Same with MOD17	Same with MOD17 for T_{min} and VPD $f(\theta) = \begin{cases} 0 & LWP \geq LWP_2 \\ \dfrac{LWP_2 - LWP}{LWP_2 - LWP_1} & LWP_1 \leq LWP < LWP_2 \\ 1 & LWP < LWP_1 \end{cases}$	Nemani et al. 2009
EC-LUE		2.14	$f(T) = \exp\left[-\left(\dfrac{T_a - T_{opt}}{T_{opt}}\right)^2\right]$ $f(\theta) = \dfrac{1}{\beta + 1} = \dfrac{LE}{LE + H}$ Same with MOD17 for T_{min} and VPD $f(T)$ of VPM model used in the old version	Yuan et al. 2007, 2014

α: quantum yield (moles CO_2 mole $PPFD^{-1}$)
α_0, α_w, and α_d: the slope between $NDVI$ and T_s, and the wet and dry edge slopes (°C $NDVI^{-1}$)
β: Bowen ratio (=H/LE)
θ: calculated soil saturation (%)
θ_{max} and θ_{min}: soil saturation of root zone at field capacity (-0.004 MPa) and permanent wilting point (−1.5 MPa) (%)
c and d: texture-specific parameters for the water stress scalar of the 3-PGS model
T_a, T_s, and T_{min}: air, surface, and daily min temperature (°C)
T_1 and T_2: biome-specific parameters for the T_{min} scalar
T_3 and T_4: min and max temperature for photosynthetic activity in the VPM model
T_{opt}: optimal temperature for photosynthetic activity (°C) (25 for EC-LUE model, monthly average air temperature when NDVI reaches its maximum in the NASA-CASA model
VPD or D: vapor pressure deficit (kPa)
D_1 and D_2: biome-specific parameters for the VPD scalar
LWP: predawn leaf water potential (−MPa), calculated from θ and soil textural parameters
LWP_1 and LWP_2: biome-specific parameters for the LWP scalar (min spring LWP and stomatal closure LWP) (−MPa)
AET and PET: actual and potential evapotranspiration
LE and H: latent and sensible heat
LSWI: land surface water index
$LSWI_{max}$: max $LSWI$ within the plant-growing season for individual pixels

7.4.5.2 Remote Sensing of Primary Production without LUE

Estimating terrestrial ecosystem primary production does not always rely on LUE. Numerous process-based models, which use remotely sensed land surface biophysical parameter products as key model inputs, but are not based on LUE theory, have been developed. The most

Ecological Characterization of Vegetation

commonly used land surface biophysical parameter is perhaps leaf area index (LAI). Running et al. (1989) conducted a pioneering study that integrated remote sensing with ecosystem models to estimate GPP. They used LAI derived from AVHRR NDVI as an input to the Forest-BGC model (Running and Coughlan 1988) and estimated both GPP and transpiration over a 28×55 km mountainous region. Nemani et al. (1993) derived an improved LAI product from Landsat Thematic Mapper (TM) imagery by correcting the background effects on NDVI using a mid-infrared reflectance. The subsequent LAI product was used as an input to the RHESSys model (Band et al. 1993) to estimate evapotranspiration and GPP at the watershed scale. Liu et al. (1997) developed the BEPS model, which estimates NPP using LAI derived from the ten-day composite AVHRR NDVI. Nemani et al. (2009) developed the TOPS model to estimate GPP and ET fluxes using the MODIS LAI product as an input. More advanced use of optical remote sensing for primary production estimation takes advantage of multi-angular remote sensing to derive more complex canopy biophysical parameters (e.g., the foliage clumping index) that can be integrated into process-based photosynthesis models to provide more detailed partitioning of solar radiation into sunlit and shaded leaves, potentially improving NPP and GPP estimates over the "big-leaf" models based on LAI (Chen et al. 2003; Chen et al. 2012). These types of models closely couple carbon and water fluxes through stomatal conductance, often by linking the Farquhar photosynthesis (Farquhar, Caemmerer, and Berry 1980) and Penman-Monteith (Monteith 1965) evapotranspiration models.

7.4.6 Uncertainties, Errors, and Accuracy

The functional products of vegetation derived from remote sensing directly, such as phenology and LUE, or indirectly, such as GPP and NPP, are theoretically invalidatable because we never have perfect reference data. The remotely sensed phenology parameters, such as the start and the end of growing season can only be derived through statistical time-series analysis, typically on a spectral vegetation index (Zhang et al. 2003). Land surface phenology is an outcome of synaptic phenomena for all vegetation within the entire pixel, which is usually quite large for the sake of frequent repeat coverage (Myneni et al. 1997; Tucker et al. 2005; Zhu et al. 2013). The synaptic seasonal timing over a sizable area cannot be accurately recorded on the ground as each individual plant may have a different phenology. Although there are now cameras mounted on flux towers which provide a much better record of the actual synaptic phenology (Richardson et al. 2009), it is nearly impossible to have a perfect spatial match between the area of the camera field of view and the satellite pixel. In addition, the direct derivation of evergreen forest phenology, which demonstrate little seasonal greenness change, remains a challenge using remotely sensed imagery (Dannenberg et al., 2013).

Similarly, the validation of GPP/NPP estimates from remote sensing over a large area is impossible because no reliable reference data is available. On the site basis, carbon fluxes from eddy-covariance flux towers are frequently used for model evaluations (Xiao et al. 2004a, 2004b; Yuan et al. 2007; Yuan et al. 2014). However, flux towers do not directly measure GPP/NPP, but net ecosystem exchange (NEE), which is the net difference between the carbon absorbed by plant photosynthesis and ecosystem respiration. Ecosystem respiration has to be subtracted from NEE in order to get GPP. Currently, there is no direct measurement for daily ecosystem respiration when photosynthesis happens, though it can be modeled based on the ecosystem respiration rate at night (Reichstein et al. 2005; Lasslop et al. 2010), adding uncertainty to the flux tower–derived GPP for model evaluation (Schaefer et al. 2012). An alternative approach for model evaluation is comparing independent estimates, such as that from atmospheric inverse modeling and that from ground inventory and analysis (Pacala et al. 2001; Piao et al. 2008). Although this does not validate any model outputs in the strict sense, it does provide more credibility to the model output. Much research and new technologies are needed in this regard in the future.

7.5 FUTURE DIRECTIONS

Optical remote sensing will continue to play a critical role in monitoring the spatial temporal dynamics of vegetation in the foreseeable future, and it will continue to provide key land surface biophysical parameter data layers to models at regional to global scales that aim to simulate and project the changes of terrestrial ecosystem functions in the future due to global environmental changes. Recently launched remote sensing missions, such as the Landsat 8 and Suomi NPP satellites by the United States and the SPOT 6 by France, will continue to provide pivotal optical remotely sensed data in the immediate future. Planning of remote sensing missions in the United States is guided by the Decadal Survey generated by the National Research Council of the National Academy of Sciences (NRC 2007). The planned mission, HyspIRI, combines a hyperspectral visible shortwave infrared imaging spectrometer with a multispectral thermal infrared spectrometer to map vegetation composition and its health. The European Space Agency's Sentinel mission, Sentinel-2, will continue to provide optical remotely sensed imagery globally as an enhanced continuity of SPOT and Landsat-type data. At the same time, high spatial resolution, optical, remotely sensed data will continue to boom in the private market. Therefore, both the quality and the quantity of optical remotely sensed data will increase in the foreseeable future.

In addition to remotely sensed data in the optical domain, radar and LiDAR remote sensing missions will increase in the future at the same time. The abundance of remotely sensed data from multiple sensors will lead to synergistic use of remotely sensed data from different sensors to extract vegetation information, particularly by fusing information from LiDAR, radar, and optical sensors (Gray and Song 2012; Nelson et al. 2009; Gao et al. 2006; Lefsky et al. 2005). Although empirical approaches will continue to be critical to understand the relationship between remotely sensed data and vegetation structure at local to regional scales, more physically based approaches that are applicable globally will continue to advance, providing increasingly higher-quality vegetation information that will enable improved understanding of the roles vegetation plays in the terrestrial ecosystem.

7.6 ACKNOWLEDGMENT

This work is partly supported by NSF Grant DEB-1313756.

REFERENCES

Adams, J. B., D. E. Sabol, V. Kapos, R. A. Filho, D. A. Roberts, M. O. Smith, and A. R. Gillespie. 1995. Classification of multispectral images based on fractions of endmembers: Application to land-cover change in the Brazilian Amazon. *Remote Sens Environ* 52:137–154.

Andersen, H. E., J. Strunk, H. Temesgen, D. Atwood, and K. Winterberger. 2011. Using multilevel remote sensing and ground data to estimate forest biomass resources in remote regions: A case study in the boreal forests of interior Alaska. *Canadian J Rem Sens*, 37(6):596–611.

Anderson, G. L., J. D. Hanson, and R. H. Haas. 1993. Evaluating Landsat Thematic Mapper derived vegetation indices for estimating above-ground biomass on semiarid rangelands. *Remote Sens Environ* 45:165–175.

Aragao, L. E. O. C., Y. E. Shimabukuro, F. D. B. Espírito-Santo, and M. Williams. 2005. Spatial validation of the Collection 4 MODIS LAI product in Eastern Amazonia. *IEEE Trans Geosci Remote Sens* 43:2526–2534.

Archibold, O. W. 1995. *Ecology of World Vegetation*. London: Chapman & Hall.

Asrar, G., M. Fuchs, E. T. Kanemasu, and J. L. Hatfield. 1984. Estimating absorbed photosynthetic radiation and leaf-area index from spectral reflectance in wheat. *Agron J* 76:300–306.

Asrar, G., R. B. Myneni, and B. J. Choudhury. 1992. Spatial heterogeneity in vegetation canopies and remote-sensing of absorbed photosynthetically active radiation—a modeling study. *Remote Sens Environ* 41:85–103.

Atkinson, P. M., M. E. J. Cutler, and H. Lewis. 1997. Mapping sub-pixel proportional land cover with AVHRR imagery. *Int J Remote Sens* 18(4):917–935. http://doi.org/10.1080/014311697218836.

Baccini, A., M. A. Friedl, C. E. Woodcock, and A. Warbington. 2004. Forest biomass estimation over regional scales using multisource data. *Geophys Res Lett* 31:L10501.

Baccini, A., S. J. Goetz, W. S. Walker, N. T. Laporte, M. Sun, D. Sulla-Menashe, J. Hackler, P. S. A. Beck, R. Dubayah, and M. A. Friedl. 2012. Estimated carbon dioxide emissions from tropical deforestation improved by carbon density maps. *Nature Clim Chang* 2:182–185.

Baccini, A., N. Laporte, S. J. Goetz, M. Sun, and H. Dong. 2008. A first map of tropical Africa's above-ground biomass derived from satellite imagery. *Environ Res Lett* 3:045011.

Band, L. E., P. Patterson, R. Nemani, and S. W. Running. 1993. Forest ecosystem processes at the watershed scale: Incorporating hillslope hydrology. *Agric Forest Meteorol* 63:93–126.

Baret, F., and G. Guyot. 1991. Potentials and limits of vegetation indexes for LAI and APAR assessment. *Remote Sens Environ* 35:161–173.

Baret, F., G. Guyot, and D. Major. 1989. Crop biomass evaluation using radiometric measurements. *Photogrammetria* 43:241–256.

Baret, F., O. Hagolle, B. Geiger, P. Bicheron, B. Miras, M. Huc, B. Berthelot, F. Nino, M. Weiss, and O. Samain. 2007. LAI, fAPAR and fCover CYCLOPES global products derived from VEGETATION: Part 1: Principles of the algorithm. *Remote Sens Environ* 110:275–286.

Baret, F., and A. Olioso. 1989. Estimation à partir de mesures de réflectance spectrale du rayonnement photosynthétiquement actif absorbé par une culture de blé. *Agronomie* 9:885–895.

Bastin, L. 1997. Comparison of fuzzy c-means classification, linear mixture modelling and MLC probabilities as tools for unmixing coarse pixels. *Int J Remote Sens* 18(17):3629–3648. http://doi.org/10.1080/014311697216847.

Bégué, A. 1993. Leaf area index, intercepted photosynthetically active radiation, and spectral vegetation indices: A sensitivity analysis for regular-clumped canopies. *Remote Sens Environ* 46:45–59.

Bergen, K. M., and I. Dronova. 2007. Observing succession on aspen-dominated landscapes using a remote sensing-ecosystem approach. *Landscape Ecol* 22(9):1395–1410.

Blackard, J. A., M. V. Finco, E. H. Helmer, G. R. Holden, M. L. Hoppus, D. M. Jacobs, A. J. Lister, G. G. Moisen, M. D. Neison, and R. Riemann. 2008. Mapping US forest biomass using nationwide forest inventory data and moderate resolution information. *Remote Sens Environ* 112:1658–1677.

Bonan, G. B. 2008a. *Ecological Climatology: Concepts and Applications*. 2nd ed. Cambridge: Cambridge University Press.

Bonan, G. B. 2008b. Forests and climate change: Forcings, feedbacks, and the climate benefits of forests. *Science* 320:1444–1449.

Brandt, J. S., T. Kuemmerle, H. Li, G. Ren, J. Zhu, and V. C. Radeloff. 2012. Using Landsat imagery to map forest change in southwest China in response to the national logging ban and ecotourism development. *Remote Sens Environ* 121:358–369.

Brown, L. J., J. M. Chen, and S. G. Leblanc. 2000. Short wave infrared correction to the simple ratio: An image and model analysis. *Remote Sens Environ* 71:16–25.

Brown, M. E., and K. M. de Beurs. 2008. Evaluation of multi-sensor semi-arid crop season parameters based on NDVI and rainfall. *Remote Sens Environ* 112(5):2261–2271.

Campbell, G. S. 1990. Derivation of an angle density function for canopies with ellipsoidal leaf angle distributions. *Agric For Meteorol* 49:173–176.

Campbell, J. B. 2002. *Introduction to Remote Sensing*. New York, NY: The Guilford Press. ISBN 1-57230-640-8.

Cao, C., J. Xiong, S. Blonski, Q. Liu, S. Uprety, X. Shao, X., Y. Bai, and F. Weng. 2013. Suomi NPP VIIRS sensor data record verification, validation, and long-term performance monitoring. *J Geophys Res-Atmos* 118:11,664–11,678.

Carlson, T. N., and D. A. Ripley. 1997. On the relation between NDVI, fractional vegetation cover, and leaf area index. *Remote Sens Environ* 62:241–252.

Chen, J. M. 1996a. Optically-based methods for measuring seasonal variation in leaf area index of boreal conifer forests. *Agric For Meteorol* 80:135–163.

Chen, J. M. 1996b. Evaluation of vegetation indices and a modified simple ratio for boreal applications. *Can J Remote Sens* 22:229–242.

Chen, J. M. 1999. Spatial scaling of a remotely sensed surface parameter by contexture. *Remote Sens Environ* 69:30–42.

Chen, J. M., and T. A. Black. 1992. Defining leaf area index for non-flat leaves. *Plant Cell Environ* 15:421–429.

Chen, J. M., T. A. Black, and R. S. Adams. 1991. Evaluation of hemispherical photography for determining plant area index and geometry of a forest stand. *Agric For Meteorol* 56:129–143.

Chen, J. M., X. Chen, W. Ju, and X. Geng. 2005b. Distributed hydrological model for mapping evapotranspiration using remote sensing inputs. *J Hydrol* 305:15–39.

Chen, J. M., and J. Cihlar. 1995. Plant canopy gap-size analysis theory for improving optical measurements of leaf area index. *Appl Optics* 34(27):6211–6222.

Chen, J. M., and J. Cihlar. 1996. Retrieving leaf area index for boreal conifer forests using Landsat TM images. *Remote Sens Environ* 55:153–162.

Chen, J. M., A. Govind, O. Sonnentag, Y. Zhang, A. Barr, and B. Amiro. 2006. Leaf area index measurements at Fluxnet Canada forest sites. *Agric For Meteorol* 140:257–268.

Chen, J. M., and S. G. Leblanc. 1997. A four-scale bidirectional reflectance model based on canopy architecture. *IEEE T Geosci Remote* 35:1316–1337.

Chen, J. M., J. Liu, S. G. Leblanc, R. Lacaze, and J.-L. Roujean. 2003. Multi-angular optical remote sensing for assessing vegetation structure and carbon absorption. *Remote Sens Environ* 84(4):516–525.

Chen, J. M., C. Menges, and S. Leblanc. 2005a. Global mapping of foliage clumping index using multi-angular satellite data. *Remote Sens Environ* 97:447–457.

Chen, J. M., G. Mo, J. Pisek, F. Deng, M. Ishozawa, and D. Chan. 2012. Effects of foliage clumping on global terrestrial gross primary productivity. *Global Biogeochem Cycles* 26:GB1019. http://doi.org/10.1029/2010GB003996.

Chen, J. M., G. Pavlic, L. Brown, J. Cihlar, S. G. Leblanc, H. P. White, R. J. Hall, D. R. Peddle, D. J. King, and J. A. Trofymow. 2002. Derivation and validation of Canada-wide coarse resolution leaf area index maps using high-resolution satellite imagery and ground measurements. *Remote Sens Environ* 80:165–184.

Chen, J. M., P. M. Rich, S. T. Gower, J. M. Norman, and S. Plummer. 1997. Leaf area index of boreal forests: Theory, techniques, and measurements. *J Geophys Res* 102(D24):29429–29443.

Chen, X., L. Vierling, D. Deering, and A. Conley. 2005c. Monitoring boreal forest leaf area index across a Siberian burn chronosequence: A MODIS validation study. *Int J Remote Sens* 26(24):5433–5451.

Chopping, M., G. G. Moisen, L. Su, A. Laliberte, A. Rango, J. V. Martonchik, and D. P. C. Peters. 2008. Large area mapping of southwestern forest crown cover, canopy height, and biomass using the NASA Multiangle Imaging Spectro-Radiometer. *Remote Sens Environ* 112:2051–2063.

Chopping, M., C. B. Schaaf, F. Zhao, Z. Wang, A. W. Nolin, G. G. Moisen, J. V. Martonchik, and M. Bull. 2011. Forest structure and aboveground biomass in the southwestern United States from MODIS and MISR. *Remote Sens Environ* 115:2943–2953.

Chuine, I., and E. G. Beaubien. 2001. Phenology is a major determinant of tree species range. *Ecology Letters* 4:500–551.

Clark, D. A., S. Brown, D. W. Kicklighter, J. Q. Chambers, J. R. Thomlinson, and J. Ni. 2001a. Measuring net primary production in forests: Concepts and field methods. *Ecol Appl* 11(2):356–370.

Clark, D. A., S. Brown, D. W. Kicklighter, J. Q. Chambers, J. R. Thomlinson, J. Ni, and E. A. Holland. 2001b. Net primary production in tropical forests: An evaluation and synthesis of existing field data. *Ecol Appl* 11(2):371–384.

Cleland, E. E., J. M. Allen, T. M. Crimmins, J. A. Dunne, S. Pau, S. E. Travers, E. S. Zavaleta, and E. M. Wolkovich. 2012. Phenological tracking enables positive species responses to climate change. *Ecology* 93(8):1765–1771.

Clevers, J. G. P. W. 1989. The applications of a weighted infrared-red vegetation index for estimating leaf area index by correcting for soil moisture. *Remote Sens Environ* 29:25–37.

Cohen, W. B., T. K. Maiersperger, T. A. Spies, and D. R. Oetter. 2001. Modelling forest cover attributes as continuous variables in a regional context with Thematic Mapper data. *Int J Remote Sens* 22(12):2279–2310.

Cohen, W. B., T. K. Maiersperger, D. P. Turner, W. D. Ritts, D. Pflugmacher, R. E. Kennedy, A. Kirschbaum, S. W. Running, M. Costa, and S. T. Gower. 2006. MODIS land cover and LAI Collection 4 product quality across nine sites in the Western Hemisphere. *IEEE Trans Geosci Remote Sens* 44:1843–1857.

Cohen, W. B., T. K. Maiersperger, Z. Yang, S. T. Gower, D. P. Turner, W. D. Ritts, M. Berterretche, and S. W. Running. 2003. Comparisons of land cover and LAI estimates derived from ETM+ and MODIS for four sites in North America: A quality assessment of 2000/2001 provisional MODIS products. *Remote Sens Environ* 88:233–255.

Cohen, W. B., and T. A. Spies. 1992. Estimating structural attributes of Douglas-fir/western hemlock forest stands from Landsat and SPOT imagery. *Remote Sens Environ* 41(1):1–17.

Cohen, W. B., T. A. Spies, R. J. Alig, D. R. Oetter, T. K. Maiersperger, and M. Fiorella. 2002. Characterizing 23 years (1972–95) of stand replacement disturbance in western Oregon forests with Landsat imagery. *Ecosystems* 5:122–137.

Cohen, W. B., T. A. Spies, and G. A. Bradshaw. 1990. Semivariograms of digital imagery for analysis of conifer canopy structure. *Remote Sens Environ* 34(3):167–178.

Cohen, W. B., T. A. Spies, and M. Fiorella. 1995. Estimating the age and structure of forests in a multi-ownership landscape of western Oregon, U.S.A. *Int J Remote Sens* 16(4):721–746.

Cohen, W. B., Z. Yang, and R. Kennedy. 2010. Detecting trends in forest disturbance and recovery using yearly Landsat time series: 2. TimeSync—Tools for calibration and validation. *Remote Sens Environ* 114(12):2911–2924.

Colditz, R. R., M. Schmidt, C. Conrad, M. C. Hansen, and S. Dech. 2011. Land cover classification with coarse spatial resolution data to derive continuous and discrete maps for complex regions. *Remote Sens Environ* 115(12):3264–3275. http://doi.org/10.1016/j.rse.2011.07.010.

Coops, N. C. 1999. Improvement in predicting stand growth of Pinus radiata (D. Don) across landscapes using NOAA AVHRR and Landsat MSS imagery combined with a forest growth process model (3-PGS). *Photogramm Eng Rem S* 65:1149–1156.

Coops, N. C., T. Hilker, F. G. Hall, C. J. Nichol, and G. G. Drolet. 2010. Estimation of light-use efficiency of terrestrial ecosystems from space: A status report. *BioScience* 60(10):788–797.

Coops, N. C., R. Waring, and J. Landsberg. 1998. Assessing forest productivity in Australia and New Zealand using a physiologically-based model driven with averaged monthly weather data and satellite-derived estimates of canopy photosynthetic capacity. *Forest Ecol Manag* 104:113–127.

Croft, H., J. M. Chen, and Y. Zhang. 2014. The applicability of empirical vegetation indices for determining leaf chlorophyll content over different leaf and canopy structures. *Ecol Complex* 17:119–130.

Croft, H., J. M. Chen, Y. Zhang, and A. Simic. 2013. Modelling leaf chlorophyll content in broadleaf and needle leaf canopies from ground, CASI, Landsat TM 5 and MERIS reflectance data. *Remote Sens Environ* 133:128–140.

Csiszar, I., W. Schroeder, L. Giglio, E. Ellicott, K. P. Vadrevu, C. O. Justice, and B. Wind. 2014. Active fires from the suomi NPP visible infrared imaging radiometer suite: Product status and first evaluation results. *J Geophys Res-Atmos*. http://doi.org/10.1002/2013JD020453.

Curran, P. J., J. L. Dungan, and H. L. Gholz. 1990. Exploring the relationship between reflectance red edge and chlorophyll content in slash pine. *Tree Physiol* 7:33–48.

Dai, Y., X. Zeng, R. E. Dickinson, I. Baker, G. B. Bonan, and M. G. Bosilovich, A. S. Denning et al. 2003. The common land model. *B Am Meteorol Soc*. http://doi.org/10.1175/BAMS-84-8-1013.

Damm, A., J. Elbers, A. Erler, B. Gioli, K. Hamdi, R. Hutjes, M. Kosvancova, M. Meroni, F. Miglietta, and A. Moersch. 2010. Remote sensing of sun-induced fluorescence to improve modeling of diurnal courses of gross primary production (GPP). *Glob Change Biol* 16(1):171–186.

Dannenberg, M., Song, C. and Hwang, T. 2013. Differences in Land Surface Phenology and Primary Productivity in the Western United States during El Niño and La Niña Events from 2000–2012. *American Geophysical Union*, 2013AGUFM.B21A0475D.

Dash, J., and P. J. Curran. 2004. The MERIS terrestrial chlorophyll index. *Int J Remote Sens* 25(23):5403–5413.

Daughtry, C. S. T., K. Gallo, and M. E. Bauer. 1983. Spectral estimates of solar radiation intercepted by corn canopies. *Agron J* 75:527–531.

Daughtry, C. S. T., K. Gallo, S. Goward, S. Prince, and W. Kustas. 1992. Spectral estimates of absorbed radiation and phytomass production in corn and soybean canopies. *Remote Sens Environ* 39:141–152.

Daughtry, C. S. T., C. L. Walthall, M. S. Kim, E. B. De Colstoun, and J. E. McMurtrey III. 2000. Estimating corn leaf chlorophyll concentration from leaf and canopy reflectance. *Remote Sens Environ* 74:229–239.

Davies, R., and M. Molloy. 2012. Global cloud height fluctuations measured by MISR on Terra from 2000 to 2010. *Geophys Res Lett* 39:L03701. http://doi.org/10.1029/2011GL050506.

Dawson, T. P., P. J. Curran, and S. E. Plummer. 1998. LIBERTY—modeling the effects of leaf biochemical concentration on reflectance spectra. *Remote Sens Environ* 65:50–60.

Deardorff, J. W. 1978. Efficient prediction of ground surface temperature and moisture, with inclusion of a layer of vegetation. *J Geophys Res* 83(C4):1889, http://doi.org/10.1029/JC083iC04p01889.

DeFries, R. S., M. C. Hansen, J. R. G. Townshend, A. C. Janetos, and T. R. Loveland. 2000. A new global 1-km dataset of percentage tree cover derived from remote sensing. *Glob Change Biol* 6:247–254.

DeFries, R. S., J. R. G. Townshend, and M. C. Hansen. 1999. Continuous fields of vegetation characteristics at the global scale at 1-km resolution. *J Geophys Res* 104(D14):16911. http://doi.org/10.1029/1999JD900057.

Delbart, N., L. Kergoat, T. Le Toan, J. Lhermitte, and G. Picard. 2005. Determination of phenological dates in boreal regions using normalized difference water index. *Remote Sens Environ* 97(1):26–38.

Demmig-Adams, B. 1990. Carotenoids and photoprotection in plants: A role for the xanthophyll zeaxanthin. *Biochim Biophys Acta* 1020:1–24.

Deng, C., and C. Wu. 2013. A spatially adaptive spectral mixture analysis for mapping subpixel urban impervious surface distribution. *Remote Sens Environ* 133:62–70.

Deng, F., J. M. Chen, S. Plummer, and M. Chen. 2006. Global LAI algorithm integrating the bidirectional information. *IEEE T Geosci Remote* 44:2219–2229.

Denslow, J. 1980. Patterns of plant species diversity during succession. *Oecologia* 46(1):18–21.

Dial, G., H. Bowen, F. Gerlach, J. Grodecki, and R. Oleszczuk. 2003. IKONOS satellite, imagery, and products. *Remote Sens Environ* 88:23–36.

Diner, D. J., J. C. Beckert, T. H. Reilly, C. J. Bruegge, J. E. Conel, R. A. Kahn, J. V. Martonchik, T. P. Ackerman, R. Davies, and S. A. Gerstl. 1998. Multi-angle Imaging SpectroRadiometer (MISR) instrument description and experiment overview. *IEEE T Geosci Remote* 36:1072–1087.

Dixon, R. K., A. M. Solomom, S. Brown, R. A. Houghton, M. C. Trexier, and J. Wisniewski. 1994. Carbon pools and flux of global forest ecosystems. *Science* 263:185–190.

Dong, J., R. K. Kaufmann, R. B. Myneni, C. J. Tucker, P. E. Kauppi, J. Liski, W. Buermann, V. Alexeyev, and M. K. Hughes. 2003. Remote sensing of boreal and temperate forest woody biomass: Carbon pools, sources and sinks. *Remote Sens Environ* 84:393–410.

Drolet, G. G., K. F. Huemmrich, F. G. Hall, E. M. Middleton, T. A. Black, A. G. Barr, and H. A. Margolis. 2005. A MODIS-derived photochemical reflectance index to detect inter-annual variations in the photosynthetic light-use efficiency of a boreal deciduous forest. *Remote Sens Environ* 98:212–224.

Drolet, G. G., E. M. Middleton, K. F. Huemmrich, F. G. Hall, B. D. Amiro, A. G. Barr, T. A. Black, J. H. McCaughey, and H. A. Margolis. 2008. Regional mapping of gross light-use efficiency using MODIS spectral indices. *Remote Sens Environ* 112:3064–3078.

Duncanson, L. I., K. O. Niemann, and M. A. Wulder. 2010. Integration of GLAS and Landsat TM data for aboveground biomass estiamtion. *Can J Remote Sens* 36:129–141.

Eastman, J. R., F. Sangermano, E. A. Machado, J. Rogan, and A. Anyamba. 2013. Global trends in seasonality of normalized difference vegetation index (NDVI), 1982–2011. *Remote Sens* 5(10):4799–4818.

Eklundh, L., K. Hall, H. Eriksson, J. Ardö, and P. Pilesjö. 2003. Investigating the use of Landsat thematic mapper data for estimation of forest leaf area index in southern Sweden. *Can J Remote Sens* 29(3):349–362.

Eklundh, L., L. Harrie, and A. Kuusk. 2001. Investigating relationships between Landsat ETMþ sensor data and leaf area index in a boreal conifer forest. *Remote Sens Environ* 78:239–251.

Farquhar, G. D., S. von Caemmerer, and J. A. Berry. 1980. A biochemical model of photosynthesic CO_2 assimilation in leaves of C3 species. *Planta* 149:78–90.

Fassnacht, K., S. T. Gower, J. M. Norman, and R. E. McMurtrie. 1994. A comparison of optical and direct methods for estimating foliage surface area index in forests. *Agric For Meteorol* 71:183–207.

Fazakas, Z., and M. N. H. Olsson. 1999. Regional forest biomass and wood volume estimation using satellite and ancillary data. *Agric For Meteorol* 98–99:417–425.

Field, C. B. 1991. Ecological scaling of carbon gain to stress and resource availability. In: H. A. Mooney, W. E. Winner, and E. J. Pell (Eds.). *Response of Plants to Multiple Stresses*. San Diego, CA: Academic Press, 35–65.

Fiorella, M., and W. Ripple. 1993a. Analysis of conifer forest regeneration using landsat thematic mapper data. *Photogramm Eng Rem S* 59(9):1383–1388.

Fiorella, M., and W. Ripple. 1993b. Determining successional stage o temperate coniferous forests with landsat satellite data. *Photogramm Eng Rem S* 59(2):239–246.

Foody, G. M. 1996. Relating the land-cover composition of mixed pixels to artificial neural network classification output. *Photogramm Eng Remote S* 62(5):491–499.

Foody, G. M., D. S. Boyd, and M. E. J. Cutler. 2003. Predictive relations of tropical forest biomass from Landsat TM data and their transferability between regions. *Remote Sens Environ* 85:463–474.

Foody, G. M., and D. P. Cox. 1994. Sub-pixel land cover composition estimation using a linear mixture model and fuzzy membership functions. *Int J Remote Sens* 15(3):619–631. http://doi.org/10.1080/01431169408954100.

Foody, G. M., M. E. Cutler, J. McMorrow, D. Pelz, H. Tangki, D. S. Boyd, and I. Douglas. 2001. Mapping the biomass of Bornean tropical rain forest from remotely sensed data. *Global Ecol and Biogeogr* 10:379–387.

Franco-Lopez, H., A. R. Ek, and M. E. Bauer. 2001. Estimating and mapping of forest stand density, volume, and cover type using the k-nearest neighbors method. *Remote Sens Environ* 77:251–274.

Frankenberg, C., J. B. Fisher, J. Worden, G. Badgley, S. S. Saatchi, J.-E. Lee, G. C. Toon, A. Bulz, M. Jung, and A. Kuze. 2011. New global observations of the terrestrial carbon cycle from GOSAT: Patterns of plant fluorescence with gross primary productivity. *Geophys Res Lett* 38:L17706.

Franklin, J., and A. H. Strahler. 1988. Invertible canopy reflectance modeling of vegetation structure in semi-arid woodland. *IEEE T Geosci Remote* 26(6):809–825.

Fridley, J. D. 2012. Extended leaf phenology and the autumn niche in deciduous forest invasions. *Nature* 485(7398):359–362.

Friedl, M. A., F. W. Davis, J. Michaelsen, and M. A. Moritz. 1995. Scaling and uncertainty in the relationship between the NDVI and land surface biophysical variables: An analysis using a scene simulation model and data from FIFE. *Remote Sens Environ* 54:233–246.

Gallo, K., C. Daughtry, and M. E. Bauer. 1985. Spectral estimation of absorbed photosynthetically active radiation in corn canopies. *Remote Sens Environ* 17:221–232.

Gamon, J. A., C. Coburn, L. B. Flanagan, K. F. Huemmrich, C. Kiddle, G. A. Sanchez-Azofeifa, D. R. Thayer, L. Vescovo, D. Gianelle, and D. A. Sims. 2010. SpecNet revisited: Bridging flux and remote sensing communities. *Can J Remote Sens* 36:376–390.

Gamon, J. A., C. B. Field, W. Bilger, O. Bjorkman, A. L. Fredeen, and J. Penuelas. 1990. Remote sensing of the xanthophyll cycle and chlorophyll fluorescence in sunflower leaves and canopies. *Oecologia* 85:1–7.

Gamon, J. A., J. Peñuelas, and C. B. Field. 1992. A narrow-waveband spectral index that tracks diurnal changes in photosynthetic efficiency. *Remote Sens Environ* 41:35–44.

Gamon, J. A., and J. S. Surfus. 1999. Assessing leaf pigment content and activity with a reflectometer. *New Phytol* 143:105–117.

Ganguly, S, R. R. Nemani, G. Zhang, H. Hashimoto, C. Milesi, A. Michaelis, W. Wang, P. Votava, A. Samanta, and F. Meiton. 2012. Generating global leaf area index from Landsat: Algorithm formulation and demonstration. *Remote Sens Environ* 122:185–202.

Ganguly, S., A. Samanta, M. A. Schull, N. V. Shabanov, C. Milesi, R. R. Nemani, Y. Knyazikhin, and R. B. Myneni. 2008b. Generating vegetation leaf area index Earth system data record from multiple sensors. Part 2: Implementation, analysis and validation. *Remote Sens Environ* 112:4318–4332.

Ganguly, S., M. A. Schull, A. Samanta, N. V. Shabanov, C. Milesi, R. R. Nemani, Y. Knyazikhin, and R. B. Myneni. 2008a. Generating vegetation leaf area index earth system data record from multiple sensors. Part 1: Theory. *Remote Sens Environ* 112:4333–4343.

Gao, F., J. Masek, M. Schwaller, and F. Hall. 2006. On the blending of the Landsat and MODIS surface reflectance: Predicting daily Landsat surface reflectance. *IEEE Trans Geogsci Rem Sens* 44(8):2207–2218.

Garbulsky, M. F., J. Peñuelas, D. Papale, and I. Filella. 2008. Remote estimation of carbon dioxide uptake by a mediterranean forest. *Glob Change Biol* 14(12):2860–2867.

Garrigues, S., R. Lacaze, F. Baret, J. T. Morisette, M. Weiss, J. E. Nickeson, R. Fernandes, S. Plummer, N. V. Shabanov, and R. B. Myneni. 2008. Validation of intercomparison of global leaf area index products derived from remote sensing data. *J Geophys Res-Biogeo* 113:G02028. http://doi.org/10.1029/2007JG000635.

Gastellu-Etchegorry, J. P., E. Martin, and F. Gascon. 2004. DART: A 3D model for simulating satellite images and studying surface radiation budget. *Int J Remote Sens* 25:73–96.

Gholz, H. L., C. C. Grier, A. G. Campbell, and A. T. Brown. 1979. *Equations for Estimating Biomass and Leaf Area of Plants in the Pacific Northwest*. Forest Research Laboratory, School of Forestry, Oregon State University.

Gobron, N., B. Pinty, and M. M. Verstraete. 1997. Theoretical limits to the estimation of the Leaf Area Index on the basis of visible and near-infrared remote sensing data. *IEEE Trans Geosc Remote Sens* 35:1438–1445.

Goel, N. S., and W. Qin. 1994. Influences of canopy architecture on relationships between various vegetation indices and LAI and FPAR: A computer Simulation. *Remote Sens Rev* 10:309–347.

Goetz, S. J., and S. D. Prince. 1999. Modeling terrestrial carbon exchange and storage: Evidence and implications of functional convergence in light-use-efficiency. *Adv Ecol Res* 28:57–92.

Goetz, S. J., S. D. Prince, S. N. Goward, M. M. Thawley, and J. Small. 1999. Satellite remote sensing of primary production: An improved production efficiency modeling approach. *Ecol Model* 122:239–255.

Gonsamo, A., J. M. Chen, and P. D'Odorico. 2013. Deriving land surface phenology indicators from CO_2 eddy covariance measurements. *Ecol Indic* 29:203–207.

Gonsamo, A., J. M. Chen, D. T. Price, W. A. Kurz, and C. Wu. 2012b. Land surface phenology from optical satellite measurement and CO2 eddy covariance technique. *J Geophys Res-Biogeo* 117(G3).

Gonsamo, A., J. M. Chen, C. Wu, and D. Dragoni. 2012a. Predicting deciduous forest carbon uptake phenology by upscaling FLUXNET measurements using remote sensing data. *Agric For Meteorol* 165:127–135.

Goward, S. N., and K. F. Huemmrich. 1992. Vegetation canopy PAR absorptance and the normalized difference vegetation index: An assessment using the SAIL model. *Remote Sens Environ* 39:119–140.

Goward, S. N., R. H. Waring, D. G. Dye, and J. L. Yang. 1994. Ecological remote-sensing at otter—satellite macroscale observations. *Ecol Appl* 4:322–343.

Gower, S. T., C. J. Kucharik, and J. M. Norman. 1999. Direct and indirect estimation of leaf area index, fAPAR, and net primary production of terrestrial ecosystems. *Remote Sens Environ* 70:29–51.

Gower, S. T., R. McMurtrie, and D. Murty. 1996. Aboveground net primary production decline with stand age: Potential causes. *Trends Ecol Evol* 11(9):378–382.

Gower, S. T., and J. M. Norman. 1991. Rapid estimation of leaf area index in conifer and broadleaf plantations. *Ecology* 72:1896–1900.

Grace, J., C. Nichol, M. Disney, P. Lewis, T. Quaife, and P. Bowyer. 2007. Can we measure terrestrial photosynthesis from space directly, using spectral reflectance and fluorescence? *Glob Change Biol* 13(7):1484–1497.

Gray, J., and C. Song. 2012. Mapping leaf area index using spatial, spectral, and temporal information from multiple sensors. *Remote Sens Environ* 119:173–183.

Grier, C. C., and R. S. Logan. 1977. Old-growth *Pseudotsuga-Menziesii* of a western Oregon watershed: Biomass distribution and production budgets. *Ecol Monogr* 47:373–400.

Guanter, L., L. Alonso, L. Gómez-Chova, J. Amorós-López, J. Vila, and J. Moreno. 2007. Estimation of solar-induced vegetation fluorescence from space measurements. *Geophys Res Lett* 34:L08401.

Guanter, L., L. Alonso, L. Gómez-Chova, M. Meroni, R. Preusker, J. Fischer, and J. Moreno. 2010. Developments for vegetation fluorescence retrieval from spaceborne high-resolution spectrometry in the O 2 -A and O 2 -B absorption bands. *J Geophys Res* 115:D19303.

Gutman, G., and A. Ignatov. 1998. The derivation of the green vegetation fraction from NOAA/AVHRR data for use in numerical weather prediction models. *Int J Remote Sens* 19(8):1533–1543.

Hall, F., D. Botkin, D. Strebel, K. Woods, and S. Goetz. 1991. Large-scale patterns of forest succession as determined by remote sensing. *Ecology* 72(2):628–640.

Hall, F. G., T. Hilker, and N. C. Coops. 2011. PHOTOSYNSAT, photosynthesis from space: Theoretical foundations of a satellite concept and validation from tower and spaceborne data. *Remote Sens Environ* 115:1918–1925.

Hall, F. G., T. Hilker, N. C. Coops, A. Lyapustin, K. F. Huemmrich, E. Middleton, H. Margolis, G. Drolet, and T. A. Black. 2008. Multi-angle remote sensing of forest light use efficiency by observing pri variation with canopy shadow fraction. *Remote Sens Environ* 112:3201–3211.

Hall, F. G., D. R. Peddle, and E. F. Ledrew. 1996. Remote sensing of biophysical variables in boreal forest stands of Picea mariana. *Int'l J Rem Sens* 17(15):3077–3081.

Hall, F. G., J. R. Townshend, and E. T. Engman. 1995. Status of remote sensing algorithms for estimation of land surface parameters. *Remote Sens Environ* 51:135–156.

Hall, R. J., R. S. Skakun, E. J. Arsenault, and B. S. Case. 2006. Modeling forest stand structure attributes using Landsat ETM+ data: Application to mapping of aboveground biomass and stand volume. *Forest Ecol Manage* 225:378–390.

Hansen, M. C., R. S. DeFries, J. R. G. Townshend, M. Carroll, C. Dimiceli, and R. A. Sohlberg. 2003. Global percent tree cover at a spatial resolution of 500 meters: First results of the modis vegetation continuous fields algorithm. *Earth Interact* 7(10):1–15. http://doi.org/10.1175/1087-3562(2003)007<0001:GPTCAA>2.0.CO;2.

Hansen, M. C., R. S. DeFries, J. R. G. Townshend, R. Sohlberg, C. Dimiceli, and M. Carroll. 2002. Towards an operational MODIS continuous field of percent tree cover algorithm: Examples using AVHRR and MODIS data. *Remote Sens Environ* 83(1–2):303–319. http://doi.org/10.1016/S0034-4257(02)00079-2.

Hansen, M. C., J. R. G. Townshend, R. S. DeFries, and M. Carroll. 2005. Estimation of tree cover using MODIS data at global, continental and regional/local scales. *Int J Remote Sens* 26(19):4359–4380. http://doi.org/10.1080/01431160500113435.

Hatfield, J., G. Asrar, and E. Kanemasu. 1984. Intercepted photosynthetically active radiation estimated by spectral reflectance. *Remote Sens Environ* 14:65–75.

Hayes, D. J., W. B. Cohen, S. A. Sader, and D. E. Irwin. 2008. Estimating proportional change in forest cover as a continuous variable from multi-year MODIS data. *Remote Sens Environ* 112(3):735–749.

Healey, S., W. Cohen, Y. Zhiqiang, and O. Krankina. 2005. Comparison of tasseled cap-based Landsat data structures for use in forest disturbance detection. *Remote Sens Environ* 97(3):301–310.

Heiskanen, J. 2006. Estimating aboveground tree biomass and leaf area index in a mountain birch forest using ASTER satellite data. *Int J Remote Sens* 27:1135–1158.

Heiskanen, J., M. Rautiainen, L. Korhonen, M. Mõttus, and P. Stenberg. 2011. Retrieval of boreal forest LAI using a forest reflectance model and empirical regressions. *Int J Appl Earth Obs* 13(4):595–606.

Helmer, E. H., S. Brown, and W. B. Cohen. 2000. Mapping montane tropical forest successional stage and land use with multi-date Landsat imagery. *Int J Remote Sens* 21(11):2163–2183.

Helmer, E. H., M. A. Lefsky, and D. A. Roberts. 2009. Biomass accumulation rates of Amazonian secondary forest and biomass of old-growth forests from Landsat time series and the Geoscience Laser Altimeter System. *J Appl Remote Sens* 3:033505.

Hilker, T., N. C. Coops, F. G. Hall, T. A. Black, B. Chen, P. Krishnan, M. A. Wulder, P. S. Sellers, E. M. Middleton, and K. F. Huemmrich. 2008. A modeling approach for upscaling gross ecosystem production to the landscape scale using remote sensing data. *J Geophys Res* 113:G03006.

Hilker, T., F. G. Hall, N. C. Coops, A. Lyapustin, Y. Wang, Z. Nesic, N. Grant, T. A. Black, M. A. Wulder, and N. Kljun. 2010. Remote sensing of photosynthetic light-use efficiency across two forested biomes: Spatial scaling. *Remote Sens Environ* 114:2863–2874.

Houghton, R. A., D. Butman, A. G. Bunn, O. N. Krankina, P. Schlesinger, and T. A. Stone. 2007. Mapping Russian forest biomass with data from satellites and forest inventories. *Environ Res Lett* 2:045032.

Hu, J., B. Tan, N. Shabanov, K. A. Crean, J. V. Martonchik, D. J. Diner, Y. Knyazikhin, and R. B. Myneni. 2003. Performance of the MISR LAI and FPAR algorithm: A case study in Africa. *Remote Sens Environ* 88:324–340.

Huang, C., S. N. Goward, J. G. Masek, N. Thomas, Z. Zhu, and J. E. Vogelmann. 2010. An automated approach for reconstructing recent forest disturbance history using dense Landsat time series stacks. *Remote Sens Environ* 114(1):183–198.

Huemmrich, K. F. 2001. The GeoSail model: A simple addition to the SAIL model to describe discontinuous canopy reflectance. *Remote Sens Environ* 75:423–431.

Huete, A. R. 1988. A soil adjusted vegetation index (SAVI). *Remote Sens Environ* 25:295–309.

Huete, A. R., K. Didan, T. Miura, E. P. Rodriguez, X. Gao, and L. G. Ferreira. 2002. Overview of the radiometric and biophysical performance of the MODIS vegetation indices. *Remote Sens Environ* 83:195–213.

Huete, A. R., and H. Q. Liu. 1994. An error and sensitivity analysis of the atmospheric- and soil-correcting variants of the NDVI for the MODIS-EOS. *IEEE T Geosci Remote* 32:897–905.

Hwang, T., C. Song, P. V. Bolstad, and L. E. Band. 2011. Downscaling real-time vegetation dynamics by fusing multi-temporal MODIS and Landsat NDVI in topographically complex terrain. *Remote Sens Environ* 115:2499–2512.

Jacquemoud, S., and F. Baret. 1990. PROSPECT: A model of leaf optical properties spectra. *Remote Sens Environ* 34:75–91.

Jakubauskas, M. 1996. Thematic Mapper characterization of lodgepole pine seral stages in Yellowstone National Park, USA. *Remote Sens Environ* 56:118–132.

Jenkins, J. C., D. C. Chojnacky, L. S. Heath, and R. A. Birdsey. 2003. National-scale biomass estimators for United States tree species. *Forest Sci* 49:12–35.

Jiang, H., J. R. Strittholt, P. A. Frost, and N. C. Slosser. 2004. The classification of late seral forests in the Pacific Northwest, USA using Landsat ETM+ imagery. *Remote Sens Environ* 91:320–331.

Jiapaer, G., X. Chen, and A. Bao. 2011. A comparison of methods for estimating fractional vegetation cover in arid regions. *Agric For Meteorol* 151(12):1698–1710. http://doi.org/10.1016/j.agrformet.2011.07.004.

Jolly, W. M., R. Nemani, and S. W. Running. 2005. A generalized, bioclimatic index to predict foliar phenology in response to climate. *Glob Change Biol* 11(4):619–632.

Jolly, W. M., and S. W. Running. 2004. Effects of precipitation and soil water potential on drought deciduous phenology in the Kalahari. *Glob Change Biol* 10:303–308.

Jonckheere, I., S. Fleck, K. Nackaerts, B. Muys, P. Coppin, M. Weiss, and F. Baret. 2004. Methods for leaf area index determination. Part I: Theories, techniques and instruments. *Agric For Meteorol* 121:19–35.

Jones, H. G., and R. A. Vaughan. 2010. *Remote Sensing of Vegetation: Principles, Techniques, and Applications.* Oxford: Oxford University Press.

Jönsson, P., and L. Eklundh. 2004. TIMESAT—A program for analyzing time-series of satellite sensor data. *Comput Geosci* 30(8):833–845.

Jordan, C. F. 1969. Derivation of leaf area index from quality of light on the forest floor. *Ecology* 50:663–666.

Justice, C. O., M. O. Román, I. Csiszar, E. F. Vermote, R. E. Wolfe, S. J. Hook, M. Friedl, Z. Wang, C. B. Schaaf, and T. Miura. 2013. Land and cryosphere products from Suomi NPP VIIRS: Overview and status. *J Geophys Res-Atmos* 118:9753–9765.

Justice, C. O., J. Townshend, E. Vermote, E. Masuoka, R. Wolfe, N. Saleous, D. Roy, and J. Morisette. 2002. An overview of MODIS Land data processing and product status. *Remote Sens Environ* 83:3–22.

Kamthonkiat, D., K. Honda, H. Turral, N. Tripathi, and V. Wuwongse. 2005. Discrimination of irrigated and rainfed rice in a tropical agricultural system using SPOT VEGETATION NDVI and rainfall data. *Int J Remote Sens* 26:2527–2547.

Kaufman, Y. J., and D. Tanre. 1992. Atmospherically resistant vegetation index (ARVI) for EOS-MODIS. *IEEE T Geosci Remote* 30:261–270.

Keeling, C. D., J. F. S. Chin, and T. P. Whorf. 1996. Increased activity of Northern vegetation inferred from atmospheric CO2 measurements. *Nature* 382:146–149.

Kennedy, R. E., W. B. Cohen, and T. A. Schroeder. 2007. Trajectory-based change detection for automated characterization of forest disturbance dynamics. *Remote Sens Environ* 110(3):370–386.

Kennedy, R. E., Z. Yang, and W. B. Cohen. 2010. Detecting trends in forest disturbance and recovery using yearly Landsat time series: 1. LandTrendr—Temporal segmentation algorithms. *Remote Sens Environ* 114(12):2897–2910.

Kimes, D. S., Y. Knyazikhin, J. L. Privette, A. A. Abuelgasim, and F. Gao. 2000. Inversion methods for physically-based models. *Remote Sens Rev* 18:381–439.

Knyazikhin, Y., J. V. Martonchik, R. B. Myneni, D. J. Diner, and S. W. Running. 1998. Synergistic algorithm for estimating vegetation canopy leaf area index and fraction of absorbed photosynthetically active radiation from MODIS and MISR data. *J Geophys Res* 103(D24):32,257–32,275.

Körner, C., and D. Basler. 2010. Phenology under global warming. *Science* 327(5972):1461–1462.

Kucharik, C. J., J. M. Norman, and S. T. Gower. 1999. Characterization of radiation regimes in nonrandom forest canopies: Theory, measurements, and a simplified modeling approach. *Tree Physiol* 19(11):695–706.

Kumar, M., and J. Monteith. 1981. Remote sensing of crop growth. In: *Plants and the Daylight Spectrum: Proceedings of the First International Symposium of the British Photobiology Society*, Leicester, 5–8 January. London, UK: Academic Press, 133–144. ISBN: 978-0-12-650980-9.

Landsberg, J. J., and R. H. Waring. 1997. A generalised model of forest productivity using simplified concepts of radiation-use efficiency, carbon balance and partitioning. *Forest Ecol Manag* 95(3):209–228.

Lasslop, G., M. Reichstein, D. Papale, A. D. Richardson, A. Arneth, A. Barr, P. Stoy, and G. Wohlfahrt. 2010. Separation of net ecosystem exchange into assimilation and respiration using a light response curve approach: Critical issues and global evaluation. *Global Change Biol* 16(1):187–208.

Latifi, H., and B. Koch. 2012. Evaluation of most similar neighbour and random forest methods for imputing forest inventory variables using data from target and auxiliary stands. *Int J Remote Sens* 33(21):6668–6694.

Law, B., P. Thornton, J. Irvine, P. Anthoni, and S. Tuyl. 2001. Carbon storage and fluxes in ponderosa pine forests at different developmental stages. *Glob Change Biol* 7:755–777.

Lawrence, R., and W. Ripple. 1999. Calculating change curves for multitemporal satellite imagery: Mount St. Helens 1980–1995. *Remote Sens Environ* 67:309–319.

Leblanc, S. G. 2002. Correction to the plant canopy gap-size analysis theory used by the Tracing Radiation and Architecture of Canopies instrument. *Appl Optics* 41(36):7667–7670.

Lefsky, M. A., D. P. Turner, M. Guzy, and W. B. Cohen. 2005. Combining lidar estimates of aboveground biomass and Landsat estimates of stand age for spatially extensive validation of modeled forest productivity. *Remote Sens Environ* 95:549–558.

Le Maire, G., C. Marsden, Y. Nouvellon, C. Grinand, R. Hakamada, J.-L. Stape, and J.-P. Laclau. 2011. MODIS NDVI time-series allow the monitoring of *Eucalyptus* plantation biomass. *Remote Sens Environ* 115:2613–2625.

Li, X. W., and A. H. Strahler. 1985. Geometric-optical modeling of a conifer forest canopy. *IEEE T Geosci Remote* GE-23(5):705–721.

Li, X. W., A. H. Strahler, and C. E. Woodcock. 1995. A hybrid geometric optical-radiative transfer approach for modeling albedo and directional reflectance of discontinuous canopies. *IEEE T Geosci Remote* 33(2):466–480.

Liesenberg, V., L. S. Galvão, and F. J. Ponzoni. 2007. Variations in reflectance with seasonality and viewing geometry: Implications for classification of Brazilian savanna physiognomies with MISR/Terra data. *Remote Sens Environ* 107:276–286.

Lieth, H. (Ed.). 1974. *Phenology and Seasonality Modeling*. New York: Springer.

Liu, J., J. M. Chen, J. Cihlar, and W. M. Park. 1997. A process-based boreal ecosystem productivity simulator using remote sensing inputs. *Remote Sens Environ* 62:158–175.

Liu, L., and Z. Cheng. 2010. Detection of vegetation light-use efficiency based on solar-induced chlorophyll fluorescence separated from canopy radiance spectrum. *IEEE J Sel Top Appl* 3(3):306–312.

Liu, L., Y. Zhang, Q. Jiao, and D. Peng. 2013. Assessing photosynthetic light-use efficiency using a solar-induced chlorophyll fluorescence and photochemical reflectance index. *Int J Remote Sens* 34(12):4264–4280.

Liu, S., T. Wang, J. Guo, J. Qu, and P. An. 2010. Vegetation change based on SPOT-VGT data from 1998 to 2007, northern China. *Environ Earth Sci* 60:1459–1466.

Liu, W., C. Song, T. A. Schroeder, and W. B. Cohen. 2008. Predicting forest successional stages using multi-temporal Landsat imagery with forest inventory and analysis data. *Int J Remote Sens* 29(13):3855–3872.

Loveland, T. R., and J. L. Dwyer. 2012. Landsat: Building a strong future. *Remote Sens Environ* 122:22–29.

Lu, D. 2005. Aboveground biomass estimation using Landsat TM data in the Brazilian Amazon. *Int J Remote Sens* 26:2509–2525.

Lucas, R. M., M. Honzák, I. Do Amaral, P. J. Curran, and G. M. Foody. 2002. Forest regeneration on abandoned clearances in central Amazonia. *Int J Remote Sens* 23(5):965–988.

Malenovský, Z., K. B. Mishra, F. Zemek, U. Rascher, and L. Nedbal. 2009. Scientific and technical challenges in remote sensing of plant canopy reflectance and fluorescence. *J Exp Bot* 60(11):2987–3004.

Markham, B. L., and D. L. Helder. 2012. Forty-year calibrated record of earth-reflected radiance from Landsat: A review. *Remote Sens Environ* 122:30–40.

Martonchik, J. V., D. J. Diner, K. A. Crean, and M. A. Bull. 2002. Regional aerosol retrieval results from MISR. *IEEE T Geosci Remote* 40:1520–1531.

Masson, V., J. L. Champeaux, F. Chauvin, C. Meriguer, and R. Lacaze. 2003. A global database of land surface parameters at 1 km resolution in meteorological and climate models. *J Climate* 16:1261–1282.

McDonald, R. I., P. N. Halpin, and D. L. Urban. 2007. Monitoring succession from space: A case study from the North Carolina Piedmont. *Appl Veg Sci* 10(2):193.

McMurtrie, R. E., D. A. Rook, and F. M. Kelliher. 1990. Modelling the yield of Pinus radiata on a site limited by water and nitrogen. *Forest Ecol Manage* 30(1–4):381–413.

Menzel, A., T. H. Sparks, N. Estrella, E. Koch, A. Aasa, R. Ahas, K. Alm-Kübler, P. Bissolli, O. Braslavská, and A. Briede. 2006. European phenological response to climate change matches the warming pattern. *Glob Change Biol* 12(10):1969–1976.

Meroni, M., M. Rossini, L. Guanter, L. Alonso, U. Rascher, R. Colombo, and J. Moreno. 2009. Remote sensing of solar-induced chlorophyll fluorescence: Review of methods and applications. *Remote Sens Environ* 113:2037–2051.

Miller, E. E., and J. M. Norman. 1971. Sunfleck theory for plant canopies 1. Lengths of sunlit segments along a transect. *Agron J* 63(5):735–738.

Miller, J. B. 1967. A formula for average foliage density. *Aust J Bot* 15:141–144.

Miller, S. D., S. P. Mills, C. D. Elvidge, D. T. Lindsey, T. F. Lee, and J. D. Hawkins. 2012. Suomi satellite brings to light a unique frontier of nighttime environmental sensing capabilities. *P Natl Acad Sci USA* 109:15706–15711.

Minocha, R., G. Martinez, B. Lyons, and S. Long. 2009. Development of a standardized methodology for quantifying total chlorophyll and carotenoids from foliage of hardwood and conifer tree species. *Can J Forest Res* 39:849–861.

Mohammed, G. H., J. Moreno, Y. Goulas, A. Huth, E. Middleton, F. Miglietta, L. Nedbal, U. Rascher, W. Verhoef and M. Drusch. 2012. *European Space Agency's Fluorescence Explorer Mission: Concept and Applications*. AGU Fall Meeting, 3–7 December, San Francisco, CA.

Monsi, M., and T. Saeki. 1953. Uber den Lichifktor in den Pflanzengesellschaften und Scine Bedeutung fur die Stoffprodcktion. *Jpn J Bot* 14:22–52.

Monteith, J. L. 1965. Evaporation and the environment. in *Proceedings of the 19th Symposium of the Society for Experimental Biology*, pp. 205–233. New York: Cambridge University Press.

Monteith, J. L. 1972. Solar radiation and productivity in tropical ecosystems. *J Appl Ecol* 9(3):747–766.

Monteith, J. L. 1977. Climate and the efficiency of crop production in Britain. *Philos T R Soc B* 281:277–294.

Moody, A., and C. E. Woodcock. 1994. Scale-dependent errors in the estimation of land-cover proportions: Implications for global land-cover dataset. *Photogra Eng Rem Sens* 60(5):585–594.

Morisette, J. T., A. D. Richardson, A. K. Knapp, J. I. Fisher, E. A. Graham, J. Abatzoglou, B. E. Wilson, D. D. Breshears, G. M. Henebry, J. M. Hanse, and L. Liang. 2009. Tracking the rhythm of the seasons in the face of global change: Phenological research in the 21st century. *Front Ecol Environ* 7(5):253–260.

Mu, Q., M. Zhao, and S. W. Running. 2011. Improvements to a MODIS global terrestrial evapotranspiration algorithm. *Remote Sens Environ* 115:1781–1800.

Mutanga, O., E. Adam, and M. A. Cho. 2012. High density biomass estimation for wetland vegetation using WorldView-2 imagery and random forest regression algorithm. *Intl J Appl Earth Obs Geoinf* 18:399–406.

Muukkonen, P., and J. Heiskanen. 2007. Biomass estimation over a large area based on standwise forest inventory data and ASTER and MODIS satellite data: A possibility to verify carbon inventories. *Remote Sens Environ* 107:617–624.

Myneni, R. B., J. Dong, C. J. Tucker, R. K. Kaufmann, P. E. Kauppi, J. Liski, L. Zhou, V. Alexeyev, and M. K. Hughes. 2001. A large carbon sink in the woody biomass of Northern forests. *Proc Nat Acad Sci USA* 98(26):14784–14789.

Myneni, R. B., S. Hoffman, Y. Knyazikhin, J. Privette, J. Glassy, Y. Tian, Y. Wang, X. Song, Y. Zhang, and G. R. Smith. 2002. Global products of vegetation leaf area and fraction absorbed PAR from year one of MODIS data. *Remote Sens Environ* 83:214–231.

Myneni, R. B., C. D. Keeling, C. J. Tucker, G. Asrar, and R. R. Nemani. 1997. Increased plant growth in the northern high latitudes from 1981 to 1991. *Nature* 386:698–702.

Myneni, R. B., S. Maggion, J. Iaquinto, J. L. Privette, N. Gobron, B. Pinty, D. S. Kimes, M. M. Verstraete, and D. L. Williams. 1995. Optical remote sensing of vegetation: Modeling, caveats, and algorithms. *Remote Sens Environ* 51:169–188.

Myneni, R. B., and D. L. Williams. 1994. On the Relationship between FAPAR and NDVI. *Remote Sens Environ* 49:200–211.

National Research Council (NRC). 2007. *Earth Science and Application from Space: National Imperatives for the Next Decade and Beyond.* Washington, DC: The National Academies Press.

Nelson, R. F., D. S. Kimes, W. A. Salas, and M. Routhier. 2000. Secondary forest age and tropical forest biomass estimation using thematic mapper imagery. *Bioscience* 50:419–431.

Nelson, R., K. J. Ranson, G. Sun, D. S. Kimes, V. Kharuk, and P. Montesano. 2009. Estimating Siberian timber volume using MODIS and ICESat/GLAS. *Remote Sens Environ* 113:691–701.

Nemani, R., H. Hashimoto, P. Votava, F. Melton, W. Wang, A. Michaelis, L. Mutch, C. Milesi, S. Hiatt, and M. White. 2009. Monitoring and forecasting ecosystem dynamics using the terrestrial observation and prediction system (TOPS). *Remote Sens Environ* 113(7):1497–1509.

Nemani, R., L. Pierce, S. Running, and L. Band. 1993. Forest ecosystem processes at the watershed scale: Sensitivity to remotely-sensed leaf area index estimates. *Int J Remote Sens* 14:2519–2534.

Ni, W., X. Li, C. E. Woodcock, J. Boujean, and R. E. Davis. 1997. Transmission of solar radiation in boreal conifer forests: Measurements and models. *J Geophy Res* 102(D24):29555–29566.

Nichol, C. J., K. F. Huemmrich, T. A. Black, P. G. Jarvis, C. L. Walthall, J. Grace, and F. G. Hall. 2000. Remote sensing of photosynthetic-light-use efficiency of boreal forest. *Agric Forest Meteorol* 101:131–142.

Nilson, T. 1971. A theoretical analysis of the frequency of gaps in plant stands. *Agr Meteorol* 8:25–38.

Norman, J. M., and G. S. Campbell. 1989. Canopy structure. In R. W. Pearcy, J. Ehlerlnger, H. A. Mooney, and P. W. Rundel (Eds.). *Plant Physiological Ecology: Field Methods and Instrumentation.* New York: Chapman and Hall, 301–325.

North, P. 1996. Three-dimensional forest light interaction model using a Monte Carlo method. *IEEE T Geosci Remote* 34:946–956.

Oker-Blom, P. 1986. Photosynthetic radiation regime and canopy structure in modeled forest stands. *Acta Forest Fennica* 197:1–44.

Oliver, C. D., and B. C. Larson. 1996. *Forest Stand Dynamics.* New York: John Wiley and Sons, Inc. ISBN 0-471-13833-9.

Oosting, H. J. 1942. An ecological analysis of the plant communities of Piedmont, North Carolina. *Am Midl Nat* 28:1–126.

Oudrari, H., J. McIntire, D. Moyer, K. Chiang, X. Xiong, and J. Butler. 2012. Preliminary assessment of Suomi-NPP VIIRS on-orbit radiometric performance, *SPIE Optical Engineering Applications, International Society for Optics and Photonics*, pp. 851011–851024.

Pacala, S. W., Hurtt, G. C., Baker, D., Peylin, P., Houghton, R. A., Birdsey, R. A., Heath, L., Sundquist, E. T., Stallard, R. F., Ciais, P., Moorcroft, P., Caspersen, J. P., Shevliakova, E., Moore, B., Kohlmaier, G., Holland, E., Gloor, M., Harmon, M. E., Fan, S.-M., Sarmiento, J. L., Goodale, C. L., Schimel, D. and Field, C. B.. 2001. Consistent land- and atmosphere-based U.S. carbon sink estimates. *Science*, 292(5525): 2316–2320.

Parry, M. L., O. F. Canziani, J. P. Palutikof, P. J. van der Linden, and C. E. Hanson (Eds.). 2007. Climate change 2007: Impacts, adaptation and vulnerability. In *Contribution of Working Group II to the Fourth Assessment Report of the Intergovernmental Panel on Climate Change*. Cambridge: Cambridge University Press, 976 p.

Peddle, D. R., F. G. Hall, and E. F. LeDrew. 1999. Spectral mixture analysis and geometric-optical reflectance modeling of boreal forest biophysical structure. *Remote Sens Environ* 67:288–297.

Peet, R. K., and N. L. Christensen. 1987. Competition and tree death. *BioScience* 37(8):586–595.

Peet, R. K., and N. L. Christensen. 1988. Changes in species diversity during secondary forest succession on the North Carolina piedmont. In H. J. During, M. J. A. Werger, and J. H. Willems (Eds.). *Diversity and Pattern in Plant Communities*. The Hague: SPB Academic Publishing.

Peñuelas, J., I. Filella, and J. A. Gamon. 1995. Assessment of photosynthetic radiation-use efficiency with spectral reflectance. *New Phytol* 131:291–296.

Peñuelas, J., T. Rutishauser, and I. Filella. 2009. Phenology feedbacks on climate change. *Science* 324(5929):887–888.

Peterson, D. L., M. A. Spanner, S. W. Running, and K. B. Teuber. 1987. Relationship of thematic mapper simulator data to leaf area index of temperate coniferous forests. *Remote Sens Environ* 22:323–341.

Pflugmacher, D., W. B. Cohen, and R. E. Kennedy. 2012. Using Landsat-derived disturbance history (1972–2010) to predict current forest structure. *Remote Sens Environ* 122:146–165.

Piao, S. L., P. Ciais, P. Friedlingstein, P. Peylin, M. Reichstein, S. Luyssaert, H. Margolis, J. Fang, A. Barr, and A. Chen. 2008. Net carbon dioxide losses of northern ecosystems in response to autumn warming. *Nature* 451:49–52.

Piao, S. L., J. Y. Fang, B. Zhu, and K. Tan. 2005. Forest biomass carbon stocks in China over the past 2 decades: Estimation based on integrated inventory and satellite data. *J Geophys Res* 110:G01006.

Piao, S. L., P. Friedlingstein, P. Ciais, N. Viovy, and J. Demarty. 2007. Growing season extension and its impact on terrestrial carbon cycle in the Northern Hemisphere over the past 2 decades. *Global Biogeochem Cycles* 21(3):GB3018.

Pickett, S. T. A., M. L. Cadenasso, and S. J. Meiners. 2013. Vegetation dynamics. In E. Van der Maarel and J. Franklin (Eds.). *Vegetation Ecology*. 2nd ed. New York: John Wiley & Sons. ISBN 978-1-4443-3889-8.

Pinty, B., N. Gobron, J.-L. Widlowski, T. Lavergne, and M. M. Verstraete. 2004. Synergy between 1-D and 3-D radiation transfer models to retrieve vegetation canopy properties from remote sensing data. *J Geophys Res* 109:D21205. http://doi.org/10.1029/2004JD005214.

Pinty, B., and M. M. Verstrate. 1992. GEMI: A non-linear index to monitor global vegetation from satellites. *Vegetation* 101:15–20.

Pisek, J., and J. M. Chen. 2007. Comparison and validation of MODIS and VEGETATION global LAI products over four BigFoot sites in North America. *Remote Sens Enviro* 109:81–94.

Potter, C. S., S. Klooster, R. Myneni, V. Genovese, P. Tan, and V. Kumar. 2003. Continental-scale comparisons of terrestrial carbon sinks estimated from satellite data and ecosystem modeling 1982–1998. *Global Planet Change* 39:201–213.

Potter, C. S., J. T. Randerson, C. B. Field, P. A. Matson, P. M. Vitousek, H. A. Mooney, and S. A. Klooster. 1993. Terrestrial ecosystem production—a process model-based on global satellite and surface data. *Global Biogeochem Cycles* 7:811–841.

Powell, S. L., W. B. Cohen, S. P. Healey, R. E. Kennedy, G. G. Moisen, K. B. Pierce, and J. L. Ohmann. 2010. Quantification of live aboveground forest biomass dynamics with Landsat time-series and field inventory data: A comparison of empirical modeling approaches. *Remote Sens Environ* 114:1053–1068.

Pregitzer, K. S., and E. S. Euskirchen. 2004. Carbon cycling and storage in world forests: Biome patterns related to forest age. *Global Change Bio* 10(12):2052–2077.

Prince, S. D. 1991. A model of regional primary production for use with coarse resolution satellite data. *Int J Remote Sens* 12(6):1313–1330.

Prince, S. D., and S. N. Goward. 1995. Global primary production: A remote sensing approach. *J Biogeogr* 22(4/5):815–835.

Qi, J., A. Chehbouni, A. R. Huete, Y. H. Kerr, and S. Sorooshian. 1994. A modified soil adjusted vegetation index. *Remote Sens Environ* 48:119–126.

Rahman, A., V. Cordova, J. Gamon, H. Schmid, and D. Sims. 2004. Potential of MODIS ocean bands for estimating CO2 flux from terrestrial vegetation: A novel approach. *Geophys Res Lett* 31:L10503.

Reichstein, M., E. Falge, D. Baldocchi, D. Papale, M. Aubinet, P. Berbigier, C. Bernhofer, N. Buchmann, T. Gilmanov, and A. Granier. 2005. On the separation of net ecosystem exchange into assimilation and ecosystem respiration: Review and improved algorithm. *Global Change Biol* 11(9):1424–1439.

Richards, J. A. 2013. *Remote Sensing Digital Image Analysis: An Introduction*. Berlin: Springer Verlag. ISBN-13: 978-3642300615.

Richardson, A. D., R. S. Anderson, M. A. Arain, A. G. Barr, G. Bohrer, G. Chen, J. M. Chen, P. Ciais, K. J. Davis, and A. R. Desai. 2012. Terrestrial biosphere models need better representation of vegetation phenology: Results from the North American Carbon Program Site Synthesis. *Glob Change Biol* 18(2):566–584.

Richardson, A. D., T. A. Black, P. Ciais, N. Delbart, M. A. Friedl, N. Gobron, D. Y. Hollinger, W. L. Kutsch, B. Longdoz, and S. Luyssaert. 2010. Influence of Spring and Autumn Phenological Transitions on Forest Ecosystem Productivity. *Philos T R Soc B* 365(1555):3227–3246.

Richardson, A. D., B. H. Braswell, D. Y. Hollinger, J. P. Jenkins, and S. V. Ollinger. 2009. Near-surface remote sensing of spatial and temporal variation in canopy phenology. *Ecol Appl* 19(6):1417–1428.

Richardson, A. D., S. P. Duigan, and G. P. Berlyn. 2002. An evaluation of noninvasive methods to estimate foliar chlorophyll content. *New Phytol* 153:185–194.

Richardson, A. D., T. F. Keenan, M. Migliavacca, Y. Ryu, O. Sonnentag, and M. Toomey. 2013. Climate change, phenology, and phenological control of vegetation feedbacks to the climate system. *Agric For Meteorol* 169:156–173.

Roberts, D. A., M. Gardner, R. Church, S. Ustin, G. Scheer, and G. O. Green. 1998. Mapping chaparral in the Santa Monica Mountains using multiple endmember spectral mixture models. *Remote Sens Environ* 65:267–279.

Rosenzweig, C., G. Casassa, D. J. Karoly, A. Imeson, C. Liu, A. Menzel, S. Rawlins, T. L. Root, B. Seguin, and P. Tryjanowski. 2007. Assessment of observed changes and responses in natural and managed systems. In: M. L. Parry, O. F. Canziani, J. P. Palutikof, P. J. van der Linden, and C. E. Hanson (Eds.). *Climate Change 2007: Impacts, Adaptation and Vulnerability. Contribution of Working Group II to the Fourth Assessment Report of the Intergovernmental Panel on Climate Change*. Cambridge: Cambridge University Press, 79–131.

Roujean, J. L., and F. M. Breon. 1995. Estimating PAR absorbed by vegetation from bidirectional reflectance measurements. *Remote Sens Environ* 51:375–384.

Rouse, J. W., R. H. Hass, J. A. Shell, and D. W. Deering. 1974. Monitoring vegetation systems in the Great Plains with ERTS-1. *Third Earth Res Technol Satell Sympos* 1:309–317.

Roy, P. S., and S. Ravan. 1996. Biomass estimation using satellite remote sensing data—an investigation on possible approaches for natural forest. *J Biosciences* 21:535–561.

Ruimy, A., B. Saugier, and G. Dedieu. 1994. Methodology for the estimation of terrestrial net primary production from remotely sensed data. *J Geophys Res-Atmos* 99:5263–5283.

Running, S. W., and J. C. Coughlan. 1988. A general model of forest ecosystem processes for regional applications I. Hydrologic balance, canopy gas exchange and primary production processes. *Ecol Model* 42:125–154.

Running, S. W., C. O. Justice, V. Salomonson, D. Hall, J. Barker, Y. J. Kaufman, A. H. Strahler, A. R. Huete, J.-P. Muller and V. Vanderbill. 1994. Terrestrial remote sensing science and algorithms planned for EOS/MODIS. *Int J Remote Sens* 15(17):3587–3620.

Running, S. W., R. R. Nemani, F. A. Heinsch, M. Zhao, M. Reeves, and H. Hashimoto. 2004. A continuous satellite-derived measure of global terrestrial primary production. *BioScience*, 54:547–560.

Running, S. W., R. R. Nemani, D. L. Peterson, L. E. Band, D. F. Potts, L. L. Pierce, and M. A. Spanner. 1989. Mapping regional forest evapotranspiration and photosynthesis by coupling satellite data with ecosystem simulation. *Ecology* 70(4):1090–1101.

Running, S. W., P. E. Thornton, R. R. Nemani, and J. M. Glassy. 2000. Global terrestrial gross and net primary productivity from the Earth Observing System. In: O. E. Sala, R. B. Jackson, H. A. Mooney, and R. W. Howarth (Eds.). *Methods in Ecosystem Science*. New York: Springer, 44–57.

Sabol, D. E., A. R. Gillespie, J. B. Adams, M. O. Smith, and C. J. Tucker. 2002. Structural stage in Pacific Northwest forests estimated using simple mixing models of multispectral images. *Remote Sens Environ* 80(1):1–16.

Sader, S. A., R. B. Waide, W. T. Lawrence, and A. T. Joyce. 1989. Tropical forest biomass and successional age class relationships to a vegetation index derived from Landsat TM data. *Remote Sens Environ* 28:143–156.

Salim, E. and Ullsten, O. 1999. Our Forests Our Future. Cambridge University Press, Cambridge, UK. ISBN: 0 521 66021 1.

Schaefer, K., C. R. Schwalm, C. Williams, M. A. Arain, A. Barr, J. M. Chen, K. J. Davis, D. Dimitrov, T. W. Hilton, and D. Y. Hollinger. 2012. A model-data comparison of gross primary productivity: Results from the North American Carbon Program site synthesis. *J Geophys Res: Biogeo* 117(G3).

Schroeder, T. A., W. B. Cohen, C. Song, M. J. Canty, and Z. Yang. 2006. Radiometric correction of multi-temporal Landsat data for characterization of early successional forest patterns in western Oregon. *Remote Sens Environ* 103(1):16–26.

Schroeder, T. A., W. B. Cohen, and Z. Yang. 2007. Patterns of forest regrowth following clearcutting in western Oregon as determined from a Landsat time-series. *Forest Ecol Manag* 243:259–273.

Schwartz, M. D. 1992. Phenology and springtime surface-layer change. *Mon Weather Rev* 120:2570–2578.

Schwarz, M. D., and N. E. Zimmermann. 2005. A new GLM-based method for mapping tree cover continuous fields using regional MODIS reflectance data. *Remote Sens Environ* 95(4):428–443. http://doi.org/10.1016/j.rse.2004.12.010.

Sellers, P. J. 1985. Canopy reflectance, photosynthesis and transpiration. *Int J Remote Sens* 6:1335–1372.

Sellers, P. J. 1987. Canopy reflectance, photosynthesis, and transpiration. 2. The role of biophysics in the linearity of their interdependence. *Remote Sens Environ* 21:143–183.

Sellers, P. J., R. E. Dickinson, D. A. Randall, A. K. Betts, F. G. Hall, J. A. Berry, G. J. Collatz, A. S. Denning, H. A. Mooney, and C. A. Nobre. 1997. Modeling the exchange of energy, water, and carbon between continents and the atmosphere. *Science* 275:502–509.

Sellers, P. J., C. Tucker, G. Collatz, S. Los, C. Justice, D. Dazlich, and D. Randall. 1994. A global 1 by 1 NDVI data set for climate studies. Part 2: The generation of global fields of terrestrial biophysical parameters from the NDVI. *Int J Remote Sens* 15:3519–3545.

Sims, D. A., and J. A. Gamon. 2002. Relationships between leaf pigment content and spectral reflectance across a wide range of species, leaf structures and developmental stages. *Remote Sens Environ* 81:337–354.

Smith, J. E., L. S. Heath, and J. C. Jenkins. 2003. Forest volume-to-biomass models and estimates of mass for live and standing dead trees of US forests. *General Technical Report NE-298*. USDA Forest Service, Northeastern Research Station.

Soenen, S. A., D. R. Peddle, R. J. Hall, C. A. Coburn, F. G. Hall. 2010. Estimating aboveground forest biomass from canopy reflectance model inversion in mountainous terrain. *Remote Sens Environ* 114(7):1325–1337.

Somers, B., G. P. Asner, L. Tits, and P. Coppin. 2011. Endmember variability in spectral mixture analysis: A review. *Remote Sens Environ* 115(7):1603–1616. http://doi.org/10.1016/j.rse.2011.03.003.

Song, C. 2004. Cross-sensor calibration between Ikonos and Landsat ETM+ for spectral mixture analysis. *IEEE Geosci Remote S* 1(4):272–276.

Song, C. 2005. Spectral mixture analysis for subpixel vegetation fractions in the urban environment: How to incorporate endmember variability? *Remote Sens Environ* 95(2):248–263.

Song, C. 2013. Optical remote sensing of forest leaf area index and biomass (invited progress report). *Progress in Physical Geography* 37(1):98–113.

Song, C., M. P. Dannenberg, and T. Hwang. 2013. Optical remote sensing of terrestrial ecosystem primary productivity. *Prog Phys Geog* 37(6):834–854.

Song, C., G. Katul, R. Oren, L. E. Band, C. L. Tague, P. C. Stoy, and H. R. McCarthy. 2009. Energy, water, and carbon fluxes in a loblolly pine stand: Results from uniform and gappy canopy models with comparisons to eddy flux data. *J Geophys Res* 114:G04021.

Song, C., T. Schroeder, and W. Cohen. 2007. Predicting temperate conifer forest successional stage distributions with multitemporal Landsat Thematic Mapper imagery. *Remote Sens Environ* 106(2):228–237.

Song, C., and C. E. Woodcock. 2002. The spatial manifestation of forest succession in optical imagery—the potential of multiresolution imagery. *Remote Sens Environ* 82:271–284.

Song, C., and C. E. Woodcock. 2003. Monitoring forest succession with multitemporal Landsat images: Factors of uncertainty. *IEEE T Geosci Remote* 41(11):2557–2567.

Song, C., C. E. Woodcock, and X. Li. 2002. The spectral/temporal manifestation of forest succession in optical imagery: The potential of multitemporal imagery. *Remote Sens Environ* 82:285–302.

Sprintsin, M., A. Karnieli, P. Berliner, E. Rotenberg, D. Yakir, and S. Cohen. 2009 Evaluating the performance of the MODIS leaf area index (LAI) product over a Mediterranean dryland planted forest. *Int J Remote Sens* 30:5061–5069.

Steininger, M. K. 1996. Tropical secondary forest regrowth in the Amazon: Age, area and change estimation with Thematic Mapper data. *Int J Remote Sens* 17(1):9–27.

Steininger, M. K. 2000. Satellite estimation of tropical secondary forest above-ground biomass: Data from Brazil and Bolivia. *Int J Remote Sens* 21:1139–1157.

Steinmetz, S., M. Guerif, R. Delecolle, and F. Baret. 1990. Spectral estimates of the absorbed photosynthetically active radiation and light-use efficiency of a winter wheat crop subjected to nitrogen and water deficiencies. *Remote Sens* 11:1797–1808.

Stenberg, P., S. Linder, H. Smolander, and J. Flower-Ellis. 1994. Performance of the LAI-2000 plant canopy analyzer in estimating leaf area index of some Scots pine stands. *Tree Physiol* 14:981–995.

Stenberg, P., M. Rautiainen, T. Manninen, P. Voipio, and H. Smolander. 2004. Reduced simple ratio better than NDVI for estimating LAI in Finnish pine and spruce stands. *Silva Fenn* 38(1):3–14.

Steven M., P. Biscoe, and K. Jaggard. 1983. Estimation of sugar beet productivity from reflection in the red and infrared spectral bands. *Int J Remote Sens* 4:325–334.

Ter-Mikaelian, M. T., and M. D. Korzukhin. 1997. Biomass equations for sixty-five North America tree species. *Forest Ecol Manage* 97:1–24.

Tian, Q., Z. Luo, J. M. Chen, M. Chen, and F. Hui. 2007. Retrieving leaf area index for coniferous forest in Xingguo County, China with Landsat ETM+ images. *J Environ Manage* 85(3):624–627.

Tomppo, E., M. Nilsson, M. Rosengren, P. P. Aalto, and R. Kennedy. 2002. Simultaneous use of Landsat-TM and IRS-1C WiFS data in estimating large area tree stem volume and above-ground biomass. *Remote Sens Environ* 82:156–171.

Tomppo, E., H. Olsson, G. Stahl, M. Nilsson, O. Hagner, and M. Katila. 2008. Combining national forest inventory field plots and remote sensing data for forest databases. *Remote Sens Environ* 112:1982–1999.

Tucker, C. J., J. E. Pinzon, M. E. Brown, D. A. Slayback, E. W. Pak, R. Mahoney, E. F. Vermote, and N. El Saleous. 2005. An extended AVHRR 8-km NDVI dataset compatible with MODIS and SPOT vegetation NDVI data. *Int J Remote Sens* 26:4485–4498.

Tucker, C. J., D. A. Slayback, J. E. Pinzon, S. O. Los, R. B. Myneni, and M. G. Taylor. 2001. Higher northern latitude normalized difference vegetation index and growing season trends from 1982 to 1999. *Int J Biometeorol* 45:184–190.

Tucker, C. J., J. R. G. Townshend, and T. E. Goff. 1985. African land-cover classification using satellite data. *Science* 227(4685):369–375.

Turner, D. P., W. B. Cohen, R. E. Kennedy, K. S. Fassnacht, and J. M. Briggs. 1999. Relationships between leaf area index and Landsat TM spectral vegetation indices across three temperate zone sites. *Remote Sens Environ* 70:52–68.

Turner, D. P., S. T. Gower, W. B. Cohen, M. Gregory, and T. K. Maiersperger. 2002. Effects of spatial variability in light use efficiency on satellite-based NPP monitoring. *Remote Sens Environ* 80(3):397–405.

Turner, D. P., S. Urbanski, D. Bremer, S. C. Wofsy, T. Meyers, S. T. Gower, and M. Gregory. 2003. A cross-biome comparison of daily light use efficiency for gross primary production. *Glob Change Biol* 9:383–396.

Uddling, J., J. Gelang-Alfredsson, K. Piikki, and H. Pleijel. 2007. Evaluating the relationship between leaf chlorophyll concentration and SPAD-502 chlorophyll meter readings. *Photosynth Res* 91:37–46.

Urban, D. L. 1990. *A Versatile Model to Simulate Forest Pattern: A User's Guide to ZELIG Version 1.0*. Charlottesville, VA: University of Virginia.

Vargas, M., T. Miura, N. Shabanov, and A. Kato. 2013. An initial assessment of Suomi NPP VIIRS vegetation index EDR. *J Geophys Res-Atmos* 118:12,301–12,316.

Verhoef, W. 1984. Light scattering by leaf layers with application to canopy reflectance modeling: The SAIL model. *Remote Sens Environ* 16:125–141.

Verrelst, J., J. Muñoz, L. Alonso, J. Delegido, J. P. Rivera, G. Camps-Valls, and J. Moreno. 2012. Machine learning regression algorithms for biophysical parameter retrieval: Opportunities for Sentinel-2 and -3. *Remote Sens Environ* 118:127–139.

Vieira, I. C. G., A. S. de Almeida, E. A. Davidson, T. A. Stone, C. J. R. de Carvalho, and J. B. Guerrero. 2003. Classifying successional forests using Landsat spectral properties and ecological characteristics in eastern Amazônia. *Remote Sens Environ* 87(4):470–481.

Wang, D., S. Liang, T. He, and Y. Yu. 2013. Direct estimation of land surface albedo from VIIRS data: Algorithm improvement and preliminary validation. *J Geophys Res-Atmos* 118:12,577–12,586.

Wang, Y., C. E. Woodcock, W. Buermann, P. Stenberg, P. Voipio, H. Smolander, T. Häme, Y. Tian, J. Hu, and Y. Knyazikhin. 2004. Evaluation of the MODIS LAI algorithm at a coniferous forest site in Finland. *Remote Sens Environ* 91(1):114–127.

Weng, Q. 2011. *Advances in Environmental Remote Sensing: Sensors, Algorithms, and Applications*. Boca Raton, FL: CRC Press. ISBN-13: 978-1-4200-9181-6.

White, M. A., K. M. de Beurs, K. Didan, D. W. Inouye, A. D. Richardson, O. P. Jensen, J. O'Keefe, G. Zhang, and R. R. Nemani. 2009. Intercomparison, interpretation, and assessment of spring phenology in North America estimated from remote sensing for 1982–2006. *Global Change Bio* 15:2335–2359.

Widlowski, J. L., M. Taberner, B. Pinty, V. Bruniquel-Pinel, M. Disney, R. Fernandes, J.-P. Gastellu-Etchegorry, N. Gobron, A. Kuusk, and T. Lavergne. 2007. Third radiation transfer model intercomparison (RAMI) exercise: Documenting progress in canopy reflectance models. *J Geophys Res* 112:D09111. http://doi.org/10.1029/2006JD007821

Wiegand, C., A. Richardson, D. Escobar, and A. Gerbermann. 1991. Vegetation indices in crop assessments. *Remote Sens Environ* 35:105–119.

Willis, C. G., B. R. Ruhfel, R. B. Primack, A. J. Miller-Rushing, and C. C. Davis. 2008. Phylogenetic patterns of species loss in Thoreau's woods are driven by climate change. *P Natl Acad Sci USA* 105(44):17029–17033.

Willis, C. G., B. R. Ruhfel, R. B. Primack, A. J. Miller-Rushing, J. B. Losos, and C. C. Davis. 2010. Favorable climate change response explains non-native species' success in thoreau's woods. *PLoS ONE* 5(1):e8878.

Wolkovich, E. M., and E. E. Cleland. 2011. The phenology of plant invasions: A community ecology perspective. *Front Ecol Environ* 9(5):287–294.

Woodcock, C. E., R. Allen, M. Anderson, A. Belward, R. Bindschadler, W. Cohen, F. Gao, S. N. Goward, D. Helder, and E. Helmer. 2008. Free access to Landsat imagery. *Science* 320:1011.

Woodcock, C. E., J. B. Collins, S. Gopal, V. D. Jakabhazy, X. Li, S. A. Macomber, S. Ryherd, V. J. Harward, J. Levitan, Y. Wu, and R. Warbington. 1994. Mapping forest vegetation using Landsat TM imagery and a canopy reflectance model. *Remote Sens Environ* 50:240–254.

Woodcock, C. E., J. B. Collins, V. D. Jakabhazy, X. Li, S. A. Macomber, and Y. Wu. 1997. Inversion of the Li-Strahler canopy reflectance model for mapping forest structure. *IEEE Trans Geosci Rem Sens* 35:405–414.

Wu, C., J. M. Chen, A. Gonsamo, D. T. Price, T. A. Black, and W. A. Kurz. 2012a. Interannual variability of net carbon exchange is related to the lag between the end-dates of net carbon uptake and photosynthesis: Evidence from long records at two contrasting forest stands. *Agric For Meteorol* 164:29–38.

Wu, C., A. Gonsamo, J. M. Chen, W. A. Kurz, D. T. Price, P. M. Lafleur, R. S. Jassal, D. Aragoni, G. Bohrer, and C. M. Gough. 2012b. Interannual and spatial impacts of phenological transitions, growing season length, and spring and autumn temperatures on carbon sequestration: A North America flux data synthesis. *Global Planet Change* 92:179–190.

Wu, Y., and A. Strahler. 1994. Remote estimation of crown size, stand density, and biomass on the oregon transect. *Ecol Appl* 4(2):299–312.

Wulder, M. A., J. G. Masek, W. B. Cohen, T. R. Loveland, and C. E. Woodcock. 2012. Opening the archive: How free data has enabled the science and monitoring promise of Landsat. *Remote Sens Environ* 122:2–10.

Xiao, X. M., D. Hollinger, J. Aber, M. Goltz, E. A. Davidson, Q. Y. Zhang, and B. Moore. 2004a. Satellite-based modeling of gross primary production in an evergreen needleleaf forest. *Remote Sens Environ* 89:519–534.

Xiao, X. M., Q. Zhang, B. Braswell, S. Urbanski, S. Boles, S. Wofsy, B. Moore III, and D. Ojima. 2004b. Modeling gross primary production of temperate deciduous broadleaf forest using satellite images and climate data. *Remote Sens Environ* 91(2):256–270.

Xin, Q., P. Olofsson, Z. Zhu, B. Tin, and C. E. Woodcock. 2013. Toward near real-time monitoring of forest disturbance by fusion of MODIS and Landsat data. *Remote Sens Environ* 135:234–247.

Xiong, X., and W. Barnes. 2006. An overview of MODIS radiometric calibration and characterization. *Adv Atmos Sci* 23:69–79.

Yang, W., B. Tan, D. Huang, M. Rautiainen, N. V. Shabanov, Y. Wang, J. L. Privette, K. F. Huemmrich, R. Fensholt, I. Sandholt, M. Weiss, D. E. Ahl, S. T. Gower, R. R. Nemani, Y. Knyazikhin, and R. B. Myneni. 2006. MODIS leaf area index products: From validation to algorithm improvement. *IEEE Trans Geosci Remote Sens* 44(7):1885–1898.

Yuan, W., W. Cai, J. Xia, J. Chen, S. Liu, W. Dong, L. Merbold, B. Law, A. Arain, and J. Beringer. 2014. Global comparison of light use efficiency models for simulating terrestrial vegetation gross primary production based on the LaThuile database. *Agr Forest Meteorol* 192:108–120.

Yuan, W., S. Liu, G. Zhou, G. Zhou, L. L. Tieszen, D. Baldocchi, C. Bernhofer, H. Gholz, A. H. Goldstein, and M. L. Goulden. 2007. Deriving a light use efficiency model from eddy covariance flux data for predicting daily gross primary production across biomes. *Agr Forest Meteorol* 143(3):189–207.

Yuan, W., S. Liu, G. Yu, J. Bonnefond, J. Chen, K. Davis, A. R. Desai, A. H. Goldstein, D. Gianelle, F. Rossi, A. E. Suyker, and S. B. Verma. 2010. Global estimates of evapotranspiration and gross primary production based on MODIS and global meteorology data. *Remote Sens Environ* 114:1416–1431.

Zarco-Tejada, P. J., J. A. J. Berni, L. Suárez, G. Sepulcre-Cantó, F. Morales, and J. R. Miller. 2009. Imaging chlorophyll fluorescence with an airborne narrow-band multispectral camera for vegetation stress detection. *Remote Sens Environ* 113(6):1262–1275.

Zarco-Tejada, P. J., A. Morales, L. Testi, and F. J. Villalobos. 2013. Spatio-temporal patterns of chlorophyll fluorescence and physiological and structural indices acquired from hyperspectral imagery as compared with carbon fluxes measured with eddy covariance. *Remote Sens Environ* 133:102–122.

Zeng, X., R. E. Dickinson, A. Walker, M. Shaikh, R. S. DeFries, and J. Qi. 2000. Derivation and evaluation of global 1-km fractional vegetation cover data for land modeling. *J Appl Meteorol* 39:826–840.

Zhang, X. Y., M. A. Friedl, C. B. Schaaf, A. H. Strahler, J. C. F. Hodges, F. Gao, B. C. Reed, and A. Huete. 2003. Monitoring vegetation phenology using MODIS. *Remote Sens Environ* 84(3):471–475.

Zhang, X. Y., and S. Kondragunta. 2006. Estimating forest biomass in the USA using generalized allometric models and MODIS land products. *Geophys Res Lett* 33:L09402.

Zhang, X. Y., D. Tarpley, and J. T. Sullivan. 2007. Diverse responses of vegetation phenology to a warming climate. *Geophys Res Lett* 34:L19405.

Zhang, Y., and R. K. Mishra. 2012. A review and comparison of commercially available pan-sharpening techniques for high resolution satellite image fusion. *Proceedings of Geoscience and Remote Sensing Symposium (IGARSS), 2012 IEEE International*. IEEE, pp. 182–185.

Zhang, Y., and C. Song. 2006. Impacts of afforestation, deforestation and reforestation on forest cover in China from 1949 to 2003. *J Forest* 104(7):383–387.

Zhao, M., F. A. Heinsch, R. R. Nemani, and S. W. Running. 2005. Improvements of the MODIS terrestrial gross and net primary production global data set. *Remote Sens Environ* 95(2):164–176.

Zhao, M., and S. W. Running. 2010. Drought-induced reduction in global terrestrial net primary production from 2000 through 2009. *Science* 329:940–943.

Zhu, Z., J. Bi, Y. Pan, S. Ganguly, A. Anav, L. Xu, A. Samanta, S. Piao, R. R. Nemani, and R. B. Myneni. 2013. Global data sets of vegetation leaf area index (LAI) 3g and fraction of photosynthetically active radiation (FPAR) 3g derived from global inventory modeling and mapping studies (GIMMS) normalized difference vegetation index (NDVI3g) for the period 1981 to 2011. *Remote Sens* 5(2):927–948.

Zhu, J., N. Tremblay, and Y. Liang. 2012b. Comparing SPAD and atLEAF values for chlorophyll assessment in crop species. *Can J Soil Sci* 92:645–648.

Zhu, Z., C. E. Woodcock, and P. Olofsson. 2012a. Continuous monitoring of forest disturbance using all available Landsat imagery. *Remote Sens Environ* 122:75–91.

Part IV

Land Use/Land Cover

8 Land Cover Change Detection

John Rogan and Nathan Mietkiewicz

ACRONYMS AND DEFINITIONS

ALI	Advanced Land Imager
ASTER	Advanced spaceborne thermal emission and reflection radiometer
AVHRR	Advanced very high resolution radiometer
ETM+	Enhanced Thematic Mapper+
EVI	Enhanced Vegetation Index
GIMMS	Global Inventory Modeling and Mapping Studies
GIS	Geographic Information System
GPS	Global Positioning System
LAI	Leaf area index
LiDAR	Light detection and ranging
LST	Land Surface Temperature
MODIS	Moderate-resolution imaging spectroradiometer
MSS	Multi-spectral scanner
NDVI	Normalized Difference Vegetation Index
OLI	Operational land imager
PAR	Photosynthetically active radiation
PV	Photosynthetic vegetation
REDD	Reducing Emissions from Deforestation in Developing Countries
SPOT	Satellite Pour l'Observation de la Terre, French Earth Observing Satellites
STAARCH	Spatial temporal adaptive algorithm for mapping reflectance change
STARFM	Spatial and temporal adaptive reflectance fusion model

8.1 INTRODUCTION

The purpose of this chapter is to explore the current trends in land cover change detection and to identify those trends that are potentially transformative to our understanding of land change, as well as identify knowledge/information gaps that should require attention in the future. The current level of understanding of the scale and pace of land cover change is inadequate (MohanRajan et al., 2024; Lv et al., 2023; Mashala et al., 2023; Parelius, 2023; Lv et al., 2022; Wang et al., 2021; Chen and Shi, 2020; Tewabe et al., 2020; Wang et al., 2020; Woodcock et al., 2020; Zhang et al., 2020; Zhang and Shi, 2020; Asokan and Anitha, 2019; Islam et al., 2018; Liping et al., 2018; Sidhu et al., 2018; Leichtle et al., 2017; Zhu, 2017; Turner et al., 2007; Frey and Smith, 2007; Hansen et al., 2013). However, it is understood that land cover change is an undisputed component of global environmental change (Kennedy et al., 2014). Land cover changes and their impacts range widely from regional temperature warming to land degradation and biodiversity loss, and from diminished food production to the spread of infectious diseases (Vitousek et al., 1997; Farrow and Winograd, 2001). Land cover change, manifested as either land cover modification and/or conversion, can occur at all spatial scales, and changes at local scales can have cumulative impacts at broader scales (Stow, 1995).

The long-standing challenge facing scientists and policymakers is the paucity of comprehensive data, at local, regional, and national levels, on the types and rates of land cover changes, and even less systematic evidence on the causes/drivers and consequences of those changes (Parelius, 2023; Lv et al., 2022; Wang et al., 2021). Such data can be generated through a dual approach (Wang et al., 2021; Chen and Shi, 2020; Islam et al., 2018; Leichtle et al., 2017; Zhu, 2017): (1) based on direct or indirect observations, for the regions and time periods for which data exist (Leichtle et al., 2017); and (2) based on projections by models (Lambin et al., 1999). A key element for the successful implementation of this dual approach is the monitoring of land cover on a systematic, operational basis (Lunetta and Elvidge, 1998; Townshend and Justice, 2002; Wulder and Coops, 2014).

In data-rich locations, such as the United States, federal resource inventory programs, such as the US Forest Service Forest Inventory and Analysis (FIA) program and the Natural Resources Conservation Service (NRCS, 2000) have provided valuable statistical information on land cover dynamics for over 35 years. These agencies provide plot-level information for remote sensing land cover mapping projects (Wulder and Coops, 2014). However, there is also a need for spatially explicit, thematically comprehensive data products that can be provided by remotely sensed data (Loveland et al., 2002). For example, The USGS Land-cover Change Program (LCCP) is designed to document the rates, causes, and consequences of land-cover change from 1973 to the present, using Landsat North American Land Characterization (NALC) data (http://landcover.usgs.gov/). The program area spans 84 ecoregions of the conterminous United States. Another example of comprehensive large-area land cover assessment is the Canadian Forest Service Earth Observation for Sustainable Development of Forests (EOSD) program (http://www.nrcan.gc.ca/), which monitors Canada's forest cover with Landsat imagery (Wood et al., 2002). Additionally, the European CORINE program (http://land.copernicus.eu/pan-european/corine-land-cover) maps land cover and land use (44 categories) using a variety of medium-resolution satellite data from 1990 to the present.

In data-poor locations, data derived from remote sensing are often the only source of information available for land cover monitoring (Wang et al., 2021; Zhang et al., 2020; Zhang and Shi, 2020; Leichtle et al., 2017; Zhu, 2017; Lambin et al., 1999). This situation places added pressure on remote sensing practitioners to produce accurate change maps using replicable methods, which cannot be verified using the traditional suite of map accuracy tools (Rogan and Chen, 2004; Dorais and Cardille, 2011). The inclusion of land cover change in international agreements such as the Kyoto Protocol under the United Nations framework convention on climate change (UNFCC), as well as the growing popularity of the United Nations Reducing Emissions from Deforestation and forest Degradation (UN-REDD and REDD+) makes it essential to advance initiatives to monitor land cover change effectively (MohanRajan et al., 2024; Lv et al., 2022; Asokan and Anitha, 2019; Zhu, 2017; DeFries and Townsend, 1999). Increased Landsat data availability (Wulder and Coops, 2014) and the growing trend in automated mapping and change detection algorithms will likely open up the current data bottleneck such that developing countries can create more precise estimates of land change (Woodcock et al., 2020; Zhu and Woodcock, 2014).

In addition to the technical advantages of remotely sensed data, the reduced data cost, increased accessibility and availability, and increased understanding of the information derived from these data, have facilitated the launch of large-area remote sensing–based monitoring programs/initiatives (Loveland et al., 2002; Eidenshink et al., 2007), as well as global-scale medium spatial resolution change map data sets (Hansen et al., 2013). Therefore, these data, in concert with enabling technologies such as global positioning systems (GPS) and geographic information systems (GIS), can form the information base upon which sound and cost-effective monitoring decisions can be made (Lunetta, 1998).

While a large body of work has accumulated regarding land cover change monitoring using remotely sensed data (e.g., see reviews by MohanRajan et al., 2024; Lv et al., 2022, 2023; Mashala et al., 2023; Parelius, 2023; Wang et al., 2020, 2021; Chen and Shi, 2020; Tewabe et al., 2020; Woodcock et al., 2020; Zhang et al., 2020; Zhang and Shi, 2020; Asokan and Anitha, 2019; Islam et al., 2018; Liping et al., 2018; Sidhu et al., 2018; Leichtle et al., 2017; Zhu, 2017; Nelson, 1983;

Land Cover Change Detection

Singh, 1989; Hobbs, 1990; Mouat et al., 1993; Stow, 1995; Coppin and Bauer, 1996; Macleod and Congalton, 1998; Ridd and Liu, 1998; Mas, 1999; Coppin et al., 2002; Civco et al., 2002; Gong and Xu, 2003; Coppin et al., 2004), little guidance exists for addressing large-area change mapping, especially in an operational context (Dobson and Bright, 1994; Loveland et al., 2002). Thus, in light of the exciting potential for future operational land cover monitoring programs, and in acknowledgment of the large amount of new, disparate methods currently employed in change detection studies in the literature, this chapter presents a review of the key requirements and chief challenges of land cover change monitoring.

A general classification of the spatial resolution of remote sensing platforms produces three categories (Rogan and Chen, 2004): (1) coarse resolution (> 250 m) (e.g., Advanced Very High Resolution Radiometer [AVHRR]); (2) medium resolution (< 250 m–> 20 m) (e.g., Landsat Multispectral Scanner [MSS]); and (3) fine resolution (< 20 m) (e.g., WorldView-2).

8.2 LAND COVER CHANGE DETECTION AND MONITORING—THEORY AND PRACTICE

Figure 8.1 presents a conceptual scheme of a forest environment and demonstrates that land cover change can result in alterations (increases or decreases) in the abundance, composition, and condition of remote sensing scene elements over various spatial and temporal resolutions (Stow et al., 1990). Conversion is shown in Figure 8.1b. In contrast, modification (Figure 8.1c–d) involves maintenance of the existing cover type in the face of changes to its scene elements (i.e., change in abundance and condition).

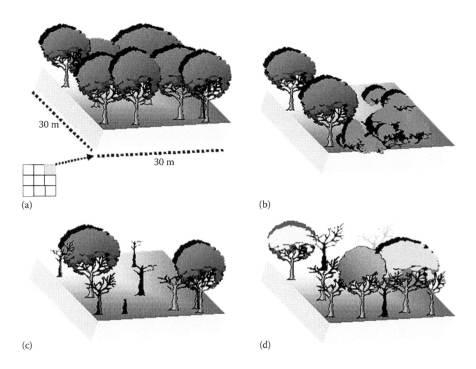

FIGURE 8.1 Conceptual scheme representing land cover changes from Time 1 (represented by (a)) to Time 2 (represented by (b), (c), and/or (d)): (b) change in composition; (c) change in abundance; and (d) change in condition of vegetation cover, which influence the spectral quantity and quality of solar-reflected radiation received by a Landsat sensor (30 m pixel).

Detecting and monitoring land cover change across large areas are two of the most important tasks that remote sensing data and technology can accomplish (Woodcock et al., 2020). Land cover change detection, one of the most common uses of remotely sensed data, is possible when changes in the surface phenomena of interest result in detectable changes in radiance, emittance (Lunetta and Elvidge, 1998), LiDAR return values (Wulder and Coops, 2014), or microwave backscatter values (Rignot and Vanzyl, 1993; Zhang and Shi, 2020; Leichtle et al., 2017), which implicitly involves spatial patterns of change (Crews-Meyer, 2002).

Khorram et al. (1999) explored the spatial context of land cover change and stated that spatial entities either: (1) become a different category; (2) expand, shrink, or change shape; (3) shift position; and (4) fragment or coalesce. These concepts are well-understood by remote sensing practitioners, and especially the resource management community, worldwide, but less so by ecology, sociology, and vulnerability communities.

However, in the last ten years a number of important developments have occurred that have helped improve the adoption of land change information by scientific communities that had not done so previously. Land change science (Turner et al., 2007) has emerged as an interdisciplinary field which seeks to understand land cover and land use (LCLU) dynamics as a coupled human-environment system. This burgeoning theoretical field claims earth observation data as a crucial component, and so has effectively exposed land cover mapping and monitoring practices to a broad audience of anthropologists, economists, and sociologists. Another important development is the opening of the Landsat archive in 2008 (Wulder and Coops, 2014). The availability of dense time series of moderate spatial resolution Landsat imagery (since 1972 to the present) has already had significant impacts on the ecology community (Kennedy et al., 2014), as temporal sequences and trajectories of importance to ecological conservation are now mostly matched by Landsat time-stacks. Overall, therefore, we can expect to see, in the near future, remotely sensed data being used to test or verify theories in a much broader array of disciplines than ever before.

Most terrestrial surfaces comprise complex configurations of land cover attributes (Turner et al., 1999). These range from being mainly "natural" to those that are largely human-dominated (Turner et al., 1999). Land cover change is viewed in terms of modifications in component attributes within either natural or human-dominated land cover, or conversions from natural to human-dominated land cover (Lambin et al., 1999). Despite the recognized importance of land cover modifications (e.g., wind or insect damage), and in contrast to conversions (i.e., forest loss due to agriculture gain), they are not as well documented at operational scales (Lambin et al., 2001). This is partly due to the fact that modifications occur at many different spatial scales and are often too subtle and cryptic to be mapped with a high level of confidence (Lv et al., 2022; Wang et al., 2020; Islam et al., 2018; Zhu, 2017; Ekstrand, 1990; Gong and Xu, 2003). Therefore, land cover modification analysis requires that a greater level of detail be accommodated in remote sensing analysis.

Macleod and Congalton (1998) listed four aspects of change detection that are important when monitoring land cover using remote sensing data (Parelius, 2023; Tewabe et al., 2020; Islam et al., 2018): (1) detecting changes that have occurred (Fung, 1990; Lunetta et al., 2002); (2) identifying the nature of the change (Hayes and Sader, 2001; Seto et al., 2002); (3) measuring the areal extent of the change (Stow et al., 1990; Rogan et al., 2003); and (4) assessing the spatial pattern of the change (Crews-Meyer, 2002). Therefore, change monitoring initiatives/programs (i.e., both current and planned) should try to accommodate these four factors, in addition to appreciating the magnitude, duration, and rate of changes that can occur (Rogan and Chen, 2004). Additionally, the burgeoning operational monitoring paradigm represents a shift away from the paradigm of the ubiquitous two-date "end-to-end" change detection approach (i.e., only two dates used in analysis), due to their greater temporal scope (Rogan and Chen, 2004).

8.3 TRENDS IN LAND COVER CHANGE DETECTION AND MONITORING

8.3.1 Historical Trends—Eight Epochs

The history of land cover change mapping and monitoring has witnessed five distinct periods, determined by the evolution of remote sensor technology, and research needs, related to resource management mandates and various scientific research interests, seen in the following list (Chen and Shi, 2020; Zhang and Shi, 2020; Asokan and Anitha, 2019; Islam et al., 2018; Sidhu et al., 2018).

1. Early case studies (late 1970s) were exploratory, and primarily focused on urban change detection (Todd, 1977).
2. Research then shifted to case-study applications (early to mid-1980s) in natural environments, based on the needs of resource management agencies and the burgeoning interest in carbon sequestration (Singh, 1989).
3. Successful applications and experience (mid- to late 1990s) led to more widespread applications of remote sensing over large areas and using a wide variety of methods (Lambin and Strahler, 1994a, b; Lambin et al., 2001).
4. Improved sensor technology facilitated the increased interest in less-researched fields, such as urban applications of remote sensing, the cryosphere, and coastal-ocean research (mid-1990s to the present) (Rashed et al., 2001), and the new approach adopted by the MODIS science team to provide image information products, such as global land cover (Friedl et al., 2010). Large-area, high spatial resolution remote sensing became possible in 1994, when the US government allowed civil commercial companies to market high spatial resolution satellite remote sensing data (i.e., 1 m and 4 m spatial resolution) (Glackin, 1998).
5. Today, a 40-year archive of Landsat imagery, a 22-year archive of AVHRR Global Inventory Modeling and Mapping Studies (GIMMS) NDVI data, and a 15-year archive of MODIS imagery and information products, coupled with an explosion in image time-series research, and increased automation have made operational regional- and global-scale land change monitoring a reality (Wulder and Coops, 2014). Table 8.1 presents a comparison of AVHRR, MODIS, and Landsat data in terms of spatial and temporal resolution. Clearly, the high temporal coverage AVHRR and MODIS data are optimal for regional-global analysis, but they can only provide this coverage at coarse spatial resolution. On the other hand, Landsat data are provided at much finer spatial scales (30 m), but they are mostly limited to local-regional coverage. However, the Global Land Survey initiative provides global Landsat coverage for five dates between the early 1970s and 2010. Spatial resolution is a key limiting factor in the ability of remote sensing imagery to resolve land cover and land cover change classes. This is because spatial scale exerts a strong influence on the ability to extract information from remotely sensed data sets and requires careful specification and analysis. As a result, the question of which remotely sensed data are appropriate for specific land cover change monitoring applications remains an open one. Obviously, the resolvability of land-cover change increases with higher spatial resolution. However, high spatial resolution imagery is not typically needed to accurately detect general land cover changes (the goal of large-area monitoring studies) in most environments (Franklin and Wulder, 2002). Studying a variety of environments, Townshend and Justice (1988) reported that spatial resolutions coarser than about 200 m undermined the reliable detection of land cover changes. Pax-Lenney and Woodcock (1997) examined the impact of coarsening the spatial resolution on the accuracy of areal estimates of agricultural fields in Egypt (30 m–120 m–240 m–480 m–960 m). Most of the coarse resolution estimates were within 10% of the original 30-m

TABLE 8.1

A Comparison of AVHRR, MODIS, and Landsat in Terms of Spatial and Temporal Resolution

Sensor/ Program	Temporal Lineage	Temporal Resolution	Geographic Coverage	Spatial Resolution	Information Content	Information
AVHRR-GIMMS	1982–2012	Bi-weekly composites	Global	1/12 degree (8 km at the equator)	NDVI	http://glcf.umd.edu/data/gimms/
MODIS	1999-present	Daily and 8-day composites	Global	250, 500, 1000 m	Multispectral/ Biophysical products	http://modis.gsfc.nasa.gov/
Landsat	1972-present	16 days	Regional	30 m Global Land Survey Global Coverage: 1970, 1990, 2000, 2005, 2010	Multispectral	http://landsat.gsfc.nasa.gov/ http://landsat.usgs.gov/science_GLS.php

estimates. Therefore, medium spatial resolution data remain the optimal choice for most land cover change studies, but more research over time will challenge this assertion in the interest of global-scale estimation and cost reduction, using coarse spatial resolution data relative to the particular application (Mashala et al., 2023; Tewabe et al., 2020; Islam et al., 2018; Leichtle et al., 2017).

8.3.2 CAUSE OF LAND COVER CHANGE

A brief survey of the number of new remote sensing journals shows that 24 journals have been launched since 2007 (an increase of 60% in a seven-year time span). The remarkable proliferation of new journals likely reflects the growing user community and wealth of new remote sensing applications, enabled by a growing time series of free data and also the increased availability of open-source software packages (e.g., Quantum GIS). Today, techniques to perform change detection have become numerous as a result of increasing versatility in manipulating digital data and growing computing power (Wang et al., 2021; Rogan and Chen, 2004). The sheer number of published articles and the importance to resource management indicates both the degree to which remote sensing is used and the proliferation of methods employed. One dimension of this proliferation is progress in developing new and improved ways of detecting change, while another dimension is the wide variety of kinds of changes being monitored (Table 8.2). Table 8.2 presents the dominant causes of multitemporal land cover change in natural and human-dominated environments, and their temporal and physical characteristics. Each change event can result in very different magnitude (i.e., small to large), duration (i.e., days to decades), and temporal rates (i.e., slow to fast) (Lv et al., 2022; Woodcock et al., 2020; Gong and Xu, 2003). Understanding the magnitude, duration, and rate of land cover disturbances has severe implications for the success of a land cover monitoring study because it permits researchers to determine the most appropriate sensor, derived data set, frequency of acquisition, level of image processing, and reproducible map legend.

It is important to note that not all land change disturbances are equally important in change detection studies, and not all disturbances may be detected as confidently as others (Gong and Xu, 2003). For example, land changes of lesser concern to forest managers include those related to inter-annual

TABLE 8.2
Causes of Land Cover Change and Their Magnitude, Duration, and Rate (after Gong and Xu, 2003)

Cause	Magnitude	Duration	Rate	References
Phenology	Small-medium	Days-months	Medium	Goodin et al. (2002); Jakubauskas et al. (2002); Zhang et al (2003)
Regeneration	Small-medium	Days-decades	Slow	Fiorella and Ripple (1993); Lawrence and Ripple (2000)
Drought	Small-medium	Months-years	Slow	Peters et al. (1993); Jacobberger-Jellison (1994)
Flooding	Medium-large	Days-weeks	Medium-fast	Blasco et al. (1992); Michener and Houhoulis (1997); Rogan et al. (2001); Zhan et al. (2002)
Wildfire	Small-large	Days-weeks	Fast	Patterson and Yool (1998); Rogan and Yool (2000)
Disease	Small-large	Days-years	Slow-medium	Wilson et al. (2002); Kelly and Meentemeyer (2002)
Insect attack	Small-large	Days-years	Slow-fast	Muchoney and Haack (1994); Chalifoux et al. (1998); Radeloff et al. (1999)
Ice storm	Small-large	Years	Medium-fast	Dupigny-Giroux et al. (2002); Millward and Kraft (2004); Olthof et al. (2004)
Mortality	Medium-large	Days-years	Slow-fast	Collins and Woodcock (1996); Allen and Kupfer (2000)
Water/nitrogen stress	Small-medium	Days-years	Slow-fast	Running and Donner (1987); Penuelas et al. (1994)
Pollution	Small-large	Years	Slow	Ekstrand (1994); Rock et al. (1988); Rees and Williams (1997); Diem (2002); Tommervik et al. (2003)
Thinning	Medium-large	Days	Fast	Olsson (1995); Nilson et al. (2001); Peddle et al. (2003a)
Clear-cutting	Large	Days	Fast	Hayes and Sader (2001)
Replanting	Small-medium	Days-decades	Fast	Coppin and Bauer (1996); Levien et al. (1999)
Mining	Large	Days-decades	Medium	Cadac (1998)
Grazing	Small-medium	Days-decades	Slow-medium	Rees et al. (2003)
Wind throw	Large	Days	Medium-fast	Mukai and Hasegawa (2000); Kundu et al. (2001); Lindemann and Baker (2002)
Erosion	Small-medium	Days-weeks	Fast	Dwivedi et al. (1997); Hong and Iisaka (1987); Michalek et al. (1993); Rosin and Hervas (2002)
Environmental quality	Small-large	Months-Years	Slow	Fung and Siu (2000)
Fragmentation	Small-large	Days	Fast	Wickham et al. (1999); Millington et al. (2003)
Conversion	Large	Years-decades	Slow-medium	Jha and Unni (1994); Loveland et al. (2002)
Desertification	Small	Years-decades	Slow	Robinove et al. (1981); Pilon et al. (1988)

variability and growth variation caused by climate variability, whereas, to global change modelers, the last type of change is of chief concern (Turner et al., 1999). A key issue in change detection is understanding how the types of change affect land cover and also how they interact with one another. For example, phenological vegetation change, which varies temporally across scales ranging from years to decades, often interacts with more temporally discrete changes, such as burn scar vegetation depletion and post-fire regeneration (Rogan et al., 2002).

8.4 LAND COVER CHANGE DETECTION APPROACHES

8.4.1 Monotemporal Change Detection—Products for Real Time and Specific Disturbance Types

Numerous land change applications using only a single image date (i.e., monotemporal change detection) (Coppin and Bauer, 1996, p. 217), which focus on a specific change event, have successfully detected a variety of land cover disturbances (Lv et al., 2023; Wang et al., 2020; Liping et al., 2018). These disturbances include water stress (Running and Donner, 1987), wildfires (Patterson and Yool, 1998; Rogan and Franklin, 2001), forest thinning (Nilson et al., 2001), forest pest damage (Leckie et al., 1988; Vogelmann and Rock, 1988; Joria and Ahearn, 1991; Franklin et al., 1994), forest mortality (Ekstrand, 1990), and the effects of pollution on vegetation vigor (Toutoubalina and Rees, 1999).

Monotemporal applications are an effective application of "swapping time for space." Applications of remotely sensed data for disturbance-specific monitoring have considerable advantages, including savings in processing time and reduced costs (Patterson and Yool, 1998). Further, end users may require a "quick look" at a particular disturbance for rapid response in the case of mudslide, wildfire, or flood events. A good example of this is the US Forest Service rapid-response wildfire detection project, which relies on MODIS active fire-detection data. However, monotemporal approaches rely heavily on assumptions about the initial state of land cover in the particular study area (Ekstrand, 1994). Indeed, an important factor in the success of these studies is that pre-change information (e.g., pre-disturbance spectral information) and stand and landscape characteristics (e.g., stratification of mixed vegetation canopies, stand-based analysis, slope, and aspect) are controlled to minimize confusion between change vs. unchanged land cover types (Ekstrand, 1990). This implies that pre-change, or pre-disturbance spectral and/or land cover information are needed to robustly resolve monotemporal disturbances using remotely sensed data (Franklin, 2001). For monotemporal (rapid response) applications coarse spatial resolution data acquired by sensors such as AVHRR, SPOT Vegetation, and MODIS data are appropriate. Image preprocessing requirements are minimal, but a spectral transformation (e.g., vegetation index) would be useful to separate the disturbance signal (e.g., wildfire or flooding) from the undisturbed background, and facilitate simple spectral change thresholding, if required.

Recent advances (MohanRajan et al., 2024; Chen and Shi, 2020; Zhang and Shi, 2020) in real-time disaster response management provide an informative application of monotemporal change detection. The International Charter Space and Major Disasters (http://www.disasterscharter.org) was founded in 1999, after the catastrophic Hurricane Mitch struck Central America. The Charter aims at providing a unified system of space data acquisition and delivery to locations affected by natural disasters and receives imagery contributions from a group of 15 international participating earth observation agencies. Additionally, the United Nations Platform for Space-based Information for Disaster Management and Emergency Response (UN-SPIDER program) was established in 2006 to serve as a gateway to space information for disaster management support (http://www.un-spider.org/). These two disaster response programs rely on high spatial resolution data to achieve their goals.

While high spatial resolution sensors cannot conveniently or cost-effectively provide wall-to-wall coverage for large-area change mapping applications due to data cost and volume, they are invaluable as a source of ground reference information for medium and coarse resolution products/applications, and for operational monitoring studies over small spatial extents (Woodcock et al., 2020). Technological advances in sensor design allow aerial photographic precision and quality in these satellite-based data and permit the investigation of thematic information at the highest order in both natural and urban/suburban landscapes. Though promising, change detection using high spatial resolution data requires further research and development (Rogan and Chen, 2004). Data costs, compared to free Landsat data, for example, are very high. Other issues include the impact of off-nadir view angles on change detection and the increasing need for object-based mapping (Stow et al., 2004). Further, geometric distortion is a vexing problem for most airborne data sets (see Franklin and Wulder, 2002).

8.4.2 Bitemporal Change Detection—Map Comparison and Disturbance Analysis

In the vast majority of land cover change studies, imagery from one date is compared to another date. Within this paradigm of analyzing images as "endpoints" there has been a tremendous variety of methods developed and used (Wang et al., 2021; Zhang et al., 2020; Sidhu et al., 2018). This proclivity of bitemporal studies has been caused by several factors: (1) there are fewer data to analyze; (2) studies have been conducted to satisfy burgeoning short-term resource management needs; (3) various researchers have needed a straightforward scenario in order to compare and evaluate a variety of change detection techniques to find an optimal method; (4) most studies have been conducted in regions of limited spatial extent and landscape heterogeneity; and (5) these studies have focused on a single disturbance event (e.g., flooding, fire, logging, or pest infestation) in environmentally (e.g., tropical forests) or politically (e.g., municipalities) important regions. Thus, while bitemporal change detection will continue to serve its purpose for a long time to come, its efficiency and consistency over large, heterogeneous areas has yet to be fully examined (Rogan et al., 2003). However, the potential for moderate spatial resolution analysis in land change monitoring is enormous (Woodcock et al., 2020).

8.4.2.1 Bitemporal Change Detection Methods

The selection of an appropriate change detection technique depends on the information requirements, data availability and quality, time and cost constraints, analysis skill, and experience (Zhang et al., 2020; Leichtle et al., 2017; Johnson and Kasischke, 1998). Table 8.3 presents a summary of a variety of land cover change detection methods and their advantages and disadvantages for operational monitoring. Twelve change detection methods are compared according to their status in terms of operational use, as well as their relative strengths and weaknesses. The chief division between the 12 methods occurs between post-classification comparison (i.e., categorical change) and the suite of existing continuous change detection techniques (e.g., image differencing).

The choice of either categorical or continuous comparison must be based on an understanding of the spectral and spatial impact of a given land cover disturbance, or range of disturbances. If land cover attributes are expected to change category (e.g., forest to urban), then post-classification comparison is suitable, if not optimal. However, in many ecosystems, complete land cover conversion rarely occurs over short time intervals (i.e., 3–5 years). In effect, modification in condition and abundance is more common than conversion (Coppin and Bauer, 1994; Rogan et al., 2002). Therefore, this makes continuous comparison a more suitable choice of change detection approach for monitoring land cover modifications, especially over relatively short time intervals (i.e., 2–5 years). When longer time periods are considered (e.g., 5–10 years), then categorical comparison may be more suitable, as actual land cover conversion may be more likely to occur. In situations where digital data are not available for earlier time periods (e.g., pre-1972) categorical comparison is the only feasible approach (e.g., a land cover map of 1775 can be compared to a 1990 land cover map) (Petit and Lambin, 2002).

TABLE 8.3
A Summary of a Variety of and Cover Change Detection Methods and Their Advantages and Disadvantages for Operational Monitoring

Change Detection Method and Status*	Advantages	Disadvantages
Post Classification Comparison (PCC) Status = I	- Provides detailed 'from-to' information	- Only complete class changes are detected
	- Can be used with different sensors, with different spatial and spectral resolutions	- Heavily dependent on the accuracy of input maps and consistency between mapping methods
	- Permits the use of data with inter-date phenological differences	- Costs often prohibitive over large areas
	- Less sensitive to radiometric/geometric errors	
Composite Analysis (CA) Status = I	- Requires only a single classification	- Can require a large number of classes and a large calibration data set
	- Can be applied to both raw and enhanced data (e.g., vegetation indices, albedo)	- Separation of spectral changes from temporal changes can be difficult
	- Makes effective use of pre-change (reference) iamge	
Image Differencing (ID) Status = I	- Can be applied to both raw and enhanced data	- Requires optimization of change/no change threshold
	- Provides detailed information on 'within class change	- Difference image interpretation can be difficult
		- Cannot differentiate spectral differences resulting from different original spectral values
		- Highly sensitive to radiometric/geometric errors
		- Does not provide 'from-to' information
Image Ratioing (IR) Status = I	- Can be applied to both raw and enhanced data	- Highly sensitive to radiometric/geometric errors
	- Can mitigate atmospheric and sun-angle effects	- Threshold optimization can be difficult, as change is nonlinearly represented
Change Vector Analysis (CVA) Status = F	- Can be applied to both raw and enhanced data	- Highly sensitive to radiometric/geometric errors
	- Provides detailed 'from-to' information	- Change-direction outputs are difficult to interpret with large number of input bands
		- Change magnitude thresholding is subjective
Multitemporal Kauth Thomas (MKT) Status = I	- Results are intuitive	- Coefficients are sensor-dependent
	- Produces suites of change, no change, and noise features	- Highly sensitive to radiometric/geometric errors
	- Standardized coefficients permit application and comparison over time and space	
Multitemporal Spectral Mixture Analysis (MSMA)	- Results are intuitive (biophysically)	- Sensitive to choice of endmember type
	- Can be used to compare fraction estimates across different sensors and platforms	

TABLE 8.3 (*Continued*)
A Summary of a Variety of and Cover Change Detection Methods and Their Advantages and Disadvantages for Operational Monitoring

Change Detection Method and Status*	Advantages	Disadvantages
Principal Components Analysis (PCA) Status = I	- Can be applied to both single-date, composite multi-date, and compsite ID data	- Components can be difficult to interpret
	- Reduces multispectral data sets into features representing change, no change, and noise	- Threshold optimization can be difficult
	- In multitemporal analysis, standardized components can minimize atmospheric and sun angle differences	- Statistically-based, so limited in space and time
		- Sensitive to disproportionate amounts of variance in the imagery
Multivariate Alteration Detection (MAD) Status = E	- Reduces multispectral data sets into features representing change, no change, and noise	- Has not been widely used
	- Can be used to compare information from different sensors	
	- Insensitive to disproportionate amounts of variance in imagery	
Multitemporal Visualization Status = I	- Simple and intuitive	- Qualitative
	- Permits inspection of three dates of imagery as RGB	- Does not provide 'from-to' information
Knowledge-based approaches Status = F	- Automatic detection of change	- Complicated approach to develop
		- Have not been widely used
Cross-Correlation Analysis (CCA) Status = F	- Allows for direct updating of land-cover maps	- Has not been widely used

* Status of the method in an operational context for land-cover change monitoring: I = Implemented in operational context, F = Feasible in an operational context, and E = Experimental

8.4.2.2 Map-Updating Approaches

Another interesting trend in bitemporal change mapping is the use of novel map-updating approaches (Chen and Shi, 2020; Asokan and Anitha, 2019; Leichtle et al., 2017; Zhu, 2017). Post-classification comparison has been implemented in hundreds of land change case studies, but it is problematic in many land change monitoring scenarios (Stow et al., 1980). Over large areas land change mapping is challenging for some of the following reasons: (1) data issues such as cost, platform continuity, and availability of aerial photographs or in situ data that inhibit comprehensive spatial and temporal coverage, and (2) cloud cover, non-stationarity in landscape features, and phenological variability further limit the usability of available imagery. In combination, these challenges make the task of re-mapping an entire landscape for a second or even third iteration very expensive and possibly unachievable at an acceptable level of map accuracy (Rogan and Chen, 2004). Actual land change due to categorical conversions (e.g., forest to urban) or within-category modifications (e.g., timber harvest) usually occupies only a small portion of a pair of 34,000 km^2 Landsat images (e.g., less than 20%) (Rogan et al., 2003) such that independent re-mapping of a landscape for a new time period is not warranted as long as there are no drastic changes to a land monitoring protocol (e.g., new map legend, change to incompatible new data sources) (Islam et al., 2018; Zhu, 2017; Rogan and Chen, 2004).

There are two main methods of map updating present in the remote sensing literature (Wang et al., 2021; Zhang and Shi, 2020; Zhu, 2017): (1) human-interpreted delineation of new changes using multitemporal data and (2) digital change detection of multitemporal imagery to detect a specific type of disturbance, such as urban sprawl or forest damage. Feranec et al. (2007) implemented a human-interpretation method of change detection with visually interpreted aerial photography to update the Coordination of Information on the Environment (CORINE) 44 category land cover map for 1990 and 2000. The 2000 land cover map was created by visually and manually editing polygons of change in the original 1990 classification with overall accuracy above 85%. Other studies have used more automated methods of pre-dating and post-dating land cover maps to monitor forest change. Wulder et al. (2008) implemented a technique to post-date a 2000 land cover map to 2003 land cover conditions to detect forest clear-cuts using the near-infrared band from Landsat TM/ETM+, SPOT-4, and ASTER data. Forest clear-cuts were detected using an ordinal ranking method that assigns pixels a value based on its reflectance relative to all other pixels. Detected clear-cuts were integrated into the pre-existing 2000 EOSD eight-category land cover product. We expect that new innovative approaches to map updating will emerge in the next decade as remote sensing practitioners merge change mapping and resource inventory in a mutually beneficial process.

8.4.3 TEMPORAL TREND ANALYSIS—AUTOMATION AND BIG DATA

Over the last four decades, voluminous amounts of digital data have been gathered from an ever-growing number of satellites and sensors continuously monitoring the earth, atmosphere, and oceans. Fortunately, the massive increase in available data has coincided with a rise in computing power, and since the widespread popularization of online mapping platforms and user-generated geographic information, often linked to the release of Google Earth in 2005, a broader user-base for the "geoweb" has developed (Tewabe et al., 2020; Wang et al., 2020; Woodcock et al., 2020; Elwood, 2011). The most significant change in the practice of land cover change mapping and monitoring has come from this "big data" paradigm, also known as "data-intensive science" (Kelling et al., 2009).

8.4.3.1 Hypertemporal Remote Sensing Data in Trend Analysis

Trend, or temporal trajectory analysis, involves the application of data acquired on a large number of observation dates (i.e., hypertemporal) (inter- and intra-annual), traditionally using coarse spatial resolution, spectrally transformed imagery (e.g., NDVI, PAR, and LAI estimates derived from AVHRR and MODIS). This topic is reviewed thoroughly by Henebry and de Beurs (2013). Once assembled, temporal-spectral profiles can be useful for describing high-frequency land cover modifications over coarse spatial scales (Eastman et al., 2009). The study of land surface phenology has witnessed a large increase in remote sensing practitioners and applications as a method for studying the patterns of plant and animal growth cycles, due to the increase in freely available information/data sets. Phenological events are sensitive to climate variation such that phenology data provide timely baseline information for documenting trends in agriculture, irrigation, forest growth rates, and detecting the impacts of climate change on multiple scales Henebry and de Beurs (2013). The increased complexity that remote sensing practitioners face when working with hypertemporal data sets is now being ameliorated through new software functionality. For example, the Earth Trends Modeler is an integrated suite of tools within IDRISI software for the analysis of image time-series data and allows the user to perform and analyze trend analysis results in both graphic and cartographic format (http://www.clarklabs.org/).

Information from trend analysis can provide information on landscape or land surface phenological variability for finer spatial resolution studies so that change related to disturbances can

Land Cover Change Detection 323

be potentially separated from climate (temperature and precipitation) variability (Borak et al., 2000). High-temporal, coarse spatial resolution imagery has also been used effectively to document the prevailing trends in vegetation phenology over large areas to guide the acquisition of medium spatial resolution imagery (i.e., to reduce commission errors caused by uneven intra-annual and inter-annual green-up) (Rees et al., 2003). As such, changes inherently linked to seasonality can potentially be separated from other land cover changes (Coppin et al., 2002). However, spatial resolution is often a limiting factor in these studies, especially when examining subtle land cover changes (Rees et al., 2003).

8.4.3.2 Challenges of Trend Analysis

One of the most challenging aspects of trend analysis is that it requires a high level of image preprocessing to account for sensor and platform differences, sensor drift, etc. (Zhang et al., 2020; Liping et al., 2018; Sidhu et al., 2018; Coppin et al., 2004). Trend analysis can be performed using coarse-to-medium spatial resolution data, although coarse resolution data are more plentiful. Substantial preprocessing is required given the large volume of data and the need for a high level of geometric and radiometric consistency. While classification is not essential, the use of image transformations to reduce data volume is size is essential. Most large-area programs utilize categorical comparison approaches to detect and monitor land cover change. While this development is noteworthy, and expected to continue, the land change science community requires information on land cover modifications, which conversion-focused programs cannot efficiently or reliably provide. However, there is potential for improvement with increased data availability and accessibility, as well as growing experience with and understanding of sensors and imagery in large-area scenarios (Franklin, 2001; Rogan and Chen, 2004).

8.4.3.3 Medium-Resolution Data for Trend Analysis

A very promising new development is the advancement of data fusion, which involves the blending of multiple co-located images to produce a hybrid information product that minimizes the limitations of each contributing dataset (Woodcock et al., 2020). A typical fusion combination merges low temporal/high spatial resolution data with high temporal/low spatial resolution data methods to extend the temporal profile of Landsat data using daily or eight-day MODIS reflectance data (Gao et al., 2006).

Medium spatial resolution data sources are considered optimal to obtain sufficient thematic detail for large-area monitoring applications. Fortunately the last decade has witnessed growth in the availability of medium spatial resolution datasets, such as the Web-Enabled Landsat Data (WELD) program (Roy et al., 2010). Since January 2008, the USGS has been providing free terrain-corrected and radiometrically calibrated Landsat data via the Internet. The WELD system is being expanded to a global scale to provide monthly and annual Landsat 30 m information for any terrestrial non-Antarctic location for six three-year epochs spaced every five years from 1985 to 2010. The WELD products are developed specifically to provide consistent data that can be used to derive land cover as well as biophysical products for assessment of land surface dynamics (Roy et al., 2010).

8.4.4 COMPARISON OF SEVERAL AUTOMATED CHANGE DETECTION APPROACHES

In recent years, much attention has been focused on automating the detection of land cover change, specifically forest disturbance, across broad landscapes and using dense image time-series stacks. Many spectral disturbance indices (Woodcock et al., 2020; Zhang et al., 2020; Sidhu et al., 2018; Healey et al., 2005; Hais et al., 2009; Mildrexler et al., 2009) and software platforms (Asner et al., 2009; Hilker et al., 2009; Huang et al., 2010; Kennedy et al., 2010) have been created to monitor forest disturbance, each with their own relative strengths and weaknesses (Table 8.4).

TABLE 8.4
A Comparison of Seven Prominent Change Detection Algorithms According to Ease of Use, Computation Time, Data Type, and Functionality

Algorithms	Ease of Use	Computation Time	Data Type	Cost	Available to Use	Highlights deforestation	Highlights degradation	Source
DI	2	NA	L	Free	Y	Y	N	Healey et al., 2005
DI'	2	NA	L	Free	Y	Y	Y	Hais et al., 2009
CLASLite	1	1	l,s,a,m	Free	Y- with permission	Y	Y	Asner et al., 2009
VCT	2	1	L,S,IRS	Free	Y	Y	Y	Huang et al., 2010
Landtrendr	3	3	L	Free	Y- requires ENVI	Y	Y	Kennedy et al., 2010
MGDI	1	NA	M	Free	N	Y	N	Mildrexler et al., 2009
STAARCH	3	NA	L,M	Free	Y	Y	Y	Hilker et al., 2009

DI = Disturbance Index; DI' = Disturbance Index Prime; MGDI = MODIS Global Disturbance Index; CLASLite = Carnegie Landsat Analysis System Lite; VCT = Vegetation Change Tracker; LandTrendr = Landsat-based Detection of Trends in Disturbance and Recovery; STAARCH = Spatial Temporal Adaptive Algorithm for mapping Reflectance Change; NA = Not Available; L = Landsat 4 and 5 Thematic Mapper (TM), Landsat 7 Enhanced Thematic Mapper Plus (ETM+); S= Satellite pour l'Observation de la Terre 4 and 5 (SPOT); A = Advanced Spaceborne Thermal Emission and Reflection Radiometer (ASTER); Moderate Resolution Imaging Spectrometer (MODIS); IRS = Indian Remote Sensing satellite; ENVI = Exelis Visual Information Solutions

8.4.4.1 Disturbance Index

Healey et al. (2005) developed a novel combination of the Tasseled Cap features (brightness (B), greenness (G), and wetness (W)) to highlight forest disturbances over single and multi-date Landsat image time series, known as the Disturbance Index (DI). The DI is a linear combination of the B, G, and W features, where each feature is rescaled to one standard deviation above or below the mean forest value of the landscape under investigation, resulting in the equation:

$$DI = B_r - (G_r + W_r)$$

where r indicates the rescaled features. The DI index is most sensitive to discrete stand-replacing disturbances, which create a strong, stable, and relatively predictable spectral signal across space and time. Alternatively, the DI index is less robust in landscapes where rapid post-disturbance succession occurs, such that the disturbance signal is weakened by increased understory vegetation growth and heterogeneity.

8.4.4.2 Disturbance Index' (DI')

Hais et al. (2009) refined the DI index to account for gradual disturbances across landscapes and forest stands exhibiting rapid succession (i.e., increased greenness) in understory vegetation. The DI' equation is as follows:

$$DI' = W_r - B_r$$

TABLE 8.5
A Comparison of Seven Prominent Change Detection Algorithms According to the Degree of Automation with Respect to Atmospheric Correction, Cloud Masking, Image Calibration, and Mosaicking

Algorithms	Atmospheric Correction	Cloud Mask	Calibration	Mosaic Multi Image
DI	N	N	Y	N
DI'	N	N	Y	N
MGDI	N	N	Y	Y
CLASlite	Y-6S	Y	Y	N
VCT	Y-LEDAPS	Y	Y	N
LandTrendr	Y-Cos(t)	Y	Y	Y
STAARCH	N	Y	Y	N

DI = Disturbance Index; DI' = Disturbance Index Prime; MGDI = MODIS Global Disturbance Index; CLASLite = Carnegie Landsat Analysis System Lite; VCT = Vegetation Change Tracker; LandTrendr = Landsat-based Detection of Trends in Disturbance and Recovery; STAARCH = Spatial Temporal Adaptive Algorithm for mapping Reflectance Change; LEDAPS = Landsat Ecosystem Disturbance Adaptive Processing System; Cos(t) = Cosine of Theta

By removing the greenness band from the original DI equation, the DI' showed a heightened sensitivity to both discrete (e.g., clear-cut, windthrow, avalanche) and gradual disturbances (e.g., defoliation, insect mortality) across space and time when compared to the DI, G, B, W, and the Normalized Difference Wetness Index (NDWI).

8.4.4.2.1 MODIS Global Disturbance Index

The MODIS Global Disturbance Index (MGDI; Mildrexler et al., 2009) is an automated change detection algorithm that fuses the MODIS Reflectance product, Land Surface Temperature (LST), and MODIS Enhanced Vegetation Index (EVI) data to detect large-area forest disturbances at global, continental, and subcontinental scales. The MGDI uses annual maximum LST composites to detect large changes in land-surface energy and links those changes to the EVI signal, thus detecting discrete disturbances. Due to the scales at which the algorithm is optimized for, disturbances such as wildland fire events, hurricane damage, large-scale windthrow, clear-cuts, and land-clearing for agriculture will be the major landscape modifiers captured over the time series.

8.4.4.3 CLASlite

CLASlite (V 3.1) is a stand-alone, fully automated software package used to map forest cover, deforestation, and forest degradation over broad spatial extents and long time series by experts and non-experts alike (Asner et al., 2009). CLASlite boasts a one-hour processing time on a standard Windows PC for a 30 m spatial resolution image across 10,000 km². CLASlite enables users to input raw data from a variety of satellite platforms (Landsat 4, 5, 7, 8; ASTER; ALI; SPOT 4, 5; MODIS) where an automation procedure atmospherically corrects, cloud masks, and classifies images across multiple dates with little user input (see Asner et al., 2009 for more details). The CLASlite algorithm utilizes a spectral mixture procedure called AutoMCU (Automated Monte Carlo Unmixing) to classify forest/non-forested areas for one or multiple image dates. Although the spectral libraries used in this procedure are optimized for tropical forests (>300,000 spectral signatures), it has also been shown to classify temperate forests with great success (see the case study later in this chapter).

8.4.4.4 Vegetation Change Tracker

The Vegetation Change Tracker (VCT; Huang et al., 2010) is an automated algorithm used to delineate forest change across 12 or more Landsat time-series stacks with little to no user parameterization for closed or near-closed forest canopies. The VCT algorithm will automatically create initial masks (i.e., clouds, cloud shadows, water) and temporally normalize for all scenes, calculate forest features, temporally interpolate masked land areas, and create a composite output image of all locations that experienced a disturbance for each time step. Additionally, the VCT algorithm calculates multiple types of change magnitude measures and tracks post-disturbance vegetation processes (i.e., succession). The VCT disturbance mapping technique is ideal for discrete, land-clearing events but works poorly for non-stand-clearing events (e.g., thinning, selective logging, insect outbreak) (MohanRajan et al., 2024; Chen and Shi, 2020; Asokan and Anitha, 2019).

8.4.4.5 LandTrendr

The Landsat-based detection of Trends in Disturbance and Recovery (LandTrendr; Wang et al., 2021; Zhang and Shi, 2020; Leichtle et al., 2017) is an algorithm that enables the user to systematically analyze a dense Landsat time-series stack to produce robust short-term disturbance and long-term vegetation trend maps. Users are able to provide dense Landsat time-series stacks into the LandTrendr, which are atmospherically corrected (Cos(t) algorithm), masked (smoke, cloud, cloud shadow, water), and temporally segmented as a means to capture landscape disturbances. Output images and figures provide a wealth of information that quantifies landscape dynamics over the time-series stack, allowing for a much more detailed assessment than bitemporal change methods can provide.

8.4.4.6 Spatial Temporal Adaptive Algorithm for Mapping Reflectance Change

The Spatial Temporal Adaptive Algorithm for mapping Reflectance Change (STAARCH; Hilker et al., 2009) blends Landsat and MODIS data to enhance the temporal resolution of Landsat (16-day) to MODIS (eight-day). The STAARCH model employs Healey et al. (2005) DI to detect landscape changes, where the DI calculation is completely automated. To aid in heterogeneous landscapes, the STAARCH model uses the minimum standard deviation of forest spectral values to increase the sensitivity of the DI to spectral forest change (i.e., disturbance). Additionally, this algorithm is able to create synthetic Landsat images for a given study area/period for each available MODIS scene used. To note, this algorithm builds upon and improves the performance of the STARFM algorithm (Gao et al., 2006).

8.4.4.7 Summary and Comparison of Automated Change Methods

To summarize the aforementioned change detection indices and algorithms it is necessary to evaluate their purposes accordingly (Tables 8.4 and 8.5). For high spatial and temporal resolution rapid change detection, it would be most advantageous to employ the CLASlite or the VCT algorithm. To evaluate longer-term environmental landscape dynamics, where computational power and time are not limiting, the LandTrendr would be the most appropriate algorithm of choice. The two disturbance indices (DI and DI') would be most efficiently utilized under the conditions where forest change detection across time would benefit from manual preprocessing steps to accommodate multi-date disparities. Additionally, the MGDI would allow for a more sophisticated approximation of landscape disturbances across a very large area. Lastly, the STAARCH algorithm allows not only for a highly accurate downscaling of MODIS to Landsat pixel scale, but accommodates an automated DI calculation; therefore, this would be the algorithm of choice if large spatial extents combined with a need for high spatial and temporal resolution is necessary. It is imperative to assess each change detection algorithm based on their strengths, weakness, and best fit for the research objectives and scales (both spatially and temporally) (Chen and Shi, 2020; Zhang and Shi, 2020; Sidhu et al., 2018).

8.5 ACCURACY ASSESSMENT—BEYOND STATISTICS

"It is extremely difficult to implement a consistent, comprehensive, quantitative accuracy assessment for large-area change maps" (Wang et al., 2021; Zhang and Shi, 2020; Sidhu et al., 2018; Loveland et al., 2002, p. 1094). Following the detection and classification/mapping of land cover change, it is preferable that the accuracy of the change maps be assessed. This topic is reviewed in detail by Olofsson et al. (2014). Accuracy assessment serves as a guide to the map quality and to reveal uncertainty and its likely implications to the end user. Accuracy assessment for change detection studies is more challenging than for single-date studies (Congalton, 1991; Khorram et al., 1999). This is because change classes usually represent a very small portion of the change image or thematic map. Additionally, when performing retrospective change detection, acquiring an adequate database of historical reference materials, such as historic aerial photographs can be very difficult, if not impossible (Biging et al., 1998). The provision of archived imagery by Google Earth provides an important component to addressing the more vexing concerns in land change accuracy assessment (Dorais and Cardille, 2011). Unfortunately, the remote sensing community has tended to focus exclusively on the calculation of map accuracy/validation statistics to demonstrate the validity of a method or the worth of a land cover map (Rogan and Chen, 2004). While having statistical information about map accuracy is very useful, it ignores many other facets of a change map that are vital to making sure that true change has been captured (Ghimire et al., 2010). These important facets include estimating the potential outcome of the mapping exercise, estimating the areal dominance of categories, and determining the desired shape, location, association, and configuration of mapped categories.

Based on ten years of experience mapping forest, wetland, and urban change in Massachusetts, the Massachusetts Forest Monitoring program (MaFoMP) (Rogan et al., 2010) developed the following list of eight steps to pursue when mapping change over a 40-year period using all available cloud-free Landsat MSS, TM, and ETM+ imagery (Woodcock et al., 2020; Liping et al., 2018; Sidhu et al., 2018; Leichtle et al., 2017; Zhu, 2017):

1. Determine optimal data needs, image processing steps based on scene model (Strahler at al., 1986; Phinn et al., 2000), and desired map legend (e.g., Anderson et al., 1976).
2. Determine optimal response design, support size, and sampling design (identify the trade-offs between support size and cost-logistical feasibility) (see Olofsson et al., 2014 for more details).
3. Qualitatively estimate the success of the mapping project based on previous experience and literature (e.g., expected outcomes, "last time we achieved 80% overall accuracy").
4. Estimate expected category area/dominance using maps from other sources or your knowledge of the study area (e.g., categories A and B should comprise over 70% of the study area, whereas categories C and D should comprise less than 2% of the study area).
5. Estimate expected category shape, location, association, and configuration (e.g., categories F and G will fall only on the coast in long linear strips, associated with ocean water).
6. Quantitatively estimate overall accuracy and per-class accuracy using validation data (should be appropriate support and sampling design). For a general purpose map, all categories should be ranked equal in importance (thus a balance must be struck between omission and commission errors) such that per-class accuracy should be equal. For a phenomenon-specific map (e.g., forest loss) certain categories should be ranked higher in importance than others such that omission errors should be avoided at all allowable costs, whereas certain levels of commission error are permissible (e.g., it is more important not to miss a rare category than it is to falsely map it). Keep in mind that resubstitution accuracy (i.e., using calibration data as validation data) can be a reasonable first-cut measure of your potential mapping success (Rogan et al., 2003).

7. Engage in post-classification editing/filtering to achieve a product that "looks right." This may make you return to your original training data and re-do the work, especially in heterogeneous locations.
8. Evaluate the map such that the end user can employ it wisely for a task that you may not have thought of (e.g., let the map user know your decisions/activities for steps 1–8).

8.6 MASSACHUSETTS CASE STUDY—CLASLITE

This case study explores the application of CLASlite (Asner et al., 2009) mapping and disturbance detection software to map forest and forest change in Massachusetts. CLASlite can operate with a variety of satellite data types, including Landsat, SPOT, ASTER, ALI, and MODIS. Landsat TM, ETM+, and OLI-TIRS data were acquired for nine individual years spanning nearly three decades (Table 8.6) across eastern Massachusetts (Figure 8.2). Four Landsat tiles were downloaded for each

TABLE 8.6
Detailed Description of Scene Date, Spatial Location, and Sensor Type Used

Acquisition Date	Landsat Scene		Landsat Sensor
	Path	Row	
August 8, 1985	12	30	TM
August 8, 1985	12	31	TM
September 1, 1985	13	30	TM
September 1, 1985	13	31	TM
August 15, 1993	12	30	TM
August 15, 1993	12	31	TM
July 5, 1993	13	30	TM
July 5, 1993	13	31	TM
August 21, 1995	12	30	TM
August 21, 1995	12	31	TM
July 15, 1999	13	30	TM
July 15, 1999	13	31	TM
July 31, 1999	12	30	ETM+
July 31, 1999	12	31	ETM+
July 23, 2002	12	30	TM
July 23, 2002	12	31	TM
July 10, 2009	12	30	TM
July 10, 2009	12	31	TM
August 18, 2009	13	30	TM
August 18, 2009	13	31	TM
August 30, 2010	12	30	TM
August 30, 2010	12	31	TM
September 6, 2010	13	30	TM
September 6, 2010	13	31	TM
July 17, 2011	12	30	TM
July 17, 2011	12	31	TM
June 16, 2011	13	30	TM
July 7, 2011	13	31	TM
August 6, 2013	12	30	OLI TIRS
August 6, 2013	12	31	OLI TIRS
September 30, 2013	13	30	OLI TIRS
September 30, 2013	13	31	OLI TIRS

Land Cover Change Detection

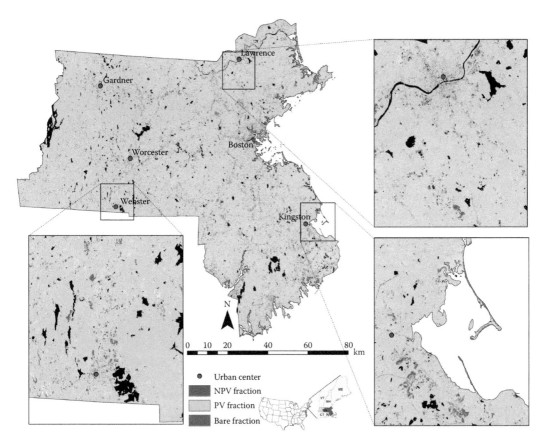

FIGURE 8.2 Study area in Central Massachusetts fraction composite image produced by CLASlite's AutoMCU with examples of rural (Webster), urban (Lawrence), and coastal (Kingston) landscapes.

respective year and georeferenced using image-to-image registration to an existing orthorectified Landsat image (http://www.landsat.org). All images were registered to an average root mean square (RMS) error of less than one pixel.

Following the manual co-registration procedure, each scene was processed for each of the nine years using CLASlite (Version 3.1; Asner et al., 2009). CLASlite is an automated change detection and mapping software optimized for tropical forests but was used here to test the feasibility across spatially heterogeneous temperate forested landscapes such as Massachusetts. CLASlite requires limited user interaction in the four main processing steps (image calibration, fraction image creation, forest cover mapping, and deforestation and disturbance delineation), which is optimal for rapid forest cover mapping spanning multiple dates.

First, all scenes were individually imported into CLASlite by specifying the required ancillary and metadata information. During image calibration, CLASlite uses 6S radiative transfer code to atmospherically correct each scene and convert the output images from radiance values to reflectance. Second, CLASlite employs a Monte-Carlo (AutoMCU; Asner and Heidebrecht, 2002) spectral decomposition algorithm to partition each scene into proportional fractional cover types of bare ground (B), photosynthetic vegetation (PV), and non-photosynthetic vegetation (NPV) for every pixel (Figure 8.2). During this stage, the user is able to specify the degree to which clouds and water bodies are masked out of the resulting image. Third, CLASlite delineates forest versus non-forest pixels via a user-defined threshold based on proportional PV against B and NPV constituents (Figure 8.3). Finally, CLASlite evaluates the fractional and reflectance images to produce disturbance and degradation classifications for each time step. As defined by Asner et al. (2009), deforestation refers to a diffuse thinning of the forest canopy, while degradation quantifies any spatial or temporal persistence of forest disturbance. In this case study CLASlite maps the location of deforestation and forest disturbance in eight eras: 1985–1993, 1993–1995, 1995–1999, 1999–2002, 2002–2009, 2009–2010, 2010–2011, and 2011–2013 (Figure 8.4).

CLASlite forest cover maps for each time period were validated using two independent approaches. The first method employed the 30 m resolution Massachusetts Forest Monitoring program (MaFoMP) land cover maps (Rogan et al., 2010) for the years 1984, 1990, 2000, and 2009 to produce a crosstabulation matrix of quantity agreement and allocation agreement with the associated CLASlite forest cover images. This assessment determined the degree to which pixels of similar land cover type (forest or non-forest) are in agreement with the 30 m MaFoMP maps (MaFoMP, 2011; Table 8.7). Errors of omission and commission were reported for each year as a percentage of all pixels in spatial and quantity agreement or disagreement to the MaFoMP map (Table 8.8). Kappa values and the Cramer's V statistic were reported for each year (Table 8.9).

Additionally, CLASlite change maps were validated using a randomly sampled collection of 200 classified pixels to compare the CLASlite delineated pixel values to high spatial resolution Google Earth imagery (Dorais and Cardille, 2011; Google Inc., 2014). The second assessment allowed for an independent evaluation of quantity and allocation pixel agreement to determine the degree to which the CLASlite outputs are correctly classifying forest versus non-forest land cover types. We used available Google Earth imagery that was closest in temporal proximity to the CLASlite-generated forest cover maps. The original fine spatial resolution data were acquired from DigitalGlobe (i.e., WorldView-2 data). Additionally, the deforestation caused by the June 2011 EF3 tornado was validated via 50 randomly sampled points using a 2011 Google Earth image captured post-tornado.

8.6.1 CLASlite Results

8.6.1.1 Forest Cover Mapping

Forest cover maps produced through an iterative thresholding procedure of the AutoMCU fraction images resulted in a 508 km^2 net reduction in forest from 1985 to 2009 (Figure 8.3). Comparatively,

TABLE 8.7
Crosstabulation Assessment Showing Pixel Agreement for Like Years in Terms of Percent of Total Pixel

	Year	Class	MAFOMP							
			1984		1990		2000		2009	
			Forest	Non-Forest	Forest	Non-Forest	Forest	Non-Forest	Forest	Non-Forest
CLASlite	1985	Forest	0.545	0.078	-	-	-	-	-	-
		Non-Forest	0.121	0.256	-	-	-	-	-	-
	1993	Forest	-	-	0.520	0.090	-	-	-	-
		Non-Forest	-	-	0.085	0.305	-	-	-	-
	2002	Forest	-	-	-	-	0.510	0.118	-	-
		Non-Forest	-	-	-	-	0.072	0.299	-	-
	2009	Forest	-	-	-	-	-	-	0.497	0.078
		Non-Forest	-	-	-	-	-	-	0.128	0.297

TABLE 8.8
Kappa (a) and Cramer's V (b) Statistics Showing the Relative Pixel Agreement Accuracy of the CLASlite Forest Cover Classification to MaFoMP Imagery across Four Time Steps

Kappa

	Year	\ MAFOMP 1984	1990	2000	2009
CLASlite	1985	0.8183	-	-	-
	1993	-	0.8275	-	-
	2002	-	-	0.8179	-
	2009	-	-	-	0.81637

Cramer's V

	Year	\ MAFOMP 1984	1990	2000	2009
CLASlite	1985	0.7839	-	-	-
	1993	-	0.7864	-	-
	2002	-	-	0.7966	-
	2009	-	-	-	0.781

TABLE 8.9
Random Sample Pixel Percent Agreement of Forest Cover Types of CLASlite Classification against High-Resolution Google Earth Imagery. Note, with Increasing Time There Is a Direct Relationship to Decreasing Forest and Increasing Non-forest Agreement

			Google Earth™								
	Year		1995		2003		2008		2010		2013	
		Class	Forest	Non-Forest	Forest	Non-Forest	Forest	Non-Forest	Forest	Non-Forest	Forest	Non-Forest
CLASlite	1995	Forest	0.638	0.064	-	-	-	-	-	-	-	-
		Non-Forest	0.037	0.25	-	-	-	-	-	-	-	-
	2002	Forest	-	-	0.613	0.032	-	-	-	-	-	-
		Non-Forest	-	-	0.048	0.296	-	-	-	-	-	-
	2009	Forest	-	-	-	-	0.608	0.322	-	-	-	-
		Non-Forest	-	-	-	-	0.032	0.317	-	-	-	-
	2010	Forest	-	-	-	-	-	-	0.585	0.032	-	-
		Non-Forest	-	-	-	-	-	-	0.037	0.335	-	-
	2013	Forest	-	-	-	-	-	-	-	-	0.5945	0.0594
		Non-Forest	-	-	-	-	-	-	-	-	0.0324	0.308

the MaFoMP maps generated a 566 km² reduction in forest from 1984 to 2009, demonstrating that CLASlite was within a 10% range of similar transitions over a similar time period. The CLASlite-generated forest cover type maps resulted in an 81% Kappa agreement with the MaFoMP maps and an average 85% accuracy when validated with randomly sampled Google Earth imagery.

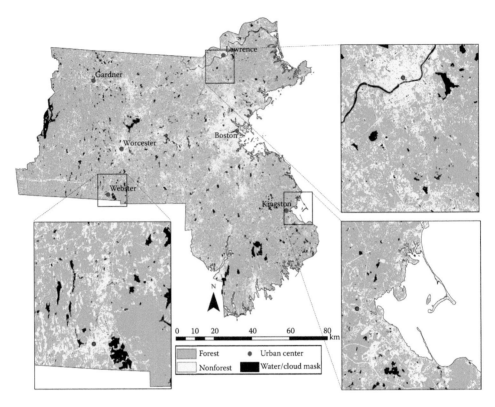

FIGURE 8.3 Statewide automated forest cover image for 2013, with examples of rural (Webster), urban (Lawrence), and coastal (Kingston) landscapes. CLASlite-produced cloud/water mask is delineated in black.

8.6.2 Deforestation and Disturbance Mapping

Between 1985 and 2013, the study area exhibited a net forest change of 2,301 km², equating to 19.5% of the study area (Table 8.10). The largest total amount of forest change was observed during 1985–1993, followed by 1995–1999 and 2002–2009, respectively (Figure 8.6), representing 13.5% of the total area that was converted from forest to non-forest (Table 8.10). A visual assessment of the deforestation and disturbance results indicated that forest change was overestimated due to subtle variation in forest phenology, though CLASlite was able to detect most major land-clearing disturbances across one to many years.

8.6.3 Gardner, Massachusetts, Forest Change

The case study located in Gardner, Massachusetts (Figure 8.4), illustrated the rural to urban land conversion, a common trend throughout the study area. Forest cover was reduced by 15.2% from 1985 (105 km²) to 2013 (84 km²). Across all years, a systematic and continuous shift from forest to non-forest cover types is revealed (Figure 8.4). CLASlite forest cover maps for 1985 report 105 km², compared to the MaFoMP maps of 106 km². Concomitantly, the 2009 CLASlite output indicated that 87 km² of forested area, while 96 km² remain according to MaFoMP. The area differences between the 2009 classifications were less than 4% of the total case study area of Gardner. Similar to the eastern Massachusetts deforestation and disturbance mapping, the amount of area affected by forest change in Gardner was overestimated. The total forest change from 1985 to 2013 was reported as being 33 km² (23%), where the greatest era of change was 1985–1993, followed by 1995–1999 and 2002–2009.

TABLE 8.10
Change Statistics per Era. Forest Change is the Sum of Disturbance and Deforestation

Era	Deforestation Total	Deforestation Percent	Disturbance Total	Disturbance Percent	Forest Change Total
1985–1993	565.37	4.81%	318.21	2.71%	883.58
1993–1995	57.1	0.49%	29.42	0.25%	86.52
1995–1999	250.76	2.13%	168.48	1.43%	419.24
1999–2002	105.49	0.90%	40.57	0.35%	146.05
2002–2009	215.35	1.83%	119.56	1.02%	334.91
2009–2010	81.14	0.69%	65.39	0.56%	146.54
2010–2011	82.88	0.71%	82.52	0.70%	165.4
2011–2013	44.97	0.38%	74.01	0.63%	118.97
Total	1403.06	11.94%	898.15	7.65%	2301.21

8.6.4 2011 Tornado Disturbance

On June 1, 2011, a 37-km long and 0.8-km wide tornado track touch-downed across south-central Massachusetts (Figure 8.5). Using the 2010–2011 CLASlite deforestation output we produced a detailed rendition of the tornado disturbed areas, encompassing 20.3 km^2 over the 60-km track (Figure 8.5). Two years post-tornado disturbance, the 2013 forest cover image reported 4.8 km^2 of forest succession along the disturbance edges, while 15.2 km^2 was still in a disturbed state. Based on 50 randomly sampled points the agreement was 93% across the tornado track.

8.7 KNOWLEDGE GAPS AND FUTURE DIRECTIONS

A remote sensing renaissance has begun. Not since the launch of Earth Resources Technology Satellite 1 in 1972 has the remote sensing community witnessed a more empowering era. Since the mid-1990s, most of the information bottlenecks to operational-style remote sensing research and application have begun to be opened wide for effective and sustainable earth observation science. The MODIS and Landsat science teams have tenaciously pushed for free, accurate data and information products that can be accessed by the rapidly growing global user community. At the same time, high spatial resolution data are available globally from a variety of private companies, most notably (for view only) the Google Earth corporation, at 1–4 m. Importantly, the fields of Landscape Ecology and Land Change Science have claimed remotely sensed data as an invaluable component of their respective scientific practice. International charters such as the UN SPIDER initiative rely completely on earth observation data to draw attention to natural and humanitarian crises. As the content of this chapter highlights, the increased availability of coarse, medium, and high spatial resolution data, and the surge in efficient automated methods place remote sensing science in a better place than it has ever been in 40 years. In the next ten years remote sensing practitioners can expect to see a multiplier effect with regard to remote sensing applications, as data, methods, and continued advocacy accumulate and expand to new fields and new problems. The following list highlights the current knowledge gaps and future directions for the remote sensing land change community (MohanRajan et al., 2024; Lv et al., 2023; Mashala et al., 2023; Parelius, 2023; Lv et al., 2022; Wang et al., 2021; Chen and Shi, 2020; Tewabe et al., 2020; Wang et al., 2020; Woodcock et al., 2020; Zhang et al., 2020; Zhang and Shi, 2020; Asokan and Anitha, 2019; Islam et al., 2018; Liping et al., 2018; Sidhu et al., 2018; Leichtle et al., 2017; Zhu, 2017).

1. Ironically, as more and more data become available, more data are needed. Referring to the Landsat program, there will be increasing demand for Landsat MSS data and also TM data that have not yet been catalogued. The collection and processing of these data from

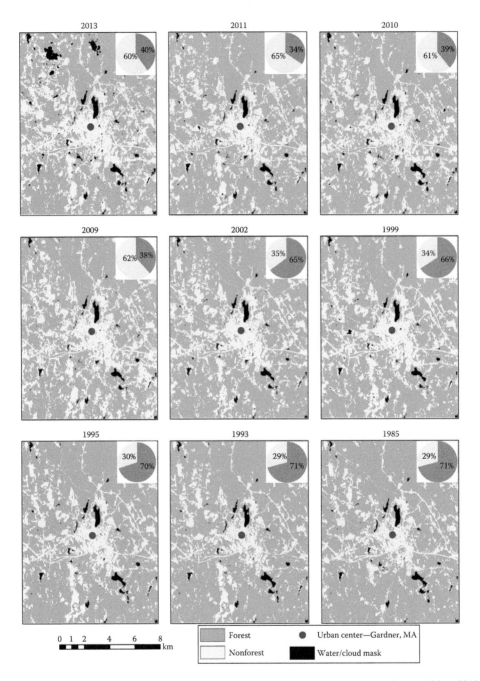

FIGURE 8.4 Forest cover temporal change sequence of Gardner, Massachusetts, from 1985 to 2013. Note, each forest cover scene has the percent proportion of pixels for forest and non-forest land cover classes.

various agencies throughout the world will greatly extend the reach of the Landsat program, especially to developing countries—the very locations where land change scientists focus their research. Additionally, the cost of high spatial resolution data is problematic; 1–4 m data are indispensable for locations where in situ data are unavailable, but these data can currently only be purchased by governments or government-affiliated research initiatives.

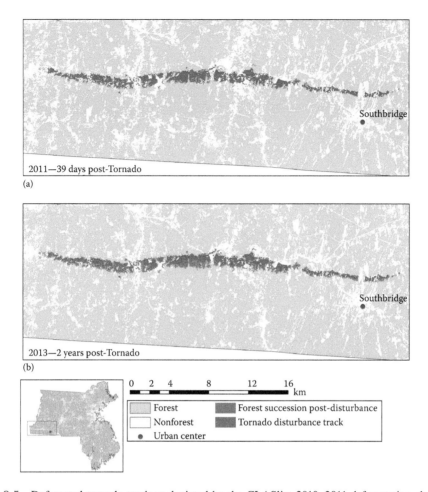

FIGURE 8.5 Deforested tornado track as depicted by the CLASlite 2010–2011 deforestation class output (a). Two years post-disturbance (b), note that successional infill (blue) has dominated the outer edges of the tornado track, while the interior of the tornado track (red) is still in a deforested state.

2. Given the importance placed currently on land cover modifications by the land change science community, it is important to distinguish their occurrence from land cover conversions. This is a difficult task because both types of change can result in similar magnitudes of reflectance in a change detection scenario. New methods are needed to ameliorate this problem, especially in developing countries where operational data availability can be scarce.
3. The remote sensing change detection community has laid a strong framework on the back of optical remote sensing imagery. While this paradigm is highly rewarding, optical data are limited in a variety of situations, especially concerning mapping in cloud-prone and data-poor locations. The next decade should hopefully see an expansion in the availability use of large-area radar and LiDAR data collections such that landscape monitoring will be as complete in Cameroon as it is in the United States.
4. All land cover change detection and monitoring relies on the availability of accurate land cover/use information for every location where remotely sensed data are captured. Unfortunately, the process of conducting change detection for a given location is hampered by the paucity of reliable ground reference, wildlife habitat, agricultural land use, and

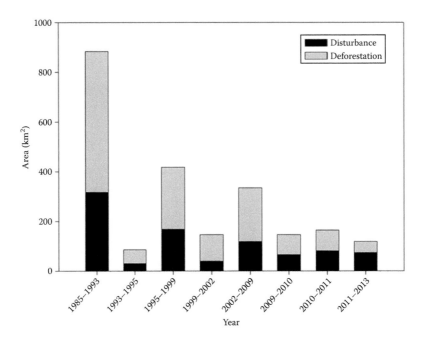

FIGURE 8.6 Area deforested in km² per time step across eastern Massachusetts.

ecological disturbance information. In the next decade, it is hoped that this knowledge gap will be at least partially filled through continued land cover/use mapping efforts, as well as map data sharing.

8.8 ACKNOWLEDGMENTS

The authors wish to thank David Wilkie and Robert Rose (Wildlife Conservation Society), who inspired the ideas that laid the basis for this chapter. Thanks, are also extended to Luisa Young (Clark University), Doug Stow (San Diego State University), and Janet Franklin (Arizona State University) for their contributions to this work.

REFERENCES

Anderson, J.R., E.E. Hardy, J.T. Roach, and R.E Witmer, 1976, A land use and land-cover classification system for use with remote sensor data. *U.S. Geological Survey Professional Paper 964*, Washington, DC.

Asner, G.P., and K.B. Heidebrecht, 2002, Spectral unmixing of vegetation, soil and dry carbon cover in arid regions: Comparing multispectral and hyperspectral observations. *International Journal of Remote Sensing*, 23, 3939–3958.

Asner, G.P., D.E. Knapp, A. Balaji, and G. Paez-Acosta, 2009, Automated mapping of tropical deforestation and forest degradation: CLASlite. *Journal of Applied Remote Sensing*, 3, 033543–033543–24.

Asokan, A., and J. Anitha, 2019, Change detection techniques for remote sensing applications: A survey. *Earth Science Informatics*, 12, 143–160. https://doi.org/10.1007/s12145-019-00380-5

Biging, G.S., D.R. Colby, and R.G. Congalton, 1998, Sampling systems for change detection accuracy assessment. In *Remote Sensing Change Detection: Environmental Monitoring Methods and Applications* (R.S. Lunetta and C.D. Elvidge, Eds.), Ann Arbor Press, Ann Arbor, MI, pp. 281–308.

Chen, H., and Z. Shi, 2020, A spatial-temporal attention-based method and a new dataset for remote sensing image change detection. *Remote Sensing*, 12(10), 1662. https://doi.org/10.3390/rs12101662

Civco, D.L., J.D. Hurd, E.H. Wilson, M. Song, and Z. Zhang, 2002, A comparison of land use and land-cover change detection methods. In *Proceedings of ASPRS-ACSM Annual Conference and FIG XXII Congress*, April 22–26, Washington, DC.

Congalton, R.G., 1991, A review of assessing the accuracy of classifications of remotely sensed data, *Remote Sensing of Environment*, 37, 35–46.

Coppin, P.R., and M.E. Bauer, 1994, Processing of multitemporal Landsat TM imagery to optimize extraction of forest cover change features. *IEEE Transactions on Geoscience and Remote Sensing*, 32, 918–927.

Coppin, P.R., and M.E. Bauer, 1996, Digital change detection in forest ecosystems with remote sensing imagery. *Remote Sensing Reviews*, 13, 207–234.

Coppin, P.R., I. Jonckheere, B. Muys, and E. Lambin, 2002, Digital change detection methods in natural ecosystem monitoring. In *Proceedings of the First Annual Workshop on the Analysis of Multi-temporal Remote Sensing Images* (L. Bruzzone and P. Smits, Eds.), World Scientific Publishing Co., Pte. Ltd., Hackensack, NJ, 440 p.

Coppin, P.R., I. Jonckheere, K. Nackaerts, B. Muys, and E. Lambin, 2004, Digital change detection methods in ecosystem monitoring: A review. *International Journal of Remote Sensing*, 25(9), 1565–1596.

Coppin, P.R., K. Nackaerts, L. Queen, and K. Brewer, 2001, Operational monitoring of green biomass change for forest management. *Photogrammetric Engineering and Remote Sensing*, 67(5), 603–611.

Crews-Meyer, K.A., 2002, Characterizing landscape dynamism using paneled-pattern metrics. *Photogrammetric Engineering and Remote Sensing*, 68(10), 1031–1040.

DeFries, R., and J. Townshend, 1999, Global land-cover characterization from satellite data: From research to operational implementation? *Global Ecology and Biogeography Letters*, 8(5), 367–379.

Dorais, A., and J. Cardille, 2011, Strategies for incorporating high-resolution Google Earth databases to guide and validate classifications: Understanding deforestation in Borneo. *Remote Sensing*, 3(6), 19.

Eastman, J.R., F. Sangermano, B. Ghimire, H. Zhu, H. Chen, N. Neeti, Y. Cai, E.A. Machado, and S.C. Crema, 2009, Seasonal trend analysis of image time series. *International Journal of Remote Sensing*, 30, 2721–2726.

Eidenshink, J., B. Schwind, K. Brewer, Z.-L. Zhu, B. Quayle, and S. Howard, 2007, A project for monitoring trends in burn severity. *Fire Ecology*, 3, 3–21.

Ekstrand, S.P., 1990, Detection of moderate damage on Norway Spruce using Landsat TM and digital stand data. *IEEE Transactions on Geoscience and Remote Sensing*, 28(4), 685–692.

Ekstrand, S.P., 1994, Assessment of forest damage with Landsat TM: Correction for varying forest stand characteristics. *Remote Sensing Environment*, 47, 291–302.

Elwood, S., 2011, Geographic information science: Visualization, visual methods, and the geoweb. *Progress in Human Geography*, 35(3), 401–408.

Farrow, A., and M. Winograd, 2001, Land use modeling at the regional scale: An input to rural sustainability indicators for Central America. In *Modeling Land Use/Cover Change from Local to Regional Scales* (A. Veldkamp and E. Lambin, Eds.). *Agricultural Ecosystems Environment*, 85, 249–268.

Feranec, J., G. Hazeu, S. Christensen, and G. Jaffrain, 2007, Corine land cover change detection in Europe (case studies of The Netherlands and Slovakia). *Land Use Policy*, 24(1), 234–247.

Franklin, S.E., 2001, *Remote Sensing for Sustainable Forest Management*, Lewis Publishers, Boca Raton, FL, 407 p.

Franklin, S.E., R.T. Gillespie, B.D. Titus, and D.B. Pike, 1994, Aerial and satellite sensor detection of Kalmia agustifolia at forest regeneration sites in central Newfoundland. *International Journal of Remote Sensing*, 15(13), 2553–2557.

Franklin, S.E., and M.A. Wulder, 2002, Remote sensing methods in medium spatial resolution satellite data land-cover classification of large areas. *Progress in Physical Geography*, 26(2), 173–205.

Frey, K.E., and L.C. Smith, 2007, How well do we know northern land cover? Comparison of four global vegetation and wetland products with a new ground-truth database for West Siberia. *Global Biogeochemical Cycles*, 21, GB1016. http://doi.org/10.1029/2006GB002706.

Friedl, M.A., D. Sulla-Menashe, B. Tan, A. Schneider, N. Ramankutty, A. Sibley, and X. Huang, 2010, MODIS collection 5 global land cover: Algorithm Refinements and characterization of datasets. *Remote Sensing of Environment*, 114, 168–182.

Ghimire, B., J. Rogan, and J. Miller, 2010, Contextual land-cover classification: Incorporating spatial dependence in land-cover classification models using random forests and the Getis statistic. *Remote Sensing Letters*: 45–54.

Glackin, D.L., 1998, International space-based remote sensing overview 1980–2007. *Canadian Journal of Remote Sensing*, 24, 307–314.

Gong, P., and B. Xu, 2003, Multi-spectral and multi-temporal image processing approaches: Part 2. Change detection. In *Methods and Applications for Remote Sensing of Forests Concepts and Case Studies* (M. Wulder and S.E. Franklin, Eds.), Kluwer Academic Publishers, Dordrecht, pp. 301–333.

Google Inc., 2014, *Google Earth™ 7.1.2.2041*. Available at: https://www.google.com/earth/

Graetz, R.D., 1990, Remote sensing of terrestrial ecosystem structure: An ecologist's pragmatic view. In *Remote Sensing of Biosphere Functioning* (R.J. Hobbs and H.A. Mooney, Eds.), Springer, New York, pp. 5–30.

Hais, M., M. Jonášová, J. Langhammer, and T. Kučera, 2009, Comparison of two types of forest disturbance using multitemporal Landsat TM/ETM+ imagery and field vegetation data. *Remote Sensing of Environment*, 113(4), 835–845.

Hansen, M.C., P.V. Potapov, R. Moore, M. Hancher, S.A. Turubanova, A. Tyukavina, D. Thau, S.V. Stehman, S.J. Goetz, T.R. Loveland, A. Kommareddy, A. Egorov, L. Chini, C.O. Justice, and J.R.G. Townshend, 2013, High-resolution global maps of 21st-century forest cover change. *Science*, 342(6160), 850–853.

Hayes, D.J., and S.A. Sader, 2001, Comparison of change-detection techniques for monitoring tropical forest clearing and vegetation regrowth in a time series. *Photogrammetric Engineering and Remote Sensing*, 67(9), 1067–1075.

Healey, S.P, W.B. Cohen, Y. Zhiqiang, and O.N. Krankina, 2005, Comparison of Tasseled Cap-based Landsat data structures for use in forest disturbance detection. *Remote Sensing of Environment*, 97(3), 301–310.

Henebry, G.M., and K.M. de Beurs, 2013, Remote sensing of land surface phenology: A prospectus. In *Phenology: An Integrative Environmental Science* (M.D. Schwartz, Ed.), 2nd ed., Springer, Heidelberg, Germany, Chapter 21, pp. 385–411.

Hilker, T., M.A. Wulder, N.C. Coops, J. Linke, G. McDermid, J.G. Masek, F. Gao, and J.C. White, 2009, A new data fusion model for high spatial- and temporal-resolution mapping of forest disturbance based on Landsat and MODIS. *Remote Sensing of Environment*, 113, 1613–1627.

Hobbs, R.J., 1990, Remote sensing of spatial and temporal dynamics of vegetation. In *Remote Sensing of Biosphere Functioning* (R.J. Hobbs and H.A. Mooney, Eds.), Ecological. Studies. 79, Springer-Verlag New York Inc., New York, pp. 203–222.

Huang, C., S.N. Goward, J.G. Masek, N. Thomas, Z. Zhu, and J.E. Vogelmann, 2010, An automated approach for reconstructing recent forest disturbance history using dense Landsat time series stacks. *Remote Sensing of Environment*, 114(1), 183–198.

Islam, K., M. Jashimuddin, B. Nath, and T.K. Nath, 2018, Land use classification and change detection by using multi-temporal remotely sensed imagery: The case of Chunati wildlife sanctuary, Bangladesh. *The Egyptian Journal of Remote Sensing and Space Science*, 21(1), 37–47. ISSN 1110-9823. https://doi.org/10.1016/j.ejrs.2016.12.005. https://www.sciencedirect.com/science/article/pii/S1110982316301594

Johnson, R.D., and E.S. Kasischke, 1998, Automatic detection and classification of land-cover characteristics using change vector analysis. *International Journal of Remote Sensing*, 19, 411–426.

Joria, P.E., and S.C. Ahearn, 1991, A comparison of the SPOT and Landsat Thematic Mapper Satellite Systems for detecting Gypsy Moth defoliation in Michigan. *Photogrammetric Engineering and Remote Sensing*, 57(12), 1605–1612.

Kelling, S., W.M. Hochachka, D. Fink, M. Riedewald, R. Caruana, G. Ballard, and G. Hooker, 2009, Data-intensive science: A new paradigm for biodiversity studies. *BioScience*, 59(7), 613–620.

Kennedy, R.E., S. Andrefouet, W.B. Cohen, C. Gomez, P. Griffiths, M. Hais, S.P. Healey, E.H. Helmer, P. Hostert, M.B. Lyons, G.W. Meigs, D. Pflugmacher, S.R. Phinn, S.L. Powell, P. Scarth, S. Susmita, T.A. Schroeder, A. Schneider, R. Sonnenschein, J.E. Vogelmann, M.A. Wulder, and Z. Zhu, 2014, Bringing an ecological view of change to Landsat-based remote sensing. *Frontiers in Ecology and the Environment*, 12(6), 339–346.

Kennedy, R.E., Z. Yang, and W.B. Cohen, 2010, Detecting trends in forest disturbance and recovery using yearly Landsat time series: LandTrendr—Temporal segmentation algorithms. *Remote Sensing of Environment*, 114(12), 2897–2910.

Khorram, S., G. Biging, N. Chrisman, D. Colby, R. Congalton, J. Dobson, R. Ferguson, M. Goodchild, J.R. Jensen, and T. Mace, 1999, *Accuracy Assessment of Land-cover Change Detection*. ASPRS Monograph Series, American Society for Photogrammetry and Remote Sensing, Bethesda, MD.

Lambin, E.F., X. Baulies, N. Bockstael, G. Fischer, T. Krug, R. Leemans, E.F. Moran, R.R. Rindfuss, Y. Sato, D.L. Skole, B.L. Turner, II, and C. Vogel, 1999, *Land-Use and land-Cover Change (LUCC) Implementation Strategy*. IGBP Report 48, IHDP Report 10. https://digital.library.unt.edu/ark:/67531/metadc12005/

Lambin, E.F., and A.H. Strahler, 1994a, Change vector analysis in multi-temporal space a tool to detect and categorize land-cover change processes using high temporal resolution satellite data. *Remote Sensing of Environment*, 48, 231–244.

Lambin, E.F., and A.H. Strahler, 1994b, Indicators of land-cover change for change-vector analysis in multi-temporal space at coarse spatial scales. *International Journal of Remote Sensing*, 15(10), 2099–2119.

Lambin, E.F., B.L. Turner, H.J. Geist, S.B. Agbola, A. Angelsen, J.W. Bruce, O.T. Coomes, R. Dirzo, G. Fischer, C. Folke, P.S. George, K. Homewood, J. Imbermon, R. Leemans, X. Li, E.F. Moran, M. Mortimore, P.S. Ramakrishnan, J.F. Richards, H. Skanes, W. Steffen, G.D. Stone, U. Svedin, T.A. Veldkamp, C. Vogel, and J. Xu, 2001, The causes of land-use and land-cover change moving beyond the myths. *Global Environmental Change*, 11, 261–269.

Leckie, D.G., P.M. Teillet, G. Fedosejevs, and D.P. Ostaff, 1988, Reflectance characteristics of cumulative defoliation of balsam fir. *Canadian Journal of Forest Research*, 18, 1008–1016.

Leichtle, T., C. Geiß, M. Wurm, T. Lakes, and H. Taubenböck, 2017, Unsupervised change detection in VHR remote sensing imagery—an object-based clustering approach in a dynamic urban environment. *International Journal of Applied Earth Observation and Geoinformation*, 54, 15–27. ISSN 1569-8432. https://doi.org/10.1016/j.jag.2016.08.010. https://www.sciencedirect.com/science/article/pii/S0303243416301490

Levien, L.M., C.S. Fischer, P.D. Roffers, B.A. Maurizi, J. Suero, and X. Huang, 1999, A machine-learning approach to change detection using multi-scale imagery. In *Proceedings of ASPRS Annual Conference Portland*, Oregon.

Liping, C., S. Yujun, and S. Saeed, 2018, Monitoring and predicting land use and land cover changes using remote sensing and GIS techniques—A case study of a hilly area, Jiangle, China. *PLoS ONE*, 13(7), e0200493. https://doi.org/10.1371/journal.pone.0200493

Loveland, T.R., T.L. Sohl, S.V. Stehman, A.L. Gallant, K.L. Sayler, and D.E. Napton, 2002, A strategy for estimating the rates of recent United States land-cover changes. *Photogrammetric Engineering and Remote Sensing*, 68, 1091–1099.

Lunetta, R.S., 1998, Project formulation and analysis approaches. In *Remote Sensing Change Detection Environmental Monitoring Methods and Applications* (R.S. Lunetta and C.D. Elvidge, Eds.), Ann Arbor Press, Chelsea, MI, 318 p.

Lunetta, R.S., 2002, Multi-temporal remote sensing analytical approaches for characterizing landscape change. In *Proceedings of the First Annual Workshop on the Analysis of Multi-Temporal Remote Sensing Images* (L. Bruzzone and P. Smits, Eds.), World Scientific Publishing Co., Pte. Ltd., Hackensack, NJ, 440 p.

Lunetta, R.S., J. Ediriwickrema, D. Johnson, J.G. Lyon, and A. McKerrow, 2002, Impacts of vegetation dynamics on the identification of land-cover change in a biologically complex community in North Carolina, USA. *Remote Sensing of Environment*, 82, 258–270.

Lunetta, R.S., and C.D. Elvidge, Eds., 1998, *Remote Sensing Change Detection Environmental Monitoring Methods and Applications*, Ann Arbor Press, Chelsea, MI, 318 p.

Lv, Z., P. Zhong, W. Wang, Z. You, and N. Falco, 2023, Multiscale attention network guided with change gradient image for land cover change detection using remote sensing images. *IEEE Geoscience and Remote Sensing Letters*, 20(Art no. 2501805), 1–5. http://doi.org/10.1109/LGRS.2023.3267879.

Lv, Z., et al., 2022, Land cover change detection with heterogeneous remote sensing images: Review, progress, and perspective, *Proceedings of the IEEE*, 110(12), 1976–1991. http://doi.org/10.1109/JPROC.2022.3219376.

Macleod, R.D., and R.G. Congalton, 1998, A quantitative comparison of change detection algorithms for monitoring eelgrass from remotely sensed data. *Photogrammetric Engineering and Remote Sensing*, 64, 207–216.

MaFoMP, 2011, *Massachusetts Forest Monitoring Program (MaFoMP), Human-Environment Regional Observatory of Central Massachusetts (HERO-CM)*. Available at: http://www.clarku.edu/department/hero/researcharea/forestchange.cfm

Mas, J.F., 1999, Monitoring land-cover changes: A comparison of change detection techniques. *International Journal of Remote Sensing*, 20, 139–152.

Mashala, M.J., T. Dube, B. T. Mudereri, K. K. Ayisi, and M. R. Ramudzuli, 2023, A systematic review on advancements in remote sensing for assessing and monitoring land use and land cover changes impacts on surface water resources in semi-arid tropical environments. *Remote Sensing*, 15(16), 3926. https://doi.org/10.3390/rs15163926

MohanRajan, S.N., A. Loganathan, P. Manoharan, et al., 2024, Fuzzy swin transformer for land use/land cover change detection using LISS-III Satellite data. *Earth Science Informatics*. https://doi.org/10.1007/s12145-023-01208-z

Mouat, D.A., G.G. Mahin, and J. Lancaster, 1993, Remote sensing techniques in the analysis of change detection. *Geocarto International*, 2, 39–50.

Nelson, R.F., 1983, Detecting forest canopy change due to insect activity using Landsat MSS. *Photogrammetric Engineering and Remote Sensing*, 49, 1303–1314.

Nilson, T., H. Olsson, J. Anniste, T. Lukk, and J. Praks, 2001, Thinning-caused change in reflectance of ground vegetation in boreal forest. *International Journal of Remote Sensing*, 22(14), 2763–2776.

NRCS (Natural Resources Conservation Service), 2000, *Summary Report: 1997 Natural Resources Inventory (Revised)*, U.S. Department of Agriculture, Washington, DC, 90 p.

Olofsson, P., G.M. Foody, M. Herold, S.V. Stehman, C.E. Woodcock and M.A. Wulder, 2014, Good practices for assessing accuracy and estimating area of land change. *Remote Sensing of Environment*, 148, 42–57.

Parelius, E.J., 2023, A review of deep-learning methods for change detection in multispectral remote sensing images. *Remote Sensing*, 15(8), 2092. https://doi.org/10.3390/rs15082092

Patterson, M.W., and S.R. Yool, 1998, Mapping fire-induced vegetation mortality using Landsat Thematic Mapper data a comparison of linear transformation techniques. *Remote Sensing of Environment*, 65, 132–142.

Pax Lenney, M., and C.E. Woodcock, 1997, The effect of spatial resolution on monitoring the status of agricultural lands. *Remote Sensing of Environment*, 61(2), 210–220.

Petit, C.C., and E.F. Lambin, 2002, Long-term land-cover changes in the Belgian Ardennes (1775–1929): Model-based reconstruction vs. historical maps. *Global Change Biology*, 8(7), 616–630.

Phinn, S.R., C. Menges, G.J.E. Hill, and M. Stanford, 2000, Optimizing remotely sensed solutions for monitoring, modeling, and managing coastal environments. *Remote Sensing of Environment*, 73, 117–132.

Rashed, T., J.R. Weeks, M.S. Gadalla, and A.G. Hill, 2001, Revealing the anatomy of cities through spectral mixture analysis of multispectral satellite imagery a case study of the greater Cairo region, Egypt. *Geocarto International*, 16, 5–15.

Rees, W.G., M. Williams, and P. Vitebsky, 2003, Mapping land-cover change in a reindeer herding area of the Russian Arctic using Landsat TM and ETM+ imagery and indigenous knowledge. *Remote Sensing of Environment*, 85, 441–452.

Ridd, M.K., and J. Liu, 1998, A comparison of four algorithms for change detection in an urban environment. *Remote Sensing of Environment*, 63, 95–100.

Rignot, E.J.M., and J.J. Vanzyl, 1993, Change detection techniques for ERS-1 SAR data. *IEEE Transactions on Geoscience and Remote Sensing*, 31(4), 1039–1046.

Rogan, J., and D. Chen, 2004, Remote sensing technology for mapping and monitoring land-cover and land use change. *Progress in Planning*, 61, 301–325.

Rogan, J., and J. Franklin, 2001, Mapping wildfire burn severity in southern California forests and shrublands using enhanced Thematic Mapper imagery. *Geocarto International*, 16(4), 89–99.

Rogan, J., J. Franklin, and D.A. Roberts, 2002, A comparison of methods for monitoring multitemporal vegetation change using Thematic Mapper imagery. *Remote Sensing of Environment*, 80, 143–156.

Rogan, J., J. Miller, D.A. Stow, J. Franklin, L. Levien, and C. Fischer, 2003, Land cover change mapping in California using classification trees with Landsat TM and ancillary data. *Photogrammetric Engineering and Remote Sensing*, 69(7), 793–804.

Rogan, J., and S.R. Yool, 2001, Mapping fire-induced vegetation depletion in the Peloncillo Mountains, Arizona and New Mexico. *International Journal of Remote Sensing*, 22, 3101–3121.

Roy, D.P., J. Ju, K. Kline, P.L. Scaramuzza, V. Kovalskyy, M.C. Hansen, T.R. Loveland, E.F. Vermote, and C. Zhang, 2010, Web-enabled Landsat data (WELD): Landsat ETM+ composited mosaics of the conterminous United States. *Remote Sensing of Environment*, 114, 35–49.

Running, S.W., and B.D. Donner, 1987, Water stress response after thinning lodgepole pine stands in Montana. In *Management of Small-Stem Stands of Lodgepole Pine*. U.S. Forest Service Int F.R.E.S. General Technical Report, INT-237, University of Montana, Montana, pp. 111–117.

Seto, K.C., C.E. Woodcock, C. Song, X. Huang, J. Lu, and R. Kaufmann, 2002, Monitoring land-use change in the Pearl River Delta using Landsat TM. *International Journal of Remote Sensing*, 23(10), 1985–2004.

Sidhu, N., E. Pebesma, and G. Câmara, 2018, Using Google Earth Engine to detect land cover change: Singapore as a use case. *European Journal of Remote Sensing*, 51(1), 486–500. http://doi.org/10.1080/22797254.2018.1451782

Singh, A., 1989, Digital change detection techniques using remotely sensed data. *International Journal of Remote Sensing*, 10, 989–1003.

Stow, D.A., 1995, Monitoring ecosystem response to global change multitemporal remote sensing analysis. In *Anticipated Effects of a Changing Global Environment in Mediterranean Type Ecosystems* (J. Moreno and W. Oechel, Eds.), Springer-Verlag, New York, pp. 254–286.

Stow, D.A., D. Collins, and D. McKinsey, 1990, Land use change detection based on multi-date imagery from different satellite sensor systems. *Geocarto International*, 5(3), 1–12.

Stow, D.A., L. Coulter, A. Johnson, and A. Petersen, 2004, Monitoring detailed land-cover changes in shrubland habitat reserves using multi-temporal IKONOS data, *Geocarto International*, 19(2), 95–102.

Stow, D.A., L. Tinney, and J. Estes, 1980, Deriving land use/land-cover change statistics from Landsat A study of prime agricultural land. In *Proceedings of the 14th International Symposium on Remote Sensing of the Environment*, Ann Arbor, MI, pp. 1227–1237.

Strahler, A.H., C.E. Woodcock, and J.A. Smith, 1986, On the nature of models in remote sensing. *Remote Sensing of Environment*, 20, 121–139.

Tewabe, T., T. Fentahun, and F. Li, 2020, Assessing land use and land cover change detection using remote sensing in the Lake Tana Basin, Northwest Ethiopia. *Cogent Environmental Science*, 6(1). http://doi.org/10.1080/23311843.2020.1778998

Todd, W.J., 1977, Urban and regional land use change detected by using Landsat data. *Journal of Research by the United States Geological Survey*, 5, 527–534.

Toutoubalina, O.V., and G.W. Rees, 1999, Remote sensing of industrial impact of Arctic vegetation around Noril'sk, northern Russia: Preliminary results. *International Journal of Remote Sensing*, 20, 2979–2990.

Townshend, J.R.G., and C.O. Justice, 1988, Selecting the spatial resolution of satellite sensors required for global monitoring of land transformations. *International Journal of Remote Sensing*, 9(2), 187–236.

Townshend, J.R.G., and C.O. Justice, 2002, Towards operational monitoring of terrestrial systems by moderate-resolution remote sensing. *Remote Sensing of Environment*, 83, 351–359.

Turner, B.L. II, E. Lambin, and A. Reenberg, 2007, The emergence of land change science for global environmental change and sustainability. *Proceedings, National Academy of Sciences of the United States of America*, 104(52), 20666–20671.

Turner, B.L., II, D. Skole, S. Sanderson, G. Fischer, L.O. Fresco, and R. Leemans, 1999, *Land-Use and Land-Cover Change Science/Research Plan*. Igbp report no. 35 and HDP report no. 7. Stockholm International Geosphere-Biosphere Programme, Stockholm.

Vitousek, P.M., H.A. Mooney, J. Lubchenco, and J.M. Melillo, 1997, Human domination of earth's ecosystems. *Science*, 277, 494–499.

Vogelmann, J.E., and B.N. Rock, 1988, Assessing forest damage in high-elevation coniferous forest in Vermont and New Hampshire using Thematic Mapper data. *Remote Sensing of Environment*, 24, 227–246.

Wang, S.W., B.M. Gebru, M. Lamchin, R.B. Kayastha, and W.-K. Lee, 2020, Land use and land cover change detection and prediction in the Kathmandu district of Nepal using remote sensing and GIS. *Sustainability*, 12(9), 3925. https://doi.org/10.3390/su12093925

Wang, S.W., L. Munkhnasan, and W. Lee, 2021, Land use and land cover change detection and prediction in Bhutan's high altitude city of Thimphu, using cellular automata and Markov chain. *Environmental Challenges*, 2, 100017. ISSN 2667-0100. https://doi.org/10.1016/j.envc.2020.100017. https://www.sciencedirect.com/science/article/pii/S2667010020300172

Wood, J.E., M.D. Gillis, D.G. Goodenough, R.J. Hall, D.G. Leckie, J.L. Luther, and M.A. Wulder, 2002, Earth observation for sustainable development of forests (EOSD): Project overview. *Proceedings of the International Geoscience and Remote Sensing Symposium (IGARSS) and 24th Symposium of the Canadian Remote Sensing Society*, June 24–28, Toronto, Canada.

Woodcock, C.E., T.R. Loveland, M. Herold, and M.E. Bauer, 2020, Transitioning from change detection to monitoring with remote sensing: A paradigm shift. *Remote Sensing of Environment*, 238, 111558. ISSN 0034-4257. https://doi.org/10.1016/j.rse.2019.111558.

Wulder, M.A., C.R. Butson, and J.C. White, 2008, Cross-sensor change detection over a forested landscape: Options to enable continuity of medium spatial resolution measures. *Remote Sensing of Environment*, 112, 796–809.

Wulder, M.A., and N.C. Coops, 2014, Make Earth observations open access. *Nature*, 513, 30–31.

Zhang, C., P. Yue, D. Tapete, L. Jiang, B. Shangguan, L. Huang, and G. Liu, 2020, A deeply supervised image fusion network for change detection in high resolution bi-temporal remote sensing images. *ISPRS Journal* of *Photogrammetry and Remote Sensing*, 166, 183–200. ISSN 0924-2716. https://doi.org/10.1016/j.isprsjprs.2020.06.003. https://www.sciencedirect.com/science/article/pii/S0924271620301532

Zhang, M., and W. Shi, 2020, A feature difference convolutional neural network-based change detection method. *IEEE Transactions on Geoscience and Remote Sensing*, 58(10), 7232–7246. https://doi.org/10.1109/TGRS.2020.2981051. https://www.sciencedirect.com/science/article/pii/S092427161730103X

Zhu, Z., 2017, Change detection using Landsat time series: A review of frequencies, preprocessing, algorithms, and applications. *ISPRS Journal of Photogrammetry and Remote Sensing*, 130, 370–384. ISSN 0924-2716. https://doi.org/10.1016/j.isprsjprs.2017.06.013.

Zhu, Z., and C.E. Woodcock, 2014, Continuous change detection and classification of land cover using all available Landsat data. *Remote Sensing of Environment*, 144, 152–171.

9 Land Use and Land Cover Mapping and Monitoring with Radar Remote Sensing

Zhixin Qi, Anthony Gar-On Yeh, Xia Li, and Qianwen Lv

ACRONYMS AND DEFINITIONS

ALOS	Advanced Land Observing Satellite
ASAR	Advanced synthetic aperture radar on board ENVISAT
AVHRR	Advanced very high resolution radiometer
COSMO-SkyMed	Constellation of Small Satellites for Mediterranean basin Observation (COSMO)-SkyMed
ENVISAT	Environmental satellite
ERS	European remote sensing satellites
ERTS	Earth Resources Technology Satellites
ETM+	Enhanced Thematic Mapper+
GIS	Geographic Information System
IKONOS	A commercial earth observation satellite, typically, collecting sub-meter to 5 m data
JERS	Japanese Earth Resources Satellite
LULC	Land use, land cover
MSS	Multi-spectral scanner
NASA	National Aeronautics and Space Administration
NDVI	Normalized Difference Vegetation Index
OLI	Operational land imager
PALSAR	Phased Array type L-band Synthetic Aperture Radar
PolSAR	Polarimetric SAR
RADARSAT	Radar satellite
SAR	Synthetic aperture radar
SEASAT	First satellite designed for remote sensing of the Earth's oceans with synthetic aperture radar (SAR)
SPOT	Satellite Pour l'Observation de la Terre, French Earth Observing Satellites
SVM	Support vector machines
TerraSAR-X	A radar Earth observation satellite with its phased array synthetic aperture radar

9.1 INTRODUCTION

Land use and land cover (LULC) mapping and monitoring is one of the most important application areas of remote sensing. Land use refers to the human activities on land that are directly related to the land (Clawson and Stewart, 1965). It usually emphasizes the functional role of land in socioeconomic activities, such as agriculture, industry, commerce, transportation, construction, and recreation. These activities are abstract and not always observable from remote-sensed images, and inference based on surrogates often has to be made to identify the land use. Land cover, on the other hand, implies the vegetation and artificial constructions covering the land surface (Burley, 1961). It encompasses natural features such as vegetation, urban areas, water, barren land, or others that are concrete and directly visible on remote-sensed images. Land cover does not describe the use of land, and the use of land may be different for lands with the same cover type. For instance, a land cover type of forest may be used for timber production, wildlife management, or recreation; it might be private land, a protected watershed, or a popular state park.

The importance of mapping, quantifying, and monitoring LULC and its change has been widely recognized in the scientific community as a key element in a variety of applications, such as ecological monitoring, habitat assessment, wildlife management, enforcement, exposure and risk assessment, global change monitoring, environmental impact assessment, state and local planning, hazardous waste remedial action, and regulatory policy development (Lo, 1998). The accurate and timely information on LULC patterns and changes has grown in importance in recent years with our increasing concern over the conflict between economic development and ecological change. The knowledge of the present distribution and area of different LULC types as well as information on their changing proportions are needed by planners, legislators, and governmental officials to determine better land-use policy, to project transportation and utility demand, to identify future development pressure points and areas, and to implement effective plans for regional development (Anderson et al., 1976).

Remote sensing technology has been employed extensively in LULC investigation because of its capability to observe land surface consistently and repetitively and its advantages of cost and time savings for large areas. A LULC classification scheme has been developed for use with remote sensing data (Anderson et al., 1976). The main characteristics of this scheme are its emphasis on resources rather than people and its capability to provide different levels of classification according to the scale and spatial resolution of the images. The development of such a classification scheme has facilitated the mapping, modeling, and measurement of many LULC applications. The scheme includes four classification levels in accordance with the image scale (Table 9.1). However, the general relationship between the classification level and the data source is not intended to restrict uses to particular scales, either in the original data source or in the final map product. For example, Level I LULC information could be not only gathered by a LANDSAT type of satellite or high-altitude imagery but also interpreted from conventional large-scale aircraft imagery or compiled by ground survey. Similarly, several Level II and III categories have been interpreted from LANDSAT data. The classification scheme for the first and second levels have been presented by Anderson et al. (1976) (Table 9.2). Levels beyond these two must be designed by users according to their needs.

TABLE 9.1
Classification Level and Corresponding Typical Data Characteristics

Classification Level	Typical Data Characteristics
I	LANDSAT (formerly ERTS) type of data
II	High-altitude data at 40,000 ft (12,400 m) or above (less than 1:80,000 scale)
III	Medium-altitude data taken between 10,000 and 40,000 ft (3,100 and 12,400 m) (1:20,000 to 1:80,000 scale)
IV	Low-altitude data taken below 10,000 ft (3,100 m) (more than 1:20,000 scale)

Source: Anderson et al. (1976)

TABLE 9.2
LULC Classification System for Use with Remote Sensing Data

Level I		Level II	
1	Urban or built-up land	11	Residential
		12	Commercial and services
		13	Industrial
		14	Transportation, communications, and utilities
		15	Industrial and commercial complexes
		16	Mixed urban or built-up land
		17	Other urban or built-up land
2	Agricultural land	21	Cropland and pasture
		22	Orchards, groves, vineyards, nurseries, and ornamental Horticultural areas
		23	Confined feeding operations
		24	Other agricultural land
3	Rangeland	31	Herbaceous rangeland
		32	Shrub and brush rangeland
		33	Mixed rangeland
4	Forest land	41	Deciduous forest land
		42	Evergreen forest land
		43	Mixed forest land
5	Water	51	Streams and canals
		52	Lakes
		53	Reservoirs
		54	Bays and estuaries
6	Wetland	61	Forested wetland
		62	Non-forested wetland
7	Barren land	71	Dry salt flats
		72	Beaches
		73	Sandy areas other than beaches
		74	Bare exposed rock
		75	Strip mines quarries, and gravel pits
		76	Transitional areas
		77	Mixed barren land
8	Tundra	81	Shrub and brush tundra
		82	Herbaceous tundra
		83	Bare ground tundra
		84	Wet tundra
		85	Mixed tundra
9	Perennial snow or ice	91	Perennial snowfields
		92	Glaciers

Source: Anderson et al. (1976).

Optical remote sensing images have been widely applied for a myriad of LULC investigation objectives. The Advanced Very High Resolution Radiometer (AVHRR) embarked on the National Oceanic Atmospheric Administration series of satellites have been predominantly used for global- to continental-scale LULC investigation because of its large swath width and twice-daily global coverage (Lambin and Ehrlich, 1997). Compared with the AVHRR, the Moderate Resolution

Imaging Spectroradiometer on the Terra and Aqua satellites has enhanced spatial, radiometric, and spectral capabilities, and has also been widely applied to large-scale LULC mapping and monitoring (Lunetta et al., 2006). However, the low resolution (250 m–1 km) of these sensors limits their ability to reveal detailed spatial distribution of LULC patterns and changes. Balancing the trade-offs involving spatial detail, areal coverage, and availability of historical data, medium-resolution images (10–90 m) obtained from Landsat 5 Thematic Mapper (TM) and Multispectral Scanner (MSS), Landsat 7 Enhanced Thematic Mapper Plus (ETM+), Landsat 8 Operational Land Imager (OLI) and Thermal Infrared Sensor (TIRS), Landsat 9 Operational Land Imager 2 (OLI-2) and the Thermal Infrared Sensor 2 (TIRS-2), Sentinel-2 MultiSpectral Instrument (MSI), and Advanced Spaceborne Thermal Emission and Reflection Radiometer are currently the most commonly used datasets for LULC mapping and monitoring (Wulder et al., 2012; Blaschke, 2016). Numerous studies have been carried out on the use of the visible to shortwave infrared bands of these datasets for forestry and agricultural land cover analysis and urban development monitoring (Adams et al., 1995; French et al., 2008). Sensors with higher spatial resolutions (0.5–6 m), including SPOT, QuickBird, IKONOS, WorldView, and Planet, have also been employed to investigate LULC with increased spatial detail (Chang et al., 2010; Pu and Landry, 2012; Wang et al., 2004).

However, optical remote sensing is limited by weather conditions. Difficulties are encountered in collecting timely LULC information in tropical and sub-tropical regions that are characterized by frequent cloud cover. Although some optical sensors such as WorldView-2 can deliver data with high temporal resolution, they cannot guarantee that the images collected at short intervals are unaffected by clouds (DigitalGlobe, 2013). Furthermore, the small coverage and high cost limit routine use of the data obtained by these sensors to investigate LULC information. Because of the shortage of cloud-free images, it might be unfeasible to collect timely LULC information using optical remote sensing. Being capable of transmitting and receiving its own electromagnetic waves with the antenna, radio detection and ranging (radar) remote sensing is nearly weather independent and can acquire imagery day and night (Figure 9.1). Furthermore, the development of synthetic aperture radar (SAR) improves the resolution beyond the limitation of physical antenna aperture. Therefore, radar remote sensing has become an effective tool for LULC mapping and monitoring in the perpetually cloud-covered tropical and equatorial regions of the world where many developing countries with the greatest need for LULC data are found. The timely information on LULC and its change is necessary for these countries to develop policies that will enable the maintenance of good balance between land development and environmental protection. In addition, compared with conventional optical remote sensing, radar remote sensing provides a different way to observe the Earth. Radar backscattering from terrain is mainly affected by (1) geometrical factors related to structural attributes of the surface and any overlying vegetation cover relative to the sensor parameters of wavelength and viewing geometry and (2) electrical factors determined by the relative dielectric constants of soil and vegetation at a given wavelength (Dobson et al., 1995). Therefore, radar can provide somewhat complementary information to optical data; and hence, classification can be significantly improved when both suites of sensors are used together.

9.2 RADAR SYSTEM PARAMETERS AND DEVELOPMENT

The primary radar system parameters influencing the intensity and patterns of radar returns from the observed objects are the frequency, polarization, and incidence angle.

Radar frequency and wavelength are interrelated, as seen in Equation 9.1:

$$c = f\lambda \tag{9.1}$$

where c is the speed of light ($3 \times 10^8 \text{ms}^{-1}$, f is the frequency and λ is the wavelength. In order to calculate λ in centimeters (cm), the value of c is given in terms of cm ($3 \times 10^{10} \text{cms}^{-1}$) and the value of frequency in terms of hertz (Hz). Radar transmits a microwave signal toward the targets and detects the backscattered portion of the signal. Microwaves are electromagnetic waves with frequencies

Land Use and Land Cover Mapping and Monitoring

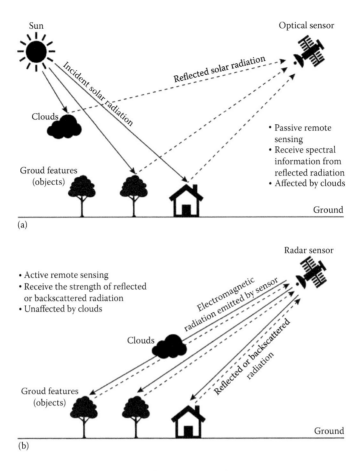

FIGURE 9.1 (a) Optical remote sensing, (b) radar remote sensing.

between 300 MHz (0.3 GHz) and 300 GHz in the electromagnetic spectrum. They are included in radio waves that are electromagnetic waves within the frequencies 30 KHz–300 GHz. Radar systems can be categorized into different bands according to the variation in frequency (Table 9.3). The definition and nomenclature for these bands were established by the US military during World War II (Waite, 1976). Although other classification systems were established outside of the United States, the system presented in Table 9.3 is the most widely used.

Polarization is a property of waves that describes the orientation of their oscillations. For transverse waves, such as many electromagnetic waves, it describes the orientation of the oscillations in the plane perpendicular to the wave's direction of travel. As shown in Figure 9.2, propagating electromagnetic radiation has three vector fields that are mutually orthogonal. The direction of propagation is one vector, and electric and magnetic fields make up the other two vector fields. Active microwave energy, as well as other frequencies of electromagnetic radiation, has a polarized component defined by the electric field vector of the radiation (Figure 9.2). Most of the radar systems are linear polarized systems that operate using horizontally (H) or vertically (V) polarized microwave radiation. For these systems, polarization refers to the orientation of the radar beam relative to the Earth's surface (Figure 9.3). If the electric vector field is parallel to the Earth's surface, the wave would be designated horizontally polarized. If it is perpendicular to the Earth's surface, the wave would be designated vertically polarized. In an active system, energy is both transmitted and received. Therefore, the linear polarization can be mixed and matched to provide the four most common linear polarization schemes, namely, HH, HV, VH, and VV (Figure 9.3).

TABLE 9.3
Radar Bands and Frequencies

Radar Frequency Band	Wavelength (cm)	Frequency Range (MHz)
P	136.00–77.00	220–390
UHF	100.00–30.00	300–1,000
L	30.00–15.00	1,000–2,000
S	15.00–7.50	2,000–4,000
C	7.50–3.75	4,000–8,000
X	3.75–2.40	8,000–12,500
Ku	2.40–1.67	12,500–18,000
K	1.67–1.18	18,000–26,500
Ka	1.18–0.75	26,500–40,000
Millimeter	<0.75	<40,000

Source: Waite (1976)

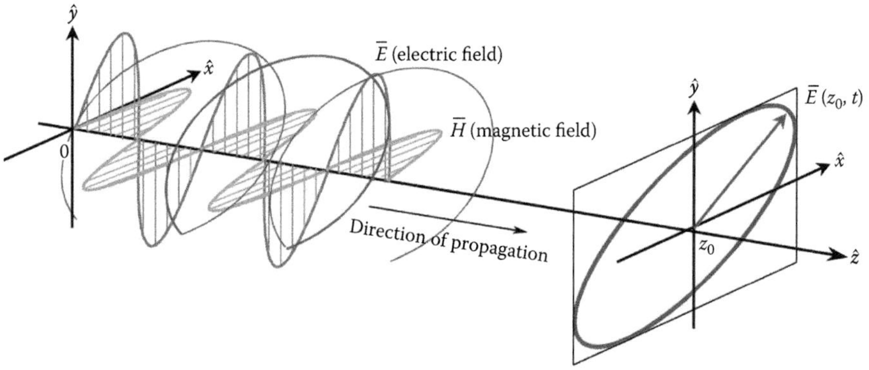

FIGURE 9.2 Components of an electromagnetic wave. The plane of polarization is defined by the electric field. Source: European Space Agency (http://earth.eo.esa.int/polsarpro/Manuals/1_What_Is_Polarization.pdf).

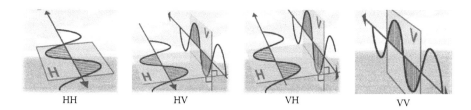

FIGURE 9.3 Linear polarization schemes. Source: European Space Agency (https://earth.esa.int/handbooks/asar/CNTR1-1-5.htm).

Incidence angle, defined as the angle between the radar line-of-sight and the local vertical (Figure 9.4) with respect to the geoid, is also a major factor influencing the radar backscatter and the appearance of objects on the imagery, caused by foreshortening or radar layover. In general, reflectivity from distributed scatters decreases with increasing incidence angles (Lewis and Henderson, 1998). Incidence angle incorporating look angle and the curvature of the Earth is shown in Figure 9.4 (a). This model assumes a level terrain on constant slope angle. In contrast, Figure 9.4

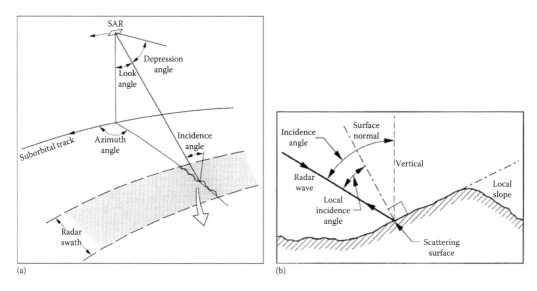

FIGURE 9.4 (a) Schematic diagrams of the system and (b) local incidence angle.
Source: Lewis and Henderson (1998).

(b) illustrates the "local incidence angle" and takes into account the local slope angle. For example, surface roughness changes as a function of the local incidence angle.

The development of radar systems has great impact on the application of radar remote sensing in LULC mapping and monitoring. Imaging radar systems could be divided into two categories: spaceborne SAR systems and airborne SAR systems. System parameters for some widely used SAR systems are summarized in Tables 9.4–9.7. Airborne imaging SARs are mainly meant for technology development as well as application developments. Early studies on the use of radar remote sensing in LULC investigation were mainly based on airborne radar imagery (Henderson, 1975; Ulaby et al., 1982). The first civilian SAR mission in space was the US SEASAT, which operated from early July to mid-September in 1978. The design lifetime of SEASAT was two years. Unfortunately, the spacecraft failed after three months of SAR operation due to problems with the power system. However, data provided by SEASAT proved to be of high quality and immense interest to the science and application communities. This interest and active research by geoscientists was further augmented in the late 1970s and 1980s by imagery from NASA's Shuttle Imaging Radar (SIR-A and SIR-B) systems along with several systematic airborne SAR projects, such as Canada's CCRS SAR-580 campaign. SIR-A and SIR-B extended the baseline established by SEASAT in the dimension of incident angle (Table 9.4). Although these SAR systems proved to be useful for LULC investigation, they were only occasionally launched to collect experimental data within a very short period. The routine investigation of LULC information using radar remote sensing has become practical after some orbital radar systems with SAR, such as ERS-1 and ERS-2, JERS-1, and RADARSAT-1, were made available for regular data collection. However, these orbital SAR systems also have limitations because only one single frequency is available. Some studies indicated that the single-frequency orbital SAR systems can be confused with separating and mapping LULC classes (Li and Yeh, 2004). To overcome the difficulty in single-frequency SAR data, some researchers utilized polarimetric SAR (PolSAR) imagery acquired by SIR-C/X-SAR or airborne SAR systems to investigate LULC information (Lee et al., 2001; Pierce et al., 1994). Their results showed that PolSAR measurements can achieve better classification results than single-polarization SAR. The use of PolSAR data in LULC mapping and monitoring has become an important research topic since PolSAR images have been made available through orbital PolSAR systems, such as ENVISAT ASAR, ALOS PALSAR, TerraSAR-X, and RADARSAT-2. In addition, constellations, such as Sentinel-1, GF-3, COSMO-SkyMed, and Radarsat Constellation Mission, composed of several satellites equipped with SAR, have been available for observing the Earth (Covello et al., 2010). The increasing availability of multi-dimensional (multi-frequency, multi-polarized, multi-temporal, and multi-incidence angle) digital radar data permits the generation of "true" color radar imagery rather than "colorized" single-channel radar imagery (Figure 9.5).

TABLE 9.4
Spaceborne SAR Systems 1978–1994

	SEASAT	SIR-A	SIR-B	Kosmos 1870	ALMAZ	ERS-1	JERS-1	SIR-C/X-SAR
Launch date	26 Jun 78	12 Nov 81	5 Oct 84	25 Jul 87	31 Mar 91	16 Jul 91	11 Feb 92	9 Apr 94, 30 Sep 94
Country	USA	USA	USA	USSR	USSR	Europe	Japan	USA
Spacecraft	SEASAT	Shuttle	Shuttle	Salyut	Salyut	ERS-1	JERS-1	Shuttle
Lifetime	3 months	2.5 days	8 days	2 years	1.5 years	9 years	6 years	10 days
Band	L	L	L	S	S	C	L	L, C; X
Frequency (GHz)	1.275	1.278	1.282	3.0	3.0	5.25	1.275	1.25, 5.3; 9.6
Polarization	HH	HH	HH	HH	HH	VV	HH	L+C: Quad/X: VV
Incident angle (degrees)	23	50	15–64	30–60	30–60	23	39	15–55
Range resolution (m)	25	40	25	30	15–30	26	18	10–30
Azimuth resolution (m)	25	40	58–17	30	15	28	18	30
Swath width (km)	100	50	10–60	20–45	20–45	100	75	15–60
Repeat cycle (days)	17, 3	nil	nil	variable	nil	3; 35; 176	44	nil

TABLE 9.5
Spaceborne SAR Systems 1994–2010

	ERS-2	RADARSAT-1	ENVISAT	PALSAR	TerraSAR-X	RADARSAT-2	TanDEM-X
Launch date	21 Apr 95	4 Nov 95	1 Mar 02	24 Jan 06	15 Jun 07	14 Dec 07	21 Jun 10
Country	Europe	Canada	Europe	Japan	Germany	Canada	Germany
Spacecraft	ERS-2	RADARSAT-1	ENVISAT	ALOS	TerraSAR-X	RADARSAT-2	TanDEM-X
Lifetime	11 years	17 years	10 years	5 years	5 years (design)	8 years (design)	5.5 years (design)
Band	C	C	C	L	X	C	X
Frequency (GHz)	5.3	5.3	5.3	1.27	9.65	5.4	9.65
Polarization	VV	HH	Quad	Quad	Quad	Quad	Quad
Incident angle (degrees)	23	10–59	15–45	8–60	20–55	10–60	20–55
Range resolution (m)	26	8–100	30–1000	7–100	1–18.5	3–100	1–18.5
Azimuth resolution (m)	28	8–100	30–1000	7–100	1–18.5	3–100	1–18.5
Swath width (km)	100	50–500	5–400	20–350	10–100	18–500	10–100
Repeat cycle (days)	35	24	35	46	11	24	11

TABLE 9.6
Spaceborne SAR Systems 2010–2020

	Sentinel-1	PALSAR-2	GF-3	COSMO-SkyMed Second Generation	Radarsat Constellation Mission (RCM)
Launch date	3 Apr 14	24 May 14	10 Aug 16	18 Dec 19	12 Jun 19
Country	Europe	Japan	China	Italy	Canada
Spacecraft	Sentinel-1A	ALOS-2	GF-3	COSMO-SkyMed Second Generation	Radarsat Constellation Mission (RCM)
Lifetime	7 year	5 year	8 year	7 year	7 year
Band	C-band	L-band	C-band	X-band	C-band
Frequency (GHz)	5.405	1.215	5.3	9.6	5.405
Polarization	Dual	Quad	Quad	Quad	Dual
Incident angle (degrees)	29.1–46.0	29–43	20–50	20–49	20–49
Range resolution (m)	5	3–10	1	1	3
Azimuth resolution (m)	5–40	10	2	1	3
Swath width (km)	80–400	70	10–650	10–100	20–350
Repeat cycle (days)	12	14	12	16	4

TABLE 9.7
Airborne SAR Systems

	AIRSAR	SAR580	EMISAR	E-SAR	PISAR	STAR-1	UAVSAR
Country	USA	Canada	Denmark	Germany	Japan	Canada	USA
Aircraft	DC-8	CV-580	Gulfstream	DO-228	Gulfstream	Cessna	Gulfstream
Band	C, L, P	X, C	C, L	X, C, L, P	X, L	X	L
Frequency (GHz)	5.3, 1.3, 0.44	9.3, 5.3	5.3, 1.25	9.6, 5.3, 1.3, 0.45	9.55, 1.27	9.6	1.26
Polarization	Quad	Quad	Quad	Single, Quad	Quad	HH	Quad
Incident angle	20–60	0–85	30–60	25–60	10–60	45–80	25–60
Range resolution (m)	7.5	6–20	2.4, 8	2–4	3	6, 12	0.5
Azimuth resolution (m)	2	< 1–10	2.4, 8	2–4	3.2	6	1.5
Swath width (km)	6–20	18–63	12, 24, 48	3	4.3, 19.3, 19.6	40, 60	16

9.3 RADAR SYSTEM PARAMETER CONSIDERATION FOR LULC MAPPING

As introduced previously, the primary radar system parameters that influence the intensity and patterns of radar returns from the observed objects are wavelength, polarization, and incidence angle. Different combinations of these parameters are normally selected to optimize an application. A number of studies have been conducted on the influence of radar frequency and polarization on LULC mapping. Several findings were made for the interaction of the target and polarized signal (Lewis and Henderson, 1998). When the plane of polarization of the transmitted microwave radiation is parallel to the dominant plane of linear features, the like-polarized radar return will be stronger than the transmitted and received signal in the orthogonal plane. For example, it can be expected that VV will have a stronger returned signal than HH if the target, such as a wheat field, has a strong vertical component. The like-polarized image (HH or VV) will have a stronger returned signal than the cross or depolarized image (HV or VH). Only the part of the transmitted signal that is depolarized has the potential of being recorded in the cross-polarized data set.

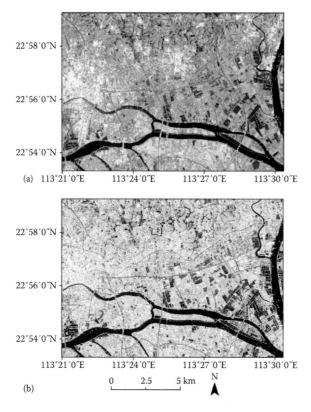

FIGURE 9.5 SAR data of the South Guangzhou area, China (a) RADARSAT-1 single-polarization data (HH) acquired on December 10, 2005; (b) RADARSAT-2 fully polarimetric data (Pauli RGB composition) acquired on March 21, 2009.

Depolarization of the transmitted radar signal is primarily a result of (1) quasi-specular reflection from corner reflectors, (2) multiple scattering from rough surfaces, and (3) multiple volume scattering due to inhomogeneities.

Land cover classification was implemented using multi-frequency (P, L, C bands) PolSAR images (Chen et al., 1996). The land cover classes included forest, water, bare soil, grass, and eight other types of crops. The radar response of crop types to frequency and polarization states were analyzed for classification based on three configurations: (1) multi-frequency and single-polarization images, (2) single-frequency and multi-polarization images, and (3) multi-frequency and multi-polarization images. The classification results showed that using partial information, P-band multi-polarization images and multi-band HH polarization images, had better classification accuracy, while a full configuration, namely, multi-band and multi-polarization, gave the best discrimination capability.

Aiming at steering the selection of optimal combinations of polarimetric SAR (PolSAR) frequency bands for different land cover classification schemes, Qi et al. (2019) investigated the capabilities of all the possible combinations of L-band ALOS PALSAR fully PolSAR data, C-band RADARSAT-2 fully PolSAR data, and X-band TerraSAR-X HH SAR data for the land cover classification, involving bare land, water, banana trees, built-up areas, forests, and crop/rangeland. The study found that (1) X-band HH SAR is not necessary for the land cover classification when C- or L-band fully PolSAR are used; (2) C-band fully PolSAR alone is adequate for classifying primitive land cover types, including water, bare land, vegetation, and built-up areas; and (3) L-band fully PolSAR alone is adequate for distinguishing between various vegetation types, such as crops, banana trees, and forests.

Lee et al. (2001) addressed the land use classification capabilities of fully polarimetric SAR versus dual-polarization and single-polarization SAR for P-, L-, and C-band frequencies. A variety of polarization combinations were investigated for application to crop and tree age classification. They found that L-band fully polarimetric SAR data was best for crop classification, but P-band was best for forest age classification, because longer wavelength electromagnetic waves provided higher penetration. For dual-polarization classification, the HH and VV phase difference was important for crop classification but less important for tree age classification. Also, for crop classification, the L-band complex HH and VV achieved correct classification rates almost as good as for full polarimetric SAR data, and for forest age classification, P-band HH and HV should be used in the absence of fully polarimetric data. In all cases, they indicated that multi-frequency fully polarimetric SAR is highly desirable. Similar results were also found by Turkar et al. (2012) in the investigation of classification capabilities of fully and partially PolSAR data for C- and L-band frequencies. They observed that L-band fully polarimetric SAR data worked better than C-band for classification of various land covers. The forest class was well classified with L-band PolSAR data, but it was poorly classified in C-band because of dominant scattering from treetops.

There are also many studies on the influence of incidence angle on the intensity and patterns of radar returns from the observed objects. The results show that the choice of optimum incidence angle varies with applications. For example, a geological application generally prefers images acquired with large incidence angles because geometric distortion is minimal and shadowing provides enhancement of topographic relief. On the other hand, a moderate incident angle is required for accurate settlement detection and urban land cover mapping. Incidence angles of less than 20–23 degrees are of minimum utility for settlement detection and urban analysis and the amount of information and accuracy of interpretation increases on the image acquired at 41 degrees but decreases on the image acquired at 51 degrees (Henderson, 1995; Henderson and Xia, 1997). Lichtenegger et al. (1991) found that SIR-A imagery was better for land-use mapping than the SEASAT imagery because of its larger incidence angle. Gauthier et al. (1998) indicated that backscatter coefficients extracted from ERS-1 SAR data over an agricultural area were found to be sensitive to incidence angle. Lang et al. (2008) found that a subtle trend of generally decreasing backscatter with increasing incidence angle, and he hypothesized that this decrease was caused by lower transmissivity of the crown layer, increased attenuation of energy from double-bounce, and multi-path scattering, and possibly increased specular reflectance of the surface layer with increasing incidence angle. Paris (1983) has also indicated that higher incidence angle increases the path length of SAR signal through the vegetation volume, resulting in higher interaction with crop canopy. Ford et al. (1986) stated that as incidence angle increases, there is sensitivity to surface roughness and decreased sensitivity to topography. The incidence angle has also proven to be sensitive to surface roughness and soil moisture (Rahman et al., 2008; Srivastava et al., 2009).

Xu et al. (2019) conducted land cover classification by using RADARSAT-2 Wide Fine Quad-Pol (FQ) images acquired at different incidence angles, including FQ8 (27.75°), FQ14 (34.20°), and FQ20 (39.95°). They found that the multi-incidence angle image produced better classification results than any of the single-incidence angle images, and the different incidence angles exhibited different superiorities in land cover classification. The difference in single-bounce scattering between trees and crops was evident in the FQ8 image, which was determined to be suitable for distinguishing between croplands and forests. The FQ8 image exhibited the largest difference in single-bounce scattering between bare lands and water and produced the least confusion between them among all the images. The single- and double-bounce scattering from urban areas and forests increased with the decrease in incidence angles. The increase in single- and double-bounce scattering from urban areas was more significant than that from forests because C-band SAR could not easily penetrate the crown layer of forests to interact with the trunks and ground. Therefore, the FQ8 image showed a slightly better performance than the other images in discriminating between urban areas and forests. As a large incidence angle resulted in a long penetration path of radar waves in the crown layer of vegetation, the FQ20 image enhanced the single- and double-bounce scattering differences between banana trees and other vegetation. Thus, the FQ20 image outperformed the other images in identifying banana trees.

Kasischke et al. (1997) have summarized the optimum SAR parameters for the use of imaging radar systems in several land surface applications (Table 9.8).

TABLE 9.8
Optimal SAR System Parameters for Monitoring Land Surface Characteristics

Application Area	Radar Frequency	Polarization	Incidence Angle[a]	Resolution[b]	Sampling Frequency[c]
Vegetation mapping	Multiple frequency data optimal—as a minimum two frequencies (one high, one low) required	Multipolarization and polarimetric data desired, especially with single-frequency systems	Both low and high desired	High resolutions desirable for mapping smaller sampling units	Low for multiple channel systems, high for single channel systems
Biomass estimation	L- or P-band optimal, as a minimum L- and C-band required	Cross-polarization data most sensitive; multipolarization data improve biomass algorithms	Low	For small forest stands, fine resolution; for larger area studies, low resolution	Low can be used—sampling at proper phenologic stage and under optimum weather conditions important
Monitoring flooded forests	L- and P-band optimal, but some sensitivity at C-band if no leaves present and HH polarization used	HH polarization most sensitive, but VV polarization can be used	Lower required	Higher resolutions may be important for mapping narrow features	High sampling frequencies usually important
Monitoring coastal/low stature wetlands	X- or C-band	HH or VV	Low	High or low, depending on ecosystem patch size	High
Monitoring tundra inundation	X- or C-band	HH or VV	Low	High or low, depending on ecosystem patch size	High
Monitoring fire-disturbed boreal forests	X- or C-band	HH or VV	Low	High or low, depending on ecosystem patch size	High
Detection of frozen/thawed vegetation	Multiple frequencies	Multiple polarizations	Low or high	High or low, depending on ecosystem patch size	High

[a] Low incidence angles = 20–40°, high incidence angles = 40–60°.
[b] High resolution = 20–40 m, low resolution > 100m.
[c] High sampling frequency = once every two weeks; low sampling frequency = once per year.

Source: Kasischke et al. (1997)

9.4 CLASSIFICATION OF RADAR IMAGERY

The classification of radar images mainly involves three steps: image preprocessing, feature extraction and selection, and selection of a suitable classifier.

9.4.1 Image Preprocessing

The preprocessing of radar images mainly includes radiometric calibration, geometric calibration, and speckle filtering. Radiometric calibration is a procedure meant to correctly estimate the target reflectance from the measured incoming radiation. Pixel values of radar images are directly related to the radar backscatter of the scene after radiometric calibration. Geometric calibration aims to tie the line/pixel positions in the image coordinates to the geographical latitude/longitude. Speckle is one of the main problems of radar image classification because a homogeneous zone on the ground still has a granular aspect and a statistical distribution with a large standard deviation (Durand et al., 1987). Image filtering is necessary for suppressing the speckle before the classification of radar images. Many adaptive filters for speckle reduction have been proposed, e.g., Lee (Lee, 1981), Lee Sigma (Lee, 1983), Frost (Frost et al., 1982), Kuan (Kuan et al., 1985), Gamma Map (Lopes et al., 1993), refined Lee (Lee et al., 1999b), Gamma WMAP (Solbo and Eltoft, 2004), and Improved Lee Sigma (Lee et al., 2009) filters.

Despite their spatial adaptive characteristic, which tends to preserve the signal's high frequency information, filter applications often give the desired speckle reduction but also an undesired degradation of the geometrical details of the investigated scene. The selection of suitable filter and window size is commonly a heuristic process, in which different filters with different window size are applied to a specific radar image and their performances are compared. Some studies have been conducted on the comparison and selection of appropriate filters. Lee et al. (1999b) stated that refined Lee filter is suitable for PolSAR images because it effectively preserves polarimetric information and retains subtle details while reducing the speckle effect in homogeneous areas. Capstick and Harris (2001) assessed the capability of Lee, Lee Sigma, Local Region, Frost, Gamma MAP, and simulated annealing filters to improve classification accuracy in agricultural applications and found the best ones are Lee-Sigma, Gamma MAP, and simulated annealing. Nyoungui et al. (2002) evaluated nine speckle reduction techniques based on their applicability to supervised land cover classification from SAR images. Issues related to suppression of speckle in a uniform area, preservation of edges, and texture preservation were pursued in these filters. The results showed that speckle suppression techniques based on the wavelet transform perform the best, followed by the modified K-nearest neighbors and Lee's local statistic filters.

9.4.2 Feature Extraction and Selection

The primary feature used in radar image classification is the intensity of each pixel that represents the proportion of microwave backscattered from that area on the ground. The pixel intensity values are often converted to a physical quantity called the backscattering coefficient or normalized radar cross-section measured in decibel (dB) units with values ranging from +5 dB for very bright objects to −40 dB for very dark surfaces. The backscatter radar intensity depends on a variety of factors: types, sizes, shapes, and orientations of the scatterers in the target area; moisture content of the target area; frequency and polarization of the radar pulses; as well as the incident angles of the radar beam. Therefore, in addition to tonal value of radar intensity images, features that can be extracted from radar images and have proved to be useful for LULC classification are mainly related to the textural, speckle, polarimetric, interferometric, multi-frequency, and multi-incidence angle information.

9.4.2.1 Textural Features

Most of the existing orbital radar systems are single-frequency types and may create confusion during the separation and mapping of LULC classes; this confusion stems from the limited information obtained by single-frequency systems (Ulaby et al., 1986). One way to compensate for the limited information from single-frequency radar data is to derive more features, such as texture, for the classification of radar images in addition to the tonal information of pixels. A variety of textural features have been extracted from radar images and proved to be useful for radar image classification. Miranda et al. (1998) performed classification of JERS-1 SAR data from the rainforest-covered area of the Uaupes River (Brazil) using the semivariogram textural classifier. It was found that the semivariogram textural classifier increased the discrimination between upland and flooded vegetation. Kurosu et al. (1999) studied texture statistics in multi-temporal JERS-1 SAR single-look imagery and demonstrated a significant improvement in the classification accuracy achieved by using the textural features as additional inputs to the classifier. Fukuda and Hirosawa (1999) derived a wavelet-based texture feature set and successfully applied it to multi-frequency PolSAR images of an agricultural area. The classification results indicated that texture was an essential key to the classification of land cover in SAR images. Simard et al. (2000) investigated some texture measures in a study to assess the map-updating capabilities of ERS-1 SAR images in urban areas. The texture measures included histogram measures, wavelet energy, fractal dimension, lacunarity, and semivariograms. The conclusion was that texture improved the classification accuracy, and the measures that performed best were mean intensity, variance, weighted-rank fill ratio, and semivariogram. Rajesh et al. (2001) compared the performance of textural features for characterization and classification of SAR images based on the sensitivity of texture measures for gray-level transformation and multiplicative noise of different speckle levels. Texture features based on gray level ran length, texture spectrum, power spectrum, fractal dimension, and co-occurrence have been considered. They found that fractal-, co-occurrence-, and texture spectrum–based features performed better, with the maximum classification accuracy for sand texture.

While the earlier studies laid the foundation for understanding the importance of textural features in radar image classification, recent advancements have further refined and expanded upon these methodologies. Lee and Pottier (2009) introduced an advanced algorithm that combined traditional textural features with spectral information, enhancing the classification accuracy of urban areas in SAR images. Lopes et al. (2012) utilized machine learning techniques, particularly support vector machines, in conjunction with textural features to classify wetlands and agricultural areas. Their approach demonstrated the potential of integrating machine learning with traditional radar image processing techniques. Zhang and Zhang (2016) developed a convolutional neural network (CNN) model that automatically extracts hierarchical textural features from SAR images, showing significant improvements in classification accuracy, especially in complex landscapes with mixed land cover types. A study by Moreira et al. (2013) explored the fusion of multi-frequency SAR data to enhance the richness of textural features, providing a more comprehensive representation of the landscape and improving the discrimination between LULC classes.

9.4.2.2 Speckle Features

It is well known that the quality of SAR data is degraded by speckle noise, superposing the true radiometric and textural information of the radar image. However, Touzi (2007) demonstrated the potential of speckle analysis in multi-polarized SAR images, revealing that the combination of speckle information from different polarizations can enhance the differentiation of certain land cover types, especially in agricultural landscapes. Esch et al. (2011) further explored that the information on the local development of speckle can be used for the differentiation of basic land cover types in a high-resolution single-polarized TerraSAR-X stripmap image. Combined with local backscatter intensity, the information on the local speckle behavior can be used for the implementation of a straightforward pre-classification of single-polarized TerraSAR-X stripmap images, showing overall accuracies of 77–86%. The results show that unsupervised speckle analysis in high-resolution SAR images supplies valuable information for a differentiation of the water, open land, woodland, and urban area. A study by Wegmuller et al. (2016) introduced a machine learning approach that integrates speckle analysis with other textural features. Their methodology demonstrated improved classification accuracies in urban areas, emphasizing the potential of combining traditional and modern techniques. Furthermore, in a recent study, Esch et al. (2018) highlighted the importance of speckle analysis in change detection using time-series SAR data. They found that understanding the temporal behavior of speckle can provide valuable insights into land cover changes, especially in regions with rapid urbanization.

9.4.2.3 Polarimetric Features

A distinctive characteristic of a PolSAR system is the utilization of polarized waves. The observed polarimetric signatures of the electric field backscattered by the scene depend strongly on the scattering properties of the image objects. In comparison with conventional single-polarization SAR, the inclusion of SAR polarimetry allows for the discrimination of different types of scattering mechanisms that leads to a significant improvement in the quality of classification results (Lee et al., 2001). However, early studies on PolSAR image classification were often conducted on the intensity images of different polarization and their simple combination, such as VH/HH and VH/VV (Wu and Sader, 1987). Recently, some polarimetric decomposition techniques have been introduced for the interpretation and classification of PolSAR images. Polarimetric decomposition techniques aim to separate a received signal by the radar as the combination of the scattering responses of simpler objects presenting an easier physical interpretation, which can be used to extract the corresponding target types in PolSAR images. Pauli decomposition is a well-known decomposition method commonly used for PolSAR data (Cloude and Pottier, 1996). In the Pauli decomposition, if the transmit and receive antennas coincide, the backscattering matrix elements can be arranged into a vector: $k = (a,b,c)/\sqrt{2} = (S_{hh} + S_{yy}, S_{hh} - S_{yy}, 2S_{hv})/\sqrt{2}$ The polarimetric parameters from the Pauli decomposition are associated for three elementary scattering mechanisms: a stands for single or odd-bounce scattering, b represents double or even-bounce scattering, and c denotes volume scattering. As shown in Figure 9.5b, Pauli RGB composition image can be formed with intensities $|a|^2$ (Blue), $|b|^2$ (Red), and $|c|^2$ (Green), which correspond to clear physical scattering mechanisms. The Pauli RGB composition image has become the standard for PolSAR image display and has often been used for visual interpretation, with the tree elements referred to as the Pauli components of the signal. The 3×3 coherency matrix T_3 is defined as the expected value of kk^{*T} (Lee and Pottier, 2009).

$$T_3 = \begin{bmatrix} T_{11} & T_{12} & T_{13} \\ T_{12}^* & T_{22} & T_{23} \\ T_{13}^* & T_{23}^* & T_{33} \end{bmatrix}$$

$$= \frac{1}{2}\begin{bmatrix} |S_{hh}+S_{vv}|^2 & (S_{hh}+S_{vv})(S_{hh}-S_{vv})^* & 2(S_{hh}+S_{vv})S_{hv}^* \\ (S_{hh}-S_{vv})(S_{hh}+S_{vv})^* & |S_{hh}-S_{vv}|^2 & 2(S_{hh}-S_{vv})S_{hv}^* \\ 2S_{hv}(S_{hh}+S_{vv})^* & 2S_{hv}(S_{hh}-S_{vv})^* & 4|S_{hv}|^2 \end{bmatrix} \quad (9.2)$$

where * denotes the conjugate and $|\ |$ denotes the module. The covariance matrix C_3 is a close relative of the coherency matrix T_3 (Lee and Pottier, 2009). They contain the same information, but this information comes in different forms. In addition to the Pauli decomposition, many other decomposition methods have been proposed to express the measured backscattering matrix as a combination of the scattering responses of simpler objects, or to separate coherency or covariance matrix as the combination of second-order descriptors corresponding to simpler or canonical objects presented as an easier physical interpretation. These widely used decomposition methods are the Huynen (Huynen, 1970), Cloude (Cloude, 1985), Barnes (Barnes, 1988), Holm (Holm and Barnes, 1988), Krogager (Krogager, 1990), Van Zyl (Rignot and Vanzyl, 1993), H/A/Alpha (Cloude

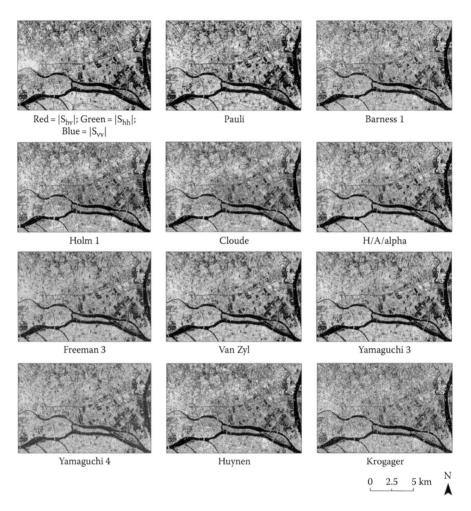

FIGURE 9.6 RGB composition images presenting different polarimetric decompositions. Source: Qi et al. (2012).

and Pottier, 1997), Freeman (Freeman and Durden, 1998), Yamaguchi (Yamaguchi et al., 2005), and Touzi (Touzi, 2007). The RGB composition images that present some of these decompositions are shown in Figure 9.6.

A number of classification methods based on decomposition results have been explored (Cloude and Pottier, 1997; Ferro-Famil et al., 2001; Lee et al., 1999a; Lee et al., 1994; Shimoni et al., 2009). The results of these methods indicated that polarimetric decomposition significantly improved the classification accuracy of PolSAR images. Qi et al. (2012) integrated polarimetric parameters extracted using different polarimetric decomposition methods into the classification of RADARSAT-2 PolSAR image. The results showed that polarimetric parameters were important in identifying different vegetation types and distinguishing between vegetation and urban/built-up.

9.4.2.4 Interferometric Features

Most SAR applications make use of the amplitude of the return signal and ignore the phase information. However, SAR interferometry uses the phase of the reflected radiation. It uses two images of the same area taken from the same or slightly different positions and finds the difference in phase

Land Use and Land Cover Mapping and Monitoring

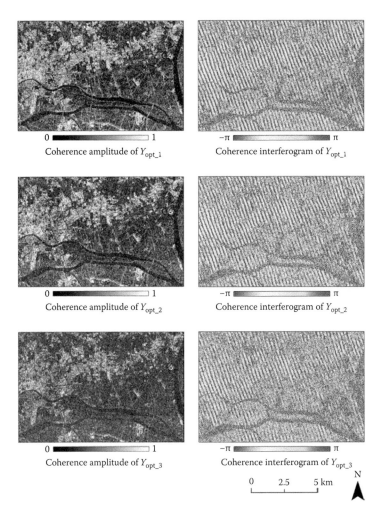

FIGURE 9.7 Polarimetric interferometric parameters extracted using PolSAR interferometry techniques for LULC classification. Source: Qi et al. (2012).

between them, producing an image known as an interferogram. The interferogram is measured in radians of phase difference and is recorded as repeating fringes which each represent a full 2π cycle (Figure 9.7). Coherence is the magnitude of an interferogram's pixels, divided by the product of the magnitudes of the original image's pixels (Figure 9.7). It is usually calculated on a small window of pixels at a time, from the complex interferogram and images. In an interferogram, the coherence serves as a measure of the quality of an interferogram, indicating a tiny and invisible change occurred in the images or the information of the surface type, such as vegetation and rocks. High coherence makes for attractive, not-noisy interferograms, whereas low coherence makes unattractive, noisy interferograms. Substantial improvements in radar image classification can be achieved by integrating interferometric information that is related to the structure and complexity of the observed objects into the classification. Askne et al. (1997) presented a model for interferometric SAR observations of forests. Such observations provide complementary information to the intensity observations and provide new information on coherence and effective interferometric tree height. This study showed the possibility of discriminating forested and non-forested areas using interferometric information. The phase stability of anthropogenic structures between SAR images has led several researchers to propose long time-scale phase correlation, or coherence, as a good measure of urban extent, and thus an appropriate tool for mapping urban change (Grey et al., 2003; Qi et al., 2012). The three optimum complex polarimetric interferometric coherences (Papathanassiou and Cloude, 2001) were extracted from two repeat-pass RADARSAT-2 PolSAR images for LULC classification by Qi et al. (2012). The polarimetric interferometric coherences were found to be very useful in reducing the confusion between urban and non-urban areas (Figure 9.7). As shown in Figure 9.7, there is a strong contrast between urban and non-urban areas in the images of polarimetric interferometric parameters. The repeat cycle of RADARSAT-2 is 22 days, which produces a very strong temporal decorrelation for non-urban areas, such as croplands and natural vegetation. Croplands and natural vegetation are significantly influenced by temporal decorrelation and lose coherence within a few days or weeks as a result of growth, movement of scatterers, and changing moisture conditions. In contrast, within urban/built-up areas, coherence remains high even between image pairs separated by a long time interval.

Zhang et al. (2021) introduced an urban flooding index using Sentinel-1 polarimetric SAR images, which showcased the potential of SAR interferometry in detecting urban flood areas (Figure 9.8). Massonnet and Feigl (2015) further showcased the significance of interferometric coherence in distinguishing between urban and agricultural terrains. Bamler and Hartl (2019) amalgamated InSAR coherence with optical data to enhance LULC classification in tropical regions, underlining the synergistic essence of these datasets.

9.4.2.5 Multi-Frequency Information

Mandy studies have been conducted on LULC classification using airborne multi-frequency SAR systems or fusing SAR data obtained by different orbital SAR systems with different frequencies (Dobson et al., 1996; Pierce et al., 1994). All these studies indicated that multi-frequency SAR data achieves higher accuracy than single-frequency SAR data in the classification. Pierce et al. (1994) performed land cover classification using SIR-C/X-SAR imagery. Using L- and C-bands alone, the single-scene classification accuracies were quite good, with each image better than 90%. With the addition of X-band data, the overall accuracies improved to 98%, due to the enhanced ability to distinguish the major tree classes. Dobson et al. (1996) combined C-band ERS-1 data with L-band JERS-1 data for land cover classification. They found that the results in a classification procedure for the composite image were superior to that obtained from either of the two sensors alone. In a more recent study, Qi et al. (2019) delved deeper into the scattering mechanisms and their implications for selecting optimal combinations of polarimetric SAR frequency bands. Their research emphasized the importance of understanding these mechanisms in achieving more accurate land cover classifications using multi-frequency SAR data.

9.4.2.6 Multiple Incidence Angle Information

Since radar backscatter from targets is affected by the incidence angle, observing targets with different incidence angle modes can provide more information to LULC classification. Cimino et al. (1986) used multiple incidence angle SIR-B data to discriminate various forest types by their relative brightness versus incidence angle signatures. The results of this study indicated that (1) different forest species, and structures of a single species, may be discriminated using multiple incidence angle radar imagery and (2) it is essential to consider the variation in backscatter due to incidence angle when analyzing and comparing data collected via varying frequencies and polarizations. Grunsky (2002) investigated the potential of the use of multi-beam RADARSAT-1 radar imagery in assisting terrain mapping over large areas. Principal components analysis was applied to ascending and descending standard beam modes with incidence angles of 20–27 degrees (S1) and 45–49 degrees (S7). The resulting components yielded imagery that highlights geomorphology, geologic structure, variation in vegetation, and an indirect measure of moisture in the study area. Xu et al. (2019) further emphasized the significance of incidence angle in fully polarimetric SAR images for land cover classification, suggesting that the incidence angle plays a crucial role in influencing the scattering mechanisms and hence the classification results.

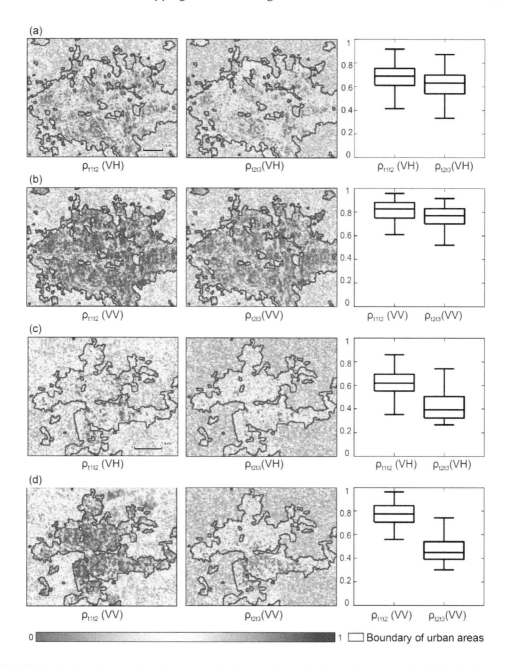

FIGURE 9.8 Interferometric coherence variation in (a) unflooded urban areas in VH polarization, (b) unflooded urban areas in VV polarization, (c) flooded urban areas in VH polarization, and (d) flooded urban areas in VV polarization (Note: ρt1t2 and ρt2t3 denote the interferometric coherence between the first and second images and that between the second and third images, respectively).

9.4.2.7 Feature Selection

Given that many potential features could be integrated into the classification of radar images, feature selection presents a problem in the classification. Using all available features in classification is improper because computation is intensive and some features may degrade classification performance (Laliberte et al., 2006). Since the recent introduction of polarimetric decomposition theorems, which have brought about abundant polarimetric parameters, the problems of feature selection have become more intractable. Therefore, the selection of suitable features is essential for successfully implementing radar image classification. Many approaches, such as principal component analysis, minimum noise fraction transform, discriminant analysis, decision boundary feature extraction, non-parametric weighted feature extraction, wavelet transform, and spectral mixture analysis may be used for feature extraction in order to reduce the data redundancy inherent in remotely sensed data or to extract specific land cover information. Most of these feature selection approaches can be found in the Weka software (Witten et al., 2011).

9.4.3 Selection of Classifiers

After the determination of features used for classification, the next step is to select a suitable classifier to implement the classification based on the selected features. A number of classifiers have been developed for the classification of remote sensing data, and each classifier has its own strengths and limitations (Lu and Weng, 2007). Many classifiers have been tested with SAR imagery, and these classifiers can be grouped as parametric and non-parametric classifiers.

9.4.3.1 Parametric Classifier

The parametric classifiers assume that the classification dataset follows a specific distribution such as Gaussian distribution, and that the statistical parameters (e.g., mean vector and covariance matrix) generated from the training samples are representative. Maximum likelihood (ML) is one of the most widely used parametric classifiers with radar images. Lim et al. (1989) introduced a ML decision rule based on the multi-variate complex Gaussian distribution of the elements of the coherent scattering matrix. In order to reduce the effects of speckle in PolSAR images, data are generally processed through incoherent averaging and are represented by coherency matrices. Lee et al. (1994) developed a Wishart classifier by introducing the ML decision rule based on the multi-variate complex Wishart distribution for the polarimetric coherency matrix. A k-mean algorithm was applied to iteratively assign the pixels of the PolSAR image to one of the classes using the ML rule. Lee et al. (1999a) proposed an unsupervised classification by combining the decomposition technique with unsupervised classification based on the Wishart classifier. The H/A/Alpha decomposition technique was used to provide an initial guess of the pixel distribution into the classes that produces a better convergence of the unsupervised classification algorithm. This unsupervised classification method has been extended to the classification of multi-frequency PolSAR data (Ferro-Famil et al., 2001).

Other commonly used parametric classifiers include maximum a posteriori (MAP), Bayesian, iterated conditional mode (ICM) contextual classifiers, fuzzy c-means clustering, and Markov random field models. Rignot et al. (1992) implemented unsupervised classification of PolSAR data by applying MAP to the covariance matrix. This method used both polarimetric amplitude and phase information, was adapted to the presence of image speckle, and did not require an arbitrary weighing of the different polarimetric channels. Vanzyl and Burnette (1992) classified polarimetric SAR images using a Bayesian classifier in which the classification was done iteratively. The results showed that only a few iterations were necessary to improve the classification accuracy dramatically. A hierarchical Bayesian classifier was developed for the classification of short vegetation using multi-frequency PolSAR data (Kouskoulas et al., 2004). It was shown that this classifier outperformed ML classifier with Gaussian assumption. Freitas et al. (2008) applied the ICM classifier to the classification of P-band PolSAR data. The ICM classifier enabled the use of contextual information to improve the classification accuracy. Based on the complex Wishart distribution of the complex covariance matrix, the fuzzy c-means clustering algorithm was used for unsupervised segmentation of multi-look PolSAR images (Du and Lee, 1996). Dong et al. (2001) implemented segmentation and classification of PolSAR data using a Gaussian Markov random field model. The model performed the classification based on image objects, which usually consist of multi-pixels, providing reliable measurement statistics and texture characteristics. Tison et al. (2004) proposed a classification method that uses a mathematical model relying on the Fisher distribution and the log-moment estimation for high-resolution SAR images over urban areas. Their contribution was the choice of an accurate model for high-resolution SAR images over urban areas and its use in a Markovian classification algorithm.

9.4.3.2 Non-parametric Classifiers

Non-parametric classifiers do not assume a particular probability density distribution of the input data, and no statistical parameters are needed to separate images. These classifiers are thus especially suitable for the incorporation of non-intensity data into the classification of SAR images. Among the most commonly used non-parametric classification classifiers for radar images are neural networks (Bruzzone et al., 2004; Hara et al., 1994; Tzeng and Chen, 1998), support vector machines (Hosseini et al., 2011; Lardeux et al., 2009; Tan et al., 2011), decision trees (Qi et al., 2012; Simard et al., 2000), and knowledge-based classifiers (Dobson et al., 1996; Pierce et al., 1994). There are several advantages of using non-parametric classifiers with SAR images. First, non-parametric classifiers may provide better classification accuracies than parametric classifiers in complex landscapes (Hosseini et al., 2011; Lardeux et al., 2009). Second, non-parametric classifiers are easy to integrate different types of data (e.g., textural, spatial, polarimetric, interferometric, GIS ancillary data) into the classification of radar images because of their non-parametric nature (Hosseini et al., 2011; Simard et al., 2000; Tan et al., 2011). Third, they are easy to use with multiple (e.g., multi-temporal) images (Bruzzone et al., 2004; Qi et al., 2012). Furthermore, each of these classifiers has its own strengths. The neutral networks have arbitrary decision boundary capability and provide fuzzy output values to take into account class mixture and the degree of membership of a pixel (Tzeng and Chen, 1998). The advantage of SVMs for data classification is their ability to be used as an efficient algorithm for non-linear classification problems, particularly in the case of extracting feature vectors from fully polarimetric SAR data (Hosseini et al., 2011). Decision tree algorithms are efficient in selecting features and implementing classification as well as provide clear classification rules that can be easily interpreted based on the physical meaning of the features used in the classification (Qi et al., 2012). This is very helpful in providing physical insight for the classification of radar images. Knowledge-based classifiers are transportable and robust because they are defined using theoretical understanding, as verified by empirical evidence, of the knowledge of the physics involved in the sensor/scene interaction (Dobson et al., 1996; Pierce et al., 1994). There are also some other non-polarimetric classifiers, such as sigma-tree structured near-neighbor classifiers (Barnes and Burki, 2006), spectral graph partitioning (Ersahin et al., 2010), and subspace (Bagan et al., 2012). Readers who want to have a detailed description of a specific classification approach should refer to cited references.

In recent years, the random forest classifier and deep learning have carved a niche for themselves in LULC classification tasks, especially when applied to SAR images. Random forest, a type of ensemble learning method, is known for its capability to handle high-dimensional data and provide accurate LULC classifications. Its application to SAR and optical imagery has yielded insights into urban growth, deforestation, and agricultural monitoring (Rodriguez-Galiano et al., 2012; Belgiu and Drăguţ, 2016). Deep learning classifiers, particularly CNNs, have been a revolutionary force in the remote sensing realm. These models have shown adeptness in extracting hierarchical features and discerning complex patterns in high-resolution imagery, including those from SAR sensors (Zhu et al., 2017). Additionally, frameworks such as U-Net and SegNet have been adapted to satellite imagery for semantic segmentation tasks, further enhancing the granularity of LULC classifications (Ronneberger et al., 2015). For multi-temporal analyses, the integration of recurrent architectures, such as Long Short-Term Memory (LSTM) networks, with CNNs offers a robust approach to capture temporal changes and dependencies in land cover types (Audebert et al., 2016).

9.5 CHANGE DETECTION METHODS FOR RADAR IMAGERY

Many change detection methods have been developed for the use of remote sensing data (Lu et al., 2004). However, there is still a general lack of the use of radar remote sensing for LULC change detection. Indeed, studies related to radar imagery change detection are fewer and more recent than optical-based ones. Change detection methods for radar imagery can be divided into two categories: (1) unsupervised change detection methods and (2) methods combining unsupervised change detection with post-classification comparison (PCC).

9.5.1 Unsupervised Change Detection Methods

Unsupervised change detection methods for radar images are usually made up of three steps: (1) image preprocessing, (2) comparing two images to generate a change magnitude map (e.g., Figure 9.9), and (3) applying threshold methods to the change magnitude map to separate "change" from "no-change." Image ratioing, statistical measure, and image differencing are the most widely utilized unsupervised change detection methods for radar imagery.

9.5.1.1 Image Ratioing

Image ratioing has been widely used in change detection with single-channel (i.e., single-frequency and single-polarization) radar images because it can effectively reduce multiplicative noise in the images (Rignot and Vanzyl, 1993). The comparison of SAR images can be carried out according to a ratio/logarithmic ratio (log-ratio) operator. An optimal threshold value is then applied to the ratio/log-ratio image (change magnitude image) to identify changes. "Trial-and-error" procedures are typically used to determine optimal threshold value (Dierking and Skriver, 2002; Singh, 1989). However, such manual operations typically turn out to be time consuming; in addition, the quality of their results critically depends on the visual interpretation of the user. To overcome this problem, many automatic thresholding algorithms have been proposed for the analysis of ratio/log-ratio images. Bazi et al. (2005) introduced a generalized Gaussian model for a log-ratio image and applied Kittler and Illingworth (K&I) thresholding algorithm to the log-ratio image to automatically detect changes. The modified K&I criterion was derived under the generalized Gaussian assumption for modeling the distributions of changed and unchanged classes. This parametric model was chosen because it is capable of better fitting the conditional densities of classes in the log-ratio image. However, in this method, changes are assumed to be on one side of the histogram of the log-ratio image, which is not true for all change detection problems. In particular, changes may be present on both sides of the histogram of the log-ratio image. Therefore, the method was further improved by combining generalized Gaussian distributions with a multiple-threshold version of K&I to detect changes on both sides of the histogram of log-ratio images (Bazi et al., 2006). Additionally, Moser and Serpico (2006) developed a K&I minimum-error thresholding algorithm to take into account the non-Gaussian distribution of the amplitude values of SAR images. This method could be applied to images acquired by two distinct sensors with different bands and polarizations. Carincotte et al. (2006) calculated change magnitude images using a log-ratio detector and used a fuzzy version of hidden Markov chains (HMCs) to classify the log-ratio images into change and no-change classes. This method took into account the fuzzy aspect of the scene behavior and change detection complexity and reached a satisfactory reliability level in the context of SAR images.

Many advanced methods have also been developed based on image ratioing to address different issues in change detection with radar images. Bujor et al. (2004) developed a method that applies log-cumulants to the detection of spatiotemporal discontinuities in multi-temporal SAR images. The contrast and the heterogeneity information was extracted by a "multi-temporal" application of the ratio of local means and by new three-dimensional texture parameters based on the log-cumulants. After that, the resulting attributes that measure the time variability or the presence of spatial features were merged. An interactive fuzzy fusion approach was proposed to provide

Land Use and Land Cover Mapping and Monitoring

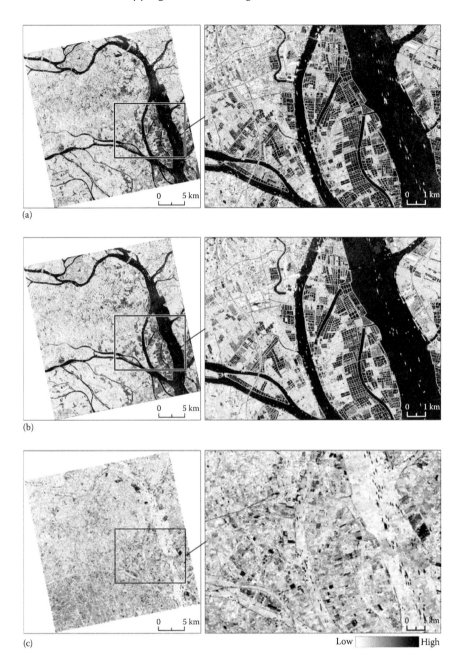

FIGURE 9.9 (a) RADARSAT-2 image acquired on March 21, 2009, (b) RADARSAT-2 image acquired on September 29, 2009, (c) change magnitude calculated using change vector analysis. Source: Qi and Yeh (2013).

end users with a simple and easily understandable tool for tuning the change-detection results. This change detection method could enable geophysicists to detect regions that contain spatial features (roads, rivers, etc.) or temporal change (flooded areas, coastline erosion, etc.). Bovolo and Bruzzone (2005) developed an approach that exploits a wavelet-based multi-scale decomposition of the log-ratio image (obtained by a comparison of the original multi-temporal SAR images) aimed at achieving different scales (levels) of representation of the change signal. Each scale was characterized by a different trade-off between speckle reduction and preservation of geometrical details. For each pixel, a subset of reliable scales was identified on the basis of a local statistic measure applied to scale-dependent log-ratio images. The final change-detection result was obtained according to an adaptive scale-driven fusion algorithm. This method could improve the accuracy and geometric fidelity of the change-detection map. Gamba et al. (2006) jointly used feature-based and pixel-based techniques to address the problem of change detection from SAR images. Image ratioing was used for deriving the first rough change map at the pixel level, and then linear features were extracted from multiple SAR images and compared to confirm pixel-based changes. This method proved to be effective in dealing with misregistration errors caused by reprojection problems or differences in the sensor's viewing geometry, which are common in multi-temporal SAR images. Bovolo and Bruzzone (2007) proposed a split-based approach (SBA) to automatic and unsupervised change detection in large-size multi-temporal SAR images. The method consisted of three steps: (1) a split of the computed ratio image into subimages, (2) an adaptive analysis of each subimage, and (3) an automatic split-based threshold-selection procedure. The SBA could detect changes in a consistent and reliable way in images of large size also when the extension of the changed area is small. Thus, it could be suitable for defining a system for damage assessment in multi-temporal SAR images.

Single-channel SAR images may result in poor discrimination between changed and unchanged areas because of the limited spectral and polarization information. Compared with single-channel SAR, multi-channel (e.g., multi-polarization and/or multi-frequency) SAR presents a great potential and is expected to provide an increased discrimination capability while maintaining the insensitivity to atmospheric and sun-illumination conditions. Change detection methods based on image ratioing have also been developed for change detection with multi-channel SAR images. Moser et al. (2007) combined the Landgrebe–Jackson expectation-maximization (LJ-EM) algorithm with a SAR-specific version of the Fisher transform to iteratively compute a scalar transformation of the multi-channel ratio image that optimally discriminates "change" from "no-change." A Markov random field (MRF) approach was also integrated into the method to take advantage of the contextual information, which is crucial to reduce the impact of speckle on the change detection results. This method can be used to generate change maps from multi-channel SAR images acquired over the same geographic region in different polarizations or at different frequencies at different times. However, its main drawback is that, even though a convergent behavior is experimentally observed, no theoretical proof of convergence is available yet. Moser and Serpico (2009) proposed an unsupervised automatic contextual change detection method for multi-channel amplitude SAR images based on image ratioing and MRFs. Each channel of the SAR amplitude ratio image was considered a separate "information source," and an additional source was derived from the spatial context; the multi-source data-fusion task was addressed together with the image thresholding task using an MRF-based approach. In order to estimate model parameters, a case-specific novel formulation of LJ-EM was developed and combined with the method of log-cumulants. This choice also overcame the convergence drawback of the approach proposed by Moser et al. (2007) because of the robust analytical properties of EM-based estimation procedures.

9.5.1.2 Statistical Measure

Inglada and Mercier (2007) developed a statistical similarity measure for change detection in multi-temporal SAR images. This measure was based on the evolution of the local statistics of the image between two dates. The local statistics were estimated by using a cumulant-based series expansion, which approximated probability density functions in the neighborhood of each pixel in the image. The degree of evolution of the local statistics was measured using the Kullback-Leibler divergence. An analytical expression for this detector was given, allowing a simple computation which depends on the four first statistical moments of the pixels inside the analysis window only. This proposed change detector outperformed the classical mean ratio detector, and the fast computation of this detector allowed a multi-scale approach in the change detection for operational use. The so-called multi-scale change profile (MCP) was introduced to yield change information on a wide range of scales and to better characterize the appropriate scale.

Chatelain et al. (2008) studied a new family of distributions constructed from multi-variate gamma distributions to model the statistical properties of multi-sensory SAR images. These distributions, referred to as multi-sensory multi-variate gamma distributions (MuMGDs), were potentially interesting for detecting changes in SAR images acquired by different sensors having different numbers of looks. This study compared different estimators for the parameters of MuMGDs. These estimators were based on the maximum likelihood principle, the method of inference function for margins, and the method of moments. The estimated correlation coefficient of MuMGDs showed interesting properties for detecting changes in radar images with different numbers of looks.

When working with multi-look fully polarimetric SAR data, an appropriate way of representing the backscattered signal consists of the covariance matrix. For each pixel, this is a 3×3 Hermitian positive definite matrix that follows a complex Wishart distribution. Based on this distribution, Conradsen et al. (2003) proposed a test statistic for equality of two such matrices and an associated asymptotic probability for obtaining a smaller value of the test statistic was derived and applied successfully to change detection in polarimetric SAR data. If used with HH, VV, or HV data only, the test statistic was reduced to the well-known test statistic for equality of the scale parameters in two gamma distributions. The derived test statistic and the associated significance measure could be applied as a line or edge detector in fully polarimetric SAR data. The new polarimetric test statistic was much more sensitive to the differences than test statistics based only on the backscatter coefficients.

9.5.1.3 Image Differencing

Image differencing is one of the most widely used unsupervised change detection methods for the use of optical remote sensing images. It has also been used for change detection with single-channel SAR images (Cihlar et al., 1992; Villasenor et al., 1993). The procedure of image differencing is similar to that of image ratioing, but the change magnitude in image differencing methods is obtained by subtracting amplitude or intensity images of two SAR images acquired at different times. A widely accepted assumption is that the distribution of pixels in the change and no-change areas in the difference image can be approximated as a mixture of Gaussian distributions (Camps-Valls et al., 2008). Thus, the EM algorithm is commonly used on change magnitude to detect changes because it finds clusters by determining a mixture of Gaussians that fit a given data set (Bruzzone and Prieto, 2000). In addition to using radar amplitude images, Grey et al. (2003) applied an image differencing technique to SAR interferometric coherence data to map urban change.

Celik (2010) proposed an unsupervised change detection algorithm for satellite images by conducting probabilistic Bayesian inferencing to perform unsupervised thresholding over subband difference images generated at the various scales and directional subbands using the dual-tree complex wavelet transform (DT-CWT) for representation. Aside from the intrascale information, the interscale information inherently provided by the DT-CWT was exploited to effectively reduce both the false-alarm and miss-detection rates. Extensive simulation results showed that this algorithm consistently performed quite well on both objective and subjective change-detection performance evaluation—under either noise-free or zero-mean Gaussian (or speckle) noise interference cases. Furthermore, the correct- and false-classification rates were almost invariant (insensitive) with respect to noise powers added to the images.

9.5.1.4 Other Methods

Other approaches have also been proposed in the literature to deal with radar change detection, including multi-temporal coherence analysis (Rignot and Vanzyl, 1993), integration of segmentation with multi-layer perceptron and Kohonen neural networks (White, 1991), case-based reasoning (Li and Yeh, 2004), maximum likelihood approach (Lombardo and Pellizzeri, 2002), radon transform and Jeffrey divergence (Zheng and You, 2013), and graph-cut and generalized Gaussian model (Zhang et al., 2013).

9.5.2 Combining Unsupervised Change Detection and Post-classification Comparison

Although unsupervised change detection approaches are relatively simple, straightforward, and easy to implement and interpret, they cannot determine types of land cover change. PCC can provide information on both the changed areas and the type of land cover change in these areas because it performs change detection by comparing separate supervised classifications of images acquired at different times (Lu et al., 2004). However, PCC is limited by the accuracies of the classification. PCC results exhibit accuracies similar to the product of the accuracies of each individual classification (Stow et al., 1980). Most orbital SAR systems are single-frequency types, and SAR images suffer from serious speckle noise caused by the SAR system's coherent nature. When applied to SAR images, PCC may yield poor results because of the poor classification accuracies caused by the speckle noise and limited spectral information. Furthermore, change detection and classification using radar images are often based on pixel-based methods, which are prone to speckle noise and difficult to use in the extraction of spatial and textural information.

A method that integrates change vector analysis (CVA) and PCC with OOIA has been developed to detect land cover changes from two repeat-pass PolSAR images (Qi and Yeh, 2012). OOIA allows for land cover change detection performed on image objects and the incorporation of textural and spatial information of the image objects. It was integrated into the method to suppress the speckle effect and extract textural and spatial information to support PolSAR image classification. In this method, two PolSAR images acquired over the same area at different times were segmented hierarchically to delineate land parcels (image objects) (Figure 9.10). The hierarchical segmentation of PolSAR images not only avoided inconsistencies in the delineation of land parcels but also delineated the changed land parcels. As shown in Figure 9.10, the changed land parcels, such as land parcels a and b_2, were delineated in the hierarchical segmentation. Afterwards, CVA was combined with PCC to detect land cover changes based on OOIA. Parcel-based CVA was performed with the features extracted from the coherency matrices of the PolSAR images to calculate the change magnitude map (Figure 9.9). The EM algorithm was applied to the change magnitude map to identify the changed land parcels. PCC that was based on a parcel-based classification approach, which integrated polarimetric decomposition, decision tree algorithms, and support vector machines, was then used to determine the type of land cover change for the changed land parcels. The combination of CVA and PCC detected different types of land cover change and also reduced the effect of the classification errors on the land cover change detection. The main advantage of this method is that it provides information on both the changed areas and the type of land cover change in these areas. Compared with conventional PCC that was based on the Wishart supervised classification, this change detection method significantly reduced overall error and false alarm rates.

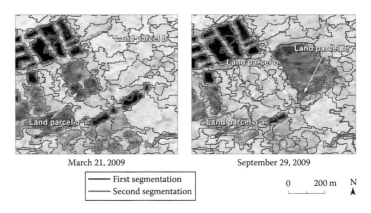

FIGURE 9.10 Hierarchical segmentation of RADARSAT-2 PolSAR images acquired at different times. Source: Qi and Yeh (2013).

9.6 APPLICATIONS OF RADAR IMAGERY IN LULC MAPPING AND MONITORING

The major applications of radar remote sensing in LULC mapping and monitoring include LULC classification and change detection, forestry inventory and mapping, crop and vegetation identification, application on urban environment, snow and ice mapping, application to wetlands, and shoreline change detection.

9.6.1 LULC Classification and Change Detection

Many studies have reported that radar remotely sensed images can provide valuable information for timely LULC classification and change detection (Table 9.9). Peng et al. (2005) found that stereo RADARSAT-1 SAR images provided valuable data sources for land cover mapping, especially in mountainous areas where cloud cover is a problem for optical data collection and topographical data are not always available. The joint use of interferometric SAR (InSAR) and textural features extracted from SAR images and its application to LULC classification has been assessed by many studies. Strozzi et al. (2000) produced two land use maps and a forest map of three different areas in Europe by using ERS-1/2 SAR interferometry. The three areas represented various geomorphological regions with different cover types. Their classification results showed that land use classification accuracies on the order of 75% are possible with, in the best case, simultaneous forest and non-forest accuracies of around 80–85%. Qi et al. (2012) implemented land cover classification by using textural, polarimetric, and interferometric information extracted from RADARSAT-2 PolSAR images. The overall accuracy of 86.64% was achieved for classifying land cover types including built-up areas, water, barren land, crop/natural vegetation, lawn, banana, and forests.

If single-frequency SAR systems are used, there is generally a considerable degree of ambiguity between different LULC types (Li et al., 2012). Combining multi-frequency SAR scenes has proved to be a valuable tool for distinguishing different LULC classes. Shimoni et al. (2009) investigated the complementarity and fusion of different frequencies (L- and P-band), PolSAR, and polarimetric interferometric (PolInSAR) data obtained by E-SAR for land cover classification. The results showed that the overall accuracy for each of the fused sets was better than the accuracy for the separate feature sets. Moreover, fused features from different SAR frequencies were complementary and adequate for land cover classification and that PolInSAR was complementary to PolSAR information and that both were essential for producing accurate land cover classification. Evans and Costa (2013) used multitemporal L-band ALOS/PALSAR, C-band RADARSAT-2 and ENVISAT/ASAR data to map ecosystems and create a lake distribution map of the Lower Nhecolandia subregion in the Brazilian Pantanal. They provided the first fine spatial resolution classification showing the spatial distribution of terrestrial and aquatic habitats for the entire subregion of Lower Nhecolandia. Holecz et al. (2009) generated land cover maps and changes over large areas by fusing single-date ALOS PALSAR Fine/Dual beam with multi-temporal ENVISAT ASAR Mode/Alternating polarization images. The results clearly demonstrated that the synergetic use enabled the reliable identification of key land cover types (in particular, cropped areas, bare soil areas, forestry, forest clear-cut, forest burnt areas, water bodies) and their evolution over time.

The use of multi-temporal SAR data can also increase the number of reliably distinguishable LULC classes. Land cover classification was conducted by using a time series of 14 ERS-1/2 SAR tandem image pairs (Engdahl and Hyyppä, 2003). A total of 14 tandem coherence images and two coherence images with a longer temporal baseline (36 and 226 days) were used in the classification. The overall accuracy for six classes, field/open land, dense forest, sparse forest, mixed urban, dense urban, and water, was found to be 90% with kappa coefficient of 0.86. Huang et al. (2008) found that multi-temporal ERS-2 SAR imagery had great potential for land use mapping or resource investigations in coastal zones under rapid development. They stated that multi-temporal SAR data can be regarded as the first choice for monitoring land uses and their dynamic changes in coastal zones that are often affected by heavy cloud or rainy weather.

TABLE 9.9
Studies on the Application of Radar Imagery in LULC Classification and Change Detection

LULC Types	Data	Accuracy	Strengths	Limitations	Reference
Bare soil, natural forest, pasture, planted forest, sugarcane plantations, soya plantations, urban or built-up areas, and water	ERS-1 SAR and Landsat-5 TM images	94.85%	Compared with using TM data alone, combining TM with SAR increases the classification accuracy for urban, pasture, and forest	Both SAR and TM images are needed	Kuplich et al. (2000)
Urban areas, water, forest, and open land	ERS-1/2 InSAR data	75%	Interferometric information is used in LULC classification	Meteorological condition effect	Strozzi et al. (2000)
Field/open land, dense forest, sparse forest, mixed urban, dense urban, and water	ERS-1/2 InSAR data	90%	Interferometric coherence carries more land cover–related information than the backscattered intensity	A large number of tandem pairs are needed	Engdahl and Hyyppä (2003)
Forest, glacial ice, grass, rock, sandy soil, shrub, snow, and water	A stereo pair of RADARSAT-1 images	83%	Suitable for land cover mapping in mountainous areas	Feature extraction and selection needs to be optimized	Peng et al. (2005)
Upland forest, lowland forest, advanced successional vegetation, intermediate successional vegetation, initial successional vegetation, degraded pasture, cultivated pasture, agroforestry, coffee plantation, infrastructure, water, and non-vegetation lowland	Landsat ETM+ and RADARSAT-1 data	72.07%	Incorporation of data fusion and textures increases classification accuracy by approximately 5.8–6.9% compared to Landsat ETM+ data	Both SAR and ETM+ images are needed	Lu et al. (2007)
Residential, road and dike, cotton field, paddy field, mixed field, orchard, forest land, river and gulf, canal and pond, aqua-farm ponds, tideland and wild land, and non-vegetated saline soil land	Multi-temporal ERS-2 SAR imagery	77.34%	Multi-temporal SAR imagery has great potential for LULC mapping or resource investigations in coastal zones under rapid development	The best time window for image acquisition should be considered	Huang et al. (2008)
Built-up areas, rural residential areas, bare land, paddy fields, vegetable land, orchards, forest, river, and fishponds	RADARSAT-1 images	75.2%	SAR provides a unique opportunity for detecting LULC changes within short intervals (e.g., monthly) in tropical and sub-tropical regions	Monthly SAR images are needed	Li et al. (2009)
Residences, roads, forests, wheat and corn fields, pastures, abandoned areas, bare soil, and rivers	E-SAR L- and P-band	75.36%	Fused features from different SAR frequencies are complementary and adequate for LULC classification	Results could be further improved by introducing spatial information into the fusion	Shimoni et al. (2009)

LULC classes	Data	Accuracy	Comments	Reference
Forest, succession, agro-pasture, water, wetland, and urban	ALOS PALSAR or RADARSAT-2 data	72.2% (L-band); 54.7% (P-band)	L-band data provides much better land cover classification than C-band data	Li et al. (2012)
Built-up, water, barren land, forest, lawn, banana, and cropland/natural vegetation	RADARSAT-2 data	86.64%	Incorporation of textural, polarimetric, and interferometric features into the classification	Qi et al. (2012)
Forests, paddy fields, croplands, lotus fields, grasslands, golf courses, parks, settlements, and water	ALOS AVNIR-2 and PALSAR images	90.34%	Combining AVNIR-2 and PALSAR data produce better accuracy	Bagan et al. (2012)
Forest woodland, open wood savanna, open grass savanna, swampy grassland, agriculture, vazantes, freshwater lake, and brackish lake	ALOS PALSAR, RADARSAT-2, and ENVISAT ASAR data	83%	Combining dual-season, C-, and L-band is essential for providing a relatively high overall accuracy of land cover classification	Evans and Costa (2013)

LULC classification with either L- or C-band is a challenge for fine LULC classification system

PolSAR images segmentation remains a challenge

Both AVNIR-2 and PALSAR data are needed

SAR data of different frequencies are needed

SAR data and optical data provide complementary information, and their combination often leads to increased classification accuracy. The use of Landsat Mss and SEASAR SAR data was evaluated in discriminating suburban and regional cover in the eastern fringe area of the Denver, Colorado, metropolitan area (Toll, 1985). The SEASAR SAR data provided a measure of surface geometry that complemented the reflective characteristics of Landsat MSS visible and near-infrared data. The integration of Landsat imagery with SAR data obtained by ERS-1, RADARSAT-1, and ALOS PALSAR for LULC classification has been carried out by many studies (Kuplich et al., 2000; Larranaga et al., 2011; Lu et al., 2007). All these studies reported a significant improvement achieved by the integration in LULC classification. Bagan et al. (2012) evaluated the potential of combined ALOS visible and near-infrared (AVNIR-2) with PALSAR fully polarimetric SAR data for land cover classification. They confirmed that, when the combined optical AVNIR-2, PALSAR, and polarimetric coherency matrix data were used, the classification accuracy of was better than that when other data combinations were used.

SAR data is also useful in the detection of LULC changes. Villasenor et al. (1993) found that the temporal changes between repeat-pass ERS-1 SAR images of the North Slope of Alaska were largely due to changes in soil and vegetation liquid water content induced by freeze/thaw events. This confirmed the viability of radar backscatter intensity comparisons using repeat-pass images as a means of change detection. Orbital SAR data has proved to be a unique opportunity for detecting land use changes within short intervals in tropical and sub-tropical regions with cloud cover (Li et al., 2009). By using object-oriented analysis with case-based reasoning, Li et al. (2009) successfully detected land use changes at monthly intervals by using multi-temporal RADARSAT-1 SAR images.

Recent advances in SAR remote sensing have proven particularly fruitful for LULC classification and change detection. For instance, Sentinel-1 SAR data has been increasingly utilized for various land cover classifications, especially in agricultural regions (Bouvet et al., 2018). The ALOS-2 PALSAR-2, with its L-band capabilities, has been pivotal for urban and rural classifications, offering precise details distinguishing natural terrains from man-made structures (Shimada et al., 2014). There's been a growing trend of integrating Sentinel-1 SAR data with optical data from satellites like Sentinel-2 for better LULC classifications, proving instrumental in diverse applications from forestry to urban sprawl (Hoekman and Reiche, 2015). Another interesting application has been in the assessment of land degradation, where SAR data can provide crucial insights into aspects like soil moisture and shifts in terrain (Lasaponara and Masini, 2016). RADARSAT-2, with its C-band capabilities, remains a preferred choice for monitoring delicate ecosystems, such as wetlands, ensuring that their degradation is noted and acted upon in a timely manner (Mahdianpari et al., 2018).

9.6.2 Forestry Inventory and Mapping

Radar remote sensing can contribute to the inventory and mapping of forest as well as to an understanding of ecosystem processes (Table 9.10). Early research to apply imaging radar to forest mapping was conducted using multiple incident angle SIR-B data (Cimino et al., 1986). The research found that different forest species might be discriminated using multiple incidence angle radar imagery, and the variation in backscatter due to incidence angle should be considered when analyzing and comparing data collected at varying frequencies and polarizations. After the availability of some orbital radar systems, such as the ERS-1/2, JERS-1, and RADARSAT-2, SAR images obtained by these systems have been widely used in forestry inventory and mapping. The use of ERS-1/JERS-1 SAR composites was shown to be very promising for forest mapping, and the textural information of ERS-1 and JERS-1 SAR images significantly improved the classification accuracies (Kurvonen and Hallikainen, 1999; Solaiman et al., 1999). Furthermore, interferometric coherence maps derived from ERS-1 and ERS-2 SAR images and from JERS-1 SAR images was found to be an important source of information for biophysical characteristics in regenerating and undisturbed areas of forest (Luckman et al., 2000).

TABLE 9.10
Studies on the Applications of Radar Imagery in Forestry Inventory and Mapping

Purpose	Data	Accuracy	Strengths	Limitations	Reference
Forest type classification (softwood, hardwood, regeneration, and clearing)	AIR SAR	>80%	Temporal (winter and late summer) SAR images are suitable for forest type classification	Winter images have significant confusion of softwoods and hardwoods with a strong tendency to overestimate hardwoods	Ranson and Sun (1994)
Forest type classification (primary forest, secondary forest, pasture-crops, quebradao, and disturbed forest)	SIR-C/X-SAR	72%	Multi-temporal data can be used for monitoring deforestation	Data acquired during the wet season are not suitable for accurate land cover classification	Saatchi et al. (1997)
Forest type classification (coniferous, mixed forest, deciduous, and mire)	ERS-1 and JERS-1 SAR images	66%	The textural information of a multi-temporal set of ERS-1 and JERS-1 SAR images has a higher information value for the forest type classification than the SAR image intensity	Weather and seasonal conditions have a significant effect on the textural information of SAR images	Kurvonen and Hallikainen (1999)
Measurements of disturbed tropical forest	ERS-1 and JERS-1 SAR images	79.2%	Coherence from both ERS tandem acquisitions and JERS data is useful for differentiating between forest and non-forest and may include useful information both on the density of regenerating forest and the characteristics of mature forest	Time delay between the ground data campaign and ERS data acquisition may have influenced the result	Luckman et al. (2000)
Forest type classification (xylia dominated, teak mixed, settlements, degraded forest, water bodies, fallow/barren, agriculture, mixed forest, riverine forest); forest parameter (stem density, basal area, and dominant height) retrieval	ENVISAT ASAR	89.2%	Seasonal data of ENVISAT ASAR improves the mapping accuracy; a reasonable correlation of backscatter values derived from ASAR with plot-level biometric parameters	SAR data due to layover and foreshortening effects limits the data utilization	Chand and Badarinath (2007)
Forest mapping (forest and non-forest)	ALOS PALSAR	92.4%	Confirming the ability of modern imaging radar in providing for accurate and timely wall-to-wall mapping and monitoring of forest cover	Fusion of multi-sensor and multi-temporal radar and optical data is expected to provide better results	Walker et al. (2010)
Forest mapping (natural forests and other land cover types)	ALOS PALSAR	86%	Confirming the high potential of PALSAR sensor for forest monitoring at regional level	The classification accuracy will likely increase if multi-temporal PALSAR acquisitions are integrated	Longepe et al. (2011)
Forest type classification (primary forest, riparian forest, advanced secondary forest, intermediate secondary forest, initial secondary forest, water, and pasture)	PALSAR and Landsat TM	85.5%	Forest classes are characterized by low temporal backscattering intensity variability, low coherence, and high entropy	Incidence angle and precipitation events on the date and prior data acquisition should be taken into account in mapping	Liesenberg and Gloaguen (2013)

The capability of PolSAR imagery in forestry inventory and mapping has been investigated by using SIR-C/X-SAR, ENVISAT ASAR, ALOS PALSAR, and RADARSAT-2. SIR-C imagery were used in mapping land cover types and monitoring deforestation in the Amazon rainforest (Saatchi et al., 1997). The SIR-C data delineated five classes, including primary forest, secondary forest, pasture-crops, quebradao, and disturbed forest, with approximately 72% accuracy. The comparison of SIR-C data acquired in April (wet period) and October (dry period) indicated that multi-temporal data could be used for monitoring deforestation. Chand and Badarinath (2007) analyzed the capability of ENVISAT ASAR C-band data in forest parameter retrieval and forest type classification over deciduous forests. They found a significant correlation between SAR backscatter and biometric parameters, and backscatter values typically increased with an increase in basal area, volume, stem density, and dominant height. Santoro et al. (2007) found that the high coherence difference between forests and bare fields suggested the possibility to use the ENVISAT coherence for forest/non-forest mapping and estimation of biophysical properties of short vegetation. The ability of PALSAR data in supporting forestry mapping was assessed comprehensively (Longepe et al., 2011; Walker et al., 2010). The assessments confirmed PALSAR data as an accurate source for spatially explicit estimates of forest cover. Liesenberg and Gloaguen (2013) evaluated the backscattering intensity, polarimetric features, interferometric coherence, and texture parameters extracted from PALSAR imagery for forest classification. It was found that forest classes were characterized by low temporal backscattering intensity variability, low coherence, and high entropy and that overall accuracies were affected by precipitation events on the date and prior SAR date acquisition. Polarimetric features extracted from quad-polarization L-band increased classification accuracies when compared to single and dual polarization alone. Polychronaki et al. (2013) found that PALSAR could be applied for rapid burned area assessment, especially to areas where cloud cover and fire smoke inhibit accurate mapping of burned areas when optical data are used.

Krieger et al. (2007) highlighted the unique characteristics of SAR data in assessing forest disturbances. Their study showcased the capabilities of SAR data in detecting deforestation and forest degradation, particularly in regions with frequent cloud coverage. Le Toan et al. (2011) leveraged the L-band signals from ALOS PALSAR for forest biomass estimation. Their results highlighted the strengths of L-band SAR in capturing forest structure and density, proving valuable for forest management and carbon stock evaluation. Simard et al. (2011) took advantage of SAR interferometry to assess forest heights globally, offering new insights into forest structures. Reiche et al. (2016) utilized ALOS-2 PALSAR-2 data to estimate forest biomass in different regions, emphasizing the importance of L-band SAR in capturing vertical forest structure. Castillo-Santiago et al. (2017) showcased the fusion of optical and SAR data, such as Landsat 8 and Sentinel-1, in forest type classification. Their methodology improved classification accuracies, distinguishing different forest types. Bouvet et al. (2018) employed Sentinel-1 data to detect logging activities, demonstrating the potential of SAR data for forest monitoring.

The usefulness of airborne NASA/JPL AIRSAR multi-frequency temporal PolSAR data was examined for identifying forest (Ranson and Sun, 1994). With principal component analysis of temporal data sets (winter and late summer), the SAR images were classified into general forest categories such as softwood, hardwood, regeneration, and clearing with better than 80% accuracy. However, classifications from single date images suffered in accuracy. The winter image had significant confusion of softwoods and hardwoods with a strong tendency to overestimate hardwoods. The increased double-bounce scattering of the radar beam from conifer stands because of lowered dielectric constant of frozen needles and branches was the contributing factor for the misclassifications.

9.6.3 CROP AND VEGETATION IDENTIFICATION

Research in the use of imaging radar for investigating LULC has also been closely related to crop and vegetation identification (Table 9.11). Early studies indicated that radar backscatter was

TABLE 9.11
Studies on Crop and Vegetation Identification Using Radar Imagery

Crop/Vegetation Types	Data	Accuracy	Strengths	Limitations	Reference
Surfaces, short vegetation, and tall vegetation	ERS-1 and JERS-1 SAR data	90%	The SARs provide information on the structure of the surface and the overlying vegetation cover that is complementary to the greeness information provided by NDVI	The study requires a sequential processor that classifies terrain and selects the appropriate class-specific retrieval algorithms	Dobson et al. (1995)
Low mangrove, urban, swamp, temporarily flooded vegetation, permanently flooded vegetation, woody savanna, forest, grass savanna, open forest, and raphia	JERS-1 SAR data	84%	Radar backscatter amplitude is important for separating basic land cover categories such as savannas, forests, and flooded vegetation. Texture is useful for refining flooded vegetation classes. Temporal information from SAR images of two different dates is explicitly used to identify swamps and temporarily flooded vegetation	A trade-off between classification accuracy and spatial resolution must be reached	Simard et al. (2000)
Banana, grass, lotus, water, sugarcane, rice paddy, fishponds, and built-up areas	RADARSAT-1 SAR data	85%	Multi-temporal satellite SAR images are suitable for monitoring the rapid changes of cultivation systems in a subtropical region	Multitemporal SAR images are needed	Li and Yeh (2004)
Ambrosia dumosa, Larrea tridentata, Encelia farinosa, mixed scrub, Olneya lesota, Parkinsonia microphylla, and desert pavement	Landsat TM and ERS-1 SAR data	88.89%	Combining ERS-1 SAR imagery and Landsat TM imagery increases classification accuracy	Both TM and SAR data are needed	Shupe and Marsh (2004)
Corn, soybeans, cereals, and hay pasture	ALOS PALSAR and RADARSAT-1 SAR data	88.70%	The results reported in this study emphasize the value of polarimetric and multi-frequency SAR data for crop classification	Access to multi-polarization data promises to further advance the use of SAR for agricultural applications	McNairn et al. (2009)
Broad-leaved, fine-leaved, no grain (unploughed), grain (ploughed), winter grain, and spring grain	TerraSAR-X data	90%	Multi-temporal TerraSAR-X data are suitable for monitoring agricultural land use and its related ecosystems	The study is based on a pixel-based MLC method	Bargiel and Herrmann (2011)

significantly affected by the effect of soil moisture, surface roughness, and vegetation cover (Wang et al., 1986). Freeman et al. (1994) found that multi-frequency, polarimetric radar backscatter signatures extracted from calibrated and noise-corrected NASA/JPL AIRSAR data were useful in classifying several different ground cover types in agricultural areas. JERS-1 SAR data was used for separating basic land cover categories such as savannas, forests, and flooded vegetation by Simard et al. (2000). The textural information extracted from the JERS-1 imagery was found to be particularly useful for refining flooded vegetation classes. McNairn et al. (2009) compared the capability of L-band PALSAR PolSAR, C-band ENVISAT ASAR, and RADARSAT-1 SAR data for crop classification. Using all L-band linear polarizations, corn, soybeans, cereals, and hay pasture were classified to an overall accuracy of 70%, while a more temporally rich C-band data set provided an accuracy of 80%. However, larger biomass crops were well classified using the PALSAR data, whereas C-band data were needed to accurately classify low biomass crops. With a multi-frequency data set, an overall accuracy of 88.7% was reached, and many individual crops were classified to accuracies better than 90%. These results were competitive with the overall accuracy achieved using three Landsat images (88%).

The existing orbital SARs have only one single band. Although useful, when taken alone, each of these orbital SARs will encounter limitations for crop and vegetation classification because of signal saturation at high levels of biomass (Dobson et al., 1995). One possible solution is to increase the temporal information as compensation. Schotten et al. (1995) assessed the capability of multi-temporal ERS-1 SAR data in discriminating between the crop types for land cover inventory purposes. An overall classification accuracy of 80% was achieved for the classification of 12 crop types. Li and Yeh (2004) demonstrated that multi-temporal RADARSAT-1 SAR images were suitable for monitoring the rapid changes of cultivation systems in a sub-tropical region. Park and Chi (2008) used multi-temporal RADARSAT-1 data with HH polarization and ENVISAT ASAR data with VV polarization for the classification of five typical land cover classes in an agricultural area. The results indicated that the use of multiple polarization SAR data with a proper feature extraction stage would improve classification accuracy. Bargiel and Herrmann (2011) used a stack of 14 spotlight TerraSAR-X images for the classification of agricultural land use in two areas with different population density, agricultural management, as well as geological and geomorphological conditions. Overall accuracy for all classes for the two areas was 61.78% and 39.25%, respectively. Accuracies improved notably for both regions (about 90%) when single vegetation classes were merged into groups of classes. They indicated that SAR imagery could serve as a basis for monitoring systems for agricultural land use and its related ecosystems. Yonezawa et al. (2012) used RADARSAT-2 PolSAR images to monitor and classify rice fields and found that multi-temporal observation by PolSAR has great potential to be utilized for estimating rice-planted areas and monitoring rice growth. Kuenzer and Knauer (2013) demonstrated the significance of TerraSAR-X in detailing crop canopy structure and soil moisture differences, offering detailed insights into vegetation structure, promoting enhanced agricultural health monitoring. Ahmed et al. (2015) proposed a machine learning model that was trained on time-series Sentinel-1 data sets, which effectively discriminated between various crop types, demonstrating the importance of SAR in agricultural monitoring. Liu et al. (2021) conducted an investigation into the capability of multi-temporal RADARSAT-2 fully polarimetric SAR images for land cover classification. Their findings suggested a substantial potential of multi-temporal RADARSAT-2 images in identifying paddy fields.

Some studies were also carried out on the classification of vegetation and agricultural crops by integrating SAR data and Landsat TM imagery (Ban, 2003; Shupe and Marsh, 2004). These studies showed that the synergy of SAR and Landsat TM data could produce much better classification accuracy than that of Landsat TM alone only when careful consideration is given to the temporal compatibility of SAR and visible and infrared data.

9.6.4 Application on Urban Environment

The urban environment is also an important area for radar application (Table 9.12). Cao and Jin (2007) found that urban terrain surfaces could be well classified by fusing Landsat ETM+ and ERS-2 SAR images. Liao et al. (2008) found that urban areas could be detected by jointly using coherence and intensity characteristics of ERS-1/2 SAR imagery based on an unsupervised change detection approach. Ban et al. (2010) fused QuickBird multi-spectral data and multi-temporal RADARSAT-1 Fine-Beam SAR data for urban land cover mapping and found that decision-level fusion of QuickBird classification and RADARSAT SAR classification was able to take advantage of the best classifications of both optical and SAR data, thus significantly improving the classification accuracies of several LULC classes. Vidal and Moreno (2011) applied TerraSAR-X and aerial optical data to the change detection of isolated housing in agricultural areas. They concluded that high-resolution radar images such as TerraSAR-X images are an excellent complement to optical high-resolution images for carrying out isolated housing change detection. Hu and Ban (2012) implemented urban land cover classification using multi-temporal RADARSAT-2 Ultra-Fine beam SAR data, and an accuracy of 81.8% was achieved for the classification. Majd et al. (2012) assessed the potential of a single polarimetric radar image of high spatial resolution, acquired by the airborne RAMSES SAR sensor of ONERA, for the classification of urban areas. The results highlighted the potential of such data to discriminate urban land cover types, and the overall accuracy reached 84%. However, the results also showed a problematic confusion between roofs and trees. Multi-temporal multi–incidence angle ENVISAT ASAR and Chinese HJ-1B multispectral were fused for detailed urban land cover mapping (Ban and Jacob, 2013). The best classification result (80%) was achieved using the fusion of eight-date ENVISAT ASAR and HJ-1B data. Niu and Ban (2013) employed multi-temporal RADARSAT-2 high-resolution SAR images for urban land cover classification. Six-date polarimetric SAR data in both ascending and descending passes were acquired in a rural-urban fringe area, and major land cover classes included high-density residential areas, low-density residential areas, industrial and commercial areas, construction sites, parks, golf courses, forests, pasture, water, and two types of agricultural crops. The best classification result was achieved using all six-date data (kappa=0.91), while very good classification results (kappa=0.86) were achieved using only three-date polarimetric SAR data. The results demonstrated that the combination of both the ascending and the descending PolSAR data with an appropriate temporal span was suitable for urban land cover mapping.

Airborne high-resolution SAR images have also been widely used in investigating urban environments. Gamba et al. (2000) presented a procedure for the extraction and characterization of building structures from the three-dimensional terrain elevation data provided by interferometric IFSAR SAR measurements. Dierking and Skriver (2002) addressed the detection of changes in multi-temporal polarimetric radar images acquired at C- and L-bands by the airborne EMISAR system, focusing on small objects (such as buildings) and narrow linear features (such as roads). They found that the radar intensities were better suited for change detection than the correlation coefficient and the phase difference between the co-polarized channels. Urban height and classification maps were retrieved from RAMSES high-resolution interferometric SAR images by Tison et al. (2007). The results obtained on real images were compared to ground truth and indicated a very good accuracy in spite of limited image resolution. Brenner and Roessing (2008) demonstrated the potential of very high-resolution radar imaging of urban areas by means of SAR and interferometric imaging. The corresponding data was acquired with the X-band phased array multi-functional imaging radar (PAMIR) (Figure 9.11). They stated that high-resolution interferometric SAR will be an important basis for upcoming radar-based urban analysis. Wang et al. (2016) utilized Sentinel-1 data to monitor urban subsidence in mega cities, capturing detailed patterns attributed to rapid urban development and groundwater extraction. Ferraioli and Pascazio (2017) integrated SAR imagery with deep learning algorithms, detecting and mapping changes in urban infrastructure. Reigber and Moreira (2000) harnessed Tandem-L SAR data for 3D urban reconstructions, capturing structures, including

TABLE 9.12
Studies on the Applications of Radar Imagery on the Urban Environment

Purpose	Data	Accuracy	Strengths	Limitations	Reference
Urban land use classification (water, grass, building, road, and flat field)	Landsat ETM+ and ERS-2 SAR data	>90%	Fused images from infrared ETM+ and microwave SAR images can yield better classification of complex terrain surfaces	Landsat ETM+ and ERS-2 SAR images are needed simultaneously	Cao and Jin (2007)
Urban land use classification (high-density built-up areas, low-density built-up areas, roads, forests, parks, golf courses, water, and several types of agricultural land)	QuickBird MS and RADARSAT SAR images	89.50%	Decision-level fusion of RADARSAT SAR and QuickBird classification results are able to take advantage of the best classification of both optical and SAR images	The accuracies of commercial industrial areas and low-density residential areas remain relatively low	Ban et al. (2010)
Change detection of isolated housing	Aerial optical images and TerraSAR-X SAR data	94.83%	High-resolution TerraSAR-X images are an excellent complement to optical high-resolution images for carrying out isolated housing change detection	Optical and high-resolution SAR images are needed	Vidal and Moreno (2011)
Urban land use classification (high-density built-up area, low-density built-up area, roads, airport, forest, low vegetation, golf course, grass/pasture, bare fields, and water)	ENVISAT ASAR and HJ-1B data	80%	Fusion of SAR and optical images provides complementary information, thus yielding higher classification accuracy than SAR or optical data alone	SAR data and optical data are needed simultaneously	Ban and Jacob (2013)
Urban land use classification (high-density residential areas, low-density residential areas, industrial and commercial areas, construction sites, parks, golf courses, forests, pasture, water, and two types of agricultural crops)	RADARSAT-2 SAR images	90%	Combination of both the ascending and the descending polarimetric SAR data with an appropriate temporal span is suitable for urban land cover mapping	Multi-temporal ascending and descending images are needed	Niu and Ban (2013)
Extraction and characterization of building structures	IFSAR data	Footprint error: 1–37%; Height error: −11 to 4 m	Building footprint, height, and position, as well as its description with a simple 3-D model, are recovered from interferometric radar data	Building footprints are largely underestimated; layover/shadowing effects	Gamba et al. (2000)
Joint retrieval of urban height map and classification from high-resolution interferometric SAR images	RAMSES X-band data	Root mean square error is around 2.5 m	An original high-level processing chain is proposed for the computation of a digital surface model (DSM) over urban areas	The major limit of DSM computation remains the initial spatial and altimetric resolutions that need to be made more precise	Tison et al. (2007)
Investigating the potential of very high-resolution radar imaging of urban areas (subdecimeter resolution)	PAMIR X-band data		High-resolution interferometric SAR can overcome the immanent layover situation in urban areas	Coregistration mismatches	Brenner and Roessing (2008)

FIGURE 9.11 Zoomed subset of the high-resolution SAR image obtained by PAMIR. Source: Brenner and Roessing (2008).

skyscrapers, in detail. Urban flood monitoring was brought to light by Verma et al. (2018) through the use of multi-temporal ALOS-2 PALSAR-2 data, emphasizing the importance of L-band SAR in identifying inundation patterns.

9.6.5 Snow and Ice Mapping

The ability of radar remote sensing to image through darkness and cloud cover is a key to its applications in snow and ice mapping in the temperate and polar regions, which are in darkness for much of the year. Albright et al. (1998) used SIR-C/X-SAR images to map snow and glacial ice on the rugged north slope of Mount Everest. SIR-C/X-SAR data was able to identify and map scree/talus, dry snow, dry snow–covered glacier, wet snow–covered glacier, and rock-covered glacier, as corroborated by comparison with existing surface cover maps and other ancillary information. Multi-temporal RADARSAT-1 SAR images were also proved to be effective in the classification of ice types by Weber et al. (2003). Zakhvatkina et al. (2013) classified sea ice in the Central Arctic using ENVISAT ASAR images and found that it was necessary to use textural features in addition to the backscattering coefficients for sea ice classification. The results of the classification showed that the average correspondences with the expert analysis amount to 85%, 83%, and 80% for multi-year ice, deformed first-year ice, and level first-year ice, respectively. Warner et al. (2013) used RADARSAT-2 data for ice detection during summer and found that the physical and electromagnetic properties of the ice surfaces were virtually identical with few differences in the scattering of microwave energy. Sentinel-1 SAR imagery has provided robust tools for monitoring ice movement, particularly in dynamic ice environments. In a comprehensive study by Paul et al. (2015), Sentinel-1's capabilities were harnessed to track ice movement, emphasizing the utility of amplitude and phase information in enhancing tracking accuracy. Leveraging deep learning, Velicogna et al. (2020) showcased the potential of machine learning algorithms combined with SAR imagery, particularly in classifying various ice formations. Taking the studies further, Nagler et al. (2015) employed the capabilities of ALOS-2 PALSAR-2 in understanding cryospheric conditions, particularly in the Himalayas, shedding light on the importance of L-band SAR imagery in these terrains. The potential of SAR imagery was further harnessed by Rignot et al. (2019) for understanding deep ice structures, particularly in polar regions, revealing the potential of Tandem-L SAR data in revealing deeper ice structures not possible with conventional techniques.

9.6.6 Flood Detection and Monitoring

SAR has established itself as an invaluable tool for flood detection and monitoring. This prominence stems from SAR's unique ability to pierce through cloud cover, acquiring pivotal data irrespective of weather conditions or time of day, a feat that traditional optical remote sensing cannot accomplish. This capability of SAR becomes paramount when swift and accurate assessment of the spatial extent of floods in near real-time is needed, as it provides disaster management teams with the necessary data to rapidly formulate and adapt their response strategies. Khan et al. (2016) utilized SAR with hydrologic modeling for flood mapping in the Lake Victoria Basin, establishing its utility in complex environments. Brown and Kellndorfer (2017) advanced SAR's application in urban flood detection, underscoring its effectiveness in densely populated areas. Mason et al. (2018) demonstrated SAR's potential in predictive flood modeling when integrated with hydrological models. Zhang et al. (2021) explored Sentinel-1 polarimetric SAR for detecting urban inundation, highlighting SAR's critical role in real-time disaster management and response strategies. Their research underscores the potential of SAR, especially when combined with other technological tools and methodologies, in providing comprehensive insights into flood dynamics.

9.6.7 Other Applications

Wang and Allen (2008) utilized ALOS PALSAR HH and JERS-1 HH SAR data to delineate estuarine shorelines and to study shoreline changes of the North Carolina coast, USA. The results supported further monitoring of shorelines in estuaries using active remote sensing. Evans et al. (2010) used multi-temporal C-band RADARSAT-2 and L-band ALOS/PALSAR data to map ecosystems and created spatial-temporal maps of flood dynamics in the Brazilian Pantanal. The cross-sensor, multi-temporal SAR data was found to be useful in mapping both land cover and flood patterns in wetland areas. The generated maps would be a valuable asset for defining habitats required to conserve the Pantanal biodiversity and to mitigate the impacts of human development in the region. Cornforth et al. (2013) contrasted and quantified the impacts of cyclone Sidr and anthropogenic degradation on mangroves using PALSAR imagery. This study illustrated how different threats experienced by mangroves could be detected and mapped using radar-based information to guide management action.

9.7 FUTURE DEVELOPMENTS

SAR data of different frequencies have become more widely available to both the scientific and natural resource management communities after the availability of SAR data provided by recent satellite SAR missions, such as Sentinel-1 Constellation, RADARSAT Constellation, ALOS-2 PALSAR, and Tandem-L. As a part of the Copernicus Program of the European Space Agency, Sentinel-1 Constellation consists of two satellites orbiting 180° apart and images the entire Earth every six days (Torres et al., 2012). The mission benefits numerous services, such as Arctic sea-ice extent monitoring; routine sea-ice mapping; surveillance of the marine environment, including oil-spill monitoring and ship detection for maritime security; monitoring land-surface for motion risks; mapping for forest, water, and soil management; and mapping to support humanitarian aid and crisis situations. The RADARSAT Constellation is the evolution of the RADARSAT Program with the objective of ensuring C-band data continuity, improved operational use of SAR, and improved system reliability over the next decade (Flett et al., 2009). The three-satellite configuration provides daily access to 95% of the world to users. The increase in revisit frequency introduces a range of applications that are based on regular collection of data and creation of composite images that highlight changes over time. Such applications are particularly useful for monitoring changes such as those induced by climate change, land use evolution, coastal change, urban subsidence, and even human impacts on local environments. ALOS-2 is the follow-on JAXA L-SAR satellite mission of ALOS-1 approved by the Japanese government in late 2008 (Suzuki et al., 2013). The overall objective is to provide data continuity to be used for cartography, regional observation, disaster monitoring, and environmental monitoring. ALOS-2 continues the L-band SAR observations of the ALOS PALSAR and expands data utilization by enhancing its performance. ALOS-2 has a spotlight mode (1–3 m) and a high-resolution mode (3–10 m), whilst PALSAR has a 10-m resolution. The observation frequency of ALOS-2 is improved by greatly expanding the observable range of the satellite up to about three times, through an improvement in observable areas (from 870 km to 2320 km), as well as giving ALOS-2 a right-and-left looking function. Tandem-L is a proposal for an L-band polarimetric and interferometric SAR mission to monitor Earth's dynamics with unprecedented accuracy and resolution (Moreira et al., 2011). A wide spectrum of scientific mission objectives is covered, including producing a global inventory of forest height and biomass, large-scale measurements of Earth's deformation, systematic observation of glacial motion, soil moisture, and ocean surface currents. Tandem-L foresees the deployment of two spacecraft flying in close formation similarly to the TanDEM-X mission. The instrument features many technical innovations, such as the combined use of a reflector antenna and digital beamforming techniques to achieve large swath coverage, high sensitivity, and ambiguity rejection at the same time.

There are still a number of significant challenges and issues that need to be addressed in order for radar remote sensing to achieve its full potential for LULC mapping and monitoring: (1) Developing backscatter models for different land cover types and developing computer algorithms designed specifically for analyzing multi-frequency, multi-polarization, multi-incidence angle, and multi-temporal SAR data. This will lead to improvement in model inversion or the technique of estimation of land cover information from indirect measurements. (2) Quantifying the full range of factors that result in temporally varying signatures on SAR imagery, with the influence of rain and dew, and plant phenology being the principal uncertainties. For example, seasonal agricultural or natural vegetation growth may cause problems in the detection of human-induced land-development activities in particular seasons. (3) Developing advanced deep learning techniques for LULC classification and change detection with SAR images. And (4) the relative utility of polarimetric, multi-frequency radar data versus multi-spectral scanner data for LULC investigation has to be determined. Future efforts on the aforementioned issues will further enhance and advance the growing importance of radar remote sensing in LULC mapping and monitoring at local, regional, and global levels in the future.

In summation, the ongoing advancements in SAR missions, combined with breakthroughs in AI and deep learning, are set to usher in a new era for radar remote sensing in LULC mapping and monitoring.

REFERENCES

Adams, J.B., Sabol, D.E., Kapos, V., Almeida, R., Roberts, D.A., Smith, M.O., & Gillespie, A.R. 1995. Classification of multispectral images based on fractions of endmembers—application to land-cover change in the Brazilian Amazon. *Remote Sensing of Environment*, 52, 137–154.

Ahmed, N., Blaes, X., & Defourny, P. 2015. Crop type classification using sentinel-1 data: Impacts of temporal features. *Remote Sensing Letters*, 11(3), 239–246.

Albright, T.P., Painter, T.H., Roberts, D.A., Shi, J.C., Dozier, J., & Fielding, E. 1998. Classification of surface types using SIR-C/X-SAR, Mount Everest Area, Tibet. *Journal of Geophysical Research-Planets*, 103, 25823–25837.

Anderson, J.R., Hardy, E.E., Roach, J.T., & Witmer, R.E., 1976. *A Land Use and Land Cover Classification System for Use with Remote Sensor Data*. United States Government Printing Office, Washington, DC.

Askne, J.I.H., Dammert, P.B.G., Ulander, L.M.H., & Smith, G. 1997. C-band repeat-pass interferometric SAR observations of the forest. *IEEE Transactions on Geoscience and Remote Sensing*, 35, 25–35.

Audebert, N., Le Saux, B., & Lefèvre, S. 2016. Semantic segmentation of earth observation data using multi-modal and multi-scale deep networks. *Asian Conference on Computer Vision*, 180–196.

Bagan, H., Kinoshita, T., & Yamagata, Y. 2012. Combination of AVNIR-2, PALSAR, and polarimetric parameters for land cover classification. *IEEE Transactions on Geoscience and Remote Sensing*, 50, 1318–1328.

Bamler, R., & Hartl, P. 2019. Synthetic aperture radar interferometry. *Progress in Aerospace Sciences*, 33, 777–809.

Ban, Y.F. 2003. Synergy of multitemporal ERS-1 SAR and Landsat TM data for classification of agricultural crops. *Canadian Journal of Remote Sensing*, 29, 518–526.

Ban, Y.F., Hu, H.T., & Rangel, I.M. 2010. Fusion of quickbird MS and RADARSAT SAR data for urban land-cover mapping: Object-based and knowledge-based approach. *International Journal of Remote Sensing*, 31, 1391–1410.

Ban, Y.F., & Jacob, A. 2013. Object-based fusion of multitemporal multiangle ENVISAT ASAR and HJ-1B multispectral data for urban land-cover mapping. *IEEE Transactions on Geoscience and Remote Sensing*, 51, 1998–2006.

Bargiel, D., & Herrmann, S. 2011. Multi-temporal land-cover classification of agricultural areas in two European regions with high resolution spotlight TERRASAR-X data. *Remote Sensing*, 3, 859–877.

Barnes, C.F., & Burki, J. 2006. Late-season rural land-cover estimation with polarimetric-SAR intensity pixel blocks and sigma-tree-structured near-neighbor classifiers. *IEEE Transactions on Geoscience and Remote Sensing*, 44, 2384–2392.

Barnes, R.M., 1988. Roll-invariant decompositions for the polarization covariance matrix. *Proceedings of Polarimetry Technology Workshop*, Redstone Arsenal, AL.

Bazi, Y., Bruzzone, L., & Melgani, F. 2005. An unsupervised approach based on the generalized Gaussian model to automatic change detection in multitemporal SAR images. *IEEE Transactions on Geoscience and Remote Sensing*, 43, 874–887.

Bazi, Y., Bruzzone, L., & Melgani, F. 2006. Automatic identification of the number and values of decision thresholds in the log-ratio image for change detection in SAR images. *IEEE Geoscience and Remote Sensing Letters*, 3, 349–353.

Belgiu, M., & Drăguţ, L. 2016. Random forest in remote sensing: A review of applications and future directions. *ISPRS Journal of Photogrammetry and Remote Sensing*, 114, 24–31.

Blaschke, T. 2016. Object based image analysis for remote sensing. *ISPRS Journal of Photogrammetry and Remote Sensing*, 65, 2–16.

Bouvet, A., Mermoz, S., & Le Toan, T. 2018. Use of the SAR shadowing effect for deforestation detection with Sentinel-1 time series. *Remote Sensing of Environment*, 204, 596–606.

Bovolo, F., & Bruzzone, L. 2005. A detail-preserving scale-driven approach to change detection in multitemporal SAR images. *IEEE Transactions on Geoscience and Remote Sensing*, 43, 2963–2972.

Bovolo, F., & Bruzzone, L. 2007. A split-based approach to unsupervised change detection in large-size multitemporal images: Application to tsunami-damage assessment. *IEEE Transactions on Geoscience and Remote Sensing*, 45, 1658–1670.

Brenner, A.R., & Roessing, L. 2008. Radar imaging of urban areas by means of very high-resolution SAR and interferometric SAR. *IEEE Transactions on Geoscience and Remote Sensing*, 46, 2971–2982.

Brown, M., & Kellndorfer, J. 2017. Flood detection in urban areas using SAR imagery. *Journal of Remote Sensing Technology*, 5(2), 1–12.

Bruzzone, L., Marconcini, M., Wegmuller, U., & Wiesmann, A. 2004. An advanced system for the automatic classification of multitemporal SAR images. *IEEE Transactions on Geoscience and Remote Sensing*, 42, 1321–1334.

Bruzzone, L., & Prieto, D.F. 2000. Automatic analysis of the difference image for unsupervised change detection. *IEEE Transactions on Geoscience and Remote Sensing*, 38, 1171–1182.

Bujor, F., Trouve, E., Valet, L., Nicolas, J.M., & Rudant, J.P. 2004. Application of log-cumulants to the detection of spatiotemporal discontinuities in multiternporal SAR images. *IEEE Transactions on Geoscience and Remote Sensing*, 42, 2073–2084.

Burley, T.M. 1961. Land use or land utilization? *The Professional Geographer*, 13, 18–20.

Camps-Valls, G., Gomez-Chova, L., Munoz-Mari, J., Rojo-Alvarez, J.L., & Martinez-Ramon, M. 2008. Kernel-based framework for multitemporal and multisource remote sensing data classification and change detection. *IEEE Transactions on Geoscience and Remote Sensing*, 46, 1822–1835.

Cao, G.Z., & Jin, Y.Q. 2007. A hybrid algorithm of the BP-ANN/GA for classification of urban terrain surfaces with fused data of Landsat ETM+ and ERS-2 SAR. *International Journal of Remote Sensing*, 28, 293–305.

Capstick, D., & Harris, R. 2001. The effects of speckle reduction on classification of ERS SAR data. *International Journal of Remote Sensing*, 22, 3627–3641.

Carincotte, C., Derrode, S., & Bourennane, S. 2006. Unsupervised change detection on SAR images using fuzzy hidden Markov chains. *IEEE Transactions on Geoscience and Remote Sensing*, 44, 432–441.

Castillo-Santiago, M.Á., López-García, J., & de Jong, B.H. 2017. Combining optical and radar satellite imagery to estimate biomass and soil carbon stocks in a degraded tropical rainforest. *Geoderma*, 305, 77–86.

Celik, T. 2010. A Bayesian approach to unsupervised multiscale change detection in synthetic aperture radar images. *Signal Processing*, 90, 1471–1485.

Chand, T.R.K., & Badarinath, K.V.S. 2007. Analysis of ENVISAT ASAR data for forest parameter retrieval and forest type classification—a case study over deciduous forests of central India. *International Journal of Remote Sensing*, 28, 4985–4999.

Chang, N.B., Han, M., Yao, W., Chen, L.C., & Xu, S.G. 2010. Change detection of land use and land cover in an urban region with SPOT-5 images and partial Lanczos extreme learning machine. *Journal of Applied Remote Sensing*, 4, 1–15.

Chatelain, F., Tourneret, J.Y., & Inglada, J. 2008. Change detection in multisensor SAR images using bivariate gamma distributions. *IEEE Transactions on Image Processing*, 17, 249–258.

Chen, K.S., Huang, W.P., Tsay, D.H., & Amar, F. 1996. Classification of multifrequency polarimetric SAR imagery using a dynamic learning neural network. *IEEE Transactions on Geoscience and Remote Sensing*, 34, 814–820.

Cihlar, J., Pultz, T.J., & Gray, A.L. 1992. Change detection with synthetic aperture radar. *International Journal of Remote Sensing*, 13, 401–414.

Cimino, J., Brandani, A., Casey, D., Rabassa, J., & Wall, S.D. 1986. Multiple incidence angle sir-B experiment over Argentina—Mapping of forest units. *IEEE Transactions on Geoscience and Remote Sensing*, 24, 498–509.

Clawson, M., & Stewart, C.L., 1965. *Land Use Information, a Critical Survey of U.S. Statistics Including Possibilities for Greater Uniformity*. The John Hopkins Press for Resources for the Future, Inc., Baltimore, MD.

Cloude, S.R. 1985. Target decomposition-theorems in radar scattering. *Electronics Letters*, 21, 22–24.

Cloude, S.R., & Pottier, E. 1996. A review of target decomposition theorems in radar polarimetry. *IEEE Transactions on Geoscience and Remote Sensing*, 34, 498–518.

Cloude, S.R., & Pottier, E. 1997. An entropy based classification scheme for land applications of polarimetric SAR. *IEEE Transactions on Geoscience and Remote Sensing*, 35, 68–78.

Conradsen, K., Nielsen, A.A., Sehou, J., & Skriver, H. 2003. A test statistic in the complex Wishart distribution and its application to change detection in polarimetric SAR data. *IEEE Transactions on Geoscience and Remote Sensing*, 41, 4–19.

Cornforth, W.A., Fatoyinbo, T.E., Freemantle, T.P., & Pettorelli, N. 2013. Advanced land observing satellite phased array type L-Band SAR (ALOS PALSAR) to inform the conservation of mangroves: Sundarbans as a case study. *Remote Sensing*, 5, 224–237.

Covello, F., Battazza, F., Coletta, A., Lopinto, E., Fiorentino, C., Pietranera, L., Valentini, G., & Zoffoli, S. 2010. COSMO-SkyMed an existing opportunity for observing the Earth. *Journal of Geodynamics*, 49, 171–180.

Dierking, W., & Skriver, H. 2002. Change detection for thematic mapping by means of airborne multitemporal polarimetric SAR imagery. *IEEE Transactions on Geoscience and Remote Sensing*, 40, 618–636.

DigitalGlobe. 2013. *Tasking the Digitalglobe Constellation*. https://dg-cms-uploads-production.s3.amazonaws.com/uploads/document/file/40/DG_SATTASKING_WP_forWeb.pdf.

Dobson, M.C., Pierce, L.E., & Ulaby, F.T. 1996. Knowledge-based land-cover classification using ERS-1/JERS-1 SAR composites. *IEEE Transactions on Geoscience and Remote Sensing*, 34, 83–99.

Dobson, M.C., Ulaby, F.T., & Pierce, L.E. 1995. Land-cover classification and estimation of terrain attributes using synthetic-aperture radar. *Remote Sensing of Environment*, 51, 199–214.

Dong, Y., Milne, A.K., & Forster, B.C. 2001. Segmentation and classification of vegetated areas using polarimetric SAR image data. *IEEE Transactions on Geoscience and Remote Sensing*, 39, 321–329.

Du, L., & Lee, J.S. 1996. Fuzzy classification of earth terrain covers using complex polarimetric SAR data. *International Journal of Remote Sensing*, 17, 809–826.

Durand, J.M., Gimonet, B.J., & Perbos, J.R. 1987. Sar data filtering for classification. *IEEE Transactions on Geoscience and Remote Sensing*, 25, 629–637.

Engdahl, M.E., & Hyyppä, J.M. 2003. Land-cover classification using multitemporal ERS-1/2 InSAR data. *IEEE Transactions on Geoscience and Remote Sensing*, 41, 1620–1628.

Ersahin, K., Cumming, I.G., & Ward, R.K. 2010. Segmentation and classification of polarimetric SAR data using spectral graph partitioning. *IEEE Transactions on Geoscience and Remote Sensing*, 48, 164–174.

Esch, T., Marconcini, M., Felbier, A., Roth, A., Heldens, W., Huber, M., . . . & Taubenböck, H. 2018. Urban footprint processor–Fully automated processing chain generating settlement masks from global data of the TanDEM-X mission. *IEEE Geoscience and Remote Sensing Letters*, 152, 288–292.

Esch, T., Schenk, A., Ullmann, T., Thiel, M., Roth, A., & Dech, S. 2011. Characterization of land cover types in terraSAR-X images by combined analysis of speckle statistics and intensity information. *IEEE Transactions on Geoscience and Remote Sensing*, 49, 1911–1925.

Evans, T.L., & Costa, M. 2013. Landcover classification of the Lower Nhecolandia subregion of the Brazilian Pantanal Wetlands using ALOS/PALSAR, RADARSAT-2 and ENVISAT/ASAR imagery. *Remote Sensing of Environment*, 128, 118–137.

Evans, T.L., Costa, M., Telmer, K., & Silva, T.S.F. 2010. Using ALOS/PALSAR and RADARSAT-2 to map land cover and seasonal inundation in the Brazilian pantanal. *IEEE Journal of Selected Topics in Applied Earth Observations and Remote Sensing*, 3, 560–575.

Ferraioli, G., & Pascazio, V. 2017. Urban area classification by multitemporal SAR images. *IEEE Journal of Selected Topics in Applied Earth Observations and Remote Sensing*, 10(6), 2730–2738.

Ferro-Famil, L., Pottier, E., & Lee, J.S. 2001. Unsupervised classification of multifrequency and fully polarimetric SAR images based on the H/A/alpha-Wishart classifier. *IEEE Transactions on Geoscience and Remote Sensing*, 39, 2332–2342.

Flett, D., Crevier, Y., & Girard, R. 2009. The radarsat constellation mission: Meeting the government of Canada's needs and requirements. *2009 IEEE International Geoscience and Remote Sensing Symposium*, 1–5, 1161–1163.

Ford, J.P., Cimino, J.B., Holt, B., & Ruzek, M.R., 1986. Shuttle imaging radar views of the earth from challenger: The SIR-B experiment. *JPL Publication 86–10, Jet Propulsion Laboratory*, Pasadena, CA, 135.

Freeman, A., & Durden, S.L. 1998. A three-component scattering model for polarimetric SAR data. *IEEE Transactions on Geoscience and Remote Sensing*, 36, 963–973.

Freeman, A., Villasenor, J., Klein, J.D., Hoogeboom, P., & Groot, J. 1994. On the use of multifrequency and polarimetric radar backscatter features for classification of agricultural crops. *International Journal of Remote Sensing*, 15, 1799–1812.

Freitas, C.D., Soler, L.D., Anna, S.J.S.S., Dutra, L.V., dos Santos, J.R., Mura, J.C., & Correia, A.H. 2008. Land use and land cover mapping in the Brazilian Amazon using polarimetric airborne P-band SAR data. *IEEE Transactions on Geoscience and Remote Sensing*, 46, 2956–2970.

French, A.N., Schmugge, T.J., Ritchie, J.C., Hsu, A., Jacob, F., & Ogawa, K. 2008. Detecting land cover change at the Jornada Experimental Range, New Mexico with ASTER emissivities. *Remote Sensing of Environment*, 112, 1730–1748.

Frost, V.S., Stiles, J.A., Shanmugan, K.S., & Holtzman, J.C. 1982. A model for radar images and its application to adaptive digital filtering of multiplicative noise. *IEEE Transactions on Pattern Analysis and Machine Intelligence*, 4, 157–166.

Fukuda, S., & Hirosawa, H. 1999. A wavelet-based texture feature set applied to classification of multifrequency polarimetric SAR images. *IEEE Transactions on Geoscience and Remote Sensing*, 37, 2282–2286.

Gamba, P., Dell'Acqua, F., & Lisini, G. 2006. Change detection of multitemporal SAR data in urban areas combining feature-based and pixel-based techniques. *IEEE Transactions on Geoscience and Remote Sensing*, 44, 2820–2827.

Gamba, P., Houshmand, B., & Saccani, M. 2000. Detection and extraction of buildings from interferometric SAR data. *IEEE Transactions on Geoscience and Remote Sensing*, 38, 611–618.

Gauthier, Y., Bernier, M., & Fortin, J.P. 1998. Aspect and incidence angle sensitivity in ERS-1 SAR data. *International Journal of Remote Sensing*, 19, 2001–2006.

Grey, W.M.F., Luckman, A.J., & Holland, D. 2003. Mapping urban change in the UK using satellite radar interferometry. *Remote Sensing of Environment*, 87, 16–22.

Grunsky, E.C. 2002. The application of principal components analysis to multi-beam RADARSAT-1 satellite imagery: A tool for land cover and terrain mapping. *Canadian Journal of Remote Sensing*, 28, 758–769.

Hara, Y., Atkins, R.G., Yueh, S.H., Shin, R.T., & Kong, J.A. 1994. Application of neural networks to radar image classification. *IEEE Transactions on Geoscience and Remote Sensing*, 32, 100–109.

Henderson, F.M. 1975. Radar for small-scale land-use mapping. *Photogrammetric Engineering and Remote Sensing*, 41, 307–319.

Henderson, F.M. 1995. An analysis of settlement characterization in Central-Europe using sir-B radar imagery. *Remote Sensing of Environment*, 54, 61–70.

Henderson, F.M., & Xia, Z.G. 1997. SAR applications in human settlement detection, population estimation and urban land use pattern analysis: A status report. *IEEE Transactions on Geoscience and Remote Sensing*, 35, 79–85.

Hoekman, D.H., & Reiche, J. 2015. A review of current applications and future potential for national to global scale monitoring using Sentinel-1. *Surveys in Geophysics*, 36, 463–489.

Holecz, F., Barbieri, M., Cantone, A., Pasquali, P., & Monaco, S. 2009. Synergetic use of multi-temporal ALOS PALSAR and ENVISAT ASAR data for topographic/land cover mapping and monitoring at national scale in Africa. *2009 IEEE International Geoscience and Remote Sensing Symposium*, 1–5, 256–259.

Holm, W.A., & Barnes, R.M., 1988. On radar polarization mixed state decomposition theorems. *Proceedings of the 1988 USA National Radar Conference*, Ann Arbor, MI.

Hosseini, R.S., Entezari, I., Homayouni, S., Motagh, M., & Mansouri, B. 2011. Classification of polarimetric SAR images using Support Vector Machines. *Canadian Journal of Remote Sensing*, 37, 220–233.

Hu, H.T., & Ban, Y.F. 2012. Multitemporal RADARSAT-2 ultra-fine beam SAR data for urban land cover classification. *Canadian Journal of Remote Sensing*, 38, 1–11.

Huang, M.X., Shi, Z., & Gong, J.H. 2008. Potential of multitemporal ERS-2 SAR imagery for land use mapping in coastal zone of Shangyu City, China. *Journal of Coastal Research*, 24, 170.

Huynen, J.R. 1970. *Phenomenological Theory of Radar Targets* (PhD dissertation). Drukkerij Bronder-offset N. V., Rotterdam.

Inglada, J., & Mercier, G. 2007. A new statistical similarity measure for change detection in multitemporal SAR images and its extension to multiscale change analysis. *IEEE Transactions on Geoscience and Remote Sensing*, 45, 1432–1445.

Kasischke, E.S., Melack, J.M., & Dobson, M.C. 1997. The use of imaging radars for ecological applications—A review. *Remote Sensing of Environment*, 59, 141–156.

Khan, S.I., Hong, Y., Wang, J., Yilmaz, K.K., Gourley, J.J., Adler, R.F., . . . & Policelli, F. 2016. Satellite remote sensing and hydrologic modeling for flood inundation mapping in Lake Victoria Basin: Implications for hydrologic prediction in ungauged basins. *IEEE Journal of Selected Topics in Applied Earth Observations and Remote Sensing*, 9(1), 386–394.

Kouskoulas, Y., Ulaby, F.T., & Pierce, L.E. 2004. The Bayesian hierarchical classifier (BHC) and its application to short vegetation using multifrequency polarimetric SAR. *IEEE Transactions on Geoscience and Remote Sensing*, 42, 469–477.

Krieger, G., Moreira, A., Fiedler, H., Hajnsek, I., Werner, M., Younis, M., & Zink, M. 2007. TanDEM-X: A satellite formation for high-resolution SAR interferometry. *IEEE Transactions on Geoscience and Remote Sensing*, 45(11), 3317–3341.

Krogager, E. 1990. New decomposition of the radar target scattering matrix. *Electronics Letters*, 26, 1525–1527.

Kuan, D.T., Sawchuk, A.A., Strand, T.C., & Chavel, P. 1985. Adaptive noise smoothing filter for images with signal-dependent noise. *IEEE Transactions on Pattern Analysis and Machine Intelligence*, 7, 165–177.

Kuenzer, C., & Knauer, K. 2013. Remote sensing of rice crop areas. *International Journal of Remote Sensing*, 34(6), 2101–2139.

Kuplich, T.M., Freitas, C.C., & Soares, J.V. 2000. The study of ERS-1 SAR and Landsat TM synergism for land use classification. *International Journal of Remote Sensing*, 21, 2101–2111.

Kurosu, T., Uratsuka, S., Maeno, H., & Kozu, T. 1999. Texture statistics for classification of land use with multitemporal JERS-1 SAR single-look imagery. *IEEE Transactions on Geoscience and Remote Sensing*, 37, 227–235.

Kurvonen, L., & Hallikainen, M.T. 1999. Textural information of multitemporal ERS-1 and JERS-1 SAR images with applications to land and forest type classification in boreal zone. *IEEE Transactions on Geoscience and Remote Sensing*, 37, 680–689.

Laliberte, A.S., Koppa, J., Fredrickson, E.L., & Rango, A. 2006. Comparison of nearest neighbor and rule-based decision tree classification in an object-oriented environment. *2006 IEEE International Geoscience and Remote Sensing Symposium*, 1–8, 3923–3926.

Lambin, E.F., & Ehrlich, D. 1997. Land-cover changes in sub-Saharan Africa (1982–1991): Application of a change index based on remotely sensed surface temperature and vegetation indices at a continental scale. *Remote Sensing of Environment*, 61, 181–200.

Lang, M.W., Townsend, P.A., & Kasischke, E.S. 2008. Influence of incidence angle on detecting flooded forests using C-HH synthetic aperture radar data. *Remote Sensing of Environment*, 112, 3898–3907.

Lardeux, C., Frison, P.L., Tison, C., Souyris, J.C., Stoll, B., Fruneau, B., & Rudant, J.P. 2009. Support vector machine for multifrequency SAR polarimetric data classification. *IEEE Transactions on Geoscience and Remote Sensing*, 47, 4143–4152.

Larranaga, A., Alvarez-Mozos, J., & Albizua, L. 2011. Crop classification in rain-fed and irrigated agricultural areas using Landsat TM and ALOS/PALSAR data. *Canadian Journal of Remote Sensing*, 37, 157–170.

Lasaponara, R., & Masini, N. 2016. SAR and optical data integration for archaeological and cultural heritage: Methods and applications. In *Remote Sensing for Archaeology and Cultural Landscapes*. Springer, Cham.

Lee, J.S. 1981. Refined filtering of image noise using local statistics. *Computer Graphics and Image Processing*, 15, 380–389.

Lee, J.S. 1983. Digital image smoothing and the sigma filter. *Computer Vision Graphics and Image Processing*, 24, 255–269.

Lee, J.S., Grunes, M.R., Ainsworth, T.L., Du, L.J., Schuler, D.L., & Cloude, S.R. 1999a. Unsupervised classification using polarimetric decomposition and the complex Wishart classifier. *IEEE Transactions on Geoscience and Remote Sensing*, 37, 2249–2258.

Lee, J.S., Grunes, M.R., & de Grandi, G. 1999b. Polarimetric SAR speckle filtering and its implication for classification. *IEEE Transactions on Geoscience and Remote Sensing*, 37, 2363–2373.

Lee, J.S., Grunes, M.R., & Kwok, R. 1994. Classification of multi-look polarimetric SAR imagery-based on complex Wishart distribution. *International Journal of Remote Sensing*, 15, 2299–2311.

Lee, J.S., Grunes, M.R., & Pottier, E. 2001. Quantitative comparison of classification capability: Fully polarimetric versus dual and single-polarization SAR. *IEEE Transactions on Geoscience and Remote Sensing*, 39, 2343–2351.

Lee, J.S., & Pottier, E., 2009. *Polarimetric Radar Imaging from Basics to Applications*. CRC Press, New York.

Lee, J.S., Wen, J.H., Ainsworth, T.L., Chen, K.S., & Chen, A.J. 2009. Improved sigma filter for speckle filtering of SAR imagery. *IEEE Transactions on Geoscience and Remote Sensing*, 47, 202–213.

Le Toan, T., Quegan, S., Davidson, M.W., Balzter, H., Paillou, P., Papathanassiou, K., . . . & Ulander, L. 2011. The BIOMASS mission: Mapping global forest biomass to better understand the terrestrial carbon cycle. *Remote Sensing of Environment*, 115(11), 2850–2860.

Lewis, A.J., & Henderson, F.M. 1998. Radar fundamentals: The Geoscience perspective. In: A.J. Lewis and F.M. Henderson (Editors), *Manual of Remote Sensing Volume2—Principles and Applications of Imaging Radar*. John Wiley & Sons, Inc., New York.

Li, G.Y., Lu, D.S., Moran, E., Dutra, L., & Batistella, M. 2012. A comparative analysis of ALOS PALSAR L-band and RADARSAT-2 C-band data for land-cover classification in a tropical moist region. *ISPRS Journal of Photogrammetry and Remote Sensing*, 70, 26–38.

Li, X., & Yeh, A.G.O. 2004. Multitemporal SAR images for monitoring cultivation systems using case-based reasoning. *Remote Sensing of Environment*, 90, 524–534.

Li, X., Yeh, A.G.O., Qian, J.P., Ai, B., & Qi, Z.X. 2009. A matching algorithm for detecting land use changes using case-based reasoning. *Photogrammetric Engineering and Remote Sensing*, 75, 1319–1332.

Liao, M.S., Jiang, L.M., Lin, H., Huang, B., & Gong, J.Y. 2008. Urban change detection based on coherence and intensity characteristics of SAR imagery. *Photogrammetric Engineering and Remote Sensing*, 74, 999–1006.

Lichtenegger, J., Dallemand, J.F., Reichart, P., Rebillard, P., & Buchroithner, M. 1991. Multi-sensor analysis for land use mapping in Tunisia. *Earth Observation Quarterly*, 33, 1–6.

Liesenberg, V., & Gloaguen, R. 2013. Evaluating SAR polarization modes at L-band for forest classification purposes in Eastern Amazon, Brazil. *International Journal of Applied Earth Observation and Geoinformation*, 21, 122–135.

Lim, H.H., Swartz, A.A., Yueh, H.A., Kong, J.A., Shin, R.T., & Vanzyl, J.J. 1989. Classification of earth terrain using polarimetric synthetic aperture radar images. *Journal of Geophysical Research-Solid Earth and Planets*, 94, 7049.

Liu, D., Qi, Z.X., Zhang, H., Li, X., Anthony Gar-on Yeh & Jiao Wang. 2021. Investigation of the capability of multitemporal RADARSAT-2 fully polarimetric SAR images for land cover classification: A case of Panyu, Guangdong province. *European Journal of Remote Sensing*, 54, 338–350.

Lo, C.P., 1998. Applications of imaging radar to land use and land cover mapping. In: F.M. Henderson and A.J. Lewis (Editors), *Maual of Remote Sensing—Principles and Applications of Imaging Radar*. John Wiley & Sons, Inc., New York.

Lombardo, P., & Pellizzeri, T.M. 2002. Maximum likelihood signal processing techniques to detect a step pattern of change in multitemporal SAR images. *IEEE Transactions on Geoscience and Remote Sensing*, 40, 853–870.

Longepe, N., Rakwatin, P., Isoguchi, O., Shimada, M., Uryu, Y., & Yulianto, K. 2011. Assessment of ALOS PALSAR 50 m orthorectified FBD data for regional land cover classification by support vector machines. *IEEE Transactions on Geoscience and Remote Sensing*, 49, 2135–2150.

Lopes, A., Nezry, E., Touzi, R., & Laur, H. 1993. Structure detection and statistical adaptive speckle filtering in SAR images. *International Journal of Remote Sensing*, 14, 1735–1758.

Lopes, F., Nezry, E., Touzi, R., & Laur, H. 2012. Structure detection and statistical adaptive speckle filtering in SAR images. *International Journal of Remote Sensing*, 14, 1735–1758.

Lu, D., Batistella, M., & Moran, E. 2007. Land-cover classification in the Brazilian Amazon with the integration of Landsat ETM plus and Radarsat data. *International Journal of Remote Sensing*, 28, 5447–5459.

Lu, D., Mausel, P., Brondizio, E., & Moran, E. 2004. Change detection techniques. *International Journal of Remote Sensing*, 25, 2365–2407.

Lu, D., & Weng, Q. 2007. A survey of image classification methods and techniques for improving classification performance. *International Journal of Remote Sensing*, 28, 823–870.

Luckman, A., Baker, J., & Wegmuller, U. 2000. Repeat-pass interferometric coherence measurements of disturbed tropical forest from JERS and ERS satellites. *Remote Sensing of Environment*, 73, 350–360.

Lunetta, R.S., Knight, J.F., Ediriwickrema, J., Lyon, J.G., & Worthy, L.D. 2006. Land-cover change detection using multi-temporal MODIS NDVI data. *Remote Sensing of Environment*, 105, 142–154.

Mahdianpari, M., Salehi, B., Mohammadimanesh, F., & Motagh, M. 2018. Very deep convolutional neural networks for complex land cover mapping using multispectral remote sensing imagery. *Remote Sensing*, 10, 1119.

Majd, M.S., Simonetto, E., & Polidori, L. 2012. Maximum likelihood classification of single high-resolution polarimetric SAR images in Urban Areas. *Photogrammetrie Fernerkundung Geoinformation*, 395–407.

Mason, D.C., Giustarini, L., Garcia-Pintado, J., & Cloke, H.L. 2018. Detection of flooded urban areas in high resolution Synthetic Aperture Radar images using double scattering. *IEEE Transactions on Geoscience and Remote Sensing*, 56(1), 184–196.

Massonnet, D., & Feigl, K.L. 2015. Radar interferometry and its application to changes in the Earth's surface. *Reviews of Geophysics*, 36, 441–500.

McNairn, H., Shang, J.L., Jiao, X.F., & Champagne, C. 2009. The contribution of ALOS PALSAR multipolarization and polarimetric data to crop classification. *IEEE Transactions on Geoscience and Remote Sensing*, 47, 3981–3992.

Miranda, F.P., Fonseca, L.E.N., & Carr, J.R. 1998. Semivariogram textural classification of JERS-1 (Fuyo-1) SAR data obtained over a flooded area of the Amazon rainforest. *International Journal of Remote Sensing*, 19, 549–556.

Moreira, A., Krieger, G., Younis, M., Hajnsek, I., Papathanassiou, K., Eineder, M., & De Zan, F. 2011. Tandem-L: A mission proposal for monitoring dynamic earth processes. *2011 IEEE International Geoscience and Remote Sensing Symposium (Igarss)*, 1385–1388.

Moreira, A., Prats-Iraola, P., Younis, M., Krieger, G., Hajnsek, I., & Papathanassiou, K.P. 2013. A tutorial on synthetic aperture radar. *IEEE Geoscience and Remote Sensing Magazine*, 1, 6–43.

Moser, G., & Serpico, S.B. 2006. Generalized minimum-error thresholding for unsupervised change detection from SAR amplitude imagery. *IEEE Transactions on Geoscience and Remote Sensing*, 44, 2972–2982.

Moser, G., & Serpico, S.B. 2009. Unsupervised change detection from multichannel SAR data by Markovian data fusion. *IEEE Transactions on Geoscience and Remote Sensing*, 47, 2114–2128.

Moser, G., Serpico, S., & Vernazza, G. 2007. Unsupervised change detection from multichannel SAR images. *IEEE Geoscience and Remote Sensing Letters*, 4, 278–282.

Nagler, T., Rott, H., Hetzenecker, M., Wuite, J., & Potin, P. 2015. The Sentinel-1 mission: New opportunities for ice sheet observations. *Remote Sensing of Environment*, 158, 348–361.

Niu, X., & Ban, Y.F. 2013. Multi-temporal RADARSAT-2 polarimetric SAR data for urban land-cover classification using an object-based support vector machine and a rule-based approach. *International Journal of Remote Sensing*, 34, 1–26.

Nyoungui, A.N., Tonye, E., & Akono, A. 2002. Evaluation of speckle filtering and texture analysis methods for land cover classification from SAR images. *International Journal of Remote Sensing*, 23, 1895–1925.

Papathanassiou, K.P., & Cloude, S.R. 2001. Single-baseline polarimetric SAR interferometry. *IEEE Transactions on Geoscience and Remote Sensing*, 39, 2352–2363.

Paris, J.F. 1983. Radar backscattering properties of corn and soybeans at frequencies of 1.6, 4.75, and 13.3 Ghz. *IEEE Transactions on Geoscience and Remote Sensing*, 21, 392–400.

Park, N.W., & Chi, K.H. 2008. Integration of multitemporal/polarization C-band SAR data sets for land-cover classification. *International Journal of Remote Sensing*, 29, 4667–4688.

Paul, F., Bolch, T., Kääb, A., Nagler, T., Nuth, C., Scharrer, K., . . . & Van Niel, T. 2015. The glaciers climate change initiative: Methods for creating glacier area, elevation change and velocity products. *Remote Sensing of Environment*, 162, 408–426.

Peng, X.L., Wang, J.F., & Zhang, Q.F. 2005. Deriving terrain and textural information from stereo RADARSAT data for mountainous land cover mapping. *International Journal of Remote Sensing*, 26, 5029–5049.

Pierce, L.E., Ulaby, F.T., Sarabandi, K., & Dobson, M.C. 1994. Knowledge-based classification of polarimetric SAR images. *IEEE Transactions on Geoscience and Remote Sensing*, 32, 1081–1086.

Polychronaki, A., Gitas, I.Z., Veraverbeke, S., & Debien, A. 2013. Evaluation of ALOS PALSAR imagery for burned area mapping in Greece using object-based classification. *Remote Sensing*, 5, 5680–5701.

Pu, R.L., & Landry, S. 2012. A comparative analysis of high spatial resolution IKONOS and WorldView-2 imagery for mapping urban tree species. *Remote Sensing of Environment*, 124, 516–533.

Qi, Z.X., & Yeh, A.G.O. 2013. Integrating change vector analysis, post-classification comparison, and object-oriented image analysis for land use and land cover change detection using RADARSAT-2 polarimetric SAR images. In: S. Timpf and P. Laube (Editors), *Advances in Spatial Data Handling*. Springer, Berlin, pp. 107–123.

Qi, Z.X., Yeh, A.G.O., and Li, X. 2019. Scattering-mechanism-based investigation of optimal combinations of polarimetric SAR frequency bands for land cover classification. *Photogrammetric Engineering & Remote Sensing*, 85, 799–813.

Qi, Z.X., Yeh, A.G.O., Li, X., & Lin, Z. 2012. A novel algorithm for land use and land cover classification using RADARSAT-2 polarimetric SAR data. *Remote Sensing of Environment*, 118, 21–39.

Rahman, M.M., Moran, M.S., Thoma, D.P., Bryant, R., Collins, C.D.H., Jackson, T., Orr, B.J., & Tischler, M. 2008. Mapping surface roughness and soil moisture using multi-angle radar imagery without ancillary data. *Remote Sensing of Environment*, 112, 391–402.

Rajesh, K., Jawahar, C.V., Sengupta, S., & Sinha, S. 2001. Performance analysis of textural features for characterization and classification of SAR images. *International Journal of Remote Sensing*, 22, 1555–1569.

Ranson, K.J., & Sun, G.Q. 1994. Northern forest classification using temporal multifrequency and multipolarimetric SAR images. *Remote Sensing of Environment*, 47, 142–153.

Reiche, J., de Bruin, S., Hoekman, D.H., & Verbesselt, J. 2016. A Bayesian approach to combine Landsat and ALOS PALSAR time series for near real-time deforestation detection. *Remote Sensing*, 8(5), 415.

Reigber, A., & Moreira, A. 2000. First demonstration of airborne SAR tomography using multibaseline L-band data. *IEEE Transactions on Geoscience and Remote Sensing*, 38(5), 2142–2152.

Rignot, E., Chellappa, R., & Dubois, P. 1992. Unsupervised segmentation of polarimetric SAR data using the covariance-matrix. *IEEE Transactions on Geoscience and Remote Sensing*, 30, 697–705.

Rignot, E., Mouginot, J., & Scheuchl, B. 2019. MEaSUREs InSAR-based Antarctica ice velocity map. *Remote Sensing*, 11(4), 423.

Rignot, E.J.M., & Vanzyl, J.J. 1993. Change detection techniques for Ers-1 Sar data. *IEEE Transactions on Geoscience and Remote Sensing*, 31, 896–906.

Rodriguez-Galiano, V.F., Ghimire, B., Rogan, J., Chica-Olmo, M., & Rigol-Sanchez, J.P. 2012. An assessment of the effectiveness of a random forest classifier for land-cover classification. *ISPRS Journal of Photogrammetry and Remote Sensing*, 67, 93–104.

Ronneberger, O., Fischer, P., & Brox, T. 2015. U-net: Convolutional networks for biomedical image segmentation. *International Conference on Medical Image Computing and Computer-Assisted Intervention*, 234–241.

Saatchi, S.S., Soares, J.V., & Alves, D.S. 1997. Mapping deforestation and land use in Amazon rainforest by using SIR-C imagery. *Remote Sensing of Environment*, 59, 191–202.

Santoro, M., Askne, J.I.H., Wegmuller, U., & Werner, C.L. 2007. Observations, modeling, and applications of ERS-ENVISAT coherence over land surfaces. *IEEE Transactions on Geoscience and Remote Sensing*, 45, 2600–2611.

Schotten, C.G.J., Vanrooy, W.W.L., & Janssen, L.L.F. 1995. Assessment of the capabilities of multitemporal Ers-1 Sar data to discriminate between agricultural crops. *International Journal of Remote Sensing*, 16, 2619–2637.

Shimada, M., Itoh, T., Motooka, T., Watanabe, M., Shiraishi, T., Thapa, R., & Lucas, R. 2014. New global forest/non-forest maps from ALOS PALSAR data (2007–2010). *Remote Sensing of Environment*, 155, 13–31.

Shimoni, M., Borghys, D., Heremans, R., Perneel, C., & Acheroy, M. 2009. Fusion of PolSAR and PolInSAR data for land cover classification. *International Journal of Applied Earth Observation and Geoinformation*, 11, 169–180.

Shupe, S.M., & Marsh, S.E. 2004. Cover- and density-based vegetation classifications of the Sonoran desert using Landsat TM and ERS-1 SAR imagery. *Remote Sensing of Environment*, 93, 131–149.

Simard, M., Pinto, N., Fisher, J.B., & Baccini, A. 2011. Mapping forest canopy height globally with spaceborne lidar. *Journal of Geophysical Research: Biogeosciences*, 116(G4).

Simard, M., Saatchi, S.S., & De Grandi, G. 2000. The use of decision tree and multiscale texture for classification of JERS-1 SAR data over tropical forest. *IEEE Transactions on Geoscience and Remote Sensing*, 38, 2310–2321.

Singh, A. 1989. Digital change detection techniques using remotely-sensed data. *International Journal of Remote Sensing*, 10, 989–1003.

Solaiman, B., Pierce, L.E., & Ulaby, F.T. 1999. Multisensor data fusion using fuzzy concepts: Application to land-cover classification using ERS-1/JERS-1 SAR composites. *IEEE Transactions on Geoscience and Remote Sensing*, 37, 1316–1326.

Solbo, S., & Eltoft, T. 2004. Gamma-WMAP: A statistical speckle filter operating in the wavelet domain. *International Journal of Remote Sensing*, 25, 1019–1036.

Srivastava, H.S., Patel, P., Sharma, Y., & Navalgund, R.R. 2009. Large-area soil moisture estimation using multi-incidence-angle RADARSAT-1 SAR data. *IEEE Transactions on Geoscience and Remote Sensing*, 47, 2528–2535.

Stow, D.A., Tinney, L.R., & Estes, J.E. 1980. Deriving land use/land cover change statistics from Landsat: A study of prime agricultural land. *Proceedings of the 14th International Symposium on Remote Sensing of Environment*. Ann Arbor, MI.

Strozzi, T., Dammert, P.B.G., Wegmuller, U., Martinez, J.M., Askne, J.I.H., Beaudoin, A., & Hallikainen, M.T. 2000. Landuse mapping with ERS SAR interferometry. *IEEE Transactions on Geoscience and Remote Sensing*, 38, 766–775.

Suzuki, S., Kankaku, Y., & Osawa, Y. 2013. ALOS-2 current status and operation plan. *Sensors, Systems, and Next-Generation Satellites Xvii*, 8889.

Tan, C.P., Ewe, H.T., & Chuah, H.T. 2011. Agricultural crop-type classification of multi-polarization SAR images using a hybrid entropy decomposition and support vector machine technique. *International Journal of Remote Sensing*, 32, 7057–7071.

Tison, C., Nicolas, J.M., Tupin, F., & Maitre, H. 2004. A new statistical model for Markovian classification of urban areas in high-resolution SAR images. *IEEE Transactions on Geoscience and Remote Sensing*, 42, 2046–2057.

Tison, C., Tupin, F., & Maitre, H. 2007. A fusion scheme for joint retrieval of urban height map and classification from high-resolution interferometric SAR images. *IEEE Transactions on Geoscience and Remote Sensing*, 45, 496–505.

Toll, D.L. 1985. Analysis of digital LANDSAT MSS and SEASAT SAR data for use in discriminating land cover at the urban fringe of Denver, Colorado. *International Journal of Remote Sensing*, 6, 1209–1229.

Torres, R., et al. 2012. GMES Sentinel-1 mission. *Remote Sensing of Environment*, 120, 9–24.

Touzi, R. 2007. Target scattering decomposition in terms of roll-invariant target parameters. *IEEE Transactions on Geoscience and Remote Sensing*, 45, 73–84.

Turkar, V., Deo, R., Rao, Y.S., Mohan, S., & Das, A. 2012. Classification accuracy of multi-frequency and multi-polarization SAR images for various land covers. *IEEE Journal of Selected Topics in Applied Earth Observations and Remote Sensing*, 5, 936–941.

Tzeng, Y.C., & Chen, K.S. 1998. A fuzzy neural network to SAR image classification. *IEEE Transactions on Geoscience and Remote Sensing*, 36, 301–307.

Ulaby, F.T., Kouyate, F., Brisco, B., & Williams, T.H.L. 1986. Textural information in SAR images. *IEEE Transactions on Geoscience and Remote Sensing*, 24, 235–245.

Ulaby, F.T., Li, R.Y., & Shanmugan, K.S. 1982. Crop classification using airborne radar and Landsat data. *IEEE Transactions on Geoscience and Remote Sensing*, 20, 42–51.

Vanzyl, J.J., & Burnette, C.F. 1992. Bayesian classification of polarimetric SAR images using adaptive a-priori probabilities. *International Journal of Remote Sensing*, 13, 835–840.

Velicogna, I., Mohajerani, Y., Landerer, F., Mouginot, J., Noel, B., Rignot, E., . . . & Wiese, D. 2020. Continuity of ice sheet mass loss in Greenland and Antarctica from the GRACE and GRACE follow-on missions. *Geophysical Research Letters*, 46(22), 12513–12520.

Verma, A.K., Bhattacharya, B.K., Kumar, R., Prasad, H.S., & Dhote, P. 2018. Rice crop parameter retrieval using multi-frequency polarimetric SAR data. *Journal of the Indian Society of Remote Sensing*, 46(5), 805–815.

Vidal, A., & Moreno, M.R. 2011. Change detection of isolated housing using a new hybrid approach based on object classification with optical and TerraSAR-X data. *International Journal of Remote Sensing*, 32, 9621–9635.

Villasenor, J.D., Fatland, D.R., & Hinzman, L.D. 1993. Change detection on Alaska North Slope using repeat-pass ers-1 SAR images. *IEEE Transactions on Geoscience and Remote Sensing*, 31, 227–236.

Waite, W.P., 1976. Historical development of imaging radar. In: A.J. Lewis (Editor), *Geoscience Applications of Imaging Radar Systems*. Remote Sensing Committee, Association of American Geographers, University of Nebraska, Nebraska, pp. 1–22.

Walker, W.S., Stickler, C.M., Kellndorfer, J.M., Kirsch, K.M., & Nepstad, D.C. 2010. Large-area classification and mapping of forest and land cover in the Brazilian Amazon: A comparative analysis of ALOS/PALSAR and Landsat data sources. *IEEE Journal of Selected Topics in Applied Earth Observations and Remote Sensing*, 3, 594–604.

Wang, J.R., Engman, E.T., Shiue, J.C., Rusek, M., & Steinmeier, C. 1986. The Sir-B observations of microwave backscatter dependence on soil-moisture, surface-roughness, and vegetation covers. *IEEE Transactions on Geoscience and Remote Sensing*, 24, 510–516.

Wang, L., Sousa, W.P., Gong, P., & Biging, G.S. 2004. Comparison of IKONOS and QuickBird images for mapping mangrove species on the Caribbean coast of Panama. *Remote Sensing of Environment*, 91, 432–440.

Wang, T., Zhang, Y., Feng, G., & Lu, P. 2016. Recent advances in SAR interferometry time series analysis for ground displacement detection. *Remote Sensing*, 8(5), 441.

Wang, Y., & Allen, T.R. 2008. Estuarine shoreline change detection using Japanese ALOS PALSAR HH and JERS-1 L-HH SAR data in the Albemarle-Pamlico sounds, North Carolina, USA. *International Journal of Remote Sensing*, 29, 4429–4442.

Warner, K., Iacozza, J., Scharien, R., & Barber, D. 2013. On the classification of melt season first-year and multi-year sea ice in the Beaufort Sea using Radarsat-2 data. *International Journal of Remote Sensing*, 34, 3760–3774.

Weber, F., Nixon, D., & Hurley, J. 2003. Semi-automated classification of river ice types on the Peace River using RADARSAT-1 synthetic aperture radar (SAR) imagery. *Canadian Journal of Civil Engineering*, 30, 11–27.

Wegmuller, U., Werner, C., Strozzi, T., & Wiesmann, A. 2016. Multi-temporal interferometric SAR aspects. *Remote Sensing*, 8, 851.

White, R.G. 1991. Change detection in SAR imagery. *International Journal of Remote Sensing*, 12, 339–360.

Witten, I.H., Frank, E., & Hall, M.A. 2011. *Data Mining: Practical Machine Learning Tools and Techniques*. Morgan Kaufmann, Burlington, MA.

Wu, S.T., & Sader, S.A. 1987. Multipolarization SAR data for surface-feature delineation and forest vegetation characterization. *IEEE Transactions on Geoscience and Remote Sensing*, 25, 67–76.

Wulder, M.A., Masek, J.G., Cohen, W.B., Loveland, T.R., & Woodcock, C.E. 2012. Opening the archive: How free data has enabled the science and monitoring promise of Landsat. *Remote Sensing of Environment*, 122, 2–10.

Xu, S., Qi, Z.X., Yeh, A.G.O., & Li, X. 2019. Investigation of the effect of the incidence angle on land cover classification using fully polarimetric SAR images. *International Journal of Remote Sensing*, 40, 1–18.

Yamaguchi, Y., Moriyama, T., Ishido, M., & Yamada, H. 2005. Four-component scattering model for polarimetric SAR image decomposition. *IEEE Transactions on Geoscience and Remote Sensing*, 43, 1699–1706.

Yonezawa, C., Negishi, M., Azuma, K., Watanabe, M., Ishitsuka, N., Ogawa, S., & Saito, G. 2012. Growth monitoring and classification of rice fields using multitemporal RADARSAT-2 full-polarimetric data. *International Journal of Remote Sensing*, 33, 5696–5711.

Zakhvatkina, N.Y., Alexandrov, V.Y., Johannessen, O.M., Sandven, S., & Frolov, I.Y. 2013. Classification of sea ice types in ENVISAT synthetic aperture radar images. *IEEE Transactions on Geoscience and Remote Sensing*, 51, 2587–2600.

Zhang, H., Qi, Z., Li, X., Chen, Y., Wang, X., & He, Y. 2021. An urban flooding index for unsupervised inundated urban area detection using sentinel-1 polarimetric SAR images. *Remote Sensing*, 13(22), 4511.

Zhang, L., & Zhang, L. 2016. Deep learning for remote sensing data: A technical tutorial on the state of the art. *IEEE Geoscience and Remote Sensing Magazine*, 4, 22–40.

Zhang, X.H., Chen, J.W., & Meng, H.Y. 2013. A novel SAR image change detection based on graph-cut and generalized gaussian model. *IEEE Geoscience and Remote Sensing Letters*, 10, 14–18.

Zheng, J., & You, H.J. 2013. A new model-independent method for change detection in multitemporal SAR images based on radon transform and Jeffrey divergence. *IEEE Geoscience and Remote Sensing Letters*, 10, 91–95.

Zhu, X.X., Tuia, D., Mou, L., Xia, G.S., Zhang, L., Xu, F., & Fraundorfer, F. 2017. Deep learning in remote sensing: A comprehensive review and list of resources. *IEEE Geoscience and Remote Sensing Magazine*, 5, 8–36.

Part V

Carbon

10 Global Carbon Budgets and the Role of Remote Sensing

R.A. Houghton

ACRONYMS AND DEFINITIONS

ALOS	Advanced Land Observing Satellite
AVHRR	Advanced very high resolution radiometer
ETM+	Enhanced Thematic Mapper+
FAO	Food and Agriculture Organization of the United Nations
GLAS	Geoscience Laser Altimeter System
JRC	Joint Research Center
LiDAR	Light detection and ranging
LULCC	Land use, land cover change
MODIS	Moderate-resolution imaging spectroradiometer
NASA	National Aeronautics and Space Administration
NDVI	Normalized Difference Vegetation Index
NOAA	National Oceanic and Atmospheric Administration
PALSAR	Phased Array type L-band Synthetic Aperture Radar
REDD	Reducing Emissions from Deforestation in Developing Countries
UNFCCC	United Nations Framework Convention on Climate Change

10.1 THE GLOBAL CARBON BUDGET

In its simplest formulation the global carbon cycle consists of four terms (atmosphere, land, ocean, and fossil fuels) (Table 10.1). The natural fluxes of carbon between land and atmosphere are 120–150 PgC/yr (1 petagram of carbon equals 10^{15} gC, or 1 billion metric tons C, or 3.67 billion metric tons CO_2), as a result of global photosynthesis and respiration (including fire). Similar fluxes of 90–120 PgC/yr occur between the ocean and atmosphere as a result of physical, chemical, and biological processes. These are not the fluxes of the global carbon *budget*, however. Instead, the global carbon budget usually refers to the anthropogenic perturbation to the global carbon budget. The fluxes resulting from anthropogenic perturbation are 1–2 orders of magnitude smaller than the natural flows (Table 10.1).

This chapter reviews the global carbon budget and the role of remote sensing—past, present, and future—in helping to define it (Ingalls et al., 2022; Wei et al., 2024; Budget, 2023; Friedlingstein et al., 2023; Liu et al., 2023; Malerba et al., 2023; Ehlers et al., 2022; Zhao et al., 2022; Harris et al., 2021; Friedlingstein et al., 2020; Pique et al., 2020; Liang et al., 2019; Cressie, 2018; Lees et al., 2018). The focus is on terrestrial ecosystems; very little is written about the role of satellites in measuring either emissions of carbon from fossil fuels or uptake of carbon by the oceans. Within the context of terrestrial ecosystems, the emphasis is on the emissions and uptake of carbon resulting from disturbance and recovery, particularly those disturbances caused by land use and land cover change (LULCC) or management.

Two broad types of explanatory mechanisms account for the loss and accumulation of carbon on land: (1) disturbances and recovery (structural mechanisms) and (2) the differential effects of

TABLE 10.1
Stocks and Flows of Carbon

Carbon stocks (PgC)	
Atmosphere	850
Land	2000
Vegetation	500
Soil	1500
Ocean	39,000
Surface	700
Deep	38,000
Fossil fuel reserves	5,000
Natural flows (PgC yr^{-1})	
Atmosphere-oceans	90
Atmosphere-land	120
The Global Carbon Budget: Anthropogenic perturbations (PgC yr^{-1} averaged over 2000–2009)	
Fossil fuels	7.8 (±0.4)
Land-use change	1.0 (±0.5)
Atmospheric increase	4.0 (±0.1)
Oceanic uptake	2.4 (±0.5)
Residual terrestrial sink	2.4 (±0.8)

environmental change (e.g., CO_2, N deposition, climate) on photosynthesis and respiration (metabolic mechanisms). The two types are not clearly distinct, e.g., a forest recovering from wood harvest may grow faster because of fertilization. Nevertheless, the distinction is useful: it is implicit or explicit in carbon models, the two mechanisms generally operate at different scales, and they are measured with different instruments. From a remote sensing perspective, the first mechanism, structural, involves changes in canopy structure (demography); the second, metabolic, may involve changes in canopy greenness.

10.1.1 THE CONTEMPORARY CARBON BUDGET

During the first decade of the 21st century, the emissions of carbon from combustion of fossil fuels averaged 7.8 (±0.4) PgC/yr (Table 10.1). These emissions are determined from (largely economic) data on the production and consumption of coal, oil, and gas (Andres et al., 2012). Another 1.0 (±0.5) PgC yr^{-1} was released to the atmosphere as a result of LULCC. That source is a net flux that includes both larger emissions and partially offsetting sinks, all attributable to land management and discussed in greater detail later. The amount of carbon accumulating in the atmosphere each year is based on measurements such as those by the National Oceanographic and Atmospheric Administration (NOAA) Friedlingstein et al., 2023; Liu et al., 2023; Cressie, 2018; Lees et al., 2018; Conway and Tans, 2012). Those accumulations averaged 4.0 (±0.1) PgC/yr over the period 2000–2009. The amount of carbon taken up by the world's oceans was 2.4 (±0.5) PgC/yr, determined by a number of global biogeochemical ocean models (Le Quéré et al., 2013). And the amount of carbon accumulating in terrestrial ecosystems, *not driven by management* (i.e., LULCC) was 2.4 (±0.8) PgC yr^{-1}. That sink is calculated by differences from the other terms in the global carbon budget. It makes the global carbon budget balance. It is commonly referred to as the residual terrestrial sink. The mechanisms driving that sink are thought to include CO_2 fertilization, N deposition, and changes in climate, but the relative contributions of these mechanisms are uncertain. Indeed, the total sink is greater than 2.4 PgC yr^{-1} because that value represents a net sink, and there are undoubtedly sources as well, such as enhanced respiration associated with permafrost thaw (Natali et al. 2014).

Keeping track of the global carbon budget annually is crucial, not only to understand how much of the carbon emitted to the atmosphere stays there, and how much accumulates on land and in the ocean (Table 10.1), but because changes in the partitioning of emissions among these reservoirs (atmosphere, land, and ocean) may provide the first indication that the global carbon cycle is changing, perhaps in response to climatic change (Friedlingstein et al., 2023; Friedlingstein et al., 2020). In particular, the fraction of carbon emissions that remains airborne has been remarkably constant (~50%) for 50 years (despite large interannual variations). As emissions have approximately doubled since the 1960s, so have the sinks on land and in the ocean, so the fraction remaining airborne has remained the same. And those sinks have kept the atmospheric increase at only half of what it would have been if all the emissions had remained in the atmosphere. In other words, the sinks have dampened global warming by about half. Whether those fractions will continue as the Earth warms is a question with both scientific and policy implications. Most of the feedback known or imaginable between carbon and climate suggest that a warming will lead to more carbon emissions (or less uptake)—i.e., that positive feedback will prevail. But there is surprisingly little evidence of a decline in land and ocean sinks yet (Lees et al., 2018), although the issue is controversial (Liu et al., 2023; Friedlingstein et al., 2020; Lees et al., 2018; Canadell et al., 2007; Knorr, 2009; Le Quéré et al. 2009; Gloor et al., 2010; Raupach et al., 2014).

10.1.2 A History of Carbon Cycle Research

Global carbon budgets were not possible to construct until after 1957, when Charles David Keeling (1928–2005) began continuous measurement of carbon dioxide concentrations at Mauna Loa, Hawaii, and the South Pole (Friedlingstein et al., 2023; Friedlingstein et al., 2020). Those measurements provided a consistent and reliable record of carbon dioxide concentrations that was required to demonstrate, first of all, the rate at which carbon dioxide was increasing in the atmosphere.

At about the same time that Keeling began his measurements, a community of climate scientists began stepping up the construction of global climate models that calculated the changes in climate expected from increased concentrations of greenhouse gases. The models were based on the physics of atmospheric circulation and the physics of radiation. To predict the rate of warming, however, and not just the equilibrium warming expected from a doubling of carbon dioxide, two additional pieces of information are required: the amount of carbon dioxide added to the atmosphere each year, and the residence time of the greenhouse gas in the atmosphere. These pieces of information enable the prediction of how rapidly the carbon dioxide concentration in the atmosphere could double, and thus how rapidly the Earth's temperature could increase in response to such a doubling (estimated at 1–4°C). Actually, the concentration of carbon dioxide does not need to double, because other greenhouse gases contribute as well, but their combined effects can be calculated in carbon dioxide equivalents. Carbon dioxide is the dominant greenhouse gas under human control. Over the last 100 years, it has accounted for more radiative forcing than all the other greenhouse gases combined and is expected to do so in the future (Ingalls et al., 2022; Friedlingstein et al., 2023).

The annual amount of carbon dioxide emitted globally from the burning of fossil fuels is obtained from statistics on oil, coal, and gas production. The residence time of carbon dioxide (and other greenhouse gases) in the atmosphere is not as readily determined. The amount of carbon dioxide emitted from fossil fuel burning that stays in the atmosphere (and for how long) can be determined from an evaluation of the global carbon budget, and that evaluation has been reconstructed annually since about 1960, based on Keeling's initial measurements of carbon dioxide at two locations, now expanded by NOAA and others to about 200 locations over the Earth (Wei et al., 2024; Friedlingstein et al., 2023; Friedlingstein et al., 2020). The atmosphere is well mixed, and all of the stations show nearly the same annual rate of growth in carbon dioxide concentration, but the spatial and seasonal variability in concentration is useful for sorting out where the emitted carbon is going (land or sea).

The problem with using the atmospheric residence time to project the rate of increase in concentration is that the residence time observed over the last decades, during which the Earth's average temperature has increased by about 0.75°C, is not necessarily the residence time on an Earth that may grow to be 1–6°C warmer over the next decades. The processes that control the uptake of carbon dioxide by the world's oceans and by terrestrial ecosystems are affected by climate (e.g., temperature and moisture) and by the concentration of carbon dioxide itself. These processes drive feedback in the carbon-climate system.

The observation that surface temperatures seem to have increased at a much slower rate since 1997 may be the result of a bias in the coverage of global temperature measurements. Correcting for the undersampling at high latitudes, where the increases in temperature have been the greatest, shows an average global rate of warming consistent with rate observed before the late 1990s (Wei et al., 2024; Ehlers et al., 2022; Liang et al., 2019; Cowtan and Way, 2014).

10.1.3 Sources and Sinks of Carbon from Land

In the 1960s and 1970s, there were no independent estimates of the net annual flux of carbon between land and atmosphere, and the net flux was calculated by difference to make the global carbon budget balance. Using that approach, the net terrestrial uptake for the period 2000–2009 averaged 2.4 (±0.8) PgC/yr (Table 10.1). And the global carbon budget is balanced.

In the early 1980s the first independent estimates of the terrestrial carbon flux were advanced, based on census data (non-spatial) concerning changes in the area of forests (Budget, 2023; Friedlingstein et al., 2023; Friedlingstein et al., 2020; Lees et al., 2018; Moore et al., 1981; Houghton et al., 2012). Deforestation was occurring in many tropical countries, and the carbon held in the trees and soils of these forests was released to the atmosphere with deforestation.

10.1.4 A Bookkeeping Model

These early analyses developed a "bookkeeping" model, which used annual rates of land-cover change and biome-averaged growth and decomposition rates per hectare to calculate annual changes in carbon pools as a result of management (Figure 10.1) (Houghton et al., 2012). For example, conversion of native vegetation to cultivated land causes 25–30% of the soil organic carbon in the top meter to be lost (Budget, 2023; Friedlingstein et al., 2023; Friedlingstein et al., 2020; Post and Kwon, 2000; Guo and Gifford, 2002; Murty et al., 2002; Don et al., 2011). This tracking approach assigns an average carbon density to the biomass and soils of a small number of ecosystem types (e.g., deciduous forest, grassland). Considerable uncertainty arises because, even within the same ecosystem type, the spatial variability in carbon density is large, partly from variations in soils and microclimate, and partly from past disturbances and recovery.

The approach in the early 1980s was not based on remote sensing. Instead, historic changes in croplands and pastures, aggregated at national or continental scales, were obtained from national and international statistics. Not being spatially explicit, the data did not specify the ecosystem type that was converted to new agricultural land. That specification required independent data, such as maps of natural ecosystems and their overlap with the distribution of croplands and pastures. Data on historical changes in land cover were reconstructed from a variety of national and international historical narratives and national land-use statistics, as well as from population data (Houghton et al., 2012).

In this highly aggregated approach, the world was divided into ten major regions, and each region was assigned two to six natural ecosystem types (Houghton et al., 2012). Since the original work in 1983, analyses with non-spatial data have been refined over the years, including lands besides forests and changes in land cover besides deforestation. The calculated flux of carbon has

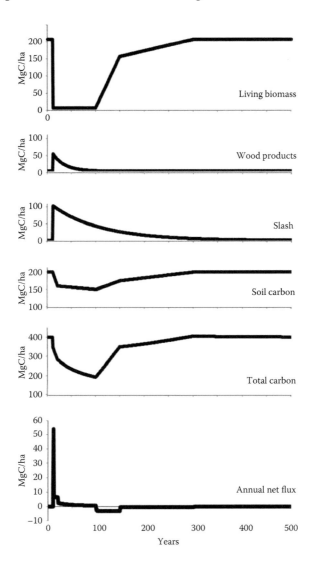

FIGURE 10.1 Idealized response curves of the bookkeeping model. The curves define the annual per-hectare changes in carbon pools in response to management (updated from Houghton et al., 2012). The bottom panel shows the annual source (sink) to the atmosphere.

been called the deforestation flux, but it is more accurately referred to as the flux from land-use and land-cover change (LULCC). It includes increases in forest area (reforestation, afforestation) as well as deforestation, and it includes losses and gains of carbon per hectare within forests as a result of wood harvest and forest growth. Ideally, the net flux would include all changes in terrestrial carbon brought about by management. During the period 2000–2009, the flux from LULCC was a net release of 1.0 PgC/yr (~ 11% of total anthropogenic carbon emissions) (Friedlingstein et al., 2023; Harris et al., 2021; Friedlingstein et al., 2020; Pique et al., 2020; Houghton et al., 2012).

Notice that adding another source of carbon to the global carbon budget leaves it unbalanced (Table 10.1). There must be another sink to compensate. Since the oceans and atmosphere are accounted for, the sink must be on land, a sink not attributable to LULCC but, instead, to environmental effects. During the period 2000–2009 this residual terrestrial sink was 2.4 (±0.8) PgC/yr.

Again, it is determined by difference, although there are a number of global dynamic vegetation models (GDVMs) that calculate an annual net uptake of similar magnitude in (unmanaged) terrestrial ecosystems (Le Quéré et al., 2013).

10.1.5 SPATIAL ANALYSES

One of the weaknesses of national (non-spatial) data on the areas of cropland and pasture is that the changes through time are net changes in area, not gross changes. Net changes in land cover underestimate gross sources and sinks of carbon that result from simultaneous clearing for, and abandonment of, agricultural lands and thus may underestimate areas of secondary forests and their carbon sinks.

Spatially explicit approaches to historic reconstructions get around this weakness. They were first developed around the year 2000. In one approach, agricultural expansion was distributed spatially on the basis of population density (Klein Goldewijk, 2001). In another, past areas were derived by hind-casting of the current distribution of agricultural lands Ramankutty and Foley (1999). The data sets of these approaches have been updated and extended to the preindustrial past (Pongratz et al., 2008; Klein Goldewijk et al., 2011). The approaches must make assumptions, just as the non-spatial approach did, about whether agricultural expansion occurs at the expense of grasslands or forests. The distinctions are important because different locations have different carbon stocks, and the carbon flux resulting from LULCC depends on both rates of land cover change and the carbon density of the lands affected. Remote sensing–based information has also been combined with regional tabular statistics to reconstruct spatially explicit land cover changes covering more than the satellite era (Thenkabail et al., 2021; Ramankutty and Foley, 1999; Klein Gooldewijk, 2001; Pongratz et al., 2008).

Two spatial data sets, along with the non-spatial data of Houghton (2013a, b, c), have been used in most of the analyses of LULCC: the SAGE data set, including cropland areas from 1700 to 1992 (Thenkabail et al., 2021; Ramankutty and Foley, 1999), and the HYDE data set, including both cropland and pasture areas (Klein Goldewijk, 2001). The difference in emissions estimates using these three data sets account for about 15% of the difference in flux estimates over the period 1850–1990 (Shevliakova et al., 2009). Other recent data sets, such as the ones compiled by Hurtt et al. (2006) and Pongratz et al. (2008), are based on combinations of SAGE, HYDE, and Houghton data sets, including updates (Houghton, 2010; Houghton et al., 2012).

The results of 13 recent analyses of LULCC (spatial and non-spatial) consistently show a net source of about 1.0 PgC/yr to the atmosphere in recent decades (Friedlingstein et al., 2023; Friedlingstein et al., 2020; Lees et al., 2018; Houghton et al., 2012). Over the longer period, 1850–2012, the annual sources and sinks of carbon from anthropogenic perturbation to the global carbon budget are shown in Figure 10.2. The emissions from fossil fuels have increased steadily through time, now accounting for ~90% of anthropogenic emissions of carbon. But before about 1900 the net emissions from LULCC were higher than fossil fuel emissions. The emissions from LULCC have not varied much from ~1 PgC/yr in the last decades.

Figure 10.2 suggests that for budget purposes there are actually five terms in the global carbon budget. The land appears twice, first as net emissions from LULCC (management) and second as net sinks attributable to processes other than management. These non-management processes are believed to result from natural and indirect anthropogenic effects, such as the effects of elevated CO_2 on plant growth, the effect of greater nitrogen (N) availability (fertilizers, N-fixation through combustion of fossil fuels), and the effects of a changing climate on the growth and respiration of vegetation and on the decay of organic matter in litter and soils. In other words, nature has had an effect on terrestrial carbon storage (and fluxes) at least as great (and in the opposite direction) as the effects of management.

It is important to note that both of these terrestrial fluxes are *net* fluxes. Management is responsible for carbon sinks in growing forests as well as for carbon sources from deforestation. And the residual terrestrial sink is composed of both increased plant growth and increased respiration and decay.

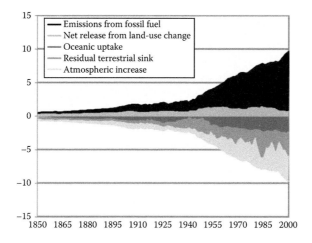

FIGURE 10.2 Annual sources (+) and sinks (-) of carbon in the global carbon budget (from Le Quéré et al., 2013).

TABLE 10.2
The Role of Forests in the Global Carbon Budget (PgC yr⁻¹). Carbon Sinks Are Negative, Sources Positive

	LULCC	Residual Terrestrial Sink	Total
Tropical forests	1.3	−1.2	0.1
Forests of the temperate and boreal zones	−0.2	−1.0	−1.2
Total	1.1	−2.2	−1.1

Source: Modified from Pan et al. (2011)

10.2 LAND USE AND LAND COVER CHANGE (LULCC), DISTURBANCES, AND RECOVERY

The terrestrial fluxes in Table 10.1 are not evenly distributed over the Earth; forests play a dominant role (Pan et al., 2011). The role of tropical forests in the global carbon cycle is nearly neutral, in part because field data from unmanaged forests suggest an increase carbon storage (Phillips et al., 2008; Gloor et al., 2012), while LULCC result in a loss of carbon of a similar magnitude (Houghton et al., 2009, 2012; Le Quéré et al., 2013) (Table 10.2). The emissions of carbon from LULCC are, themselves, uncertain because of differing estimates of deforestation rates and carbon densities (Friedlingstein et al., 2023; Zhao et al., 2022; Harris et al., 2021; Thenkabail et al., 2021; Friedlingstein et al., 2020; Pique et al., 2020; Baccini et al., 2012; Harris et al., 2012; Houghton et al., 2012).

Outside the tropics, the net carbon balance of forests is reasonably well documented because analyses are based on data from systematic forest inventories. Nevertheless, the question remains: how much of the observed carbon sink is attributable to recovery from a past disturbance, as opposed to an environmentally enhanced rate of growth (Williams et al., 2012; Zhang et al., 2012; Fang et al., 2014)?

Table 10.2 also shows that the residual terrestrial sink is roughly the same in tropical forests and in temperate and boreal forests (~1.0 PgC yr⁻¹). The major difference between the regions is the

FIGURE 10.3 Landsat images showing clearings (light blue) within a forested landscape (red) in the state of Rondonia, Brazil. Resolution: 30 m.

effect of LULCC—high rates of deforestation (and thus emissions) in the tropics versus a small sink outside the tropics where regrowth is slightly greater than emissions. In tropical forests the sink from environmental effects offsets the emissions from LULCC, for a total balance close to zero. Outside the tropics both LULCC and environmental effects result in sinks. Overall, the global terrestrial carbon sink is in northern mid-latitude forests.

10.2.1 Use of Satellite Data

The use of satellite data to help determine the sources and sinks of carbon from disturbance and recovery is discussed here (Section 10.3). Section 2.5 will address the use of satellite data to help locate and identify other metabolic changes in terrestrial carbon storage.

Satellite data at moderate spatial resolution (30–100 m) have been used to document land cover change since the mid-1980s, but only at local or regional scales. Recently Landsat data have been used to determine global rates of forest loss and gain (Hansen et al., 2013) (Figure 10.3). Furthermore, satellite data have recently begun to be used to document the carbon density of aboveground woody vegetation (Ingalls et al., 2022; Wei et al., 2024; Friedlingstein et al., 2020; Saatchi et al., 2011; Baccini et al., 2012).

10.2.1.1 Rates of Change in Forest Area

A major uncertainty in estimating the net flux of carbon from LULCC has always been, and remains, rates of change in the areas of forests (deforestation and reforestation). Satellites were first recognized as being useful for documentation of these changes in 1984 (Woodwell et al., 1984). One of the earliest studies using Landsat data was of deforestation rates in Amazonia (Skole and Tucker, 1993).

With a time series of satellite data on land cover change, it is possible to estimate the changes (Wei et al., 2024; Budget, 2023; Friedlingstein et al., 2023; Liu et al., 2023; Malerba et al., 2023). In general, satellite data alleviate the concerns of bias, inconsistency, and subjectivity in country reporting (Grainger, 2008). Depending on the spatial and temporal resolution, satellite data can also distinguish between gross and net losses of forest area. However, increases in forest area are more difficult to define with satellite data than deforestation because the growth of trees is a more gradual process. Furthermore, although satellite data are good for measuring losses of forest area, identifying the types of land use that follow deforestation (e.g., croplands, pastures, shifting cultivation) requires repeated looks at the land. Exceptions include the regional studies by Morton et al. (2006) and Galford et al. (2008).

Satellite-based methods for measuring changes in forest area include both high-resolution, sample-based methods and wall-to-wall mapping analyses. Sample-based approaches employ systematic or stratified random sampling to quantify gains or losses of forest area at national, regional, and global scales (Achard et al., 2002, 2004; Hansen et al., 2010, 2013). Systematic sampling provides a framework for forest area monitoring. The UN-FAO Forest Resource Assessment Remote Sensing Survey uses samples at every latitude/longitude intersection to quantify biome and global-scale forest change dynamics from 1990 to 2005 (Food and Agriculture Organization and Joint Research Centre) (FAO/JRC, 2012). Other sampling approaches stratify by intensity of change, thereby reducing sample intensity. Achard et al. (2002) provided an expert-based stratification of the tropics to quantify forest cover loss from 1990 to 2000 using whole Landsat image pairs. Hansen et al. (2010, 2011, 2013) employed Moderate Resolution Imaging Spectrometer (MODIS) data as a change indicator to stratify biomes into regions of homogeneous change for Landsat sampling.

Sampling methods such as those described earlier provide regional estimates of forest area and change with uncertainty bounds, but they do not provide a spatially explicit map of forest extent or change. Wall-to-wall mapping does. While coarse-resolution data sets (> 4 km) have been calibrated to estimate wall-to-wall changes in area (DeFries et al., 2002), recent availability of moderate spatial resolution data (< 100 m), typically Landsat imagery (30 m), allows a more finely resolved approach. Historical methods rely on photointerpretation of individual images to update forest cover on annual or multi-year bases, such as with the Forest Survey of India (Global Forest Survey of India, 2008) or the Ministry of Forestry Indonesia products (Government of Indonesia/World Bank, 2000). Advances in digital image processing have led to an operational implementation of mapping annual forest-cover loss, for example, with the Brazilian PRODES (INPE, 2010) and the Australian National Carbon Accounting products (Caccetta et al., 2007). These two systems rely on cloud-free data to provide single-image/observation updates on an annual basis. Persistent cloud cover has limited the derivation of products in regions such as the Congo Basin and Insular Southeast Asia (Hansen et al., 2011). For such areas, Landsat data can be used to generate multi-year estimates of forest-cover extent and loss (Hansen et al., 2013; Broich et al., 2011a). For regions experiencing forest change at an agro-industrial scale, MODIS data provide a capability for integrating Landsat-scale change to annual time-steps (Broich et al., 2011b).

In general, moderate spatial resolution imagery is limited in tropical forest areas by data availability. Currently Landsat is the only source of data at moderate spatial resolution available for tropical monitoring, but to date an uneven acquisition strategy among bioclimatic regimes limits the application of generic biome-scale methods with Landsat. No other system has the combination of (1) global acquisitions, (2) historical record, (3) free and accessible data, and (4) standard terrain-corrected imagery, along with robust radiometric calibration, that Landsat does. Future improvements in moderate spatial resolution monitoring can be obtained by increasing the frequency of data acquisition.

The primary weakness of satellite data is that they are not available before the satellite era (Landsat began in 1972). Long time series are required for estimating legacy emissions of past land-use activity. Although maps, at varying resolutions, exist for many parts of the world, spatial data on land cover and land cover change became available at a global level only after 1972, at best. In fact, there are many gaps in the coverage of the Earth's surface before 1999 when the first global acquisition strategy for moderate spatial resolution data was undertaken with the Landsat Enhanced Thematic Mapper Plus sensor (Arvidson et al., 2001). The long-term plan of Landsat ETM+ data includes annual global acquisitions of the land surface, but cloud-cover and phenological variability limit the ability to provide annual global updates of forest extent and change. The only other satellite system that can provide global coverage of the land surface at moderate resolution is the ALOS PALSAR radar instrument, which also includes an annual acquisition strategy for the global land surface (Rosenqvist et al., 2007). However, large-area forest change mapping using radar data has not yet been implemented.

10.2.1.2 Uncertainties

Since 1990 forest areas have been reported at five-year intervals by the United Nations' (UN's) Food and Agriculture Organization (FAO). Those estimates (FAO, 2010) have been used frequently by analyses calculating the carbon fluxes from LULCC. These FAO data rely on reporting by individual countries. They are more accurate for some countries than for others and are not without inconsistencies and ambiguities (Grainger, 2008). Revisions in the reported rates of deforestation from one five-year FRA assessment to the next may be substantial due to different methods or data being used.

To estimate the uncertainty in reported rates of deforestation, the FAO began sampling with data from Landsat in the early 1990s to determine the area of forest and changes in that area (FAO, 1996). But the satellite-based estimates are not consistent with the country-based estimates (Liu et al., 2023; Ehlers et al., 2022; Zhao et al., 2022). The FRA2010 (FAO, 2010), based on data reported by countries, reports a declining rate of tropical deforestation, while the FAO/JRC (2011) study, based on a sampling with Landsat, reports an increasing rate for the years 2000–2005.

As mentioned, the country-based estimates of deforestation from the FAO (2010) are different from estimates determined with Landsat data FAO/JRC (2011). There are at least two explanations, besides the explanation generally given, that data from satellites are more consistent and objective. First of all, changes in forest cover as measured from satellites include both natural and anthropogenic disturbances, as caused, for example, by wildfire and clearing for agriculture, respectively. They may also include clear-cut harvests. Census data for agriculture, on the other hand, include only the conversion of forest to cropland or pasture, not natural disturbances, and not harvested forests, which are still defined as forest. These differences in observation and definition raise the important issue of attribution (Section 10.3.8): remote sensing observes all disturbances, not simply those attributable to agricultural conversion and forest management. Everything else being equal, estimates of disturbance observed by remote sensing should be higher than estimates based on management.

Another possible explanation is that census data may include the deforestation of small land parcels, much less than one hectare. Such small clearings may be missed even with Landsat data of 30 m resolution. In the Democratic Republic of the Congo small clearings added 35% to the rate of deforestation obtained by a more traditional analysis of change with Landsat (van der Werf et al., 2010). Rates of degradation and associated carbon emissions are even more uncertain, as they are not as easily observed with satellite data (Huang and Asner, 2010).

And, finally, one other approach for estimating deforestation rates should be mentioned: satellite detection of forest fires (van der Werf et al., 2010). The approach provides an estimate of forest loss as long as deforestation is accompanied with burning. The approach does not identify LULCC if fire is absent, for example, harvest of wood. Nor does it distinguish between intentional deforestation fires and escaped wildfires. The approach combines estimates of burned area (Giglio et al., 2010) with complementary observations of fire occurrence (Giglio et al., 2003). It makes assumptions about how much fire is for clearing. At the province or country level, clearing rates calculated this way capture up to about 80% of the variability and also 80% of the total clearing rates found by other approaches (Hansen et al., 20013, 2011; INPE, 2010). Two advantages of the fire-counting approach are (1) that it allows for an estimate of interannual variability in LULCC emissions and (2) that the emissions of carbon monoxide from burning, routinely monitored by satellites, provide a much larger departure from background conditions than emissions of CO_2 (e.g., van der Werf et al., 2008).

10.2.1.3 Biomass Density

The second type of information required for calculating the emissions of carbon from LULCC is the carbon density of the forests being deforested or harvested (Malerba et al., 2023; Ehlers et al., 2022; Zhao et al., 2022; Harris et al., 2021; Friedlingstein et al., 2020).

As mentioned, non-spatial analyses assigned average carbon densities to the vegetation and soils of a small number of natural ecosystems found in each of the ten major regions. Until recently, data on the distribution of carbon density were not adequate for finer spatial detail. A study of Amazonia, for example, showed that none of the seven different maps of biomass density were in agreement as to the total biomass of the region or even where the largest and smallest densities were to be found (Houghton et al., 2001).

With respect to calculating carbon emissions, the spatial co-occurrence of both forest loss and carbon density is especially important and only available with spatially detailed data. Average carbon densities and average rates of forest loss over large regions may yield accurate estimates of carbon emissions if the disturbances are distributed randomly. But if disturbances, particularly LULCC, affect forests with carbon densities that are systematically different from the mean carbon density, that difference will bias emissions estimates. One way to counter that bias is to co-locate changes in areas with carbon densities—at the spatial resolution of disturbance.

10.2.1.4 Spatial Analyses

Recently, new satellite techniques have been applied to estimate aboveground carbon densities (Ingalls et al., 2022; Cressie, 2018; Lees et al., 2018; Goetz and Dubayah, 2011). Examples of mapping aboveground carbon density over large regions include work with MODIS (Houghton et al., 2007), multiple satellite data (Saatchi et al., 2007, 2011), radar (Treuhaft et al., 2009), and LiDAR (Baccini et al., 2012) (see Goetz et al., 2009; Goetz and Dubayah, 2011 for reviews). While the accuracy of fine-scale satellite-based estimates may be lower than site-based inventory measurements (inventory data are generally used to calibrate satellite algorithms), the satellite data are far less intensive to collect, can cover a wide spatial area, and thus can better capture the spatial and temporal variability in aboveground carbon density. By matching carbon density to the forests actually being deforested, this approach has the potential to increase the accuracy of flux estimates, especially in tropical areas where variability of carbon density is high, and data availability is poor. Recently published maps of forest biomass are in greater agreement and at finer spatial resolution than previous maps, but differences still remain (Malerba et al., 2023; Pique et al., 2020; Liang et al., 2019; Cressie, 2018; Lees et al., 2018; Mitchard et al., 2013).

One method used to determine the carbon densities of the forests being deforested or those in close proximity to those being deforested is the approach by Pique et al. (2020). They used a 500 m x 500 m grid of aboveground biomass density determined from MODIS data, calibrated with circa 5.5 million GLAS shots, which, in turn, were calibrated with field measurements at more than 400 locations in the tropics (Figure 10.4) (Baccini et al., 2012). An advantage of the approach is that the GLAS shots and field plots were at similar scales. A second approach might use not the average aboveground carbon density of a MODIS pixel but the sample point data from GLAS estimates to determine the aboveground biomass density in the vicinity of the deforestation or degradation. This approach would not assume an average aboveground carbon density, but it would miss many of the forests being deforested.

Neither approach yields a carbon density at the resolution of forest loss (30–60 m with Landsat); thus there is still the potential for bias if deforestation or degradation takes place in forest patches systematically different from the mean density of 500 m x 500 m cells. This mismatch in scale is one of the largest sources of uncertainty (bias).

It is interesting to note, however, that the relative error of the first approach is highest in regions with low biomass density and lowest in dense humid forest characterized by high biomass density

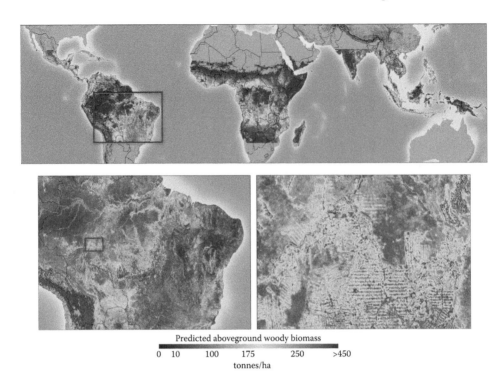

FIGURE 10.4 Aboveground carbon density in woody vegetation throughout the tropics. Resolution: 35% (from Baccini et al., 2012).

(Baccini et al., 2012). This observation suggests that errors in biomass density will contribute relatively little to the error in flux estimates calculated for deforestation.

10.2.1.5 Measurement of *Changes* in Carbon Density

If one can measure aboveground biomass density from satellites, then it should be possible to estimate *changes* in biomass density "directly" with time-series data (Figure 10.5). Some of the changes will result from changes in forest area, as measured by the first approach. But some changes in biomass density may occur without a change in cover type. Such gains and losses in carbon density that exceed the gains and losses from changes in forest area are presumably a measure of growth and degradation (Ingalls et al., 2022; Liu et al., 2023; Friedlingstein et al., 2020; Liang et al., 2019).

The traditional approach (changes in area, with average densities assigned) estimates only those density changes related to outright clearing (and recovery). The second, more direct approach estimates changes from both clearing *and* any other factors affecting carbon density. Because the two approaches observe different processes, the difference defining changes in biomass density are attributable to degradation (and growth) and/or environmental change (e.g., CO_2, climate, N deposition).

Measuring *changes* in aboveground carbon density is just beginning, but it provides a method for estimating carbon sources and sinks that is more direct than identifying disturbance first and then assigning a carbon density or change in carbon density (Houghton and Goetz, 2008). The direct measurement of change in density will still require models and ancillary data for full carbon accounting (i.e., changes in soil, slash, and wood products) to yield the total flux of carbon (Section 10.3.6). Furthermore, estimation of change, by itself, does not distinguish between deliberate

Global Carbon Budgets and the Role of Remote Sensing 407

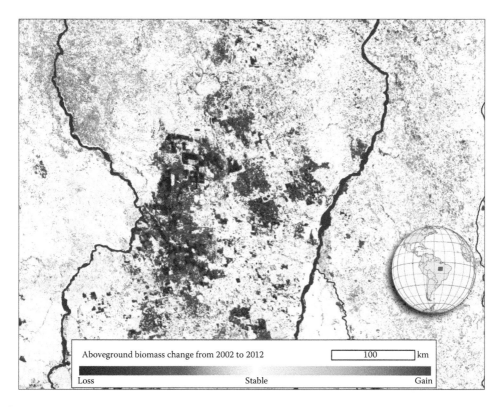

FIGURE 10.5 A 190 km x 215 km region northeast of the Xingu basin in Brazil, showing gains and losses of aboveground live woody biomass density (Mg/ha) between 2002 and 2012.

LULCC activity and indirect anthropogenic or natural drivers (attribution), because deforestation, as discussed earlier, may result from either management or natural disturbance. Nevertheless, estimation of change in aboveground carbon density has clear potential for improving calculations of sources and sinks of carbon.

Direct measurement of biomass density can provide a continuous range of biomass densities, and thus might be expected to yield a more precise estimate of carbon *change* through time than estimates based on changes in area (Houghton and Goetz, 2008). But the change would have to be greater than the uncertainty surrounding any one measurement. Thus, *trends* in MODIS-based biomass density may be more compelling than a single change between two years even if the years are far apart.

Whether measuring changes in forest area (and assigning carbon density) or measuring changes in aboveground carbon density directly, most pixels will probably appear unchanged because rates of growth in mature forests are slow relative to (1) the time interval of observation and (2) the error associated with measurement. However, one might expect two other, less common outcomes. First, both approaches might yield a downward trend in carbon density (deforestation and/or degradation) or an upward trend (growth). In these cases, both area changes and density changes would be consistent, although not necessarily equal.

The direct measurement of density change might be expected to indicate greater carbon sinks than the combined approach because global lands, in general, are a net carbon sink despite the fact that managed lands, globally, are a net source (Table 10.1) (Le Quéré et al., 2013). To the extent the terrestrial carbon sink is in aboveground biomass, it should be observed; and the location of changes in carbon *not* attributable to disturbance and recovery would be most instructive for locating the residual terrestrial sink (Ingalls et al., 2022; Ehlers et al., 2022; Friedlingstein et al., 2020).

In sum, direct measurement of change in aboveground density has the advantage of by-passing the classification step for identifying type of change. But the trade-off is that, without an understanding of LULCC, the observed changes cannot be readily attributed to cause (i.e., anthropogenic or natural, harvest or clearing).

10.3 THE POLICY REALM: ISSUES INHERENT IN ESTIMATING THE FLUX OF CARBON FROM LULCC, WITH AN EXAMPLE USING RED, REDD, AND REDD+

RED, REDD, and REDD+ are policy mechanisms proposed for reducing emissions of greenhouse gases under the UN Framework Convention on Climate Change (UNFCCC). **RED** refers to **R**educing **E**missions from **D**eforestation. If a developing country can demonstrate that it had reduced its emissions of carbon from deforestation, it is eligible for carbon credits. The mechanism was expanded (**REDD**) to include a second "D"—forest Degradation. A demonstrated reduction in emissions from forest degradation would also qualify for additional carbon credits. REDD was subsequently expanded to **REDD+**, which adds conservation, the sustainable management of forests, and the enhancement of forest carbon stocks. These three REDD mechanisms are more an application of carbon science to policy than they are a component of the global carbon budget. Nevertheless, the (reduced) emissions of carbon associated with these mechanisms provide a context for discussing issues related to measuring the role of land in the global carbon budget (Ingalls et al., 2022; Liu et al., 2023; Friedlingstein et al., 2020; Pique et al., 2020). The following paragraphs discuss these issues.

RED. Reduced Emissions from Deforestation is the simplest of the three mechanisms, conceptually and practically. But one still has to

1. Agree on a definition for deforestation,
2. Assign a carbon density to the area deforested, and
3. Decide whether to count committed or actual emissions (see Section 10.3.3).
   REDD. When forest degradation (the second "D") is added, accounting is more difficult.
   The same three requirements that apply to RED also apply to REDD. In addition, one must:
4. Agree on a definition for degradation (what thresholds for changes in carbon density?) and
5. Decide whether to count gross emissions or net emissions.

REDD+. For REDD+, emissions and sinks are both counted, emissions from deforestation and degradation, and sinks from the enhancement of forest carbon stocks. From the perspective of the atmosphere, reduced emissions are equivalent to increased stocks. Growth, as well as degradation, is counted. In addition to deforestation and forest degradation, both reforestation and afforestation are also counted, as they represent means for the enhancement of forest carbon stocks.

Nine issues are discussed briefly in the following sections.

10.3.1 Definitions

Two recent studies used different estimates of change in forest area to estimate very different (a factor of three) emissions of carbon from tropical deforestation (Liu et al., 2023; Zhao et al., 2022; Harris et al., 2021; Friedlingstein et al., 2020). Baccini et al. (2012) used FAO rates of deforestation; Harris et al. (2012) used Hansen's rates of forest loss. As discussed earlier, the two estimates of change in forest area are not measures of the same process. *Deforestation*, as defined by the FAO and the Intergovernmental Panel on Climate Change (IPCC), is the permanent conversion of forest cover to another cover. *Forest loss* includes lands deforested, but it also includes temporary losses of

Global Carbon Budgets and the Role of Remote Sensing

forest cover from fires or logging (not defined as deforestation by the IPCC) (see Houghton, 2013a). The difference is important in the UNFCCC intent for accounting for carbon credits and debits. The intent is to reward carbon management, but not to reward natural effects (i.e., attribution). The difference is also important in accounting for differences among estimates of forest loss and estimates of carbon emissions.

10.3.2 Assigning a Carbon Density to the Areas Deforested

This issue is addressed in detail in Section 10.2.1.3.

10.3.3 Committed versus Actual Emissions (Legacy Effects)

In the process of deforestation and forest degradation, only some of the carbon initially held in the forest is released to the atmosphere in the year of the activity (Figure 10.6). Some of the wood may be removed from the forest and converted to wood products with average life spans of a year (fuelwood) to centuries (lumber used in buildings). And some wood may accumulate on the forest floor as woody debris. These pools of woody material will decay over decades and release carbon to the atmosphere well after the actual activity (legacy flux). The same legacy effects pertain to growing forests, only in reverse. That is, harvested forests that are allowed to recover will accumulate carbon for centuries, or until they are harvested again.

Counting the carbon emissions from deforestation and forest degradation can be done in at least two ways. If the total change in carbon density is counted in the year of the activity, the emissions are referred to as committed emissions. If the changes in living and dead biomass, woody debris,

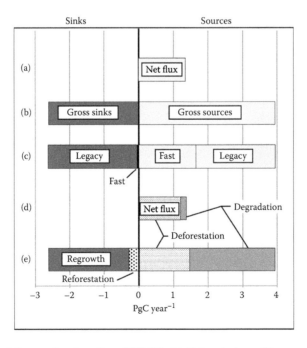

FIGURE 10.6 Annual fluxes of carbon from LULCC. (a) Net emissions; (b) gross sources and sinks; (c) gross fluxes divided into fast (year of disturbance) and legacy (from previous disturbances); (d) net emissions divided into those from deforestation and degradation; (e) gross uptake divided between growth in existing forests and growth of new forests, and gross emissions divided between deforestation and degradation (from Houghton et al., 2012).

harvested products, and soils are tracked through time, then the resulting net emissions reflect actual emissions. Committed emissions are more easily computed. But they cannot be verified by independent measurements of carbon flux, for example, based on forest inventory measurements, eddy covariance measurements, or inverse calculations based on variations of atmospheric CO_2 concentrations. They are easily calculated but not verifiable by independent methods. Remote sensing can help with measuring the sink in growing biomass, but the sources of carbon from dead biomass, woody debris, wood products, and soil are not observable from space.

Two other issues follow from these legacy effects: gross versus net emissions and initial conditions.

10.3.4 Gross and Net Emissions of Carbon from LULCC

Gross emissions refer to the releases of carbon from living biomass, dead biomass, woody debris, wood products, and soils (Ingalls et al., 2022; Ehlers et al., 2022; Pique et al., 2020; Liang et al., 2019). Gross uptake (or sinks) of carbon refers to the accumulation of carbon in living and dead biomass and soils as a forest grows. Together, gross emissions and gross sinks yield the net flux of carbon (Figure 10.6).

Obviously, gross emissions are greater than net emissions except in the case of deforestation, when they are equal. Gross emissions may be much greater than the net emissions if the gross emissions and gross sinks offset each other in time, that is, if the emissions from decaying wood are balanced by the uptake of carbon in recovering forests.

With deforestation, net and gross emissions are the same (there is no regrowth). But with forest degradation, either net or gross emissions may be counted. Forest degradation is often followed by or offset by forest regrowth, which raises the question: is regrowth a part of degradation, or should only the gross emissions be counted? If the emphasis is strictly on emissions, then using gross emissions is defensible, using estimates of either committed or actual emissions. But increasing carbon sinks on land is equivalent, in terms of carbon, to reducing emissions and, besides, offers another management opportunity. That is, an emphasis on reducing gross emissions may miss a larger potential for increasing sinks, just as reducing withdrawals from a bank account is only one way to achieve a higher balance. And if reducing emissions and increasing sinks are equivalent, then the emphasis shifts from gross to net emissions, where net emissions are defined as the sum of gross emissions and gross uptake. Net emissions are what affect the atmospheric concentrations of CO_2. Again, remote sensing may help monitor growth of aboveground forest biomass but not the decay of dead wood, wood products, or soil (Ingalls et al., 2022; Malerba et al., 2023; Ehlers et al., 2022; Zhao et al., 2022; Harris et al., 2021).

10.3.5 Initial Conditions

Legacy effects (growing forests and accumulated pools of decaying wood) have a large effect on calculated emissions and sinks. If one wants to know the emissions in the year 2000, for example, the history of disturbance before 2000 is important. It determines the areas of forest recovering from disturbance as well as the magnitude of carbon pools decaying. Without accounting for that history, one misses the sinks in secondary forests and the sources of carbon from landfills, for example.

There are three ways to handle legacy effects. The simplest way is to count committed emissions. That accounting ignores legacy effects. For actual emissions, however, one must either reconstruct the history of LULCC and disturbance for the years before 2000 (in this example) or determine the age structure of forests in 2000 and the pools of wood in products and woody debris. These two pools (secondary forests and decay pools) determine the current sources and sinks of carbon as well as future sources and sinks. Soils, also, may be either losing or gaining carbon from earlier disturbances. Remote sensing can help with the first and third approaches; the second (reconstruction of history) requires a historic approach.

Global Carbon Budgets and the Role of Remote Sensing

10.3.6 Full Carbon Accounting

Full carbon accounting refers to the changes in all pools of carbon, not just those in aboveground biomass (Ingalls et al., 2024; Friedlingstein et al., 2023; Pique et al., 2020; Liang et al., 2019; Cressie, 2018; Lees et al., 2018). The net and gross emissions of carbon from disturbance and recovery also include the sources of carbon from burning on site; from the decay of stumps, roots, and plant material left on site; and from the decay of soil organic matter if the soils are cultivated (Figure 10.7). A full global accounting must also consider the decay of wood products removed from the forest. Gross rates of carbon uptake result from the accumulation of carbon in growing forests (vegetation and soils). Remote sensing may help with measuring changes in aboveground living biomass (either gains or losses) but not with the other pools.

The good news is that living biomass is estimated to account for 75–90% of the total net flux of carbon from LULCC (Houghton, 2003). Soils accounted for most of the rest. Estimates of dead and fallen aboveground biomass density (including coarse wood debris) may be obtained from the literature, documented for both natural forests and changes as a result of disturbance (e.g., wood harvest, fire, shifting cultivation) (e.g., Harmon et al., 1990; Delaney et al., 1998; Chambers et al., 2000; Idol et al., 2001). Distinguishing among types of disturbance helps define the amount of coarse woody debris. For example, harvests and fires remove some wood; storms and insects kill trees but leave the wood. Clearing forests for pastures in Amazonia leaves more debris (dead biomass) than clearing for cultivated agriculture (Morton et al., 2008).

As a first approximation, belowground biomass can be estimated as a fraction of aboveground biomass (e.g., 21%, Houghton et al., 2001). Alternative estimates may be obtained from documented relationships between forest types or climatic variables and belowground biomass (Cairns et al., 1997; Mokany et al., 2006; Cheng and Niklas, 2007).

Data on harvested wood products and wood removed from deforested sites may be compiled at the country level from FAO Production Yearbooks and FAOstat (2011). Those data can be used

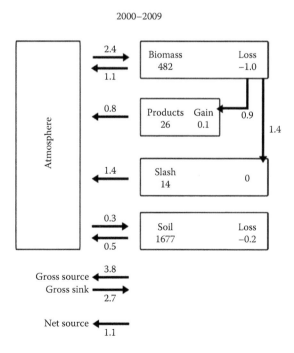

FIGURE 10.7 Average annual fluxes of carbon and changes in terrestrial pools of carbon as a result of global LULCC (from Houghton, 2013c).

together with case studies in the literature covering different types of disturbance and different ecosystems to simulate harvests for the years before satellite data.

A significant fraction of soil organic carbon is lost with cultivation and may accumulate again if croplands are abandoned and forests return. Changes vary with land management and type of disturbance, but cultivation seems to produce a consistent change (a loss of 25–30% of the upper 1 m of soil) (Post and Kwon, 2000; Guo and Gifford, 2002; Murty et al., 2002; Don et al., 2011).

10.3.7 Accuracy and Precision

In general, the errors are smallest for deforestation and committed emissions (based on aboveground biomass). These are the observations most directly obtained from satellites. The errors increase for degradation, for actual (delayed) emissions, for net emissions, and for pools of carbon other than aboveground biomass (e.g., belowground carbon, harvested products, decay pools [slash, logging debris]). The errors are largest for distinguishing between managed and natural effects (attribution), especially when considering sinks.

10.3.8 Attribution

Separating sinks that result from natural and indirect processes from sinks that result from management (i.e., attribution) is difficult at any scale, from plot level to satellite. And how should emissions or sinks of carbon from natural processes be counted? For example, if unmanaged forests burn, should the emissions count as a carbon debit? It is perhaps better to change the focus from attribution to a focus on those management practices that lead to the greatest net sinks (or lowest net sources), regardless of attribution? With such an approach, all carbon accumulation would be counted as a credit; all emissions would be counted as debits. And the net flux, whatever the cause, defines the credit/debit. This may seem unfair for a country whose forests burn (because of drought), but that country is credited in subsequent years as those forests recover. In choosing the most appropriate policy, it is perhaps more important to bear in mind the net effect on the atmosphere than the causes (anthropogenic or natural) of the emissions.

10.3.9 Uncertainties

The two primary pieces of information for determining changes in terrestrial carbon attributable to LULCC (rates of LULCC and carbon density) contribute about equally to the uncertainty of flux estimates. Before the use of satellite data, the uncertainties contributed by these two variables were each about ±0.3 PgC/yr (Houghton et al., 2012). With the use of satellite data for both rates of land cover change and carbon density, these uncertainties should be much lower (Baccini et al., 2012; Harris et al., 2012). Use of satellite data have helped reduce uncertainties substantially.

Nevertheless, there are other issues that contribute to making the terrestrial net emissions considerably more uncertain. Some of these uncertainties may be reduced in the future through new satellite data or new analyses with satellite data.

The issues include more the subtle forms of LULCC than clear-cutting and the clearing of forests for croplands and pastures. They involve distinguishing among the land uses following forest clearing (croplands, pastures, shifting cultivation, plantations, etc.). There are also issues about accounting for land degradation, agricultural management, and fire management. Satellite data may help identify some of these more subtle effects.

And there are also processes not yet well accounted for in analyses of global LULCC; e.g., losses of carbon from wetlands, especially the draining and burning of peatlands in Southeast Asia; the sources and sinks of carbon associated with settled lands (urban and ex-urban); woody encroachment (or transitions between woody and herbaceous vegetation types); and lateral transport of carbon resulting from erosion and redeposition (Houghton et al., 2012).

And last, but not least, is the issue of environmental effects. The actual fluxes of carbon from LULCC are affected by variations in temperature and moisture (including longer growing seasons) and long-term trends in CO_2 concentrations and nitrogen loading. These environmental changes affect the annual sources and sinks from managed lands and also the magnitude of the residual terrestrial sink in lands not managed (next section). Not only do the actual fluxes of carbon vary as a result of environmental variation, but also the estimates calculated by different models vary depending on the models. In particular, estimates of the flux of carbon from LULCC vary as much or more from the way environmental effects are modeled than they do from data on LULCC alone (Gasser and Ciais, 2013; Houghton, 2013b; Pongratz et al., 2014).

10.4 THE RESIDUAL TERRESTRIAL SINK

The residual terrestrial sink is the net flux of carbon from the sum of all processes not accounted for in analyses of LULCC. In the ideal case, where all effects of management are included in analyses of LULCC, the residual flux would be the result of natural and indirect effects on terrestrial carbon storage. But as discussed in Section 10.3.9, many of the more subtle forms of management are not included in analyses of LULCC. Furthermore, it is unclear whether some of the processes that affect carbon storage on land are the result of management or environmental change. Is woody encroachment the effect of grazing or fire management, or is it the effect of climate change or of increased CO_2 in the atmosphere? Are the emissions from burning tropical peatlands the result of management or El Niño? The answer is probably that both management and environmental change are involved. And rather than debate the role of intention in these processes, it is more important to try to quantify the net and gross fluxes for each. Such an approach would yield a graph similar to Figure 10.2 but with a greater number of smaller bands of terrestrial fluxes, not just the two (anthropogenic and natural) that appear in Figure 10.2. Each of the bands appearing now would be broken into individual processes. And as scientists estimate the net flux of carbon for more and more processes, the residual (unexplained) sink (or source) should get smaller and smaller.

The residual terrestrial sink is defined by difference. It is the sink required to balance the global carbon budget. But there may be several ways to quantify it more directly, using remote sensing. One way was discussed in Section 10.2.1.5. The argument there was that changes in aboveground biomass density observed in locations without LULCC might reveal areas of carbon decline or accumulation not caused directly by human activity.

There are at least two additional ways that satellite data might be used to constrain the global carbon budget. One is with repeated measurements of CO_2 in the atmospheric column, and one is with measurement of canopy photosynthesis.

10.4.1 THE ORBITING CARBON OBSERVATORY (OCO)

Repeated measurements of CO_2 in the column of air over the Earth's surface should add a spatial dimension to the observed seasonal oscillation of CO_2 concentrations: that is, low concentrations in late summer in the Northern Hemisphere after the growing season in which photosynthesis has exceeded respiration; and high concentrations in early spring after the season in which respiration has exceeded photosynthesis (Houghton, 1987). With many more observations of this oscillation, one may be able to deduce sources and sinks of carbon over the surface of the Earth. With frequent enough sampling, day-night changes as well as seasonal changes might be documented.

The OCO-2 satellite was designed to make measurements of total CO_2 in air over the Earth's surface, with repeat coverage every 16 days (https://directory.eoportal.org/web/eoportal/satellite-missions/o/oco-2) (Table 10.3). The approach for calculating the net flux for any area is similar to the inverse calculations based on ~200 air sampling stations (Conway and Tans, 2012), but with much higher temporal frequency and greater spatial coverage. GOSAT offers the same approach but with coarser spatial resolution.

TABLE 10.3

The Orbiting Carbon Observatory (OCO-2)

Science objective: determine the global geographic distribution of CO_2 sources and sinks.

Measurement approach: the mission will not directly measure CO_2 sources and sinks. Rather, it will use variations in the column-averaged CO_2 mole fraction of air in data assimilation models to infer sources and sinks. The CO_2 data are derived from spectrometers measuring the intensity of sunlight reflected from the presence of CO_2 in a column of air.

Orbit: the OCO-2 will fly in a near-polar, sun-synchronous orbit, viewing the same location on earth once every 16 days.

Launch date: July 2014

Reference: Boesch, H., D. Baker, B. Connor, D. Crisp, and C. Miller. 2011. Global characterization of CO_2 column retrievals from shortwave-infrared satellite observations of the Orbiting Carbon Observatory-2 mission. *Remote Sensing* 3(2): 270–304; doi: http://dx.doi.org/10.3390/rs3020270

The global coverage of these satellites means that the data can be used to deduce sources and sinks of carbon, not only from land but from the ocean as well. And the terrestrial sources and sinks will include all mechanisms, including both management (LULCC) and natural effects (residual terrestrial sink). However, the fossil fuel contribution must be factored out, and the transport or mixing of air must be accounted for.

10.4.2 Satellite Monitoring of Vegetation Activity (Greenness)

Some of the most interesting continental and global measurements of terrestrial metabolism are those of vegetation activity, greenness, or photosynthesis. The most common index is the Normalized Difference Vegetation Index (NDVI), as measured with Advanced Very High Resolution Radiometer (AVHRR) or MODIS (Ingalls et al., 2022; Ehlers et al., 2022; Friedlingstein et al., 2020; Pique et al., 2020; Liang et al., 2019). Such measurements have been correlated with the annual growth of CO_2 in the atmosphere (Myneni et al., 1997).

A major limitation of the approach is that it cannot close the carbon budget; it misses respiration. Thus, the flux is a gross flux, calculated with algorithms to be gross primary production or net primary production. Respiration is indeed related to photosynthesis, nearly balancing it in most cases. But the imbalance determines whether the net flux is a source or a sink. For example, Myneni et al. (1995) found, surprisingly, that the years with the greatest growth rates of CO_2 coincided with the greenest years. The observation seems contrary to what would be expected if moisture limited photosynthesis. But the finding does not contradict that expectation. Rather, respiration is even more sensitive to moisture than photosynthesis, such that respiration exceeds photosynthesis in moist (green) years, increasing the net emissions from land.

Although seasonal and interannual variations in NDVI fail to capture the respiratory fluxes of carbon, and thus may not be of direct use to carbon budgeting, they are very important for observing physiological responses of photosynthesis to variations in climate. For example, NDVI from AVHRR and MODIS has been used to look for year-to-year variation and trends in greenness (plant productivity) at high latitudes (e.g., Beck and Goetz, 2011) (Figure 10.8). An initial greening of tundra and boreal forests over the 1980s subsequently reversed in many boreal forest areas to declining productivity ("browning") during the 1990–2000s, while tundra areas continued to systematically increase in greenness. The browning observed in northern forests is believed to result from an increase in summer droughts (high vapor pressure deficits) associated with a warming climate (Goetz et al., 2011).

Other examples of physiological responses to variations in moisture come from the tropics. Some analyses with satellite data suggested that tropical forests in Amazonia were light limited rather than moisture limited because productivity was apparently greater during the dry season (less cloudy)

Global Carbon Budgets and the Role of Remote Sensing

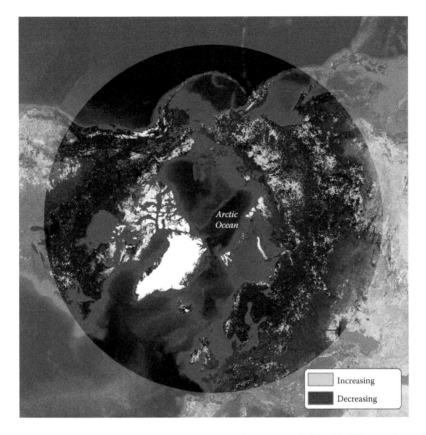

FIGURE 10.8 Circumpolar trends in photosynthetic activity as recorded by NDVI over the period 1982–2005 (modified from Goetz et al., 2011)

(Nemani et al., 2003). Recent work has shown, however, that the increase in dry season greenness was an artifact of variations in sun-sensor geometry. Correcting for bidirectional reflectance eliminates the seasonal changes in surface reflectance, suggesting that the forests of the region may not be light limited (Morton et al., 2014).

10.5 CONCLUSIONS

Over the years, remote sensing has played an ever increasing role in helping to evaluate the global carbon budget, and that trend will continue (Ingalls et al., 2024; Wei et al., 2024; Budget, 2023; Friedlingstein et al., 2023; Liu et al., 2023; Malerba et al., 2023; Ehlers et al., 2022; Zhao et al., 2022; Harris et al., 2021; Friedlingstein et al., 2020; Pique et al., 2020; Liang et al., 2019; Cressie, 2018; Lees et al., 2018). This review suggests that data from remote sensing are used in at least five ways to help constrain the terrestrial component of the budget. The primary role of remote sensing before 2010 was in measuring rates of change in the areas of forest. Although such data were spatial, they were used non-spatially in bookkeeping models to calculate sources and sinks of carbon with empirical response curves that assigned carbon densities to vegetation and soil of different types of ecosystems and different land uses. The net emissions from LULCC defined one term in the global carbon budget.

When satellites were also used to measure the aboveground carbon density of forests, the co-location of deforestation rates and carbon densities enabled a more accurate assignment of carbon density to the forests deforested and degraded, thus yielding a more accurate estimate of the emissions of carbon from disturbance and recovery (Baccini et al., 2012; Harris et al., 2012). Uncertainties remain because definitions of deforestation vary among analyses, because different methods are used to estimate emissions, and because other components of an ecosystem besides its aboveground biomass density contribute to carbon dynamics (Houghton, 2013a).

An active area of current research focuses on a third approach: using satellite data to estimate *changes* in aboveground carbon density. The approach is more comprehensive than the first (based on identifying areas disturbed) because some changes in carbon density occur without a change of cover type. Because the direct measurement of change is more sensitive to changes within a cover type (forest degradation and growth), it may yield an estimate closer to the *total net* change in forest carbon, including changes attributable to disturbance and recovery but also changes attributable to environmental change (e.g., enhanced or retarded growth). The magnitude and geographic distribution of the differences between the two approaches may suggest the explanatory mechanisms, including feedback to climate change, and where further research should focus.

Satellite data (e.g., NDVI) have been used for decades to identify and measure changes in photosynthetic activity. Although the data do not enable construction of a strict carbon budget, because respiration must be approximated, they have been vital in revealing physiological responses to environmental trends and variability, particularly drought.

Starting in 2014, the Orbiting Carbon Observatory will be providing data on spatial and temporal variations in CO_2 concentrations over the Earth's surface. Those data will be used, in a fifth approach, to infer the net sources and sinks of carbon from all processes, including both terrestrial ecosystems and the oceans.

This brief summary of satellites in the global carbon budget suggests a few observations. First, there are cross-cutting themes that repeatedly arise in all methods employing satellite data. The most common themes are net versus gross fluxes, attribution, and full carbon accounting. A major justification for attribution is to separate the processes that affect carbon storage into those that can be managed from those that cannot. There's a difference, of course, between understanding the sources and sinks of carbon and being able to manage them. In the end, whatever reduces emissions or increases sinks is helpful for mitigating climate change.

Second, there is a trade-off between comprehensive coverage (full carbon accounting of all processes) and attribution. Those approaches that are most comprehensive, including OCO as well as direct measurement of changes in terrestrial carbon density, yield net fluxes but do not distinguish among specific processes. Measures of greenness are specific to photosynthesis but do not account for changes in all the other pools of carbon. Measures of disturbance and recovery account for a portion of the terrestrial carbon budget but miss fluxes in undisturbed lands.

Data from satellites require some form of modeling and/or ancillary data to yield estimates of changes in carbon storage or fluxes. For example, changes in terrestrial pools of carbon belowground or on the surface need to be estimated or modeled with data from field measurements. Data from the OCO will require models of atmospheric transport to infer sources and sinks and will need ancillary data to partition those fluxes into fossil fuel, terrestrial, and oceanic components. Thus satellite data and modeling work in combination to evaluate some of the terms in the global carbon budget. No single satellite provides the data necessary to evaluate even one term.

Finally, it is important to note that satellite observations are of generally short-term processes, days to months. These observations are sensitive to metabolic and physical processes, but are not necessarily sufficient to predict, or even record, the long-term effects of climate change on land and oceanic carbon storage. Monitoring for the understanding of longer-term effects requires a commitment to long-term data acquisition and data continuity (Tollefson, 2013; Wunsch et al., 2013).

10.6 ACKNOWLEDGMENT

Support for this chapter was from NASA's Carbon Monitoring System (CMS) program, and that support is gratefully acknowledged.

REFERENCES

Achard, F., H. D. Eva, P. Mayaux, H.-J. Stibig, and A. Belward. 2004. Improved estimates of net carbon emissions from land cover change in the tropics for the 1990s. *Global Biogeochemical Cycles* 18: GB2008. http://doi.org/10.1029/2003GB002142.

Achard, F., H. D. Eva, H.-J. Stibig, P. Mayaux, J. Gallego, T. Richards, and J.-P. Malingreau. 2002. Determination of deforestation rates of the world's humid tropical forests. *Science* 297:999–1002.

Andres, R. J., T. A. Boden, F.-M. Bréon, P. Ciais, S. Davis, D. Erickson, J. S. Gregg, A. Jacobson, G. Marland, J. Miller, T. Oda, J. G. J. Olivier, M. R. Raupach, P. Rayner, and K. Treanton. 2012. A synthesis of carbon dioxide emissions from fossil-fuel combustion. *Biogeosciences* 9:1845–1871. http://doi.org/10.5194/bg-9-1845-2012.

Arvidson, T., J. Gasch, and S. N. Goward. 2001. Landsat 7's long-term acquisition plan—an innovative approach to building a global imagery archive. *Remote Sensing of Environment* 78:13–26.

Baccini, A., S. J. Goetz, W. S. Walker, N. T. Laporte, M. Sun, D. Sulla-Menashe, J. Hackler, P. S. A. Beck, R. Dubayah, M. A. Friedl, S. Samanta, and R. A. Houghton. 2012. Estimated carbon dioxide emissions from tropical deforestation improved by carbon-density maps. *Nature Climate Change* 2:182–185. http://doi.org/10.1038/nclimate1354.

Beck, P. S. A., and S. J. Goetz. 2011. Satellite observation of changes in high latitude vegetation productivity changes between 1982 and 2008: Ecological variability and regional differences. *Environmental Research Letters* 6:045501. http://doi.org/10.1088/1748-93266/4/045501.

Boesch, H., D. Baker, B. Connor, D. Crisp, and C. Miller. 2011. Global characterization of CO_2 column retrievals from shortwave-infrared satellite observations of the Orbiting Carbon Observatory-2 mission. *Remote Sensing* 3(2):270–304. http://doi.org/10.3390/rs3020270.

Broich, M., M. C. Hansen, P. Potapov, B. Adusei, E. Lindquist, and S. V. Stehman. 2011b. Time-series analysis of multi-resolution optical imagery for quantifying forest cover loss in Sumatra and Kalimantan, Indonesia. *International Journal of Applied Earth Observation and Geoinformation* 13:277–291.

Broich, M., M. C. Hansen, F. Stolle, P. Potapov, B. A. Margono, and B. Adusei. 2011a. Remotely sensed forest cover loss shows high spatial and temporal variation across Sumatra and Kalimantan, Indonesia 2000–2008. *Environmental Research Letters* 6(1). http://doi.org/10.1088/1748-9326/6/1/014010.

Budget, G. 2023. *Global Carbon Budget 2023*. Global Carbon Budget.

Caccetta, P. A., S. L. Furby, J. O'Connell, J. F. Wallace, and X. Wu. 2007. Continental monitoring: 34 years of land cover change using Landsat imagery. In: *32nd International Symposium on Remote Sensing of Environment, June 25–29, 2007*. San José, Costa Rica.

Cairns, M. A., S. Brown, E. H. Helmer, and G. A. Baumgardner. 1997. Root biomass allocation in the world's upland forests. *Oecologia* 111:1–11.

Canadell, J. G., C. Le Quéré, M. R. Raupach, C. B. Field, E. T. Buitenhuis, P. Ciais, T. J. Conway, N. P. Gillett, R. A. Houghton, and G. Marland. 2007. Contributions to accelerating atmospheric CO_2 growth from economic activity, carbon intensity, and efficiency of natural sinks. *Proceedings of the National Academy of Sciences of the United States of America* 104:18866–18870.

Chambers, J. Q., N. Higuchi, J. P. Schimel, L. V. Ferreira, and J. M. Melack. 2000. Decomposition and carbon cycling of dead trees in tropical forests of the central Amazon. *Oecologia* 122:380–388.

Cheng, D.-L. and K. J. Niklas. 2007. Above- and below-ground biomass relationships across 1534 forested communities. *Annals of Botany* 99(1):95–102.

Conway, T. J., and P. P. Tans. 2012. *Trends in Atmospheric Carbon Dioxide*. http://www.esrl.noaa.gov/gmd/ccgg/trends.

Cowtan, K., and R. G. Way. 2014. Coverage bias in the HadCRUT4 temperature series and its impact on recent temperature trends. *Quarterly Journal of the Royal Meteorological Society*. http://doi.org/10.1002/qj.2297

Cressie, N. 2018. Mission CO2ntrol: A statistical scientist's role in remote sensing of atmospheric carbon dioxide. *Journal of the American Statistical Association* 113(521):152–168. http://doi.org/10.1080/01621459.2017.1419136.

DeFries, R. S., R. A. Houghton, M. C. Hansen, C. B. Field, D. Skole, and J. Townshend. 2002. Carbon emissions from tropical deforestation and regrowth based on satellite observations for the 1980s and 90s. *Proceedings of the National Academy of Sciences of the United States of America* 99:14256–14261.

Delaney, M., S. Brown, A. E. Lugo, A. Torres-Lezama, and N. Bello Quintero. 1998. The quantity and turnover of dead wood in permanent forest plots in six life zones of Venezuela. *Biotropica* 30:2–11.

Don, A., J. Schumacher, and A. Freibauer. 2011. Impact of tropical land-use change on soil organic carbon stocks—a meta-analysis. *Global Change Biology* 17:1658–1670. http://doi.org/10.1111/j.1365-2486.2010.02336.x

Ehlers, D., C. Wang, J. Coulston, Y. Zhang, T. Pavelsky, E. Frankenberg, C. Woodcock, and C. Song. 2022. Mapping forest aboveground biomass using multisource remotely sensed data. *Remote Sensing* 14(5):1115. https://doi.org/10.3390/rs14051115

Fang, J., T. Kato, Z. Guo, Y. Yang, H. Hu, H. Shen, X. Zhao, A. Kishimodo, Y. Tang, and R. A. Houghton. 2014. Environmental change enhances tree growth: Evidence from Japan's forests. *Proceedings of the National Academy of Sciences of the United States of America*, in press.

FAO. 1996. Forest Resources Assessment 1990: Survey of tropical forest cover and study of change processes. *FAO Forestry Paper 130*. Rome.

FAO. 2010. Global forest resources assessment 2010. *FAO Forestry Paper 163*. Rome.

FAO/JRC. 2011. Global forest land-use change from 1990 to 2005. *Initial Results from a Global Remote Sensing Survey*. http://www.fao.org/forestry/fra/remotesensingsurvey/en/

FAOstat. 2011. http://faostat.fao.org/site/377/default.aspx#ancor.

Friedlingstein, P., M. O'Sullivan, M. W. Jones, R. M. Andrew, D. C. E. Bakker, J. Hauck, P. Landschützer, C. Le Quéré, I. T. Luijkx, G. P. Peters, W. Peters, J. Pongratz, C. Schwingshackl, S. Sitch, J. G. Canadell, P. Ciais, R. B. Jackson, S. R. Alin, P. Anthoni, L. Barbero, N. R. Bates, M. Becker, N. Bellouin, B. Decharme, L. Bopp, I. B. M. Brasika, P. Cadule, M. A. Chamberlain, N. Chandra, T.-T.-T. Chau, F. Chevallier, L. P. Chini, M. Cronin, X. Dou, K. Enyo, W. Evans, S. Falk, R. A. Feely, L. Feng, D. J. Ford, T. Gasser, J. Ghattas, T. Gkritzalis, G. Grassi, L. Gregor, N. Gruber, Ö. Gürses, I. Harris, M. Hefner, J. Heinke, R. A. Houghton, G. C. Hurtt, Y. Iida, T. Ilyina, A. R. Jacobson, A. Jain, T. Jarníková, A. Jersild, F. Jiang, Z. Jin, F. Joos, E. Kato, R. F. Keeling, D. Kennedy, K. Klein Goldewijk, J. Knauer, J. I. Korsbakken, A. Körtzinger, X. Lan, N. Lefèvre, H. Li, J. Liu, Z. Liu, L. Ma, G. Marland, N. Mayot, P. C. McGuire, G. A. McKinley, G. Meyer, E. J. Morgan, D. R. Munro, S.-I. Nakaoka, Y. Niwa, K. M. O'Brien, A. Olsen, A. M. Omar, T. Ono, M. Paulsen, D. Pierrot, K. Pocock, B. Poulter, C. M. Powis, G. Rehder, L. Resplandy, E. Robertson, C. Rödenbeck, T. M. Rosan, J. Schwinger, R. Séférian, T. L. Smallman, S. M. Smith, R. Sospedra-Alfonso, Q. Sun, A. J. Sutton, C. Sweeney, S. Takao, P. P. Tans, H. Tian, B. Tilbrook, H. Tsujino, F. Tubiello, G. R. van der Werf, E. van Ooijen, R. Wanninkhof, M. Watanabe, C. Wimart-Rousseau, D. Yang, X. Yang, W. Yuan, X. Yue, S. Zaehle, J. Zeng, and B. Zheng. 2023. Global carbon budget 2023. *Earth System Science Data*, 15:5301–5369. https://doi.org/10.5194/essd-15-5301-2023.

Friedlingstein, P., M. O'Sullivan, M. W. Jones, R. M. Andrew, J. Hauck, A. Olsen, G. P. Peters, W. Peters, J. Pongratz, S. Sitch, C. Le Quéré, J. G. Canadell, P. Ciais, R. B. Jackson, S. Alin, L. E. O. C. Aragão, A. Arneth, V. Arora, N. R. Bates, M. Becker, A. Benoit-Cattin, H. C. Bittig, L. Bopp, S. Bultan, N. Chandra, F. Chevallier, L. P. Chini, W. Evans, L. Florentie, P. M. Forster, T. Gasser, M. Gehlen, D. Gilfillan, T. Gkritzalis, L. Gregor, N. Gruber, I. Harris, K. Hartung, V. Haverd, R. A. Houghton, T. Ilyina, A. K. Jain, E. Joetzjer, K. Kadono, E. Kato, V. Kitidis, J. I. Korsbakken, P. Landschützer, N. Lefèvre, A. Lenton, S. Lienert, Z. Liu, D. Lombardozzi, G. Marland, N. Metzl, D. R. Munro, J. E. M. S. Nabel, S.-I. Nakaoka, Y. Niwa, K. O'Brien, T. Ono, P. I. Palmer, D. Pierrot, B. Poulter, L. Resplandy, E. Robertson, C. Rödenbeck, J. Schwinger, R. Séférian, I. Skjelvan, A. J. Smith, A. J. Sutton, T. Tanhua, P. P. Tans, H. Tian, B. Tilbrook, G. van der Werf, N. Vuichard, A. P. Walker, R. Wanninkhof, A. J. Watson, D. Willis, A. J. Wiltshire, W. Yuan, X. Yue, and S. Zaehle. 2020. Global carbon budget 2020. *Earth System Science Data* 12:3269–3340. https://doi.org/10.5194/essd-12-3269-2020.

Galford, G. L., J. F. Mustard, J. Melillo, A. Gendrin, C. C. Cerri, and C. E. P. Cerri. 2008. Wavelet analysis of MODIS time series to detect expansion and intensification of row-crop agriculture in Brazil. *Remote Sensing of Environment* 112:576–587.

Gasser, T., and P. Ciais. 2013. A theoretical framework for the net land-to-atmosphere CO_2 flux and its implications in the definition of "emissions from land-use change. *Earth System Dynamics* 4:171–186. http://doi.org/10.5194/esd-4-171-2013.

Giglio, L., J. Descloitres, C. O. Justice, and Y. J. Kaufman. 2003. An enhanced contextual fire detection algorithm for MODIS. *Remote Sensing of Environment* 87:273–282. http://doi.org/10.1016/S0034-4257(03)00184-6.

Giglio, L., J. T. Randerson, G. R. van der Werf, P. S. Kasibhatla, G. J. Collatz, D. C. Morton, and R. S. DeFries. 2010. Assessing variability and long-term trends in burned area by merging multiple satellite fire products. *Biogeosciences* 7:1171–1186. http://doi.org/10.5194/bg-7-1171-2010.

Global Forest Survey of India. 2008. *State of the Forest Report 2005*. Dehradun, India: Forest Survey of India, Ministry of Environment and Forests.

Gloor, M., L. Gatti, R. Brienen, T. R. Feldpausch, O. L. Phillips, J. Miller, J. P. Ometto, H. Rocha, T. Baker, B. de Jong, R. A. Houghton, Y. Malhi, L. E. O. C. Aragão, J.-L. Guyot, K. Zhao, R. Jackson, P. Peylin, S. Sitch, B. Poulter, M. Lomas, S. Zaehle, C. Huntingford, P. Levy, and J. Lloyd. 2012. The carbon balance of South America: A review of the status, decadal trends and main determinants. *Biogeosciences* 9:5407–5430.

Gloor, M., J. L. Sarmiento, and N. Gruber. 2010. What can be learned about carbon cycle climate feedbacks from the CO_2 airborne fraction? *Atmospheric Chemistry and Physics* 10:7739–7751.

Goetz, S. J., A. Baccini, N. T. Laporte, T. Johns, W. Walker, J. Kellndorfer, R. A. Houghton and M. Sun. 2009. Mapping and monitoring carbon stocks with satellite observations: A comparison of methods. *Carbon Balance and Management* 4(2). http://doi.org/10.1186/1750-0680-4-2.

Goetz, S., and R. Dubayah. 2011. Advances in remote sensing technology and implications for measuring and monitoring forest carbon stocks and change. *Carbon Management* 2(3):231–244.

Goetz, S. J., H. E. Epstein, U. S. Bhatt, G. J. Jia, J. O. Kaplan, H. Lischke, Q. Yu, A. Bunn, A. H. Lloyd, D. Alcaraz-Segura, P. S. A. Beck, J. C. Comiso, M. K. Raynolds, and D. A. Walker. 2011. Recent changes in Arctic vegetation: Satellite observations and simulation model predictions. In: G. Gutman and A. Reissell (Eds.), *Arctic Land Cover and Land Use in a Changing Climate* (pp. 9–36). Amsterdam: Springer-Verlag. http://doi.org/10.1007/978-90-481-9118-5_2.

Government of Indonesia/World Bank. 2000. *Deforestation in Indonesia: A Review of the Situation in 1999*. Jakarta: Government of Indonesia/Work Bank.

Grainger, A. 2008. Difficulties in tracking the long-term trend of tropical forest area. *Proceedings of the National Academy of Sciences of the United States of America* 105(2):818–823.

Guo, L. B., and R. M. Gifford. 2002. Soil carbon stocks and land use change: A meta analysis. *Global Change Biology* 8:345–360.

Hansen, M. C., P. V. Potapov, R. Moore, M. Hancher, S. A. Turubanova, A. Tyukavina, D. Thau, S. V. Stehman, S. J. Goetz, T. R. Loveland, A. Kommareddy, A. Egorov, L. Chini, C. O. Justice, and J. R. G. Townshend. 2013. High-resolution global maps of 21st-century forest cover change. *Science* 342:850–853.

Hansen, M. C., S. V. Stehman, and P. V. Potapov. 2010. Quantification of global gross forest cover loss. *Proceedings of the National Academy of Sciences of the United States of America* 107:8650–8655.

Hansen, M. C., et al. 2011. Continuous fields of land cover for the conterminous United States using Landsat data: First results from the Web-Enabled Landsat Data (WELD) project. *Remote Sensing Letters* 2:279–288.

Harmon, M. E., W. K. Ferrell, and J. F. Franklin. 1990. Effects on carbon storage of conversion of old-growth forests to young forests. *Science* 247:699–702.

Harris, N. L., S. Brown, S. C. Hagen, et al. 2012. Baseline map of carbon emissions from deforestation in tropical regions. *Science* 336:1573–1576.

Harris, N. L., D. A. Gibbs, A. Baccini, et al. 2021. Global maps of twenty-first century forest carbon fluxes. *Nature Climate Change* 11:234–240. https://doi.org/10.1038/s41558-020-00976-6.

Houghton, R. A. 1987. Biotic changes consistent with the increased seasonal amplitude of atmospheric CO_2 concentrations. *Journal of Geophysical Research* 92:42234230.

Houghton, R. A. 2003. Revised estimates of the annual net flux of carbon to the atmosphere from changes in land use and land management 1850–2000. *Tellus* 55B:378–390.

Houghton, R. A. 2010. How well do we know the flux of CO_2 from land-use change? *Tellus B* 62(5):337–351. http://doi.org/10.1111/j.1600-0889.2010.00473.x.

Houghton, R. A. 2013a. The emissions of carbon from deforestation and degradation in the tropics: Past trends and future potential. *Carbon Management* 4(5):539–546.

Houghton, R. A. 2013b. Keeping management effects separate from environmental effects in terrestrial carbon accounting. *Global Change Biology* 19:2609–2612.

Houghton, R. A. 2013c. Role of forests and impact of deforestation in the global carbon cycle. In: F. Achard and M. C. Hansen (Eds.), *Global Forest Monitoring from Earth Observation* (pp. 15–38). Boca Raton: CRC Press.

Houghton, R. A., D. Butman, A. G. Bunn, O. N. Krankina, P. Schlesinger, and T. A. Stone. 2007. Mapping Russian forest biomass with data from satellites and forest inventories. *Environmental Research Letters* 2:045032. http://doi.org/10.1088/1748-9326/2/4/045032.

Houghton, R. A., M. Gloor, J. Lloyd, and C. Potter. 2009. The regional carbon budget. In M. Keller, M. Bustamante, J. Gash, and P. Silva Dias (Eds.), *Amazonia and Global Change* (pp. 409–428), Geophysical Monograph Series 186. Washington, DC: American Geophysical Union.

Houghton, R. A., and S. J. Goetz. 2008. New satellites help quantify carbon sources and sinks. *Eos* 89(43):417–418.

Houghton, R. A., J. I. House, J. Pongratz, G. R. van der Werf, R. S. DeFries, M. C. Hansen, C. Le Quéré, and N. Ramankutty. 2012. Carbon emissions from land use and land-cover change. *Biogeosciences* 9:5125–5142. http://doi.org/10.5194/bg-9-5125-2012.

Houghton, R. A., K. T. Lawrence, J. L. Hackler, and S. Brown. 2001. The spatial distribution of forest biomass in the Brazilian Amazon: A comparison of estimates. *Global Change Biology* 7:731–746.

Huang, M., and G. P. Asner. 2010. Long-term carbon loss and recovery following selective logging in Amazon forests. *Global Biogeochemical Cycles* 24:GB3028. http://doi.org/10.1029/2009GB003727.

Hurtt, G. C., S. Frolking, M. G. Fearon, B. Moore, E. Shevliakova, S. Malyshev, S. W. Pacala, and R. A. Houghton. 2006. The underpinnings of land-use history: Three centuries of global gridded land-use transitions, wood harvest activity, and resulting secondary lands. *Global Change Biology* 12:1–22.

Idol, T. W., R. A. Filder, P. E. Pope, and F. Ponder. 2001. Characterization of coarse woody debris across a 100 year chronosequence of upland oak-hickory forest. *Forest Ecology and Management* 149:153–161.

Ingalls, T. C., J. Li, Y. Sawall, R. E. Martin, D. R. Thompson, and G. P. Asner. 2022. Imaging spectroscopy investigations in wet carbon ecosystems: A review of the literature from 1995 to 2022 and future directions. *Remote Sensing of Environment* 305:114051. ISSN 0034-4257. https://doi.org/10.1016/j.rse.2024.114051. https://www.sciencedirect.com/science/article/pii/S0034425724000622

Instituto Nacional de Pesquisas Espaciais (INPE). 2010. *Deforestation Estimates in the Brazilian Amazon*. São José dos Campos: INPE. http://www.obt.inpe.br/prodes/.

Klein Goldewijk, K. 2001. Estimating global land use change over the past 300 years: The HYDE database. *Global Biogeochemical Cycles* 15:417–433.

Klein Goldewijk, K., A. Beusen, G. van Drecht, and M. de Vos. 2011. The HYDE 3.1 spatially explicit database of human-induced global land-use change over the past 12,000 years. *Global Ecology and Biogeography* 20:73–86.

Knorr, W. 2009. Is the airborne fraction of anthropogenic CO_2 emissions increasing? *Geophysical Research Letters* 36:L21710. http://doi.org/10.1029/2009GL040613.

Lees, K. J., T. Quaife, R. R. E. Artz, M. Khomik, and J. M. Clark. 2018. Potential for using remote sensing to estimate carbon fluxes across northern peatlands—A review. *Science of the Total Environment* 615:857–874. ISSN 0048-9697. https://doi.org/10.1016/j.scitotenv.2017.09.103. https://www.sciencedirect.com/science/article/pii/S0048969717324464

Le Quéré, C., et al. 2013. The global carbon budget 1959–2011. *Earth System Science Data* 5:165–185.

Le Quéré, C., M. R. Raupach, J. G. Canadell, G. Marland, L. Bopp, P. Ciais, P., T. J. Conway, S. C. Doney, R. A. Feely, P. Foster, P. Friedlingstein, K. Gurney, R. A. Houghton, J. I. House, C. Huntingford, P. E. Levy, M. R. Lomas, J. Majkut, N. Metzl, J. P. Ometto, G. P. Peters, I. C. Prentice, J. T. Randerson, S. W. Running, J. L. Sarmiento, U. Schuster, S. Sitch, T. Takahashi, N. Viovy, G. R. van der Werf, and F. I. Woodward. 2009. Trends in the sources and sinks of carbon dioxide. *Nature GeoScience* 2:831–836.

Liang, S., D. Wang, T. He, and Y. Yu. 2019. Remote sensing of earth's energy budget: Synthesis and review. *International Journal of Digital Earth* 12(7):737–780. http://doi.org/10.1080/17538947.2019.1597189

Liu, Z., Z. Deng, S. Davis, et al. 2023. Monitoring global carbon emissions in 2022. *Nature Reviews Earth & Environment* 4:205–206. https://doi.org/10.1038/s43017-023-00406-z.

Malerba, M. E., M. D. D. P. Costa, D. A. Friess, L. Schuster, M. A. Young, D. Lagomasino, O. Serrano, S. M. Hickey, P. H. York, M. Rasheed, J. S. Lefcheck, B. Radford, T. B. Atwood, D. Ierodiaconou, and P. Macreadie. 2023. Remote sensing for cost-effective blue carbon accounting. *Earth-Science Reviews* 238:104337. ISSN 0012-8252. https://doi.org/10.1016/j.earscirev.2023.104337. https://www.sciencedirect.com/science/article/pii/S0012825223000260

Mitchard, E. T. A., S. S. Saatchi, A. Baccini, G. P. Asner, S. J. Goetz, N. L. Harris, and S. Brown. 2013. Uncertainty in the spatial distribution of tropical forest biomass: A comparison of pan-tropical maps. *Carbon Balance and Management* 8:10. http://www.cbmjournal.com/content/8/1/10

Mokany, K., R. J. Raison, and A. S. Prokushkin. 2006. Critical analysis of root:shoot ratios in terrestrial biomes. *Global Change Biology* 12:84–96.

Moore, B., R. D. Boone, J. E. Hobbie, R. A. Houghton, J. M. Melillo, B. J. Peterson, G. R. Shaver, C. J. Vorosmarty, and G. M. Woodwell. 1981. A simple model for analysis of the role of terrestrial ecosystems in the global carbon budget. In: B. Bolin (Ed.), *Carbon Cycle Modelling* (pp. 365–385, SCOPE 16). New York: John Wiley and Sons.

Morton, D. C., R. S. DeFries, J. T. Randerson, L. Giglio, W. Schroeder, and G. R. van der Werf. 2008. Agricultural intensification increases deforestation fire activity in Amazonia. *Global Change Biology* 14(10):2262–2276.

Morton, D. C., R. S. DeFries, Y. E. Shimabukuro, L. O. Anderson, E. Arai, F. del Bon Espirito-Santo, R. Freitas, and J. Morisette. 2006. Cropland expansion changes deforestation dynamics in the southern Brazilian Amazon. *Proceedings of the National Academy of Sciences of the United States of America* 103(39):14637–14641. http://doi.org/10.1073/pnas.0606377103.

Morton, D. C., J. Nagol, C. C. Carabajal, J. Rosette, M. Palace, B. D. Cook, E. F. Vermote, D. J. Harding, and P. R. North. 2014. Amazon forests maintain consistent canopy structure and greenness during the dry season. *Nature* 506(7487):221–224. http://doi.org/10.1038/nature13006.

Murty, D., M. F. Kirschbaum, R. E. McMurtrie, and H. McGilvray. 2002. Does conversion of forest to agricultural land change soil carbon and nitrogen? A review of the literature. *Global Change Biology* 8:105–123.

Myneni, R. B., C. D. Keeling, C. J. Tucker, G. Asrar, and R. R. Nemani. 1997. Increased plant growth in the northern high latitudes from 1981–1991. *Nature* 386:698–701.

Myneni, R. B., S. O. Los, and G. Asrar. 1995. Potential gross primary productivity of vegetation from 1982–1990. *Geophysical Research Letters* 22:2617:2620.

Natali, S. M., E. A. G. Schuur, E. E. Webb, C. E. Hicks Pries, and K. G. Crummer. 2014. Permafrost degradation stimulates carbon loss from experimentally warmed tundra. *Ecology* 95(3):602–608.

Nemani, R. R., et al. 2003. Climate-driven increases in global terrestrial net primary production from 1982–1999. *Science* 300:1560–1563.

Pan, Y., R. A. Birdsey, J. Fang, R. Houghton, P. E. Kauppi, W. A. Kurz, O. L. Phillips, A. Shvidenko, S. L. Lewis, J. G. Canadell, P. Ciais, R. B. Jackson, S. W. Pacala, A. D. McGuire, S. Piao, A. Rautiainen, S. Sitch, and D. Hayes. 2011. A large and persistent carbon sink in the world's forests. *Science* 333:988–993.

Phillips, O. L., S. L. Lewis, T. R. Baker, K.-J. Chao, and N. Higuchi. 2008. The changing Amazon forest. *Philosophical Transactions of the Royal Society*, Series B, 363:1819–1828.

Pique, G., R. Fieuzal, A. A. Bitar, A. Veloso, T. Tallec, A. Brut, M. Ferlicoq, B. Zawilski, J. Dejoux, H. Gibrin, and E. Ceschia. 2020. Estimation of daily CO2 fluxes and of the components of the carbon budget for winter wheat by the assimilation of Sentinel 2-like remote sensing data into a crop model. *Geoderma* 376:114428. ISSN 0016-7061. https://doi.org/10.1016/j.geoderma.2020.114428. https://www.sciencedirect.com/science/article/pii/S0016706119321998

Pongratz, J., C. H. Reick, R. A. Houghton, and J. I. House. 2014. Terminology as a key uncertainty in net land use and land cover change carbon flux estimates. *Earth System Dynamics* 5:177–195.

Pongratz, J., C. Reick, T. Raddatz, and M. Claussen. 2008. A reconstruction of global agricultural areas and land cover for the last millennium. *Global Biogeochemical Cycles* 22:GB3018. http://doi.org/10.1029/2007GB003153.

Post, W. M., and K. C. Kwon. 2000. Soil carbon sequestration and land-use change: Processes and potential. *Global Change Biology* 6:317–327.

Ramankutty, N., and J. A. Foley. 1999. Estimating historical changes in global land cover: Croplands from 1700 to 1992. *Global Biogeochemical Cycles* 13:997–1027.

Raupach, M. R., M. Gloor, J. L. Sarmiento, J. G. Canadell, T. L. Frölicher, T. Gasser, R. A. Houghton, C. Le Quéré, and C. M. Trudinger. 2014. The declining uptake rate of atmospheric CO_2 by land and ocean sinks. *Biogeosciences*, in press.

Rosenqvist, A., M. Shimada, N. Ito, and M. Watanabe. 2007. ALOS PALSAR: A pathfinder mission for global-scale monitoring of the environment. *IEEE Transactions on Geoscience and Remote Sensing* 45:3307–3316.

Saatchi, S. S., N. L. Harris, S. Brown, M. Lefsky, E. T. A. Mitchard, W. Salas, B. R. Zutta, W. Buermann, S. L. Lewis, S. Hagen, S. Petrova, L. White, M. Silman, and A. Morel. 2011. Benchmark map of forest carbon stocks in tropical regions across three continents. *Proceedings of the National Academy of Sciences of the United States of America* 108:9899–9904.

Saatchi, S. S., R. A. Houghton, R. C. dos Santos Alvala, J. V. Soares, and Y. Yu. 2007. Distribution of aboveground live biomass in the Amazon basin. *Global Change Biology* 13:816–837.

Shevliakova, E., S. Pacala, S. Malyshev, G. Hurtt, P. C. D. Milly, J. Caspersen, L. Sentman, J. Fisk, C. Wirth, and C. Crevoisier. 2009. Carbon cycling under 300 years of land use change: Importance of the secondary vegetation sink. *Global Biogeochemical Cycles* 23:1–16.

Skole, D. L., and C. J. Tucker. 1993. Tropical deforestation and habitat fragmentation in the Amazon: Satellite data from 1978 to 1988. *Science* 260:1905–1910.

Thenkabail, P. S., Teluguntla, P. G., Xiong, J., Oliphant, A., Congalton, R. G., Ozdogan, M., Gumma, M. K., Tilton, J. C., Giri, C., Milesi, C., Phalke, A., Massey, R., Yadav, K., Sankey, T., Zhong, Y., Aneece, I., and Foley, D. 2021. Global cropland-extent product at 30-m resolution (GCEP30) derived from Landsat satellite time-series data for the year 2015 using multiple machine-learning algorithms on Google Earth Engine cloud. *U.S. Geological Survey Professional Paper 1868*:63 p. https://doi.org/10.3133/pp1868.

Tollefson, J. 2013. Budget crunch hits Keeling's curves. *Nature* 503:321–322.

Treuhaft, R. N., B. D. Chapman, J. R. dos Santos, F. G. Gonçalves, L. V. Dutra, P. M. L. A. Graça, and J. B. Drake. 2009. Vegetation profiles in tropical forests from multibaseline interferometric synthetic aperture radar, field, and lidar measurements. *Journal of Geophysical Research* 114:D23110. http://doi.org/10.1029/2008JD011674.

van der Werf, G. R., J. Dempewolf, S. N. Trigg, J. T. Randerson, P. S. Kasibhatla, L. Giglio, D. Murdiyarso, W. Peters, D. C. Morton, G. J. Collatz, A. J. Dolman, and R. S. DeFries. 2008. Climate regulation of fire emissions and deforestation in equatorial Asia. *Proceedings of the National Academy of Sciences of the United States of America* 105:20350–20355.

van der Werf, G. R., J. T. Randerson, L. Giglio, G. J. Collatz, M. Mu, P. S. Kasibhatla, D. C. Morton, R. S. DeFries, Y. Jin, and T. T. van Leeuwen. 2010. Global fire emissions and the contribution of deforestation, savanna, forest, agricultural, and peat fires (1997–2009). *Atmospheric Chemistry and Physics* 10:11707–1735. http://doi.org/10.5194/acp-10-11707-2010.

Wei, W., R. Hao, L. Ma, et al. 2024. Characteristics of carbon budget based on energy carbon emissions and vegetation carbon absorption. *Environmental Monitoring and Assessment* 196:134. https://doi.org/10.1007/s10661-024-12295-w.

Williams, C. A., G. J. Collatz, J. Masek, and S. Goward. 2012. Carbon consequences of forest disturbance and recovery across the conterminous United States. *Global Biogeochemical Cycles* 26:GB1005. http://doi.org/10.1029/2010GB003947.

Woodwell, G. M., J. E. Hobbie, R. A. Houghton, J. M. Melillo, B. Moore, A. Park, B. J. Peterson and G. R. Shaver. 1984. Measurement of changes in the vegetation of the earth by satellite imagery. In: G. M. Woodwell (Ed.), *The Role of Terrestrial Vegetation in the Global Carbon Cycle: Measurement by Remote Sensing* (pp. 221240, SCOPE 23). Chichester: John Wiley and Sons.

Wunsch, C., R. W. Schmitt, and D. J. Baker. 2013. Climate change as an intergenerational problem. *Proceedings of the National Academy of Sciences* 110:4435–4436.

Zhang, F., J. M. Chen, Y. Pan, R. A. Birdsey, S. Shen, W. Ju, and L. He. 2012. Attributing carbon changes in conterminous U.S. forests to disturbance and non-disturbance factors from 1901 to 2010. *Journal of Geophysical Research: Biogeosciences* 117(G2):18 p.

Zhao, J., D. Liu, Y. Cao, L. Zhang, H. Peng, K. Wang, H. Xie, and C. Wang. 2022. An integrated remote sensing and model approach for assessing forest carbon fluxes in China. *Science of the Total Environment* 811:152480. ISSN 0048-9697. https://doi.org/10.1016/j.scitotenv.2021.152480. https://www.sciencedirect.com/science/article/pii/S0048969721075586

11 Aboveground Terrestrial Biomass and Carbon Stock Estimations from Multi-Sensor Remote Sensing

Wenge Ni-Meister

ACRONYMS AND DEFINITIONS

AGB	Aboveground biomass
ALOS	Advanced Land Observing Satellite
ASAR	Advanced synthetic aperture radar on board ENVISAT
ASTER	Advanced spaceborne thermal emission and reflection radiometer
AVHRR	Advanced very high resolution radiometer
CHM	Canopy height model
ENVISAT	Environmental satellite
ERS	European remote sensing satellites
ETM+	Enhanced Thematic Mapper+
EVI	Enhanced Vegetation Index
FAO	Food and Agriculture Organization of the United Nations
GHG	Greenhouse gas
GLAS	Geoscience Laser Altimeter System
GPS	Global Positioning System
ICESat	Instrument on board the Ice, Cloud, and land Elevation
IKONOS	A commercial earth observation satellite, typically, collecting sub-meter to 5 m data
JERS	Japanese Earth Resources Satellite
LAI	Leaf area index
LiDAR	Light detection and ranging
MODIS	Moderate-resolution imaging spectroradiometer
NASA	National Aeronautics and Space Administration
NDVI	Normalized Difference Vegetation Index
NIR	Near-infrared
NOAA	National Oceanic and Atmospheric Administration
PALSAR	Phased Array type L-band Synthetic Aperture Radar
RADAR	Radio detection and ranging
RADARSAT	Radar satellite
REDD	Reducing Emissions from Deforestation in Developing Countries
SAR	Synthetic aperture radar
SPOT	Satellite Pour l'Observation de la Terre, French Earth Observing Satellites
SRTM	Shuttle Radar Topographic Mission
SWIR	Shortwave infrared
SVM	Support vector machines

TerraSAR-X	A radar Earth observation satellite with its phased array synthetic aperture radar
TIR	Thermal infrared
UNFCCC	United Nations Framework Convention on Climate Change
USDA	United States Department of Agriculture

11.1 INTRODUCTION

Recent global observation systems provide measurements of horizontal and vertical vegetation structure of ecosystems which will be critical for estimating global carbon storage and assessing ecosystem response to climate change and natural and anthropogenic disturbances. Remote sensing overcomes the limitations associated with sparse field surveys; it has been used extensively as a basis for inferring forest structure and aboveground biomass (AGB) over large areas (Sainuddin et al., 2024; Sanam et al., 2024; Singh et al., 2024; Ma et al., 2024; Datta et al., 2023; Tian et al., 2023; Zhang et al., 2023; Musthafa and Singh, 2022; Pötzschner et al., 2022; Wang et al., 2022; Zeng et al., 2020; Zhang et al., 2022; Das et al., 2021; Ronoud et al., 2021; Abbas et al., 2020; Bispo et al., 2020; Chapungu et al., 2020; Sinha et al., 2020; Ghosh and Behera, 2018; Kumar and Mutanga, 2017). This chapter summarizes recent progress on AGB estimate using remote sensing technology, including the strength and weakness of using optical passive, radar, and LiDAR remote sensing and the fusion of multi-sensor for AGB estimates. It lays out the potential of remote sensing in AGB and carbon storage estimates at large scales for meeting the requirements of the United Nations Framework Convention on Climate Change (UNFCCC) Measuring, Reporting and Verification. The purpose of this chapter is to review recent progress on AGB estimates using remote sensing data.

11.1.1 Importance of the Terrestrial Ecosystem Carbon and Carbon Changes Estimates

Vegetation biomass is a crucial ecological variable for understanding the evolution and potential future changes of the climate system. Global carbon stored in vegetation is comparable in size to atmospheric carbon and plays an important role in the global carbon cycle (Houghton, 2005). Changes of forest biomass in time can be used as an essential climate variable (ECV), because it is a direct measure of sequestration or release of carbon between terrestrial ecosystems and the atmosphere. During productive seasons, forests take up carbon dioxide (CO_2) from the atmosphere and store it as plant biomass, while they release CO_2 to the atmosphere during deforestation, decomposition, and biomass burning. Changes in the amount of vegetation biomass due to deforestation significantly affect the global atmosphere by acting as a net source of carbon. The Global Climate Observing System (GCOS) recognizes AGB and associated carbon stocks of the world's forests as an Essential Climate Variable (ECV) (Hollman et al., 2013).

However the terrestrial carbon cycle is the most uncertain component of the global carbon cycle (Heimann and Reichstein, 2008). Large uncertainties in terrestrial carbon cycle arise from inadequate data on the current state of the land surface vegetation structure and the carbon density of forests. Consequently, there is an urgent need for improved data sets that characterize the global distribution of AGB, especially in the tropics. Therefore, a global assessment of biomass and its dynamics is an essential input to climate change prognostic models and mitigation and adaptation strategies.

11.1.2 Importance of Tropical Rainforests in Carbon Storage

Tropical forests are disappearing rapidly due to land conversion, selective cutting, and fires (Zhang et al., 2023; Wang et al., 2022; Abbas et al., 2020; Kumar and Mutanga, 2017). The single biggest direct cause of tropical forest loss is due to the conversion of forests to cropland and pasture.

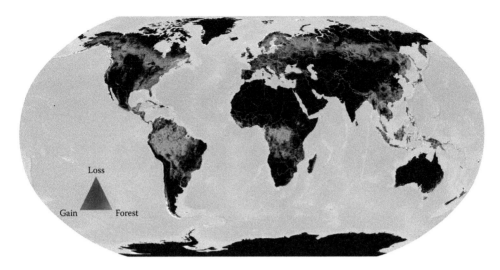

FIGURE 11.1 Global distribution of forest cover change, circa 1990–2000. The false-color composite was aggregated from 30-m to 5-km grid cells. Forest loss is represented in red, forest gain in blue, and persistent forest in green. Colors are stretch in the proportion of 1 (forest): 4 (gain): 4 (loss). (For interpretation of the references to color in this figure legend, the reader is referred to the web version of this article.) (Source: Kim et al., 2014).

Humans harvest timber for construction and fuel, and wildfires pose a big threat to Amazon forests. The tropics exhibit a large trend of forest loss, increasing by 2101 km²/year, half of which occurred in South American rainforests (Hansen et al., 2013), with recent report of reduced rates of forest loss from a high of over 40,000 km²/year in 2003 to 2004 and a low of under 20,000 km²/year in 2010 to 2011 in Central America. However, this decreasing rate of loss was counterbalanced by increased forest loss from other tropical forest regions (Figure 11.1).

Land use, land use change and forestry (LULUCF) is the second-largest source of anthropogenic greenhouse gas (GHG) emissions, dominated by tropical deforestation (Canadell et al., 2007). The loss of Amazon forest releases huge amounts of carbon into the atmosphere. Tropical deforestation contributes about one-eighth to one-fifth of total anthropogenic CO_2 emissions to the atmosphere (Houghton et al., 2012; Houghton, 2005). However, the magnitude of these emissions has remained poorly constrained. Emissions from land use change remains one of the most uncertain components of the global carbon cycle. Global carbon emission estimates using different approaches and the uncertainties associated with each approach ranges from 10% to 34% (Houghton et al., 2012). A recent estimate of gross carbon emissions across tropical regions between 2000 and 2005 was 0.81 petagrams of carbon per year (Harris et al., 2012), which was only 25–50% of recently published estimates (Baccini et al., 2012; Pan et al., 2011; FAO, 2010). Huge discrepancies exist in the carbon emission estimates.

The lack of reliable estimates of forest carbon storage and rates of deforestation and forestation result in the uncertainties of terrestrial carbon emissions estimates (Houghton et al., 2009; Houghton, 2005; Houghton et al., 2012). Estimates of the biomass storage disagree with biomass obtained from large-scale wood-volume inventories (Houghton et al., 2001). Large uncertainties in the carbon stocks estimates contribute to the broad range of possible emissions of carbon from tropical deforestation and degradation (Houghton, 2005).

Reducing emissions from deforestation and forest degradation in developing countries (REDD) launched by the United Nations Framework Convention on Climate Change provides positive incentives to individuals, communities, projects, and government agencies in developing countries to reduce GHG emissions from forests through monetary compensation. REDD was extended as

REDD+ to include conservation, sustainable management of forests, and the enhancement of forest carbon stocks. As a mechanism under the multi-lateral climate change agreement, REDD+ is a vehicle to financially reward developing countries for their verified efforts to reduce emissions and enhance removals of GHG through a variety of forest management options.

Efforts to mitigate climate change through the Reduced Emissions from Deforestation and Degradation (REDD) depend on mapping and monitoring of tropical forest carbon stocks and emissions over large geographic areas. There are many challenges to making REDD work, and mapping forest carbon stocks and emissions at the high resolution demanded by investors and monitoring agencies remains a technical barrier. Foremost among the challenges is quantifying nations' carbon emissions from deforestation and forest degradation, which requires information on forest clearing and carbon storage (Datta et al., 2023; Pötzschner et al., 2022; Ronoud et al., 2021; Sinha et al., 2020).

11.1.3 Summary of Methods Used to Estimate Terrestrial Biomass and Carbon Stocks

Vegetation AGB is defined as the mass per unit area (Mg/ha) of live or dead plant organic matter. Forest biomass consists of AGB and belowground biomass. AGB represents all living biomass above the soil, including stem, stump, branches, bark, seeds, and foliage, while belowground biomass consists of all living roots, excluding fine roots (less than 2 mm in diameter) (FAO, 2010). Because AGB is relatively easy to measure and accounts for the majority of the total accumulated biomass in a forest ecosystem, AGB is usually estimated in many studies as to forest biomass. At the level of individual plants and forest stand levels, aboveground and belowground biomass are different but share strikingly similar scaling exponents (Figure 11.2) (Cheng and Niklas, 2007). Belowground biomass is often estimated based on AGB. This review mainly focuses on AGB estimates.

Biomass estimate methods range from simple to more complex methods (Singh et al., 2024; Musthafa and Singh, 2022; Bispo et al., 2020). The biome-averaged method is to estimate the biome-averaged AGB first, and the spatial distribution of biomass is mapped based on biome type. A more complex method is to develop species- and site-specific allometric models depending on bole diameter at breast height (dbh; cm) or diameter and tree height. The plot estimates of national forest inventories are commonly aggregated to represent forest biomass at national or regional scales (Brown et al., 1989; Jenkins, 2003; Gibbs et al., 2007; Goetz et al., 2009).

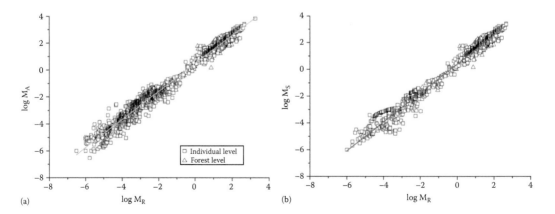

FIGURE 11.2 Log-log bivariate plots of above- vs. belowground (root) biomass (M_A vs. M_R) and stem vs. root biomass (M_S vs. M_R) at the level of individual plants (n=1406) and Chinese forest samples (n=1534). Solid lines are RMA regression curves; dashed lines denote +/-2 s.d. (Cheng and Niklas, 2007).

11.1.4 The Role of Remote Sensing in Terrestrial Ecosystem Carbon Estimates

Recent global observation systems provide measurements of horizontal and vertical vegetation structure of ecosystems, which will be critical for estimating global carbon storage and assessing ecosystem response to climate change and natural and anthropogenic disturbances. Remote sensing overcomes the limitations associated with sparse field surveys, it has been used extensively as a basis for inferring forest structure and AGB over large areas (Sainuddin et al., 2024; Tian et al., 2023; Das et al., 2021; Sinha et al., 2020; Kumar and Mutanga, 2017). Although no sensor has been developed that is capable of providing direct measures of vegetation biomass, the radiometry is sensitive to vegetation structure (crown size, tree density, height), texture, and shadow, which are correlated with AGB. Three types of remote sensing data are often used, which are:

- Optical remote sensing,
- Radar (radio detection and ranging, microwave) data, and
- LiDAR (light detection and ranging) data.

Optical spectral reflectances are sensitive to vegetation structure (leaf area index, crown size, and tree density), texture, and shadow, which are correlated with AGB. Radar data are directly related to AGB through measuring dielectric and geometrical properties of forests (Le Toan et al., 2011). LiDAR remote sensing is promising in characterizing vegetation vertical structure and height, which are then associated to ABG (Ma et al., 2024; Zhang et al., 2023; Wang et al., 2022; Zhang et al., 2022; Drake et al., 2002a, 2002b). Vegetation structure characteristics measured from satellite data are linked to field-based AGB estimates, and their relationships are used to map large-scale AGB from satellite data (see Table 11.1 for details). Recently, remote sensing has been extensively used as a robust tool in delivering forest structure and AGB because it provides a practical means of acquiring spatially distributed forest biomass from local to continental areas (Houghton, 2005; Lu, 2006; Goetz and Dubayah, 2011; Zhang and Kondragunta, 2006).

11.1.5 Specific Topics Covered in This Chapter

Significant progress has been made in recent years regarding the large-area application of spaceborne remote sensing for the mapping of terrestrial ecosystem carbon stocks, which manifested in the release of several regional- to continental-scale maps of AGB. This chapter reviews recent progress of terrestrial AGB and carbon stock estimations from remote sensing (Singh et al., 2024; Ma et al., 2024; Musthafa and Singh, 2022; Zeng et al., 2020; Ronoud et al., 2021; Ghosh and Behera, 2018; Kumar and Mutanga, 2017). It focuses on not only the current state of remote sensing of biomass using one particular sensor but also recent progress on biomass mapping through the fusion of multi-sensors. First, we brief the traditional method of AGB estimates, then summarize what types of remote sensing data are being used for biomass estimates followed by a summary of research methods. Later sections provide recent progresses on biomass estimates using optical, radar, and LiDAR sensors and fusion of multi-sensors. Finally, we discuss the strengths and potential improvement of remote sensing approaches for mapping terrestrial ecosystem biomass and carbon stocks and point out future research directions.

11.2 CONVENTIONAL METHODS OF CARBON STOCKS ESTIMATES

Table 11.1 lists the strengths and limitations of conventional methods for carbon stock estimates. The direct method is to harvest trees, dry, and weigh the biomass. It is the most accurate method; however, it is the most labor-intensive. For a small area, the most direct way to measure the carbon stored in aboveground living forest biomass is to harvest all trees, dry them, and weigh the biomass. The dry biomass can be converted to carbon content by taking half of the biomass weight (carbon

TABLE 11.1
Strengths and Limitations of Conventional Methods to Estimate AGB and Forest Carbon Stocks

Methods	Descriptions	Input Parameters	Strengths	Limitations	References
Direct measurement	• Harvest all trees • Dry them • Weigh the biomass	n/a	• Very accurate	• Very small areas	Brown et al. (1989)
Biome average	• Estimate average forest carbon stocks for each biome based on inventory data • Map carbon stocks based on land cover types	• Land cover types • Averaged biomass for each biome	• Easy and quick • Globally consistent • Low cost	• Low accuracy • Lost local variations	FAO (2010)
Species-based allometric method	• Use allometric relationships to estimate AGB based on DBH	• DBH • Species	• Easy to implement	• Low accuracy if the allometric relationship is not local	Jenkins et al. (2004)
Woody volume and density-based allometric method	• Use generalized allometric relationships for all species stratified by broad forest types or ecological zones	• DBH • Tree height • Wood density • Forest types (dry or wet forest)	• Quite accurate • Effective for tropical forests	• Need extra wood density and tree height measurement	Brown (2002), Chave et al. (2005), Chave et al. (2014)

content ≈ 50% of biomass). This method is destructive, expensive, extremely time consuming, and impractical for any large regions.

No methodology can directly measure forest carbon stocks across a landscape. Different methods are used to approximate large-scale carbon stocks ranging from simple, empirical to more complex, physically based methods. At the national level, the Intergovernmental Panel on Climate Change (IPCC) proposed different tiers of carbon stocks quality, ranging from Tier 1 (simplest to use; globally available data) up to Tier 3 (high-resolution methods specific for each country and repeated through time) (Gibbs et al., 2007).

11.2.1 BIOME AVERAGE METHODS

The simplest one is to use the biomass average for each biome to approximate a nation's carbon stocks (IPCC's Tier 1) (Gibbs et al., 2007; Houghton et al., 2001). Biome averages are compiled based on tree harvesting measurements and analysis of forest inventory data archived by the United Nations Food and Agricultural Organization (FAO) (Gibbs et al., 2007).

This method has both strengths and limitations. Biomes account for major bioclimatic gradients such as temperature, precipitation, and geologic substrate; it is a quick and easy way to estimate forest carbon stocks based on biomes. Besides, biome averages are free and easily accessible to map global forest carbon systematically. It provides a starting point for a country to access their carbon emission from disturbance. However, biome averages were generally focused on mature stands and were based on a few plots that may not adequately represent the biome or region. Further, forest carbon stocks vary significantly with slope, elevation, drainage class, soil type, and land-use history within each biome, therefore an average value cannot adequately represent the variation for an entire forest category or country. Finally, the carbon stock estimates over disturbed areas could also be biased, as the carbon stocks for the new growth systematically differ from the biome-average values (Houghton et al., 2001).

11.2.2 Allometric Biomass Methods

Another commonly used approach is the allometric-based biomass and carbon stocks estimates (IPCC's Tier 2 or 3). It depends on forest inventory measurements to develop allometric relationships between tree diameters at breast height (DBH) alone or in combination with tree height with AGB. Ground-based DBH and height measurements in large areas are converted to forest carbon stocks using allometric relationships. Many allometric equations for estimating AGB have been published (Chave et al., 2014; Chave et al., 2005; Brown, 2002; Brown et al., 1989; West et al., 1999; Jenkins, 2003; Jenkins et al., 2004). Two allometric-based approaches are commonly used to estimate AGB.

- **Species-Based Allometric Method**

The first one is a species-based approach to estimate biomass based on a given tree DBH (Tian et al., 2023; Pötzschner et al., 2022; Zeng et al., 2020; Ronoud et al., 2021; Chapungu et al., 2020; Kumar and Mutanga, 2017). It requires the measurement of the diameter for each individual tree and allometric equations for each individual tree species. For example, Jenkins et al. (2004) developed a set of generalized allometric regression models to predict AGB in tree components for all tree species in the United States. It is used by the USDA Forest Service, Forest Inventory and Analysis (FIA) program to estimate US national forest carbon estimates. This approach provides a nationally consistent method for estimation of biomass and C stocks at large scales and requires only a single field-based variable—tree diameter at breast height (DBH;1.37 m)—as input.

- **Woody Volume and Density-Based Allometric Method**

The second approach is more generalized, using woody volume and wood density to calculate biomass (Brown, 2002; Chave et al., 2005; Chave et al., 2014). Developing allometric equations for each individual species can be very difficult. However, grouping all species together and using generalized allometric relationships, stratified by broad forest types or ecological zones, is highly effective, particularly for the tropics because DBH alone explains more than 95% of the variation in aboveground tropical forest carbon stocks, even in highly diverse regions (Brown, 2002).

Chave et al. (2005) developed generalized allometric equations for the pan-tropics based on an exceptionally large dataset of 2410 trees across a wide range of forest types. They included wood density and tree height within their models and proposed a global forest classification system that contains three climatic categories (dry, moist, and wet) to account for climatic constraints determining the AGB variation. Chave et al. (2014) updated their allometric equations and developed a single model using trunk diameter, total tree height, and wood specific gravity across tropical vegetation types, with no detectable effect of region or environmental factors. The new allometric models should contribute to improving the accuracy of biomass assessment protocols in tropical vegetation types and to improving accuracy of carbon stock estimates for tropical forests.

Studies show that the most important parameters in estimating biomass (in decreasing order of importance) were diameter, wood density, tree height, and forest type (classified as dry, moist, or wet forest). Including tree height reduced the standard error of biomass estimates from 19.5% to 12.5% (Chave et al., 2005). Tree biomass estimation was significantly improved by including

TABLE 11.2
Characteristics of Satellite Sensors Used to Estimate AGB

Sensor Characteristics			Sensor	Spectral Range	Spatial Resolution	Spatial Coverage	Temporal Resolution	Temporal Coverage
Active	LiDAR	Ground LiDAR	• EVI, DEWL	1064 nm, 1548 nm	Site level	Site level	Discontinuous	2000s
		Small footprint LiDAR	• Optech ALTM 3100C • Leica ALS50-II • Riegl LMS-Q140i-60	1064 nm	Foot-meter scale	Local	Discontinuous	1988–present
		Medium footprint LiDAR	• LVIS	1064 nm	15 m–25 m	Regional	Discontinuous	1999–present
		Large footprint LiDAR	• GLAS	1064 nm	60 m–90 m	Global	Discontinuous	2003–2009
	Radar	P-band	• Biomass	200 m	<50m	Semi-global No Europe/US	25–45 days	Launched in 2024
		L-band	• ALOS-PALSAR • ALOS-PALSAR(2)	15–30 cm	7–89 m	Global	46 days	2006–2011 (PALSAR) 2014–present (PALSAR2)
		X-/C-band	• ERS • ENVISAT • RADARSAT • TerraSAR-X	2.5–7.5 cm	• ERS: 30 m • ENVISAT: 30 m–90 m • RADARSAT: 1–100 m • TerraSAR-X: 1–16 m	Global	3 days, 35 days, and 336 days	• 1995–present • ERS: 1991–2011 • ENVISAT: 2002–2012 • RADARSAT:1995–present • TerraSAR-X: 2007–present

Passive	Multi-spectral/Hyperspatial	• IKONOS • QuickBird • Orbit View	VIS-NIR	1–5 m	Global	No regular repeat cycle	2000–present
	Multi-spectral High Spatial	• LandSat • SPOT HRV • ASTER	VIS-TIR	30 m	Global	16 days	1972–present
	Multi-spectral Coarse Resolution	• MODIS • AVHRR	VIS-TIR	1 km	Global	Daily	2000–present
	Multi-spectral and multi-angular	• MISR	VIS-NIR	1 km	Global	Daily	1999–present
	Hyperspectral	• AVRIS • Hyperion	VIS-IR	4–20 m, 30 m	Global	Discontinuous	2000–present

ASTER = Advanced Spaceborne Thermal Emission and Reflection Radiometer
AVHRR = Advanced Very High Resolution Radiometer
AVIRIS = Airborne Visible/Infrared Imaging Spectrometer
ETM = Enhanced Thematic Mapper
DWEL = Dual-Wavelength Echidna® LiDAR
EVI = Echidna® Validation Instrument
GLAS = Geoscience Laser Altimeter System, onboard the Ice, Cloud, and land Elevation Satellite (ICESat)
LVIS = Land Vegetation and Ice Sensor
MODIS = Moderate Resolution Imaging Spectroradiometer
NIR = Near infrared
PALSAR = Phased Array-type L-band Synthetic Aperture Radar
SPOT HRV = Le Syst'eme Pour l'Observation de la Terre High Resolution Visible
SIR-C/X-SAR = Spaceborne Imaging Radar-C/X-band Synthetic Aperture Radar
TIR = Thermal infrared

wood density (Brown et al., 1989) and tree height (Tian et al., 2023; Wang et al., 2022) in the allometric models in addition to tree diameter. However, measuring height (H) and wood density (q) requires additional work, increasing project time and costs. This approach is not often used, as it requires additional height measurements for each individual tree.

Despite the difficulty, more and more studies demonstrated the importance of these parameters for biomass estimates. For example, studies by Feldpausch et al. (2012); and Feldpausch et al. (2011) demonstrate incorporating height in biomass estimates for the pan-tropical region improves biomass estimates by lowering biomass estimates. For tropical forests, carbon storage can be overestimated by 35 PgC if height is ignored. The study by Domke et al. (2012) for the United States also demonstrates similar results. Domke et al. (2012) compared estimates of carbon stocks using Jenkin's and a tree height–based approach—the component ratio method (CRM) (Woodall et al., 2011) for the 20 most abundant tree species in the 48 states of the United States and found the method incorporating height decreased national carbon stocks estimates by an average of 16% for the species. These results implicate that tree height, an important allometric factor, needs to be included in future forest biomass estimates to reduce error in estimates of tropical carbon stocks and emissions due to deforestation and to improve accuracy of national and global forest carbon.

11.3 REMOTE SENSING DATA

A variety of remote sensing systems have been used to estimate AGB estimates: passive optical remote sensing, radar, and LiDAR (Sanam et al., 2024; Tian et al., 2023; Zeng et al., 2020; Bispo et al., 2020). Table 11.3 summarizes the characteristics of satellite sensors used to estimate AGB and carbon storage.

11.3.1 Passive Optical Remote Sensing Data

Passive remote sensors measure different wavelengths of reflected solar radiation, providing two-dimensional information that can be indirectly linked to biophysical properties of vegetation and AGB and carbon stocks (Tian et al., 2023; Pötzschner et al., 2022; Ronoud et al., 2021; Chapungu et al., 2020; Ghosh and Behera, 2018; Kumar and Mutanga, 2017). Several optical satellite instruments are available for mapping AGB and carbon stocks at different spatial scales. The spatial resolutions of these satellite data range from meter to kilometer scales. Those data spans from 1970s to current and some recent satellite data are collected on a daily scale. The most popular optical remote sensing satellite data being used to map AGB are multispectral satellite data at various spatial resolutions.

NOAA's Advanced Very High Resolution Radiometer (AVHRR) and NASA's Moderate Resolution Imaging Spectroradiometer (MODIS) data are promising data in producing biomass at continental and global scales (Baccini et al., 2008; Blackard et al., 2008; Dong et al., 2003; Zhang and Kondragunta, 2006; Baccini et al., 2004). AVHRR has provided global observations at 1-km scale every one or two days since 1979. MODIS aboard the Aqua and Terra satellites have imaged the entire globe approximately every two days at resolutions of 250–500 m, dating as far back as 2000. These datasets are used alone or fused with other remote sensing data to provide AGB and carbon stock estimates at large scales.

Landsat Thematic Mapper (TM), Enhanced Thematic Mapper Plus (ETM+), and Advanced Spaceborne Thermal Emission and Reflection Radiometer (ASTER) provide biomass estimates at local and regional scales at high spatial resolution (Zhang and Kondragunta, 2006; Pflugmacher et al., 2014; Muukkonen and Heiskanen, 2005). Landsat provides four decades of imagery of the entire globe at 30-m spatial resolution, the longest continuous record of space-based moderate-resolution land remote sensing data freely available to the public. With the advantages of being free and long-term data records, methods of using spectral information or more complicated methods using both spectral and temporal information or fusion with other remote sensing data have been developed to estimate AGB estimates. Landsat images are invaluable data sources to AGB and

TABLE 11.3
Strengths and Limitations of Using Different Remote Sensing Data to Estimate AGB and Forest Carbon Stocks

Remote Sensing Data Types	Measured Forest Structure Parameters Inputs	Methods	Strengths	Limitations	References
Passive optical remote sensing	• Reflectances • Spectral indices • Tree shadows • Height for sparse canopy • Stand age • Land cover types	• Linear regression • Non-parametric method	• High spatial imaging capability • Consistent at all scales • Free for most imageries except for very high spatial data	• Saturation at high biomass	• Baccini et al.(2004, 2008) • Dong et al. (2003) • Chopping et al. (2009) • Zhang and Kondragunta (2006)
RADAR	• Radar signals • Woody volume • Crown center height	• Linear regression • Non-parametric method • Physical models	• Accurate at low biomass • High spatial imaging capability • Free data	• Saturation at high biomass • Impact from underness topography/roughness and soil wetness	• Le Toan et al. (1992, 2004, 2011) • Askne and Santoro (2005) • Carreiras et al. (2012) • Cartus et al. (2012) • Chowdhury et al. (2014)
LiDAR	• Tree height • Height metrics • Foliage profiles • Crown sizes	• Linear regression • Non-parametric method	• Most accurate • Free data	• Sparse samplings for spaceborne LiDAR data • Small regions for small footprint LiDAR data	• Asner et al. (2012) • Blair et al. (1999) • Drake et al. (2002a,) • Dubayah and Drake (2000) • Garcia et al. (2010)

carbon stock estimates. ASTER, an imaging instrument onboard Terra launched in December 1999, images the Earth at 15-m resolution in visible to near infrared spectrum, which is the most sensitive to vegetation properties. Other passive optical systems such as multiangular data from MISR on board Terra and airborne/spaceborne hyperspectral data from AVIRIS and EO1 sensors are also used for biomass estimates (Chopping et al., 2009; Anderson et al., 2008).

11.3.2 Radar Data

Radar data physically measure biomass through the interaction of the radar waves with tree scattering elements. The widely used active radar data for biomass estimates are from spaceborne synthetic aperture radar (SAR) sensors, such as the L-band Advanced Land Observing Satellite (ALOS), Phased Array type L-band Synthetic Aperture Radar (PALSAR), the C-band European remote sensing satellite (ERS)/SAR, RADARSAT/SAR or Environmental Satellite (ENVISAT)/Advanced Synthetic Aperture Radar (ASAR) and the X-band TerraSAR-X instrument, which transmit microwave energy at wavelengths from 3.0 cm (X-band) to 23.6 cm (L-band).

The European remote sensing satellite (ERS) and ENVISAT operated by the European Space Agency (ESA) have collected C-band SAR data since 1991. The Canadian RADARSAT has been collecting C-band data since 1995. The German TerraSAR-X has been in space since 2010. Those data have been used to estimate AGB with low density. The L-band Phased Array type L-band Synthetic Aperture Radar (PALSAR) was launched by JAXA (Japan Aerospace Exploration Agency). ALOS/PALSAR was operated in-orbit from January 2006 until April 2011. It showed great potential for forestry applications in the boreal regions due to its high signal/noise ratio, high resolution (~20 m), provision of cross-polarized data, and because data were being systematically collected across the Northern Hemisphere. ALOS2 was launched in 2014, and PALSAR-2 has updated features of PALSAR. A spaceborne P-band SAR, which would be less affected by saturation at higher biomass levels, is planned to launch in the coming years in the frame of the Earth Explorer Program of the ESA. Many airborne L-band and P-band data were also collected for biomass estimates. The major advantage of all SAR systems is their weather- and daylight-independency.

11.3.3 LIDAR DATA

LiDAR is an active remote sensing system based on laser ranging that measures the distance between a sensor and target surface. Vegetation LiDAR systems typically emit at wavelengths between 900 and 1064 nm and record the time during which the emitted laser pulse is reflected off an object and returns to the sensor. The time-return interval is used to calculate the range (distance) between the sensor and the object. LiDAR provides direct and indirect measurements of vegetation structure, which can be used to estimate global carbon storage. Recent advances in LiDAR technology have made LiDAR data widely available to study vegetation structure characteristics and forest biomass.

LiDAR systems are classified as small-footprint LiDAR (laser footprint at less than 1 m scale) and large footprint LiDAR (laser footprint at 10 m or greater) based on the size of the laser footprint or discrete return and full waveform recording based on how the laser energy is recorded (Wulder et al., 2012; Dubayah and Drake, 2000). Discrete return systems record single or multiple returns from a given laser pulse. As the laser signal is reflected back to the sensor, large peaks (i.e., bright returns) represent discrete objects in the path of the laser beam and are recorded as discrete points. Most small-footprint LiDAR systems record discrete energy returns. In contrast, full waveform–recording LiDAR systems digitize the entire reflected energy from a return, resulting in complete sub-meter vertical vegetation profiles. The waveform is a function of canopy height and vertical distribution of foliage, as it is made up of the reflected energy from the surface area of canopy components, such as foliage, trunks, twigs, and branches, at varying heights within the large footprint. The total waveform is therefore a measure of both the vertical distribution of vegetation surface area and the distribution of the underlying ground height. Waveform recording instruments are mainly large-footprint LiDAR systems; however, recent advances made full waveform instruments with increasingly smaller footprint sizes available.

Small-footprint multiple return LiDAR data have been collected in many regions of the globe, and more recently small footprint scanning waveform systems have become operational. Such small footprint airborne LiDAR systems are available on a commercial basis and are now used at the operational level in forest resource inventories (Næsset and Gobakken, 2008). At the standard level, ground-based LiDAR data, such as EVI, were collected and used for AGB estimates (Ni-Meister et al., 2010; Strahler et al., 2008). With many ground LiDAR systems, complex and detailed vegetation structure data have been recorded over various study sites.

The Geoscience Laser Altimeter System (GLAS) was a large footprint spaceborne full waveform profiling LiDAR carried on the Ice, Cloud, and land Elevation Satellite (ICESat) for 2003–2009. GLAS was the first spaceborne LiDAR and global measurement of canopy height was one of the science objectives of the ICESat mission (Zwally et al., 2002). The size and shape of the GLAS footprints vary from 50 to 65 m in diameter and from elliptical to circular, depending on the date of the acquisition. The pulses are spaced approximately 172 m apart.

Airborne data have also been collected using a Scanning LiDAR Imager of Canopies by Echo Recovery (SLICER) with a 15-m footprint and the Laser Vegetation Imaging Sensor

(LVIS) with a 20/25-m footprint over several large areas for improved vegetation structure characterization since 1998 (Blair et al., 1999). This large-footprint LiDAR system records full waveform laser energy returns. These global, regional, and local LiDAR data can provide the detailed vegetation structure and biomass maps necessary for carbon models and ecosystem processes studies.

11.4 RESEARCH APPROACHES/METHODS

Many methods are adopted to convert field measured AGB at the local scale to large scale based on remote sensing measurements or extrapolating from small-scale LiDAR and field measurements to large-scale maps of AGB (Sainuddin et al., 2024; Datta et al., 2023; Pötzschner et al., 2022; Ronoud et al., 2021; Chapungu et al., 2020; Sinha et al., 2020; Ghosh and Behera, 2018). Common methods include linear statistical models, support vector machines, nearest neighbor–based methods, random forest, and Gaussian processes (e.g., Figure 11.3). The most common approach is line statistical regression (Fassnacht et al., 2014), then non-parametric nearest neighbor, machine learning (Carreiras et al., 2012; Zhao et al., 2011) and random forest (Baccini et al., 2012), and Gaussian processes (Zhao et al., 2011) (See Figure 11.3 for a summary by Fassnacht et al., 2014). Some physically based or semi-empirical models have also been used (Saatchi et al., 2007a, 2007b).

- **Nonparametric Methods**

With recent advancement in geospatial statistical methods and ongoing technology improvement in performing expensive statistical computations, the nonparametric method appears more prevalent in more recent studies (Datta et al., 2023; Wang et al., 2022; Baccini et al., 2008, 2004). These methods perform recursive partitioning of data sets, make no assumptions regarding the distribution and correlation of the input data, effectively solve complex non-linear relationships between the response and predictor variables, and show great advantages for nonlinear problems and often perform better than standard linear regression models.

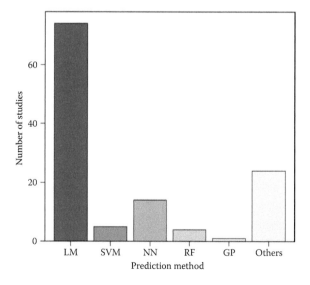

FIGURE 11.3 Frequency distribution of the prediction methods used for AGB estimates. (lin = linear regression model, SVM = support vector machines, NN = nearest neighbor–based methods, RD = random forest, GP = Gaussian processes) (Fassnacht et al., 2014).

- **Tree-Based Models**

One nonparametric approach, tree-based models (Breiman et al., 1984), is a fundamental tool in data mining. It performs recursive partitioning of data sets to capture nonlinear relationships between the response and predictor variables for predicting a categorical (classification tree) or continuous (regression tree) outcome. This method has been previously used in remote sensing field to predict for classification and continuous variables (Baccini et al., 2004; Baccini et al., 2008). Tree-based models are known for their simplicity and efficiency when dealing with domains with a large number of variables and cases. However, it can also lead to poor decisions in lower levels of the tree due to the unreliability of estimates based on small samples of cases.

A commonly used tree-based model in AGB estimate is random forest (Breiman, 2001; Breiman et al., 1984). Random forest constructs a multitude of decision trees at a training time in which different bootstrap samples of the data are used to estimate each tree and outputting the class that is the mode of the classes output by individual trees. The resulting model is more accurate and less sensitive to noise in input data relative to conventional tree-based modeling algorithms.

The use of random forest for biomass estimate demonstrates the advantages of the nonparametric statistical method. For example, Baccini et al. (2008) compared the performance between random forest with more traditional multiple regression analysis, and they found that the traditional regression explained variance is 71% compared to 85% from random forest of their AGB in their study region. LiDAR data, in combination with a random forest algorithm and a large number of reference sample units on the ground often yields the lowest error for biomass predictions and becomes very popular in most research efforts on biomass estimates. There are, however, limitations to the random forest model in the prediction phase. The model tends to over-predict low biomass values and under-predict high biomass values. This trend is intrinsic of regression tree–based models whose predictions are the average of the values within the terminal node. Different methods are used in different remote sensing field for biomass estimates.

11.5 REMOTE SENSING–BASED ABOVEGROUND BIOMASS ESTIMATES

Different remote sensing datasets were used to estimate AGB (Ma et al., 2024; Musthafa and Singh, 2022; Zeng et al., 2020; Bispo et al., 2020; Chapungu et al., 2020; Sinha et al., 2020; Ghosh and Behera, 2018; Kumar and Mutanga, 2017). Table 11.3 lists the strengths and limitations of using different remote sensing data to estimate AGB and forest carbon stocks. Details are discussed in the following sections.

11.5.1 Optical Remote Sensing

Optical remote sensing data have been extensively used to map AGB (Sanam et al., 2024; Singh et al., 2024; Zhang et al., 2023; Zeng et al., 2020; Ronoud et al., 2021). One simple method to map AGB is to use remotely sensed land cover classification maps where each class is assigned an average value of biomass density based on literature estimates or forest inventories. The IPCC tier-I approach was applied to the study area using their prescribed forest carbon density values combined with land cover data generated from the globally available land cover dataset, Global Land Cover 2000. Land cover data were reclassified as forest or non-forest, using all forest classes of GLC2000. Aboveground carbon densities were assigned to each land cover class using IPCC values.

The other more commonly used method is determination of relationships between in situ biomass density and remote sensing characteristics/signals that can be consistently mapped over large regions (Saatchi et al., 2007). This approach has the advantage of providing spatially consistent and continuous values of the amount of biomass present at any given location. The suite of freely available optical satellite sensors, such as Landsat, AVHRR, and MODIS, have been used extensively to map AGB based on statistical relationships between ground-based measurements and satellite-observed surface reflectance and vegetation indices (Lu, 2006). Spectral reflectances of optical remote sensing are the simplest variables in biomass estimates. Vegetation indices are particularly useful in biomass observations because it enhances green vegetation signals and minimizes the impacts from surface and atmospheric effects. Alternatively, tree canopy attributes such as leaf area index (LAI), tree cover, crown size, density, and tree shadow fraction derived from optical satellite data are considered to be effective proxies of AGB. Tree shadow fraction is an indicator of vertical vegetation structure and can be an indicator of biomass.

At continental- and global-scale biomass mapping, the coarse spatial resolution optical sensors, such as the NOAA AVHRR (1.1 km) and MODIS (250 m–1 km) have been useful for forest biomass estimates due to the good trade-off between spatial resolution, image coverage, and frequency in data acquisition (Lu, 2006). Dong et al. (2003) used the normalized difference vegetation index (NDVI) estimate provided by the AVHRR sensor to estimate forest biomass at a continental scale. A regression model was developed to relate AGB to latitude and the inverse of the AVHRR NDVI. Their results were encouraging for a study at this scale but were ultimately unreliable for small-area, high-accuracy forest inventories required by small property owners seeking to quantify their forests.

Recent studies using MODIS data based on random forest (Baccini et al., 2004; Baccini et al., 2008) found that the short-wave infrared (SWIR) bands (MODIS bands 6 [1628–1652 nm] and 7 [2105–2155 nm] are particularly sensitive to forest structural parameters (crown size and tree density), texture, and shadow, which are correlated with AGB. They have found a negative relationship between AGB and SWIR reflectance. They argue that SWIR signal is a strong indicator of tree shadows, which is related to stand age structure. Generally, the structure of young forests is often characterized by a single canopy layer, high density, relatively few canopy gaps, and trees of roughly the same size. Conversely, older forests are characterized by a mix of tree ages and sizes, canopy gaps, and multiple canopy layers resulting in increases of the shadow component, thus decreases of SWIR reflectance. Baccini et al. (2008) report a high accuracy, with the map explaining 82% of the variance in AGB for 10% of field plots held back for validation, with a root mean squared error (RMSE) of 50.5 Mg ha^{-1}. However, many other studies using MODIS data have various success (Anaya et al., 2009; Blackard et al., 2008). The main limitation is that MODIS signals are not very sensitive to high biomass values (Zheng et al., 2007; Lu, 2006; Anaya et al., 2009).

For quantifying biomass at local to regional scales, data provided by finer spatial resolution instruments, such as Landsat TM (Lu et al., 2005) and Advanced Spaceborne Thermal Emission and Reflection Radiometer (ASTER) (Muukkonen and Heiskanen, 2005) are required. Typically, finer spatial resolution satellite data has been used as an intermediate step when relating ground reference data with coarser spatial resolution data, usually by regression techniques. For example, Muukkonen and Heiskanen (2005) used stand-wise forest inventory data and moderate-resolution ASTER data to estimate biomass with coarse-resolution MODIS data for a large area with good accuracy. The demonstrated approach can be used as a cost-effective tool to produce preliminary biomass estimates for large areas where more accurate national or large-scale forest inventories do not exist.

The Landsat series of satellites has proven to be a successful venture, providing decades of free-access moderate-resolution multispectral imagery (Datta et al., 2023; Pötzschner et al., 2022; Das et al., 2021; Ronoud et al., 2021; Sinha et al., 2020). To estimate forest biomass, many of the studies used band combinations of the Landsat data and vegetation indices in a regression with a variety of standard field variables, including mean height, Lorey's mean height (mean stand height weighted by basal area per tree), maximum height, crown width, and others. These efforts met with varying degrees of success. Das et al. (2021) and Ronoud et al. (2021) reported greater success using Landsat to map biomass than radar data. Landsat data, in the form of a canopy density product, was an important predictor for the AGB of forests in Africa (Avitabile et al., 2012) and in the Amazon (Saatchi et al., 2007). Canopy density metrics works well for open canopies (i.e., primarily during early-successional stages of forest development). Biomass differences between forests with closed canopies are not captured. Dong et al. (2019) employed a feed-forward neural network to model forest biomass and was successful in extracting forest biomass with high levels of accuracy. Ronoud et al. (2021) also note a key issue in remote sensing of biomass: the inability of models to transfer from study site to study site. Empirical models built from satellite imagery rarely transfer from one study area to another, even if the study sites are composed of similar forest species and climatic conditions. Small forest plots are not represented well by image pixels larger than their spatial extent (Lu, 2006), and complex biophysical environments are not well represented at the scale of Landsat data.

Recent advances of mapping disturbance using Landsat data leads to a new approach to map AGB dynamics using forest disturbance and recover history maps derived from Landsat (Tian et al., 2023; Zhang et al., 2022; Sinha et al., 2020; Ghosh and Behera, 2018; Kumar and Mutanga, 2017; Pflugmacher et al., 2012; Pflugmacher et al., 2014; Powell et al., 2010; Main-Knorn et al., 2013). With recently developed algorithms that characterize trends in disturbance (e.g., year of onset, duration, and magnitude) and post-disturbance regrowth, the new method improved Landsat-based mapping of current biomass across large regions. The new approach includes information on vegetation trends prior to the date to enhance Landsat's spectral relationships with biomass. The method was tested in various US forests in Oregon, Arizona, Minnesota, Montana, as well as in Europe, using Landsat-based disturbance and recovery (DR) metrics. They found that the new method substantially improved predictions of AGB compared to models based on only single-date reflectance. Conversely, they also found that their method performed significantly better in estimating AGB dead than LiDAR models; and single-date Landsat data failed completely.

Chopping et al. (2009) investigated the usability of Multi-angle Imaging Spectro-Radiometer (MISR) on board the Terra satellite to measure woody biomass and other forest parameters for large parts of Arizona and New Mexico. The advantages of MISR over active or other passive sensors are: (1) timely and extensive estimates of forest biomass and (2) other parameters at low cost.

Gonzalez et al. (2010) used QuickBird's panchromatic band to automatically detect tree crowns and then used regression techniques to estimate biomass from the diameter of each tree crown. They found that the QuickBird imagery resulted in higher error and lower total biomass estimates than the LiDAR data due to the shadowing that interfered with the crown detection algorithm. The cost of acquiring the images from these sensors is prohibitive for most research purposes. While the spatial resolution offered by these sensors is excellent for crown delineation, care must be taken with shadowing and other effects of sun angle and tree height, further reducing the utility of these data for small-area forest quantification (Gleason and Im, 2011).

11.5.2 Radar

Radar signals are sensitive to dielectric and geometrical properties of forests and are thus directly related to measurements of AGB. The ability of radar sensors to measure biomass mainly depends on how deep the radar signals can penetrate the canopy. The longer the wavelength is, the deeper the penetration is. The L- and P-band backscatter, particularly HV and HH polarized backscatter, is strongly dependent on biomass amount (Le Toan et al., 2011; Ranson and Sun, 1994; Imhoff, 1995; Saatchi et al., 2012). P-band backscatter shows stronger dependence on biomass than L-band backscatter. The radar backscatter increases approximately linearly with increasing biomass until it is saturated at a certain biomass level that varies with the radar wavelength (Imhoff, 1995). The biomass level for backscatter saturation is about 200 tons/ha at P-band, 100 tons/ha at L-band, and 30–50 tons/ha at X- and C-bands (Le Toan et al., 2004, 2011).

The observed relationship between radar backscatter and biomass can be physically illustrated using electromagnetic scattering models (Sun and Ranson, 1995). HV backscatter is dominated by volume scattering from the woody elements in the trees so that HV is strongly related to AGB. For the HH and VV polarizations, ground conditions can affect the biomass-backscatter relationship, because HH backscatter comes mainly from trunk-ground scattering, while VV backscatter results from both volume and ground scattering.

Application of the radar biomass estimation at a continental or globe scale is best at 1.0 ha scale (100 m × 100 m pixel size). At this scale, the distribution of AGB over the landscape is both stationary and normal and the radar resolution is large enough to reduce the speckle noise and the geolocation error between radar pixel and the plot location. Errors associated with the biomass estimation from radar backscatter or height measurements at this scale can be reduced to acceptable levels (10–20%) for mapping the AGB globally (Saatchi et al., 2011; Saatchi et al., 2012).

Synthetic aperture radar (SAR) sensors on board several satellites (ERS-1, JERS-1, Envisat, RADARSAT) with C- and X-band were used to quantify forest carbon stocks in relatively homogeneous or young forests, but the signal tends to saturate at fairly low biomass levels (~50–100 t C/ha) (Le Toan et al., 2004, 2011). Mountainous or hilly conditions also increase errors. Several studies have used the phased array type L-band SAR (PALSAR) on board the Japanese Advanced Land Observing Satellite (ALOS), launched in 2005, to estimate biomass and carbon stocks in sparse canopies from African savanna woodlands to boreal forests (Peregon and Yamagata, 2013; Mermoz et al., 2014; Cartus et al., 2012; Carreiras et al., 2012). Those studies found that ALOS/PALSAR data can successfully map AGB in sparse canopies when aggregating the ALOS biomass maps at a large scale (county or hectare scale). Synergistic use of L- and X-band SAR can provide large-scale AGB (Englhart et al., 2011). They combined multi-temporal TerraSAR-X X-band and ALOS PALSAR L-band to estimate large-scale biomass for tropical forests with $r^2=0.53$ with an RMSE of 79 t/ha.

Many studies have demonstrated that radar backscattering only works best to estimate biomass for sparse canopy. As an alternative to SAR backscatter intensity, recent advancement in interferometric radar analysis techniques, such as polarimetric and interferometric radar (PolInSAR), have shown great potential to predict biomass (Askne and Santoro, 2005). These interferometric techniques allow for a characterization of the vertical forest structure and thus a more immediate estimation of forest biophysical attributes. Coherence saturation levels are generally higher than those reported for backscatter intensity. Under favorable conditions, correlations exist for values of up to 250–300 tons/ha (Santoro et al., 2007; Chowdhury et al., 2014). The backscattering intensity for C- and X-band are not very good for forest biomass estimation. But the InSAR coherence, and the phase center height of X-band InSAR can be used for the purpose. However, the potential to implement such experimental techniques across large areas depends on suitable configurations of future spaceborne SAR missions. With the advancement in interferometric radar analysis techniques, radar data has great potential for global biomass estimates due to its independence from clouds and therefore the possibility to obtain continuous global coverage.

11.5.3 LIDAR

Use of LiDAR to estimate forest biomass has accelerated rapidly in recent years. Observation from both discrete- or full-return LiDAR can be translated into various forest structure metrics such as maximum canopy height and multi-strata heights aboveground as well as characteristic height at which different proportions of the total reflected energy are returned to the sensor. The various derived metrics can be related to AGB, typically via correlative model with associated field measurements (Wulder et al., 2012; Goetz and Dubayah, 2011).

Many studies have demonstrated the strong relationship between AGB and LiDAR measured height metrics, ranging from boreal conifers to equatorial rain forests. LiDAR has been widely used to map AGB using different LiDAR systems. LiDAR is recognized as the state-of-the-art remote sensing technology for mapping AGB because it is much less sensitive to the saturation problem, compared to conventional remote sensing optical and radar data. We summarize recent progress on LiDAR-based biomass mapping activities from the following two perspectives:

- **Small-Footprint Discrete-Return LiDAR**

AGB has been estimated successfully with remote sensing, especially using small-footprint discrete-return LiDAR data (Næsset and Gobakken, 2008; Næsset and Nelson, 2007; García et al., 2010; Nelson, 1988; Nelson et al., 2004). Nelson et al. (2004) demonstrated that tree height obtained from airborne LiDAR is a good predictor of biomass for large area averages. Næsset and Goabakken (2008) found that LiDAR tree height and forest density were able to explain 88% and 85% of the variability in aboveground and belowground biomass, respectively, for 1395 sample plots in the coniferous boreal zone of Norway. These studies often use LiDAR data alone, or in combination with passive optical or radar data.

Most studies were conducted based on regression equations relating vegetation biomass to LiDAR-derived variables across different scales from individual tree to plot and stand scales. The plot-based approach commonly involves field-measured biomass regressed against derived statistics from plot-level LiDAR data. The LiDAR statistics can be from the individual returns or from the height of the canopy (also called canopy height model [CHM]). This approach adopts distributional metrics such as the mean canopy height and the standard deviation of the canopy height derived from the CHM or the raw returns. These metrics are then used in conjunction with regression equations to predict forest properties (Nelson et al., 2004; García et al., 2010). However, many recent studies used LiDAR return intensities rather than height metrics to estimate biomass. García et al. (2010) found that several biomass estimation models based on LiDAR intensity or height combined with intensity data provide better biomass estimate than using height metrics alone.

- **Large-Footprint Full-Waveform LiDAR**

Large-footprint full-waveform LiDAR systems have been shown to provide accurate estimates of AGB in tropical and temperate deciduous, conifer, and mixed forests over a wide range of conditions. Over the past decade, several airborne Land Vegetation Ice System (LVIS), and Scanning LiDAR Imager of Canopies by Echo Recovery (SLICER) LiDAR systems have demonstrated the ability to retrieve AGB over various biomes ranging from boreal conifers to equatorial rain forests (Lefsky et al., 2005b; Dubayah et al., 2010; Drake et al., 2002b; Drake et al., 2002a; Anderson et al., 2006; Anderson et al., 2008). Most studies adapted stepwise multiple regressions to predict ground-based measures of stand structure from both conventional canopy structure indices, including mean and maximum canopy surface height as well as canopy cover, and indices derived from the canopy height profile (CHP), or vegetation height metrics: RH100, RH75, RH50, and RH25 are defined as the relative height (RH), relative to the ground elevation, at which 100%, 75%, 50%, and 25%, respectively, of the accumulated full-waveform energy occurs (Blair et al., 1999).

The Geoscience Laser Altimeter System (GLAS), onboard the Ice, Cloud and land Elevation Satellite (ICESat), is a full-waveform digitizing LiDAR system with a nominal footprint size of ~65 m that acquires information on topography and the vertical structure of the vegetation (Zwally et al., 2002; Carabajal and Harding, 2005; Harding and Carabajal, 2005). A series of studies using GLAS data have successfully demonstrated the capabilities of GLAS data for estimating forest biomass on ground plots in tropical, temperate, and conifer forests (Boudreau et al., 2008; Nelson et al., 2009; Lefsky et al., 2005a; Baccini et al., 2012).

One major limitation of current spaceborne LiDAR systems (i.e., ICESat GLAS) is the lack of imaging capabilities and the fact that it provides sparse sampling information on the forest structure. To overcome this problem, it has been fused with other data to map large-scale AGB. Boudreau et al. (2008) and Nelson et al. (2009) used a multiphase sampling approach to relate GLAS waveforms to airborne profiling LiDAR measurements and profiling LiDAR measurement to field estimates of total aboveground dry biomass in Québec, Canada, and Siberia, Russia. Some combine optical remote sensing images with GLAS data to map biomass at large scales (Baccini et al., 2008; Baccini et al., 2012). Another issue of ICESat data is that the LiDAR waveform mixes LiDAR energy returns from both vegetation and underneath topography. To mitigate this problem, researchers have limited their analyses to areas with < 10 DEG slope (Nelson et al., 2009). Lefsky et al. (2005a, 2005b) uses waveform shapes to remove the impact of underneath topography on waveform. Yang et al. (2011) developed a physical approach to remove the underneath topography effect. It is important to evaluate the accuracy, precision, and sources of un-certainty involved in using GLAS for large-scale biomass estimation in different regions of the world.

Full-waveform instruments such as GLAS (and LVIS and SLICER) must use high pulse energies in order to penetrate dense canopy and detect the ground surface. As a result of the high pulse energies, the pulse rate must be low, which limits the spatial sampling and resolution of these instruments. Furthermore, the width of the pulse "acts as a low pass filter, thereby smoothing the waveform and limiting the vertical resolution of the canopy features. This also broadens the return from the ground and reduces its amplitude, thus making its detection more difficult.

11.5.4 Multi-Sensor Fusion

The use of LiDAR data, particularly spaceborne data, is limited by its sparse spatial sampling (Zhang et al., 2023; Wang et al., 2022; Zeng et al., 2020; Bispo et al., 2020; Chapungu et al., 2020; Sinha et al., 2020; Ghosh and Behera, 2018; Kumar and Mutanga, 2017). Both radar and passive optical remote sensing provide large scales of imaging capability. However, both optical and SAR estimates of AGB are limited by a loss of sensitivity with increasing biomass, commonly known as "saturation." A promising development is to combine radar/passive optical data with LiDAR to develop models that improve biomass estimates by exploiting the strengths of each sensor. The fusion of metrics from multiple sensors has produced biomass models with high accuracy. While results have been variable, multi-sensor fusion can produce models with accuracy levels similar to or better than those of LiDAR alone (see Table 11.4 for a summary).

Many studies investigate if additional hyperspectral signature from hyperspectral data or radar and optical imaging capability beside LiDAR measurements improve biomass estimates (Sun et al., 2011; Anderson et al., 2008; Swatantran et al., 2011; Gonzalez et al., 2010). The results vary, but most studies found that LiDAR provides the best biomass estimates and additional optical passive or radar data do not improve biomass estimates.

However, another series of multi-sensor fusion study for AGB is fusion of airborne LiDAR, spaceborne radar, Landsat, and field data to map AGB at large scales through two stages of upscaling: scaling from field measurements to airborne LiDAR scale, then from airborne LiDAR scale to spaceborne radar scale (Asner, 2009; Asner et al., 2012; Asner and Mascaro, 2014; Asner et al., 2010; Nelson et al., 2009). Baccini et al. (2008) generated AGB estimates of tropical Africa from MODIS data using GLAS height metrics (average height and height of median energy or HOME metrics). Asner et al. (2010, 2012) uses airborne LiDAR and Landsat data together with field data to map AGB and carbon at high spatial scale in the Amazon. Nelson et al. (2009) combined field data, airborne LiDAR, and spaceborne GLAS data to map AGB at large scales in boreal forests.

Most recent development on biomass and carbon estimates using remote sensing data is large regional mapping of AGB through multi-sensor fusion. Those activities include the fusion of LiDAR and multispectral data (Baccini et al., 2012; Asner, 2009; Asner et al., 2012) with radar data (Saatchi et al., 2011). Two independent studies have produced pan-tropical maps of AGB at 500 m and 1 m spatial resolution (Baccini et al., 2012; Saatchi et al., 2011). These two maps have been widely used by sub-national and national-level activities in relation to Reducing Emissions from Deforestation and forest Degradation (REDD+).

Both maps use similar input data layers and are driven by the same spaceborne LiDAR dataset providing systematic forest height and canopy structure estimates, but they use different ground datasets for calibration and different spatial modeling methodologies. Field data were upscaled to GLAS footprint level (70 m) over a broad range of conditions in tropical Africa, America, and Asia based on the statistical relationships between LiDAR metrics and filed AGB, then GLAS footprint

TABLE 11.4
Capabilities to Estimate AGB and Forest Carbon Stocks through Multi-Sensor Fusion

Multi-Sensors	Study Area	Biomass Parameters	Method	Resolution	Accuracy	References
GLAS/ICESat, MODIS, SRTM, and QSCAT	Tropical forests: • Latin America • Sub-Saharan Africa • SE Asia	Lorey's height	Maximum entropy	1 km, 10 km, 100 km	• 1 km scale: +/−6–53% • 10 km scale: +/−5% • 100 km scale: +/−1%	Saatchi et al. (2011)
GLAS/ICESat, MODIS, and SRTM	Pan-tropical forest	• Waveform metrics • Surface reflectance • Temperature • Topography	Random forest	500 m	• Tropical America: +/−8.4/117.7=7% • Africa: +/−8.4/64.5=13% • Asia: +/−3.0/46.5=6%	Baccini et al. (2012)
Airborne LiDAR, GLAS/ICESat, and MODIS	Colombia and Peru	• MCH • Surface reflectance • Temperature • Topography	Random forest	1.1 km	RMSE Colombia: +/−15.7 Mg C/ha Peru: +/−17.6 Mg C/ha	Barccini et al. (2008)
GLAS, MODIS	South-central Siberia	• GLAS waveform metrics • MODIS land cover	Neural network	500 m	< 100 slope: +/−11.8/163.4=7% > 100 slope: +/−12.4/171.9=7%	Nelson et al. (2009)
Airborne LiDAR GLAS/ICESat Landsat ETM+ SRTM	Quebec, Canada 1.3M km^2	• GLAS waveform metrics • Land cover	Regression	30 m	• Carbon density: +/−2.2/39=6% • Total carbon: +/−0.3/4.9=6% • R2=0.56–0.65	Boudreau et al. (2008)
GLAS/ICESat Landsat	CA	• Tree height • LAI	Regression	30m	RMSE:40–150MgC/ha Relative error: +/−40%	Zhang et al. (2022)
Airborne LiDAR Landsat	Peruvian Amazon 43 Mha	• MCH • Forest cover	Regression	0.1 ha and 5 ha	At 0.1 ha and 5 ha: RMSE=23 and 5 Mg C/ha	Asner et al. (2010)
LVIS and AVIRIS	Bartlett Forest	RH50 AVIRIS NMF	Stepwise regression	20 m	• RMSE improved from 0.55 to 0.51 when combined Mg/ha • Adjusted R2 from 027,0.3 to 0.39. • Fusion reduced error by 5–8%	Anderson et al. (2008)

Multi-Sensors	Study Area	Biomass Parameters	Method	Resolution	Accuracy	References
LVIS and AVIRIS	Sierra Nevada, CA	• RH100, RH75, RH50, RH25 • NDWI, DGVI, CC	Regression	• 20 m • Species based	• $R^2=0.84$, RMSE=58.78 Mg/ha • No significant improvement fusing AVIRIS and LiDAR compared to LiDAR alone	Swatantran et al. (2011)

AVIRIS MNF: AVIRIS minimum noise fraction transform (MNF) rotation
CC: Canopy cover
DGVI: First/second derivative of red edge normalized to 626–795 nm baseline
Lorey's height: Basal area weighted height of all trees >10 cm in diameter)
MCH: Mean canopy vertical height profiles, the distance from ground (digital terrain models) to the approximate centroid of the tree crowns
NDVI: Normalized difference of vegetation index
NDWI: Normalized difference of water index
RH100: Relative height (RH) to the ground elevation at which 100% of the accumulated full waveform energy occurs
RH75: Relative height (RH) to the ground elevation at which 75% of the accumulated full waveform energy occurs
RH50: Relative height (RH) to the ground elevation at which 50% of the accumulated full waveform energy occurs
RH25: Relative height (RH) to the ground elevation at which 25% of the accumulated full waveform energy occurs
SRTM: Shuttle Radar Topography Mission

biomass was scaled to 500 m wall-to-wall biomass map through a random forest machine learning using MODIS BRDF, surface temperature, and SRTM digital elevation data (Baccini et al., 2012).

Saatchi et al. (2011) calibrated ICESat/GLAS Lorey's height (basal area weighted height of all trees > 10 cm in diameter) to AGB using field data collected from 4079 in situ inventory plots across

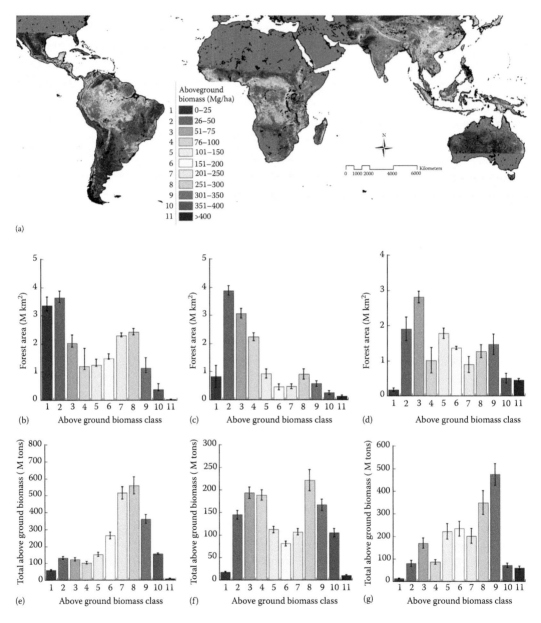

FIGURE 11.4 Distribution of forest AGB (Saatchi et al., 2011). (A) Forest AGB is mapped at 1 km spatial resolution. The study region was bounded at 30° north latitude and 40° south latitude to cover forests of Latin America and sub-Saharan Africa and from 60° to 155° east and west longitude. The map was colored on the basis of 25–50 Mg ha^{-1} AGB classes to better show the overall spatial patterns of forest biomass in tropical regions. Histogram distributions of forest area (at 10% tree cover) for each biomass class were calculated by summing the pixels over Latin America in B, Africa in C, and Asia in D. Similarly, total AGB for each class was computed by summing the values in each region with distributions provided for Latin America in E, Africa in F, and Asia in G. All error bars were computed by using the prediction errors from spatial modeling.

three tropical continents. These AGB estimates were extrapolated from inventory plots (0.25 ha) to the entire landscape at 1 km scale based on spatial imagery from multiple sensors (moderate resolution imaging spectroradiometer [MODIS], shuttle radar topography mission [SRTM], and quick scatterometer [QSCAT]) using a data fusion model based on the maximum entropy (MaxEnt) approach. This "benchmark" map of biomass carbon stocks over 2.5 billion ha of forests on three continents, encompassing all tropical forests, for the early 2000s (see Figure 11.4).

A recent study compared these two maps and found significant difference in their AGB estimates over a wide variety of forest cover types and scales; however, at the country level there is general agreement, with much of the country-level difference explained by the choice of different allometric equations (Mitchard et al., 2013). These two maps were also compared to a high-resolution, locally calibrated map (Asner, 2009). A further limitation present in both studies is the lack of local wood density or diameter-height calibration. Both are known to vary considerably across the landscape but using constant wood density and/or diameter-height relationship smooths out the variations of AGB estimates. This has an important implication for REDD+—it appears we have the algorithms and tools to estimate biomass stocks with some certainty.

11.6 SUMMARY

A variety of remote sensing data types, including optical, LiDAR, and radar (mostly Synthetic Aperture Radar [SAR]) are used to estimate biomass (Tian et al., 2023; Wang et al., 2022; Bispo et al., 2020). The most frequently applied sensors were discrete-return airborne LiDAR, spaceborne multispectral, and airborne or spaceborne radar systems (Figure 11.5) (Fassnacht et al., 2014).

Several studies were conducted for an analysis of reported biomass accuracy estimates using different remote sensing platforms (airborne and spaceborne) and sensor types (optical, radar, and LiDAR) (Zolkos et al., 2013; Goetz and Dubayah, 2011). These studies reported that LiDAR is significantly better at estimating biomass than passive optical or radar sensors used alone (Figure 11.6). AGB models developed from airborne LiDAR metrics are significantly more accurate than those

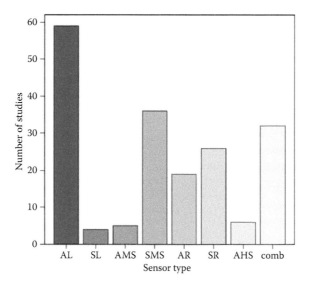

FIGURE 11.5 Frequency distribution of the data sources (sensors) for AGB estimates. (AL = airborne LiDAR, SL = spaceborne LiDAR, AMS = airborne multispectral, SMS = spaceborne multispectral, AR = airborne radar, SR = spaceborne radar, AHS = airborne hyperspectral, comb = studies using data from at least two sensors) (Fassnacht et al., 2014).

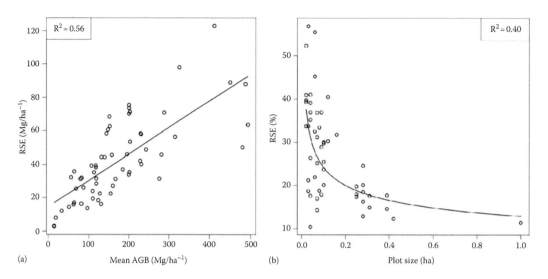

FIGURE 11.6 LiDAR model RSE vs. mean field-estimated AGB for (A) 51 LiDAR-only studies and (B) RSE(%) variability with plot size for 48 studies (Zolkos et al., 2013).

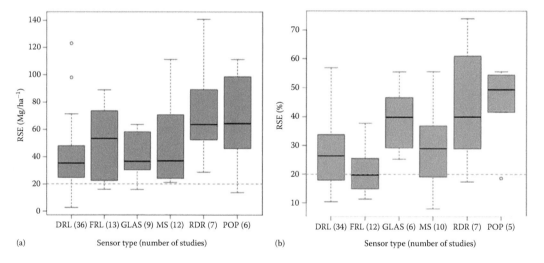

FIGURE 11.7 (A) RSE of remote sensing–AGB regression models, with dotted horizontal line at RSE = 20 Mg ha^{-1}, and (B) RSE(%) (RSE standardized by mean AGB from field measurements) categorized by sensor type, with dotted horizontal line at RSE = 20% of mean AGB. The number of studies for each type is indicated in parentheses. Not all studies reported both mean AGB and RSE values, hence the slight disparity in sample sizes between mean RSE and AGB by sensor type. DRL = discrete-return LiDAR, FRL = full-return LiDAR, MS = multi-sensor, RDR = radar, POP = passive optical (Zolkos et al., 2013).

using radar or passive optical data. The LiDAR model error is positively correlated with the magnitude of AGB, various at higher biomass, and decreases with plot size (Figure 11.7). The fusion of LiDAR with other sensors does not always improve biomass estimates. The spatial extent of airborne LiDAR is typically restricted to relatively small areas (tens of km^2) and is also often integrated with imaging sensors for larger-area mapping. Airborne LiDAR metrics–produced AGB models were significantly more accurate than those based on the spaceborne GLAS instrument due to its sparse samplings. The sparse sampling density of GLAS requires fusion with image data for any AGB mapping application, with associated losses in model accuracy.

Previous studies have reported that error varies with forest type, with higher accuracies for biomass estimates in coniferous stands compared to hardwood stands (Nelson et al., 2004; Ni-Meister et al., 2010). Studies also reported that model errors tend to decrease with increasing plot size (Frazer et al., 2011). Large plot size lowers between-plot variance and has greater spatial overlap and is more resilient to GPS positional errors (Frazer et al., 2011).

Zolkos et al. (2013) reported that the error from LiDAR and multi-sensor models, but not radar or passive optical alone, may satisfy Measurement, Reporting and Verification (MRV) guidelines, particularly in tropical forests. The best LiDAR model and fusion of LiDAR and imaging satellite data at large spatial extents have demonstrated accuracies that may be suitable for carbon accounting purposes at the project level.

11.7 CONCLUSIONS AND FUTURE DIRECTIONS

Tremendous progress has been made to estimate AGB in the last decade or so (Sainuddin et al., 2024; Sanam et al., 2024; Singh et al., 2024; Ma et al., 2024; Datta et al., 2023; Tian et al., 2023; Zhang et al., 2023; Musthafa and Singh, 2022; Pötzschner et al., 2022; Wang et al., 2022; Zeng et al., 2020; Zhang et al., 2022; Das et al., 2021; Ronoud et al., 2021; Abbas et al., 2020; Bispo et al., 2020; Chapungu et al., 2020; Sinha et al., 2020; Ghosh and Behera, 2018; Kumar and Mutanga, 2017). With the development of new LiDAR technology, a number of investigators have developed innovative approaches to fuse passive optical imagery or radar imaging data with to spatially extend point-based estimates of biophysical parameters derived from LiDAR to develop high-quality, wall-to-wall AGB maps with unprecedented accuracy and spatial resolution. Particularly, the synergy of spaceborne large-footprint LiDAR (ICESAT GLAS) and medium-resolution optical data, primarily from the Moderate Resolution Imaging Spectrometer (MODIS), has been exploited to map canopy height and biomass at regional to continental scales. Combining with high-quality forest loss maps, these high-quality carbon stock maps are being used to estimate carbon emissions due to forest cover change at regional and continental scales (Harris et al., 2012; Baccini et al., 2012). Those large-scale maps of biomass, carbon, and carbon emissions can be extremely useful for REDD and global carbon monitoring programs and has the potential to substantially reduce uncertainty in global carbon exchanges and net carbon budgets.

To develop accurate and consistent biomass maps is still challenging. One issue is a lack of large-scale densely sampled LiDAR data at continental and global scales. We call an urgent need for a LiDAR mission to quantify forest carbon store and carbon change at global scale. Currently a laser-based instrument called the Global Ecosystem Dynamics Investigation (GEDI) LiDAR is being developed for the International Space Station, which will provide a unique 3D view of Earth's forest structure, as the valuable information for global carbon estimates. Through combining high vertical and spatial resolution, photon-counting systems might overcome the limitations of full-waveform low detector sensitivity and restricted vertical and spatial resolution. However much vegetation structure properties can be retrieved from the future spaceborne LiDAR missions, ICESat-II, with a 10 kHz, 532 nm micropulse photon counting laser altimeter, still needs further investigation. With recent advancement in polarimetric and interferometric radar (PolInSAR), the fusion of LiDAR and PolInSAR may have great potential to provide accurate AGB estimates at continental and global scales. However, implementation of such experimental techniques across large areas heavily depends on suitable configurations of future spaceborne SAR missions.

With increasing use of remote sensing technology to map AGB and carbon stocks at large scales, yet their calibration will still rely on the accuracy of ground-based carbon storage estimation. Accuracy of aboveground estimates using remote sensing data depends heavily on the accuracy of the allometric equations chosen. Different allometric equations used to calibrate the remote sensing data resulted in different carbon estimates. Recent studies suggested that regional variation allometric equations was an important source of variation in tree AGB (Feldpausch et al., 2012; Goodman, 2014). With the recently updated allometric equations for tropical forests (Chave et al., 2014), remote sensing products could be improved.

There is an urgent need for improved data sets that characterize the global distribution of AGB, especially in the tropics. For the UN Framework Convention on Climate Change to implement a Reduced Emissions from Deforestation and Degradation (REDD+) scheme, more accurate and precise country-based carbon inventories are needed. With recent progress made on biomass and carbon store estimates at continental scales and recently published, global, high-resolution (30 m) forest cover change maps (Hansen et al., 2013), accurate estimates of global carbon store and carbon emission estimates are possible in the near future.

11.8 ACKNOWLEDGMENT

This review was partially supported by NASA under the contract number NNX10AG28G.

REFERENCES

Abbas, S., M. S. Wong, J. Wu, N. Shahzad & S. M. Irteza (2020) Approaches of satellite remote sensing for the assessment of above-ground biomass across tropical forests: Pan-tropical to national scales. *Remote Sensing* 12(20), 3351. https://doi.org/10.3390/rs12203351

Anaya, J. A., E. Chuvieco & A. Palacios-Orueta (2009) Aboveground biomass assessment in Colombia: A remote sensing approach. *Forest Ecology and Management* 257, 1237–1246.

Anderson, J., M. E. Martin, M. L. Smith, R. O. Dubayah, M. A. Hofton, P. Hyde, B. E. Peterson, J. B. Blair & R. G. Knox (2006) The use of waveform lidar to measure northern temperate mixed conifer and deciduous forest structure in New Hampshire. *Remote Sensing of Environment* 105, 248–261.

Anderson, J., L. Plourde, M. Martin, B. Braswell, M. Smith, R. Dubayah, M. Hofton & J. Blair (2008) Integrating waveform lidar with hyperspectral imagery for inventory of a northern temperate forest. *Remote Sensing of Environment* 112, 1856–1870.

Askne, J. & M. Santoro (2005) Multitemporal repeat pass SAR interferometry of boreal forests *IEEE Transactions on Geoscience and Remote Sensing* 43, 10.

Asner, G. P. (2009) Tropical forest carbon assessment: Integrating satellite and airborne mapping approaches. *Environmental Research Letters* 4, 034009.

Asner, G. P., J. K. Clark, J. Mascaro, G. A. Galindo García, K. D. Chadwick, D. A. Navarrete Encinales, G. Paez-Acosta, E. Cabrera Montenegro, T. Kennedy-Bowdoin, Á. Duque, A. Balaji, P. von Hildebrand, L. Maatoug, J. F. Phillips Bernal, A. P. Yepes Quintero, D. E. Knapp, M. C. García Dávila, J. Jacobson & M. F. Ordóñez (2012) High-resolution mapping of forest carbon stocks in the Colombian Amazon. *Biogeosciences* 9, 2683–2696.

Asner, G. P. & J. Mascaro (2014) Mapping tropical forest carbon: Calibrating plot estimates to a simple LiDAR metric. *Remote Sensing of Environment* 140, 614–624.

Asner, G. P., G. V. N. Powellb, J. Mascaroa, D. E. Knappa, J. K. Clarka, J. Jacobsona, T. Kennedy-Bowdoina, A. Balajia, G. Paez-Acostaa, E. Victoriac, L. Secadad, M. Valquid & R. F. Hughese (2010) High-resolution forest carbon stocks and emissions in the Amazon. *PNAS* 107(38), 16738–16742. https://www.pnas.org/doi/full/10.1073/pnas.1004875107

Avitabile, V., A. Baccini, M. A. Friedl & C. Schmullius (2012) Capabilities and limitations of Landsat and land cover data for aboveground woody biomass estimation of Uganda. *Remote Sensing of Environment* 117, 366–380.

Baccini, A., M. A. Friedl, C. E. Woodcock & R. Warbington (2004) Forest biomass estimation over regional scales using multisource data. *Geophysical Research Letters* 31.

Baccini, A., S. J. Goetz, W. S. Walker, N. T. Laporte, M. Sun, D. Sulla-Menashe, J. Hackler, P. S. A. Beck, R. Dubayah, M. A. Friedl, S. Samanta & R. A. Houghton (2012) Estimated carbon dioxide emissions from tropical deforestation improved by carbon-density maps. *Nature Climate Change* 2, 182–185.

Baccini, A., N. Laporte, S. J. Goetz, M. Sun & H. Dong (2008) A first map of tropical Africa's above-ground biomass derived from satellite imagery. *Environmental Research Letters* 3, 045011.

Bispo, Polyanna da Conceição, Pedro Rodríguez-Veiga, Barbara Zimbres, Sabrina do Couto de Miranda, Cassio Henrique Giusti Cezare, Sam Fleming, Francesca Baldacchino, Valentin Louis, Dominik Rains, Mariano Garcia, et al. (2020) Woody aboveground biomass mapping of the Brazilian savanna with a multi-sensor and machine learning approach. *Remote Sensing* 12(17). 2685. https://doi.org/10.3390/rs12172685

Blackard, J., M. Finco, E. Helmer, G. Holden, M. Hoppus, D. Jacobs, A. Lister, G. Moisen, M. Nelson & R. Riemann (2008) Mapping U.S. forest biomass using nationwide forest inventory data and moderate resolution information. *Remote Sensing of Environment* 112, 1658–1677.

Blair, J. B., D. L. Rabine & M. A. Hofton (1999) The Laser Vegetation Imaging Sensor (LVIS): A medium-altitude, digitization-only, airborne laser altimeter for mapping vegetation and topography. *ISPRS Journal of Photogrammetry and Remote Sensing* 54, 8.

Boudreau, J., R. Nelson, H. Margolis, A. Beaudoin, L. Guindon & D. Kimes (2008) Regional aboveground forest biomass using airborne and spaceborne LiDAR in Québec. *Remote Sensing of Environment* 112, 3876–3890.

Breiman, L. (2001) Random forests. *Machine Learning* 45, 5–32.

Breiman, L., J. H. Friedman, R. A. Olshen & C. J. Stone (1984) *Classification and Regression Trees*. Wadsworth, Belmont, CA.

Brown, S. (2002) Measuring carbon in forests: Current status and future challenges. *Environmental Pollution* 116, 9.

Brown, S., A. J. Gillespie & A. E. Lugo (1989) Biomass estimation methods for tropical forests with applications to forest inventory data. *Forest Science* 35, 21.

Canadell, J. G., C. Le Quere, M. R. Raupach, C. B. Field, E. T. Buitenhuis, P. Ciais, T. J. Conway, N. P. Gillett, R. A. Houghton & G. Marland (2007) Contributions to accelerating atmospheric CO2 growth from economic activity, carbon intensity, and efficiency of natural sinks. *Proceedings of the National Academy of Sciences of the United States of America* 104, 18866–18870.

Carabajal, C. C. & D. J. Harding (2005) ICESat validation of SRTM C-band digital elevation models. *Geophysical Research Letters* 32.

Carreiras, J. M. B., M. J. Vasconcelos & R. M. Lucas (2012) Understanding the relationship between aboveground biomass and ALOS PALSAR data in the forests of Guinea-Bissau (West Africa). *Remote Sensing of Environment* 121, 426–442.

Cartus, O., M. Santoro & J. Kellndorfer (2012) Mapping forest aboveground biomass in the Northeastern United States with ALOS PALSAR dual-polarization L-band. *Remote Sensing of Environment* 124, 466–478.

Chapungu, L., L. Nhamo & R. C. Gatti (2020) Estimating biomass of savanna grasslands as a proxy of carbon stock using multispectral remote sensing. *Remote Sensing Applications: Society and Environment* 17, 100275. ISSN 2352-9385. https://doi.org/10.1016/j.rsase.2019.100275. https://www.sciencedirect.com/science/article/pii/S2352938519300953

Chave, J., C. Andalo & S. Brown (2005) Tree allometry and improved estimation of carbon stocks and balance in tropical forests. *Oecologia* 145.

Chave, J., M. Rejou-Mechain, A. Burquez, E. Chidumayo, M. S. Colgan, W. B. Delitti, A. Duque, T. Eid, P. M. Fearnside, R. C. Goodman, M. Henry, A. Martinez-Yrizar, W. A. Mugasha, H. C. Muller-Landau, M. Mencuccini, B. W. Nelson, A. Ngomanda, E. M. Nogueira, E. Ortiz-Malavassi, R. Pelissier, P. Ploton, C. M. Ryan, J. G. Saldarriaga & G. Vieilledent (2014) Improved allometric models to estimate the aboveground biomass of tropical trees. *Global Change Biology*. https://doi.org/10.1111/gcb.12629

Cheng, D. L. & K. J. Niklas (2007) Above- and below-ground biomass relationships across 1534 forested communities. *Annals of Botany* 99, 95–102.

Chopping, M., A. Nolin, G. G. Moisen, J. V. Martonchik & M. Bull (2009) Forest canopy height from the Multiangle Imaging SpectroRadiometer (MISR) assessed with high resolution discrete return lidar. *Remote Sensing of Environment* 113, 2172–2185.

Chowdhury, T. A., C. Thiel & C. Schmullius (2014) Growing stock volume estimation from L-band ALOS PALSAR polarimetric coherence in Siberian forest. *Remote Sensing of Environment* 155, 129–144.

Das, B., R. Bordoloi, S. Deka, A. Paul, P. K. Pandey, L. B. Singha, O. P. Tripathi, B. P. Mishra & M. Mishra (2021). Above ground biomass carbon assessment using field, satellite data and model based integrated approach to predict the carbon sequestration potential of major land use sector of Arunachal Himalaya, India. *Carbon Management* 12(2), 201–214. http://doi.org/10.1080/17583004.2021.1899753

Datta, D., M. Dey, P. K. Ghosh, S. Neogy & A. K. Roy. 2023. Coupling multi-sensory earth observation datasets, in-situ measurements, and machine learning algorithms for total blue C stock estimation of an estuarine mangrove forest. *Forest Ecology and Management* 546, 121345. ISSN 0378-1127.

https://doi.org/10.1016/j.foreco.2023.121345. https://www.sciencedirect.com/science/article/pii/S03781 12723005790

Domke, G. M., C. W. Woodall, J. E. Smith, J. A. Westfall & R. E. McRoberts (2012) Consequences of alternative tree-level biomass estimation procedures on U.S. forest carbon stock estimates. *Forest Ecology and Management* 270, 108–116.

Dong, J., R. K. Kaufmann, R. B. Myneni, C. J. Tucker, P. E. Kaupp, J. Liski, W. Buermann, V. Alexeyev & M. K. Hughes (2003) Remote sensing estimates of boreal and temperate forest woody biomass: Carbon pools, sources, and sinks. *Remote Sensing of Environment* 84, 8.

Dong, L., Tang, S., Min, M., Veroustraete, F., & Cheng, J. (2019) Aboveground forest biomass based on OLSR and an ANN model integrating LiDAR and optical data in a mountainous region of China. *International Journal of Remote Sensing*, 40(15), 6059–6083. https://doi.org/10.1080/01431161.2019.1587201

Drake, J. B., R. O. Dubayah, D. B. Clark, R. G. Knox, J. B. Blair, M. A. Hofton, R. L. Chazdon, J. F. Weishampel & S. D. Prince (2002b) Estimation of tropical forest structural characteristics using large-footprint lidar. *Remote Sensing of Environment* 79, 15.

Drake, J. B., R. Dubayah, R. G. Knox, D. B. David, B. Clark & J. B. Blaire (2002a) Sensitivity of large-footprint lidar to canopy structure and biomass in a neotropical rainforest. *Remote Sensing of Environment* 81, 15.

Dubayah, R. O. & J. Drake (2000) Lidar remote sensing for forestry. *Journal of Forestry* 98, 12.

Dubayah, R. O., S. L. Sheldon, D. B. Clark, M. A. Hofton, J. B. Blair, G. C. Hurtt & R. L. Chazdon (2010) Estimation of tropical forest height and biomass dynamics using lidar remote sensing at La Selva, Costa Rica. *Journal of Geophysical Research-Biogeosciences* 115.

Englhart, S., V. Keuck & F. Siegert (2011) Aboveground biomass retrieval in tropical forests—The potential of combined X- and L-band SAR data use. *Remote Sensing of Environment* 115, 1260–1271.

FAO (2010) *Food and Agriculture Organization of the United Nations Global Forest Resources Assessment 2010 FAO Forestry Paper 163*. https://www.fao.org/forest-resources-assessment/past-assessments/fra-2010/en/

Fassnacht, F. E., F. Hartig, H. Latifi, C. Berger, J. Hernández, P. Corvalán & B. Koch (2014) Importance of sample size, data type and prediction method for remote sensing-based estimations of aboveground forest biomass. *Remote Sensing of Environment* 154, 102–114.

Feldpausch, T. R., L. Banin, O. L. Phillips, T. R. Baker, S. L. Lewis, C. A. Quesada, E. Affum-Baffoe, E. J. M. M. Arets, N. J. Berry, M. Bird, E. S. Brondizio, P. de Camargo, J. Chave, G. Djagbletey, T. F. Domingues, M. Drescher, P. M. Fearnside, M. B. França, N. M. Fyllas, G. Lopez-Gonzalez, A. Hladik, N. Higuchi, M. O. Hunter, Y. Iida, K. A. Salim, A. R. Kassim, M. Keller, J. Kemp, D. A. King, J. C. Lovett, B. S. Marimon, B. H. Marimon-Junior, E. Lenza, A. R. Marshall, D. J. Metcalfe, E. T. A. Mitchard, E. F. Moran, B. W. Nelson, R. Nilus, E. M. Nogueira, M. Palace, S. Patiño, K. S. H. Peh, M. T. Raventos, J. M. Reitsma, G. Saiz, F. Schrodt, B. Sonké, H. E. Taedoumg, S. Tan, L. White, H. Wöll & J. Lloyd (2011) Height-diameter allometry of tropical forest trees. *Biogeosciences* 8, 1081–1106.

Feldpausch, T. R., J. Lloyd, S. L. Lewis, R. J. W. Brienen, M. Gloor, A. Monteagudo Mendoza, G. Lopez-Gonzalez, L. Banin, K. Abu Salim, K. Affum-Baffoe, M. Alexiades, S. Almeida, I. Amaral, A. Andrade, L. E. O. C. Aragão, A. Araujo Murakami, E. J. M. M. Arets, L. Arroyo, G. A. Aymard C, T. R. Baker, O. S. Bánki, N. J. Berry, N. Cardozo, J. Chave, J. A. Comiskey, E. Alvarez, A. de Oliveira, A. Di Fiore, G. Djagbletey, T. F. Domingues, T. L. Erwin, P. M. Fearnside, M. B. França, M. A. Freitas, N. Higuchi, E. H. C, Y. Iida, E. Jiménez, A. R. Kassim, T. J. Killeen, W. F. Laurance, J. C. Lovett, Y. Malhi, B. S. Marimon, B. H. Marimon-Junior, E. Lenza, A. R. Marshall, C. Mendoza, D. J. Metcalfe, E. T. A. Mitchard, D. A. Neill, B. W. Nelson, R. Nilus, E. M. Nogueira, A. Parada, K. S. H. Peh, A. Pena Cruz, M. C. Peñuela, N. C. A. Pitman, A. Prieto, C. A. Quesada, F. Ramírez, H. Ramírez-Angulo, J. M. Reitsma, A. Rudas, G. Saiz, R. P. Salomão, M. Schwarz, N. Silva, J. E. Silva-Espejo, M. Silveira, B. Sonké, J. Stropp, H. E. Taedoumg, S. Tan, H. ter Steege, J. Terborgh, M. Torello-Raventos, G. M. F. van der Heijden, R. Vásquez, E. Vilanova, V. A. Vos, L. White, S. Willcock, H. Woell & O. L. Phillips (2012) Tree height integrated into pantropical forest biomass estimates. *Biogeosciences* 9, 3381–3403.

Frazer, G. W., S. Magnussen, M. A. Wulder & K. O. Niemann (2011) Simulated impact of sample plot size and co-registration error on the accuracy and uncertainty of LiDAR-derived estimates of forest stand biomass. *Remote Sensing of Environment* 115, 636–649.

García, M., D. Riaño, E. Chuvieco & F. M. Danson (2010) Estimating biomass carbon stocks for a Mediterranean forest in central Spain using LiDAR height and intensity data. *Remote Sensing of Environment* 114, 816–830.

Ghosh, S. M. & M. D. Behera (2018) Aboveground biomass estimation using multi-sensor data synergy and machine learning algorithms in a dense tropical forest. *Applied Geography* 96, 29–40. ISSN

0143-6228. https://doi.org/10.1016/j.apgeog.2018.05.011. https://www.sciencedirect.com/science/article/pii/S0143622818303114

Gibbs, H. K., S. Brown, J. O. Niles & J. A. Foley (2007) Monitoring and estimating tropical forest carbon stocks: Making REDD a reality. *Environmental Research Letters* 2.

Gleason, C. J. & J. Im (2011) A review of remote sensing of forest biomass and biofuel: Options for small-area applications. *GIScience & Remote Sensing* 48, 141–170.

Goetz, S. J., A. Baccini, N. T. Laporte, T. Johns, W. Walker, J. Kellndorfer, R. A. Houghton & M. Sun (2009) Mapping and monitoring carbon stocks with satellite observations: A comparison of methods. *Carbon Balance Manag* 4, 2.

Goetz, S. J. & R. Dubayah (2011) Advances in remote sensing technology and implications for measuring and monitoring forest carbon stocks and change. *Carbon Management* 2, 14.

Gonzalez, P., G. P. Asner, J. J. Battles, M. A. Lefsky, K. M. Waring & M. Palace (2010) Forest carbon densities and uncertainties from Lidar, QuickBird, and field measurements in California. *Remote Sensing of Environment* 114, 1561–1575.

Goodman, R. C. (2014) The importance of crown dimensions to improve tropical tree biomass estimates. *Ecological Applications* 24, 9.

Hansen, M. C., P. V. Potapov, R. Moore, M. Hancher, S. A. Turubanova, A. Tyukavina, D. Thau, S. V. Stehman, S. J. Goetz, T. R. Loveland, A. Kommareddy, A. Egorov, L. Chini, C. O. Justice & J. R. Townshend (2013) High-resolution global maps of 21st-century forest cover change. *Science* 342, 850–853.

Harding, D. J. & C. C. Carabajal (2005) ICESat waveform measurements of within-footprint topographic relief and vegetation vertical structure. *Geophysical Research Letters* 32.

Harris, N. L., S. Brown, S. C. Hagen, S. S. Saatchi, S. Petrova, W. Salas, M. C. Hansen, P. V. Potapov & A. Lotsch (2012) Baseline map of carbon emissions from deforestation in tropical regions. *Science* 336, 1573–1576.

Heimann, M. & M. Reichstein (2008) Terrestrial ecosystem carbon dynamics and climate feedbacks. *Nature* 451, 289–292.

Hollman, R., C. J. Merchant, R. Saunders, C. Downy, M. Buchwitz, A. Cazenave, E. Chuvieco, P. Defourny, G. D. Leeuw, R. Forsberg, T. Holzer-Popp, F. Paul, S. Sandven, S. Sathyendranath, M. V. Roozendael & W. Wagner (2013) The ESA climate change initiative—satellite data records for essential climate variables. *Bulletin of American Meteorology Society* 10, 12.

Houghton, R. A. (2005) Aboveground forest biomass and the global carbon balance. *Global Change Biology* 11, 945–958.

Houghton, R. A. (2007) Balancing the global carbon budget. *Annual Review of Earth and Planetary Sciences* 35, 35.

Houghton, R. A., F. Hall & S. J. Goetz (2009) Importance of biomass in the global carbon cycle. *Journal of Geophysical Research* 114.

Houghton, R. A., J. I. House, J. Pongratz, G. R. V. D. Werf, R. S. DeFries, M. C. Hansen, C. L. Quere & N. Ramankutty (2012) Carbon emissions from land use and land-cover change. *Biogeosciences* 9, 18.

Houghton, R. A., K. T. Lawrance, J. L. Hackler & S. Brown (2001) The spatial distribution of forest biomass in Brazilian Amazon: A comparison of estimates. *Global Climate Biology* 4, 16.

Imhoff, M. (1995) Radar backscatter and biomass saturation: Ramifications for global biomass inventory. *IEEE Transactions on Geoscience and Remote Sensing* 33, 9.

Jenkins, J. C. (2003) National-scale biomass estimators for United States tree species. *Forest Science* 49, 24.

Jenkins, J. C., D. C. Chojnacky, L. S. Heath & R. A. Birdsey (2004) Comprehensive database of diameter-based biomass regressions for North American tree species. *United States Department of Agriculture Forest Service, General Technical Report NE-319*. https://doi.org/10.2737/NE-GTR-319

Kim, D. K., J. O. Sexton, P. Noojipady, C. Huang, A. Anand, S. Channan, M. Feng & J. R. Townshend (2014). Global, Landsat-based forest-cover change from 1990 to 2000. *Remote Sensing of Environment*. Available online 26 September 2014. ISSN 0034-4257. http://doi.org/10.1016/j.rse.2014.08.017.

Kumar, L. & O. Mutanga (2017) Remote sensing of above-ground biomass. *Remote Sensing* 9(9), 935. https://doi.org/10.3390/rs9090935

Le Toan, T., A. Beaudoin, J. Riom & D. Guyon (1992) Relating forest biomass to SAR data. *IEEE Transactions on Geoscience and Remote Sensing* 30, 8.

Le Toan T., S. Quegan, M. Davidson, H. Balzter, P. Paillou, K. Papathanassiou, S. Plummer, S. Saatchi, H. Shugart & L. Ulander (2011) The BIOMASS mission: Mapping global forest biomass to better understand the terrestrial carbon cycle. *Remote Sensing of Environment* 115, 11.

Le Toan T., S. Quegan, I. Woodward, M. Lomas, N. Delbart & G. Picard (2004) Relating radar remote sensing of biomass to modelling of forest carbon budgets. *Journal of Climate Change* 67, 4.

Lefsky, M. A., D. J. Harding, M. Keller, W. B. Cohen, C. C. Carabajal, F. D. Espirito-Santo, M. O. Hunter & R. de Oliveira (2005a) Estimates of forest canopy height and aboveground biomass using ICESat. *Geophysical Research Letters* 32.

Lefsky, M. A., A. T. Hudak, W. B. Cohen & S. A. Acker (2005b) Geographic variability in lidar predictions of forest stand structure in the Pacific Northwest. *Remote Sensing of Environment* 95, 532–548.

Lu, D. S. (2006) The potential and challenge of remote sensing-based biomass estimation. *International Journal of Remote Sensing* 27, 1297–1328.

Lu, D. S., M. Mateus Batistella & E. Moran (2005) Satellite estimation of aboveground biomass and impacts of forest stand structure. *Photogrammetric Engineering & Remote Sensing* 71, 9.

Ma, T., C. Zhang, L. Ji, Z. Zuo, M. Beckline, Y. Hu, X. Li & X. Xiao (2024) Development of forest aboveground biomass estimation, its problems and future solutions: A review. *Ecological Indicators* 159, 111653. ISSN 1470-160X. https://doi.org/10.1016/j.ecolind.2024.111653. https://www.sciencedirect.com/science/article/pii/S1470160X24001109

Main-Knorn, M., W. B. Cohen, R. E. Kennedy, W. Grodzki, D. Pflugmacher, P. Griffiths & P. Hostert (2013) Monitoring coniferous forest biomass change using a Landsat trajectory-based approach. *Remote Sensing of Environment* 139, 277–290.

Mermoz, S., T. Le Toan, L. Villard, M. Réjou-Méchain & J. Seifert-Granzin (2014) Biomass assessment in the Cameroon savanna using ALOS PALSAR data. *Remote Sensing of Environment* 155 109–119.

Mitchard, E. T., S. S. Saatchi, A. Baccini, G. P. Asner, S. J. Goetz, N. L. Harris & S. Brown (2013) Uncertainty in the spatial distribution of tropical forest biomass: A comparison of pan-tropical maps. *Carbon Balance Manag* 8, 10.

Musthafa, M. & G. Singh (2022) Improving forest above-ground biomass retrieval using multi-sensor L- and C- band SAR data and multi-temporal spaceborne LiDAR data. *Frontiers in Forests and Global Change* 5, 822704. http://doi.org/10.3389/ffgc.2022.822704

Muukkonen, P. & J. Heiskanen (2005) Estimating biomass for boreal forests using ASTER satellite data combined with standwise forest inventory data. *Remote Sensing of Environment* 99, 434–447.

Muukkonen, P. & J. Heiskanen (2007) Biomass estimation over a large area based on standwise forest inventory data and ASTER and MODIS satellite data: A possibility to verify carbon inventories. *Remote Sensing of Environment* 107, 617–624.

Næsset, E. & T. Gobakken (2008) Estimation of above- and below-ground biomass across regions of the boreal forest zone using airborne laser. *Remote Sensing of Environment* 112, 3079–3090.

Næsset, E. & R. Nelson (2007) Using airborne laser scanning to monitor tree migration in the boreal–alpine transition zone. *Remote Sensing of Environment* 110, 357–369.

Nelson, R. (1988) Estimating forest biomass and volume using airborne laser data. *Remote Sensing of Environment* 24, 21.

Nelson, R. (2010) Model effects on GLAS-based regional estimates of forest biomass and carbon. *International Journal of Remote Sensing* 31, 1359–1372.

Nelson, R., K. J. Ranson, G. Sun, D. S. Kimes, V. Kharuk & P. Montesano (2009) Estimating Siberian timber volume using MODIS and ICESat/GLAS. *Remote Sensing of Environment* 113, 691–701.

Nelson, R., A. Short & M. Valenti (2004) Measuring biomass and carbon in Delaware using an airborne profiling LIDAR. *Scandinavian Journal of Forest Research* 19, 12.

Ni-Meister, W., S. Y. Lee, A. H. Strahler, C. E. Woodcock, C. Schaaf, T. A. Yao, K. J. Ranson, G. Q. Sun & J. B. Blair (2010) Assessing general relationships between aboveground biomass and vegetation structure parameters for improved carbon estimate from lidar remote sensing. *Journal of Geophysical Research-Biogeosciences* 115.

Pan, Y., R. A. Birdsey, J. Fang, R. Houghton, P. E. Kauppi, W. A. Kurz, O. L. Phillips, A. Shvidenko, S. L. Lewis, J. G. Canadell, P. Ciais, R. B. Jackson, S. W. Pacala, A. D. McGuire, S. Piao, A. Rautiainen, S. Sitch & D. Hayes (2011) A large and persistent carbon sink in the world's forests. *Science* 333, 988–93.

Peregon, A. & Y. Yamagata (2013) The use of ALOS/PALSAR backscatter to estimate above-ground forest biomass: A case study in Western Siberia. *Remote Sensing of Environment* 137, 139–146.

Pflugmacher, D., W. B. Cohen & R. E. Kennedy (2012) Using Landsat-derived disturbance history (1972–2010) to predict current forest structure. *Remote Sensing of Environment* 122, 146–165.

Pflugmacher, D., W. B. Cohen, R. E. Kennedy & Z. Yang (2014) Using Landsat-derived disturbance and recovery history and lidar to map forest biomass dynamics. *Remote Sensing of Environment* 151, 124–137.

Pötzschner, F., M. Baumann, N. I. Gasparri, G. Conti, D. Loto, M. Piquer-Rodríguez & T. Kuemmerle (2022) Ecoregion-wide, multi-sensor biomass mapping highlights a major underestimation of dry forests carbon stocks. *Remote Sensing of Environment* 269, 112849. ISSN 0034-4257. https://doi.org/10.1016/j.rse.2021.112849. https://www.sciencedirect.com/science/article/pii/S0034425721005691

Powell, S. L., W. B. Cohen, S. P. Healey, R. E. Kennedy, G. G. Moisen, K. B. Pierce & J. L. Ohmann (2010) Quantification of live aboveground forest biomass dynamics with Landsat time-series and field inventory data: A comparison of empirical modeling approaches. *Remote Sensing of Environment* 114, 1053–1068.

Ranson, K. J. & G. Sun (1994) Mapping biomass in a Northern forest using multifrequency SAR data. *IEEE Transactions on Geoscience and Remote Sensing* 32, 8.

Ronoud, G., P. Fatehi, A. A. Darvishsefat, E. Tomppo, J. Praks & M. E. Schaepman (2021) Multi-sensor aboveground biomass estimation in the broadleaved hyrcanian forest of Iran. *Canadian Journal of Remote Sensing* 47(6), 818–834. http://doi.org/10.1080/07038992.2021.1968811

Saatchi, S. S., N. L. Harris, S. Brown, M. Lefsky, E. T. A. Mitchard, W. Salas, B. R. Zutta, W. Buermann, S. L. Lewis, S. Hagen, S. Petrova, L. White, M. Silman & A. Morel (2011) Benchmark map of forest carbon stocks in tropical regions across three continents. *Proceedings of the National Academy of Sciences of the United States of America* 108, 9899–9904.

Saatchi, S. S., R. A. Houghton, R. C. Dos Santos Alvala, J. V. Soares & Y. Yu (2007a) Distribution of aboveground live biomass in the Amazon. *Global Change Biology* 14, 12.

Saatchi, S. S., R. A. Houghton, R. C. Dos Santos AlvalÁ, J. V. Soares & Y. Yu (2007b) Distribution of aboveground live biomass in the Amazon basin. *Global Change Biology* 13, 816–837.

Saatchi, S., L. Ulander, M. Williams, S. Quegan, T. LeToan, H. Shugart & J. Chave (2012) Forest biomass and the science of inventory from space. *Nature Climate Change* 2, 826–827.

Sainuddin, F. V., G. Malek, A. Rajwadi, et al. 2024. Estimating above-ground biomass of the regional forest landscape of Northern Western Ghats using machine learning algorithms and multi-sensor remote sensing data. *Journal of the Indian Society of Remote Sensing*. https://doi.org/10.1007/s12524-024-01836-y

Sanam, H., A. A. Thomas, A. P. Kumar, et al. (2024) Multi-sensor approach for the estimation of aboveground biomass of mangroves. *Journal of the Indian Society of Remote Sensing*. https://doi.org/10.1007/s12524-024-01811-7

Santoro, M., A. Shvidenko, I. McCallum, J. Askne & C. Schmullius (2007) Properties of ERS-1/2 coherence in the Siberian boreal forest and implications for stem volume retrieval. *Remote Sensing of Environment* 106, 154–172.

Singh, R. K., C. M. Biradar, M. D. Behera, A. J. Prakash, P. Das, M. R. Mohanta, G. Krishna, A. Dogra, S. K. Dhyani & J. Rizvi (2024) Optimising carbon fixation through agroforestry: Estimation of aboveground biomass using multi-sensor data synergy and machine learning. *Ecological Informatics* 79, 102408. ISSN 1574-9541. https://doi.org/10.1016/j.ecoinf.2023.102408. https://www.sciencedirect.com/science/article/pii/S1574954123004375

Sinha, S., S. Mohan, A. K. Das, L. K. Sharma, C. Jeganathan, A. Santra, S. S. Mitra & M. S. Nathawat (2020) Multi-sensor approach integrating optical and multi-frequency synthetic aperture radar for carbon stock estimation over a tropical deciduous forest in India. *Carbon Management* 11(1), 39–55. http://doi.org/10.1080/17583004.2019.1686931

Strahler, A. H., D. L. B. Jupp, C. E. Woodcock, C. B. Schaaf, T. Yao, F. Zhao, X. Yang, J. Lovell, D. Culvenor, G. Newnham, W. Ni-Miester & W. Boykin-Morris (2008) Retrieval of forest structural parameters using a ground-based lidar instrument (Echidna®). *Canadian Journal of Remote Sensing* 34, 14.

Sun, G. & K. J. Ranson (1995) A three dimensional radar backscattering model for forest canopies. *IEEE Transactions on Geoscience and Remote Sensing* 33, 11.

Sun, G., K. J. Ranson, Z. Guo, Z. Zhang, P. Montesano & D. Kimes (2011) Forest biomass mapping from lidar and radar synergies. *Remote Sensing of Environment* 115, 2906–2916.

Swatantran, A., R. Dubayah, D. Roberts, M. Hofton & J. B. Blair (2011) Mapping biomass and stress in the Sierra Nevada using lidar and hyperspectral data fusion. *Remote Sensing of Environment* 115, 2917–2930.

Tian, Lei, Xiaocan Wu, Yu Tao, Mingyang Li, Chunhua Qian, Longtao Liao & Wenxue Fu (2023) Review of remote sensing-based methods for forest aboveground biomass estimation: Progress, challenges, and prospects. *Forests* 14(6), 1086. https://doi.org/10.3390/f14061086

Wang, Xiaoyi, Caixia Liu, Guanting Lv, Jinfeng Xu & Guishan Cui (2022) Integrating multi-source remote sensing to assess forest aboveground biomass in the Khingan Mountains of North-Eastern China using machine-learning algorithms. *Remote Sensing* 14(4), 1039. https://doi.org/10.3390/rs14041039

West, G. B., J. H. Brown & B. J. Enquist (1999) A general model for the structure and allometry of plant vascular systems. *Nature* 400, 4.

Woodall, C. W., L. S. Heath, G. M. Domke & M. C. Nichols (2011) Methods and equations for estimating aboveground volume, biomass, and carbon for trees in the U.S. forest inventory, 2010. *United States Department of Agriculture, Forest Service, General Technical Report NRS-88*. United States Department of Agriculture.

Wulder, M. A., J. C. White, R. F. Nelson, E. Naesset, H. O. Orka, N. C. Coops, T. Hilker, C. W. Bater & T. Gobakken (2012) Lidar sampling for large-area forest characterization: A review. *Remote Sensing of Environment* 121, 196–209.

Yang, W. Z., W. Ni-Meister & S. Lee (2011) Assessment of the impacts of surface topography, off-nadir pointing and vegetation structure on vegetation lidar waveforms using an extended geometric optical and radiative transfer model. *Remote Sensing of Environment* 115, 2810–2822.

Zeng, N., H. He, X. Ren, L. Zhang, Y. Zeng, J. Fan, Y. Li, Z. Niu, X. Zhu & O. Chang (2020) The utility of fusing multi-sensor data spatio-temporally in estimating grassland aboveground biomass in the three-river headwaters region of China. *International Journal of Remote Sensing* 41(18), 7068–7089. http://doi.org/10.1080/01431161.2020.1752411

Zhang, Fanyi, Xin Tian, Haibo Zhang & Mi Jiang (2022). Estimation of aboveground carbon density of forests using deep learning and multisource remote sensing. *Remote Sensing* 14(13), 3022. https://doi.org/10.3390/rs14133022

Zhang, X. & S. Kondragunta (2006) Estimating forest biomass in the USA using generalized allometric models and MODIS land products. *Geophysical Research Letters* 33.

Zhang, Y., N. Wang, Y. Wang & M. Li (2023) A new strategy for improving the accuracy of forest aboveground biomass estimates in an alpine region based on multi-source remote sensing. *GIScience & Remote Sensing* 60(1). http://doi.org/10.1080/15481603.2022.2163574

Zhao, K., S. Popescu, X. Meng, Y. Pang & M. Agca (2011) Characterizing forest canopy structure with lidar composite metrics and machine learning. *Remote Sensing of Environment* 115, 1978–1996.

Zheng, D., L. S. Heath & M. J. Ducey (2007) Forest biomass estimated from MODIS and FIA data in the Lake States: MN, WI and MI, USA. *Forestry* 80, 265–278.

Zolkos, S. G., S. J. Goetz & R. Dubayah (2013) A meta-analysis of terrestrial aboveground biomass estimation using lidar remote sensing. *Remote Sensing of Environment* 128, 289–298.

Zwally, H. J., B. Schutz, W. Abdalati, J. Abshire, C. Bentley, A. Brenner, J. Bufton, J. Dezio, D. Hancoc, D. Harding, T. Herring, B. Minster, K. Quinn, S. Palm, J. Spinhirne & R. Thomas (2002) ICESat's laser measurements of polar ice, atmosphere, ocean, and land. *Journal of Geodynamics* 34, 11.

Part VI

Summary and Synthesis of Volume IV

12 Forests, Biodiversity, Ecology, LULC, and Carbon

Prasad S. Thenkabail

ACRONYMS AND DEFINITIONS

ACD	Aboveground carbon density
AGB	Aboveground biomass
ALOS	Advanced Land Observing Satellite
ALS	Airborne laser scanning
ANN	Artificial Neural Networks
APAR	Absorbed photosynthetically active radiation
ASAR	Advanced synthetic aperture radar on board ENVISAT
AVHRR	Advanced very high resolution radiometer
CHM	Canopy height model
CHRIS	Compact High Resolution Imaging Spectrometer
COSMO-SkyMed	Constellation of Small Satellites for Mediterranean basin Observation (COSMO)-SkyMed
DEM	Digital Elevation Model
DTM	Digital terrain model
ENVISAT	Environmental satellite
EO	Earth Observation
ERS	European remote sensing satellites
FLUXNET	A network of micrometeorological tower sites to measure carbon dioxide, water, and energy balance between terrestrial systems and the atmosphere
GEO	Group on Earth Observation
GHG	Greenhouse gas
GIMMS	Global Inventory Modeling and Mapping Studies
GLAS	Geoscience Laser Altimeter System
GPP	Gross primary productivity
HVI	Hyperspectral vegetation indices
ICESat	Instrument on board the Ice, Cloud, and land Elevation
IKONOS	A commercial earth observation satellite, typically, collecting sub-meter to 5 m data
JERS	Japanese Earth Resources Satellite
LAI	Leaf area index
LiDAR	Light detection and ranging
LSP	Land surface phenology
LUC	Land use classes
LUE	Light use efficiency
LULC	Land use, land cover
LULCC	Land use, land cover change
MLS	Mobile laser scanning
MODIS	Moderate-resolution imaging spectroradiometer

NASA	National Aeronautics and Space Administration
NDVI	Normalized Difference Vegetation Index
NOAA	National Oceanic and Atmospheric Administration
NPP	NPOESS Preparatory Project
PALSAR	Phased Array type L-band Synthetic Aperture Radar
PAR	Photosynthetically active radiation
PolSAR	RADARSAT-2 polarimetric SAR
PRI	Photochemical reflectance index
PROBA	Project for On Board Autonomy
PROSPECT	Radiative transfer model to measure leaf optical properties spectra
PV	Photosynthetic vegetation
RADARSAT	Radar satellite
REDD	Reducing Emissions from Deforestation in Developing Countries
SAR	Synthetic aperture radar
SAVI	Soil-adjusted vegetation index
SRTM	Shuttle Radar Topographic Mission
STAARCH	Spatial temporal adaptive algorithm for mapping reflectance change
SVM	Support vector machines
TerraSAR-X	A radar Earth observation satellite with its phased array synthetic aperture radar
TLS	Terrestrial laser scanning
TRMM	Tropical Rainfall Measuring Mission
UAS	Unmanned Aircraft System
UAV	Unmanned Aerial Vehicle
UNFCCC	United Nations Framework Convention on Climate Change
VHRI	Very high resolution imagery

This chapter provides a summary of each of the 11 chapters in Volume IV of the six-volume *Remote Sensing Handbook*. The topics covered include (Figure 12.0): (a) forests, (b) biodiversity, (c) ecology, (d) land use/land cover, and (e) carbon. Under each of these broad topics, there are one or more chapters. For example, there are four chapters under forests. In a nutshell, these chapters provide a complete and comprehensive overview of these critical topics, capture the advances of the last 60+ years, and provide a vision for further development in the years ahead. By reading this summary chapter, a reader can have a quick understanding of what is in each of the chapters of Volume IV, see how the chapters interconnect and intermingle, and get an overview of the importance of various chapters in developing complete and comprehensive knowledge of remote sensing for land resources. These chapters together not only capture the advances of the last 60+ years but also provide a vision for the future.

12.1 TROPICAL FOREST CHARACTERIZATION USING MULTI-SPECTRAL IMAGERY

Forest carbon (C) estimates vary widely (Chapters 1–4) because of knowledge gaps, data and methods used, and rapid changes in tropical land use that may account for the "missing sink" of carbon in the global C budget (Gao et al., 2024). Depending on changes in land use and global climate, tropical forests can alternate between sources and sinks of atmospheric C, leading to uncertainty in future trends in forest C fluxes. The long-term net flux of carbon between terrestrial ecosystems and the atmosphere has been dominated by two factors: (1) changes in the total area of forests and (2) per-hectare changes in forest biomass resulting from management and regrowth (Chapters 1–4). Apart from regional-level uncertainties in tropical forest C fluxes, uncertainties also exist in the regenerative capacity of forests and in harvest and management policies (Chapters 1–4).

Forests, Biodiversity, Ecology, LULC, and Carbon

**Chapter 12: Summary Chapter for
Remote Sensing Handbook (Second Edition, Six Volumes): Volume IV**

Volume IV: Forests, Biodiversity, Ecology, LULC and Carbon

Chapter 1: Tropical Forest Characterization using Multispectral Imagery
Chapter 2: LiDAR and Radar Forest Informatics
Chapter 3: Hyperspectral and LiDAR for Forest Quantities
Chapters 4 : Tree and Stand Height from Optical Remote Sensing
Chapters 5 to 7: Biodiversity Studies, Habitat Mapping, and Ecological Studies from Space
Chapters 8 and 9: Land Cover Change Detection using Optical and Radar Data
Chapters 10 and 11: Global Forest Biomass and Carbon Budgets from Remote Sensing

FIGURE 12.0 Overview of the chapters in Volume IV of the *Remote Sensing Handbook* (Second Edition).

The need to remove uncertainties and errors in estimates of tropical forest C storage and fluxes is more urgent than ever before. Under the United Nations Framework Convention on Climate Change (UNFCCC), countries regularly report that the state of their forest resources and emerging mechanisms, such as Reducing Emissions from Deforestation and Degradation (REDD+), are likely to require temporally and spatially fine-grained assessments of carbon stocks (Friedlingstein et al., 2020).

In Chapter 1, Dr. E. H. Helmer et al. focus on characteristics of tropical forests that are relevant to REDD+ and studied with various types of multi-spectral imagery. These parameters include aboveground live tree biomass (AGLB; e.g., Figure 12.1) or height, age (to estimate rates of C accumulation in regrowth), degradation, and forest type. In the chapter they use coarse-, medium-, and high-resolution imagery to quantify and model these parameters. They highlight the use of remote sensing methods, such as:

1. **Forest type or tree species community mapping,** which can be critical to REDD+ and is achieved over a range of resolutions with various classification methods. The detail at which these classes are mapped and accuracies at which they are mapped depend on the resolution of the imagery, the methods used, and the richness of data (e.g., how frequent the temporal images are or whether they include climate phenological extremes).
2. **Forest degradation studies,** which are conducted using multi-spectral imagery with pixel sizes less than 10 m or with Spectral Mixture Analysis (SMA) of multi-spectral imagery with coarser spatial resolution. SMA decomposes green vegetation (GV), non-photosynthetic vegetation (NPV), soil, and shade. Normalized Difference Fraction Index (NDFI) for forest degradation studies values range from 1 for intact forests and −1 for bare soil. Forest degradation is of many types that include roads and trails for selective logging, slash and burn agriculture, etc. that are detected using fine-resolution imagery.
3. **AGLB has been modeled and mapped (e.g., Figure 12.1) using spectral indices or bands of various multispectral sensors**. C stored in tree biomass is relatively accurate when summed over large areas, even though pixelwise estimates are somewhat uncertain because models underestimate AGLB at high biomass and overestimate it at low biomass.
4. **When image time series span the age range of regrowth forests, spectral data can precisely estimate forest height or AGLB.** Stand age can also be determined, which is required for estimating rates of C removal from the atmosphere in tree biomass.
5. **Fine spatial resolution imagery is used to characterize the distribution of canopy crown sizes,** which is then used to estimate AGLB.

There are emerging technologies driving paradigm changes in the nexus of ecology, remote sensing, and analytics, such as near-surface remote sensing and wireless sensor networks, arising from the eScience paradigm, which offers unique opportunities to integrate field observations at hypertemporal and -spatial resolutions (Sanchez-Azofeifa et al., 2017). Characterization and quantification of forest characteristics are advanced as illustrated for cross-scale phenology monitoring in a Central Amazon tropical evergreen forest by integrating PlanetScope with BRDF-adjusted MODIS (Wang et al., 2020). Without the cross-scale and without integrating multiple satellite data, there can be significant uncertainties in measured forest quantities. For example, tree heights of tropical forests in Ghana measured and compared using terrestrial laser scanning, airborne LiDAR scanning, stereo-photogrammetry (with imagery acquired by a RGB camera mounted on Uncrewed Aerial

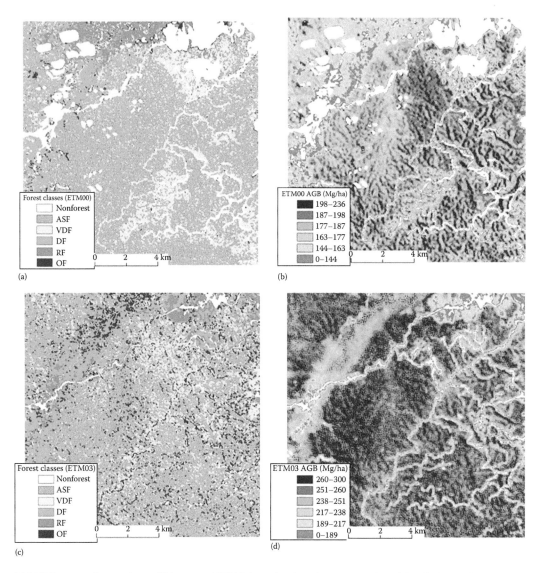

FIGURE 12.1 Comparison of bitemporal ETM data showing the subsets of (a) ETM 2000 land cover classification with the (b) ETM 2000 aboveground biomass (AGB) estimates (b) and the (c) ETM 2003 land cover map, as well as the AGB predictions of the radiometrically calibrated (d) ETM 2003 image.

Source: Wijaya et al., 2010.

Systems [UAS]), and reference to ground methods showed significant differences (Laurin et al., 2019). Recent studies have repeatedly shown the need for integrating multi-scale remote sensing to attain maximum accuracies, greatest precision of different forest quantities, and to limit uncertainties and errors. Nevertheless, this will also depend on the access to data and resources available. In assessing forest diversity and structure Ganivet and Bloomberg (2019) established the use of small (e.g., 20 × 50 m) plots with a 10-cm diameter at breast height (DBH) minimum measurement. In terms of remotely sensed techniques: (1) if funding is sufficient, airborne imagery seems the best regarding the quality of information (i.e., hyperspectral and hyperspatial imagery, LiDAR), and (2) if funding is limited, a cost-effective alternative providing reasonably accurate estimates would be the use of high-resolution satellite imagery such as WorldView. For studies where freely accessible data is the only possible option, Sentinel-2 can be used, although it is relatively coarser in terms of quality (Li et al., 2020). Recent forest studies are increasingly using machine learning (ML) and cloud computing. Jackson and Adam (2022) mapped deforestation and forest degradation due to selective logging using WorldView-3 multispectral data using random forest (RF) and support vector machine (SVM) with radial basis function kernel classifiers. Using the same approaches adopted in multiple forest remote sensing studies, Gupta and Sharma (2022) incorporated a continuous coverage of multi-spectral optical and synthetic aperture radar (SAR) along with sparsely global ecosystem dynamics investigation (GEDI) spaceborne Light Detection and Ranging (LiDAR) data in ML models for mapping mixed tropical forest canopy height in Gujarat, India. The combination of spectral data with geometric information from LiDAR improves the classification of tree species in a complex tropical forest, and these results can serve to inform management and conservation practices of these forest remnants (Pereira et al., 2023). Availability of time-series data from multiple remote sensing platforms (e.g., ground-based, airborne, spaceborne) along with advanced ML methods and cloud-computing platforms like GEE has enabled near real-time mapping of tropical forest disturbance (Kilbride et al., 2023) and in quantifying other forest characteristics.

12.2 LIDAR AND RADAR FOR FOREST INFORMATICS

LiDAR and radar are both active sensors with their own light sources and hence have the ability to acquire data during the day or night as well as offer better cloud penetration. Juha Hyyppä et al., in Chapter 2, provide an exhaustive state of the art on forest informatics assessed, modeled, and mapped by collecting 3D information using LiDAR and radar.

LiDAR acquires data by illuminating the surface with a laser in: (1) 600–1000 nm—inexpensive, but unsafe for the eyes; (2) 1550 nm—safe for the eyes but less accurate (note: this wavelength is widely applied and there are no accuracy problems); (3) 1064 nm—safe for the eyes but greater attenuation in water; (4) 532 nm—bathemetric applications such as to survey underwater terrain. The most common types of LiDAR data acquisition platforms are:

1. Terrestrial (ground-based) laser scanning (TLS), or terrestrial LiDAR
2. Mobile laser scanning (MLS), or mobile LiDAR
3. Airborne laser scanning (ALS), or airborne LiDAR
4. Spaceborne LiDAR missions (SLM). NASA's Geoscience Laser Altimeter System (GLAS) on board NASA's Ice, Cloud and land Elevation Satellite (ICESat) is the first spaceborne LiDAR. ICESat/GLAS acquired data from 2003 to 2009.

Radar data is collected in various bands by a number of spaceborne synthetic aperture radars (SAR) over many years as outlined here:

1. X-band (frequency: 12.5–8 GHz; wavelength: 2.4–3.75 cm). Satellites include TerraSAR-X, TanDEM-x, and COSMO-SkyMed. Data is heavily used in military reconnaissance and surveillance.

2. C-band (frequency: 8–4 GHz; wavelength: 3.75–7.5 cm). Radarsat, ERS. Sea-ice surveillance. Penetration of vegetation is limited to top layers.
3. S-band (frequency: 4–2 GHz; wavelength: 7.5–15 cm). It is used for meteorological applications (e.g., rainfall measurement).
4. L-band (frequency: 2–1 GHz; wavelength: 15–30 cm). Satellites include ALOS, PALSAR. It can penetrate vegetation, ice, and glacier studies.
5. P-band (frequency: 1–0.3 GHz; wavelength: 30–100 cm). Satellites include the ESA Explorer 7 (435 MGz). It has high biomass penetration.

An important advance is the ability to assess forest biomass and, in turn, C stocks and fluxes using data from the ESA's Explorer 7 that acquires data in P-band synthetic aperture polarimetric radar operating at 435 MHz. The early days of radar were limited to 2D application with the real advantage of cloud penetration, unlike optical sensors.

But as outlined by Juha Hyyppä et al. in Chapter 2, both LiDAR and radar provide point clouds (echo) that allow for creating 3D maps of trees and other vegetation. The forest informatics derived by the LiDAR and radar 3D point clouds (e.g., Figure 12.2) include parameters such as tree location, tree height, diameter at breast height (DBH), species, age, basal area, crown area, volume, biomass, and leaf area index (LAI). These forest variables from 3D point clouds are obtained through two approaches: (1) area-based approaches (ABAs) and (2) individual/single-tree detection approaches (ITDs). Regression models, neural networks, random forest are some of the methods used in ABAs and ITDs. The 3D data further help derive digital terrain model (DTM), digital surface model (DSM), and canopy height model, normalized digital surface model (CHM/nDSM).

Extensive discussion on how to derive forest informatics using LiDAR and Radar 3D point clouds, including methods and approaches used, strengths, and limitations, are presented and discussed in Chapter 2.

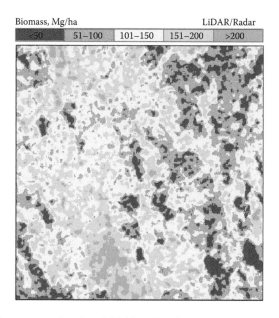

FIGURE 12.2 Forest biomass mapping from LiDAR and radar synergies. Biomass map from SRTM phase center height and PALSAR data developed from regression model using random biomass samples from LVIS-derived reference map. The image was smoothed using a 5-by-5 window (pixel size of 15 m).

Source: Sun et al., 2011.

The radar and LiDAR data used along with optical remote sensing data help address multiple issues in forest studies like penetrating cloud and haze, ability to provide height data, and increasing accuracies of forest classifications and their biophysical and biochemical quantification.

For example, the polarimetric radar, LiDAR, and near-IR passive optical sensing platforms in conjunction with physics-based models accurately estimate forest aboveground biomass (AGB) (Benson et al., 2021). Padalia et al. (2023) integrated GEDI LiDAR, Landsat-8, and ALOS-2 SAR data for improved AGB estimates of forests. Increasing use of artificial intelligence in determining forest informatics is evident in recent literature. Lee et al. (2020) estimated forest AGB from optical and radar satellite images using artificial neural networks (ANN). For this purpose, the eight input neurons of the forest-related layers, based on remote sensing data, were prepared: normalized difference vegetation index (NDVI), normalized difference water index (NDWI), NDVI texture, NDWI texture, average canopy height, standard deviation canopy height, and two types of coherence maps were created using the Kompsat-3 optical image, L-band ALOS PALSAR-1 radar images, digital surface model (DSM), and digital terrain model (DTM) (Lee et al., 2020). The fusion of hyperspectral data and LiDAR data and their integration with ML and DL algorithms can play an essential role in assessing biophysical and biochemical variables of forest species (Sharma et al., 2022). Fagua et al. (2019) integrated LiDAR and SAR data with multi-spectral Landsat data to improve the prediction and mapping of forest canopy height (CH) at high spatial resolution (30 m) in tropical forests in South America. Numerous recent studies have illustrated the strength of ML/DL/AI methods and tools in forest studies. Liu et al. (2021) developed a deep learning–based algorithm (Deep-RBN) that combined the fully connected network (FCN) deep learning algorithm with the optimized radial basis neural network (RBN) algorithm for forest structural parameter estimation using airborne LiDAR data.

12.3 HYPERSPECTRAL IMAGER (HSI) AND LIDAR DATA IN THE STUDY OF FOREST BIOPHYSICAL, BIOCHEMICAL, AND STRUCTURAL PROPERTIES

When remote sensing data are acquired in narrow bands and contiguously over a wavelength range representing a spectrum, it is called hyperspectral data. The number of bands itself is not as critical a factor, so 20 or 30, or 100 or 200, narrowbands (typically, ≤ 10 nm bandwidth) over 400–2500 nm is quite commonly used as hyperspectral data. When hyperspectral data are collected in an image format using ground-based, airborne, or spaceborne sensors, such data are called HSI. In contrast, LiDAR is an active sensor based on emitted laser pulses, which provides three-dimensional information in the form of laser point clouds, as enumerated in Chapter 3 by Dr. Asner et al., providing information on:

1. Biophysical properties
2. Biochemical properties
3. Canopy physiological properties
4. Canopy structural and carbon properties

HSI is ideal for the study of biophysical, biochemical, and physiological properties of vegetation (e.g., Figure 12.3). LiDAR is ideal for characterizing the structural and architectural properties of vegetation and for advancing biomass assessments. Dr. Gregory Asner et al. in Chapter 3 discuss all of these and show us the advances one can make in studying these features. Biophysical variables include LAI (m^2/m^2), biomass (kg/m^2), equivalent water thickness (EWT; mm), and leaf mass area (LMA; g m^{-2}). Biochemical properties include nitrogen (N), cellulose, lignin, pigments (e.g., chlorophyll a, b, total; anthocyanins; carotenoids), photosynthetically active radiation (PAR), absorbed photosynthetically active radiation (APAR), and light use efficiency (LUE). These are studied using HSI-derived hyperspectral vegetation indices (HVIs; see Thenkabail et al., 2014, 2013), radiative transfer models such as PROSPECT (Jacquemoud et al., 2009), and empirical spectroscopic algorithms (Asner and Martin, 2008). Specific HVIs can be applied to study specific properties, e.g., biomass or LAI using narrow spectral bands centered at 680 nm and 910 nm, or LUE using Photochemical Reflectance Index (PRI; 531 and 570 nm). In contrast, LiDAR is often used to study tree heights, AGB of vegetation, forest structure, and aboveground carbon density (ACD). Research is still in progress on establishing the accuracies of LiDAR-estimated ACD with that of plot-based measurements. An important part of Chapter 3 is the illustration and enumeration of advances one can make in improved understanding of structural, architectural, biophysical, biochemical, and ACD characteristics of forests by integrating HSI and LiDAR data.

The forest biophysical and biochemical trait accuracies are improved by advanced remote sensing data, methods, and approaches. For example, acquiring hyperspectral data combined with high-density, three-dimensional (3D) information from LiDAR, both acquired from an uncrewed aerial system (UAS), to help studies pertaining to physiological activities of forest canopy for carbon accumulation estimation as well as precision forestry applications such as nutrition diagnosis, water regulation, and subsequent productivity enhancement of forest systems (Shen et al., 2020). Imaging spectroscopy data (284 bands, 380 nm–2500 nm) acquired by the Airborne Prism Experiment (APEX) spectrometer with a spatial resolution of 3 m × 3 m, and airborne discrete return LiDAR data with an average point density of 23 points/m^2 substantially improved forest canopy N and Phosphorus (P) concentrations in a study in lowland forests located in the humid temperate climate zone (Ewald et al., 2018). Generally, the fusion of hyperspectral (spectral) and LiDAR (structural) data is ideal for gathering biophysical and biochemical traits of plants and helps improve tree species classification (Shi et al., 2018). New forest mapping methods determine forest quantities using varied remote sensing data and ML and AI methods. Aksoy et al. (2023) used random forest (RF), eXtreme Gradient Boosting (XGBoost) ML models as well as neural networks (NN) AI-based models to estimate the stem volume (V), AGB, basal area (B), tree height (H), stem diameter (D), and forest stand age (A) and found: (1) the XGBoost ML algorithm outperformed RF 1–3% in the R^2 metric and (2) the NN model was superior in the estimation of V, AGB, and B parameters.

FIGURE 12.3 Demonstration of a virtual active hyperspectral LiDAR in automated point cloud classification. Hyperspectrally classified point cloud visualized from two viewing directions. The background points were left out, and only the needle (green) and trunk (brown) points are plotted. The red arrow points toward the point of measurement.

Source: Suomalainen et al., 2011.

12.4 TREE AND STAND HEIGHTS FROM OPTICAL REMOTE SENSING

Over the years, remote sensing has been used to map and monitor forests in terms of their cover, type, distribution, species dominance, deforestation, tree crown, biomass, LAI, stand area, and a host of other parameters, including forest health and change over time. But almost all of these remote sensing measurements have been two-dimensional. The National Aeronautics and Space Administration (NASA) created the first 3D global map of forest heights using data from the Geoscience Laser Altimeter System (GLAS) on the Ice, Cloud, and land Elevation Satellite (GLAS/ICESat), MODIS, and TRMM (Lefsky, 2010). Greater accuracies and lesser uncertainties in forest biomass and C estimates are feasible through improved accuracies in the measurement of tree heights. Traditionally, tree heights are measured by plot sampling in the field. This is extremely tedious, difficult in complex forests due to inaccessibility, and resource prohibitory to cover the forests of the world, which make up about 30% of the terrestrial area. Further, since rapid changes occur in forests, particularly due to anthropogenic activities, repeated measurement of these changes are required.

Chapter 4 by Dr. Sylvie Durrieu, Dr. Cédric Vega, et al. shows us the approaches and methods of estimating tree heights through 3D vertical measurement using digital photogrammetry and, more recently (and increasingly), through LiDAR remote sensing. The main advantage of LiDAR is that it sees all the vegetation within the plot, whereas field-plot data may describe only a sample of trees in a plot for the sake of cost-effectiveness, and digital photogrammetry points see only dominant vegetation. Dr. Sylvie Durrieu, Dr. Cédric Vega, et al. present and discuss in detail:

1. The principles of height measurement using LiDAR and photogrammetry.
2. Stand-level height assessments made through area-based approaches that include: (1) field inventory at the plot level of forest parameters, (2) extracting LiDAR point clouds of the forest parameters for inventoried plots, (3) establishing empirical models linking LiDAR data with field-plot data for each forest parameter, and (4) extrapolating the empirical models over the entire forest, leading to large-area inventories of forest parameters. The process involves developing models relating LiDAR or photogrammetric 3D data *versus* field-plot data of forest characteristics (e.g., tree height, basal area) and applying the same over larger forest areas. For greater accuracies, large-area inventories require segmenting forest types into distinct categories and developing the stand-level models for each of these forest categories separately.
4. Individual tree height assessment through: (1) raster-based approaches (e.g., canopy height modeling, detecting tree apices, measuring tree crowns), (2) point-based approaches, and (3) hybrid approaches. For tree height modeling using a raster-based approach the tree crown data may or may not be required, but canopy height modeling (CHM) and detecting tree apices are required. The point-based approach detects not only the dominant trees (apices) but also over-topped trees taking advantage of the ALS 3D point cloud (e.g., Figure 12.4). Hybrid approaches, combining raster-based and point-based approaches, have shown improved individual-tree extraction.

Optical images from various satellites remain the backbone of forest trait studies. Often high-resolution optical stereo images are used for tree height estimation. Wang et al. (2021) produced the forest tree height maps and AGB products based on data from Ziyuan-3 satellite (ZY-3) stereo images combined with a digital elevation model (DEM) obtained from Advanced Land Observing Satellite (ALOS) and Sentinel-2 multi-spectral images. Terrestrial laser scanning (TLS), airborne LiDAR Scanning (LiDAR), and stereo-photogrammetry and field-plot reference collected through ground surveys form the essential components of tree height and AGB estimations (Laurin et al., 2019). They found that tree heights were overestimated by ground methods and underestimated by TLS, ALS, and UAV. They also found that TLS provides robust estimates, including diameter at breast height (DBH), simultaneous use of different methods helps correct estimates, and height uncertainty

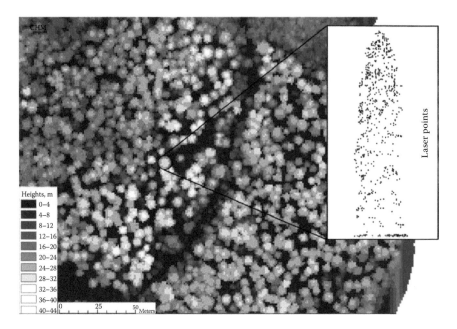

FIGURE 12.4 Single-tree biomass modeling using airborne laser scanning (ALS). Example of ALS points inside one tree canopy segment and of CHM with 0.5-m grid size.

Source: Kankare et al., 2013.

causes 6–10% error in AGB. Yang et al. (2022) proposed a Sentinel-2 optical data fused with LiDAR canopy height data as best for AGB estimations. Lee and Lee (2018) showed various ways of generating a forest height map using limited data. In the absence of LiDAR and X- and P-band SAR data, they determined forest height data using common remote data, such as discrete-return LiDAR data, SRTM, satellite L-band SAR data, and Sentinel-2 optical data. Illarionova et al. (2022) used airplane-based LiDAR data and trained a deep neural network to predict the vegetation height.

12.5 STUDY OF BIODIVERSITY FROM SPACE

Chapter 5 by Dr. Thomas Gillespie et al. provides a lucid outline on how to use remote sensing data from space in biodiversity studies. They approach this by looking at three categories of biodiversity assessment using optical, radar, LiDAR, and thermal remote sensing data from space. The three categories are:

1. Mapping
2. Modeling
3. Monitoring

Mapping presentation and discussions include vegetation categories and invasive species. Mapping of vegetation categories and broad habitats can typically be performed using 30-m Landsat or better resolution. However, accurate and detailed mapping of species or individual trees requires hyperspectral data (5 m or better) from sensors such as QuickBird, GeoEye, and IKONOS and/or hyperspectral data. Ability to map animals from space is limited due to lack of coverage of frequent images at sufficiently high spatial resolution and data from specific spectral bands such as thermal, which can detect body heat from animals if data is within a meter or so.

Forests, Biodiversity, Ecology, LULC, and Carbon

Spatial modeling for biodiversity studies such as understanding species richness, ecosystem richness for habitats, and carrying capacity will be a powerful tool. This would involve using a wide array of spatial data, such as precipitation, land use/cover, soils, and elevation, and then performing spatial models for planning and decision-making processes (e.g., which lands to conserve, where the richest habitats for biodiversity are).

Monitoring biodiversity is important for understanding factors such as habitat loss or degradation, productivity changes, and for assessing development and conservation.

Many biodiversity studies, like identifying species or trees, require hyperspatial data that is also hyperspectral. Other biodiversity studies, like habitat mapping, require more temporal data at moderate resolutions, like Landsat 30 m or MODIS time series (e.g., Figure 12.5). Monitoring animals will require hyperspatial data that is possibly thermal as well. In Chapter 5, Dr. Thomas Gillespie et al. provide the remote sensing data characteristics needed for biodiversity studies and present the state of the art in mapping, modeling, and monitoring biodiversity.

Essential Biodiversity Variables (EBVs) have been suggested to harmonize biodiversity monitoring worldwide, and many of these can be monitored from space using Earth observation (EO) data. A priority list of biodiversity metrics to observe from space are provided by Skidmore et al. (2021). Wang and Gamon (2019) summarize the pros and cons of different methods in remote sensing of

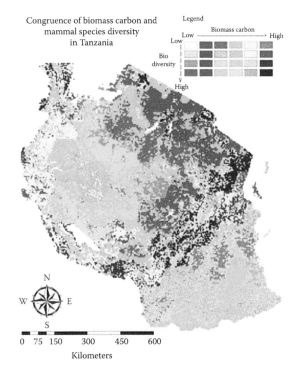

FIGURE 12.5 A framework for integrating biodiversity concerns into national REDD+ programs. Example national-scale map for Tanzania displaying congruence values between carbon and biodiversity at the scale of a 5-km grid and across all vegetation types. Map generated using freely available land cover data from MODIS, mammal data from the freely available African mammal databank (African Mammals Databank [AMD]) (IEA, 1998), and African carbon data provided by UNEP-WCMC, based on multiple sources (Gardner et al., 2012). This kind of simple overlay map can help in identifying those areas of both high opportunity (strong positive correlation in carbon and biodiversity values) and risk (low in carbon but high in biodiversity) in the REDD+ planning process.

Source: Gardner et al., 2012.

plant biodiversity and outline the major gaps in global biodiversity monitoring systems. Fusing LiDAR and hyperspectral remote sensing data has greater potential of predicting taxonomic, functional, and phylogenetic tree diversity in temperate forests (Kamoske et al., 2022). Hyperspectral data is specifically valuable in advancing biodiversity studies of specific categories. For example, Papp et al. (2021) characterized and mapped of invasive species and achieved at 92 accuracies using support vector machine (SVM) and 99% accuracies using artificial neural network (ANN) classification algorithms. The models were trained using highly accurate field reference data.

12.6 MULTI-SCALE HABITAT MAPPING AND MONITORING USING SATELLITE DATA AND ADVANCED IMAGE ANALYSIS TECHNIQUES

There are wide-ranging habitats on the planet that house plants and animals, such as forests, savannas, deserts, and wetlands. The ability of EO data to map and study habitats varies widely depending on the detail with which a habitat needs to be mapped and the basic characteristics of remote sensing data, like their spatial, spectral, and temporal resolutions. If the need of mapping is to discern a single species of tree or shrub or grass, the requirement of spatial resolution could be submeter to a few meters. If the need of habitat mapping is to get a broad understanding of the density of forest cover, then coarse resolution, like 30 m or 250 m, may suffice. However, if the goal of the habitat mapping is to get a broad understanding of habitat land cover over vast areas, even 1-km data that is more temporally rich maybe needed.

In Chapter 6, Dr. Stefan Lang et al. provide habitat mapping protocols using a wide array of EO data. The biodiversity of each habitat defines the richness of plant and animal species contained in these habitats. They begin with providing the importance of habitat studies by referring to Group on Earth Observation (GEO), identifying biodiversity as one of the nine societal beneficial areas. Central to their chapter are the strategy, approaches, and methods adopted in two major European Union projects: MS.MONINA (Multi-scale Service for Monitoring NATURA 2000 Habitats of European Community Interest) and BIO_SOS (Biodiversity Multi-Source Monitoring System) (Spanhove et al., 2014). For example, as highlighted in Chapter 6, these projects use:

1. Very high spatial resolution imagery (VHRI) from sensors like WorldView-2 for fine-scale habitat mapping
2. Imaging spectroscopy (hyperspectral) data from CHRIS/PROBA or Hyperion for plant biophysical and biochemical properties
3. LiDAR data from ICESat/GLAS to determine tree height and 3D biomass
4. X-band radar for from fine-resolution sensors like TerraSAR-X to differentiate plant species
5. Thermal VHRI can be used to even count the number of cows in rangelands

However, in order to map fine details of habitats, such as individual species, one may require a combination of hyperspatial, hyperspectral, and other data (e.g., bathymetry) analyzed with an ensemble of algorithms (e.g., Figure 12.6).

Chapter 6 provides a sensor suitability table showing what forest, grassland, heathland, and wetland habitat variables are mapped and at what detail by various low-, medium-, very high-, hyperspectral-, laser-, and microwave sensors.

Habitat mapping often uses land cover maps that are coarse resolution (e.g., MODIS 250 m or higher) and only provide a broad guidance. For specific monitoring and management of fine-scale habitats (Landsat 30 m or better), habitat maps are required. Mondal et al. (2020) developed a multi-scale reporting framework for the United Nations' Sustainable Development Goal 15 (SDG 15) (Mondal et al., 2020) using MODIS (250–1000 m), Landsat (30 m), and Sentinel (10 m) data. They infer: (1) coarse-scale satellite data (≥ 1 km) might not capture subtle forest degradation, (2)

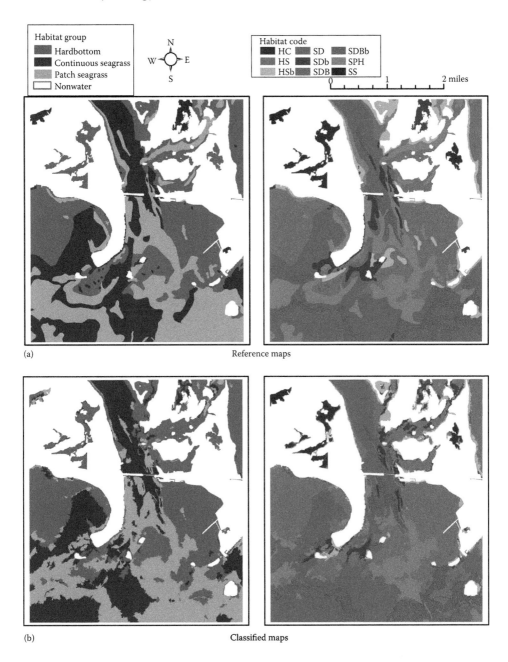

FIGURE 12.6 Habitat maps from (a) reference data and (b) classification results of the fused dataset (hyperspectral imagery, aerial photography, and bathymetry data) and ensemble analysis of random forest (RF), support vector machines (SVMs), and k-nearest neighbor (k-NN). Six code-level habitats were observed: HC (soft coral, hard coral, sponge, and algae hardbottom), HS (hardbottom with perceptible seagrass [<50%]), SD (moderate to dense, continuous beds of seagrass), SDB (moderate to dense nearly continuous beds [seagrass > 50%], with blowouts and/or sand or mud patches), SPH (dense patches of seagrass [> 50%] in a matrix of hardbottom), and SS (sparse continuous beds of seagrass).

Source: Zhang, 2014.

high-resolution data (≤ 10 m) must be used to assess current forest condition, and (3) a multi-step framework should be employed for accurate reporting of SDG 15 (Mondal et al., 2020). Oeser et al. (2019) highlight the considerable potential of Landsat-based spectral temporal metrics for assessing wildlife habitat. Jung et al. (2020) provides a global, spatially explicit characterization of 47 terrestrial habitat types, as defined in the International Union for Conservation of Nature (IUCN) habitat classification scheme, which is widely used in ecological analyses, including for quantifying species' area of habitat. They produced this novel habitat map for the year 2015 by creating a global decision tree that intersects the best currently available global data on land cover, climate, and land use.

12.7 ECOLOGICAL CHARACTERIZATION OF VEGETATION USING MULTI-SENSOR REMOTE SENSING

Launch of optical sensors, starting with the Soviet Union's Sputnik and NOAA AVHRR, changed our view of the world and how we study Planet Earth. In Chapter 7, Dr. Conghe Song et al. provide an exhaustive series of optical satellites, their brief history and characteristics, and their value in studying vegetation. These satellites provide data in distinct wavebands and have unique spectral, spatial, radiometric, and temporal coverage of the entire planet. Hence, quantifying, modeling, and mapping of vegetation from remote sensing became widespread, especially with the launch of the first Landsat in 1972. Dr. Conghe Song et al. group vegetation characterization into two broad categories:

1. Vegetation structure that includes:
   a. Vegetation cover
   b. Forest successional stages
   c. Leaf area index (LAI)
   d. Biomass and net primary productivity (NPP)
2. Vegetation functions that include:
   a. Land surface phenology (LSP)
   b. Fraction of absorbed photosynthetically active radiation (fAPAR)
   c. Chlorophyll
   d. Light use efficiency (LUE)
   e. Gross Primary Productivity (GPP)/NPP

Chapter 7 shows us:
1. Vegetation cover modeling using various techniques that include statistical regression with vegetation indices, classifications, and spectral mixture analysis (SMAs)
2. Forest successional stages characterized by physically based models, empirical models, and change detection approaches
3. LAI algorithms based on VIs and radiative transfer models.
4. Biomass and NPP through regression models, k-NN algorithms, machine learning algorithms, and biophysical approaches
5. LSP through VIs
6. fAPAR through empirical models involving VIs and biophysical models
7. Chlorophyll assessment through VIs and radiative transfer models
8. LUE through PRI
9. GPP and NPP based on LUE, and other process-based models using remotely sensed data as inputs

As we can clearly see, this is an exhaustive list of vegetation parameters that are used in a wide array of global and local studies, such as primary productivity of GPP and NPP, understanding ecosystems, assessing degradation, and changes over space and time (e.g., Figure 12.7). These products

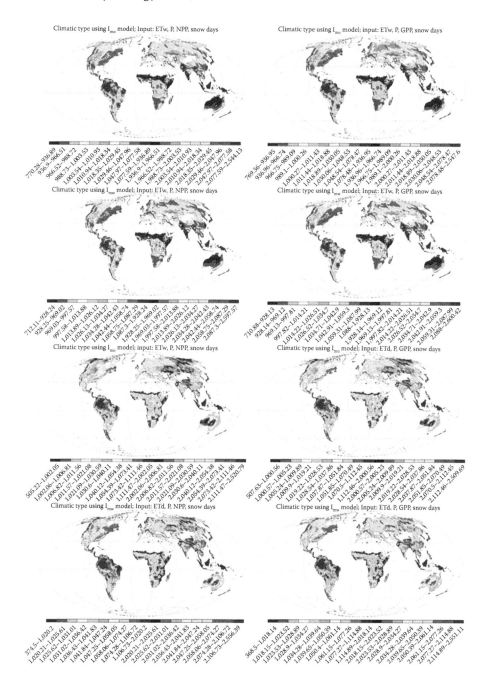

FIGURE 12.7 Terrestrial Earth couple climate–carbon spatial variability and uncertainty. Climatic types influenced by NPP MODIS (left-hand) and GPP MODIS (right-hand).

Source: Alves et al., 2013.

are often the most accurate data on vegetation and their characteristics that feed into global change models and climate models.

Multi-sensor data from a wide array of satellites will help overcome data gaps, provide more frequent temporal data, and help address the needs of vegetation characterization at various spatial, spectral, temporal, and radiometric scales. For example: (1) high spatial scales are needed to study individual plant species or individual plants of plant communities; (2) high spectral data is needed to establish relationships of specific biophysical and biochemical quantities, such as plant pigments, chlorophyll, plant water content, and lignin; (3) high temporal data is needed to understand specific events like drought intensity and progression at various times throughout the growing season; and (4) high radiometric data is needed to overcome sensor data saturation at high biomass or growth levels. Two factors need to be noted when using data from within and between sensors. First, ensure that all sensors are calibrated over time to ensure any sensor degradation is well accounted for. Second, inter-sensor relationships are developed to ensure that the relationships across sensors are well understood. Campbell et al. (2020) showed that structural/spectral characteristics of woodlands that pose challenges to mapping mortality can be overcome by fusing UAS imagery, airborne LiDAR, and Landsat time series. Chen et al. (2022) demonstrated the ability of multi-sensor remote sensing to significantly improve forest AGB. This they demonstrated using satellite LiDAR data from GEDI and ICESat-2, and images of ALOS-2 yearly mosaic L-band SAR (synthetic aperture radar), Sentinel-1 C-band SAR, Sentinel-2 MSI, and ALOS-1 DSM that were combined for pixel- and object-based forest AGB mapping in a vital heterogeneous mountainous forest. These multi-sensor studies have become common given the strong inter-sensor calibration products available, such as the Harmonized Landsat-8 and Sentinel-2 (HLS) (Masek et al., 2022; Masek et al., 2021; Claverie et al., 2018).

12.8 LAND COVER CHANGE DETECTION

Remote sensing is an ideal way to observe, quantify, and monitor land cover and land cover changes (LCLCC) over space and time. The advantage remote sensing offers is repeated coverage, synoptic views over large areas, global coverage, and the ability to study LCLCC in different resolutions or scales and using consistent data over time. Remote sensing does not directly provide information on land use. But it is inferred from land cover. So, land cover and land use (LCLU) studies are widespread using a plethora of remotely sensed data. Many types of LCLU applications are possible through remote sensing. These applications involve forests, grasslands, croplands, and so forth. The degree of detail one can study in LCLU and LCLU change (LCLUC) will depend on the characteristics of remotely sensed data—their spatial, spectral, radiometric, and temporal resolution. Also, the degree of detail depends on methods, techniques, and approaches used in classifying and synthesizing. All these factors influence the details at which LCLU and LULUC is mapped, and their accuracies achieved (e.g., Figure 12.8).

Chapter 8 by John Rogan and Nathan Mietkiewicz provides a background on the importance of land cover change detection studies, outlines the theory and practice, enumerates on trends of change detection studies, outlines and discusses methods and approaches, and provides an assessment of map accuracy strategies. They show us how the 40+ years of spaceborne remote sensing archives, from various sensors such as 1–10 km AVHRR, 250–1000 m MODIS, and 30 m Landsat, are helping us study and understand land cover trends in any part of the world. For example, there is now a monthly continuous record of AVHRR Global Inventory Modeling and Mapping Studies (GIMMS) NDVI data for 40 years (1981–2022) (Pinzon et al., 2023), and the 10+ years' record of Landsat-8 and -9, Sentinel-2, and NASA's Harmonized Landsat Sentinel-2 (HLS) Landsat product (HLSL30) for 2013–present and HLS S2 product (HLSS30) for 2015–present (Masek et al., 2022). Such long-term records have enabled land cover studies in various spatial, spectral, temporal, and radiometric resolutions over several decades.

Chapter 8 reviews the three types of change detection approaches: (1) monotemporal change detection, where only a single image of an area is involved; (2) bitemporal change detection, where two images on two distinct dates of an area are involved; and (3) temporal trend analysis, where continuous series of images (e.g., monthly maximum value composites over one or more years) of an area are involved. The authors compare several automated change detection approaches that include:

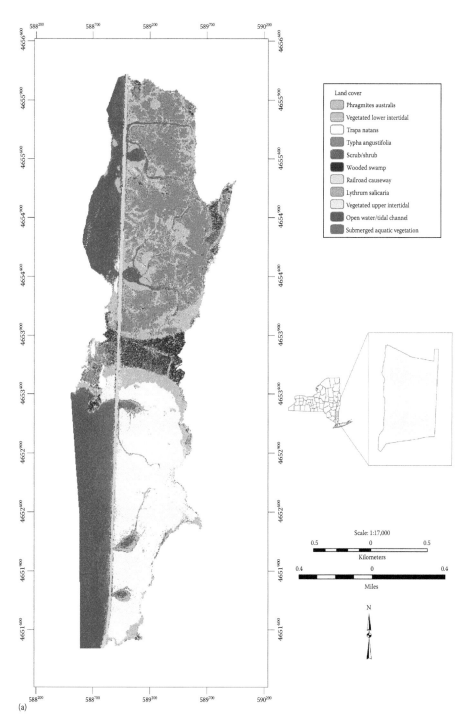

FIGURE 12.8 Tivoli Bays land cover map produced using IKONOS data with two methods: (a) Method 1 relied solely on the four spectral bands (blue, green, red, and near infrared) of the IKONOS image; (b) Method 2 used a maximum-likelihood classification of the four spectral bands, supplemented by local texture information (variance) calculated in a moving 3-by-3 pixel (Method 1) or 5-by-5 pixel (Method 2) window, superposed separately on each band of the IKONOS image. Ultimately, eight bands were used in the classifications for Method 2.

Source: Laba et al., 2010.

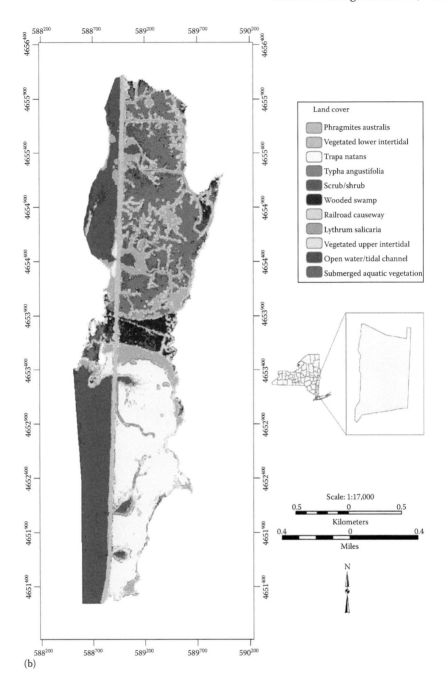

FIGURE 12.8 (*Continued*)

(1) disturbance index, (2) MODIS global disturbance index (Gao et al., 2021), (3) CLASlite (a forest cover automated change detection algorithm) (Redowan et al., 2020), (4) vegetation change tracker, and (5) Spatial Temporal Adaptive Algorithm for mapping Reflectance Change (STAARCH) (Li et al., 2020). They demonstrate the CLASlite method of change detection for three study areas (rural, urban, and coastal) using Landsat images for eight eras (1985–1993, 1993–1995, 1995–1999, 1999–2002, 2002–2009, 2009–2010, 2010–2011, and 2011–2013) in central Massachusetts. The CLASlite

method partitions each scene into proportional fractional cover types of bare ground (B), photosynthetic vegetation (PV), and non-photosynthetic vegetation (NPV) for every pixel. They (Redowan et al., 2020) use very high spatial resolution data (sub-meter to 5 m) obtained from Google Earth as reference data to establish forest versus non-forest class accuracies using classic error matrices.

The increase in number of EO satellites and frequencies of data acquisitions has changed the way change detection is conducted. Zhu (2017) outlined four aspects of change detection: (1) frequencies (e.g., 2–3 days with the availability of HLS data, Masek et al., 2022; Masek et al., 2021; Masek et al., 2018), (2) preprocessing (e.g., atmospheric correction, cloud and cloud shadow detection, and composite/fusion/metrics techniques), (3) algorithms (e.g., thresholding, differencing, segmentation, trajectory classification, statistical boundary, and regression), and (4) applications (e.g., change target and change agent detection). Nevertheless, land cover and land cover change are best studied using multiple sensor data. De Alban et al. (2018) showed that the combined Landsat and L-band SAR data, specifically Japan Earth Resources Satellite (JERS-1) and Advanced Land Observing Satellite-2 Phased Array L-band Synthetic Aperture Radar-2 (ALOS-2/PALSAR-2) to map and quantify land use/cover change transitions between 1995 and 2015 in the Tanintharyi Region of Myanmar using random forests ML classifier produced best overall classification outperforming the accuracies obtained from individual sensors. Polykretis et al. (2020) explored the use of five spectral indices (normalized difference vegetation index [NDVI], soil adjusted vegetation index [SAVI], albedo, bare soil index [BSI], tasseled cap greenness [TCG], and tasseled cap brightness [TCB]) in change vector analysis (CVA) in Crete Island, Greece, and found NDVI and albedo were found to provide superior results against the other combinations. The web-based remote sensing platform Google Earth Engine (GEE) is increasingly used in change detection studies (Thenkabail et al., 2021). Nevertheless, there is a paradigm shift away from change detection, typically using two points in time, to monitoring or an attempt to track change continuously in time (Woodcock et al., 2020). This has been made possible by freely available and web-accessible remote sensing data at various resolutions and in frequent time periods from Landsats, Sentinels, MODIS, and even from sub-meter to 5-m very high spatial resolution imagery (VHRI) from satellites like Planet Labs' Doves and Super Doves.

12.9 RADAR REMOTE SENSING IN LAND USE AND LAND COVER MAPPING AND CHANGE DETECTION

Radar remote sensing has some unique features, such as cloud penetration and all-day imaging ability (since it is an active sensor), when compared with optical remote sensing. Radar data is acquired over 0.3–100 cm wavelength, in one or more of the four polarizations (HH, HV, VH, and VV), three modes (SpotLight, StripMap, and ScanSAR), various incident angles, repeat frequency, and resolutions (range and azimuth). Radar data is also processed at various levels: slant range data, ground range data, and geocoded and orthorectified data. How the radar data is acquired and processed is important in determining what applications these data are used for. An application such as change detection and interferometry requires radar data acquisition to have identical parameters (e.g., orbit, incidence angle, and polarization). In Chapter 9, Dr. Zhixin Qi and Dr. Anthony Gar-On Yeh discuss a number of applications of radar data that include land use and land cover (LULC) classification, forest inventory and mapping, crop and vegetation identification, urban environments, snow and ice, and a number of others. These studies are reported based on data acquired in various frequencies and wavelengths by a wide array of spaceborne SAR sensors, such as Envisat ASAR, TanDEM-X, TerraSAR-X, RADARSAT constellation, ERS, JERS, ALOS PALSAR, and Cosmo-Sky-MED. Chapter 9 highlights the following strengths of SAR data in various applications:

1. LULC classification: wide array of SAR data has been used in classifying forest types (e.g., primary, secondary, slash and burn agriculture, regrowth or regenerative forests, plantations) or LULC classes (e.g., cropped areas, bare soil areas, forestry, forest clear-cut,

forest burnt areas, water bodies) (Rajat et al., 2023). A large number of LULC classes, as mentioned, are, for example, mapped when a single-date ALOS PALSAR fine-/dual-beam is combined with a multi-temporal ENVISAT ASAR. Some studies have shown overall classification accuracies as high as 87% for several land cover classes such as built-up areas, water, barren land, crop/natural vegetation, lawn, banana fields, and forests by combining the SAR textural, polarimetric, and interferometric information extracted from RADARSAT-2 PolSAR images. Combining multiple frequency SAR scenes (e.g., L-band, P-band, PolSAR, PolInSAR), and fusing them provided greater accuracies in LULC classification than using single-frequency SAR images. Numerous studies have reported significant improvement in classification accuracies when radar data is combined with optical data (Jafarzadeh and Attarchi, 2023).
2. Forest species identification has been successfully performed using SAR data with multiple incident angles and variation of backscatter coefficient in various incident angle.
3. Crop classification: crops such as corn, soybeans, cereals, and hay pasture are classified with 70–89% accuracy with SAR data (Mirzaei et al., 2023); with accuracies increasing with increasing number of temporal coverage and multiple frequencies.
4. Biophysical characterization of forests has been conducted using interferometric coherence maps derived from ERS-1 and ERS-2 SAR images and from JERS-1 SAR images.
5. Urban applications: increased accuracies in urban mapping were possible when very high resolution optical imagery (e.g., QuickBird) was combined with SAR data.

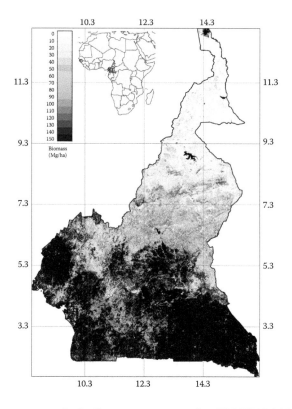

FIGURE 12.9 Biomass assessment in the Cameroon savanna using ALOS PALSAR data. Pixels saturate at 150 Mg.ha^{-1}. Dense forest classes were masked out using the GlobCover 2009 land cover map (Mermoz et al., 2014). The figure shows the north-south AGB gradient.

Source: Mermoz et al., 2014.

Radar data has also been used extensively for AGB estimations and for carbon stock assessments (e.g., Figure 12.9). However, radar data has high noise and large geometric distortion relative to optical imagery; hence, it requires its own specialized algorithms to process data. This is nowhere as developed as for optical sensors.

Radar data can provide complementary/supplementary information to optical remote sensing to advance our understanding and better map, model, and monitor land themes. So, wherever feasible, it is better to use optical and radar data to complement/supplement information of each sensor type.

The SAR images have great significance for change detection by providing data that can capture all-weather and all-time polarization information. The commonly available information in satellite SAR is amplitude and phase information, and change detection techniques have been developed based on each technology (Baek and Jung, 2019): (1) amplitude change detection (ACD) and (2) coherence change detection (CCD). Convolutional neural network (CNN) methods for SAR image change detection are increasingly implemented (Li et al., 2019). The role of Sentinel-1 SAR (amplitude) backscattered signal variations for change detection analyses when a natural (e.g., fire, flash flood) or human-induced (e.g., disastrous war situations, large-area forest clearance) events were established using random forest ML algorithms, which quickly helped assess binary decisions (changed/unchanged) (Mastro et al., 2022). In Sentinel-1 SAR images, forest clearings exhibit reduced backscatter as well as increased interferometric coherence (Durieux et al., 2019). Mastro et al. (2022) and Durieux et al. (2019) leveraged Sentinel-1 SAR data to monitor deforestation and implied that the SAR data has potential to achieve better performance in forest disturbance monitoring than Global Forest Watch (Zhang et al., 2020), the current Landsat-based gold standard. Nevertheless, there are numerous studies (e.g., Kluczek et al., 2024; Luo et al., 2023) that have shown the combined use of multiple satellite data types (e.g., optical, radar, LiDAR, hyperspectral) to best study change detection.

12.10 GLOBAL CARBON BUDGETS AND REMOTE SENSING

The global carbon budget consists of four terms: atmosphere, land, oceans, and fossil fuels, as presented by Dr. Richard Houghton in Chapter 10. The long-term net flux of carbon between terrestrial ecosystems and the atmosphere has been dominated by two factors: (1) changes in the total area of forests and (2) per-hectare changes in forest biomass resulting from management and regrowth. Apart from regional-level uncertainties, the carbon flux of tropical forests is greatly influenced by uncertainty in the regenerative capacity of forests and in harvest and management policies.

Chapter 10 highlights the need to keep track of global carbon budget annually to determine how much carbon is emitted to the atmosphere, how much is absorbed by land and oceans, and how much stays in the atmosphere. Currently, of the 32 billion tons of C emitted into the atmosphere each year due to anthropogenic activity, tropical forests sequester about 4.25 billion tons, soils and other vegetation another 4.25 billion tons, and oceans 8.5 billion tons, leaving the residual 15 billion tons in the atmosphere (Lewis et al., 2009). Also, land use change, mainly from deforestation in the tropics, is responsible for estimated net emissions of about 6 billion tons of greenhouse gases—greater than the emissions from all the world's planes, ships, trucks, and cars (Lewis et al., 2009). Dr. Richard Houghton points out that the fraction of C that remains in the atmosphere has been remarkably constant over the last 50 years; with the increase in emissions compensated by an increase in sinks on land and in the oceans. Estimates of C storage in terrestrial ecosystems are still very approximate. For example, Lewis et al. (2009) reports a wide range (0.29–0.66 Mg C ha^{-1}yr^{-1}) of tropical forest C storage. The Wet Tropical Asian Bioregion forests, for example, contain high C density of up to 500 Mg/ha (Lasco, 2004) but are changing rapidly due to selective logging, forest conversion to agriculture resulting in C density of less than 40 Mg/ha, and conversion to plantations (agroforests), which are responsible for at least a 50% decline in forest C density (Gómez and Ellis, 2023).

In order to track carbon sources and sinks from the land, first, Chapter 10 discusses the "bookkeeping model" of the early days, which uses annual rates of land cover change and standard growth and decomposition rates per hectare to calculate annual changes in carbon pools as a result of management. This highly aggregated approach did not use remote sensing as input but was based on statistics available from national and international sources, which were subjective and approximate.

The current approach to global carbon budgets enumerated by Dr. Houghton in Chapter 10 is based on "two broad types of explanatory mechanisms [that] account for the loss and accumulation of carbon on land: (1) disturbances and recovery (structural mechanisms) and (2) the differential effects of environmental change (e.g., CO_2, N deposition, climate) on photosynthesis and respiration (metabolic mechanisms)." This approach uses significant remote sensing. The flux of carbon from land use and land cover change (LULCC) is based on disturbances and recovery, especially from medium resolution (30–100 m). Analyses include:

1. Rates of change in forest area
2. Biomass density
3. Measurement of changes in carbon density

Houghton presents an example of estimating the flux of carbon from LULCC, taking the UN Framework Convention on Climate Change (UNFCCC) Reducing Emissions from Deforestation, forest Degradation, and Conservation (REDD+) (Morita and Matsumoto, 2023). Under this mechanism, when a country reduces emissions, it is eligible for carbon credits. The nine issues inherent in estimating the flux of carbon from LULCC outlined by Dr. Richard Houghton in Chapter 10 are:

1. Definitions
2. Assigning a carbon density to the deforested areas
3. Committed versus actual emissions
4. Gross and net emissions of carbon from LULCC
5. Initial conditions
6. Full carbon accounting
7. Accuracy and precision
8. Attrition
9. Uncertainties

Opportunities to significantly improve estimates of C storage and flux through improved estimates of land use classes (LUC) and modeling are possible with the evolution in spaceborne hyperspectral, hyperspatial, and advanced multi-spectral sensors, as a result of improvements in the spatial, spectral, radiometric, and temporal properties as well as in the optics and signal-to-noise ratio of data (e.g., Figure 12.10). High spatial resolution allows location, while high spectral resolution allows identification of features. Hyperspectral remote sensing sensors allow direct measurement of canopy chemical content (e.g., chlorophyll, N), forest species, chemistry distribution, timber volumes, and water, and improved biophysical and yield characteristics. Thenkabail et al. (2004a) demonstrated an increased accuracy of about 30% in LUC and biomass when 30 hyperspectral wavebands were used relative to six non-thermal Landsat TM bands. Hyperspatial data have demonstrated the ability to extract individual tree crowns from 1-m panchromatic data. Agroforest successional stages have been mapped, and their varying carbon sink strengths assessed using IKONOS (Thenkabail et al., 2004b). In contrast, forest structure variables (e.g., biomass, LAI) are poorly predicted by the older generation sensors. One also has to look at the new Orbiting Carbon Observatory-2 (OCO-2) launched in 2014 to study CO_2 in the column of air over the earth's surface, which will further advance our understanding of CO_2 sources and sinks.

Compared to ground-based observations, satellite remote sensing has been providing more and more accurate and higher-resolution global GHG detection, and as a result the European Union,

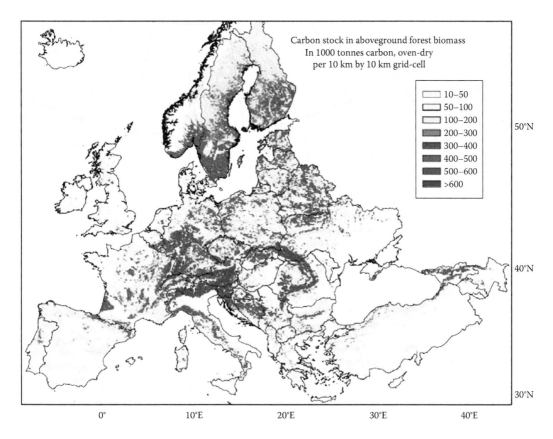

FIGURE 12.10 Total carbon stock of aboveground forest biomass for the European Union countries, calculated separately for broadleaves and conifers with a spatial resolution of 500 m MODIS data. Aggregated biomass conversion and expansion factors (BCEFs) were used to convert the remote sensing–based growing stock classification results to carbon stock of the aboveground forest biomass (Gallaun et al., 2010).

Source: Gallaun et al., 2010.

the United States, Japan, and Canada are vigorously developing MVS (monitoring and verification support) capabilities for accounting GHG emissions using satellite remote sensing (Liangyun et al., 2022). There are 18 GHG satellites (launched or soon to be launched) with resolutions ranging from 25 m to 100 km and include satellites like SCIAMACHY, GOSAT, GOSAT-2, GOSAT-GW, OCO-2, TanSAT, Sentinel-5P, Sentinel-5, FY-3D, GF-5, OCO-3, Microcarb, MethaneSAT, Metop-SGA, FengYun-3G, GEOCARB, DQ-01, and CO2M. This will help reduce uncertainties in the global carbon budgets released yearly (Friedlingstein et al., 2022; Friedlingstein et al., 2020). The satellite-based GHG estimates also make it feasible for more objective and globally verifiable data, methods, and products. For example, the satellite remote technology has led to a new era of observations and provides multi-scale information on essential climate variables (ECVs) that is independent of in situ measurements and model simulations (Zhao et al., 2023).

12.11 REMOTE SENSING OF GLOBAL TERRESTRIAL CARBON

Methods for measuring carbon stocks and fluxes include satellite remote sensing, forest inventory, soil inventory, eddy flux, flask measurements, ecosystem modeling, and biome modeling (Bates et al., 2008). Micrometeorological eddy covariance studies and studies using forest inventory plots have yielded conflicting results regarding the sink strength of mature tropical forests in the Amazon (Robinson et al., 2009). All these methods vary in complexity, precision, accuracy, and costs.

However, satellite remote sensing offers the most distinct advantages in consistency of data, synoptic coverage, global reach, cost per unit area, repeatability, precision, and accuracy (Meng et al., 2009). Opportunities to significantly advance C storage and flux estimates through improved LUC estimates and modeling exist with the evolution in spaceborne hyperspectral, hyperspatial, and advanced multispectral sensors (e.g., Figure 12.11a) as a result of improvements in the spatial, spectral, radiometric, and temporal properties as well as in optics and the signal-to-noise ratio of data. High spatial resolution allows location, while high spectral resolution allows identification of features. Hyperspectral remote sensing sensors allow direct measurement of canopy chemical content (e.g., chlorophyll, nitrogen), forest species, chemistry distribution, timber volumes, and water (Asner and Martin, 2008), and improved biophysical and yield characteristics (Thenkabail et al., 2004a, 2004b). Thenkabail et al. (2004c) demonstrated an increased accuracy of about 30% in LUC and biomass when 30 hyperspectral wavebands are used relative to six non-thermal Landsat TM bands. Hyperspatial data have demonstrated the ability to extract individual tree crowns from 1-m panchromatic data. Agroforest successional stages have been mapped and their varying carbon sink strengths assessed using IKONOS (Thenkabail et al., 2004a). In contrast, forest structure variables (e.g., biomass, LAI) are poorly predicted by the older-generation sensors.

A model can be developed for characterizing forest structure, biomass yield, and C storage based on hyperspectral, hyperspatial, and advanced multispectral data (e.g., Figure 12.11a) and the field-plot data by discerning a large number, say K, of LUC. Letting τ_k, k = 1......K, represent the total C in the kth LUC. The proposed model:

$$\tau k = \lambda k A k + \varepsilon k \tag{12.1}$$

is a practical model for regional estimates of C storage by LUC, which becomes possible as we can discern a sufficiently narrow number of LUC using advanced remote sensing. In this model, λk is average C per hectare in LUC_k, A_k is the total land area in LUC_k, and εk is the departure of τk from its expected value. The historical CGTS field-plot data, supplemented with additional above- and belowground sampling, to evaluate λk for a suite of LUCs of importance in the WTAB region. Remotely sensed data will be used to evaluate A_k at various scales or pixel resolutions, radiometry, bandwidth, and time of acquisition.

The previous model attempts to capitalize on having a sufficiently large number of LUC so that each is relatively homogeneous with respect to C and biomass variability. For example, Thenkabail et al. (2004a) has demonstrated the potential for such fine-resolution classification by LUC. The temporal change in τk is given by:

$$\frac{d\tau k}{dt} = \frac{d(\lambda k A k + \varepsilon k)}{dt} = \frac{d\lambda k}{dt} A k + \lambda k \frac{dA k}{dt} + \frac{d\varepsilon k}{dt} \tag{12.2}$$

If the instantaneous rate of change in the previous model is integrated into annual changes, the resulting discretized model of change in C storage and biomass is:

$$\Delta \tau k = \Delta \lambda k + \lambda k \Delta A k + \Delta \varepsilon k \tag{12.3}$$

For many LUC, e.g., mature upland forests, $\Delta \lambda k$ is anticipated to be small so that the model for $\Delta \tau_k$ will be dominated by the $\lambda k \Delta A k$ term. Moreover, $\Delta A k$ is discernible from satellite data, whereas regional estimates of λk will be determined by fieldwork. For LUCs that cannot be so finely discerned as to support an assertion that $\Delta \lambda \approx 0$, then a spatially averaged value, say $\overline{\Delta \lambda k}$, will be used in (12.3).

Chapter 11 by Dr. Wenge Ni-Meister has exclusive focus on AGB assessment and C stock estimations using multi-sensor remote sensing. Recent global observation systems provide measurements of horizontal and vertical vegetation structure of ecosystems, which will be critical for estimating global carbon storage and assessing ecosystem response to climate change and natural and anthropogenic disturbances. Remote sensing overcomes the limitations associated with sparse field surveys; it has been used extensively as a basis for inferring forest structure and AGB over large areas. They provide a systematic approach wherein they discuss AGB estimates based on:

1. Optical remote sensing
2. Radar remote sensing
3. LiDAR remote sensing

Optical passive remote sensing data are sensitive to vegetation structure (LAI, crown size, and tree density), texture, and shadow, which are correlated with AGB. Radar data are directly related to AGB through measuring dielectric and geometrical properties of forests (Le Toan, et al., 2011). LiDAR remote sensing is promising in characterizing vegetation vertical structure and height, which are then associated to ABG. Vegetation structure characteristics measured from satellite data are linked to field-based AGB estimates, and their relationships are used to map large-scale AGB from satellite data.

Many studies have demonstrated that LiDAR is significantly better at estimating biomass than passive optical or radar sensors used alone. LiDAR directly measures horizontal and vertical vegetation structure characteristics; AGB models developed from LiDAR structure metrics are significantly more accurate than those using radar or passive optical data. The use of LiDAR data, particularly spaceborne data, is limited by its sparse spatial sampling. Both radar and passive optical remote sensing provide a large scale of imaging capability. However, both optical and SAR estimates of AGB are limited by a loss of sensitivity with increasing biomass. Innovative approaches have been developed to fuse passive optical imagery or radar imaging data with spatially extended point-based estimates of biophysical parameters derived from LiDAR to develop high-quality, wall-to-wall AGB maps with unprecedented accuracy and spatial resolution. Particularly, the synergy of spaceborne large-footprint LiDAR (ICESat/GLAS) and medium-resolution optical data, primarily from the Moderate Resolution Imaging Spectrometer (MODIS) (Qin et al., 2024), has been exploited to map AGB and C storage at regional to continental scales. A laser-based instrument being developed for the International Space Station will provide a unique 3D view of Earth's forests, helping to fill in missing information about their role in the carbon cycle (e.g., Figure 12.11b).

Such advances in remote sensing of carbon stock assessment in various landscapes such as forests, agroforests (e.g., Figure 12.11c), and other land cover/land use categories will help in setting up operational global carbon monitoring frameworks such as one under the United Nations Framework Convention on Climate Change (UNFCCC) mechanism for Reducing Emissions from Deforestation in Developing Countries (REDD) (Morita and Matsumoto, 2023). An important advance presented and discussed in Chapter 11 is the use of multi-sensor data-fusion approaches to increase accuracies of biomass quantification and carbon stock estimations. Other aspects of biomass and carbon estimations using remote sensing are discussed in this and other chapters of this volume.

Terrestrial carbon fluxes remain the one of the largest uncertainties in the global carbon cycle. A review of 50 years of terrestrial carbon cycles and its advances using remote sensing highlights (Xiao et al., 2019) revealed: (1) the broad wavelength range (visible, infrared, and microwave) of the electromagnetic spectrum was used to estimate C fluxes and/or stocks; (2) a historical overview of the key milestones in remote sensing of the terrestrial carbon cycle; (3) platforms/sensors, methods, findings, and challenges in remote sensing of carbon fluxes; (4) platforms/sensors, methods, findings, and challenges in passive optical, microwave, and LiDAR remote sensing of biomass carbon stocks as well as remote sensing of soil organic carbon; (5) progress in remote sensing of disturbance impacts on the carbon cycle; and (6) uncertainty and validation of the resulting C flux and

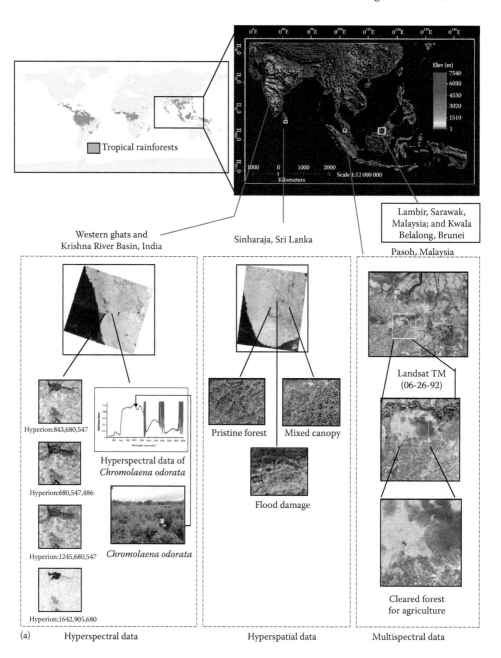

FIGURE 12.11A Tropical forests. Snapshots of hyperspectral, hyperspatial, and advanced multi-spectral data are illustrated for certain random benchmark study areas in Wet Tropical Asian Bioregion (WTAB) in Sri Lanka, India, Malaysia, and Brunei.

stock estimates. Zeng et al. (2020) estimated the global terrestrial C fluxes of 1999–2019 by upscaling terrestrial net ecosystem exchange, gross primary production, and ecosystem respiration from FLUXNET 2015. Remote sensing data were derived from the Copernicus Global Land Service. Terrestrial vegetation, as the key component of the biosphere, has had a greening trend since the beginning of this century (Zhang et al., 2019). However, their remote sensing–based, data-driven model, which was calibrated based on the global eddy flux data set (FLUXNET2015) and Moderate

Forests, Biodiversity, Ecology, LULC, and Carbon

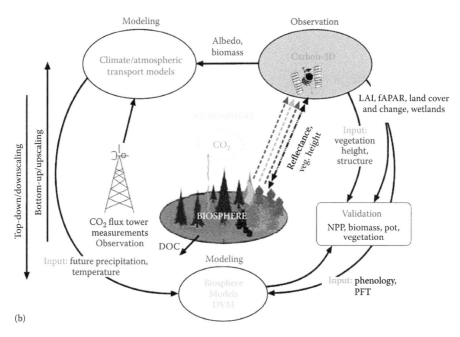

(b)

FIGURE 12.11B Role of carbon-3D in the observation and modeling strategy of the carbon cycle.

Source: Hese et al., 2005.

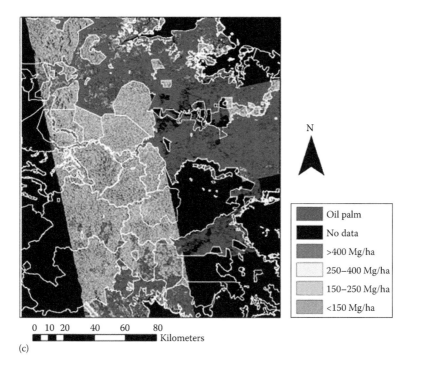

(c)

FIGURE 12.11C Biomass map of Sabah Malaysian Borneo. Biomass values were generated only from non-precipitation-affected imagery. Oil palm areas were derived from all images and are pictured in red. Biomass values have been grouped for all high biomass areas (> 250Mg ha-1), moderate biomass areas (two groups from 50 to 250 Mg ha–1), and severely degraded areas (< 50 Mg ha–1).

Source: Morel et al., 2011.

Resolution Imaging Spectroradiometer vegetation index data showed no proportional increase of terrestrial gross carbon sequestration from the greening Earth.

12.12 ACKNOWLEDGMENTS

I would like to thank the lead authors and co-authors of each of the chapters for providing their insights and edits of my chapter summaries. Any use of trade, firm, or product names is for descriptive purposes only and does not imply endorsement by the US government.

REFERENCES

Aksoy, S. et al., 2023. Forest biophysical parameter estimation via machine learning and neural network approaches. *IGARSS 2023–2023 IEEE International Geoscience and Remote Sensing Symposium*, Pasadena, CA, pp. 2661–2664. http://doi.org/10.1109/IGARSS52108.2023.10282899.

Alves, M.C., Carvalho, L.G., and Oliveira, M.S. 2013. Terrestrial Earth couple climate–carbon spatial variability and uncertainty. *Global and Planetary Change*, 111, 9–30. ISSN 0921-8181. http://doi.org/10.1016/j.gloplacha.2013.08.009.

Asner, G.P., and Martin, R.E. 2008. Spectral and chemical analysis of tropical forests: Scaling from leaf to canopy levels. *Remote Sensing of Environment*, 112(10), 3958–3970.

Baek, W.-K., and Jung, H.-S. 2019. A review of change detection techniques using multi-temporal synthetic aperture radar images. *Korean Journal of Remote Sensing*, 35(5_1), 737–750. https://doi.org/10.7780/KJRS.2019.35.5.1.10

Bates, B.C., Kundzewicz, Z.W., Wu, S., and Palutikof, J.P., Eds. 2008. Climate change and water. *Technical Paper of the Intergovernmental Panel on Climate Change*. IPCC Secretariat, Geneva, 210 p.

Benson, M.L., Pierce, L., Bergen, K., and Sarabandi, K. 2021. Model-based estimation of forest canopy height and biomass in the Canadian boreal forest using radar, LiDAR, and optical remote sensing. *IEEE Transactions on Geoscience and Remote Sensing*, 59(6), 4635–4653. https://doi.org/10.1109/TGRS.2020.3018638

Campbell, M.J., Dennison, P.E., Tune, J.W., Kannenberg, S.A., Kerr, K.L., Codding, B.F., and Anderegg, W.R.L. 2020. A multi-sensor, multi-scale approach to mapping tree mortality in woodland ecosystems. *Remote Sensing of Environment*, 245, 111853. ISSN 0034-4257. https://doi.org/10.1016/j.rse.2020.111853. https://www.sciencedirect.com/science/article/pii/S0034425720302236

Chen, L., Ren, C., Bao, G., Zhang, B., Wang, Z., Liu, M., Man, W., and Liu, J. 2022. Improved object-based estimation of forest aboveground biomass by integrating LiDAR data from GEDI and ICESat-2 with multi-sensor images in a heterogeneous mountainous region. *Remote Sensing*, 14(12), 2743. https://doi.org/10.3390/rs14122743

Claverie, M., Ju, J., Masek, J.G., Dungan, J.L., Vermote, E.F., Roger, J.-C., Skakun, S.V., and Justice, C. 2018. The harmonized Landsat and Sentinel-2 surface reflectance data set. *Remote Sensing of Environment*, 219, 145–161.

De Alban, J.D.T., Connette, G.M., Oswald, P., and Webb, E.L. 2018. Combined Landsat and L-band SAR data improves land cover classification and change detection in dynamic tropical landscapes. *Remote Sensing*, 10(2), 306. https://doi.org/10.3390/rs10020306.

Durieux, A.M.S., Calef, M.T., Arko, S., Chartrand, R., Kontgis, C., Keisler, R., and Warren, M.S. 2019. Monitoring forest disturbance using change detection on synthetic aperture radar imagery. *Proceedings of SPIE*, 11139, Applications of Machine Learning, 1113916. https://doi.org/10.1117/12.2528945

Ewald, M., Aerts, R., Lenoir, J., Fassnacht, F.E., Nicolas, M., Skowronek, S., Piat, J., Honnay, O., Garzón-López, C.X., Feilhauer, H., Van De Kerchove, R., Somers, B., Hattab, T., Rocchini, D., and Schmidtlein, S. 2018. LiDAR derived forest structure data improves predictions of canopy N and P concentrations from imaging spectroscopy. *Remote Sensing of Environment*, 211, 13–25. ISSN 0034-4257. https://doi.org/10.1016/j.rse.2018.03.038. https://www.sciencedirect.com/science/article/pii/S0034425718301391

Fagua, J.C., Jantz, P., Rodriguez-Buritica, S., Duncanson, L., and Goetz, S.J. 2019. Integrating LiDAR, multispectral and SAR data to estimate and map canopy height in tropical forests. *Remote Sensing*, 11(22), 2697. https://doi.org/10.3390/rs11222697

Friedlingstein, P., O'Sullivan, M., Jones, M.W., Andrew, R.M., Gregor, L., Hauck, J., Le Quéré, C., Luijkx, I.T., Olsen, A., Peters, G.P., Peters, W., Pongratz, J., Schwingshackl, C., Sitch, S., Canadell, J.G., Ciais, P., Jackson, R.B., Alin, S.R., Alkama, R., Arneth, A., Arora, V.K., Bates, N.R., Becker, M., Bellouin, N., Bittig, H.C., Bopp, L., Chevallier, F., Chini, L.P., Cronin, M., Evans, W., Falk, S., Feely, R.A., Gasser, T., Gehlen, M., Gkritzalis, T., Gloege, L., Grassi, G., Gruber, N., Gürses, Ö., Harris, I., Hefner, M., Houghton, R.A., Hurtt, G.C., Iida, Y., Ilyina, T., Jain, A.K., Jersild, A., Kadono, K., Kato, E., Kennedy, D., Klein Goldewijk, K., Knauer, J., Korsbakken, J.I., Landschützer, P., Lefèvre, N., Lindsay, K., Liu, J., Liu, Z., Marland, G., Mayot, N., McGrath, M.J., Metzl, N., Monacci, N.M., Munro, D.R., Nakaoka, S.-I., Niwa, Y., O'Brien, K., Ono, T., Palmer, P.I., Pan, N., Pierrot, D., Pocock, K., Poulter, B., Resplandy, L., Robertson, E., Rödenbeck, C., Rodriguez, C., Rosan, T.M., Schwinger, J., Séférian, R., Shutler, J.D., Skjelvan, I., Steinhoff, T., Sun, Q., Sutton, A.J., Sweeney, C., Takao, S., Tanhua, T., Tans, P.P., Tian, X., Tian, H., Tilbrook, B., Tsujino, H., Tubiello, F., van der Werf, G.R., Walker, A.P., Wanninkhof, R., Whitehead, C., Willstrand Wranne, A., Wright, R., Yuan, W., Yue, C., Yue, X., Zaehle, S., Zeng, J., and Zheng, B. 2022. Global carbon budget 2022. *Earth System Science Data*, 14, 4811–4900. https://doi.org/10.5194/essd-14-4811-2022.

Friedlingstein, P., O'Sullivan, M., Jones, M.W., Andrew, R.M., Hauck, J., Olsen, A., Peters, G.P., Peters, W., Pongratz, J., Sitch, S., Le Quéré, C., Canadell, J.G., Ciais, P., Jackson, R.B., Alin, S., Aragão, L.E.O.C., Arneth, A., Arora, V., Bates, N.R., Becker, M., Benoit-Cattin, A., Bittig, H.C., Bopp, L., Bultan, S., Chandra, N., Chevallier, F., Chini, L.P., Evans, W., Florentie, L., Forster, P.M., Gasser, T., Gehlen, M., Gilfillan, D., Gkritzalis, T., Gregor, L., Gruber, N., Harris, I., Hartung, K., Haverd, V., Houghton, R.A., Ilyina, T., Jain, A.K., Joetzjer, E., Kadono, K., Kato, E., Kitidis, V., Korsbakken, J.I., Landschützer, P., Lefèvre, N., Lenton, A., Lienert, S., Liu, Z., Lombardozzi, D., Marland, G., Metzl, N., Munro, D.R., Nabel, J.E.M.S., Nakaoka, S.-I., Niwa, Y., O'Brien, K., Ono, T., Palmer, P.I., Pierrot, D., Poulter, B., Resplandy, L., Robertson, E., Rödenbeck, C., Schwinger, J., Séférian, R., Skjelvan, I., Smith, A.J.P., Sutton, A.J., Tanhua, T., Tans, P.P., Tian, H., Tilbrook, B., van der Werf, G., Vuichard, N., Walker, A.P., Wanninkhof, R., Watson, A.J., Willis, D., Wiltshire, A.J., Yuan, W., Yue, X., and Zaehle, S. 2020. Global carbon budget 2020. *Earth System Science Data*, 12, 3269–3340. https://doi.org/10.5194/essd-12-3269-2020.

Gallaun, H., Zanchi, G., Nabuurs, G.-J., Hengeveld, G., Schardt, M., and Verkerk, P.J. 2010. EU-wide maps of growing stock and above-ground biomass in forests based on remote sensing and field measurements. *Forest Ecology and Management*, 260(3), 252–261. ISSN 0378-1127. http://doi.org/10.1016/j.foreco.2009.10.011.

Ganivet, E., and Bloomberg, M. 2019. Towards rapid assessments of tree species diversity and structure in fragmented tropical forests: A review of perspectives offered by remotely-sensed and field-based data. *Forest Ecology and Management*, 432, 40–53. ISSN 0378-1127. https://doi.org/10.1016/j.foreco.2018.09.003. https://www.sciencedirect.com/science/article/pii/S0378112718307102.

Gao, Y., Quevedo, A., Szantoi, Z., and Skutsch, M. 2021. Monitoring forest disturbance using time-series MODIS NDVI in Michoacán, Mexico. *Geocarto International*, 36(15), 1768–1784. https://doi.org/10.1080/10106049.2019.1661032.

Gao, Y., Wang, Y., Chen, L., Wu, F., and Yu, G. 2024. Achieving accurate regional carbon-sink accounting and its significance for "missing" carbon sinks. *The Innovation*, 5(1), 100552. ISSN 2666-6758. https://doi.org/10.1016/j.xinn.2023.100552. https://www.sciencedirect.com/science/article/pii/S2666675823001807.

Gardner, T.A., Burgess, N.D., Aguilar-Amuchastegui, N., Barlow, J., Berenguer, E., Clements, T., Danielsen, F., Ferreira, J., Foden, W., Kapos, V., Khan, S.M., Lees, A.C., Parry, L., Roman-Cuesta, R.M., Schmitt, C.B., Strange, N., Theilade, I., and Vieira, I.C.G. 2012. A framework for integrating biodiversity concerns into national REDD+ programmes. *Biological Conservation*, 154, 61–71. ISSN 0006-3207. http://doi.org/10.1016/j.biocon.2011.11.018.

Gupta, R., and Sharma, L.K. 2022. Mixed tropical forests canopy height mapping from spaceborne LiDAR GEDI and multisensor imagery using machine learning models. *Remote Sensing Applications: Society and Environment*, 27, 100817. ISSN 2352-9385. https://doi.org/10.1016/j.rsase.2022.100817. https://www.sciencedirect.com/science/article/pii/S2352938522001252.

Hese, S., Lucht, W., Schmullius, C., Barnsley, M., Dubayah, R., Knorr, D., Neumann, K., Riedel, T., and Schröter, K. 2005. Global biomass mapping for an improved understanding of the CO2 balance—the Earth observation mission Carbon-3D. *Remote Sensing of Environment*, 94(1), 94–104. ISSN 0034-4257. http://doi.org/10.1016/j.rse.2004.09.006.

IEA. 1998. *AMD African Mammals Databank—A Databank for the Conservation and Management of the African Mammals Vol 1 and 2*. Report to the Directorate-General for Development (DGVIII/A/1) of the European Commission. Project No. B7–6200/94–15/VIII/ENV. Bruxelles, pp. 1174.

Illarionova, S., Shadrin, D., ignatiev, V., Shayakhmetov, S., Trekin, S., and Oseledets, I. 2022. Estimation of the canopy height model from multispectral satellite imagery with convolutional neural networks. *IEEE Access*, 10, 34116–34132. http://doi.org/10.1109/ACCESS.2022.3161568.

Jackson, C.M., and Adam, E. 2022. A machine learning approach to mapping canopy gaps in an indigenous tropical submontane forest using WorldView-3 multispectral satellite imagery. *Environmental Conservation*, 49(4), 255–262. http://doi.org/10.1017/S0376892922000339.

Jacquemoud, S., Verhoef, W., Baret, F., Bacour, C., Zarco-Tejada, P.J., Asner, G.P., François, C., and Ustin, S.L. 2009. Prospect + sail: A review of use for vegetation characterization. *Remote Sens Environ*, 113, S56–S66.

Jafarzadeh, J., and Attarchi, S. 2023. Increasing the spatial accuracy of the land use map using fusion of optical and radar images of sentinel and Google earth engine. *ISPRS Annals of the Photogrammetry, Remote Sensing and Spatial Information Sciences*, X-4/W1–2022, 321–326, https://doi.org/10.5194/isprs-annals-X-4-W1-2022-321-2023, 2023.

Jung, M., Dahal, P.R., Butchart, S.H.M., et al.2020. A global map of terrestrial habitat types. *Scientific Data*, 7, 256. https://doi.org/10.1038/s41597-020-00599-8.

Kamoske, A.C., Dahlin, K.M., Read, Q.D., Record, S., Stark, S.C., Serbin, S.P., and Zarnetske, P.L. 2022. Towards mapping biodiversity from above: Can fusing lidar and hyperspectral remote sensing predict taxonomic, functional, and phylogenetic tree diversity in temperate forests? *Global Ecology and Biogeography*, 31(7), 1440–1460. Willey Online Library. https://doi.org/10.1111/geb.13516.

Kankare, V., Räty, M., Yu, X., Holopainen, M., Vastaranta, M., Kantola, T., Hyyppä, J., Alho, P., and Viitala, R. 2013. Single tree biomass modelling using airborne laser scanning. *ISPRS Journal of Photogrammetry and Remote Sensing*, 85, 66–73. ISSN 0924-2716. http://doi.org/10.1016/j.isprsjprs.2013.08.008.

Kilbride, J.B., Poortinga, A., Bhandari, B., Thwal, N.S., Quyen, N.H., Silverman, J., Tenneson, K., Bell, D., Gregory, M., Kennedy, R., et al. 2023. A near real-time mapping of tropical forest disturbance using SAR and semantic segmentation in Google earth engine. *Remote Sensing*, 15(21), 5223. https://doi.org/10.3390/rs15215223.

Kluczek, M., Zagajewski, B., and Kycko, M. 2024. Combining multitemporal optical and radar satellite data for mapping the tatra mountains non-forest plant communities. *Remote Sensing*, 16, 1451. https://doi.org/10.3390/rs16081451.

Laba, M., Blair, B., Downs, R., Monger, B., Philpot, W., Smith, S., Sullivan, P., and Baveye, P.C. 2010. Use of textural measurements to map invasive wetland plants in the Hudson River National Estuarine Research Reserve with IKONOS satellite imagery. *Remote Sensing of Environment*, 114(4), 876–886. ISSN 0034-4257. http://doi.org/10.1016/j.rse.2009.12.002.

Lasco, R.D. 2004. Forest carbon budgets in Southeast Asia following harvesting and land cover change. *Science in China*. Series 3. 45(Suppl.), 55–64.

Laurin, G.V., Ding, J., Disney, M., Bartholomeus, H., Herold, M., Papale, D., and Valentini, R. 2019. Tree height in tropical forest as measured by different ground, proximal, and remote sensing instruments, and impacts on above ground biomass estimates. *International Journal of Applied Earth Observation and Geoinformation*, 82, 101899. ISSN 1569-8432. https://doi.org/10.1016/j.jag.2019.101899. (https://www.sciencedirect.com/science/article/pii/S0303243419300844)

Lee, W.J., and Lee, C.W. 2018. Forest canopy height estimation using multiplatform remote sensing dataset. *Journal of Sensors*, 2018(Article ID 1593129), 9. https://doi.org/10.1155/2018/1593129

Lee, Y.-S., Lee, S., Baek, W.-K., Jung, H.-S., Park, S.-H., and Lee, M.-J. 2020. Mapping forest vertical structure in Jeju Island from optical and radar satellite images using artificial neural network. *Remote Sensing*, 12(5), 797. https://doi.org/10.3390/rs12050797

Lefsky, M.A. 2010. A global forest canopy height map from the moderate resolution spectroradiometer and the geoscience laser altimeter system. *Geophysical Research Letters*, 37(15), 1–5. http://doi.org/10.1029/2010GL043622

Le Toan, T., Quegan, S., Davidson, M.W.J., Balzter, H., Paillou, P., Papathanassiou, K., Plummer, S., Rocca, F., Saatchi, S., Shugart, H., and Ulander, L. 2011. The BIOMASS mission: Mapping global forest biomass to better understand the terrestrial carbon cycle. *Remote Sensing of Environment*, 115, 2850–2860.

Lewis, S., Lopez-Gonzalez, G., Sonké, B. et al. 2009. Increasing carbon storage in intact African tropical forests. *Nature*, 457, 1003–1006. https://doi.org/10.1038/nature07771.

Li, W., Niu, Z., Shang, R., Qin, Y., Wang, L., and Chen, H. 2020. High-resolution mapping of forest canopy height using machine learning by coupling ICESat-2 LiDAR with Sentinel-1, Sentinel-2 and Landsat-8 data. *International Journal of Applied Earth Observation and Geoinformation*, 92, 102163. ISSN 1569-8432. https://doi.org/10.1016/j.jag.2020.102163. https://www.sciencedirect.com/science/article/pii/S030324342030026X.

Li, Y., Peng, C., Chen, Y., Jiao, L., Zhou, L., and Shang, R. 2019. A deep learning method for change detection in synthetic aperture radar images. *IEEE Transactions on Geoscience and Remote Sensing*, 57(8), 5751–5763. http://doi.org/10.1109/TGRS.2019.2901945.

Liangyun, L.I.U., Liangfu, C.H.E.N., Yi, L.I.U., et al. 2022. Satellite remote sensing for global stocktaking: Methods, progress and perspectives [J]. *National Remote Sensing Bulletin*, 26(2), 243–267. https://doi.org/10.11834/jrs.20221806.

Liu, H., et al., 2021. Deep learning in forest structural parameter estimation using airborne LiDAR data. *IEEE Journal of Selected Topics in Applied Earth Observations and Remote Sensing*, 14, 1603–1618. http://doi.org/10.1109/JSTARS.2020.3046053.

Luo, W., Ma, H., Yuan, J., Zhang, L., Ma, H., Cai, Z., and Zhou, W. 2023. High-accuracy filtering of forest scenes based on full-waveform LiDAR data and hyperspectral images. *Remote Sensing*, 15, 3499. https://doi.org/10.3390/rs15143499

Masek, J.G., Ju, J., Claverie, M., Skakun, S., Roger, J.C., Vermote, E., Franch, B., Yin, Z., and Dungan, J.L. 2022. *Harmonized Landsat Sentinel-2 (HLS) Product User Guide Product Version 2.0*.

Masek, J.G., Ju, J., Roger, J.C., Skakun, S., Claverie, M., and Dungan, J. 2018. Harmonized landsat/sentinel-2 products for land monitoring. *IGARSS 2018–2018 IEEE International Geoscience and Remote Sensing Symposium*, Valencia, Spain, pp. 8163–8165. http://doi.org/10.1109/IGARSS.2018.8517760.

Masek, J.G., Ju, J., Roger, J.C., Skakun, S., Vermote, E., Claverie, M., Dungan, J., Yin, Z., Freitag, B., and Justice, C. 2021. *HLS Sentinel-2 MSI Surface Reflectance Daily Global 30m v2.0*. Distributed by NASA EOSDIS Land Processes DAAC. https://doi.org/10.5067/HLS/HLSS30.002.

Mastro, P., Masiello, G., Serio, C., and Pepe, A. 2022. Change detection techniques with synthetic aperture radar images: Experiments with random forests and sentinel-1 observations. *Remote Sensing*, 14(14), 3323. https://doi.org/10.3390/rs14143323

Meng, Q., Cieszewski, C., and Madden, M. 2009. Large area forest inventory using Landsat ETM+: A geostatistical approach. *ISPRS Journal of Photogrammetry and Remote Sensing*, 64(1), 27–36.

Mermoz, S., Toan, T.L., Villard, L., Réjou-Méchain, M., and Seifert-Granzin, J. 2014. Biomass assessment in the Cameroon savanna using ALOS PALSAR data. *Remote Sensing of Environment*, Available online 15 May 2014. ISSN 0034-4257. http://doi.org/10.1016/j.rse.2014.01.029.

Mirzaei, A., Bagheri, H., and Khosravi, I. 2023. Enhancing crop classification accuracy through synthetic SAR-optical data generation using deep learning. *ISPRS International Journal of Geo-Information*, 12, 450. https://doi.org/10.3390/ijgi12110450

Mondal, P., McDermid, S.S., and Qadir, A. 2020. A reporting framework for sustainable development goal 15: Multi-scale monitoring of forest degradation using MODIS, landsat and sentinel data. *Remote Sensing of Environment*, 237, 111592. ISSN 0034-4257. https://doi.org/10.1016/j.rse.2019.111592. https://www.sciencedirect.com/science/article/pii/S0034425719306121

Morel, A.C., Saatchi, S.S., Malhi, Y., Berry, N.J., Banin, L., Burslem, D., Nilus, R., and Ong, R.C. 2011. Estimating aboveground biomass in forest and oil palm plantation in Sabah, Malaysian Borneo using ALOS PALSAR data. *Forest Ecology and Management*, 262(9), 1786–1798. ISSN 0378-1127. http://doi.org/10.1016/j.foreco.2011.07.008.

Morita, K., and Matsumoto, K. 2023. Challenges and lessons learned for REDD+ finance and its governance. *Carbon Balance Manage*, 18, 8. https://doi.org/10.1186/s13021-023-00228-y.

Padalia, H., Prakash, A., and Watham, T. 2023. Modelling aboveground biomass of a multistage managed forest through synergistic use of Landsat-OLI, ALOS-2 L-band SAR and GEDI metrics. *EcologicalInformatics*, 77, 102234. ISSN 1574-9541. https://doi.org/10.1016/j.ecoinf.2023.102234. https://www.sciencedirect.com/science/article/pii/S1574954123002637

Papp, L., van Leeuwen, B., Szilassi, P., Tobak, Z., Szatmári, J., Árvai, M., Mészáros, J., and Pásztor, L. 2021. Monitoring invasive plant species using hyperspectral remote sensing data. *Land*, 10(1), 29. https://doi.org/10.3390/land10010029

Pereira Martins-Neto, R., Tommaselli, A.M.G., Imai, N.N., Honkavaara, E., Miltiadou, M., Saito Moriya, E.A., and David, H.C. 2023. Tree species classification in a complex Brazilian tropical forest using hyperspectral and LiDAR data. *Forests*, 14(5), 945. https://doi.org/10.3390/f14050945

Pinzon, J.E., Pak, E.W., Tucker, C.J., Bhatt, U.S., Frost, G.V., and Macander, M.J. (2023). *Global Vegetation Greenness (NDVI) from AVHRR GIMMS-3G+, 1981–2022*. ORNL DAAC, Oak Ridge, TN. https://doi.org/10.3334/ORNLDAAC/2187

Polykretis, C., Grillakis, M.G., and Alexakis, D.D. 2020. Exploring the impact of various spectral indices on land cover change detection using change vector analysis: A case study of crete Island, Greece. *Remote Sensing*, 12(2), 319. https://doi.org/10.3390/rs12020319

Qin, Y., Xiao, X., Tang, H., Dubayah, R., Doughty, R., Liu, D., Liu, F., Shimabukuro, Y., Arai, E., Wang, X., and Moore III, B. 2024. Annual maps of forest cover in the Brazilian Amazon from analyses of PALSAR and MODIS images. *Earth System Science Data*, 16, 321–336. https://doi.org/10.5194/essd-16-321-2024.

Rajat, P., Avtar, R., Malik, R., Musthafa, M., Rathore, V.S., Kumar, P., and Singh, G. 2023. Forest plantation species classification using Full-Pol-Time-Averaged SAR scattering powers. *Remote Sensing Applications: Society and Environment*, 29, 100924. ISSN 2352-9385. https://doi.org/10.1016/j.rsase.2023.100924. https://www.sciencedirect.com/science/article/pii/S235293852300006X.

Redowan, M., Phinn, S.R., Roelfsema, C.M., and Aziz, A.A. 2020. CLASlite unmixing of Landsat images to estimate REDD+ activity data for deforestation in a Bangladesh forest. *Journal of Applied Remote Sensing*, 14(2), 024505. https://doi.org/10.1117/1.JRS.14.024505

Robinson, D.T., Brown, D.G., and Currie, W.S. 2009. Modelling carbon storage in highly fragmented and human-dominated landscapes: Linking land-cover patterns and ecosystem models. *Ecological Modelling*, 220(9–10), 1325–1338.

Sanchez-Azofeifa, A., Guzman, J.A., Campos, C.A., Castro, S., Garcia-Millan, V., Nightingale, J., and Rankine, C. 2017. Twenty-first Century remote sensing technologies are revolutionizing the study of tropical forests. *Biotropica*, 49(5), 604–619. Wiley online library. https://doi.org/10.1111/btp.12454.

Sharma, L., Gupta, R., and Verma, J.K. 2022. Efficacy of advanced remote sensing (hyperspectral and LiDAR) in enhancing forest resources management. *Source Title: Research Anthology on Ecosystem Conservation and Preserving Biodiversity*. IGI Global, p. 20. https://doi.org/10.4018/978-1-6684-5678-1.ch083

Shen, X., Cao, L., Coops, N.C., Fan, H., Wu, X., Liu, H., Wang, G., and Cao, F. 2020. Quantifying vertical profiles of biochemical traits for forest plantation species using advanced remote sensing approaches. *Remote Sensing of Environment*, 250, 112041. ISSN 0034-4257. https://doi.org/10.1016/j.rse.2020.112041. https://www.sciencedirect.com/science/article/pii/S0034425720304119

Shi, Y., Skidmore, A.K., Wang, T., Holzwarth, S., Heiden, U., Pinnel, N., Zhu, X., and Heurich, M. 2018. Tree species classification using plant functional traits from LiDAR and hyperspectral data. *International Journal of Applied Earth Observation and Geoinformation*, 73, 207–219. ISSN 1569-8432. https://doi.org/10.1016/j.jag.2018.06.018. https://www.sciencedirect.com/science/article/pii/S030324341830504X.

Skidmore, A.K., Coops, N.C., Neinavaz, E. et al. 2021. Priority list of biodiversity metrics to observe from space. *Nature Ecology and Evolution*, 5, 896–906. https://doi.org/10.1038/s41559-021-01451-x.

Spanhove, T., Vanden Borre, J., Corbane, C., Buck, O., and Lang, S. 2014. *User Needs, Possibilities and Limitations of Remote Sensing for Natura 2000 Monitoring at Multiple Scales—Results from the European MS*. MONINA project.

Sun, G., Ranson, K.J., Guo, Z., Zhang, Z., Montesano, P., and Kimes, D. 2011. Forest biomass mapping from lidar and radar synergies. *Remote Sensing of Environment*, 115(11), 2906–2916. ISSN 0034-4257. http://doi.org/10.1016/j.rse.2011.03.021.

Suomalainen, J., Hakala, T., Kaartinen, H., Räikkönen, E., and Kaasalainen, S. 2011. Demonstration of a virtual active hyperspectral LiDAR in automated point cloud classification. *ISPRS Journal of Photogrammetry and Remote Sensing*, 66(5), 637–641. ISSN 0924-2716. http://doi.org/10.1016/j.isprsjprs.2011.04.002.

Thenkabail, P.S., Enclona, E.A., Ashton, M.S., Legg, C., and De Dieu, M.J. 2004a. Hyperion, IKONOS, ALI, and ETM+ sensors in the study of African rainforests. *Remote Sensing of Environment*, 90(1), 23–43.

Thenkabail, P.S., Enclona, E.A., Ashton, M.S., Legg, C., and Jean De Dieu, M. 2004b. Hyperion, IKONOS, ALI, and ETM+ sensors in the study of African rainforests. *Remote Sensing of Environment*, 90, 23–43.

Thenkabail, P.S., Enclona, E.A., Ashton, M.S., Legg, C., and Jean De Dieu, M., 2004d. Hyperion, IKONOS, ALI, and ETM+ sensors in the study of African rainforests. *Remote Sensing of Environment*, 90, 23–43.

Thenkabail, P.S., Enclona, E.A., Ashton, M.S., and Van Der Meer, V. 2004c. Accuracy assessments of hyperspectral waveband performance for vegetation analysis applications. *Remote Sensing of Environment*, 91(2–3), 354–376.

Thenkabail, P.S., Gumma, M.K., Teluguntla, P., and Mohammed, I.A. 2014. Hyperspectral remote sensing of vegetation and agricultural crops. Highlight article. *Photogrammetric Engineering and Remote Sensing*, 80(4), 697–709.

Thenkabail, P.S., Mariotto, I., Gumma, M.K., Middleton, E.M., Landis, D.R., and Huemmrich, F.K. 2013. Selection of hyperspectral narrowbands (HNBs) and composition of hyperspectral twoband vegetation indices (HVIs) for biophysical characterization and discrimination of crop types using field reflectance and Hyperion/EO-1 data. *IEEE Journal of Selected Topics in Applied Earth Observations and Remote Sensing*, 6(2), 427–439. http://doi.org/10.1109/JSTARS.2013.2252601

Thenkabail, P.S., Smith, R.B., and De-Pauw, E. 2000b. Hyperspectral vegetation indices for determining agricultural crop characteristics. *Remote sensing of Environment*, 71, 158–182.

Thenkabail, P.S., Stucky, N., Griscom, B.W., Ashton, M.S., Diels, J., Van Der Meer, B., and Enclona, E. 2004a. Biomass estimations and carbon Stock calculations in the oil palm plantations of African derived savannas using IKONOS data. *International Journal of Remote Sensing*, 25(23), 5447–5472.

Thenkabail, P.S., Teluguntla, P.G., Xiong, J., Oliphant, A., Congalton, R.G., Ozdogan, M., Gumma, M.K., Tilton, J.C., Giri, C., Milesi, C., Phalke, A., Massey, R., Yadav, K., Sankey, T., Zhong, Y., Aneece, I., and Foley, D. 2021. Global cropland-extent product at 30m resolution (GCEP30) derived from Landsat satellite time-series data for the year 2015 using multiple machine-learning algorithms on Google Earth Engine cloud: U.S. *Geological Survey Professional Paper 1868*, 63 p. https://doi.org/10.3133/pp1868. https://lpdaac.usgs.gov/news/release-of-gfsad-30meter-cropland-extent-products/IP-119164.

Wang, J., Yang, D., Detto, M., Nelson, B.W., Chen, M., Guan, K., Wu, S., Yan, Z., and Wu, J. 2020. Multi-scale integration of satellite remote sensing improves characterization of dry-season green-up in an Amazon tropical evergreen forest. *Remote Sensing of Environment*, 246, 111865. ISSN 0034-4257. https://doi.org/10.1016/j.rse.2020.111865. https://www.sciencedirect.com/science/article/pii/S0034425720302352

Wang, R. and Gamon, J.A. 2019. Remote sensing of terrestrial plant biodiversity. *Remote Sensing of Environment*, 231, 111218. ISSN 0034-4257. https://doi.org/10.1016/j.rse.2019.111218. https://www.sciencedirect.com/science/article/pii/S0034425719302317

Wang, Y., Zhang, X., and Guo, Z. 2021. Estimation of tree height and aboveground biomass of coniferous forests in North China using stereo ZY-3, multispectral Sentinel-2, and DEM data. *Ecological Indicators*, 126, 107645. ISSN 1470-160X. https://doi.org/10.1016/j.ecolind.2021.107645. https://www.sciencedirect.com/science/article/pii/S1470160X21003101

Wijaya, A., Liesenberg, V., and Gloaguen, R. 2010. Retrieval of forest attributes in complex successional forests of Central Indonesia: Modeling and estimation of bitemporal data. *Forest Ecology and Management*, 259(12), 2315–2326. ISSN 0378-1127. http://dx.doi.org/10.1016/j.foreco.2010.03.004.

Woodcock, C.E., Loveland, T.R., Herold, M., and Bauer, M.E. 2020. Transitioning from change detection to monitoring with remote sensing: A paradigm shift. *Remote Sensing of Environment*, 238, 111558. ISSN 0034-4257. https://doi.org/10.1016/j.rse.2019.111558. https://www.sciencedirect.com/science/article/pii/S0034425719305784

Xiao, J., Chevallier, F., Gomez, C., Guanter, L., Hicke, J.A., Huete, A.R., Ichii, K., Ni, W., Pang, Y., Rahman, A.F., Sun, G., Yuan, W., Zhang, L., and Zhang, X. 2019. Remote sensing of the terrestrial carbon cycle: A review of advances over 50 years. *Remote Sensing of Environment*, 233, 111383. ISSN 0034-4257. https://doi.org/10.1016/j.rse.2019.111383. https://www.sciencedirect.com/science/article/pii/S003442571930402X.

Yang, Q., Su, Y., Hu, T., Jin, S., Liu, X., Niu, C., Liu, Z., Kelly, M., Wei, J., and Guo, Q. 2022. Allometry-based estimation of forest aboveground biomass combining LiDAR canopy height attributes and optical spectral indexes. *Forest Ecosystems*, 9, 100059. ISSN 2197-5620. https://doi.org/10.1016/j.fecs.2022.100059. https://www.sciencedirect.com/science/article/pii/S2197562022000598

Zeng, J., Matsunaga, T., Tan, Z.H. et al. Global terrestrial carbon fluxes of 1999–2019 estimated by upscaling eddy covariance data with a random forest. *Scientific Data*, 7, 313. https://doi.org/10.1038/s41597-020-00653-5

Zhang, C. 2014. Applying data fusion techniques for benthic habitat mapping and monitoring in a coral reef ecosystem. *ISPRS Journal of Photogrammetry and Remote Sensing*, Available online 27 June 2014. ISSN 0924-2716. http://doi.org/10.1016/j.isprsjprs.2014.06.005.

Zhang, D., Wang, H., Wang, X., and Lü, Z. 2020. Accuracy assessment of the global forest watch tree cover 2000 in China. *International Journal of Applied Earth Observation and Geoinformation*, 87, 102033.

Zhang, Y., Song, C., Band, L.W., and Sun, G. 2019. No proportional increase of terrestrial gross carbon sequestration from the greening earth. *Journal of Geophysical Research: Biogeosciences*, 124(8), 2540–2553. https://doi.org/10.1029/2018JG004917

Zhao, S., Liu, M., Tao, M., Zhou, W., Lu, X., Xiong, Y., Li, F., and Wang, Q. 2023. The role of satellite remote sensing in mitigating and adapting to global climate change. *Science of the Total Environment*, 904, 166820. ISSN 0048-9697. https://doi.org/10.1016/j.scitotenv.2023.166820. https://www.sciencedirect.com/science/article/pii/S0048969723054451

Zhu, Z. 2017. Change detection using landsat time series: A review of frequencies, preprocessing, algorithms, and applications. *ISPRS Journal of Photogrammetry and Remote Sensing*, 130, 370–384. ISSN 0924-2716. https://doi.org/10.1016/j.isprsjprs.2017.06.013. https://www.sciencedirect.com/science/article/pii/S092427161730103X.

Index

Note: Page numbers in *italics* indicate a figure and page numbers in **bold** indicate a table on the corresponding page.

3D airborne laser scanning (ALS), 58
  area-based approaches, *152*
  vegetation height, 137–140, **138**
3D data, light detection and ranging (LiDAR) and radar techniques, 52–54, **54**, 126, 458–463
  AGB, *see* aboveground biomass (AGB)
  ALS, 58
    vegetation height, 137–140, **138**
  biodiversity, *see* biodiversity from space
  canopy height model, 62
  canopy structure, 108–110
  data sources, 98–99
  DBH measurements, **68**
  diameter, 67–69
  DSM, 62
  DTM, 61–62
  forestry data, 62–64
  HIS, *see* hyperspectral imaging (HSI)
  individual tree height derivation, *66*, 66–67
  InSAR, 57
  ITD, 65–66
  light penetration, 110
  MLS, 59–60
  multispectral point clouds, *59*
  multi-scale habitat mapping and monitoring, *see* multi-scale habitat mapping and monitoring
  optical remote sensing, *see* optical remote sensing, tree and stand heights
  point height metrics, 62, **63**
  SAR, 55–56, 83
  SAR radargrammetry, 57–58
  space-borne LiDAR, 54–55
  species assemblages, 194
  stem curve derivation, 67–69
  TLS, 58–59

A

aboveground biomass (AGB), 432
  ALS and, 58
  belowground and, 411
  biophysical approaches and, 274
  carbon changes estimates, 424
  carbon stocks, 426, 427–428, **428**
    allometric biomass methods, 429–432
    biome average methods, 428–429
    satellite sensors, characteristics of, **430–431**
  change, *407*
  estimate terrestrial biomass, 426
  forest cover change, global distribution of, *425*
  frequency distribution, *435*
  vs. mean field-estimated, *446*
  machine learning algorithms and, 273
  nonparametric methods, 434–435
  remote sensing–based
    distribution of forest, *444*
    frequency distribution, *445*
    LiDAR, 108–109, **108**, 439–441
    multi-sensor fusion, 441–445, **442–443**
    optical remote sensing, 436–438
    radar signals, 438–439
  remote sensing data, 432
    forest carbon stocks, **433**
    LiDAR, 434–435
    passive optical, 432–433
    radar data, 433–434
  RSE of remote sensing, *446*
  SAR, estimation with, 55–56
  terrestrial ecosystem carbon, 424
  terrestrial ecosystem carbon estimates, 427
  tree-based models, 436
  tropical rainforests in carbon storage, 424–426
aboveground carbon density (ACD), 477
  changes in, measuring, 406–407, 415–416
  LiDAR and, 109, 464
  mapping, 405, *406*
accuracy assessment, change detection, 327–328
advanced topographic laser altimeter system (ATLAS), 99, 136, 142
advanced very high resolution radiometer (AVHRR), 251, **252**, 253, 315, **316**
  LULC investigation and, 345–346
  passive optical remote sensing data, 432
advanced wide field sensor (AWiFS), 10, 30
airborne laser scanning (ALS), 52, **54**
  and aerial stereo-photogrammetry, 71
  CHM could, 146
  vs. DP, 147, **148–149**, 150
  in forest inventory, 58
  the individual tree level, *75*
  ITDs, 65–66
  multi-epoch data, 72
  multispectral, 58
  Optech Titan data, 58
  photogrammetric 3D data, 147–152
  point height metrics, 62
  Riegl VUX-1HA, 74
  SAR interferometry, 57
  single-photon, 58
  3D, 74
    vegetation height, 137–140, **138**
  3D point clouds, 145
  time series, *167*
  Toposys-I Falcon, 74
  tree detection, 65
  tree height estimates, 66, 67
airborne LiDAR, 49, 58, 62, 99
  biodiversity, 203

491

Landsat imagery with, 274
light penetration, 110
multi-sensor fusion, 441
tropical forest biomass, 5
vegetation height, 140
allometric biomass methods, 429–432
alpha diversity, 197
Amazon rainforest, 374
area-based approaches (ABAs), 462
  accurate training plot–level data, 63–64
  individual tree–based features, 64
  regression models, 64
  RF, 64
artificial neural networks (ANN), 463, 468
augmented reality (AR), 78, *445*
automated change detection approaches, 323, **324**, **325**
  CLASlite, 325
  comparison, 326
  disturbance index, 324, 324–325
  LandTrendr, 326
  LCLCC, 472
  MGDI, 325
  STAARCH, 326
  VCT, 326
automatic cloud cover assessment (ACCA), 11–12
automatic time-series analysis (ATSA), 14
autonomous big data, 80–81

B

Bayesian spectral mixture analysis (BSMA), 263
belowground biomass, 5, 270–271, 411, 426, 440
beta diversity, 197
biodiversity
  climate change regulation, 126
  EO-based, 227–228
    forest habitats, 229, 231
    grassland habitats, *231*, 231–232
    heathland habitats, 232
    increasing resolution, influence of, **230**
    land cover, habitats and indicators, 228–229, *229*
    types, 229
    wetland habitats, 233
  fauna, 156
  habitat monitoring, 213
  land cover changes, 311
  monitoring service, 236–238
  species richness, 28
biodiversity from space, 191–192, 466–468
  LiDAR, 203
  measurement
    animals, 194
    individual trees, 193, *194*
    species, 192–193, *193*
    species assemblages, 194
    vegetation types, 192–193
  modeling
    land cover and diversity, 196–197
    multiple sensors and diversity, 198
    species distribution modeling, 194–196, *195–197*
    spectral indices and diversity, 197–198
  monitoring, 198–199, **199**
    protected areas, 199–200, *200*
    urban areas, 200–201, *201*

policies, 214–215, *215*
radar sensors, 202–203
satellites and sensors, 203
spectral sensors, 201–202
study of, 466–468, *467*
biodiversity surrogates (BS), 233, 235
biomass, 271–272, 459, 462
  AGB, 426
  allometric-based, 429–432
  ALOS PALSAR data, *476*
  belowground, 5, 270–271, 411, 426, 440
  bookkeeping model, 398–400
  biophysical approaches, 274
  canopy structure and, 108–110
  crops, 376
  density, 405
  forest inventory, 49
  kNN, 272
  LiDAR, 439–441
  machine learning algorithms, 273–274
  multi-sensor fusion, 441–445
  optical remote sensing data, 436–438
  radar data, 433–434, 438–439
  rainforest, 98
  regression, 272–273
  Sabah Malaysian Borneo, map of, *483*
  SAR interferometry, 57
  3D SAR techniques, 83
  tree heights, 127
  tropical forests, 16
    age with one image epoch, 20–21
    accumulation in forest regrowth, 17–18
    from high-resolution multispectral imagery, 16–17
    with one image epoch, 18–20
  uncertainties, errors, and accuracy, 274–275
  vegetation, 424
biomass conversion and expansion factors (BCEFs), *479*
biome average methods, 428–429
biophysical approaches, 274
bitemporal change detection, 319
  map-updating approaches, 321–322
  methods, 319
bookkeeping model, 398–400, *399*

C

canopy height model (CHM), 60, 62, 65
  ALS, 158, *167*
  DP, 146, 147, 151
  hole-filling algorithms, 159
  LM, 160
  modeling and optimization, 159–160, *160*
  point-based approach, 163
  time series, *167*
  tree and stand heights, 465
carbon, 102, *see also* global carbon budgets
  average annual fluxes, *409*, *411*
  canopy physiology, 105
  changes estimates, 424
  forest-bound, 50
  in live biomass, 6
  offset project, 4

Index

sources and sinks, 398
stocks estimates, 427–432
stocks/canopy, 22, 426
stocks and fluxes, 71, 128
storage capacity, 31, 424–426
terrestrial ecosystem, 427
carbon density
   areas deforested, 409
   biomass density, 405
   changes in, 406–408, *407*
   spatial variability in, 398
   woody vegetation, 402, *406*
   uncertainties, 412
carbon dioxide (CO_2), 105, 424
carbon stocks, 426, 427–428, **428**
   AGB, 447
   allometric biomass methods, 429–432
   biome average methods, 428–429
   forest biomass, *479*
   LiDAR technology, 434
   tropical rainforests, 424–426
   passive remote sensors, 432–433
   persistent reduction in, 22
   satellite sensors, characteristics of, **430–431**
carbon storage, 31, 413, 432
   LiDAR, 434
   LULCC, 401
   residual terrestrial sink, 413
   tropical rainforests in, 424–426
Carnegie landsat analysis system lite (CLASlite), 328–330
   2011 Tornado Disturbance, 333
   automated change detection approaches, 325
   cloud/water mask, *332*
   crosstabulation assessment, **330**
   deforestation, 332, **333**
   deforested tornado track, *335*
   disturbance mapping, 332
   forest change, 332
   forest cover mapping, 330–331, *332*
   Gardner, 332, *334*
   Google Earth imagery, **331**
   MaFoMP, **331**
   Massachusetts, 332, *334*, *336*
case-based reasoning, 367
change vector analysis (CVA), 368, 475
China-Brazil earth resources satellite series (CBERS), 10
classifiers, 362
   ensemble, 104
   non-parametric, 363
   parametric, 362–363
   spatial and contextual, 23
cloud screening, 10–11, *11*
   manual and semi-automated approaches, 11
clumping index, 265–270
commercial high-resolution satellites, 259, **259–260**
committed *vs.* actual emissions, 409–410
component ratio method (CRM), 432
constellation of small satellites for mediterranean basin observation (COSMO)-SkyMed, 71, 225
crop and vegetation identification, 374–376, **375**
crosstabulation assessment, **330**
crowdsourcing, 77–78, 82

D

DBSCAN clustering, 65
deep learning networks (DNNs), 71
deforestation, 398, 407, 408
   Amazon rainforest, 374, 402
   biomass density, 405–406
   carbon emissions, 128
   definition, 408–409
   CLASlite, 325, 332, **333**, 335
   five-year FRA assessment, 404
   greenhouse gas emissions, 4
   SPOT program, 258
deforested tornado track, *335*
deformation, ecosystem structure and dynamics of ice (DESDynI), 165
density-based allometric method, 429, 432
diameter, 26, 67–69
   allometric-based biomass, 429
   area-based model implementation, 156
   AGLB, 16
   forest measurements, 50–52, **51**
   harvester laser scanning, 80
   LiDAR, 99, 140, 434
   tree heights, 127
diameter at breast height (DBH), 64, 130, 461
   allometric-based biomass, 429
   autonomous perception data, 79
   forest inventory, 72
   forest measurements, 50
   laser scanning, 80
   stem curve derivation, 67, **68**
   single-scan methods, **68**
   TLS, 59
   tropical forest, 461, 462
digital elevation model (DEM), 12, 465
digital photogrammetric (DP), 126, 131
   optical remote sensing, tree and stand heights *vs.* ALS, 147, **148–149**, 150
   area-based model implementation, 152
   CHM, 147
   development of, 144
   LiDAR systems, 141
   principle and brief history, 143–145
   stereoscopic measurement, *143*
   3D point, 145–146, *146*
   time series, *167*
digital stereo-photogrammetry, 71
digital surface model (DSM), 139, 166, 168, 462, 472
   DTM calculation, 61
   forest cuttings, 76
   matching strategy, *146*
   MICMAC software, *146*
   3D data, 60
   processing, 62
digital terrain model (DTM), 48, 74, 139, 462, 463
   ALS, 145, 166
   CHM, 62
   errors, 140
   filtering techniques, 61–62
   forest cuttings, 76
   quality, 140
   3D data, 60
   TIN densification method, 61

disturbance index (DI), 324–326
disturbance mapping, 332
diversity
    land cover, 196–197
    multiple sensors, 198
    spectral indices, 197–198
drone-based laser scanning, 59–60
drone laser scanning, 78–79
droughts, tropical forests, 31–32
dual-tree complex wavelet transform (DT-CWT), 367

E

Earth observation (EO), 212
    biodiversity and habitat mapping, 227–228
        forest habitats, 229, 231
        grassland habitats, *231*, 231–232
        heathland habitats, 232
        increasing resolution, influence of, **230**
        land cover, habitats and indicators, 228–229, *229*
        types, 229
        wetland habitats, 233
    habitat delineation and characterization, **213**
    mapping approaches, 226
ecosystem services (ES), 216
electromagnetic spectrum (EMS), 222
empirical-based approaches, 264
    multi-temporal change detection, 265
    space-for-time substitution, 264–265
    time-series analysis, 265
enhanced vegetation index (EVI), 276, 279, 325, 434
enzyme ribulose-1,5-bisphosphate carboxylase-oxygenase (RuBisCO), 105, 106
EO data for habitat monitoring (EODHaM), 238
equivalent water thickness (EWT), 101, 103
essential biodiversity variables (EBVs), 467
essential climate variable (ECV), 424, 479
EU Biodiversity Strategy 2020, *215*
EU Habitats Directive, 216–217
European nature information system (EUNIS), 228, 238
European remote sensing satellite (ERS), 433–434
European space agency (ESA), 98, 212, 260, 433, 434

F

feature extraction and selection, 355
    feature selection, 361–362
    interferometric coherence variation, *361*
    interferometric features, 358–360, *359*
    multi-frequency information, 360
    multiple incidence angle information, 360
    polarimetric features, 357–358, *358*
    speckle features, 356–357
    textural features, 355–356
filtering techniques, 61–62
flood
    detection and monitoring, 380
    urban, monitoring, 379
        vs. unflooded, *361*
    vegetation, 376
flooding index, 360
Fmask and time series approaches, 12, *13*
forest biomass mapping
    accumulation, 17–18
    age with one image epoch, 20–21
    high-resolution multispectral imagery, 16–17
    with one image epoch, 18–20, *19*
    regrowth rates, 17–18
    sensitivity, 19
forest carbon offsets, 4–5
forest carbon stocks, **433**
forest change, 332
forest cover mapping, 330
forest cuttings, 76, *77*
forest degradation, 4, 16, 459
    CLASlite, 325
    derived indices, 23
    interpreting and combining subpixel endmember fractions, 23
    mining detection, 22
    pixel level, 21–22
    road and trail detection, 21–22
    SAR, 374
    subpixel level with spectral mixture analysis, 22–23
    tropical forests, 21
forest fires, 31–32
forest habitats, 229, 231
forest inventory, 48, 82, 143, 475
    allometric-based biomass, 429
    ALS, 58, 150
    biome averages, 428
    carbon stocks, 479
    DTM, 61
    large-scale strategic planning, 49, **50**
    LiDAR, 69
    operative forest management, 49, **50**
    pre-harvest planning, 49, **50**
    3D data, 52
    SAR, 55
forest inventory and analysis (FIA), 264, 273, 312, 429
forest management and ecology, 127–129
forest measurements, 50–52, **51**
forestry inventory and mapping, 372–374, **373**
forest succession, 263
    empirical-based approaches, 264–265
    physical-based models, 263–264
fraction of absorbed photosynthetically active radiation (FPAR), 279–280
    application, **283**
    with biophysical models, 281–282
    with empirical models, **280**, 280–281, *281*
full carbon accounting, *411*, 411–412
fully connected network (FCN), 463
fuzzy c-means algorithm, 261–262

G

gap filling, 15–16
Gaussian filtering, 159
general habitat categories (GHC), 228, 238
geographic information system (GIS), 49, 312
    forest polygons, 72
    software and tools, 218
    Web-, 238
geoscience laser altimeter system (GLAS), 17, 99, 434
    ICESat, 225, 274, 444
    large-footprint full-waveform LiDAR systems, 440–441

Index

LiDAR system, 137
multi-sensor fusion, 441
space-borne LiDAR, 54–55
spatial analyses, 405
tree and stand heights, 465
global carbon budgets, 395–396, 477–479, *479*
accuracy, 412
attribution, 412
bookkeeping model, 398–400, *399*
carbon density, 409
committed *versus* actual emissions, 409–410
contemporary, 396–397
full carbon accounting, *411*, 411–412
gross and net emissions, 410
history, 397–398
initial conditions, 410
LULCC, *see* land use and land cover (LULC)
precision, 412
RED, 408–413
REDD, 408–413
REDD+, 408–413
residual terrestrial sink, 413
OCO, 413–414, **414**
vegetation activity (greenness), 414–415, *415*
sources and sinks, 398
spatial analyses, 400, *401*
stocks and flows, **396**
uncertainties, 412–413
global change
disturbance, drought and fire, 31–32
flooding, 32
International Year of Biodiversity, *214*
landslides, 32
satellite EO-based mapping, **213**
storms, 32
subcontinental to global scales, 30–31
global earth observation system of systems (GEOSS), 212
global ecosystem dynamics investigation (GEDI), 55, 126, 140, 472
large-area inventories, 157
LiDAR data, 99, 137, 203, 463
urban areas, remote sensing of, 201
vegetation height errors, 141, 142
global inventory monitoring and modeling system (GIMMS), 277
global navigation satellite system (GNSS), 52, 59, 77
global positioning systems (GPS), 135, 312, 447
global terrestrial carbon, 479–484, *482*, *483*
global vegetation models (GVM), 128
graph-cut and generalized Gaussian model, 367
grassland habitats, *231*, 231–232
greenhouse gases (GHGs), 3, 425–426, 478
carbon offsets, 4
forest regrowth, 17
inventories and forest carbon offsets, 4–5
REDD+, 24
green vegetation (GV), 22–23, 264
gross primary productivity (GPP), 278, 279, 288, 470
primary production based LUE, 288–289, **290**
primary production without LUE, 290–291
uncertainties, errors, and accuracy, 291
ground-based LiDAR, *see* terrestrial laser scanning (TLS)
ground measurement techniques, 266–267
group on earth observations (GEO), 212, 468

H

HabDir, 216–217, 228, 233, 236
habitat mapping, 227–228
forest habitats, 229, 231
grassland habitats, *231*, 231–232
heathland habitats, 232
increasing resolution, influence of, **230**
land cover, habitats and indicators, 228–229, *229*
physical and mappable expression, *229*
quality, 233–234, *234*
types, 229
wetland habitats, 233
hand-held laser scanning, 59–60
harvester laser scanning, 80, *81*
heathland habitats, 232
hidden Markov chains (HMCs), 364
hierarchical clustering, 65
high-resolution (HR) images, 218, 219, 232
hyperspectral imaging (HSI)
data fusion, benefits, *111*, 111–113, *112*
data sources, 97–98, **98**
forest biophysical, biochemical, and structural properties, 463–464, *464*
remote sensing of forests
biochemical properties, **101**, 102–105, *104*
biophysical properties, 100–102, **101**
canopy physiology, 105–108, *107*
hyperspectral infrared imager (HyspIRI), 98, 292

I

IKONOS, 259
image differencing, 367
image preprocessing, 355
image ratioing, 364–366
image spatial resolution, 29–30
increasing resolution, influence of, **230**
individual tree detection approaches (ITDs)
with ALS, 65–66
clustering techniques, 65
forestry data, 62–64
segmentation, 65
individual tree height derivation, *66*, 66–67
individual tree inventories, 72–76, *73*, *75*
in-situ monitoring LiDAR (IML) system, 82
instrument on board the ice, cloud, and land elevation (ICESat), 99, 126, 136, 137
large-area inventories, 157
LiDAR data, 434
space-borne LiDAR, 54–55
intergovernmental panel on climate change (IPCC), 428, 436
multispectral imagery, 5–6
vegetation phenology, 275
inverse distance weighted (IDW), 159
iPad Pro, 60, 68

K

Kittler and Illingworth (K&I) thresholding, 364
K-means clustering, 65, 162
K-nearest neighbors (KNN)-bagging approach, 158, 272, 273

L

land, vegetation, and ice sensor (LVIS), 99, 225, 435, 440
land cover, habitats and indicators, 228–229, *229*
land cover and diversity, 196–197
land cover and land cover changes (LCLCC), 472–475, *473*, *474*
land cover change detection, 311–313
    accuracy assessment, 327–328
    advantages and disadvantages, **320–321**
    automated change detection approaches, 323, **324**, **325**
        CLASlite, 325
        comparison, 326
        disturbance index, 324–325
        LandTrendr, 326
        MGDI, 325
        STAARCH, 326
        VCT, 326
    AutoMCU, *329*
    AVHRR, MODIS, and Landsat, **316**
    bitemporal change detection, 319
        map-updating approaches, 321–322
        methods, 319
    cause, 316–318, **317**
    CLASlite, 328–330
        2011 Tornado Disturbance, 333
        cloud/water mask, *332*
        crosstabulation assessment, **330**
        deforestation, 332, **333**
        deforested tornado track, *335*
        disturbance mapping, 332
        forest change, 332
        forest cover mapping, 330
        Gardner, 332, *334*
        Google Earth imagery, **331**
        MaFoMP, **331**
        Massachusetts, 332, *334*, *336*
    conceptual scheme, *313*
    history of, 315–316
    monotemporal change detection, 318–319
    temporal trend analysis, 322
        challenges of, 323
        hypertemporal remote sensing data, 322–323
        medium-resolution data, 323
    theory and practice, 313–314
Landgrebe–Jackson expectation-maximization (LJ-EM) algorithm, 366
landsat-based detection of trends in disturbance and recovery (LandTrendr), 265, 326
Landsat program, 253–256, **256–257**
Landsat TM/ETM+ images, 11–14
land surface phenology (LSP), 275–276
    metrics derivation and validation, 279
    vegetation index for, 276–279, *278*
land use, land use change and forestry (LULUCF), 425
land use and land cover (LULC), 343–346
    annual fluxes of carbon, *409*
    applications, 380
    carbon, 395
    change detection, 369–372, **370–371**
    classification, 369–372, **370–371**
    classification level, **344**
    classification system, **345**
    classifiers, 362
        non-parametric, 363
        parametric, 362–363
    crop and vegetation identification, 374–376, **375**
    feature extraction and selection, 355
        feature selection, 361–362
        interferometric coherence variation, *361*
        interferometric features, 358–360, *359*
        multi-frequency information, 360
        multiple incidence angle information, 360
        polarimetric features, 357–358, *358*
        speckle features, 356–357
        textural features, 355–356
    flood detection and monitoring, 380
    forestry inventory and mapping, 372–374, **373**
    future developments, 380–381
    gross and net emissions, 410
        full carbon accounting, *411*, 411–412
        initial conditions, 410
    image preprocessing, 355
    radar system parameters
        electromagnetic wave, *348*
        frequencies, **348**
        linear polarization schemes, *348*
        local incidence angle, *349*
        mapping, 351–354, *352*
        optical remote sensing, *347*
        optimal SAR system parameters, **354**
        polarization, 347
        radar remote sensing, *347*
        spaceborne SAR systems, **350–351**
    satellite data, 402
        biomass density, 405
        carbon density, changes in, 406–408, *407*
        rates of change in forest area, 402–404
        spatial analyses, 405–406, *406*
        uncertainties, 404
    snow and ice mapping, 379–380
    typical data characteristics, **344**
    unsupervised change detection methods, 364
        image differencing, 367
        image ratioing, 364–366
        vs. post-classification, 368
        RADARSAT-2 image, *365*
        statistical measure, 366–367
    urban environment, 377–379, **378**, *379*
land vegetation ice system (LVIS), 440, *462*
    large-footprint full-waveform LiDAR systems, 440
    LiDAR data sources, 99, 225, 435
large-footprint full-waveform LiDAR, 440–441
laser scanning (LS), 52, 79
laser vegetation imaging sensor (LVIS), 99, 225
leaf area index (LAI), 107, 250, 265
    ALS, 58
    biophysical properties, 100–102
    black spruce forest, *103*
    definitions, 266–267
    ground measurement techniques, 266–267
    LiDAR and radar point clouds, 462
    LUE, 291
    MODIS, 82
    optical remote sensing, 437
    products and characteristics, **271**
    radiative transfer models, 270
    retrieval, 267–270, **268–269**

spectral vegetation indices, 268–270
tree and stand heights, 465
uncertainties, errors, and accuracy, 274–275
leaf chlorophyll content, 282–283
  laboratory extraction, 283
  with radiative transfer models, 284–285
  with SPAD meter, 283–284
  with spectral vegetation indices (VIs), 284
leaf mass area (LMA), 102, 105, 113
LiDAR Phenology (LiPhe), 82, *83*
light detection and ranging (LiDAR) and radar, remote sensing of forests, 48–49
  aboveground biomass (AGB), 434–435
  automizing field inventories, 77–80
  biodiversity from space, 203
  biomass, 108–110
  canopy structure, 108–110
  change detection, 72–76
  characteristics, **53**
  data fusion, benefits, *111*, 111–113, *112*
  data quality, 99–100
  data sources, 98–99
  features of, 52
  forest biophysical, biochemical, and structural properties, 463–464, *464*
  forest cuttings, 76, *77*
  forest informatics, 461–463, *462*
  forest inventory, 49–50, **50**
  forest measurements, 50–52, **51**
  harvester laser scanning, 80, *81*
  individual tree inventories, 72–76, *73*, *75*
  large-area mapping, improvement, 71–72
  light penetration, 110
  mature tropical forests, cross-sectional views, *110*
  multi-scale habitat mapping and monitoring, 222–225, **224–225**
  optical remote sensing, tree and stand heights
    accuracy of vegetation height measurements, **138**
    error, tree height measurement, *141*
    principle and brief history, 135–137, *136*
    spaceborne LiDAR systems, 140–142
    3D ALS data, 137–140
  plot concept, 69–71, *70*
  PLSS, 80, 82
  processing, 60–61
  radar backscatter coefficient, 52
  radio waves, 52
  3D data, 52–54, **54**
    ALS, 58
    canopy height model, 62
    DBH measurements, **68**
    diameter, 67–69
    DSM, 62
    DTM, 61–62
    forestry data, 62–64
    individual tree height derivation, *66*, 66–67
    InSAR, 57
    ITD, 65–66
    MLS, 59–60
    multispectral point clouds, *59*
    point height metrics, 62, **63**
    SAR, 55–56
    SAR radargrammetry, 57–58
    space-borne LiDAR, 54–55
    stem curve derivation, 67–69
    TLS, 58–59
  structural properties, **108**
  waveform and discrete return measurements, **108**, *109*
light use efficiency (LUE), 106, 291, 463, 464
  biophysical basis of, 285–286
  photosynthesis-light response curve, *286*
  remote sensing, 286–288, *287*
Li-Strahler model, 263
local maxima (LM), 160, 161

M

machine learning algorithms, 273–274
mapping tropical forest types with multispectral imagery
  associations and land cover, *25*
  coarse spatial scale, 29
  finely scaled habitats
    characteristics, 24
    edges/linear features, 24
    object-oriented classification, 26
  forest associations, 24
  forest formations, 24
  image spatial resolution, 29–30
  with medium-resolution imagery
    ancillary data and machine learning, 27
    mono-and bi-dominant tree floristic classes, 28
    physiognomic/floristic classes, 27–28
  REDD+, 24
  remote tree species identification, 26
  species richness, endemism and functional traits, 28–29
map-updating approaches, 321–322
Markov random field (MRF), 14, 366
Massachusetts forest monitoring program (MaFoMP), 327, 330–332
maximum likelihood approach, 367
mean canopy profile height (MCH), 109
mean-shift clustering, 65
millennium ecosystem assessment, 233
mining detection, 22
mini-UAV-based airborne laser scanning, 78
mixture models with spectral vegetation indices, 262
mobile laser scanning (MLS), 52, **54**
  drone-based scanning, 59–60
  drone laser scanning, 78–79
  hand-held laser scanning, 59–60
  mini-UAV-based airborne laser scanning, 78
  phone-based laser scanning, 59
  SLAM, 78
  stem mapping, 68
  terrestrial laser scanning, 59–60
  vehicle-based scanning, 59–60
moderate resolution imaging spectroradiometer (MODIS), 10, 82, 202, 204, **316**, 333
  biomass, 274
  cloud screening, 11
  cloud/shadow contamination, 14–15
  LAI, 272, 275, 291
  LCLCC, 472, 474
  LUE, 286
  machine learning algorithms, 273–274
  medium-resolution data for trend analysis, 323

multi-sensor fusion, 441
NPP, *471*
passive remote sensors measure, 432
PRI, 287–288
satellite data, 403
spatial analyses, 407
in spatial and temporal resolution, **316**
STAARCH, 326
tree and stand heights, 465
tropical forest age with one image epoch, 20–21
tropical forest disturbance, drought, and fire, 31
vegetation activity, 414
vegetation, optical sensors, 251–253, **254**
MODIS global disturbance index (MGDI), 325, 326
monotemporal change detection, 318–319
multi-angle imaging spectroradiometer (MISR), 21
biomass, 274
passive remote sensors, 433
optical remote sensing, 438
optical sensors for vegetation mapping, 251–253
multi-epoch ALS, 72, 74, 76
multi image matches for auto correlation methods (MICMAC), 145, *146*
multiple endmember spectral mixture analysis (MESMA), 263
multiple sensors and diversity, 198
multi-scale change profile (MCP), 367
multi-scale habitat mapping and monitoring, 468–470, *469*
   biodiversity monitoring service, 236–238, *237*, *238*
   EO-based biodiversity and habitat mapping, 227–228
      forest habitats, 229, 231
      grassland habitats, *231*, 231–232
      heathland habitats, 232
      increasing resolution, influence of, **230**
      land cover, habitats and indicators, 228–229, *229*
      types, 229
      wetland habitats, 233
   policy framework
      biodiversity, 214–215, *215*
      EU Habitats Directive, 216–217
      global change, 212–214, **213**
      International Year of Biodiversity, *214*
      MS.MONINA, three-level information service concept, *217*
      state of ecosystems, 216
   quality, pressures, and changes
      habitat quality, 233–234, *234*
      identifying, 234–236
   satellite sensor capabilities, 217–218
      advancement of imaging technology, *226*, 226–227, *227*
      LiDAR, 222–225, **224–225**
      radar, 222–225, **224–225**
      revisiting time, 218, 225
      spatial resolution, 218, 219, 222
      spectral resolution, 218
multi-sensor fusion, 441–445, 441–445, **442–443**
multi-sensory multi-variate gamma distributions (MuMGDs), 367
multi-spectral scanner (MSS), 14, 254, 255, 265
multi-temporal change detection, 265

N

national forest inventory (NFI), 49, 166
   ALS, 72
   area-based model implementation, 156
   forest management and ecology, 128
   large-area inventories, 157–158
national oceanic and atmospheric administration (NOAA), 397
   carbon budget, 396, 397
   LSP, 277
   optical remote sensing, 436
   species distribution modelling, 196
   vegetation mapping, 251, **252**, 253
Natura 2000, 216, 225
near infrared (NIR), 136–137
net ecosystem productivity (NEP), 278
net primary productivity (NPP), 288
   biomass, 270–272
   primary production based LUE, 288–289, **290**
   primary production without LUE, 290–291
   MODIS, *471*
   uncertainties, errors, and accuracy, 291
   vegetation mapping, 250, 470
neural networks (NN), 64, 273
non-parametric classifiers, 363
non-photochemical quenching (NPQ), 105, 106
non-photosynthetic vegetation (NPV), 22, 23, 104, 264, 329
normalized difference fraction index (NDFI), 23, 459
normalized difference infrared index (NDII), 277
normalized difference moisture index (NDMI), 20
normalized difference vegetation index (NDVI), 251, 463
   circumpolar trends in, *415*
   filling cloud and scan-line gaps, 15
   graphical representations, *281*
   HSI data, 101, 113
   vs. FPAR, *281*
   land cover, *197*
   LSP, 277–279
   mixture models with spectral vegetation indices, 262
   optical remote sensing , 437
   spectral indices, 197–198
   tropical forest, 29
   vegetation activity, 414, *415*
normalized difference water index (NDWI), 277, 463

O

operational land imager (OLI), 328, 346
   cloud and cloud shadow screening, 11–14
   Landsat program, 255
optical remote sensing, tree and stand heights, 126
   AGB, 436–438
   area-based approaches
      general presentation, 150–152
      goodness-of-fit statistics, **154**
      HL values, *155*
      implementation, 152–157
      maximum LiDAR height, *151*
      principle, *152*
   determinants, 127
   DP
      vs. ALS, 147, **148–149**, 150

Index

principle and brief history, 143–145
stereoscopic measurement, *143*
3D point, 145–146, *146*
field measurements, **127–134**, 129, 135
forest management and ecology, 127–129
large-area inventories, 157–158
LiDAR
 accuracy of vegetation height measurements, **138**
 error, tree height measurement, *141*
 principle and brief history, 135–137, *136*
 spaceborne LiDAR systems, 140–142
 3D ALS data, 137–140
model extrapolation, 157–158
point-based approaches, *130*, 162–165, *163*, **164**
raster-based approaches
 CHM modeling and optimization, 159–160, *160*
 strengths and limitations, **164**
 tree crowns, 161–162
 treetop detection, 160–161, *161*
vertical tree, *130*
orbiting carbon observatory (OCO), 413–414, **414**

P

parametric classifier, 362–363
partial least squares regression (PLSR), 102, 104
passive optical remote sensing data
 aboveground biomass (AGB), 432–433
permanent laser scanning systems (PLSS), 80, 82
phased array multi-functional imaging radar (PAMIR), 377, *379*
PhillyTreeMap, 77
phone-based laser scanning, 59–60
photochemical reflectance index (PRI), 106, *107*
 LUE, 286–288
physical and mappable expression, *229*
point height metrics, 62, **63**
polarimetric interferometric (PolInSAR), 135, 369, 439
polar orbiting platforms (POES), 251, **252**

Q

quality indices (QI), 212

R

radar backscatter coefficient, 48, 52, 56
RADARSAT-2, *352*, 353, *365*
 crop and vegetation identification, 376
 hierarchical segmentation of, *368*
 LULC, 369, 372
 radar system parameters, 349
 snow and ice mapping, 380
radar sensors, 202–203
 aboveground biomass (AGB), 433–434
 forest informatics, 461–463, *462*
 land use and land cover mapping and change detection, 475–477, *476*
 multi-scale habitat mapping and monitoring, 222–225, **224–225**
radar system parameters
 electromagnetic wave, *348*
 frequencies, **348**
 linear polarization schemes, *348*
 local incidence angle, *349*
 mapping, 351–354, *352*
 optical remote sensing, *347*
 optimal SAR system parameters, **354**
 polarization, 347
 radar remote sensing, *347*
 spaceborne SAR systems, **350–351**
radiative transfer models, 270, 284–285
radiative transfer model to measure leaf optical properties spectra (PROSPECT), 284, 463
 HSI data, 102, *103*
radio waves, 52, 347
radon transform and Jeffrey divergence, 367
random forest (RF), 64, 273, 461, 464
raster-based approaches, tree and stand heights
 CHM modeling and optimization, 159–160, *160*
 tree crowns, 161–162
 treetop detection, 160–161, *161*
reducing emissions from deforestation (RED), 408
reducing emissions from deforestation and forest degradation (REDD+), 4, 33, 408
 forest carbon offsets, 4–5
 forest degradation, 22
 forest type, 24
 GHGs, 4–5
reducing emissions from deforestation in developing countries (REDD), 408, 426
regression models, 17, 64, 100, 262, 437, 462
 vegetation, optical sensors, 261
relative heights (RH), 141, 142, 440
residual terrestrial sink, 413
 OCO, 413–414, **414**
 vegetation activity (greenness), 414–415, *415*
resolution with average estimation errors (RMSE), 64
 ALS, 139–142, 150
 crowdsourcing, 78, 79
 diameter and stem curve derivation, 68–69
 InSAR, 57–58
 LiDAR plots, 70
 optical remote sensing, 437
 machine learning algorithms, 273
 MAE, 80
 regression, 272
 tree height derivation, 66–67
return beam vidicon (RBV), 254, 255
road and trail detection, 21–22

S

SAR interferometry (InSAR), 57
satellite-based forest mapping, 71–72
Satellite Pour l'Observation de la Terre, French Earth Observing Satellites (SPOT), 20–22
 biophysical approaches, 274
 cloud and cloud shadow screening, 14
 MODIS, 14–15
 vegetation, optical sensors, 257–258, **258**
 VGT sensors, *278*
scan-line gaps, 15–16
scanning LiDAR imager of canopies by echo recovery (SLICER), 137, 434, 440
Sentinel-2 imagery, 14
shortwave infrared (SWIR), 10, 225
 biochemical properties, 103

HSI data, 102
LSP, 276
optical remote sensing, 437
mapping tropical forest types, 24
SPOT, 257
tropical forest age with one image epoch, 20
shuttle radar topography mission (SRTM), 445, *462*, 466
InSAR, 57
biomass map from, *462*
multi-sensor fusion, 444, 445
radar, 202
signal-to-noise (SNR), 99–100
small-footprint discrete-return LiDAR, 440
snow and ice mapping, 379–380
societal benefit areas (SBAs), 212, 366
solar-induced fluorescence (SIF), 106, 107
space-borne LiDAR
GEDI, 55
GLAS, 55–56
spaceborne LiDAR systems, 140–142, *141*
space-borne synthetic aperture radar (SAR), 53, **54**, **346**, 445, 447
ABA, 64
crop and vegetation identification, 376
ERS-1 and ERS-2 tandem mission, 56
forestry inventory and mapping, 372
high-resolution, *379*
image differencing, 367
image preprocessing, 355
image ratioing, 364, 366
interferometry, 56, 358–360
large-area mapping, 71
LULC, 349, 350, 369
multi-frequency information, 360
multiple incidence angle information, 360
non-parametric classifiers, 363
parameters, **354**
parametric classifiers, 362–363
pixel level information, 55
polarimetry, 56, 357
radar cross-section, 56
snow and ice mapping, 379–380
statistical measure, 366–367
strengths of, 475–477
South Guangzhou data, *352*
speckle features, 356–357
radar pulse, 56
3D techniques, 83
textural features, 356
urban environment, application 377, 379
space-for-time substitution, 264–265
spatial analyses, global carbon budget, 400, *401*
spatial resolution, 218
sensor's suitability, **220–221**
sun-source systems, 222
WorldView-2 data, *219*, 222
spatial temporal adaptive algorithm for mapping reflectance change (STAARCH), 326, 474
spatial wavelet analysis (SWA), 161
species assemblages, thematic maps of, 194
species-based allometric method, 429
species distribution modeling, 194–196, *195–197*
spectral indices and diversity, 197–198

spectral mixture analysis (SMA), 262–264, 459
subpixel level with, 22–214
spectral resolution, 218
spectral sensors, 201–202
spectral vegetation indices (VIs), 284
split-based approach (SBA), 366
stem curve derivation, 67–69
sun-source systems, 222
Suomi national polar-orbiting partnership (Suomi NPP), *251*, 253, **255**, 292
support vector machines (SVM), *435*, 461, 468
grassland classification, *231*
sustainable forest management (SFM), 49, 72, *75*
synthetic aperture radar (SAR), 48, 439
radargrammetry, 57–58
3D, 71, 83

T

TanDEM-X mission (TDX), 57
temporal trend analysis, 322
challenges of, 323
hypertemporal remote sensing data, 322–323
medium-resolution data, 323
TerraSAR-X radargrammetry (TSX), 57–58
Terrascan software, 61
terrestrial biomass, 426
terrestrial ecosystem carbon
carbon changes estimates, 424
remote sensing in, 427
terrestrial laser scanning (TLS), 52, **54**, 59–60
in forest field inventories, 59
thematic mapper (TM), 255, 264, 291, 432
time-series analysis, 12–14, 265, 265
tree and stand heights, 465–466, *466*
tree-based models, 436
triangulated irregular network (TIN), 61, 62
iterative approach, 140
tropical forests with multispectral imagery, 3–4
area, age, height, 17–18, *18*
AWiFS, 10
biomass
accumulation, 17–18
age with one image epoch, 20–21
high-resolution multispectral imagery, 16–17
with one image epoch, 18–20, *19*
regrowth rates, 17–18
characterization, 458–461, *46*
clouds
Fmask and time series approaches, *13*
gap filling, 15–16
Landsat TM, 11–14
MODIS, 14–15
MSS, 14
OLI imagery, 11–14
scan-line gaps, 15–16
screening, 10–11, *11*
Sentinel-2 imagery, 14
SPOT, 14
degradation, 21
derived indices, 23
interpreting and combining subpixel endmember fractions, 23
mining detection, 22

Index

pixel level, 21–22
road and trail detection, 21–22
subpixel level with spectral mixture analysis, 22–23
global change
    disturbance, drought and fire, 31–32
    flooding, 32
    landslides, 32
    storms, 32
    subcontinental to global scales, 30–31
mapping types, *see* mapping tropical forest types with multispectral imagery
REDD+ programs, *see* REDD+
roles, 5–6
"stock-difference" method, 5
SWIR, 10
types, **6–9**, 6–10

U

uncertainties
    LULCC, 404–405, 412–413
    vegetation structure, 274–275
United Nations framework convention on climate change (UNFCC), 128, 312, 424
unmanned aerial vehicles (UAVs), 111, 145, 194
    LiDAR data, 139, 165
unsupervised change detection methods, 364
    image differencing, 367
    image ratioing, 364–366
    vs. post-classification, 368
    RADARSAT-2 image, *365*
    statistical measure, 366–367
urban environment, 377–379, **378**, *379*

V

vegetation, optical sensors, 250–251, *251*
    AVHRR, 251, **252**
    biomass, 271–272
        biophysical approaches, 274
        kNN, 272
        machine learning algorithms, 273–274
        regression, 272–273
        uncertainties, errors, and accuracy, 274–275
    clumping index, 265–270
    commercial high-resolution satellites, 259, **259–260**
    ecological characterization, 470–472, *471*
    forest succession, 263
        empirical-based approaches, 264–265
        physical-based models, 263–264
    FPAR, 279–280
        application, **283**
        with biophysical models, 281–282
        with empirical models, **280**, 280–281, *281*
    future direction, 260

GPP/NPP, 288
    primary production based LUE, 288–289, **290**
    primary production without LUE, 290–291
    uncertainties, errors, and accuracy, 291
IKONOS, 259
LAI, 265
    definitions, 266–267
    ground measurement techniques, 266–267
    products and characteristics, **271**
    retrieval, 267–270, **268–269**
Landsat program, 253–256, **256–257**
leaf chlorophyll content, 282–283
    laboratory extraction, 283
    with radiative transfer models, 284–285
    with SPAD meter, 283–284
    with spectral vegetation indices (VIs), 284
LUE
    biophysical basis of, 285–286
    photosynthesis-light response curve, *286*
    remote sensing, 286–288, *287*
MISR, 251–253
MODIS, 251–253, **254**
NOAA, 251, **252**
POES satellites, 251, **252**
SPOT, 257, **258**
Suomi NPP, 253, **255**
vegetation cover, 261
    fuzzy classification, 261–262
    mixture models with spectral vegetation indices, 262
    regression, 261
    SMA, 262–263
vegetation phenology, 275–276
    LSP studies, 276–279, *278*
VIIRS, 253, **255**
vegetation change tracker (VCT), 265, 265, 326
vegetation cover, 261
    fuzzy classification, 261–262
    mixture models with spectral vegetation indices, 262
    regression, 261
    SMA, 262–263
vegetation phenology, 275–276
    LSP studies, 276–279, *278*
vehicle-based laser scanning, 59–60
very high resolution (VHR), 218, 219
visible infrared imaging radiometer suite (VIIRS), 202, 253, **255**
visible to near-infrared (VNIR), 10, 11, 105, *111*, 113
visible to shortwave-infrared (VSWIR), 97, 113

W

wetness brightness difference index (WBDI), 20
wide field sensor (WiFS), 29–30
woody volume, 429, 432